Understanding Biophotonics

Understanding Biophotonics

Fundamentals, Advances, and Applications

edited by

Kevin K. Tsia

Pan Stanford Publishing

Published by

Pan Stanford Publishing Pte. Ltd.
Penthouse Level, Suntec Tower 3
8 Temasek Boulevard
Singapore 038988

Email: editorial@panstanford.com
Web: www.panstanford.com

British Library Cataloguing-in-Publication Data
A catalogue record for this book is available from the British Library.

Understanding Biophotonics: Fundamentals, Advances, and Applications

ISBN 978-981-4411-77-6 (Hardcover)
ISBN 978-981-4411-78-3 (eBook)

Printed in the USA

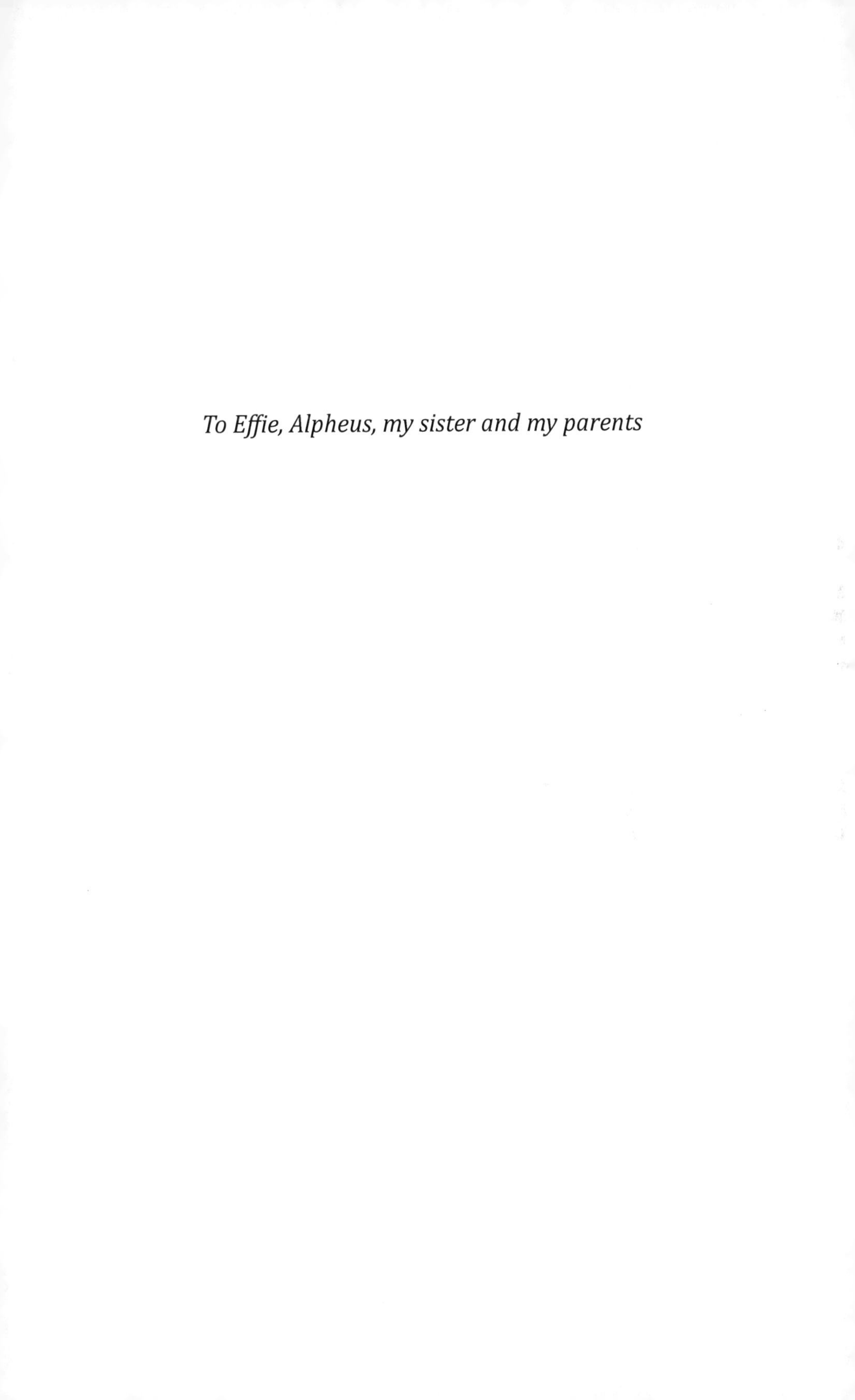

To Effie, Alpheus, my sister and my parents

Contents

Preface xix

1 Essential Basics of Photonics **1**
Kevin K. Tsia
1.1 Introduction 1
1.2 Wave Nature of Light 4
1.2.1 Electromagnetic (EM) Waves 5
1.2.1.1 EM spectrum 6
1.2.1.2 Basic formalism 7
1.2.1.3 Energy, power, and intensity 10
1.3 Polarization 14
1.4 Phase Velocity and Group Velocity 18
1.5 Coherence of Light 19
1.5.1 Temporal Coherence 20
1.5.2 Spatial Coherence 21
1.6 Interference 22
1.6.1 Essential Concepts of Interference 23
1.6.2 Interferometers 26
1.6.3 Significance of Phase Difference in Interference 26
1.6.4 Relationship between Coherence and Interference 28
1.7 Diffraction 29
1.7.1 Near-Field and Far-Field Diffraction 31
1.7.2 Diffraction and Fourier Optics 33
1.7.3 Resolution in an Optical Imaging System 35
1.7.4 Diffraction Grating 41
1.8 Gaussian Beam Optics 42
1.9 Optical Waveguides and Optical Fibers 46

1.10 Photons 52
1.10.1 Dual Nature of Light 52
1.10.2 Photon Flux 53

2 Essential Basics of Light–Matter Interaction in Biophotonics 57
Kevin K. Tsia
2.1 Introduction 57
2.2 Classical Interpretations 58
2.3 Optical Anisotropy and Birefringence 63
2.4 Light Absorption 69
2.4.1 Absorption Properties of Biological Cells and Tissues 72
2.5 Dispersion 78
2.6 Light Scattering 82
2.6.1 Rayleigh Scattering 86
2.6.2 Mie Scattering 87
2.7 Light Transport in Tissue 87
2.7.1 Radiative Transport Theory 89
2.7.2 Diffusion Approximation 91
2.7.3 Numerical Simulation: Monte Carlo (MC) Method 93
2.8 "Optical Window" for Biological Tissue 94
2.8.1 Examples: NIR Spectroscopy and Tomography 97
2.9 Quantum Approach to Light–Matter Interaction 101
2.9.1 General Concepts 101
2.9.2 Energy Diagrams 103
2.9.3 Atomic Transitions through Radiative Processes 108
2.10 Fluorescence Basics 115
2.10.1 Fluorophores 125
2.10.1.1 Intrinsic fluorophores 126
2.10.1.2 Extrinsic fluorophores 126
2.10.1.3 Near-infrared (NIR) fluorophores 130
2.10.1.4 Quantum dots (QDs) 131
2.10.1.5 Fluorescent proteins 134
2.11 Nonlinear Optics 135
2.11.1 General Descriptions of Optical Nonlinearity 137
2.11.2 Frequency Mixing 139

2.11.3 Intensity-Dependent Modification of Material Properties 144
2.11.3.1 Optical Kerr effect 144
2.11.3.2 Multi-photon absorption 146
2.11.4 Raman Effects 147
2.11.4.1 Spontaneous Raman scattering 147
2.11.4.2 Coherent Raman scattering 154
2.11.5 Applications of Nonlinear Optics in Biophotonics 156
2.12 More on Light–Tissue/Cell Interaction 166
2.12.1 Photochemical Effect: Photodynamic Therapy 169
2.12.2 Thermal Effects 172
2.12.3 Photoablation 177
2.12.3.1 UV versus IR photoablation 181
2.12.4 Plasma-Induced Ablation and Photodisruption 183

3 Multiphoton Microscopy **199**
Shuo Tang
3.1 Introduction 199
3.2 Principles and Instrumentation 201
3.2.1 One-Photon versus Two-Photon Fluorescence 201
3.2.2 MPM Contrast Signals 202
3.2.3 MPM Instrumentation 204
3.3 MPM Imaging of Cells and Tissues 207
3.4 Combined MPM/OCT System 209
3.4.1 MPM/OCT System Configuration 209
3.4.2 Imaging with the Combined MPM/OCT System 212
3.5 Multiphoton Endomicroscopy 214
3.6 Summary 217

4 Biological Fluorescence Nanoscopy **221**
David Williamson, Astrid Magenau, Dylan Owen, and Katharina Gaus
4.1 Introduction 221
4.2 Resolution and Diffraction 222
4.3 Extending Resolution 225

4.4 Super-Resolution Ensemble Nanoscopy: STED, GSD, SSIM 229
4.5 Super-Resolution Single Molecule Nanoscopy: (F)PALM and STORM 237
4.5.1 Probe Selection for PALM/STORM 242
4.5.2 Dual-Color PALM/STORM 245
4.5.3 Live-Cell PALM 248
4.5.4 3D PALM/STORM 249
4.6 Conclusions and Future Directions 251

5 Raman Spectroscopy of Single Cells **259**
Thomas Huser and James Chan
5.1 Introduction 260
5.2 The Principles of Raman Scattering 262
5.3 Single-Cell Analysis by Raman Scattering 274
5.3.1 Micro-Raman Spectroscopy Systems 274
5.3.1.1 Confocal detection 274
5.3.1.2 Laser wavelength 276
5.3.1.3 Optical excitation and detection geometries 278
5.4 Data Processing and Analysis of Raman Spectral Data 283
5.4.1 Smoothing 284
5.4.2 Background Subtraction 284
5.4.3 Principal Component Analysis (Multivariate Analysis Methods) 286
5.5 Laser Tweezers Raman Spectroscopy 289
5.5.1 Different Optical Trapping Geometries 290
5.5.2 Biological and Biomedical Applications of LTRS 291
5.5.3 Microdevices with Raman Tweezers Systems 294
5.5.4 Multifocal Raman Tweezers 297
5.5.5 Multimodal LTRS System for Measuring Mechanical and Chemical Properties of Single Cells 298
5.6 Outlook 300

6 High-Speed Calcium Imaging of Neuronal Activity Using Acousto-Optic Deflectors **309**
Benjamin F. Grewe
6.1 Introduction 310

6.2 The AOD Two-Photon Microscope Design 313
6.3 Laser Scanning with Acousto-Optic Devices 315
6.3.1 Operation Modes of Acousto-Optic Deflectors (AODs) 315
6.3.2 Scanning Resolution of Acousto-Optic Devices 317
6.3.3 Specifications of Acousto-Optic Deflectors 318
6.4 Dispersive Effects on Femtosecond Laser Pulses 320
6.4.1 Spatial Dispersion Compensation 320
6.4.2 Temporal Dispersion Compensation 322
6.5 Example of Application: High-Speed Population Imaging in Mouse Barrel Cortex 323
6.5.1 The Random-Access Pattern Scanning (RAPS) Approach 324
6.5.2 Automated Extraction of Spiking Activity from Fluorescent Data Traces 326
6.6 Discussion 328
6.7 Future Directions 329
6.8 Final Conclusions 330

7 Intravital Endomicroscopy **335**
Gangjun Liu and Zhongping Chen
7.1 Introduction 336
7.2 Development of Intravital Endomicroscopy 338
7.2.1 Light-Guiding Media 339
7.2.2 Beam Scanning Mechanisms 346
7.2.2.1 External scanning mechanism 347
7.2.2.2 Distal-end scanning mechanism 349
7.2.3 Focusing Optics 353
7.3 Embodiment of Introvital Endoscopic System 354
7.3.1 OCT Endoscopic Imaging 355
7.3.2 MPM Endoscopic Imaging 361
7.3.3 Multimodal Endoscopic Imaging 367
7.4 Summary 372

8 Nanoplasmonic Biophotonics **385**
Luke Lee
8.1 Conventional Biophotonics versus Nanoplasmonic Biophotonics 385

8.2 Fundamentals of Nanoplasmonics 390
8.2.1 History of Nanoplasmonic Materials 390
8.2.2 Overview of the Electrodynamics of Nanoplasmonics 390
8.2.3 Dielectric Constant of a Noble Metal 393
8.2.4 The Rigorous Field Distributions of a Metallic Sphere from Mie Theory 394
8.2.5 The Simplified Field Distributions of a Small Metallic Sphere 395
8.2.6 Extinction, Scattering, and Absorption Cross Sections 396
8.2.7 Complex Nanoplasmonic Geometries 397
8.2.8 Strongly Coupled Nanoplasmonic Structures 399
8.3 Nanoplasmonic Molecular Ruler 401
8.4 Plasmon Resonance Energy Transfer 403
8.5 Metal-Enhanced Fluorescence 405
8.6 Surface-Enhanced Raman Spectroscopy 407
8.7 Plasmonic Enhanced Photothermal Effect 413
8.8 Nanoplasmonic Trapping 418
8.9 Future Outlook 421
8.9.1 Nanoplasmonics and Fluidic Integration for Point-of-Care Diagnosis 422
8.9.2 Multiplex Gene Manipulation Using Tunable Optical Antennae 424
8.10 Appendix 425
8.10.1 Mie Theory 425
8.10.2 Far-Field Scattering of Metallic Spheres 426
8.10.3 Theory of Plasmonic Enhanced Photothermal Effects 427

9 Label-Free Detection and Measurement of Nanoscale Objects **441**
Ş. K. Özdemir, L. He, W. Kim, J. Zhu, F. Monifi, and L. Yang
9.1 Introduction and Overview 442
9.2 Quality Factor and Mode Volume of a Resonator 444
9.3 Surface Plasmon Resonance Sensors 446
9.4 Mechanical Resonators 451
9.5 Photonic Crystal Resonators 454

9.6 Whispering Gallery Mode Optical Resonators 459
9.7 Spectral Shift Method 466
9.7.1 Mode Splitting Method 471
9.8 Sensing Using WGM Active Microresonators and Microlasers 482
9.8.1 Reactive Shift in Active Resonators 486
9.8.2 Mode Splitting in Active Resonators 487
9.8.2.1 Below lasing threshold 487
9.8.2.2 Above the lasing threshold: WGM microlaser 489
9.9 Conclusions and Outlook 493

10 Optical Tweezers **507**
R. W. Bowman and M. J. Padgett
10.1 Introduction 507
10.2 Optical Forces 509
10.2.1 Nonconservative forces 511
10.2.2 Rayleigh Scattering: Very Small Particles 512
10.2.3 Mie Scattering: Intermediate-Sized Particles 513
10.3 System Designs for Optical Tweezers 513
10.3.1 Dual Traps 515
10.3.2 Time-Shared Traps 515
10.3.3 Holographic Optical Tweezers 515
10.3.4 Laser Sources 518
10.3.5 Objective Lenses and Beam Shape 519
10.4 Counterpropagating Optical Traps 520
10.5 Force and Position Measurement 522
10.5.1 Laser Position Detection 522
10.5.2 Video Particle Tracking 524
10.5.3 Particle Dynamics and Calibration 527
10.6 Conclusions 529

11 Coherent Nonlinear Microscopy with Phase-Shaped Beams **537**
Varun Raghunathan and Eric Olaf Potma
11.1 Introduction 537
11.2 Focal Fields in a Microscope 539
11.3 Phase Shaping of Focal Fields 543

11.3.1 Dynamic Diffractive Optical Elements 543
11.3.2 Fixed Diffractive Optical Elements 548
11.3.3 Beam Interference 551
11.4 Nonlinear Optics in a Microscope 553
11.5 Applications of Beam Shaping in Nonlinear Microscopy 557
11.5.1 Interface-Specific Imaging 557
11.5.2 Resolution Enhancement with Annular Pupils 559
11.5.3 Nanoscale Phase Resolution of Focal Volume 561
11.6 Conclusions 565

12 Supercontinuum Sources for Biophotonic Applications 571
J. R. Taylor
12.1 Introduction 571
12.2 Early Optical Fiber-Based Sources 574
12.3 Contributing Physical Processes 577
12.3.1 Dispersion 578
12.3.2 Optical Solitons: Self-Phase Modulation and Dispersion 579
12.3.3 Modulational Instability and Four-Wave Mixing 581
12.3.4 Soliton Instabilities and Dispersive Wave Interaction 582
12.4 Modern Supercontinuum Sources 584
12.4.1 Femtosecond Pulse Pumped Supercontinua 585
12.4.2 Picosecond Pulse Pumped Supercontinua 588
12.4.3 CW Pumped Supercontinua 594
12.5 Application and Future Developments 598

13 Novel Sources for Optical Coherence Tomography Imaging and Nonlinear Optical Microscopy 609
Kenneth Kin-Yip Wong
13.1 Introduction 610
13.2 Background 611
13.2.1 Optical Coherence Tomography 611

13.2.2 Fourier Domain Optical Coherence Tomography (FD-OCT) 612
13.2.3 Alternative OCT Systems 613
13.3 Coherent Anti-Stokes Raman Scattering (CARS) Systems 614
13.3.1 Fiber-Based CARS Systems 615
13.4 Enabling Optical Amplifier Technologies 615
13.4.1 Erbium-Doped Fiber Amplifier (EDFA) 616
13.4.2 Fiber Raman Amplifier (FRA) 618
13.4.3 Semiconductor Optical Amplifiers (SOAs) 620
13.4.4 OPAs as Amplifiers 622
13.4.4.1 Fiber-optical parametric oscillator (FOPO) 626
13.5 Recent Research Efforts 628
13.5.1 OCT Applications 628
13.5.1.1 Fourier domain mode locking (FDML) 628
13.5.1.2 Hybrid FDML 629
13.5.1.3 Swept-pump FOPO/FOPA 632
13.5.1.4 FOPA in phase-conjugated OCT (PC-OCT) 633
13.5.2 CARS Applications 633

14 Optical Scanning Holography and Sectional Image Reconstruction for Biophotonics 649
Edmund Y. Lam
14.1 Digital Holographic Microscopy 650
14.2 Optical Scanning Holography 651
14.2.1 The Physical System 652
14.2.2 Mathematical Representation 653
14.2.3 Diffraction Tomography 656
14.3 Sectional Image Reconstruction 659
14.3.1 Identifying the Impulse Response 659
14.3.2 Conventional Method in Sectioning 659
14.3.3 Sectioning Using the Wiener Filter 660
14.3.4 Sectioning Using the Wigner Distribution 661
14.3.5 Sectioning Using a Random-Phase Pupil 661
14.3.6 Sectioning Using Inverse Imaging 662

14.4 Experiments 664
14.4.1 Reconstruction Experiment with a Synthetic Object 664
14.4.2 Reconstruction Experiment with a Biological Specimen 665
14.5 Concluding Remarks 668

15 Subcellular Optical Nanosurgery of Chromosomes 673
Linda Shi, Veronica Gomez-Godinez, Norman Baker, and Michael W. Berns
15.1 Introduction 674
15.2 Robotic Laser Microscope Systems Design 675
15.2.1 Robotic Laser Microscope Using a Picosecond Green Laser (RoboLase II) 675
15.2.2 Robotic Laser Microscope with a Femtosecond Near-Infrared (NIR) Laser 678
15.3 Chromosome Studies 680
15.3.1 DNA Damage Responses of Mitotic Chromosomes 680
15.3.2 Chromosome Tips (Telomeres) Regulate Cytokinesis 685
15.4 Conclusions 689

16 Optical Transfection of Mammalian Cells 693
Maciej Antkowiak, Kishan Dholakia, and Frank Gunn-Moore
16.1 Introduction 693
16.2 Biophysical Mechanisms of Laser-Mediated Membrane Permeabilization and Cell Transfection 694
16.2.1 Laser-Mediated Membrane Poration 694
16.2.2 Mechanisms of Optoinjection and Phototransfection 696
16.3 Optical Transfection with Various Types of Lasers 700
16.3.1 Continuous-Wave Lasers 703
16.3.2 Nanosecond Pulsed Lasers 705
16.3.3 Femtosecond Pulsed Lasers 708
16.3.4 Nanoparticle-Mediated Techniques 710

16.4 Technical Advancements Toward High-Throughput, in vivo, and Clinical Applications 713
16.4.1 Reliability, Ease of Use, and Throughput 713
16.4.2 Towards in vivo Applications 718
16.5 Conclusions 723

Index 735

Preface

Here is what I told a five-year-old kid last summer when he asked me what I do for a living: "I teach and play tricks with light in the university for better health." He stared at me with a puzzled look in his eyes and said, "You are not a doctor, right? But it sounds like fun!" Yes indeed. This is precisely the "fun" part since I have regarded myself working in the field of biophotonics. While it does not have a rigorous definition, *biophotonics* is generally conceived to bear a fundamental concept: to understand and manipulate how light interacts with biological matter, from molecules and cells to tissues, and even whole organisms. Looking at the tip of the iceberg, we can see that light can now be used to probe biomolecular events with impressively high sensitivity and specificity, such as gene expression and protein–protein interaction One can also visualize the spatial and temporal distribution of biochemical constituents with light and thus the corresponding physiological dynamics in living cells, tissues, and organisms in real time. To make it more fun, light can be utilized to alter the properties and behaviors of biological matter, such as to damage cancerous cells by laser surgery or therapy. Light can also be used to manipulate the neuronal signaling in a brain network. Indeed, fueled by the innovations in photonic technologies in the past half-century (particularly the laser development), biophotonics continues to play a ubiquitous role in revolutionizing the basic life science studies as well as biomedical diagnostics and therapies.

Biophotonics always requires integration of knowledge from multiple disciplines to solve a specific biomedical or life science problem. To correct the kid's question, certainly not only relying on the expertise of the (medical) doctors, biophotonics very often entails synergism among physicists, biologists, chemists, engineers, and healthcare professionals. Such diversity, however, often sets a barrier that impedes knowledge exchange across disciplines and thus the development of the field. It is not uncommon to see a baffled look in the eyes of a medical doctor (the same look that the kid gave me) when an engineer introduces and explains a brand new technology with engineering jargon, or likewise to get the same perplexed stare from the engineer when a biomedical scientist describes the pressing difficulties in his or her research using the language of biology.

To be better aligned with the momentous efforts to advance current biophotonic technologies, particularly cellular and molecular optical imaging, for translational research and clinical practices, it is always essential for a biomedical researcher to grasp the essence of the state-of-the-art technologies. Likewise, it is equally crucial for an engineer to understand the actual biomedical problems to be solved. Indeed, closing this knowledge gap now becomes more important than ever, particularly because of the booming interest in emerging technologies, such as laser nanosurgery and nanoplasmonics, which could create new insights for understanding, monitoring, and even curing diseases on the molecular basis—thereby possibly setting up the next high tide.

The past two decades have witnessed an assortment of comprehensive references and handbooks of biophotonics. In view of this literature, this book, instead of being exhaustive, particularly serves as a condensed reference that presents the essential basics of optics and biophotonics to the newcomers from either the engineering or biomedical discipline (senior undergraduates or postgraduate researchers) who are interested in this multidisciplinary field. With contributions from leading experts, the book also highlights the major current advancements in preclinical diagnostics using optical microscopy and spectroscopy, including multiphoton microscopy, super-resolution microscopy, and endomicroscopy. It also introduces a number of emerging techniques and toolsets for biophotonics applications, such as nanoplasmonics, microresonators for molecular detection, and subcellular optical nanosurgery.

I would like to sincerely express my gratitude to all the contributing authors for their excellent work and patience over the course of chapter preparation and editing. And I would also like to give my special thanks to my research postgraduate students, particularly Terence Wong and Andy Lau, who spent a great deal of time in preparing the professional drawings of the figures and in the editing of the book—particularly in Chapters 1 and 2, one of the key features of the book. Although the field is fast changing, I hope that the book, coming in this International Year of Light, will serve as a good trigger or a pointer for the readers to embrace the continuing fun of biophotonics and contribute to the field in the future.

Kevin K. Tsia
Spring 2015

Chapter 1

Essential Basics of Photonics

Kevin K. Tsia
Department of Electrical and Electronic Engineering,
The University of Hong Kong, Pokfulam Road, Hong Kong
tsia@hku.hk

1.1 Introduction

Biophotonics entails how we play tricks with a myriad of light properties in order to help perform different diagnostic and therapeutic tasks in basic life science research and clinical applications. Before we go into learning how to make good use of light as a versatile tool in biophotonics, understanding *what light is* and *how light interacts with matter* should be regarded and emphasized as the foundation of biophotonics. While there are numerous comprehensive references and handbooks of optics/photonics [e.g., 1–5], this chapter of the book is geared toward highlighting the fundamental concepts that are most relevant to biophotonics. Hence, it serves as a condensed review for the readers who already have a background in photonics. On the other hand, this chapter, without tedious mathematics, can help newcomers to photonics to grasp the most essential concepts, which should be sufficient to comprehend the overall picture of

Understanding Biophotonics: Fundamentals, Advances, and Applications
Edited by Kevin K. Tsia

ISBN 978-981-4411-77-6 (Hardcover), 978-981-4411-78-3 (eBook)
www.panstanford.com

biophotonics. The next chapter covers the essential concepts of light–matter interactions.

Generally speaking, different optical phenomena are often described by three main models in the field of optics/photonics. They are

(i) *Wave optics.* Light is treated as a form of waves propagating in a medium (e.g., air, glass, skin tissues). It can be used to explain many common light phenomena, such as polarization, interference, and diffraction.
(ii) *Quantum optics.* Light is treated as moving discrete particles (quanta), named *photons.* This quantum approach is particularly useful to explain light–matter interaction, light generation, photo-detection, and nonlinear optical effects.
(iii) *Ray (geometrical) optics.* Light is treated as rays/lines. It is regarded as an approximate theory of light that applies only when the wavelength of light is significantly smaller than the dimensions of the apparatuses in the system. It is a common tool for studying imaging problems, e.g., how an illuminated object can be seen by our eyes or be captured by a camera.

On many occasions, the choice of these three models depends on whether the model is adequate (and easy) to explain the phenomena involved in a particular situation. For instance, only the quantum approach can explain the working principle of laser, while both ray optics and wave optics fail to do so. Another example is to explain the imaging principle of an optical microscope. We can simply use the ray optics approach to trace the light rays in the microscope and hence to find out the final image position. However, the image resolution particularly cannot be fully evaluated by this approach and is mostly explained by the wave optics approach, namely diffraction theory (Fig. 1.1), which will be introduced in a later section. Hence, practical imaging problem incorporates both models most of the time. Table 1.1 shows the adequacy of these models in explaining some optical phenomena and related problems.

As discussed in the later sections, the study of light propagation in biological tissues also does not rely on merely a single model. Specifically, the transport dynamics of light in tissue is usually modeled in a form of propagating waves. In addition, the *photon*

Table 1.1 Adequacy of the three main models—ray optics, wave optics, and quantum optics—to explain the common optical phenomena. Being an approximate theory, ray optics is mostly sufficient in explaining the simple optical effects in which only the geometry of the light path is relevant, e.g., reflection, refraction, and total internal reflection (which can partially explain the waveguiding effect). Wave optics in general is adequate to explain most of the light propagation characteristics (e.g., interference, diffraction), except light generation and detection, which can only be fully explained by quantum optics (i.e., considering how the atoms/molecules interact with photon)

Optical phenomena	Ray optics	Wave optics	Quantum optics
Reflection and refraction	√	√	√
Diffraction and interference	×	√	√
Nonlinear optics	×	√	√
Optical waveguiding (optical fibers)	√/×	√	√
Photo-detection	×	×	√
Light generation (e.g., lasers, LEDs)	×	×	√

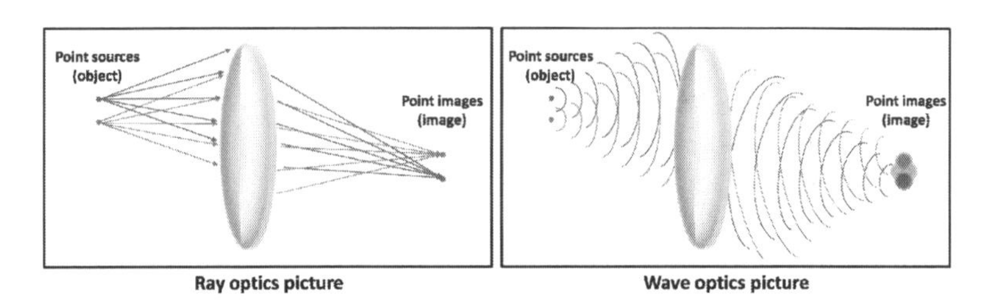

Figure 1.1 Two different models describing a simple imaging problem of two point sources: (left) ray optics and (right) wave optics. Note that ideal *point source* is physically unattainable. Nevertheless, it can in practice be viewed as the light radiation from a single molecule or nanometer-sized particle. In ray optics, the light rays adequately outline the image-formation trajectories, which help locate the image position (see the figure on the left). In wave optics, not only the image position can be located, but the image quality can also be explained. In this example (figure on the right), the final image of the two closely packed point sources would appear as two blurred spots—this is an effect of diffraction (a wave phenomenon).

picture typically comes into play in an ad hoc manner to explain light absorption, fluorescence, and nonlinear light scattering.

1.2 Wave Nature of Light

The wave nature of light was first formally argued by Christiaan Huygens in the 17th century. The wave theory had, however, been overlooked until the early 19th century, when Thomas Young observed the light interference phenomenon by a double-slit experiment and Augustin-Jean Fresnel presented a rigorous wave theory treatment to explain the phenomena of diffraction and interference. Later, James Clerk Maxwell, one of the greatest physicists of all time, put together the prior fundamental concepts of electricity and magnetism, which were once thought to be unrelated, and elegantly reformulated them into a set of four equations—now known as the celebrated *Maxwell's equations*—revealing the interrelationship between the electric and magnetic fields. Even more intriguingly, he was able to show, from the four equations, that the time-varying electric and magnetic fields propagate in a medium in a form of a wave, called the *electromagnetic wave* (Fig. 1.2). The most significant finding is that the speed of the electromagnetic wave evaluated from this theory is consistent with the measured speed of light—a concrete proof to support the wave theory of light!

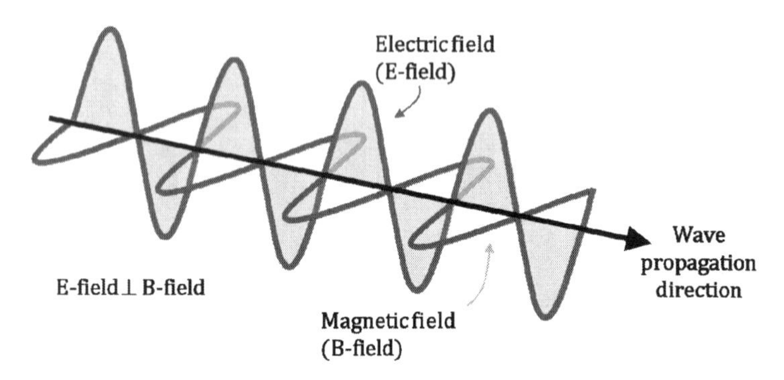

Figure 1.2 Simple representation of the light wave, i.e., electromagnetic (EM) wave, which consists of two orthogonal oscillating components: (i) electric field (**E** field), and (ii) magnetic field (**B** field).

Heinrich Rudolf Hertz later experimentally verified the existence of electromagnetic waves in 1888, eight years after Maxwell died.

1.2.1 *Electromagnetic (EM) Waves*

Light can be described as a form of wave traveling at the speed of light $c = 3 \times 10^8$ m/s in vacuum. In other media, such as water, glass, or biological tissues, the speed of light is different. Fundamentally, the speed of light depends on how the atoms or molecules in the medium interact with light. It can be characterized by refractive index (n), which is an intrinsic and dimensionless parameter of a medium, defined as the ratio of the speed of light in vacuum (c) to that in the medium (υ):

$$n = \frac{c}{\upsilon} \tag{1.1}$$

Table 1.2 shows the refractive indexes of a list of common materials, including various biological tissues and tissue components.

In its simplest form, the light wave can be viewed as a sinusoidal wave (harmonic wave) that is of infinite extent in space (Fig. 1.2). For a wave propagating at a speed of υ along the z direction, the spatial

Table 1.2 Refractive indexes of different materials, including biological tissues. Note that the refractive indexes correspond to the values measured in the visible light range [2, 6]

Materials	Refractive index
Free space (vacuum)	1 (exactly)
Air (room temp, 1 atm)	1.000293
Water	1.33
Cornea (human)	1.376
Aqueous humor	1.336
Enamel (teeth)	1.62
Gray matter	1.36
Glass	1.52
Salt (NaCl)	1.54
Sapphire	1.78
Gallium phosphide (GaP)	3.5
Gallium arsenide (GaAs)	3.9
Silicon	4.2

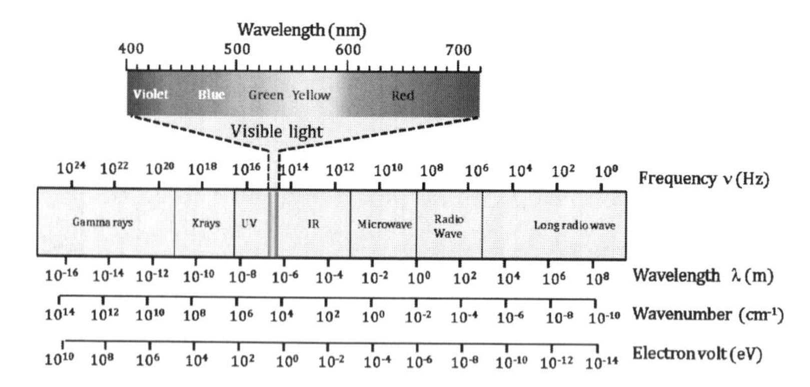

Figure 1.3 Electromagentic (EM) spectrum. Four different units are commonly adopted to specify the frequency (Hz, or cm^{-1}), wavelength (m), or the photon energy (eV).

period is known as wavelength λ (unit: m), whereas the temporal period T (unit: s) is simply related to wavelength by $T = \lambda/\upsilon$. The frequency of the wave (unit: s^{-1} or Hz) equals to the inverse of period, i.e., $f = 1/T = \upsilon/\lambda$. The amplitude of the wave relates to the power (in watts, W), energy (in joule, J), or intensity (W/m^2) of the light (discussed in Section 1.10.2).

1.2.1.1 EM spectrum

According to Maxwell's equations, light is an electromagnetic field consisting of the oscillating electric (**E**) and magnetic (**B**) fields, which altogether propagate as a wave, called electromagnetic (EM) wave, at the speed of light. In general, an EM wave is categorized into different classes according to the wavelength, frequency, wavenumber (a reciprocal of the wavelength [unit: m^{-1}]) or photon energy (unit eV, electron-volt) in the EM spectrum (Fig. 1.3). The wavelength span of the EM spectrum is impressively wide-ranging, from ~1 pm (10^{-12} m), which corresponds to gamma ray, all the way to ~1 km (10^3 m), which belongs to long radio waves. In general, light is loosely defined to cover the wavelength range from ultraviolet (UV: 200–400 nm), visible (400–700 nm), near-infrared (NIR: 700–2000 nm) to mid- and far-infrared (10–100 μm). This is regarded as the *optical wave* regime. It will be discussed in Chapter

2 that light with different wavelengths in the EM spectrum can interact with matter in totally different manners. In the other words, the optical properties of any materials, including biomolecules, cells, and tissues, can vary drastically at different wavelengths. This leads to the fact that light or laser sources with different wavelengths can have their unique applications. For instance, NIR light (e.g., neodymium-doped yttrium aluminum garnet [Nd:YAG] laser, titanium sapphire [Ti:S] laser) is generally favorable for deep-tissue imaging, whereas mid-IR lasers (e.g., CO_2 laser at 10.6 μm, or erbium-doped yttrium aluminium garnet [Er:YAG] laser at 2.94 μm) have been used for skin resurfacing or other therapeutic treatments. Of course, choices of light source depend on not only the wavelength but also other parameters such as light power.

It would be useful here to clarify the common practice of using the wavelength or frequency unit. In mid-IR and NIR, wavelength is defined in micrometers (μm, 10^{-6} m). In some applications, especially in IR spectroscopy, it is common to use the wavenumber in cm^{-1} to specify the wavelength (frequency) of light. In UV and visible light, one usually uses nanometers (nm, 10^{-9}) to define wavelength. On some occasions, light can be described in the unit of photon energy (in eV), which will be discussed later.

1.2.1.2 Basic formalism

As described earlier, the light wave (i.e., EM wave) can be modeled as a simple harmonic (sinusoidal) function. Indeed, it is one possible solution of the wave equation derived from Maxwell's equations. More precisely, Maxwell's equations tell us that the EM wave consists of two oscillating (e.g., sinusoidal) fields (i.e., **E** field and **B** field), which are always perpendicular to each other during the propagation, as shown in Fig. 1.2. Considering that the EM wave propagates along the z direction (Fig. 1.4), the time-varying **E** field can be expressed as

$$\mathbf{E}(z, t) = \mathbf{E}_0 \cos(kz - \omega t) \tag{1.2}$$

where ω is the angular frequency $\omega = 2\pi\nu$, in which ν is frequency. Equation 1.2 is often called the *wave function*. $\mathbf{E}_0$ is the amplitude of the electric field, which is a *vector* quantity specifying its oscillation

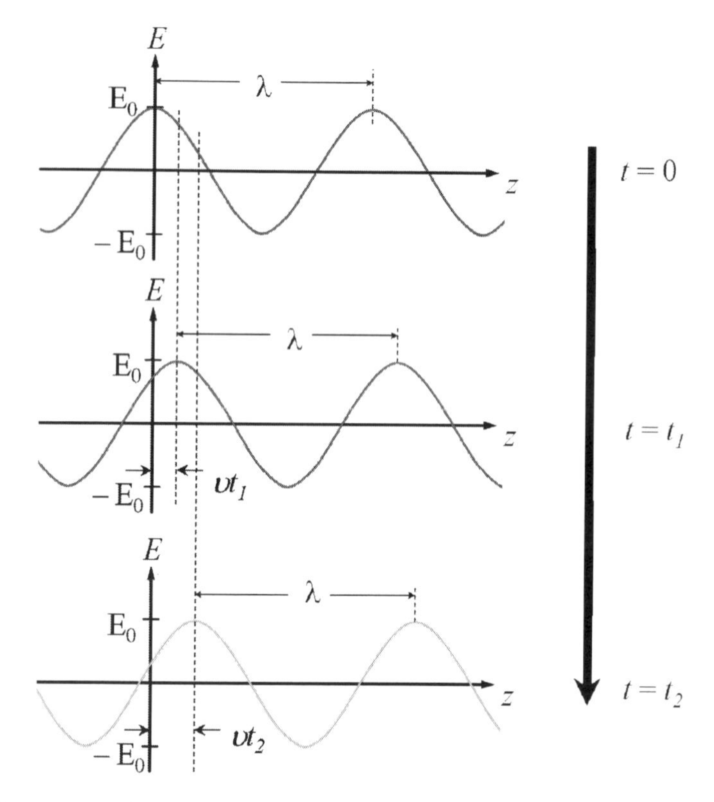

Figure 1.4 The propagation evolution of the wavefunction (Eq. 1.2), showing the series of snapshots of the **E** field at different time points: $t = 0$ (top), $t = t_1$ (middle), and $t = t_2$ (bottom), where $t_1 < t_2$. The speed of the wave is $\upsilon = c/n = \omega/k$ (see Eq. 1.3).

direction, or the *polarization* direction of light (will be discussed in the next section).

k is an imperative parameter, called *k number or wavenumber*. It is defined as

$$k = \frac{2\pi n}{\lambda} = \frac{\omega n}{c}. \tag{1.3}$$

In general, this is a vector quantity as well that specifies the propagation direction of the EM wave. It is called *wave-vector* or **k** *vector*. For any arbitrary propagation directions, the wave function in Eq. 1.2 can be rewritten as

$$\mathbf{E}(\mathbf{r}, t) = \mathbf{E}_0 \cos(\mathbf{k} \cdot \mathbf{r} - \omega t) \tag{1.4}$$

where $\mathbf{r} = x\hat{\mathbf{i}} + y\hat{\mathbf{j}} + z\hat{\mathbf{k}}$ is the position vector and $\hat{\mathbf{i}}$, $\hat{\mathbf{j}}$, and $\hat{\mathbf{k}}$ are the three orthogonal unit vectors in the Cartesian coordinate system. Note that the magnitude of $\mathbf{k}$, i.e., $|\mathbf{k}| = k = 2\pi n/\lambda$ (i.e., Eq. 1.4).

In most cases, it is mathematically convenient to express the wave function in a complex representation, rather than a real-quantity representation (Eqs. 1.2 and 1.4), i.e.,

$$E(\mathbf{r}, t) = \mathbf{E}_0 e^{+i(\mathbf{k}\cdot\mathbf{r}-\omega t)}. \tag{1.5}$$

We can see that the relationship between the real field (Eq. 1.4) and the complex field representation (Eq. 1.5) can be expressed as follows:

$$\begin{aligned}\mathbf{E}(\mathbf{r}, t) &= \frac{1}{2}\left(E(\mathbf{r}, t) + E^*(\mathbf{r}, t)\right) \\ &= \frac{1}{2}\left(\mathbf{E}_0 e^{+i(\mathbf{k}\cdot\mathbf{r}-\omega t)} + \mathbf{E}_0^* e^{-i(\mathbf{k}\cdot\mathbf{r}-\omega t)}\right) = \frac{1}{2}\left(\mathbf{E}_0 e^{i(\mathbf{k}\cdot\mathbf{r}-\omega t)} + \text{c.c.}\right)\end{aligned} \tag{1.6}$$

where c.c. is complex conjugate. In fact, Eq. 1.5 can be regarded as a *plane wave* in a three-dimensional space—a wave with constant amplitude (hence constant phase, i.e., $(\mathbf{k}\cdot\mathbf{r} - \omega t) = \text{constant}$) along a plane at a particular time instant, oscillating sinusoidally in a direction perpendicular to the propagation direction (i.e., $\mathbf{k}$ vector) (Fig. 1.5). This is a commonly used model to describe the propagation characteristics of light in many media.

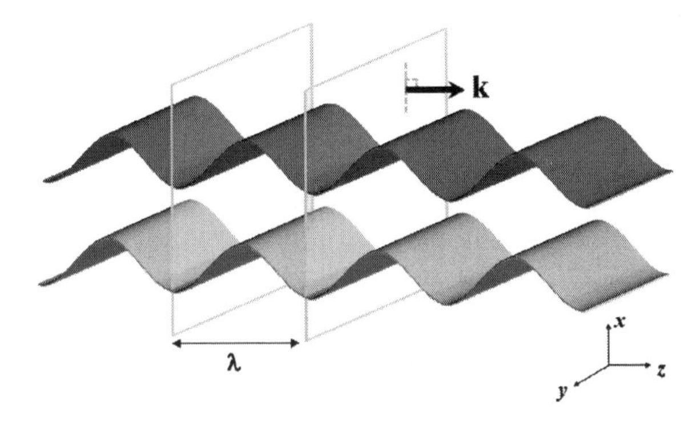

Figure 1.5 Graphical representation of a plane wave propagating in the z direction. The phase (or equivalently the amplitude) of the wave across the entire x–y plane at any z position is constant.

Following the same arguments, the expression of the associated **B** field in a plane EM wave can be written as

$$\mathfrak{B}(\mathbf{r}, t) = \mathbf{B}_0 e^{i(\mathbf{k}\cdot\mathbf{r}-\omega t)} \tag{1.7}$$

where $\mathbf{B}_0$ is the amplitude of the **B** field. The magnetization due to **B** field vanishes at optical frequencies (typically UV to NIR). As a result, the **B** field in light (EM wave) has in most cases negligible effects on light–matter interactions (except for ferromagnetic materials). Indeed, it should be noted that most interactions between light and biological cells/tissues are primarily determined by the electrical responses of the cells/tissues, called polarization (be cautious that this is *not* light polarization, to be discussed in the next section). This refers to how susceptible the charges in a molecule or an atom are redistributed in response to an external electric field. As a result, we mainly focus on using the **E** field of light to characterize light wave properties and light–matter interactions.

1.2.1.3 Energy, power, and intensity

In practice, the most commonly measurable quantity of light is its *energy* (unit: joule, J), or *power* (unit: watt, W). On the basis of electromagnetic theory, we can represent the energy of an EM wave by a vector quantity specifying the direction and magnitude of the energy flow, which is called Poynting vector

$$\mathbf{S}\,(\mathbf{r}, t) = c^2\varepsilon_0\mathbf{E}_0 \times \mathbf{B}_0 \cos^2\,(\mathbf{k}\cdot\mathbf{r} - \omega t) \tag{1.8}$$

where ε_0 is the permittivity of free space (unit: F/m). Here, **S** specifies the *instantaneous* energy per unit time per unit area, i.e., $\mathrm{W/m^2}$, and its direction is perpendicular to both the **E** field and the **B** field (**S** is proportional to the cross-product of $\boldsymbol{E}$ and $\boldsymbol{B}$). Therefore, the wavefront propagation (**k** vector) is parallel to the energy flow direction (Poynting vector). This condition is, however, not always true, especially in an antisotropic medium, where **k** and **S** are indeed not parallel. While this is out of the scope of this text, interested readers can refer to Refs. [2, 7] for more details.

From Eq. 1.8, **S** is an extremely fast-varying function at optical frequencies. For instance, the oscillation frequency of **S** can be as high as 10^{15}Hz (i.e., a period of femtosecond) for red light ($\lambda = 600$ nm). To date, there is no photo-detector which has high enough

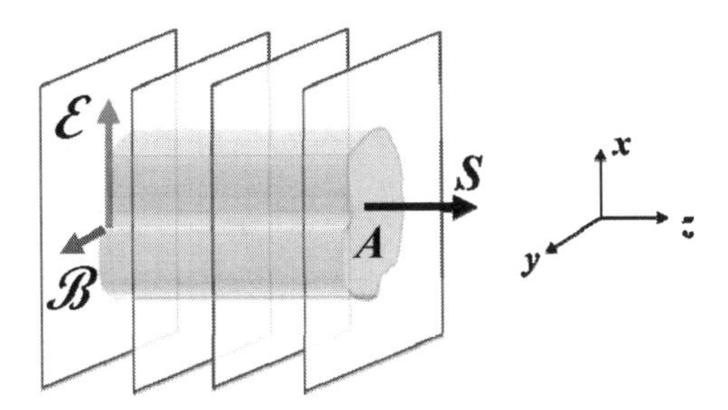

Figure 1.6 Defining optical power of a plane wave, with an optical intensity $I \equiv \langle S \rangle_T$, enclosing an area A. The power is simply $P = IA$. The planes in the figure represent the wavefronts of the plane wave.

speed to *directly* capture the signal with such a rapid oscillation. It is thus more practical to measure an averaged value of **S** over a time interval. On the basis of Eq. 1.7, one can evaluate the time-averaged **S**, i.e., $\langle S \rangle_T$, as follows:

$$\langle S \rangle_T = \frac{1}{T_0} \int_0^{T_0} |\mathbf{S}(\mathbf{r}, t)| dt, \tag{1.9}$$

where T_0 is the averaging time, which is typically much longer than one cycle (or period) of the wave. On the basis of EM theory as well as the fact that the time-averaged $\cos^2$ function is $\left\langle \cos^2 (\mathbf{k} \cdot \mathbf{r} - \omega t) \right\rangle_T = 1/2$, it can be shown that

$$\langle S \rangle_T = \frac{c\varepsilon_0 |\mathbf{E}_0|^2}{2}. \tag{1.10}$$

Eq. 1.10 shows that $\langle S \rangle_T$ is directly proportional to $|\mathbf{E}_0|^2$[a]. In most cases, people term $\langle S \rangle_T$ as average *intensity* (or *irradiance*), denoted as $I \equiv \langle S \rangle_T$. The average power across a given area A (unit: W) can be expressed as

$$P = \int_A \langle S \rangle_T \, dA \tag{1.11}$$

which can simply be represented as $P = IA$ if it is a plane wave with a uniform amplitude across the entire space (Fig. 1.6). Note that in general the intensity in Eq. 1.10 should be expressed as $\langle S \rangle_T = nc\varepsilon_0 |\mathbf{E}_0|^2/2$ in a medium with a refractive index n.

[a] Equation 1.10 is derived by using the relationship $|\mathbf{E}_0| = c|\mathbf{B}_0|$ for a plane wave. For detailed derivation, refer to [1, 2].

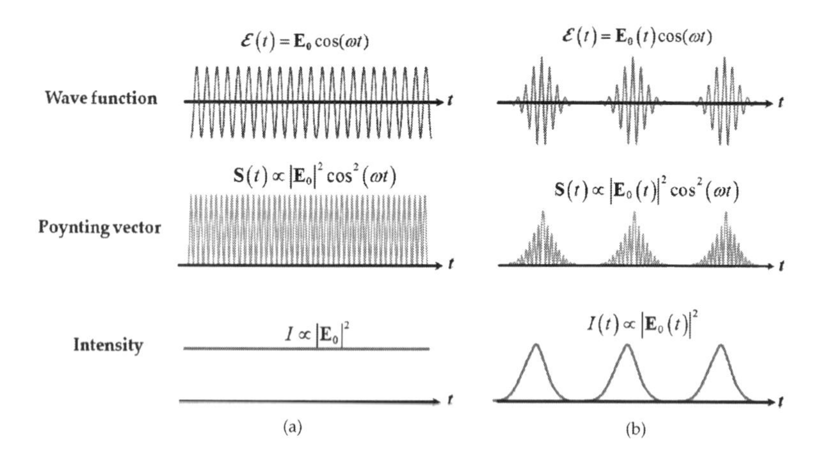

Figure 1.7 Relationship among wavefunction, Poynting vector, and intensity (or power) in (a) the CW and (b) the pulse cases. Note that $|\mathbf{E}_0(t)|$ is the amplitude envelope of the pulse.

The wave function we have considered so far is a continuous wave (CW), which maintains its oscillation with constant amplitude indefinitely (constant $\mathbf{E}_0$). The definitions of power and intensity in this case are straightforward. That is, the time-averaged power and intensity are simply constant regardless the value of the averaging time T_0 (see Eqs. 1.9 and 1.10 and Fig. 1.7a). In contrast, it is worth pointing out another common example: measuring the power and the intensity of a train of light wave pulses (or wave packets). It is particularly relevant to characterize the power of pulsed lasers. In the case of a pulse, the amplitude $\mathbf{E}_0(t)$ is a time-varying quantity instead of a constant. The power or intensity is thus also time-varying. Following Eq. 1.9 by taking the averaging time T_0 longer than one cycle but shorter than the pulse duration, one can indeed show that the power or intensity essentially follows the envelope of the $\mathbf{E}$ field, i.e., $\propto |\mathbf{E}_0(t)|^2$(Fig. 1.7b), similar to Eq. 1.10. Rigorous mathematical treatments of the power and intensity of a pulse can be found in Ref. [2].

Consider a simplified case in which the envelope of the pulse is in rectangular shape with a width τ (Fig. 1.8). Very often, the pulse laser source has a fixed repetition rate R or period $T = 1/R$. In such a case, there are usually *two ways* to describe to power (or intensity)

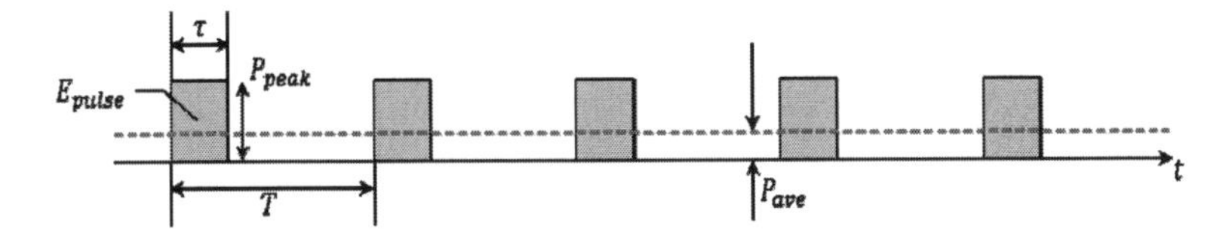

Figure 1.8 Definitions of the peak power (P_{peak}), the average power (P_{ave}), and the pulse energy (E_{pulse}) of a pulse train with pulse width τ and period T.

of the pulses: *average* power P_{ave} (or intensity I_{ave}) and *peak* power P_{peak} (or intensity $I_{\text{peak}} = P_{\text{peak}}/$ area). One can relate P_{ave} with P_{peak} as follows:

$$P_{\text{ave}} = P_{\text{peak}}\tau R = E_{\text{pulse}} R \tag{1.12}$$

where E_{pulse} is the energy of the individual pulse. In practice, the pulses come in different shapes. The representative examples are Gaussian pulses and hyperbolic secant pulses (Fig. 1.9). For these well-defined pulse functions, the pulse widths are typically defined as full width at half maximum (FWHM) of the peak power P_{peak}, as indicated in Fig. 1.9. Although accurate calculation of the pulse energy should take a full integration of the entire pulse, Eq. 1.12 serves as a good and quick estimate in most occasions.

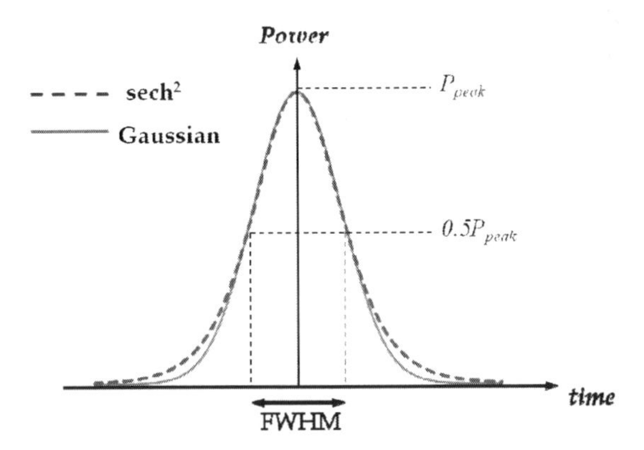

Figure 1.9 Intensity profiles of a Gaussian pulse and a hyperbolic secant sech^2 pulse. Compared with a Gaussian pulse with the same FWHM and same P_{peak}, the sech^2 pulse has stronger wings.

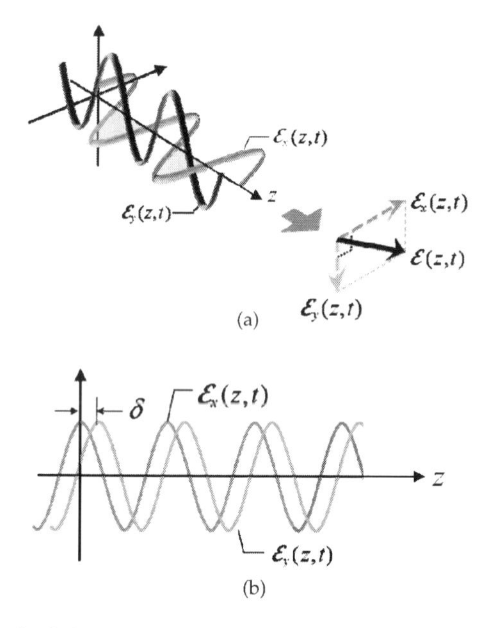

Figure 1.10 The light polarization state can be described by decomposing the **E** field into two orthogonal plane wave components (see (a)). The light polarization essentially describes the **E** field vector sum of the two orthogonal components: $E(z,t) = E_x(z,t) + E_y(z,t)$ at any time and position. Note that, in general, a relative phase difference δ appears between two components (see (b), in which the orthogonal components are overlaid).

1.3 Polarization

The orientation of the **E** field oscillation, i.e., $\mathbf{E}_0$, in a light wave is of great importance in terms of the light propagation characteristics as well as light–matter interaction. If the **E** field of a light wave oscillates in a particular and well-defined orientation, the light is said to be *polarized*. For example, light is said to be *linearly polarized* when its **E** field oscillates along one linear direction only (see Fig. 1.11a). In other words, if the **E** field oscillates randomly in all perpendicular planes with respect to the propagation direction **k**, the light is *randomly polarized* (or *unpolarized*) (see Fig. 1.12).

In general, any polarization orientation (or *polarization state*) can be described by decomposing the **E** field into two orthogonal

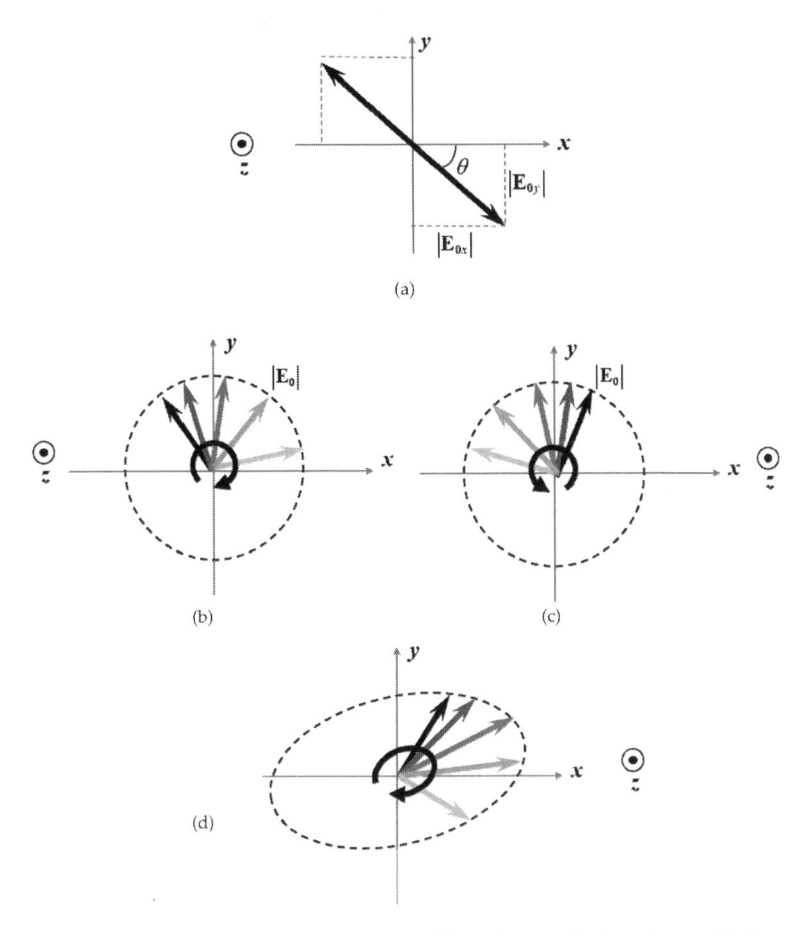

Figure 1.11 (a) Linear polarized light, (b) right-circularly polarized light, (c) left-circularly polarized light, and (d) elliptically polarized light viewed by facing against the propagation direction z (out of paper).

plane wave components in which a relative phase difference δ appears between two waves (Fig. 1.10a). Considering that the light wave is propagating along the z direction, one of the plane wave components, with the **E** field $E_x(z, t)$ oscillating in the x direction, can be expressed as

$$\mathcal{E}_x(z, t) = \mathbf{E}_{0x} e^{i(kz-\omega t)}, \tag{1.13}$$

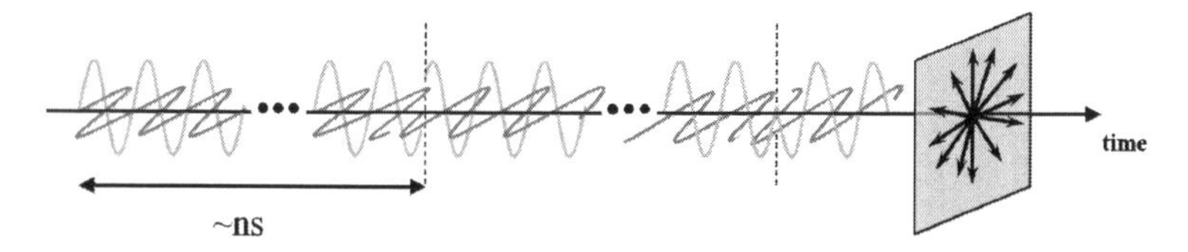

Figure 1.12 Graphical representation of randomly polarized light, e.g., natural light.

whereas another orthogonal component with the **E** field $E_y(z, t)$ oscillating in the y direction can be written as

$$\mathcal{E}_y(z, t) = \mathbf{E}_{0y} e^{i(kz-\omega t+\delta)} \tag{1.14}$$

where $\mathbf{E}_{0x}$ and $\mathbf{E}_{0y}$ are the amplitudes of the x and y components. Hence, the resultant light wave is the vector sum of the two orthogonal components at any position and time (Fig. 1.10):

$$E(z, t) = E_x(z, t) + E_y(z, t). \tag{1.15}$$

The key parameter that determines the polarization state is the phase difference δ. In general, polarization can be classified into three different types:

(i) *Linear polarization* (Fig. 1.11a). When the two orthogonal components are in phase or 180° out of phase, i.e., $\delta = 0$ or $m\pi$ (m is an integer), the light is said to be linearly polarized. The polarization direction makes an angle

$$\theta = \tan^{-1} \frac{|\mathbf{E}_{0y}|}{|\mathbf{E}_{0x}|} \tag{1.16}$$

with respect to the x direction. The resultant amplitude has a magnitude of

$$|\mathbf{E}_0| = \sqrt{|\mathbf{E}_{0x}|^2 + |\mathbf{E}_{0y}|^2} \tag{1.17}$$

(ii) *Circular polarization* (Fig. 1.11b,c). When the two orthogonal components have a quarter-wave relative phase difference, i.e., $\delta = \pm\pi/2 + 2m\pi$ (m is an integer) *and* both of them have the same amplitude, i.e.,$|\mathbf{E}_0| = |\mathbf{E}_{0x}| = |\mathbf{E}_{0y}|$, the light is said to be circularly polarized as the resultant **E** field vector moves in a rotary fashion (in a circular trajectory) when propagating. The rotation can be in a clockwise or anticlockwise depending on the sign of δ Fig. 1.11b,c).

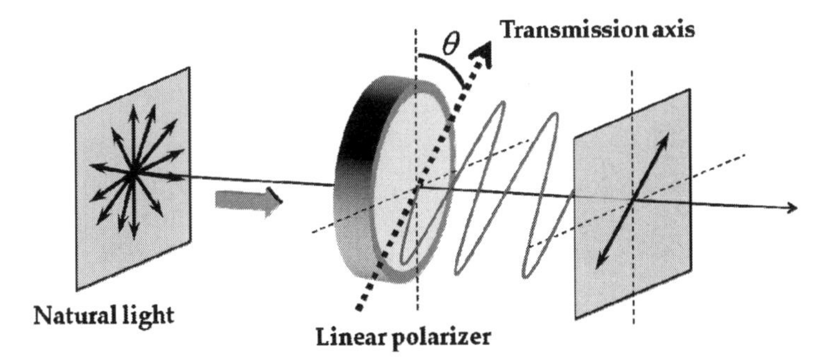

Figure 1.13 A linear polarizer is used to generate a linearly polarized light with the polarization direction aligning with the transmission axis of the polarizer. When a dichroic material is used as linear polarizer, this transmission axis is typically the direction along which the **E** field of the randomly polarized input light has the lowest absorption.

(iii) *Elliptical polarization* (Fig. 1.11d). This is the generalized description of light polarization in which the relative phase difference δ and the individual amplitudes $|\mathbf{E}_{0x}|$ and $|\mathbf{E}_{0y}|$ can be any arbitrary values. The resultant **E** field vector also moves in a rotary manner, but in an elliptical trajectory.

Some light sources generate randomly polarized light. Examples are natural sunlight, light-emitting diodes (LEDs), and incandescent lamps. Randomly polarized light can also be understood in terms of the vector sum of the two orthogonal components. However, the relative phase difference between these two components varies rapidly (~ns) and randomly. The orientation of the resultant **E** field vector, and thus the polarization, cannot be well defined (Fig. 1.12).

In practice, linearly polarized light can be generated by passing a randomly polarized light through a *linear polarizer* (Fig. 1.13), which can be made of a material only selectively absorbing one of the two orthogonal linear polarization states of the incident light—an effect called *dichroism*. One popular example is the polaroid sheet, which can be used for making "3D glasses" for viewing 3D movies [3]. Other materials can also be used to make polarizer, such as birefringent materials (see Section 2.3). In contrast, circularly and elliptically polarized light can be generated by using a device called phase retarder (mostly made of birefringent materials), which can control the phase difference δ between two orthogonal **E** field components. A detailed description can be is given in Section 2.3.

Manipulation of light polarization has been found versatile in many applications in biophotonics. One well-known example is *polarization microscopy*, which employs polarized light for enhancing image contrast of the cellular or tissue structures, as opposed to randomly polarized light, e.g., incandescent light sources [8]. Polarization microscopy is particularly useful for revealing the fine anisotropic structure of a specimen (i.e., with directionally dependent structural or morphological properties). Such anisotropic materials exhibit an optical phenomenon called *birefringence*, which is discussed in Section 2.3.

1.4 Phase Velocity and Group Velocity

Consider the monochromatic (single frequency) light wave described in Eq. 1.5. Its propagation speed (i.e., the speed of light) can be expressed as

$$\upsilon_p = \frac{\omega}{k}. \tag{1.18}$$

This is called *phase velocity*. This is the speed at which the phase of the light wave propagates in space (Fig. 1.14) and is equivalent to the definition stated in Eq. 1.1, i.e., $\upsilon_p = c/n$. And in reality, the refractive index n is a wavelength-dependent parameter—an effect called dispersion. Therefore, phase velocity can be used to correctly describe the speed of light *only if* the wave is monochromatic. Light at different wavelengths would have different phase velocities. It becomes clear when we consider the following situation: if a light source generates a composite wave that has two frequencies ω_1 and ω_2, and thus two wavenumbers k_1 and k_2, what is the speed of light in this case? ω_1/k_1? Or ω_2/k_2? Certainly, neither one is appropriate to define the speed of such a composite light wave. Therefore, another definition of speed of light, called *group velocity*, is used particularly for a light wave with multiple or a band of frequencies. It is expressed as

$$\upsilon_g = \frac{\partial \omega}{\partial k}. \tag{1.19}$$

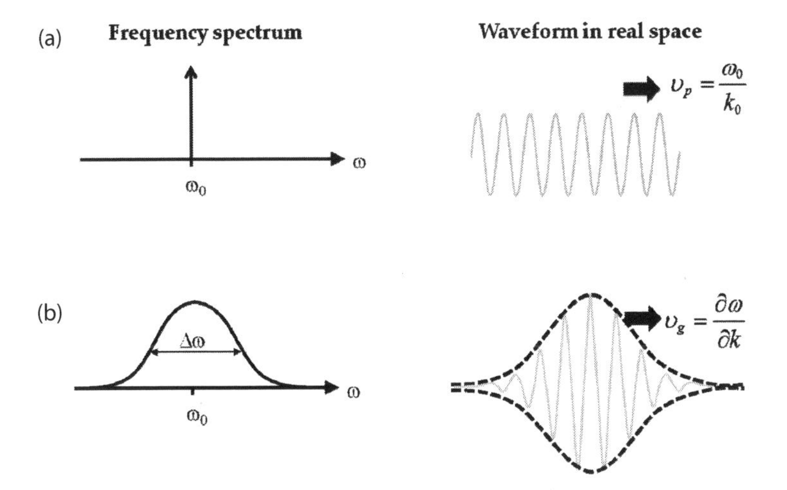

Figure 1.14 Defining (a) the phase velocity of a monochromatic wave and (b) the group velocity of a wave packet (pulse), i.e., the speed at which the pulse envelope is traveling.

The physical interpretation of group velocity can be understood by considering a pulse light source that generates a light wave packet (temporal pulse), as shown in Fig. 1.14. By the *Fourier transform* definition, any wave packet function in time can be represented in its frequency spectrum. In the other words, a wave packet implies that it consists of a band of frequencies in its spectrum with a bandwidth of Δw (see Fig. 1.14b). Group velocity describes the speed of the overall envelope of the wave packet, instead of the speeds of individual frequency components in the spectrum. The concept of group velocity is particularly important for studying the light propagation characteristics in a medium with *dispersion* (discussed in Section 2.5).

1.5 Coherence of Light

So far, we have considered that a light wave can be represented by a wavefunction that has both a well-defined amplitude and a phase in time as well as across a certain space. However, in practice, the light wave indeed possesses some sorts of randomness

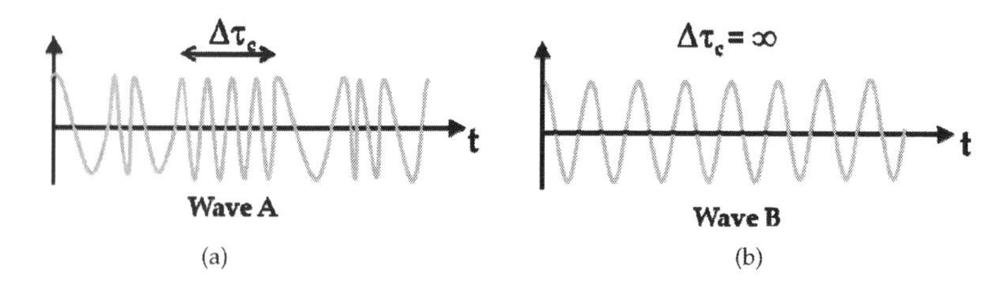

Figure 1.15 (a) A wave with low temporal coherence (wave A). (b) A wave, which is monochromatic, with a *perfect* temporal coherence (wave B).

in its amplitude and phase. The resultant wave, thus, has a stochastic dependence on both time and position that is essentially unpredictable unless statistical analyses are performed [2]. This leads to an important concept called *coherence*, which is used to evaluate the "randomness" of the light wave, or more particularly *the randomness of the phase of the wave*. Coherence of light is typically divided into two classifications: temporal coherence and spatial coherence.

1.5.1 *Temporal Coherence*

Consider the temporal fluctuation of two different waves at a fixed location, as shown in Fig. 1.15. The phase and amplitude of wave A vary more randomly, and thus it is hard to predict their values beyond a certain time period. In contrast, wave B appears to be more "stable" and resembles a well-defined sinusoidal variation. Hence, the time window within which the phase can be predicted for wave B is much wider than for wave A. In this case, wave B is said to have higher temporal coherence. To be more quantitative, the aforementioned "time window" can be defined as *coherence time* ($\Delta\tau_c$), *the temporal interval over which we can reasonably predict the phase of the wave at a given point in space.* The coherence time can be quantitatively evaluated by the autocorrelation function of the wavefunction [2].

Coherence time can also be translated to a coherence length scale (along the propagating direction) simply by

$$\Delta l_c = c\Delta\tau_c. \tag{1.20}$$

This is called coherence length (Δl_c), which is the extent in space (along the propagating direction) over which the wave is "nicely sinusoidal," so that its phase can be predicted reliably.

Again, Fourier analysis tells us that information within any time window can always be represented in its frequency (wavelength) domain. Therefore, the coherence time is always associated with a finite bandwidth in the frequency $\Delta\nu$ spectrum, or equivalently in wavelength spectrum $\Delta\lambda$. They simply follow an inversely proportional relationship:

$$\Delta\nu \approx \frac{1}{\Delta\tau_c}, \tag{1.21}$$

or in terms of wavelength,

$$\Delta\lambda \approx \frac{\lambda^2}{c\Delta\tau_c} = \frac{\lambda^2}{\Delta l_c}. \tag{1.22}$$

Hence, the broader the light source's bandwidth, the shorter its coherence time. Table 1.3 lists some of the representative light sources and their corresponding bandwidths and coherence lengths. Equation 1.21 or 1.22 is of great importance in applications involving interference mechanism, such as optical coherence tomography (OCT) used in ophthalmology and dermatology [9].

1.5.2 *Spatial Coherence*

Instead of considering a fixed location over time, spatial coherence measures how well the phase of the wave can be predicted over a space at a fixed time point. Some literature uses the term *lateral spatial coherence* to differentiate it from the concept of coherence length (along *longitudinal*, i.e., propagation, direction), which is related to temporal coherence. Figure 1.16 shows a comparison between a highly spatially coherent wave and a spatially incoherent

Table 1.3 The bandwidths and the corresponding coherence lengths of typical light sources

Light source	Center wavelength (λ_o)	Bandwidth (Hz)	Δl_c
Filtered sunlight (400–700 nm)	~550 nm	2.98×10^{14}	1000 nm
Near infrared LED	~1 mm	1.5×10^{13}	20 mm
He–Ne gas laser	633 nm	1×10^{6}	400 m

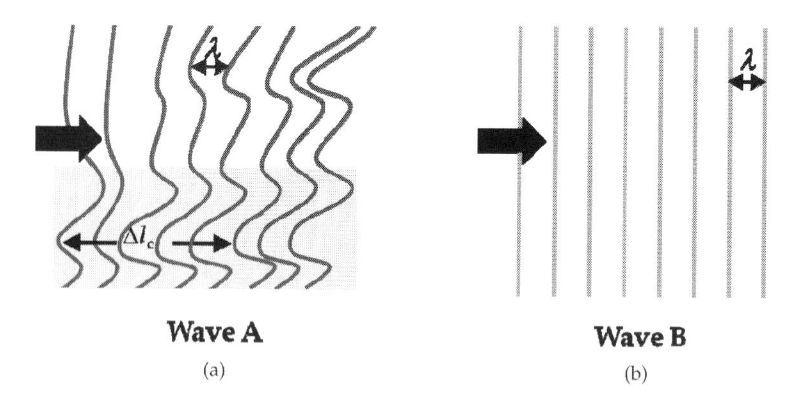

Figure 1.16 (a) A wave showing low spatial coherence (wave A). (b) A wave (plane wave) showing a perfect spatial coherence (wave B). The shaded area in (a) represents the approximate area over which we can reasonably predict the phase of the wave at a given time instant. Note that Δl_c is the coherence length in the longitudinal direction.

wave. From the figure, we can tell that a spatially coherent wave allows us to draw a set of well-defined wavefronts in a space.

Control of light wavefronts is of profound interest in many biomedical applications, such as digital holographic microscopy for quantitative cell imaging [10], and high-resolution deep biological tissue imaging by optical phase conjugation, a technique enabling reconstruction of wavefronts propagated in tissue [11].

The degree of coherence (for both temporal and spatial) can be evaluated by its ability to produce interference fringes—a measure of how well the two (or more) waves can be destructively interfered as well as constructively interfered. The concepts of interference are discussed in the next section.

1.6 Interference

Interference occurs when two or more *(coherent or partially coherent)* waves superimpose together in time and in space. The overall *optical intensity* of such superposition is not simply the sum of the individual waves' intensities, but it can instead be greater or less than the sum, or can even be zero. This effect primarily depends

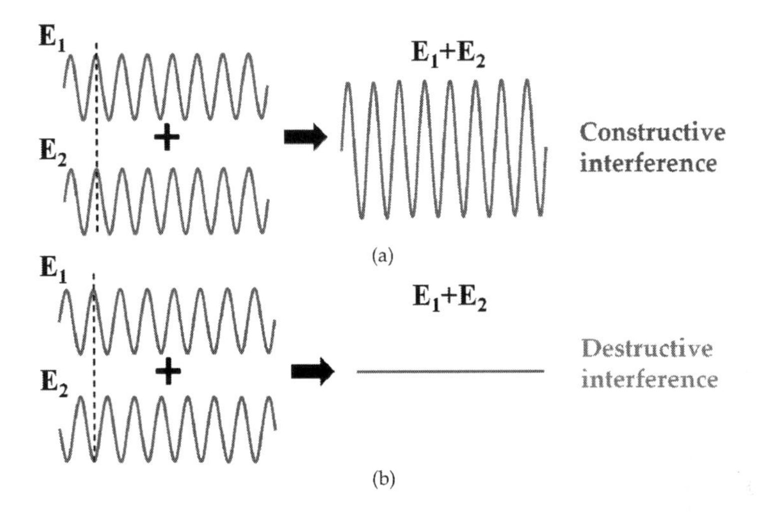

Figure 1.17 (a) Constructive interference. (b) Destructive interference.

on the *phase of the waves*. If two overlapping waves are in phase they are said to be *constructively interfered*, whereas if they are 180° out of phase they are *destructively interfered* (Fig. 1.17).

1.6.1 *Essential Concepts of Interference*

In general, when two waves, with optical intensities I_1 and I_2, respectively, are superposed, the total intensity of the resultant wave I is given by

$$I = I_1 + I_2 + 2\sqrt{I_1 I_2}\cos(\delta) \quad (1.23)$$

where δ is the relative phase difference between the two waves. Note that the dependence of I, I_1, and I_2 on position $\mathbf{r}$ has been skipped for convenience. It is clear from Eq. 1.23 that the total intensity does not necessarily equal to the sum of the individual intensities. The difference is chiefly determined by the interference term (i.e., the third term in Eq. 1.23), which can be positive (constructive interference) or negative (destructive interference). Consider two extreme cases:

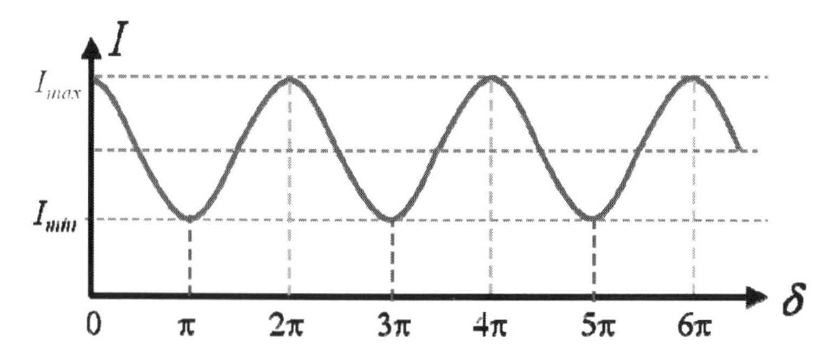

Figure 1.18 Interference profile as a function of the phase difference between the two waves (Eq. 1.23).

- when $\delta = 0$ or in general $2m\pi$ (m is integer), i.e., the waves are in phase, the intensity reaches its maximum: $I = I_{\max} = I_1 + I_2 + 2\sqrt{I_1 I_2}$
- when $\delta = (2m + 1)\pi$, i.e., the waves are 180° out of phase, the intensity comes to a minimum: $I = I_{\min} = I_1 + I_2 - 2\sqrt{I_1 I_2}$

Figure 1.18 shows a plot of the sinusoidal dependence of the overall intensity on the phase difference δ. So, how do we introduce such a phase difference in practice? Or why do we need to generate the phase difference, and hence the interference?

Let us consider two identical waves that have the same frequency and same field amplitude. The only difference between them is that the sources start at different position, so that source A is l_A away from the detector P, where the superposed intensity of the two waves is measured. And source B is l_B away from the detector. The schematic of this example is shown in Fig. 1.19a. From Eq. 1.2, we learned that the phase of a wave is $(kz - \omega t)$. As a result, at a fixed time instant, say $t = 0$ for simplicity, the phase shift of wave from the source A to the detector is $\varphi_A = kl_A$, while the phase shift of wave B is $\varphi_B = kl_B$. Because of the path length difference between wave A and B, i.e., $l_A - l_B$, there is an associated phase difference δ between the two waves, which is given by

$$\delta = \varphi_A - \varphi_B = k(l_A - l_B) \tag{1.24}$$

We can observe from the above equation that the phase difference δ, and thus the interference effect, can be manipulated by adjusting the relative "path length delay," i.e., $l_A - l_B$, between

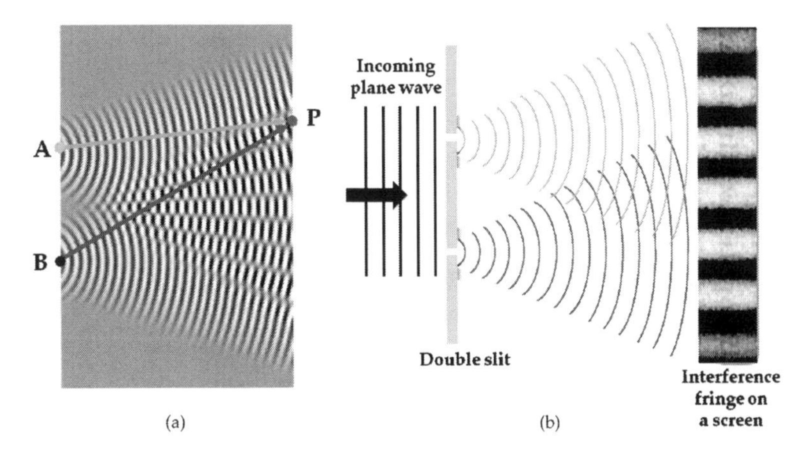

Figure 1.19 (a) Interference of two waves generated from two point sources A and B. The resultant intensity at point P is primarily governed by the path length difference, PA – PB = $l_A - l_B$, which gives rise to the phase difference δ (see Eqs. 1.23 and 1.24). (b) Young's double-slit experiment (essentially similar to (a)). The two point sources can be realized in practice by introducing two tiny slits. An interference fringe pattern is observed in the far field by a viewing screen.

the two waves. It is essentially similar to the well-known Young's double-slit experiment, in which an interference fringe pattern is generated in the far field [1]. It is generated by illuminating a plane wave on two tiny slits, which act like two point sources, such that two different waves emerging from the two slits interfere with each other(Fig. 1.19b).

It should be remembered that $k = nk_0$ is the wavenumber in the given medium (with refractive index n) and k_0 is the free-space wavenumber. Therefore, the phase difference can also be affected by different media in which the light waves propagate. Hence, it should be more precise to define a so-called optical path length (OPL) to take the refractive index of the medium into account. OPL is simply defined by the product of the physical length l and the refractive index n, i.e., OPL $= nl$. As a result, Eq. 1.24 can be rewritten as $\delta = k_0(nl_A - nl_B) = k_0(OPL_A - OPL_B) = k_0\Lambda$, where OPL_A and OPL_B are the OPL of paths A and B, respectively. Λ is defined as the optical path length difference.

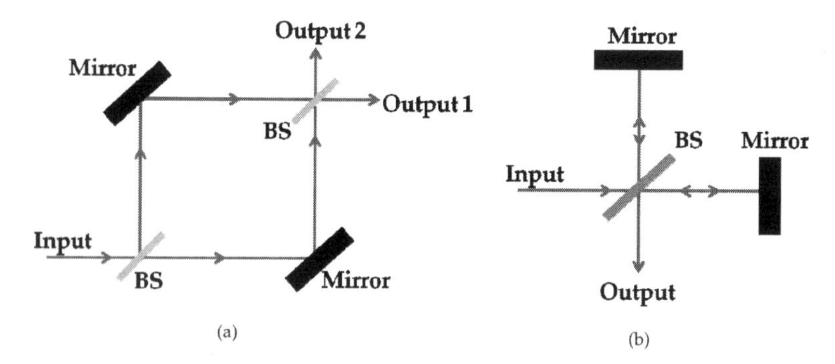

Figure 1.20 (a) Mach–Zehnder interferometer (MZI). (b) Michelson interferometer (MI). Two distinct differences between the two interferometers can be seen: (i) MZI requires two beam splitters (BS), one for splitting the beam into two paths and the other for combining the beams. MI uses only one BS to split the input beam and recombine the beams. (ii) In the MZI, there are two interference outputs (Output 1 and 2), which are separate from the input. This is in contrast to the case of the MI, in which only one of the outputs is separate from the input (going downwards in (b)). Another output, which goes back to the input (not shown), would usually be blocked by an optical isolator [2].

1.6.2 *Interferometers*

In practice, interference is generated by an instrument called *interferometer*, which can come with different configurations depending on the applications. The two most commonly used configurations are (i) the Mach–Zehnder interferometer (Fig. 1.20a) and (ii) the Michelson interferometer (Fig. 1.20b). There is one common feature shared by the two interferometers, and this is indeed the central idea of generating the interference effect in many applications: *splitting the original light wave into two arms followed by recombining them*. The phase difference, and thus the interference effect, can be manipulated by altering the optical path length difference between two arms or introducing different materials in the individual arms (with different n).

1.6.3 *Significance of Phase Difference in Interference*

The phase difference of the interfering waves is the key factor governing the interference effect. In the other words, interference

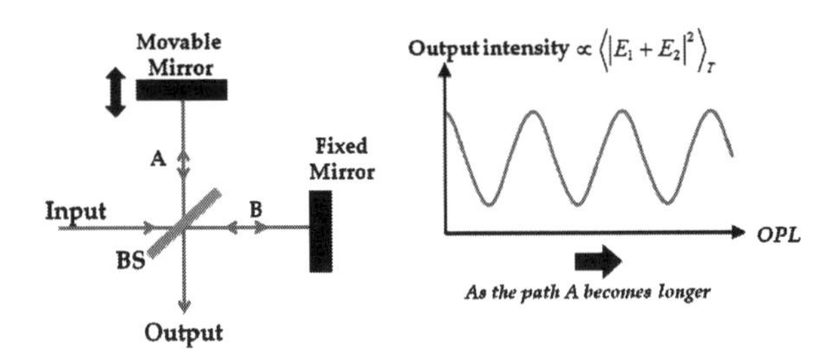

Figure 1.21 An example of using a Michelson interferometer to measure indirectly the field oscillation of a monochromatic input wave. By moving the mirror in path A while keeping the path length of path B fixed (left), an interferogram can be observed at the output (right). This operation is essentially equivalent to measuring the autocorrelation between the two monochromatic waves (E_1 and E_2), from which the field amplitude and oscillation can be evaluated [2]. Such a setup represents a basic configuration for many applications, such as FROG, Fourier transform spectroscopy, and OCT.

can help one extract the phase information of the waves—an intriguing capability that makes interference a versatile tool in numerous applications. As mentioned in Section 1.2.2, there is no photo-detector to date that is able to *directly* measure the field amplitude oscillation, and thus the phase, of the light wave at optical frequencies. To this end, interference serves as an ideal method to *indirectly* measure the phase and field oscillation. The general concept is depicted in Fig. 1.21. Consider that a monochromatic wave is split into two paths (arms) in a Michelson interferometer. The two waves (E_1 in path A and E_2 in path B) are reflected by the mirrors in the individual arms and recombined, or thus interfered, after the beam splitter. Here the detector at the output port is not fast enough to capture the field oscillation and thus instead can only measure the average value of the intensity over a certain time period T (can be nanoseconds for fast detectors or milliseconds for the slower ones), which is proportional to $\langle |E_1 + E_2|^2 \rangle_T$.[a] When one of the path lengths is varied, the relative delay and thus the phase shift between E_1 and E_2 are introduced. As a result,

[a]Mathematically, the time average value can be expressed in $\langle |E_1 + E_2|^2 \rangle_T = \frac{1}{T} \int_T |E_1 + E_2|^2 \, dt$.

the measured average intensity will be changed accordingly and follows a sinusoidal interference fringe pattern as a function of the path length difference (following Eq. 1.23). Such fringe pattern, typically called *interferogram*, reveals the phase information of the ideal monochromatic wave. In fact, it is essentially the central idea of performing an *autocorrelation* (or *optical autocorrelation*) process [2]. It is particularly useful for characterizing the properties of an optical pulse, e.g., the phase, amplitude, pulse width, and frequency spectrum[a] [9]. Note that the setup shown in Fig. 1.21 is oversimplified, especially for the case of characterizing the phase information of the arbitrary pulse shape. In this case, we have to rely on a more sophisticated version of the interferometer plus advanced signal processing to retrieve the complete information of the phase and amplitude of the pulse. The most common technique is called frequency-resolved optical gating (FROG) [12].

In addition, optical autocorrelation based on interferometry also serves as the basic concept for realizing *Fourier transform spectroscopy*, a common tool for metrology and molecule identification [13]. The setup shown in Fig. 1.21 also represents the original configuration of OCT, which is perhaps one of most successful optical tissue imaging modalities being used clinically nowadays, particularly for ophthalmology and intravascular assessment in cardiology [9].

1.6.4 *Relationship between Coherence and Interference*

Interference and coherence are closely related to each other. We can see from Fig. 1.22 that if the light wave is incoherent (i.e., the phase variation is randomized as described in Section 1.2.5), no clear interference fringe will be visible. It applies to both temporally and spatially incoherent light (Figs. 1.15 and 1.16). Indeed, the *fringe visibility* (or the contrast) of the interference pattern can be used as a measure of the degree of coherence and is given by

$$V = \frac{I_{\max} - I_{\min}}{I_{\max} + I_{\min}} \tag{1.25}$$

[a] In fact, the Fourier transform of the field autocorrelation (i.e., the interferogram) is the frequency spectrum of the light wave, as stated by the Wiener–Khinchin theorem [9].

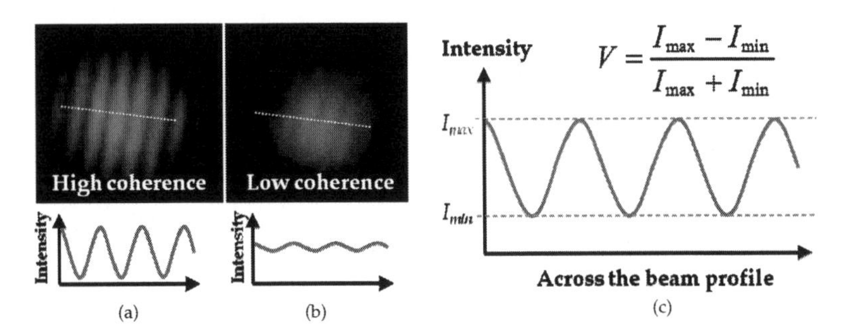

Figure 1.22 Interference patterns generated by spatially overlapping two different light beams: (a) two highly coherent light beams, and (b) two beams with low coherence. The bottom plots show the intensity profiles of the line cuts (dotted lines in the top images). (c) Definition of fringe visibility V.

where I_{max} and I_{min} are the maximum and minimum intensity in the interference fringe pattern. When $V = 1$, the light is perfectly coherent. On the other hand, when $V = 0$, the light is incoherent. For $0 < V < 1$, the light is said to be partially coherent.

The interference effect has been utilized in numerous applications in biophotonics. One well-known example is phase contrast microscopy [8]. A lot of the biological cells are transparent and thus generate very small image contrast under visible light illumination. Phase contrast microscopy makes use of the fact light undergoes different phase shifts as it propagates in different parts of the cells. Therefore, if we could interfere the light with a reference light that does not pass through the cell, the phase shift information can be transformed to intensity variation, enhancing the image contrast (Fig. 1.23). More detailed principles of phase contrast microscopy can be found in Ref. [8]. Other optical imaging modalities based on the concepts of interference are digital holographic microscopy [14] and OCT [9].

1.7 Diffraction

The plane wave is an idealized model that is adequate to explain the general characteristics of the light wave propagation. Nevertheless, unlike the plane waves that infinitely extend in space and have

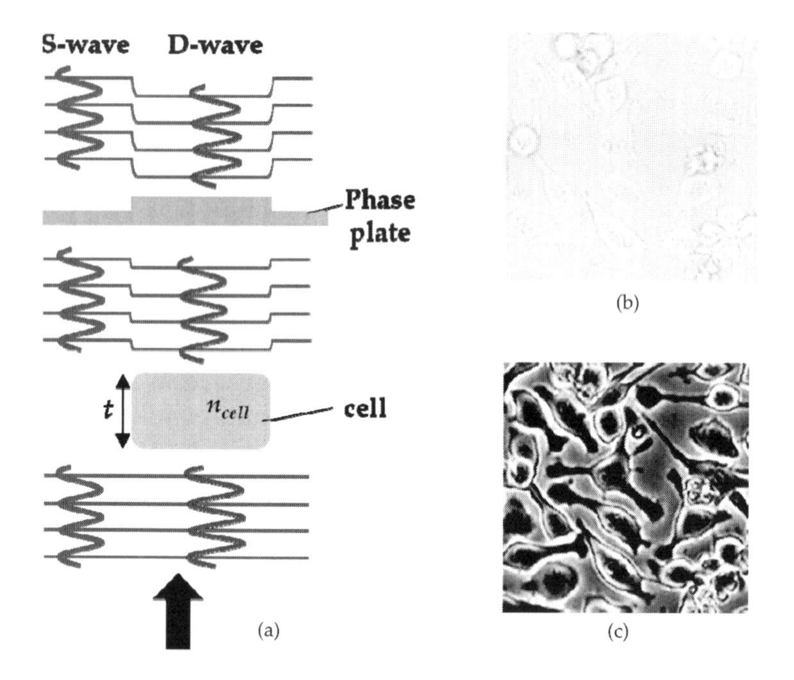

Figure 1.23 (a) A crude picture illustratrating the basic concept of phase-contrast microscopy. We can treat the light wave passing through the biological specimen (e.g., transparent/unstained cell) as two different components: the light wave propagating through the cell is called D wave. The other part of the wave that does not interact with the cell is called S wave. Hence, there is a phase difference between D and S waves, $\delta = k_0 n_{\text{cell}} t - k_0 t$, where n_{cell} and t are the refractive index and the thickness of the cell, respectively. Typically, this phase difference in most of the biological cells is around $\pi/2$ at visible wavelengths. Therefore, if we place a component, called phase plate [8], which introduces an additional $\pi/2$ between the two waves, we can achieve a $\sim \pi$ phase shift. By bringing the two waves together (through focusing the light waves by the lenses) onto the image plane, destructive interference occurs. In the other words, the phase difference information across the cell is now transformed into intensity variation. The image contrast can thus be enhanced. It can be clearly observed in the comparison between the images captured by (b) bright field microscopy and (c) phase-contrast microscopy. (Image source: http://www.microscopyu.com).

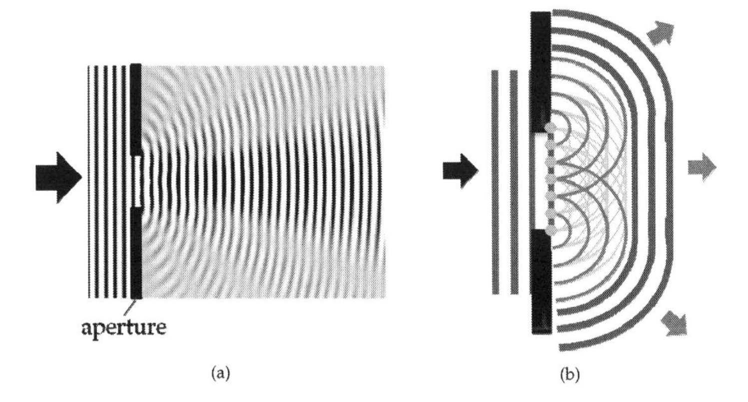

Figure 1.24 (a) Diffraction of a plane wave after it passes through an aperture. (b) Explaining diffraction by the Huygens–Fresnel principle: every point of a wavefront can be treated as a point source of a spherical wave (see the dots at the aperture). The amplitude of the light wave at any point beyond is the superposition of all these point sources.

perfectly flat wavefronts, light in practice can be of *finite spatial extent* and *of any arbitrary wavefont shape*. One example is to use an aperture or obstacle to limit the passage of the light wave (Fig. 1.24a). The unobstructed wave distribution after passing through the aperture tends to spread out and the wavefronts tend to bend around the corner. This phenomenon is called *diffraction.*

Diffraction of the light wave can be well explained by the Huygens–Fresnel principle: *Every point of a wavefront can be treated as a point source of a spherical wave. The amplitude of the light wave at any point beyond is the superposition of all these point sources* (see Fig. 1.24b). Therefore, diffraction can simply be regarded as an interference effect of numerous sources. Indeed, diffraction is fundamentally not different from interference. In general, whenever a wave encounters an obstacle, the emergent wave shows a different wave distribution, called diffraction pattern, which is of immense importance in optical imaging problems.

1.7.1 *Near-Field and Far-Field Diffraction*

Detailed evaluation of the diffraction pattern requires complex and tedious mathematics, and thus is out of scope of this book. Interested

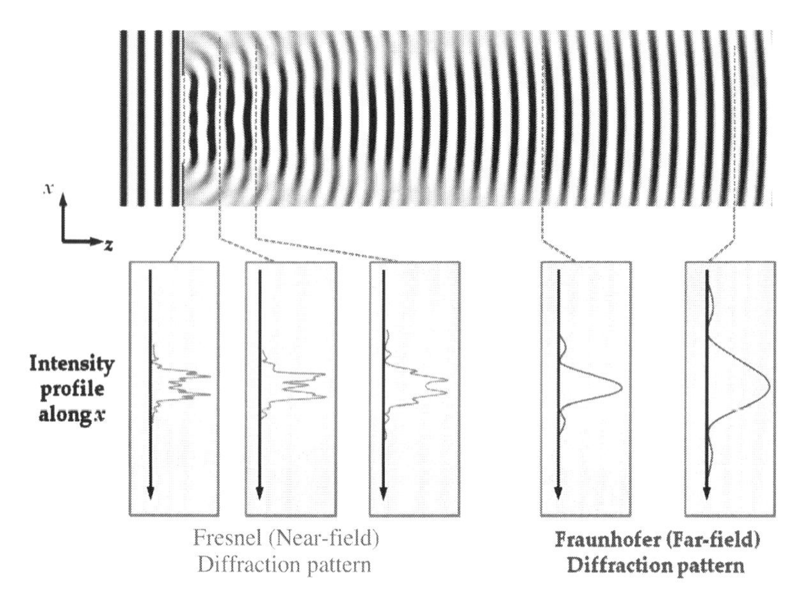

Figure 1.25 Fresnel (near-field) and Fraunhofer (far-field) diffraction patterns of a plane wave incident onto an aperture. The intensity profiles along x at different z locations behind the aperture are also shown.

readers could refer to the references which present the rigorous diffraction theories, e.g., Ref. [15]. The diffraction patterns highly depend on its distance from the aperture, the shape and the size of the aperture, and the wavelength of the light. Nevertheless, we can in general classify two diffraction regimes (Fig. 1.25):

(i) *Near-field diffraction* (or Fresnel diffraction). When the plane of observation is very close to the aperture, or the wavelength of the light is comparable to the size of the aperture, the diffraction pattern exhibits complicated features.

(ii) *Far-field diffraction* (or Fraunhofer diffraction). When the plane of observation is moved further away or the wavelength of the light is much less than the aperture size, the complex structure of the diffraction pattern is smoothened and more spread out.

As a general rule of thumb, the assumption of far-field diffraction is valid as long as the following condition is satisfied:

$$L > \frac{d^2}{\lambda} \tag{1.26}$$

where L is the distance between the aperture (or obstacle) and the plane of observation. d is the aperture size and λ is the wavelength.

1.7.2 *Diffraction and Fourier Optics*

Understanding diffraction often inevitably requires the knowledge of Fourier analysis, a specialized field in optics called *Fourier optics* [15]. The principle of Fourier optics describes that any arbitrary wave in space can always be regarded as a superposition of plane waves propagating along all directions, each of which has a different weighting factor (Fig. 1.26). Each plane wave along a particular direction corresponds to one of the Fourier (or called spatial frequency) components of the original wave. It means that we can always analyze the light wave distribution in either the spatial domain or the spatial frequency domain (spatial frequency directly relates to the plane wave's propagation angle). This is exactly in the same way of analyzing the temporal signal (e.g., sinusoidal signal) in its frequency spectrum. As a result, *space* and *spatial frequency* basically form a Fourier transform pair, similar to the Fourier transform pair of *time* and *frequency*. Except now, the Fourier analysis of light wave distribution can be multi-dimensional, i.e., two-dimensional or even three-dimensional.

The relationship between diffraction and Fourier analysis can readily be appreciated by the fact that the far field diffraction pattern is essentially the Fourier transform of the input light distribution. For examples, when a square aperture is introduced at the input, the far-field diffraction pattern resembles a two-dimensional sinc^2 function (Fig. 1.27(a)), which is the squared modulus of the Fourier transform of a two-dimensional *rect* function,[a] i.e., the shape of the

---

[a]The 2D sinc^2 function is expressed as

$$I_{\text{sqaure}}\left(x', y'\right) \propto \left(\text{sinc}^2 qx'\right)\left(\text{sinc}^2 qy'\right) = \left(\frac{\text{sin}\pi qx'}{\pi qx'}\right)^2\left(\frac{\text{sinc}\pi qy'}{\pi qy'}\right)^2$$

where $q = (\pi D/(\lambda \text{L})$. D is the length of the square side and L is the distance between the aperture and the diffraction pattern. The Airy disk function is, on the other hand, given as

$$I_{\text{Airy}}\left(r\right) \propto \left[\frac{2J_1(2\pi Rr/\lambda L)}{2\pi Rr/\lambda L}\right]^2$$

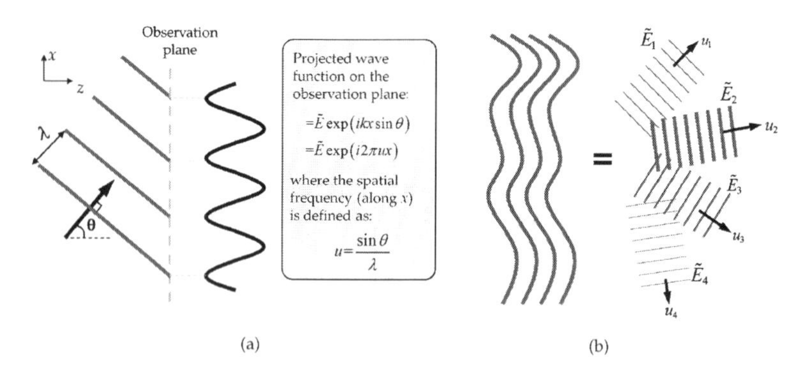

Figure 1.26 (a) A plane wave propagating at the angle θ with respect to the z direction. The spatial frequency of the wave is defined as the frequency of the projected wave function, with a field amplitude $\bar{\mathrm{E}}$, on the observation plane along the x direction, i.e. $u = \sin\theta\lambda$. (b) A simple picture showing the central concept of Fourier optics, i.e., any arbitrary wave can be treated and analyzed as a superposition of plane waves, each of which has its own spatial frequency u_i and field amplitude $\bar{\mathrm{E}}_i$, $i = 1, 2, 3,$ and 4 in the figure.

aperture; a circular aperture gives an *Airy disk* far-field diffraction pattern (Fig. 1.27b), which is indeed the squared modulus of the Fourier transform of a uniform circular disk function, i.e., the shape of the aperture.

Another intriguing phenomenon worth mentioning is the capability of a simple thin convex lens to perform a Fourier transform operation. Again, it can be proved by the diffraction theory that the light field at the back focal plane of the convex lens is essentially the Fourier transform of the light field at the front focal plane (Fig. 1.28a). Based on this observation, we can further notice if this Fourier-transformed light field is located at the front focal plane of a second lens (Fig. 1.28b), we can retrieve back the original (but inverted) light field at the back focal plane of this second lens because of the Fourier transform property—performing the Fourier transform twice on a function $g(x)$ will yield $g(-x)$. This lens-pair system called 4-f system is particularly useful for *spatial frequency filtering* (or simply spatial filtering) [2]. The phenomenon

where $r = \sqrt{x'^2 + y'^2}$ is the radial distance from the origin on the diffraction pattern plane. R is the radius of the circular aperture. J_1 is called *Bessel function* of the first kind of order [15].

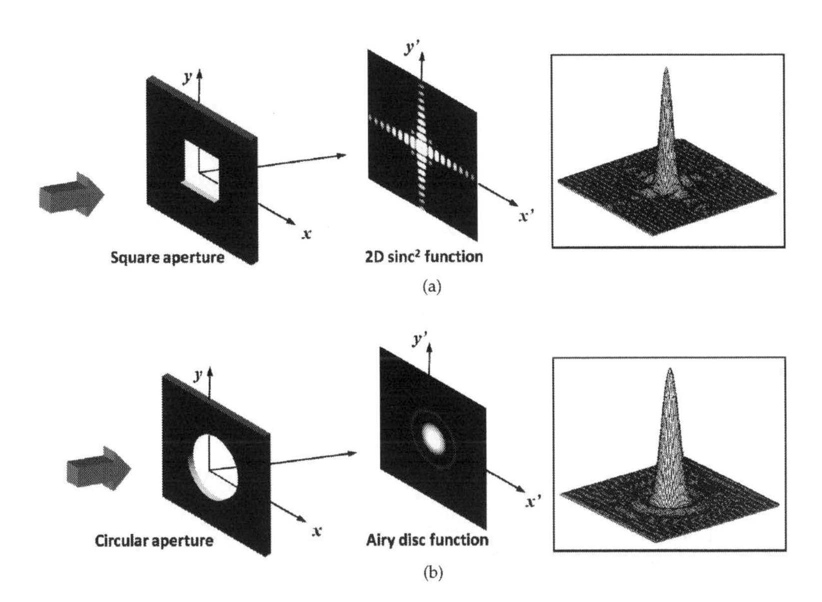

Figure 1.27 Far-field diffraction patterns of two different aperture shapes: (a) square and (b) circular. The square aperture gives the 2D sinc^2 function, which is the Fourier transform of the square *rect* function (i.e., the shape of the square aperture). On the other hand, the circular aperture gives the Airy disc function, which is the Fourier transform of a circular *rect* function. The 3D plots of the two diffraction patterns are shown in the right column.

of diffraction has profound impact on a wide variety of areas in optics. In particular, two representative examples are described as follows: (i) diffraction-limited resolution in optical imaging, and (ii) diffraction grating.

1.7.3 *Resolution in an Optical Imaging System*

In an ideal optical imaging system (e.g., as simple as a thin convex lens), the image is always the *perfect replica* of the object (apart from the image magnification). For instance, a point object always yields a point image using a thin lens (refer back to Fig. 1.1). However, the point object in real case will yield an image of a blurred spot, instead of a perfectly sharp point (Fig. 1.1). Such blurring effect is ultimately attributed to the diffraction of light wave. Specifically, diffraction theory shows that the *blurred spot image* is a Fourier transform

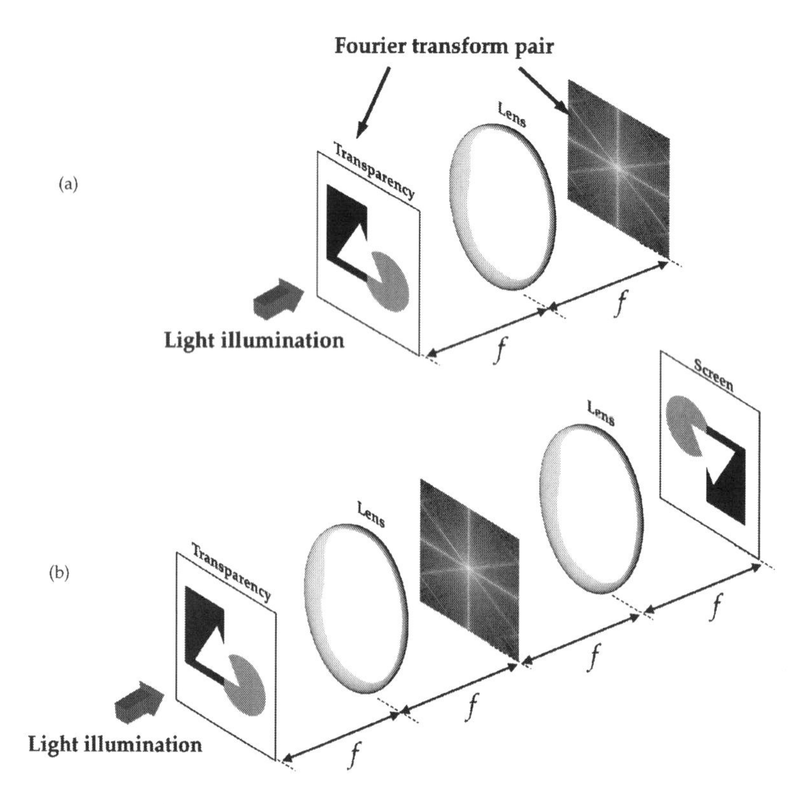

Figure 1.28 (a) A thin convex lens works as a Fourier transformer. The light field pattern (diffraction pattern) at the back focal plane of the lens (i.e., a distance of focal length f behind the lens) forms a Fourier transform pair with the input light field at the front focal plane (i.e., a distance of f in front of the lens). (b) A schematic of a $4f$ system. When the Fourier transformed light field in (a) is located at the front focal plane of a second lens, the original but inverted light field at the back focal plane of the second lens can be retrieved. It is essentially the consequence of the Fourier transform property. Performing the Fourier transform twice on a function $g(x)$ will yield $g(-x)$.

of the aperture function representing the lens. For an example, a circular lens (i.e., can be treated as a circular aperture) will yield a blurred point image resembling an Airy disk pattern (similar to Fig. 1.27). The distribution of the blurred spot can be described by a *point spread function*, an important metric of evaluating the resolution of an imaging system.

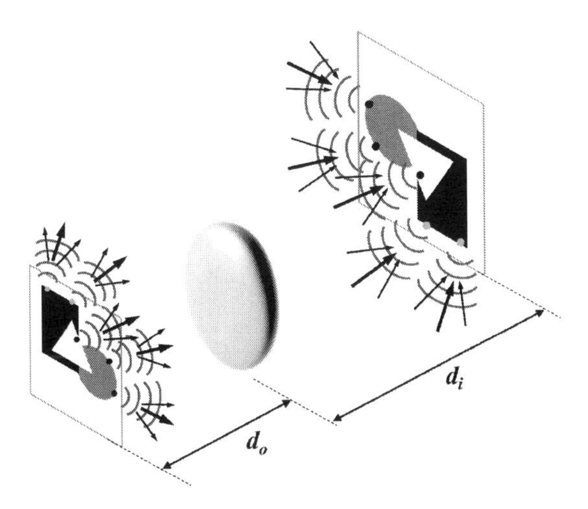

Figure 1.29 The concept of optical imaging. By considering an object (self-illuminating or being illuminated) consisting of numerous point sources, each of which radiates the light wave (the wave can be simply the scattered wave for illuminated object), we can visualize the image by focusing these waves emerging from the point sources to the corresponding points on the image plane. In the case of a simple lens with a focal length f, the relationship between the image distance d_i and the object distance d_o is written as $1/f = 1/d_o + 1/d_i$.

Any object, either self-illuminating or being illuminated, can be regarded as the entity consisting of numerous point sources. An imaging system is to capture the waves emerging from these sources and to map them back to the individual points which altogether constitute an image (Fig. 1.29). It is equivalent, using the language of ray optics, to say that in order to achieve image formation, all the light rays emerging from the point sources of the object will be "re-routed" and converged back to the corresponding points on the image plane. That is the fundamental concept applied in all the ray-tracing techniques for studying the imaging problem (e.g., imaging by a simple convex lens [Fig. 1.30]).

In practice, the image is essentially the superposition of all the blurred spots, instead of perfect replica of the point sources. The ultimate resolution is typically defined by the smallest separation between the two blurred spots such that we can still resolve the two spots. An example of a two-point image is shown in Fig. 1.31.

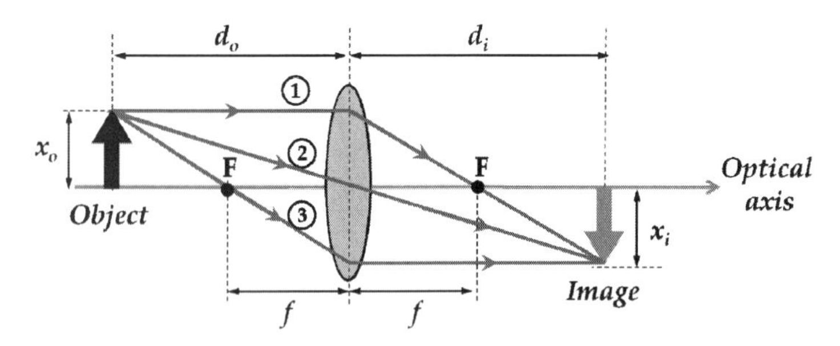

Figure 1.30 A ray-tracing diagram of thin-lens imaging. Imaging condition tells us that the light rays emerging from any point of the object (the tip of the arrow) will eventually converge back to another point on a plane (image plane), where the image is formed. This ray–optic approach is in accordance to the concept based on the wave approach (in Fig. 1.29). There are a set of rules in ray-tracing technique to be followed: The ray propagating in parallel with the optical axis will be bent by the lens and passes through the focal point F (ray 1). The reverse light path is also true (ray 3). Also, the ray passing through the center of the lens will pass straight through the lens (ray 2). Using at least two of these three rays to perform ray tracing, we are able to locate the image and thus evaluate the image properties, e.g., image magnification $M = x_i/x_o = d_i/d_o$.

The fact that the blurring is caused by diffraction, the associated resolution is termed as *diffraction-limited resolution* (Δx) and is typically expressed as (according to Rayleigh criterion[a])

$$\Delta x = 0.61\frac{\lambda}{NA}, \tag{1.27}$$

where λ is the wavelength and *NA* is called the numerical aperture of the imaging lens (Fig. 1.32)—a product of the sine of the half-angle of the light cone ($\sin\alpha$) acceptable by the lens and the refractive index of the medium between the lens and the object (n), i.e., NA $= n\sin\alpha$. Each blurred spot shown in Fig. 1.31 is called the point spread function (psf), which is useful for quantifying the optical resolution of an imaging system, e.g., microscope. The psf of the thin-lens (circular in shape) is the Airy disk function (Fig. 1.31).

[a] Rayleigh criterion tells us that two point sources are regarded as resolvable on the image plane when the central peak of the Airy disk from one of the point sources coincides with the first minimum of the Airy disk from the other point source. (see Fig. 1.31b). This definition is named after *Lord Rayleigh* (1842–1919) who further refined the diffraction-limited resolution theory developed by German physicist *Ernst Abbe* in 1873.

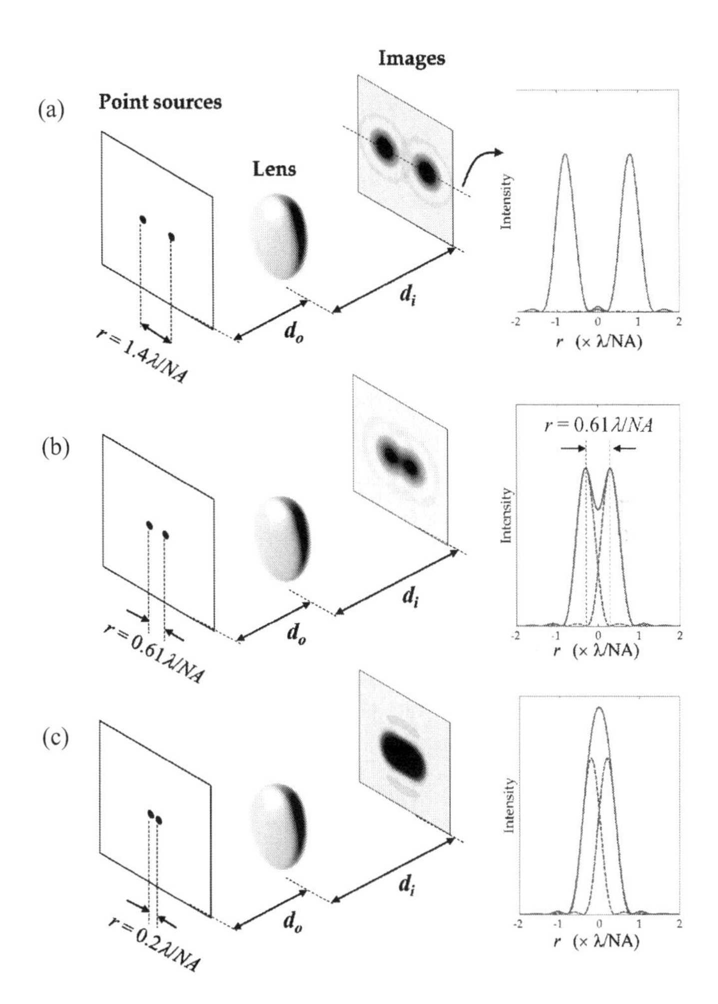

Figure 1.31 Defining optical resolution of an imaging system (e.g., a thin lens): the minimum separation between two point sources so that the two blurred spots (called point spread function [psf]) can still be distinguished. The right column shows the line-cut intensity profiles of the individual psfs (dashed lines) and the summed psf (solid lines). In (a), the separation $r = 1.4\ \lambda/\text{NA} > \Delta x$, the two spots can clearly be discerned. In (b), $r = 0.61\lambda/\text{NA} = \Delta x$, we still can barely see the two spots. This is the Rayleigh criterion for defining the diffraction-limited resolution: two point sources are regarded as resolvable on the image plane when the central peak of the Airy disk from one of the point sources coincides with the first minimum of the Airy disk from the other point source (right inset of (b)). In (c), $r = 0.2\lambda/\text{NA} < \Delta x$, the overall intensity does not show two distinct peaks and the two point sources are said to be unresolvable.

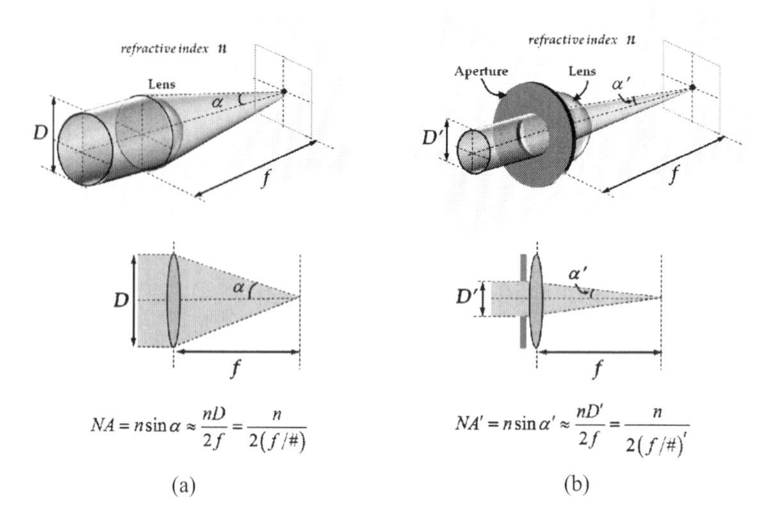

Figure 1.32 (a)–(b) Defining the numerical aperture (NA) of a lens: NA $= n\sin\,\alpha \;\approx nD/2f \;=\; n/2(f/\#)$, where α is the half-angle of the light cone accepted by the lens, n the refractive index of the medium, and D the aperture size of the lens. $f/\# = f/D$ is called the f number of the lens, a parameter that is extensively used in photography. A comparison between (a) and (b) illustrates the effect on the NA of the same lens introduced by adding an aperture right behind the lens. Clearly, varying the aperture size D' directly affects the acceptance angle α', which in turn changes the NA′ and thus the image resolution (as given by Eq. 1.27).

Equation 1.27 essentially describes a pivotal concept in optical imaging: any lens-based optical imaging system is, because of light diffraction, unable to distinguish features which are closer together than roughly half of the wavelength of light—an important theory developed by Ernst Abbe back in 1873, and refined by Lord Rayleigh [16]. Such diffraction barrier can however be overcome by a number of specialized techniques together which are now generalized as *super-resolution* imaging. These techniques are able to reveal features with the dimensions well below the diffraction limit (down to nanometer scale). This is particularly useful in fluorescence microscopy for cellular and molecular imaging [16]. The Nobel Prize in Chemistry 2014 was awarded to Erik Betzig, Stefan W. Hell and W. E. Moerner for recognizing their pioneering

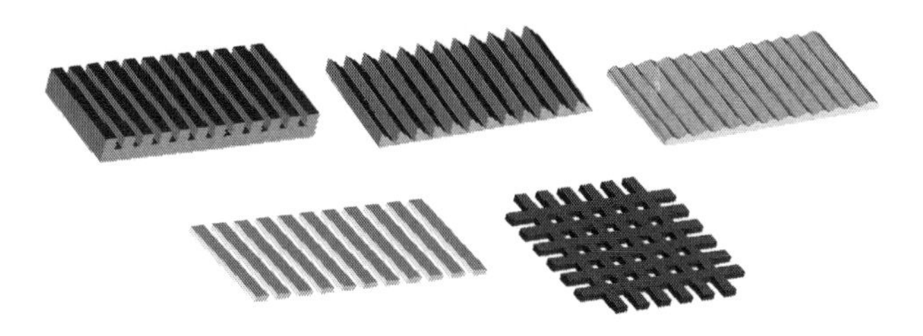

Figure 1.33 Examples of diffraction grating structures. Note that the periodicity of the grating is typically on the order of the illumination wavelength in order to ensure the diffraction to be observable (observing the lower diffraction orders; see Eq. 1.28).

research in super-resolution microscopy (and sometimes called nanoscopy). The key relevant techniques are discussed in Chapter 4.

1.7.4 *Diffraction Grating*

A diffraction grating is a periodic array of diffracting elements (in 1D, 2D or even 3D), which can appear in different forms, e.g., multiple slits, corrugated surface structures (can be blazed or sinusoidal), or an array of materials with alternating refractive indexes and so on (Fig. 1.33). In any case, the key effect of the diffraction grating is to introduce periodic modulation of the phase or/and the amplitude of the incident waves. Each of the diffracting elements (e.g., slits, scratches) can be regarded as a point source emitting a wave which propagates in all directions. The light observed from a particular direction, θ_m, is a superposition of all the diffracting elements, which interfere altogether. In general, when the path length difference d between the light waves from neighboring elements is equal to an integer multiple of the wavelength, λ, the waves will be constructively interfered (i.e., in phase), as depicted in Fig. 1.34. This condition can be expressed as

$$d = \Lambda \sin(\theta_m) = m\lambda \tag{1.28}$$

where Λ is the groove period of the grating, θ_m is the diffracted angle of the light, and m is an integer called *diffraction order*. Hence, a monochromatic light wave will be diffracted and split into distinct directions with the propagation directions governed by Eq. 1.28, which is called the *grating equation*. Equation 1.28 assumes the light

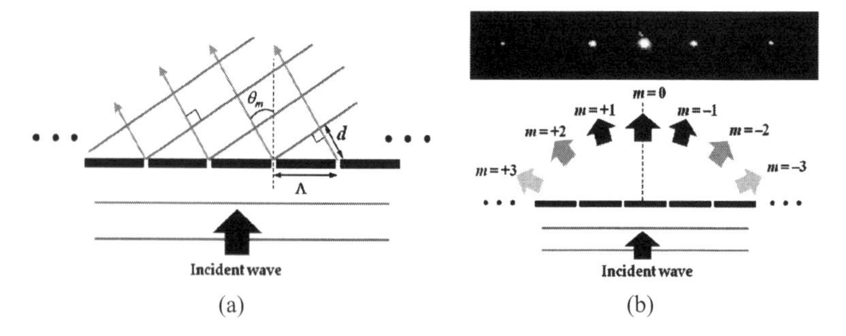

Figure 1.34 (a) Working principle of a diffraction grating. A monochromatic light can only be diffracted by a diffraction grating into a discrete set of directions or angles θ_m. These conditions are satisfied when the path length difference d between the neighboring diffractive elements (e.g., the slits in this figure) equals to an integer multiple of wavelength λ, i.e., $d = m\lambda$, where m is an integer and is called diffraction order. From the geometry depicted in the figure, $d = \Lambda \sin\theta_m = m\lambda$, where Λ is the grating period. (b) Different diffraction orders of a diffraction grating. The upper inset shows the observed diffracted light spots after passing through a grating.

is incident onto the grating perpendicularly, i.e., the incident angle $\theta_i = 0°$. Consider an arbitrary incident angle, the grating equation in Eq. 1.28 can be generalized as

$$\Lambda\,[\sin(\theta_m) - \sin(\theta_i)] = m\lambda \tag{1.29}$$

This equation applies to both reflection-type and transmission-type diffraction gratings (Fig. 1.35).

The fact that the diffracted angle depends on the wavelength (Eq. 1.28 or 1.29) makes the diffraction gratings useful for numerous spectroscopic applications. The wavelength (or frequency) spectrum of any light wave can be analyzed by the diffraction pattern generated a diffraction grating (Fig. 1.36). Indeed, Fig. 1.36 shows a generic configuration of virtually all *optical spectrometers* (or called optical spectrum analyzers) available nowadays.

1.8 Gaussian Beam Optics

Light waves in many occasions have finite spatial extent, such as a beam of laser light. In order to accurately study and predict the

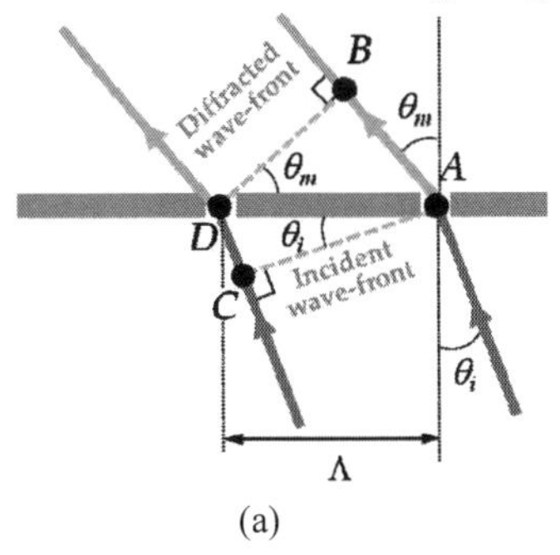

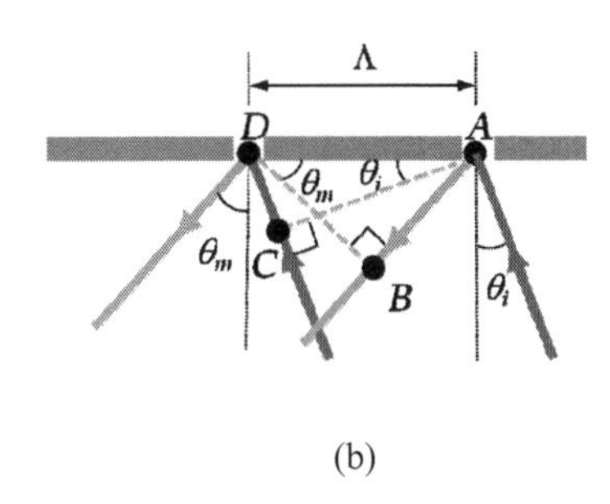

Figure 1.35 (a) Transmission diffraction grating. (b) Reflection diffraction grating. In the case of oblique incidence at an angle θ_i, the diffraction angle θ_m can be found by considering the length difference $AB - CD = m\lambda$, which will yield Eq. 1.29. This is based on the fact that points A and C are on the same incident wavefront, whereas points B and D are on the same diffracted wavefront.

light propagation characteristics, ones often take the "finite size" of the light wave into account, instead of using the idealized plane waves which infinitely extend in space, or using the ray optics model which is merely an approximate theory as mentioned earlier. One commonly used model is *Gaussian beam*, i.e., the intensity profile of light beam along the transverse axis exhibits a bell-shaped curve which is symmetrical around the central peak (Fig. 1.37a). It maintains such Gaussian intensity profile at any location along the beam propagation axis. Only the beam radius varies during propagation. Indeed, this can readily be proved by Fresnel diffraction theory [2]. The beam shape even remains Gaussian after passing through some simple optical elements, e.g., aberration-free lenses and mirrors [2]. Moreover, the facts that the light beams of many lasers are in the Gaussian shapes as well as that Gaussian beam is simply one possible solution of wave equation makes this model popular in many applications.

Mathematically, the intensity profile of a Gaussian beam $I(r, z)$ is written as (Fig. 1.37b)

$$I(r, z) = I_0 \left(\frac{w_0}{w(z)} \right)^2 \exp\left(-\frac{2r^2}{w(z)^2} \right) \tag{1.30}$$

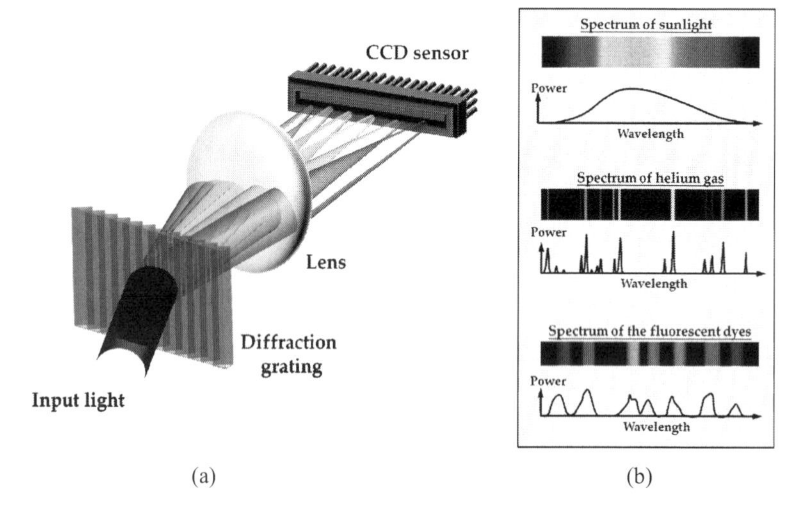

Figure 1.36 (a) A generic schematic of an optical spectrometer, which mainly consists of a diffraction grating spreading out the input spectrum in space (see the "rainbow"). A lens is used to focus the rainbow onto a photodetector array or image sensor (e.g., charge-coupled device [CCD]), which captures the entire spectrum. Some examples of the input spectra (e.g., sunlight, helium gas, and fluorescent dyes) are shown in (b). The spectrum is essentially the intensity or power profile detected across the sensor area.

where I_0 is the peak intensity at $z = 0$ on the longitudinal axis and $r = 0$ on the transverse axis. The beam size at any z is typically characterized by the *beam radius* $w(z)$. It is defined as the transverse distance at which the intensity of the beam drops to $1/e^2$ (13.5%) of its central peak intensity, i.e., when $r = w(z)$ (see Fig. 1.37a). w_0 is the beam radius when $z = 0$. The beam radius is given by

$$w(z)^2 = w_0^2\left[1 + \left(\frac{z}{z_0}\right)^2\right] \tag{1.31}$$

where z_0 is known as the Rayleigh range (or sometimes called collimation range) (Fig. 1.38). It is the distance over which $w(z) = \sqrt{2}w_0$. The Rayleigh range is given by

$$2z_0 = \frac{2\pi n w_0^2}{\lambda} \tag{1.32}$$

where n is the refractive index of the medium and λ is the wavelength of the light. The overall picture of Gaussian beam

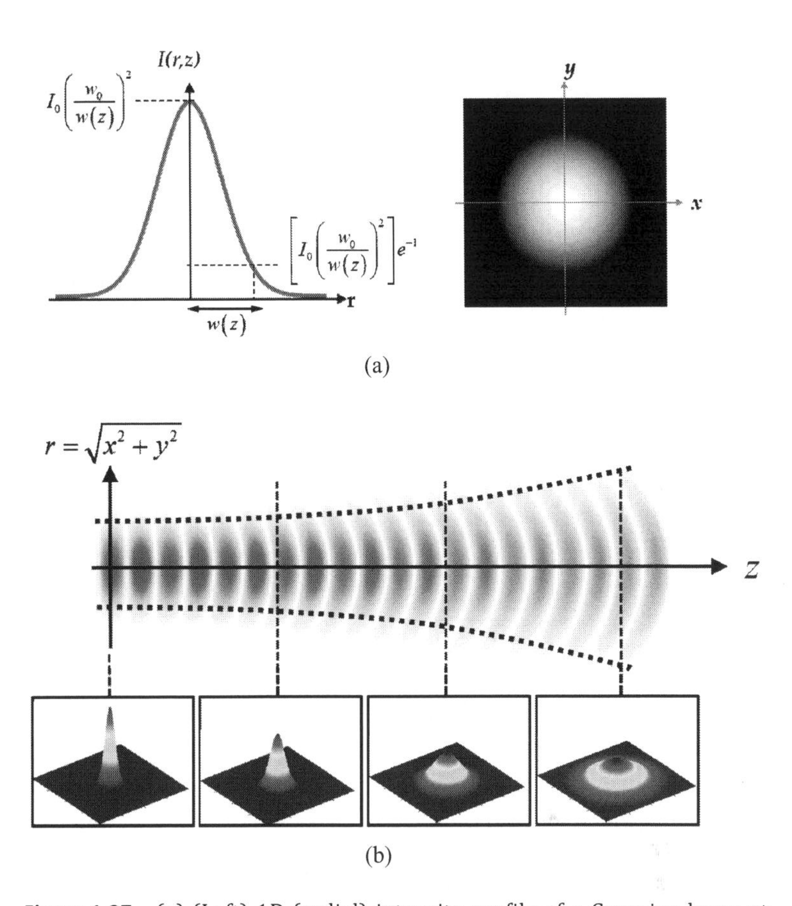

Figure 1.37 (a) (Left) 1D (radial) intensity profile of a Gaussian beam at any position z. (Right) 2D intensity profile of a Gaussian beam. (b) The **E** field pattern of the complete Gaussian beam. Note that the intensity drops and the beam size $w(z)$ increases as it propagates in the z direction (see the bottom 4 insets).

propagation can clearly be described by Eqs. 1.30 and 1.31: the beam tends to diverge with its peak intensity decreased as it is propagating along z. Notably, the beam radius is at its minimum or *beam waist* when $z = 0$, i.e., $w(0) = w_0$ (Fig. 1.38). From Eq. 1.32, *the smaller the beam waist, the shorter the Rayleigh range* (i.e., the "faster" the beam diverges), *and vice versa*.

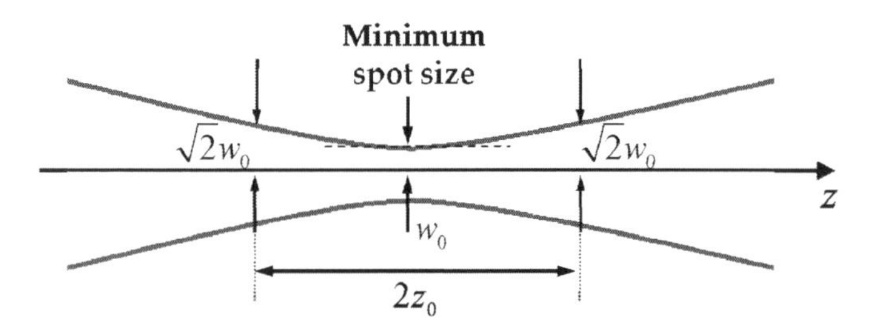

Figure 1.38 Defining the beam waist w_0 (i.e., minimum spot size at $z = 0$) and the Rayleigh range z_0 of a Gaussian beam.

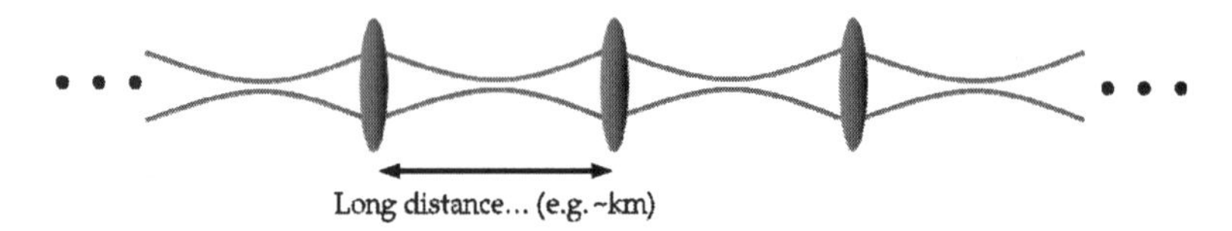

Figure 1.39 An example of implementing long-haul light transmission in free space by the cascaded lenses, each of which refocuses the diverging beam.

1.9 Optical Waveguides and Optical Fibers

As a result of light diffraction, a light beam with a finite size tends to diverge as it propagates in free space. In order to efficiently route the light beam from one location to another, one can use multiple lenses to repetitively refocus the beam such that the most of the power can be captured at the destination and will not be lost because of beam spreading (Fig. 1.39). While this method is cumbersome particularly in the case requiring (i) a long-haul transmission ($>>$ km) or (ii) a complex light routing scheme within in a very compact area (e.g., with a thumb size), *optical waveguide* represents an excellent platform to confine the light wave propagation in a much more effective and flexible manner.

Optical waveguiding can simply be understood by using a ray optics picture. The central idea is to trap or confine the light wave in a physical structure (called waveguide core), which is surrounded or sandwiched by a cladding medium (Fig. 1.40a). This trapping mechanism is done by *total internal reflection* (TIR) which occurs only if

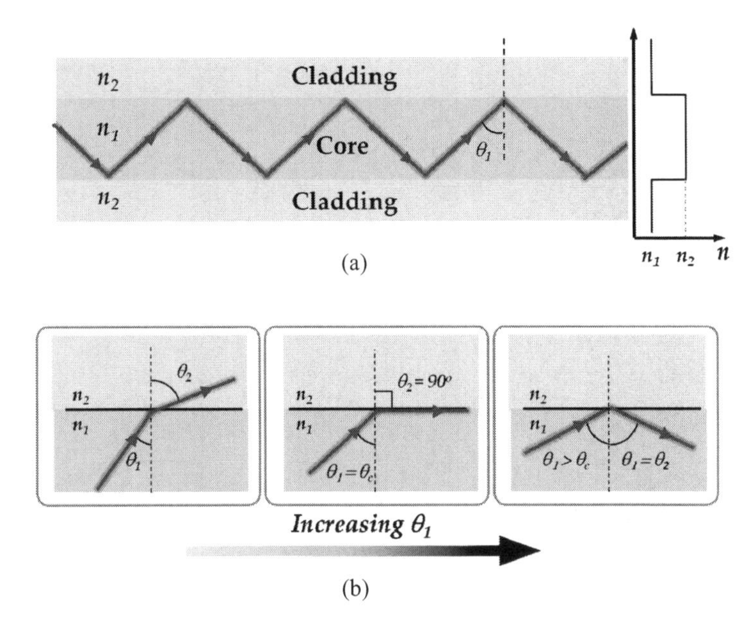

Figure 1.40 (a) Ray optics approach explaining waveguiding in an optical waveguide, which consists of a core (refractive index n_1) sandwiched by a cladding (refractive index n_2). In order to achieve waveguiding, two conditions have to be satisfied: (i) n_2 has to be greater than n_1 (see the index profile on the right), and (ii) the incident angle upon the core-cladding interface has to be greater than the critical angle θ_c (see (b)) so that total internal reflection (TIR) occurs. In this way, the light ray is bounced back and forth and "guided" within the waveguide core. (b) Schematic of describing Snell's law (law of refraction). When $n_2 > n_1$ and $\theta_1 < \theta_c$, the light will pass through the interface and bends away from the normal according the Snell's law. When $\theta_1 = \theta_c$, the refracted light is bended at 90° propagating along the interface (Eq. 1.34). When $\theta_1 > \theta_c$, TIR occurs.

(i) the refractive index of the core (n_1) is larger than that of the cladding medium (n_2) and
(ii) the angle of incidence at the core-cladding interface (θ_1) has to be greater than the critical angle (θ_c) (see Fig. 1.40a).

If $\theta_1 < \theta_c$, the light will be leaked out and be lost to the cladding at a refracted angle (θ_2) which is governed by Snell's law (Fig. 1.40b), i.e.,

$$n_1 \sin\theta_1 = n_2 \sin\theta_2 \tag{1.33}$$

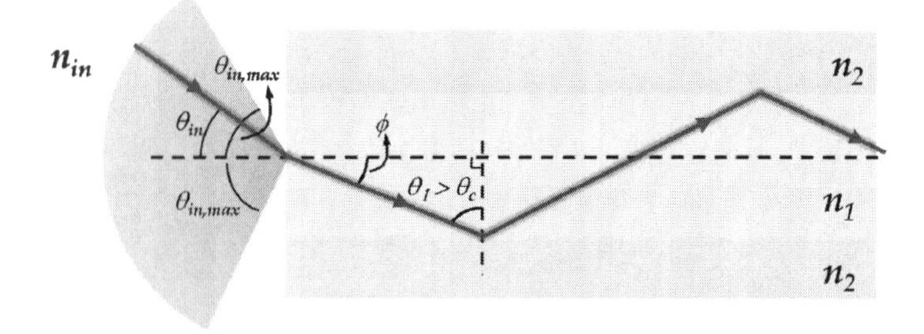

Figure 1.41 The input coupling condition for waveguiding. There is a maximum entrance angle $\theta_{\text{in,max}}$ such that the light ray can still be confined within the core by TIR, i.e., when $\theta_{\text{in}} = \theta_{\text{in,max}}$, we can have $\theta_1 = \theta_c$. Hence, any input angle within the cone angle $2\theta_{\text{in,max}}$ (shaded cone) can satisfy the propagation angle $\theta_1 > \theta_c$, i.e., the waveguiding condition is satisfied.

From Eq. 1.33, the refracted angle $\theta_2 = 90^\circ$ when the incident angle θ_1 equals to a so-called critical angle θ_c, which is given by

$$\theta_c = \sin^{-1}\left(\frac{n_2}{n_1}\right). \tag{1.34}$$

Consider that the light enters the waveguide core from a medium with a refractive index of n_{in}, one can relate the condition of Eq. 1.34 to the maximum entrance angle $\theta_{\text{in,max}}$ such that the light wave can still be confined within the core by the following relationship:

$$NA \equiv n_{\text{in}} \sin\theta_{\text{in,max}} \quad n_1 \cos\theta_c = \sqrt{n_1^2 - n_2^2} \tag{1.35}$$

where NA is the numerical aperture of the waveguide, which relates to the maximum acceptance angle (Fig. 1.41). The larger the NA, the larger the acceptance cone angle. Hence, the NA of the waveguide is typically used for evaluating how efficient the light can be coupled into the waveguide.

The ray optics picture depicted in Fig. 1.41 appears to allow light rays at any angle $\theta_1 > \theta_c$ to be confined and to propagate along the waveguide core. However, when we associate the wave picture to the light ray, in particular the phase of the wave, we will observe that only the waves at a *discrete set of angles* greater than the critical angle θ_c are permitted to propagate along the core. Each allowed angle represents a particular propagation mode, called *guided mode* or *waveguide mode*. The explanation is depicted in Fig. 1.42. We here

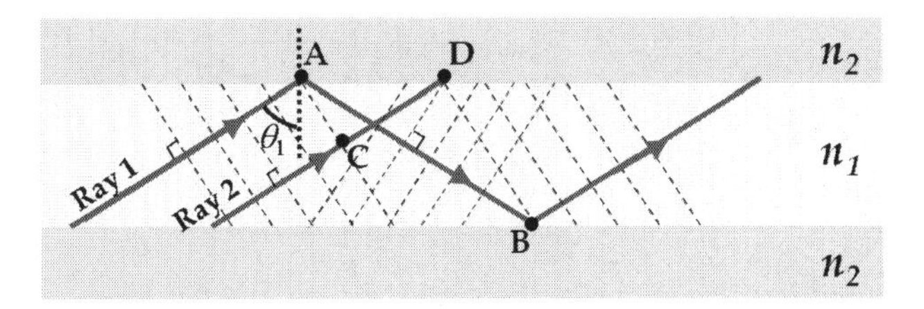

Figure 1.42 Explaining the existence of a discrete set of waveguide modes. Only light rays (modes) with a certain discrete set of propagation angles θ_1 can satisfy the condition AB – CD $= m\lambda$, where m is an integer.

consider ray 1 and ray 2, both of which belong to the same wave. Note that points A and C are on the same wave front. They thus have to be in phase. The ame argument applies to point B and D. Therefore, the necessary condition for wave propagation (guiding) in the waveguide is that the path length difference (l_1 – l_2) between the distance AB and CD has to be an integer multiple of wavelength λ/n_1—a condition of constructive interference! This condition is written as

$$l_1 - l_2 = \frac{m\lambda}{n_1} \tag{1.36}$$

where m is an integer. Therefore, only those waves (modes), which have the discrete set of angles θ_1, can satisfy the condition of Eq. 1.36. For the waveguides which can only support one mode, they are called *single-mode* waveguides (Fig. 1.42). In contrast, for those which can support more than one mode, they are called *multi-mode* waveguides (Fig. 1.43). The number of guiding modes is primarily governed by the geometry and the size of the waveguide. A detailed wave theory of waveguide modes can be found in Refs. [7, 17].

In principle, any material combination which could achieve the aforementioned *refractive index contrast* (between the core and the cladding) can be used for making optical waveguides. One well-known example is *optical fiber*—a flexible and handy waveguide, which is typically not much wider than a human hair. Optical fiber typically follows the cylindrical geometry (Fig. 1.44) and is mostly made of silica glass, polymer or other flexible materials. Typically the index difference between core and cladding is $<1\%$, which is

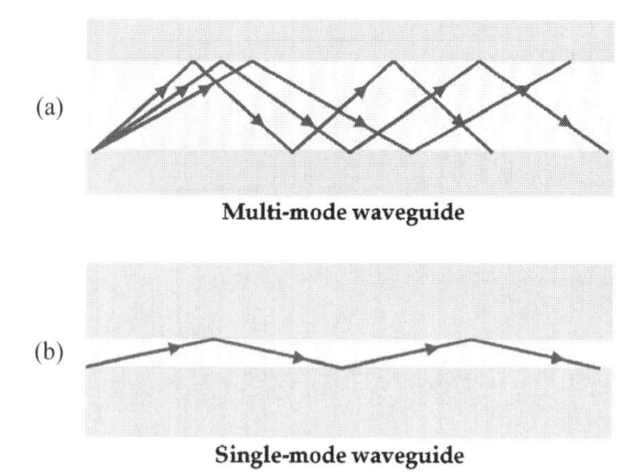

Figure 1.43 (a) Multi-mode waveguide and (b) single-mode waveguide. Typically a multimode waveguide has a bigger core size than a single-mode waveguide.

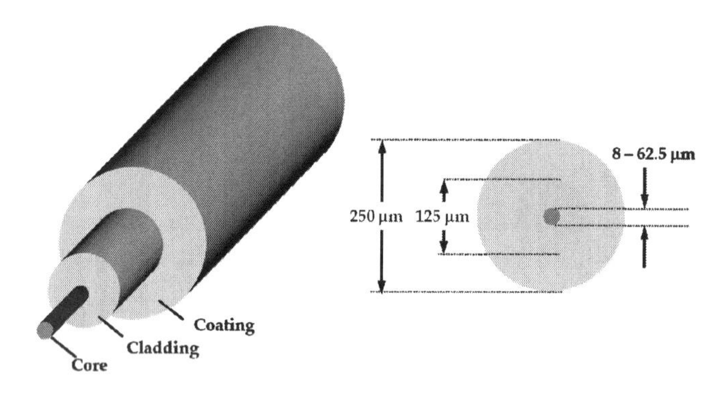

Figure 1.44 Common structure of an optical glass fiber: a fiber core surrounded by a cladding. A coating is used to protect the inner fiber structures. Typically the refractive index difference between the core and the cladding is $<1\%$. The typical dimensions of the glass fiber are shown on the cross-sectional view (right).

already large enough for efficient waveguiding. The utility of optical glass fiber becomes dramatically widespread after Charles K. Kao first promoted the idea in the 1960s that the optical glass fibers can be made with very low optical attenuation ($<<20$ dB/km)—a significant discovery in the last century which triggers the rapid

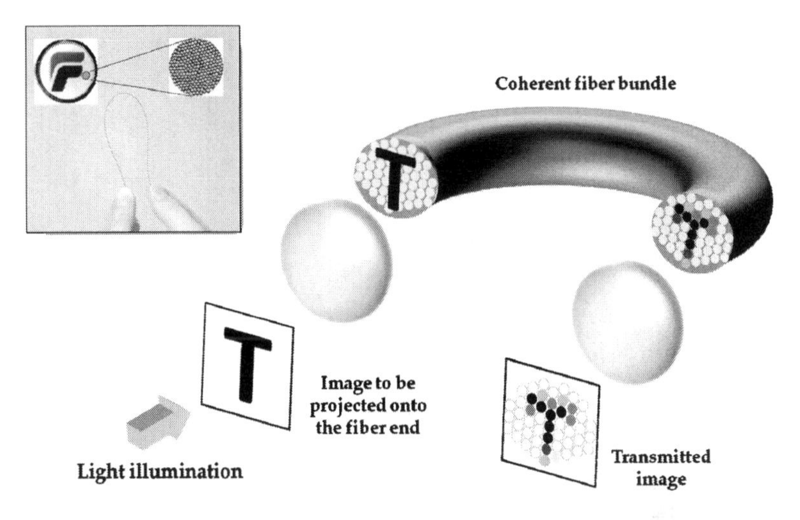

Figure 1.45 Imaging using a coherent fiber bundle. The image is projected onto one end of the bundle and is transmitted through it. The transmitted image can thus be observed from the other end. Note that the transmitted image is pixelated due to the individual fibers of the bundle. In practice, the imaging fiber bundle consists of ~100–100,000 fibers depending on the application (see left upper inset). (Source: www.fujikura.com)

development of telecommunication, and grants him the 2009 Nobel Prize in Physics.

The applications of fibers also extend to biomedicine, e.g., fiber-optic sensors [19], and fiber-optic endoscopy [20, 21]. In particular, a bundle of fibers would be employed in many cases, especially for imaging. When a bundle of fibers is bound together and the individual fibers are carefully and neatly aligned and are arranged relative to one another, it is able to transmit images from one end to another (Fig. 1.45). Typical number of fibers are around ~1000–10,000. The light from different locations of the imaged object is coupled into the individual fibers, which can be regarded as the individual pixels of the final captured image. Such bundles are called *coherence fiber bundles* (Fig. 1.45). Owing to its great flexibility, it can be used for endoscopic imaging for internal body examination *in vivo*. On the other hand, single-fiber-based endoscopic imaging, which is even more compact and flexible, is also possible. It can be achieved by scanning the single fiber across the imaging field in order to reconstruct the whole image [21]. Chapter 7 will discuss this appealing technology in more detail.

1.10 Photons

While the classical EM theory is capable of explaining a great deal of optical effects, it however fails to account for some of the phenomena, especially how light can be generated and absorbed in a medium. An example is *blackbody radiation*, an optical phenomenon which was once a paradox when people attempted to use classical EM theory to explain it in the late 19th century. Until 1901, Max Planck introduced an idea that light energy comes with a set of discrete or quantized values which is proportional to the light frequency ν. The quantized nature of light was later substantiated by Albert Einstein who was able to explain the photoelectric effect in 1905 by considering the quanta of light called photons, the elementary *stable* particles which are *massless* and *chargeless* and appear only at a speed of c. The energy of a photon that has a frequency ν in free space can be written as

$$E = \frac{hc}{\lambda} = h\nu, \tag{1.37}$$

where h is called Planck's constant with $h = 6.626 \times 10^{-34}$J·s.

Note that the photon energy is sometimes defined in the unit of electron volt (eV), i.e., E (in eV) $= E$ (in J)$/e$, where e is the electron charge ($e = 1.602 \times 10^{-19}$ C). Thus, a photon with an energy of 1 eV $= 1.602 \times 10^{-19}$ J. Using this relation and Eq. (37), we can arrive at a commonly used expression which relates the energy (in eV) and wavelength (in μm):

$$E\,(\text{in eV}) = \frac{1.24}{\lambda(\text{in }\mu\text{m})}. \tag{1.38}$$

For example, a NIR photon at 1.24 μm has a photon energy of 1 eV (Fig. 1.3).

1.10.1 *Dual Nature of Light*

The particle picture of light, i.e., photons, does not conflict with the wave model of light. Indeed, they both represent the nature of light—a peculiar but fundamental concept in quantum mechanics, called *wave–particle duality*. Such dualism describes that light exhibits both the wave and particle properties. And this can also be generalized to a similar duality in all matters [1, 2]. Wave–particle duality addresses the inability of classical "particle-only" and "wave-only" concepts to fully describe the behavior of quantum-scale objects, such as photons and electrons. Wave model alone cannot be used to explain light absorption and generation. On the other hand, double-slit interference would become perplexing if one considers that the input light consists of *only* one photon [1, 2, 23, 24]. Figure 1.46 shows an elegant experiment which demonstrates the wave–particle duality of light. A detailed description of this duality is out of the scope of this book. The interested reader can refer to Refs. [1, 2, 22].

In 1924, Louis-Victor de Broglie hypothesized that matter (including photons) exhibits a wavelike nature. His hypothesis simply relates wavelength λ and momentum p by

$$p = \frac{h}{\lambda} = \frac{h\nu}{c} \tag{1.39}$$

The picture of photon momentum becomes more important when light scattering or diffraction in general in a medium. Light scattering alters the photon propagation direction, which creates a momentum change (momentum in general is a vectorial quantity). Such momentum change implies that the light can exert a mechanical force on microscopic particles—an interesting phenomenon called optical trapping. This effect has been utilized to manipulate the motion of biological cells or microparticles with light. This technique is called optical tweezers, which will be covered in detail in Chapter 10.

1.10.2 *Photon Flux*

In many phenomena, we deal with streams of photons instead of a single photon unless extremely weak light/illumination or localized light–atom interaction is considered. Governed by the principle of quantum mechanics, the photon number of the light (e.g., a laser beam) in general is random and follows different probabilistic laws (e.g., see the detected photons in Fig. 1.46). The most common one

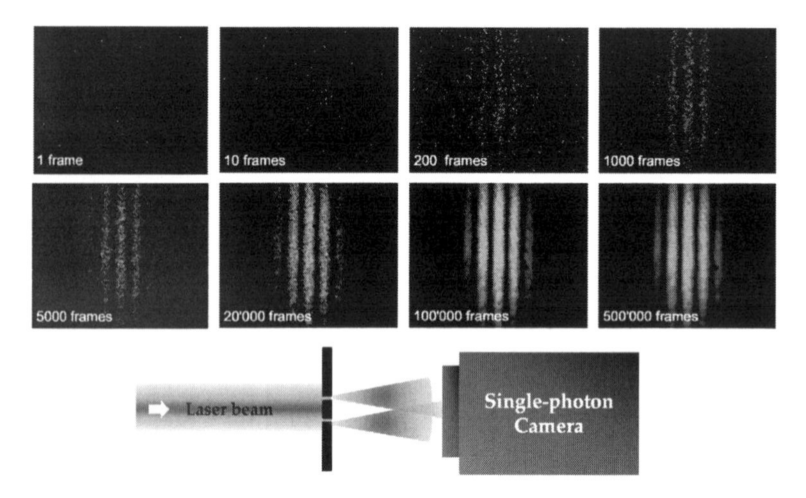

Figure 1.46 An experiment detecting the extremely weak light diffracted from a double slit on a using a camera, which is sensitive enough to detect individual photons. Each dot on the image frame corresponds to the individual event of a detected photon. Although single frames show an apparently random distribution of the detected photon, progressively large number of events (frames) reveals the classical fringe pattern of a double-slit interference experiment. This is an elegant experiment showing the "transition" of particle nature of photons to its wave nature, proving the "particle-wave duality" of light [23, 24].

is that the photon number of the coherent light (such as laser) follows Poisson statistics [2]. Different probability distribution is determined by the light generation mechanism from different light source [2].

Apart from knowing the detailed probabilistic statistics, we are often simply interested in measuring the number of photons in the light, or more precisely the *photon flux density* (i.e., the number of photons per unit time per unit area), which can be related to the classical measure of optical intensity (W/cm^2). Consider a monochromatic light with a frequency ν and optical intensity of $I(\mathbf{r})$. It carries a photon flux density:

$$\Phi(\mathbf{r}) = \frac{I(\mathbf{r})}{h\nu} \tag{1.40}$$

In practice, if the light is quasimonochromatic (has a finite frequency bandwidth) with a central frequency ν_0, we can define its *mean photon flux density* as

$$\Phi_0(\mathbf{r}) = \frac{I(\mathbf{r})}{h\nu_0} \tag{1.41}$$

As an example, the mean photon flux densities of bright sunlight and a 10 mW laser beam with a spot diameter of 20 μm at $\lambda = 600$ nm are 10^{14} and 10^{22} photons/s-cm^2, respectively. The *mean photon flux* over an area (i.e., photons/s) can, on the other hand, be given by

$$\Phi = \int_A \Phi_0(\mathbf{r})dA = \frac{P}{h\nu_0} \tag{1.42}$$

where P is the optical power of the light over the area A:

$$P = \int_A I(\mathbf{r})dA \tag{1.43}$$

The quantum picture of light, i.e., photons, becomes more important when it is necessary to explain light–matter interaction, such as light scattering, absorption, fluorescence, and even lasing. This will be discussed in the next chapter.

References

1. Hecht, E. (2001). *Optics* (4th ed.). Addison-Wesley.
2. Saleh, B. E. A., and Teich, M. C. (2007). *Fundamentals of Photonics* (2nd ed.). Wiley-Interscience.
3. Bass, M., DeCusatis, C., Enoch, J., Lakshminarayanan, V., Li, G., MacDonald, C., Mahajan, V., and Stryland, E. V. (2009). *Handbook of Optics*. Vol. 1: *Geometrical and Physical Optics, Polarized Light, Components and Instruments* (3rd ed.). McGraw-Hill Professional.
4. Bass, M., DeCusatis, C., Enoch, J., Lakshminarayanan, V., Li, G., MacDonald, C., Mahajan, V., and Stryland, E. V. (2009). *Handbook of Optics*. Vol. 2: *Design, Fabrication and Testing, Sources and Detectors, Radiometry and Photometry* (3rd ed.). McGraw-Hill Professional.
5. Bass, M., DeCusatis, C., Enoch, J., Lakshminarayanan, V., Li, G., MacDonald, C., Mahajan, V., and Stryland, E. V. (2009). *Handbook of Optics*. Vol. 4: *Optical Properties of Materials, Nonlinear Optics, Quantum Optics* (3rd ed.). McGraw-Hill Professional.
6. Dinh, T. V. (2010). B*iomedical Photonics Handbook* (2nd ed.). CRC.
7. Liu, J. M. (2005). *Photonic Devices*. Cambridge University Press.

8. Murphy, D. B., and Davidson, M. W. (2012). *Fundamentals of Light Microscopy and Electronic Imaging* (2nd ed.). Wiley-Blackwell.
9. Brezinski, M. E. (2006). *Optical Coherence Tomography: Principles and Applications*. Academic Press.
10. Popescu, G. (2011). *Quantitative Phase Imaging of Cells and Tissues*. McGraw-Hill Professional.
11. Ackermann, G. K., and Eichler, J. (2007). *Holography: A Practical Approach*. Wiley-VCH.
12. Trebino, R. (2002). *Frequency-Resolved Optical Gating: The Measurement of Ultrashort Laser Pulses*. Springer.
13. Smith, B. C. (1995). *Fundamentals of Fourier Transform Infrared Spectroscopy*. CRC Press.
14. Ferraro, P., Wax, A., and Zalevsky, Z. (2011). *Coherent Light Microscopy: Imaging and Quantitative Phase Analysis*. Springer.
15. Goodman, J. (2004). *Introduction to Fourier Optics* (3rd ed.). Roberts and Company Publishers.
16. Pawley, J. (2006). *Handbook of Biological Confocal Microscopy* (3rd ed.). Springer.
17. Hecht, J. (2005). *Understanding Fiber Optics* (5th ed.). Prentice Hall.
18. Okamoto, K. (2005). *Fundamentals of Optical Waveguides* (2nd ed.). Academic Press.
19. Marazuela, M. D., and Moreno-Bondi, M. C. (2002). Fiber-optic biosensors: an overview. *Anal. Bioanal. Chem.*, **372**:664–682.
20. Flusberg, B. A., Cocker, E. D., Piyawattanametha, W., Jung, J. C., Cheung, E. L. M., and Schnitzer, M. J. (2005). Fiber-optic fluorescence imaging. *Nat. Methods*, **2**:941–950.
21. Lee, C. M., Engelbrecht, C. J., Soper1, T. D., Helmchen, F., and Seibel, E. J. (2010). Scanning fiber endoscopy with highly flexible, 1 mm catheterscopes for wide-field, full-color imaging. *J. Biophoton.*, **3**:385–407.
22. Eisberg, R., and Resnick, R. (1985). *Quantum Physics of Atoms, Molecules, Solids, Nuclei, and Particles* (2nd ed.). Wiley.
23. Weis, A., and Wynands, R. (2003). Three demonstration experiments on the wave and particle nature of light. *Physik und Didaktik*, **1**(2):67–73.
24. Dimitrova, T. L., and Weis, A. (2008). The wave–particle duality of light: a demonstration experiment. *Am. J. Phys.*, **76**:137–142.

Chapter 2

Essential Basics of Light–Matter Interaction in Biophotonics

Kevin K. Tsia
Department of Electrical and Electronic Engineering, The University of Hong Kong, Pokfulam Road, Hong Kong
tsia@hku.hk

2.1 Introduction

The interaction of light with matter, particularly biological cells and tissues, is of enormous interest in the field of biophotonics. Simply put, shining light onto the biological cells/tissues can help us to reveal a wealth of information about the cells/tissues, including the structural morphology, the molecular or chemical constituents, and even the physiological and biochemical dynamics involved in the cells/tissues. These capabilities are particularly useful for *biomedical diagnostics* in both basic life science research and clinical practice. Moreover, shining light onto the cells/tissues can also alter their properties and/or the behaviors in order to, say, damage the cancerous cells by laser surgery/therapy and to manipulate the cellular signaling in, e.g., a neural network by optical stimulation.

Understanding Biophotonics: Fundamentals, Advances, and Applications
Edited by Kevin K. Tsia

ISBN 978-981-4411-77-6 (Hardcover), 978-981-4411-78-3 (eBook)
www.panstanford.com

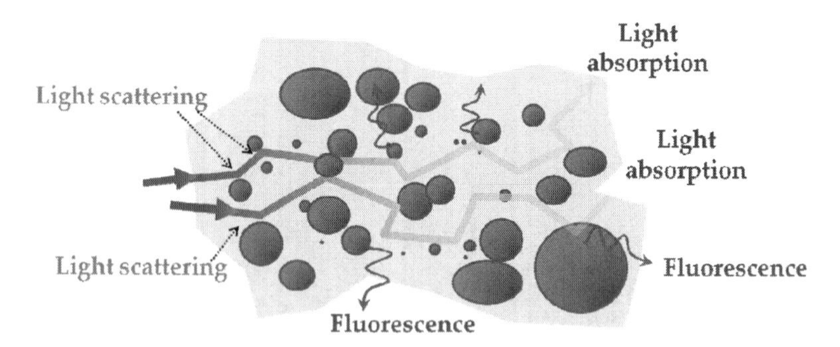

Figure 2.1 Three main phenomena occur when light propagates in a biological tissue: (i) light absorption, in which the light intensity is attenuated; (ii) light scattering, during which the light is deviated from its original pathway by the scatterers in the tissues (represented by circles in the figure); and (iii) fluorescence, which is the light radiation process emitting light at a different (longer) wavelength.

These areas fall into the category of *biomedical therapeutics* in biophotonics. Light–matter interaction is also essential for understanding the working principles of many essential optical devices and components, such as lasers and photodetectors. These are often the key elements which constitute many important systems for both the diagnostic and therapeutic applications.

When light propagates and interacts in a biological material, three main phenomena occur (Fig. 2.1): *light absorption*, *light scattering*, and *fluorescence*. The occurrences of these effects chiefly depend on the optical properties of the material (i.e., how the material responses to light) and the characteristics of the light source. In general, there are two fundamental approaches to explain light–matter interactions: (i) classical harmonic oscillator model and (ii) quantum mechanical model. These will be elucidated in more detail in this chapter.

2.2 Classical Interpretations

When a material is illuminated by an incident harmonic light wave (i.e., an EM wave), its internal electric charge distribution is disturbed. More specifically, the charges experience a time-varying force which is proportional to the strength of the oscillating **E** field in the EM wave. For instance, in polar molecules, the dipoles[a] in the molecules tend to align with the **E**-field orientation. On the other hand, in nonpolar molecules, the surrounding electron cloud will be distorted. A dipole is thus induced and is aligned with the **E** field. In both cases, the medium is said to be polarized and the resultant dipoles in the two cases can be generalized to an effect called *electronic polarization* (Fig. 2.2a,b). Similar processes can also be applied to atoms in a molecule. The atoms or the ions can be driven into vibration by the incoming **E** field. And such vibration can exist in different manners, say, stretching, rotating, bending, etc. (Fig. 2.2c). Since the nuclei are much more massive than the electron clouds, the induced vibration behavior is different from the aforementioned electronic contribution. Such effect is typically named as *molecular polarization*.

In general, such polarization **P**, which is proportional to the input **E** field, is given as

$$\mathbf{P} = \varepsilon_0 \chi \mathbf{E} \tag{2.1}$$

where ε_0 is the permittivity of the free space; χ is called *electric susceptibility*, which characterizes how easy the material is in response to the incoming **E** field; and χ is directly related to the refractive index of the material n by

$$n = \sqrt{1 + \chi} \tag{2.2}$$

Therefore, refractive index essentially describes how the material responds to a light wave.

Electric susceptibility can be evaluated by using the *classical harmonic oscillator* model, which treats the atom or molecule as an oscillator being driven by an external time-varying force (i.e., from the **E** field of the light) (Fig. 2.3). The driven oscillation amplitude of the oscillator depends on several parameters: (i) the **E**-field amplitude, (ii) the restoring force of the oscillator, (iii) the friction of the oscillator, which results in a damping effect during oscillation, and (iv) the driving frequency ω, i.e., the frequency or wavelength of the light. Hence, the oscillation amplitude can be evaluated by

---

[a] A pair of electric charges of equal magnitude but opposite sign is called a dipole (see Fig. 2.2a).

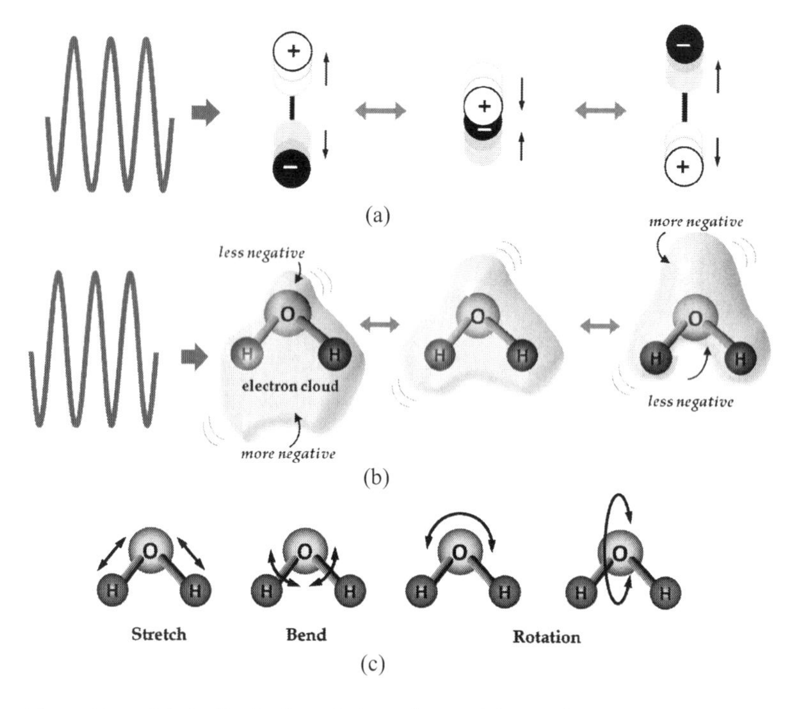

Figure 2.2 (a) A dipole (consisting of a positive and a negative charge) driven into an oscillation by an incoming EM wave. (b) A crude picture showing that the EM wave drives and distorts the electron cloud of a water molecule (H_2O), which is a polar molecule. The distortion in effect creates a dipole oscillation, resembling the case in (a). The molecule is thus said to be polarized and this effect is called *electronic polarization*. (c) The EM wave can also drive the molecule into different forms of vibrational motions, e.g., stretch, bend, and rotation, making it polarized. This effect is called *molecular polarization*. Note that the electron cloud is omitted for clarity in (c).

the classical equation of motion, which is governed by the Newton's Second Law [1–3]. Without going into the detailed derivation, we here show the *electric susceptibility*, which is proportional to the oscillation amplitude and is frequency dependent, as

$$\chi(\omega) = \frac{\chi_0}{\omega_0^2 - \omega^2 - i\Gamma\omega} \tag{2.3}$$

where χ_0 is a constant related to a parameter called the plasma frequency of the material [1–3], Γ is the damping rate of the

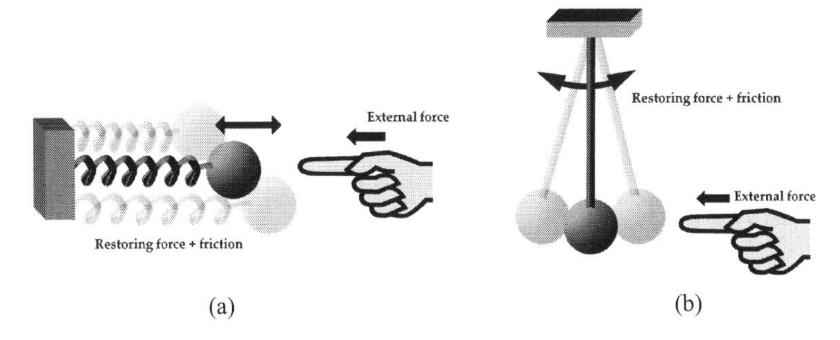

Figure 2.3 Examples of the classical harmonic oscillator: (a) spring–mass system and (b) pendulum system. In both systems, the mass will be driven in a harmonic (sinusoidal) oscillation when an external force is applied. The oscillation is a result of that restoring force set by the spring constant of the spring (or the string–mass in the pendulum) counteracts the external force. The friction introduces damping during the oscillation.

oscillation, and ω_0 is the resonance frequency of the oscillator (i.e., the chemical bond of the molecule) at which the light energy is most efficiently absorbed by (or transferred to) the oscillator, i.e., the strongest light absorption occurs. It is clear from Eq. 2.3 that $\chi(\omega)$ is a complex quantity. Its real $\chi'(\omega)$ and imaginary $\chi''(\omega)$ parts are plotted in Fig. 2.4.

So, *what is the physical meaning of a complex* $\chi(\omega)$? On the basis of Eq. 2.2, we know that the refractive index can also be a complex value and is frequency dependent. It can be denoted as $n(\omega) = \eta(\omega) + i\kappa(\omega)$, where $\eta(\omega)$ and $\kappa(\omega)$ are the real and imaginary parts of refractive index, respectively. As a result, this implies that the wavenumber k is complex as well (see Eq. 1.3 in Chapter 1). It becomes clear when we look at the wavefunction $\mathbf{E}(z, t)$ with a complex $k(\omega) = k'(\omega)+ik''(\omega)$:

$$\begin{aligned}\mathbf{E}(z,t) &= \mathbf{E_0}\exp\left[i(kz-\omega t)\right]\\ &= \mathbf{E_0}\exp\left[i\left((k'+ik'')z-\omega t\right)\right]\\ &= \mathbf{E_0}\exp\left(-k''z\right)\exp\left[i\left(k'z-\omega t\right)\right]\end{aligned} \quad (2.4)$$

Equation 2.4 tells that the light wave attenuates at an exponential decay rate of k'' as it propagates at a phase velocity of ω/k'. Figure 2.5 depicts the wave propagation characteristics in this situation. As k, n, and χ are interrelated to each other, one can expect that the

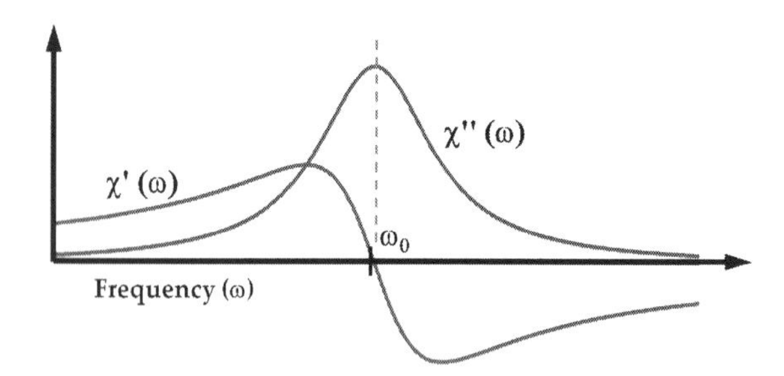

Figure 2.4 The real part $\chi'(\omega)$ and the imaginary part $\chi''(\omega)$ of the electric susceptibility. Note that the complex refractive index n and the wavenumber k follow the same wavelength dependence as the electric susceptibility.

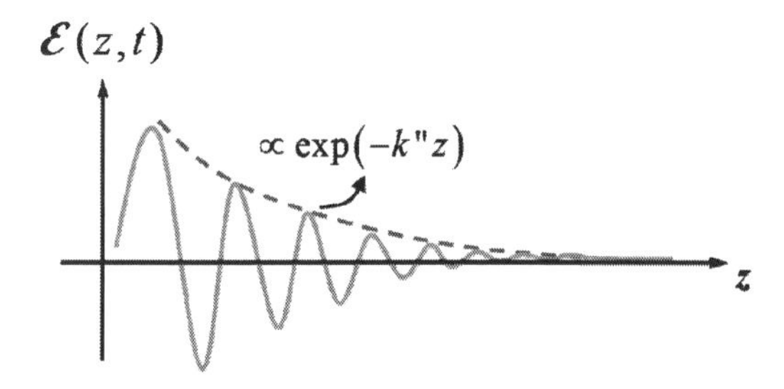

Figure 2.5 The wavefunction with a complex wavenumber k. The decaying feature (dashed line) originates from the exponential decaying term with the imaginary part k'' (see Eq. (4)).

frequency dependence of k follows the same trend of χ, as shown in Fig. 2.4. As a result, the physical meaning of a complex $\chi(\omega)$ is that

- the imaginary part $\chi''(\omega)$ describes the energy loss of the wave and this is called *light absorption*.
- the real part $\chi'(\omega)$ describes the phenomenon called *dispersion*—the refractive index, and thus the phase velocity are frequency dependent.

The wavelength dependence of the refractive indexes of the selected materials is plotted in Fig. 2.6.

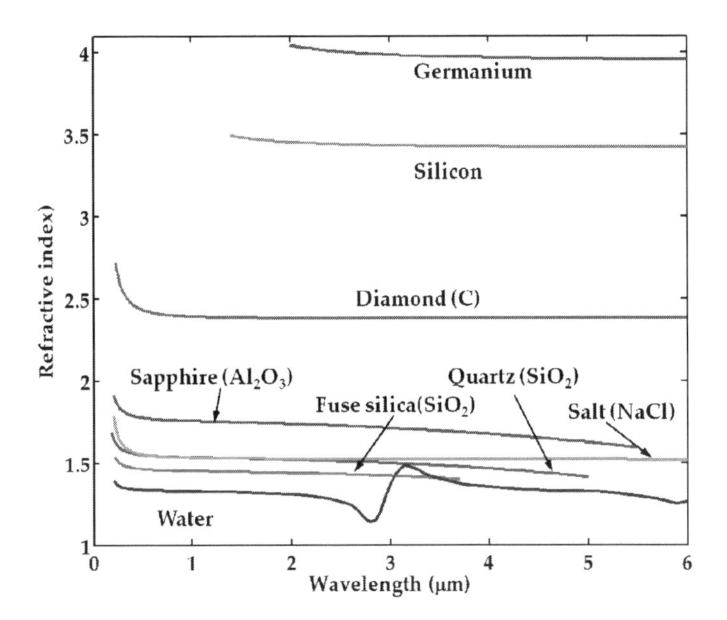

Figure 2.6 Refractive indexes of typical materials: semiconductor crystals (germanium and silicon), diamond, sapphire, silicon dioxide (in the form of quartz and fused silica), salt, and water [2, 4]. Note that the refractive index variations of these materials are relatively small from the visible to IR (~6 μm) wavelengths. It indicates that this range is outside the resonance band (Fig. 2.4) where the material is strongly absorptive. The exception in the figure is water, which has a strong resonance (i.e., strong absorption) at ~3 μm. It shows the characteristic asymmetric shape, as shown in Fig. 2.4.

2.3 Optical Anisotropy and Birefringence

The electric susceptibility and thus the refractive index of a medium we have been considering so far do not depend on *light polarization* and the *light propagation direction*. This type of medium is said to be *isotropic*. Examples are gas, liquids, and amorphous solids. The molecules in the isotropic media are located and oriented randomly in space (Fig. 2.7a). So, *macroscopically*, the optical properties make no difference when the light interacts with them from different directions or with different light polarization states.

In contrast, if the molecules in the medium are arranged and oriented in some kinds of ordered manners, such medium is

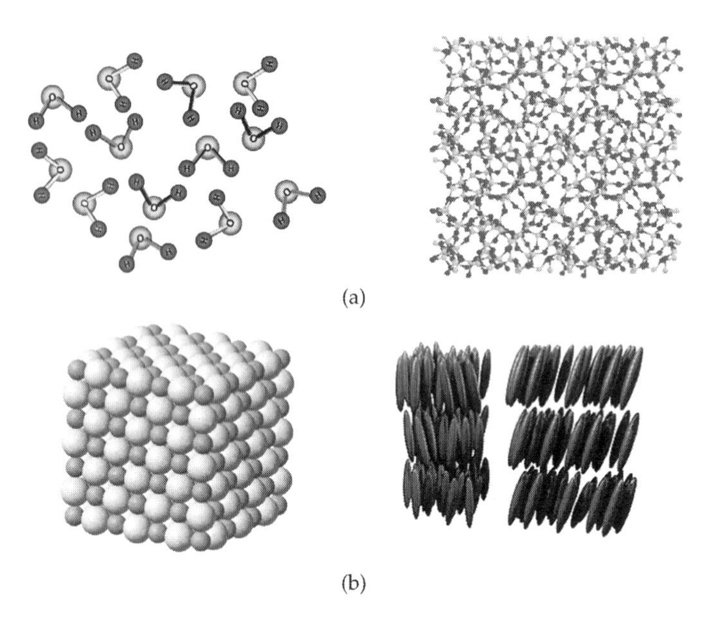

Figure 2.7 (a) Isotropic media: (left) water (H_2O) and (right) fused silica glass (SiO_2). (b) Anisotropic media: (left) salt crystal (NaCl), and (right) liquid crystal. Source: en.wikipedia.org/ wiki/Liquid_crystal and www.research.ibm.com/amorphous/

anisotropic. Examples are crystals and liquid crystals (Fig. 2.7b). Anisotropic medium exhibits different refractive indexes along different directions or equivalently when the incoming light has different polarization states (e.g., linear polarization at different polarization angles). This phenomenon can readily be explained by the classical harmonic oscillator model that the **E** field of the light wave "sees" different molecular arrangements or organizations in the medium when the light wave comes from different directions. Therefore, the driven oscillatory motions of the molecules, and thus the refractive index (or equivalently the susceptibility), will be different according to the orientation of the **E** field (i.e., light polarization state) and the direction of the light wave propagation (Fig. 2.8a,b).

In order to describe the anisotropy mathematically, the *electronic polarization* or *molecular polarization* **P** in Eq. 2.1 has to be generalized to have the x, y, and z components in a three-

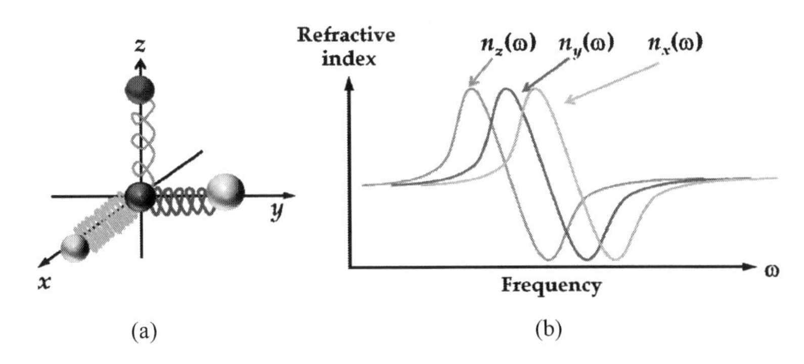

Figure 2.8 (a) A graphical representation of the harmonic oscillator model describing an anisotropic medium. If we choose the appropriate coordinate system (x, y, z), we can identify three independent refractive indexes, n_x, and n_y, and n_z along the three orthogonal directions [2, 4]. The refractive indexes can be evaluated by the harmonic oscillator model, i.e., considering the three harmonic oscillators (with different "spring constants" and "masses") oriented themselves orthogonal to each other. In general, n_x, and n_y and n_z can be different, as shown in (b), each of which shows its own characteristic wavelength dependence (similar to Fig. 2.4). Note that only the real parts of the refractive indexes are shown in (b).

dimensional space. Moreover, the susceptibility is now characterized by a 3 × 3 matrix of nine coefficients $\{\chi_{ij}\}$ called *susceptibility tensor*, where $i, j = 1, 2, 3$ indicate the x, y, and z components, respectively. Hence, Eq. 2.1 in a generalized case can be expressed in the following matrix representation:

$$\mathbf{P} = \begin{bmatrix} \mathbf{P}_x \\ \mathbf{P}_y \\ \mathbf{P}_z \end{bmatrix} = \varepsilon_0 \begin{bmatrix} \chi_{11} & \chi_{12} & \chi_{13} \\ \chi_{21} & \chi_{22} & \chi_{23} \\ \chi_{31} & \chi_{32} & \chi_{33} \end{bmatrix} \begin{bmatrix} \mathbf{E}_x \\ \mathbf{E}_y \\ \mathbf{E}_z \end{bmatrix} \tag{2.5}$$

where $[\mathbf{E}_x\ \mathbf{E}_y\ \mathbf{E}_z]$ is the electric field vector. Similarly, the refractive index can be expressed in such matrix form. In fact, one can always find a new coordinate system such that only 3 (instead of 9) independent refractive indexes, n_x, n_y, and n_z, are required to fully describe the anisotropy of the medium. Such a coordinate system defines three *principal axes* of the medium (the x, y, and z axes indicated in Fig. 2.8)[a]. Interested readers can refer to more detailed mathematical treatments related to optical anisotropy [e.g., 5, 6].

[a]Eq. 2.5 shows that there are 9 different (independent) variables of susceptibility. However, it is possible to find a coordinate system in which this susceptibility tensor

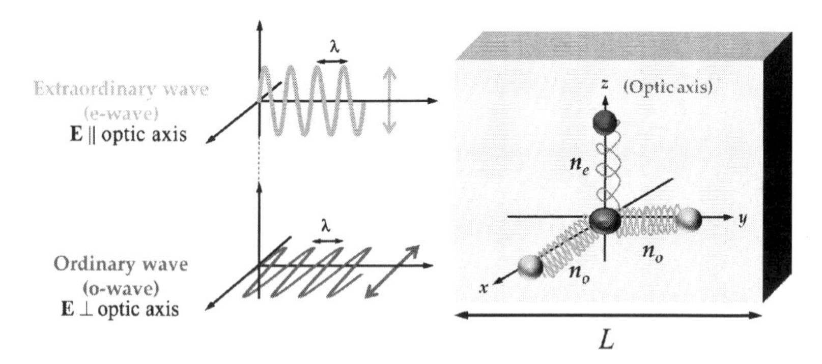

Figure 2.9 Explaining the birefringence effect. An anisotropic medium is uniaxial or birefringent when the refractive indexes $n_x = n_y = n_o$, and $n_z = n_e$. This can be understood by using the same oscillator model illustrated in Fig. 2.8. Instead of having three different refractive indexes, the picture here shows that the oscillators in the x and y directions exhibit the same spring constant and mass, leading to only two independent refractive indexes. The index difference, or the degree of birefringence $\Delta n = |n_e - n_o$. One special case considered here is when the input light propagation direction is perpendicular to the optic axis, the e wave and o wave will respectively "see" entirely the refractive index n_e and n_o. A phase difference (retardation) between the e wave and o wave is thus expected and can be easily evaluated by Eq. 2.7.

In particular, the medium is said to be *uniaxial* when the refractive indexes $n_x = n_y = n_o$, and $n_z = n_e$. The suffixes *o* and *e* stand for *ordinary* and *extraordinary*, respectively. When the polarization of the light wave is in the x–y plane (i.e., perpendicular to z), the wave is called ordinary wave (o wave). On the other hand, when the light polarization contains the z component, the wave is called extraordinary wave (e wave). The z axis in this case is defined as *optic axis*—the propagation of light along the optic axis is independent of its polarization states (Fig. 2.9).

---

and the corresponding refractive index tensors become "diagonalized," i.e.,

$$n^2 = \begin{bmatrix} n_x^2 & 0 & 0 \\ 0 & n_y^2 & 0 \\ 0 & 0 & n_z^2 \end{bmatrix}$$

Remember that χ and n are related by Eq. 2.2. Details of the diagonalization can be found in Refs. [5, 6].

Such uniaxial material, which displays two different refractive indexes, is also said to be *birefringent*. The degree of birefringence Δn can be evaluated by

$$\Delta n = |n_e - n_o|. \qquad (2.6)$$

Note that birefringence is also a wavelength-dependent parameter because of dispersion. When a light wave is incident on a birefringent medium at an angle oblique to its optic axis, it would be split into two orthogonally polarized waves (one is o wave while another is e wave), each of which propagates along a different direction. This is an effect called *double refraction*. When the incident light is perpendicular to the optic axis, the o wave and the e wave would propagate co-linearly in the birefringent material (Fig. 2.9). However, because of the fact that the o wave has the refractive index of n_o whereas the e wave has refractive index of n_e, these two orthogonally polarized light waves experience different phase shifts. If the length of the birefringent medium is L, the difference in phase shift between e wave and o wave is

$$\Delta\phi = \left|\frac{2\pi}{\lambda}n_e L - \frac{2\pi}{\lambda}n_o L\right| = \frac{2\pi}{\lambda}|n_e - n_o|\,L = \frac{2\pi}{\lambda}\Delta n L \qquad (2.7)$$

Recall from Chapter 1 that the polarization state of a light wave is primarily governed by the phase difference between the two orthogonal **E**-field components (in this case, they are e wave and o wave). Hence, the birefringent material in this situation acts as a phase retarder which can modify the polarization states of light with difference birefringence Δn or/and medium length L. Such phase retarder is sometimes called *wave plate*. Examples are *half-wave plate*, which introduces a 180° relative phase shift between e wave and o wave; and *quarter-wave plate*, which introduces a 90° phase shift (Fig. 2.10).

Indeed, a lot of biological cells and tissues are structurally anisotropic, such as lipid bilayers of the cell membrane, collagen, actin, and myosin filaments in the muscle cells, as well as epithelial and nervous tissues (Fig. 2.11). Hence, one could utilize birefringence as a useful tool to reveal the anisotropic organization of the cells and tissues. The *polarized light microscope* is a key

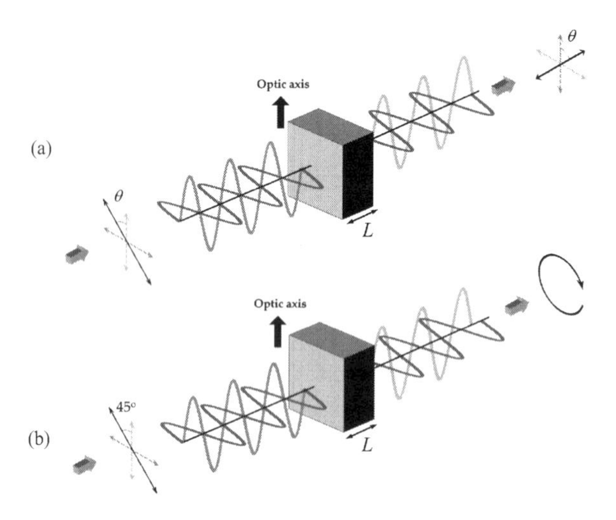

Figure 2.10 (a) Half-wave plate, which introduces a 180° relative phase shift between e wave and o wave, i.e., $\Delta\varphi = 2\pi\,\Delta nL/\lambda = 180°$. It is mainly used for manipulating the orientation of a linearly polarized light. In brief, if an input linearly polarized light makes an angle of θ with respect to the optic axis, the half-wave plate will rotate the polarization angle by 2θ. (b) Quarter-wave plate, which introduces a 90° relative phase shift between e wave and o wave, i.e., $\Delta\varphi = 2\pi\,\Delta nL/\lambda = 90°$. It is particularly used to transform a linearly polarized light (oriented at 45° with respect to the optics axis to a circularly polarized light, and vice versa.

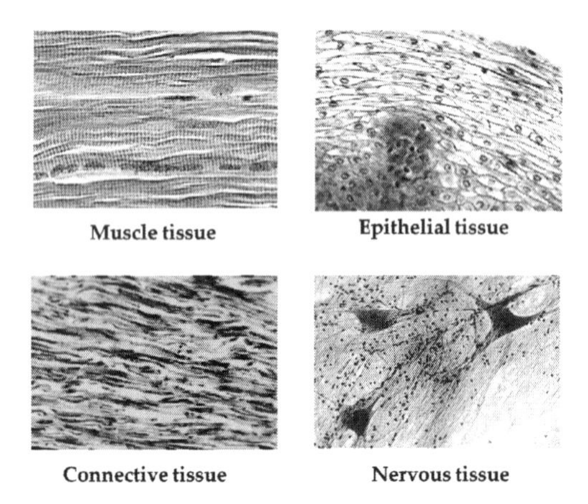

Figure 2.11 Typical biological tissues. They all are anisotropic or more specifically birefringent as the cellular/molecular arrangements have the preferential orientations.

instrument used for this purpose. It employs polarized light to interact with the biological specimens and makes use of the birefringence of the specimens, which introduces phase retardation, to provide the image contrast based on the anisotropy. Hence polarized light microscopy has successfully been used for studying the structures and dynamics of numerous birefringent biological specimens, e.g., the mitosis dynamics of spindle microtubules and the packing arrangement of DNA molecules in the chromosomes [7, 8].

2.4 Light Absorption

As discussed earlier, light absorption occurs when the energy is transferred from the light to the molecules or atoms in the material. More specifically, when the light of a given frequency matches the natural vibrational frequencies of either the electron cloud motions (electronic contribution) or the atomic/molecular motions (molecular contribution), the light energy would be greatly absorbed and is then transformed into the vibrational motion of the electrons clouds or the atoms/molecules. This situation is called *resonance* and such frequency is named as *resonance frequency*. As discussed in Section 2.1, the imaginary part of the susceptibility $\chi''(\omega)$ or refractive index, which describes light absorption, shows a spectral peak at such resonance frequency (see Fig. 2.4). In reality, a lot of materials contain multiple resonances, corresponding to different forms of electronic and molecular vibrations. The overall $\chi''(\omega)$ is the sum of all the contributions from these resonances: multiple absorption peaks can be seen (Fig. 2.12). As a rule of thumb, it is generally true that absorption of UV and visible light mostly corresponds to the electronic contribution whereas absorption of IR light corresponds to the molecular contribution.

On the basis of Eq. 2.4, the light absorption (i.e., attenuation) can be expressed in terms of the optical intensity which is proportional to the modulus square of **E**-field amplitude (Eq. 1.9 in Chapter 1), i.e., $I \propto |\mathbf{E}|^2$. Therefore, the light intensity loss as a function of propagation distance z is given by

$$I(z) = I(0)\exp\left[-2k''z\right] = I(0)\exp\left[-\mu_a z\right] \qquad (2.8)$$

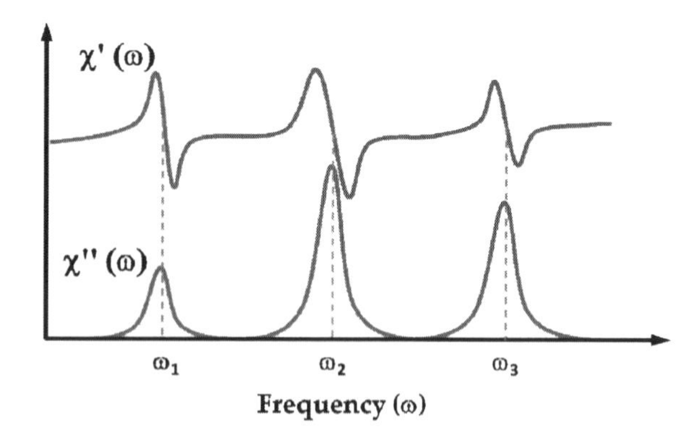

Figure 2.12 Typical materials contains multiple resonances, corresponding to different molecular and electronic vibrations. The overall complex χ is the sum of all the contributions from these resonances; see ω_1, ω_2, and ω_3 in this example.

which is known as the *Beer–Lambert law*, or *Beer's law* (Fig. 2.13). $I(0)$ is the optical intensity at $z = 0$ and $\mu_a = 2k''$ is called the absorption coefficient (unit: m^{-1} or more commonly in cm^{-1} for biological specimens). Here we can define two commonly used parameters for characterizing the optical loss of a given specimen: *transmittance* (T) and *absorbance* (A). Transmittance is defined as

$$T = \frac{I(z)}{I(0)} \tag{2.9}$$

In the logarithmic scale (in decibel [dB]), it is given as

$$T_{\log} = 10\log_{10}\frac{I(z)}{I(0)} = 10\log_{10}(\exp[-\mu_a z]) = -4.343\mu_a z \tag{2.10}$$

Absorbance A or sometimes called *optical density* (O.D.) is defined as

$$A = -\log_{10}(T) \tag{2.11}$$

From Eqs. 2.10 and 2.11, we see that O.D. of 3 means $T_{\log}$ of −30 dB, i.e., the output optical intensity (or power) is 1000 times less than the input at $z = 0$.

Another way to interpret the absorption coefficient μ_a is to use the concept of an *absorption cross section* of the molecules σ_a (unit: m^2). More precisely, this absorption cross-section should be referred to a certain part of a molecule (e.g., particular chemical bonds,

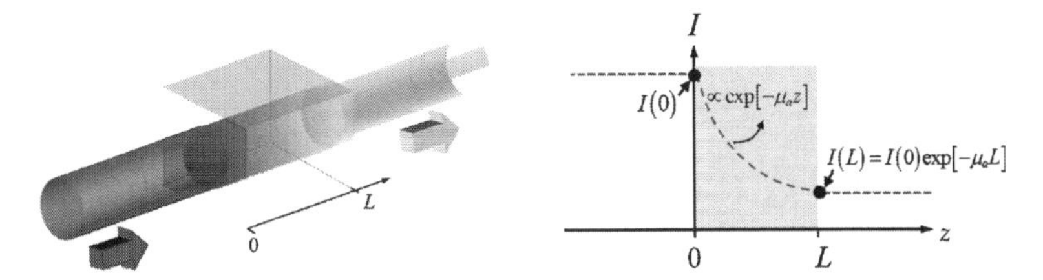

Figure 2.13 Light absorption described by Beer's law. Considering a material with a length of L and an absorption coefficient μ_a (left), the input light intensity will be attenuated according to Beer's law (exponential decay as a function of distance) (right).

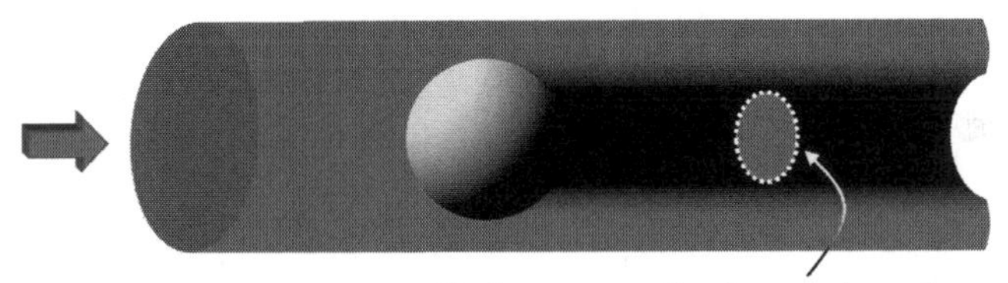

Figure 2.14 Defining absorption cross-section, σ_a. We can regard a chromophore as an idealized sphere that blocks the incoming light and casts a shadow, which constitutes absorption. The size of absorption shadow is defined as the absorption cross-section.

or functional groups in the entire molecule), called *chromophore*, which absorbs the certain wavelengths and transmits others. By treating a chromophore as an idealized sphere, we can imagine that this sphere blocks the incoming light and casts a shadow, which constitutes absorption. We can then define the area of the shadow effectively to be the absorption cross-section σ_a (Fig. 2.14). If there are N_a chromophores per unit volume (i.e., density of chromophores [m^{-3}]), the absorption coefficient (m^{-1}) can be related to the absorption cross-section (m^2) by[a]

$$\mu_a = N_a \sigma_a. \tag{2.12}$$

Equation 2.11 assumes that σ_a is independent of the relative orientations of the incident light and the chromphores. It also

[a]Note that in many biomedical applications (e.g., imaging of biological cells), the commonly used length scale unit is *centimeter* instead of *meter*, and so are N_a (cm^{-3}), σ_a (cm^2), and μ_a(cm^{-1}).

assumes that the distribution of chromophores is uniform and homogeneous in space. On many occasions, it is more convenient to specify the concentration c (unit: mol/l) rather than the density of the chromophore. The conversion between the two unit is given as

$$N\left(\text{cm}^{-3}\right) \cdot \left(\frac{1 \text{ mol}}{6.023 \times 10^{23}}\right) \cdot \left(\frac{1000 \text{ cm}^3}{\text{liter}}\right) = c\left(\frac{\text{mol}}{\text{liter}}\right) \tag{2.13}$$

Note that 1 mole of molecules (unit: mol) corresponds to the molecule number of 6.023×10^{23}, which is known as *Avogadro's number* N_A. By substituting Eq. 2.12 into Eq. 2.8 and using the conversion stated in Eq. 2.13, the absorbance A in Eq. 2.11 can be rewritten in the following form

$$\begin{aligned} A &= -\log_{10}\left(\frac{I(z)}{I(0)}\right) \\ &= -\log_{10}\left(\exp -(N_a \cdot \sigma_a \cdot z)\right) \\ &= \left\{6.023 \times 10^{20} \cdot \sigma_a \cdot \log_{10}\left(\exp(1)\right)\right\} \cdot c \cdot z \\ &= \varepsilon \cdot c \cdot z \end{aligned} \tag{2.14}$$

where z is the length of the specimen. We define ε as the molar extinction coefficient (unit: $\text{cm}^{-1}\ \text{mol}^{-1}$ liter[a]), which is another parameter to measure the "absorbing power" of the chromophore:

$$\varepsilon = 6.023 \times 10^{20} \cdot \sigma_a \cdot \log_{10}\left(\exp(1)\right) = 2.616 \times 10^{20} \cdot \sigma_a \tag{2.15}$$

2.4.1 *Absorption Properties of Biological Cells and Tissues*

Light absorption is one of the most important intrinsic properties of the cells and tissues that can be harnessed in both diagnostic and therapeutic applications:

- **Diagnostic applications:** The resonance frequencies of a molecule are well defined and could serve as the spectral "fingerprint" to help identify the chemical constituents in the cells and tissues. They are useful for, e.g., tumor detection and other physiological assessments (e.g., pulse oximetry)

---

[a] In the older literature, the unit of the extinction coefficient is defined as $\text{cm}^2\ \text{mol}^{-1}$, which make the quoted value 1000 times larger than that with the unit of $\text{cm}^{-1}\text{mol}^{-1}$ liter.

- **Therapeutic applications:** Absorption is the primary mechanism that allows light from a light source (mostly laser) to produce physical or biochemical effects on tissue for various treatment purposes, e.g., Lasik (laser-assisted in situ keratomileusis) eye surgery, laser hair removal, tattoo removal, photodynamic therapy (PDT). It will be further discussed in Section 2.12.

As mentioned earlier, light absorption has two contributions: electronic and molecular. Typically, electronic contribution constitutes the absorption of UV and visible light whereas molecular contribution typically leads to absorption of near IR (NIR) to IR light. Here is a description of the common chromophores in biological cells and tissues covering the absorption spectrum from UV to IR:

UV light absorption

- *Amino acids and proteins:* The side chains of some typical amino acids, e.g., Glu, Gln, Asn, His, Arg, and Asp, have transitions around 210 nm. In particular, the aromatic ring structures in the amino acids, e.g., phenylalanine, tyrosine, and tryptophan, have the strong absorption peaks in the 260 to 280 nm range and some higher peaks at shorter wavelengths (Fig. 2.15). These features distinguish them

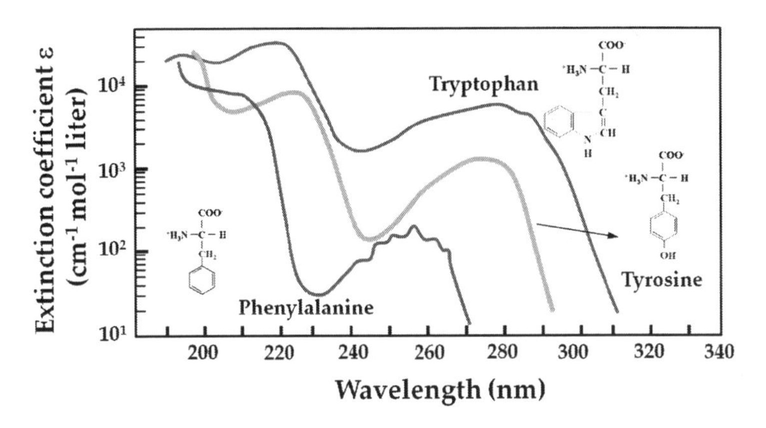

Figure 2.15 Absorption spectra of three standard amino acids: tryptophan, tyrosine, and phenylalanine. Reproduced with permission from [9].

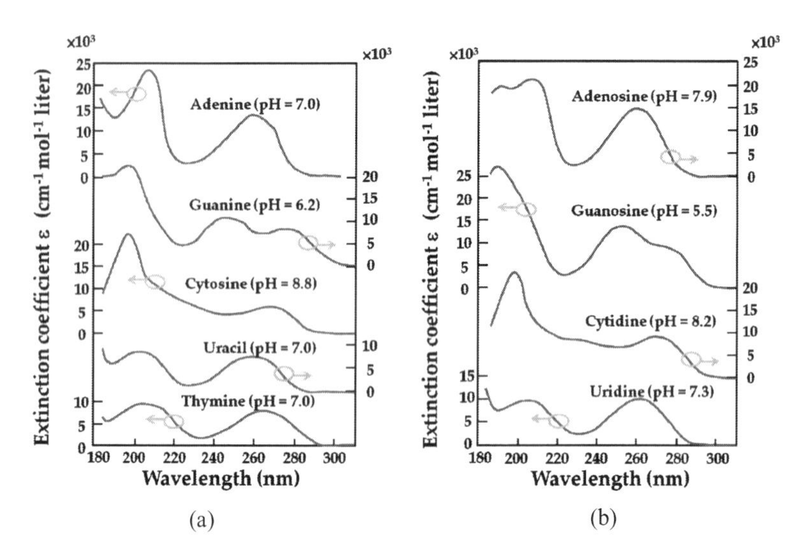

Figure 2.16 Absorption spectra of (a) the bases and (b) the ribonucleosides, which are the building blocks for the DNA and RNA. Note that the absorption spectra are pH-value dependent. Reproduced with permission from [10].

from the other amino acids. The fact that tryptophan is generally the most absorptive one makes it the common measure of protein concentration.

- *Nucleic acids (ribonucleic acid [RNA] and deoxyribonucleic acid [DNA])*: The four nucleotide bases of the DNA (adenine [A], guanine [G], cytosine [C], and thymine [T]) have the absorption peaks in the range of 260–270 nm (Fig. 2.16a). However, the absorption falls drastically (by 3–4 orders of magnitude) as the wavelength is beyond the visible range. RNA exhibits similar absorption properties, having the absorption peaks at ~260 nm (Fig. 2.16b).
- *Coenzymes (nicotinamide adenine dinucleotide [NAD^+])*: NAD^+ is one of the most important coenzymes in the biological cells. It is essential for metabolism which is responsible for cellular energy production. During metabolism, NAD^+ undergoes a redox reaction, which forms NADH. Measurements of NAD^+ and NADH absorption can be used to monitor the metabolism occurring in biological cells and

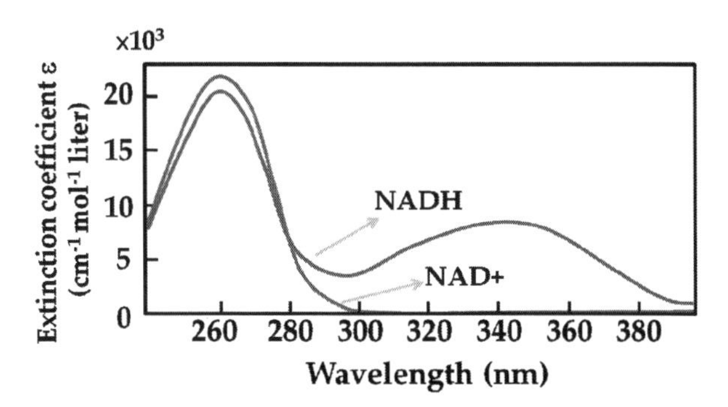

Figure 2.17 Absorption spectra of NADH and NAD+. Reproduced with permission from [9].

tissues, especially useful for drug discovery. NADH has the absorption peaks at 340 nm and 260 nm, while NAD^+ absorbs only at 260 nm (Fig. 2.17).

Visible to NIR light absorption

- *Hemogoblin:* It is an iron-containing oxygen-transport protein in the red blood cells (RBCs). About 98% of the oxygen is carried by hemogoblin molecules in the blood whereas the remaining 2% is dissolved in the blood plasma. There are two main types of hemogoblin in normal blood: (1) *Oxyhemoglobin* (HbO_2) is formed during respiration when the oxygen atoms bind to hemoglobin in RBCs. (2) *Deoxyhemoglobin* (*Hb*) is the form of hemoglobin without the bound oxygen. Under certain pathologic conditions, small concentrations of *methemoglobin (Methb) or carboxyhemoglobin (COHb)* would also appear in blood. It is a common practice to use *oxygen saturation* (SpO_2) to evaluate the oxygen content in blood. It is defined as the ratio of oxyhaemoglobin over all types of haemoglobin. Oxygen saturation is an important indicator of oxygen delivery and utilization as well as metabolic activity. The absorption spectra of different types of hemogoblin show distinct differences in the visible to NIR wavelength range. For examples, absorption peaks of HbO_2 appear at 418, 542, 577, and 925 nm while those of Hb appear at 550, 758, and

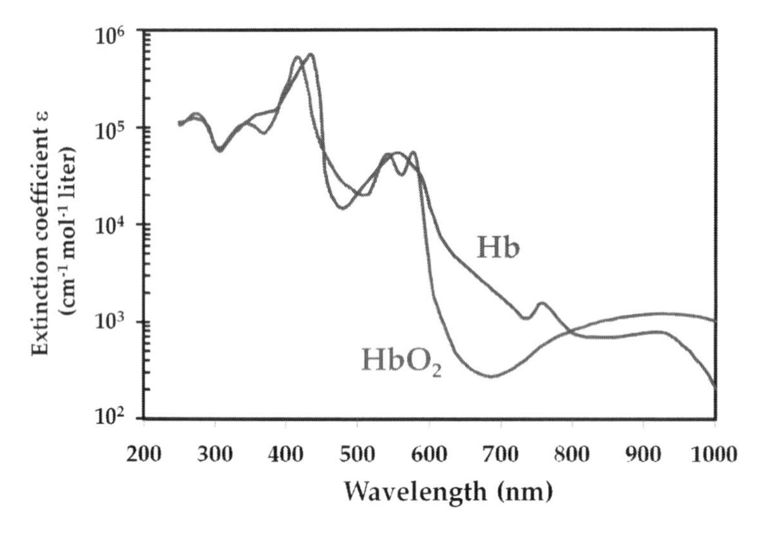

Figure 2.18 Absorption spectra of oxyhemogoblin (HbO_2) and deoxyhemogoblin (Hb) [9].

910 nm (Fig. 2.18). As a result, by performing the multiple light absorption measurements of the pulsatile arterial blood at different wavelengths, one can easily extract the value of SpO_2. This is the basic principle of *pulse oximetry*, a non-invasive optical technique to measure the oxygen content in blood [11].

- *Melanins*: They are the biopolymers which mainly give rise to the skin pigmentation and are the dominant chromophores in the skin. Melanin is also involved in skin diseases such as malignant melanoma. Most of light absorption of epidermis is usually dominated by melanin absorption. In the visible range, melanin exhibits a strong absorption, which decreases at longer (NIR) wavelengths (Fig. 2.19). There are two types of melanins: eumelanin (black-brown) and pheomelanin (red-yellow).

NIR to IR light absorption

- *Water*: Typically, water content in humans constitutes ~60% of total body weight. Because of its high concentration in most biological tissues, water is considered to be

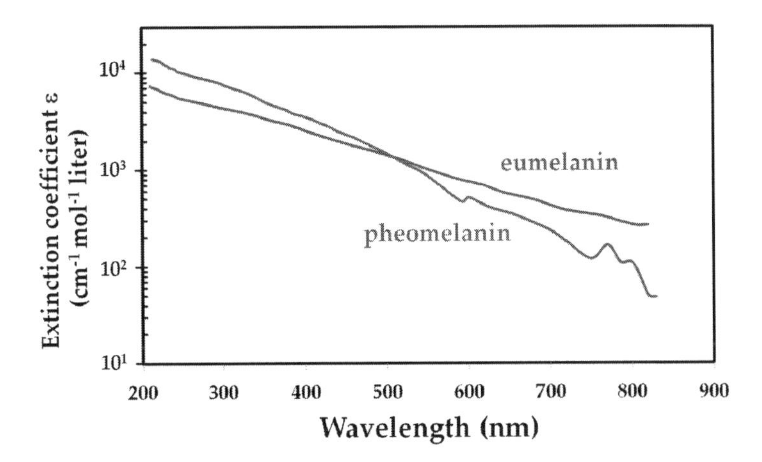

Figure 2.19 Absorption spectra of eumelanin and pheomelanin. Based on Refs. [12, 13].

one of the most important chromophores in tissue spectroscopic measurements. Water has a complex absorption spectrum ranging from UV to IR. Its strong IR absorption, particularly peaks at ~3 μm, ~1.9 μm, and ~1.4 μm, originates from different forms of molecular vibrational resonances (the –OH bonds). It exhibits low absorption in the visible range, which accounts for the transparent nature of water. The absorption rises significantly as the wavelength moves into the UV range because of the electronic contribution (Fig. 2.20).

As discussed, IR absorption of the biomolecules (e.g., proteins and lipids) originates from the different forms of the molecular vibrations. The fact that these absorption peaks (i.e., resonance frequencies and their *overtones*[a] of the molecule) are well defined and unique makes them useful for identifying the chemical constituents in the cells and tissues. The technique which is used to perform such spectral measurement is called *IR spectroscopy* or *NIR*

[a]While IR absorption (loosely from ~3 to ~10 μm) corresponds to the *fundamental* resonance frequencies of the molecules, *overtone* vibrational frequencies (overtone is referred to any resonant frequency above the fundamental frequency) of the molecules typically constitute NIR absorption (from ~800 to ~1600 nm).

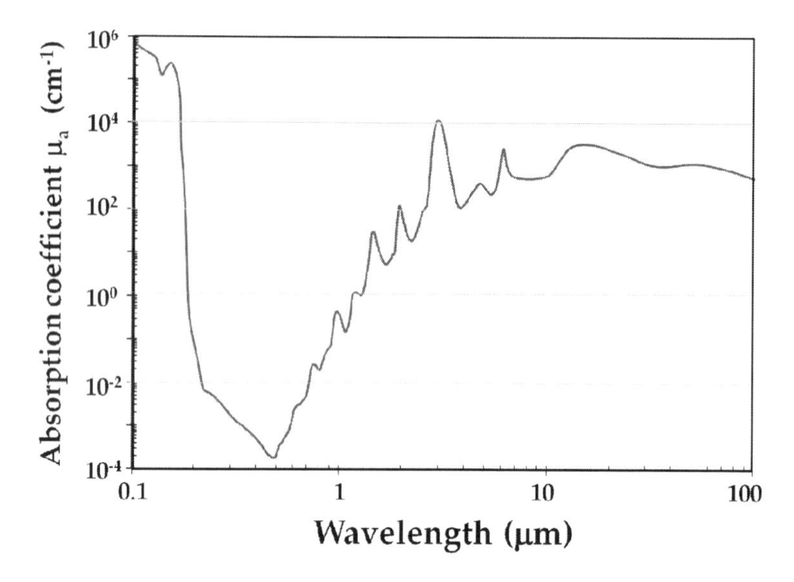

Figure 2.20 Absorption spectrum of water [4].

spectroscopy [14–16]. Some representative vibrational frequencies are listed in Table 2.1.

2.5 Dispersion

From Section 2.2, we know that the refractive index is in general frequency dependent $n(\omega)$—an optical effect called *dispersion*, or called *chromatic dispersion*. This explains why a prism can spread out a beam of white light into a rainbow, which is because of that different colors are refracted at different angles based on the Snell's law (Fig. 2.21a). Dispersion also explains the *chromatic aberration* of the lenses, which needs to be corrected in high-quality optical imaging (Fig. 2.21b).

Dispersion is particularly important for describing the optical pulse propagation in the dispersive media. In Chapter 1, we know that an optical pulse consists of a band of frequencies in its frequency spectrum. When the pulse is propagating in a dispersive medium, each frequency component experiences different phase velocity because of dispersion. Hence, the individual frequency

Table 2.1 Molecular vibrational frequencies of typical chemical bonds [17]. The frequency is in cm^{-1}, which is commonly used in IR spectroscopy

Bond	Vibrational motion	Functional group	Frequency cm^{-1}
O–H	stretch	alcohols, phenols	3640–3200
N–H	stretch	amines, amides	3400–3250
O–H	stretch	carboxylic acids	3300–2500
C–H	stretch	alkynes (–C≡C–H)	3330–3270
C–H	stretch	aromatics	3100–3000
C–H	stretch	alkanes	3000–2850
C–H	stretch	aldehydes (H–C=O)	2830–2695
S-H	stretch	thoils	2600–2550
C≡N	stretch	nitriles	2260–2210
-C≡C-	stretch	alkynes	2260–2100
C=O	stretch	carbonyls, esters, ketones	1760–1665
-C=C-	stretch	alkenes	1680–1640
N–H	bend	primary amines	1650–1580
C–C	stretch (in–ring)	aromatics	1600–1400
N–O	asymmetric stretch	nitro compounds	1550–1475
C–H	bend	alkanes	1470–1450
C–H	rock	alkanes	1370–1350
N–O	symmetric stretch	nitro compounds	1360–1290
C–N	stretch	aromatic amines	1335–1250
C–O	stretch	carboxylic acids, esters, ethers	1320–1000
C–N	stretch	aliphatic amines	1250–1020
=C–H	bend	alkenes	1000–650
O–H	bend	carboxylic acids	950–910
N–H	wag	primary, secondary amines	910–665
C–Cl	stretch	alkyl halides	850–550
C–H	rock	alkanes	725–720
C-S	stretch	disulfides	705–570
S-S	stretch	disulfides, aryldisulfides	620–430

components are separated in time. In the other words, the *instantaneous* frequency changes (or "sweeps") in time within the pulse. The consequence of this effect is *pulse broadening* or *chirping* (Fig. 2.22).

Recall that the speed of an optical pulse should be described by group velocity, *not* by phase velocities of the individual frequencies. The defining group velocity of an optical pulse thus becomes particularly essential in the dispersive media. In the presence of

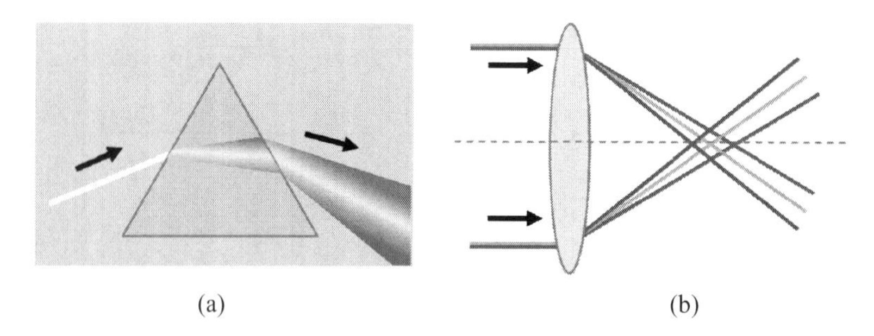

(a) (b)

Figure 2.21 Examples of the dispersion effect: (a) creating rainbow using a prism; (b) chromatic aberration of a simple lens which leads to different focal points for different colors and thus degrades the image quality.

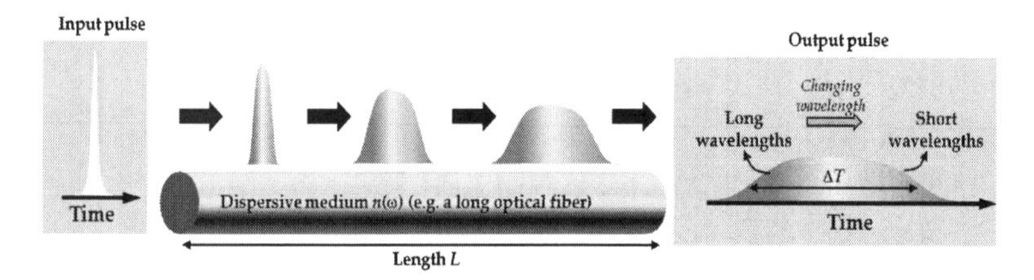

Figure 2.22 Pulse broadening or chirping in a dispersive medium. Given a pulse with a bandwidth of $\Delta\lambda$, the pulse broadening ΔT after traveling for a distance L is given by $\Delta T = DL\Delta\lambda$ (Eq. 2.20), where D is the group velocity dispersion (GVD) of the medium.

dispersion, Eq. 1.3 in Chapter 1 should be frequency dependent, i.e.,

$$k(\omega) = \frac{\omega}{c} n(\omega) \tag{2.16}$$

Equation 2.16 is also known as *dispersion relation*. As described in Eq. 1.19 in Chapter 1, the group velocity is defined as $\upsilon_g = \partial\omega/\partial k$. Using Eq. (16), the group velocity can be rewritten as

$$\upsilon_g = \frac{c}{n + \omega\frac{\partial n}{\partial \omega}} \equiv \frac{c}{n_g} \tag{2.17}$$

where n_g is known as group index

$$n_g = n + \omega\frac{\partial n}{\partial \omega}^{a} \tag{2.18}$$

[a] On the basis of the sign of the frequency derivative of refractive index, $\partial n/\partial\omega$, we can define two regimes of dispersion: *normal dispersion* when $\partial n/\partial\omega$ is positive (or D is

Consider an optical pulse traveling at a speed of υ_g in a dispersive medium with a length of L, the total time taken to travel through the medium is $T = L/\upsilon_g$. If the pulse has a frequency bandwidth $\Delta\omega$, the pulse broadening ΔT is given by

$$\Delta T = \frac{\partial T}{\partial \omega}\Delta\omega = \frac{\partial}{\partial \omega}\left(\frac{L}{\upsilon_g}\right)\Delta\omega = L\frac{\partial^2 k}{\partial \omega^2}\Delta\omega = L\beta_2\Delta\omega \qquad (2.19)$$

where β_2 is the second derivative of $k(\omega)$ and is called *group velocity dispersion (GVD) parameter* (unit: s^2/m). It describes the "dispersive power" of the medium. If we write Eq. 2.19 in terms of wavelength, i.e., for a given bandwidth of $\Delta\lambda$, ΔT would be

$$\Delta T = \frac{dT}{d\lambda}\Delta\lambda = \frac{d}{d\lambda}\left(\frac{L}{\upsilon_g}\right)\Delta\lambda = LD\Delta\lambda \qquad (2.20)$$

where D is the GVD parameter in terms of wavelength and is given by

$$D = -\frac{2\pi c}{\lambda^2}\beta_2 \qquad (2.21)$$

Equation 2.19 or 2.20 gives a useful estimation of pulse broadening in the presence of dispersion. The larger the frequency bandwidth (i.e., ultrashort pulses, e.g., femtosecond pulses) or the higher the GVD parameter (i.e., highly dispersive), the wider the pulse broadening. This relation is useful for the applications in which the pulse broadening needs to be carefully manipulated. For instance, in multiphoton microscopy, broadening of the femtosecond pulses due to the dispersion introduced by the optical components in the microscope should be minimized or compensated (Chapter 3).

It is worthwhile to emphasize two important points related to dispersion: (i) The frequency spectrum of the pulse is actually mapped into the time domain. Hence, dispersion can be exploited as a mechanism in which the frequency spectrum can be captured in real time and thus is useful for high-speed optical spectroscopy—a technique called *optical time-stretch* or *dispersive Fourier transform (DFT)* [18, 19]. (ii) There is a fundamental connection between dispersion and absorption (as shown in Figs. 2.4 and 2.12) described

negative) and *anomalous dispersion* when $\partial n/\partial\omega$ is negative (or D is positive). The propagation characteristics of an optical pulse depend largely on the signs of $\partial n/\partial\omega$ or equivalently D. Further discussion can be found in Refs. [1, 2, 5, 6].

by the *Kramers–Kronig relations*, which states that dispersive media inevitably exhibit optical loss [2, 3]. This explains why the amplitude of the broadening pulse is attenuated during propagation (Fig. 2.22).

2.6 Light Scattering

Light scattering is a physical process in which the light path is deviated from a straight trajectory by one or more localized non-uniformities (or *scatterers*) in the medium. Scatterers can be particles, bubbles, droplets, density fluctuations in fluids, defects in crystalline solids, or surface roughness. In the biological tissue, light scattering generally arises from the presence of heterogeneities such as physical inclusions or non-uniform distribution of refractive indexes of different cellular components, such as cell membranes and organelles.

Microscopically, light scattering can be understood by that when the charged particles in a medium are driven into oscillatory motion (at nonresonance frequencies) by the incident light, they almost instantaneously re-emit (as opposed to absorb) light of the same frequency as the incident wave. This is known as *elastic scattering*—no net energy exchange between the light and the medium (remember photon energy is proportional to the light frequency, i.e., $E = h\nu$). In contrast, if the input energy is partially absorbed, the energy of the re-radiation would be less than the input energy, i.e., the frequency of the re-emitted light will be *down-shifted* relative to that of the incident light. This is known as *inelastic scattering*—a net energy exchange occurs.

In either case, the re-radiation, i.e., scattering, direction depends on various parameters such as (i) the size, the shape, and the orientation of the scatterers, (ii) the refractive index contrast between the scatterers and the ambient environment, (iii) the light wavelength, and (iv) the light polarization state. Similar to the case of light absorption, we can define a scattering cross-section σ_S to quantify the "scattering power" of the scatterer. By treating a scatterer as an idealized sphere, we can imagine this sphere scatters the incoming light and casts a shadow. The size of this shadow is

Figure 2.23 Defining scattering cross-section, σ_s. Similar to Fig. 2.14, we can regard a scatterer as an idealized sphere which scatters the incoming light and effectively casts a shadow. The size of the shadow is defined as the scattering cross-section.

defined as σ_s (Fig. 2.23). Similar to light absorption, if the density of the scatterers (i.e., the number of scatterers per unit volume) is N_s, we can define a scattering coefficient μ_s (unit: m^{-1} or cm^{-1}) which is given by

$$\mu_s = N_s\sigma_s. \tag{2.22}$$

μ_s can be regarded as another optical loss factor in the medium apart from the absorption coefficient μ_a. Different types of tissues or even the same type of tissues but in different forms (e.g., normal versus cancerous) can give rise different μ_s. As a result, this serves as a useful parameter for quantitative measurements in medical diagnostics. The scattering coefficients of the brain and skin tissues are shown in Figs. 2.24 and 2.25. Some other representative tissues are listed in Table 2.2.

It is common to describe the angular distribution of scattered light by considering that there is an associated probability for a photon to be scattered at an angle θ. If the probabilities of light scattering in all directions are equal, the scattering is *isotropic*. Otherwise, it is said to be *anisotropic*. The anisotropy of scattering can be measured by a parameter called g factor, which represents the average value of the cosine of the scattering angle in all directions,

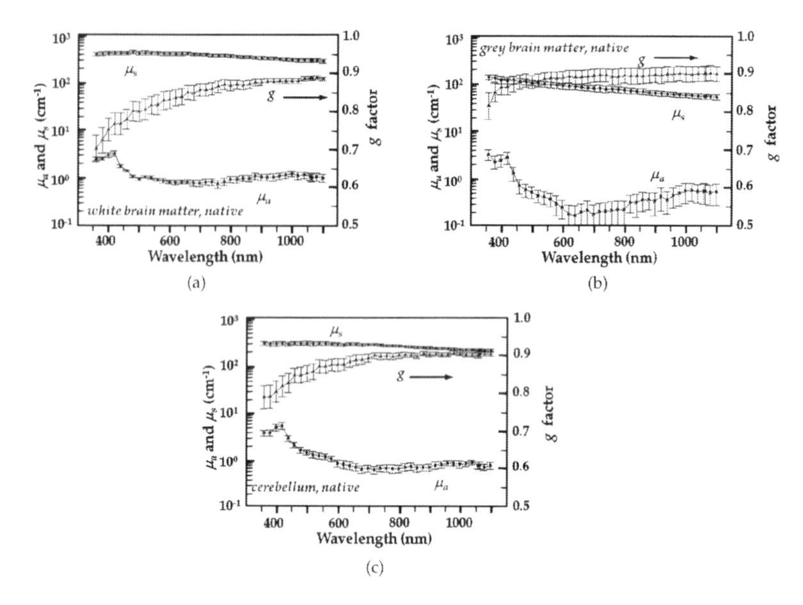

Figure 2.24 Scattering and absorption coefficients (cm^{-1}) of different brain tissues: (a) white brain matter, (b) grey brain matter, and (c) cerebellum. The anisotropy parameter, called g factor, is also plotted in the figures. On the basis of the figures, the reduced scattering coefficient μ_s' can be calculated by Eq. 2.33 [20].

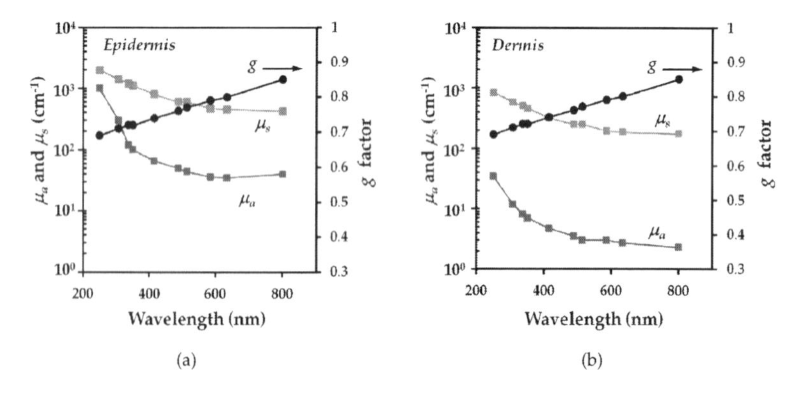

Figure 2.25 Scattering and absorption coefficients (cm^{-1}) of the skin tissues: (a) epidermis and (b) dermis. The anisotropy parameter, called g factor, is also plotted in the figures. Reproduced with permission from [22].

Table 2.2 Scattering coefficients (cm^{-1}) of the selected tissues [22]

Tissues	Wavelength (nm)	$\mu_s(cm^{-1})$	g factor
Lung	635	324	0.75
	1064	39	0.91
Bladder, mucous	1064	7.5	0.85
Bladder, wall	1064	54.3	0.85
Bladder, integral	1064	116	0.9
Heart, myocardial	1060	177.5	0.96
Heart, epicardial	1060	127.1	0.93
Heart, aneurysm	1060	137	0.98
Heart, trabecula	1064	424	0.97
Liver	635	313	0.68
	1064	356	0.95
Colon, muscle	1064	238	0.93
Colon, submucous	1064	117	0.91
Colon, mucous	1064	39	0.91
Colon, integral	1064	261	0.94
Spleen	1064	137	0.9

i.e.,

$$g = \langle \cos(\theta) \rangle \tag{2.23}$$

where $\langle \cdot \rangle$ represents the average value [9, 21]. Note that when the light is scattered by a particle so that its trajectory is deflected by an angle θ, $\cos(\theta)$ represents the forward component of a newly scattered trajectory (Fig. 2.26). There are three limiting cases of g: (i) when $g = 1$, the light is completely forward scattered; (ii) when $g = 0$, the light scattering is isotropic; (iii) when $g = -1$, the light is completely backward scattered. In biological tissue, g falls within the range $0.69 \leq g \leq 0.99$, which means the scattering is preferentially forward (see Ref. [22], Figs. 2.24 and 2.25, and Table 2.2).

Elastic scattering can be classified into three main types in terms of the scatterers' size relative to the wavelength: (i) *Rayleigh scattering*, where the wavelength is much smaller than the scatterers; (ii) *Mie scattering*, where the size of the scatterer is comparable to the wavelength; and (iii) *geometric scattering*, where the scatterer is much larger than the wavelength. In the third case, the scattered light trajectory can be simply evaluated

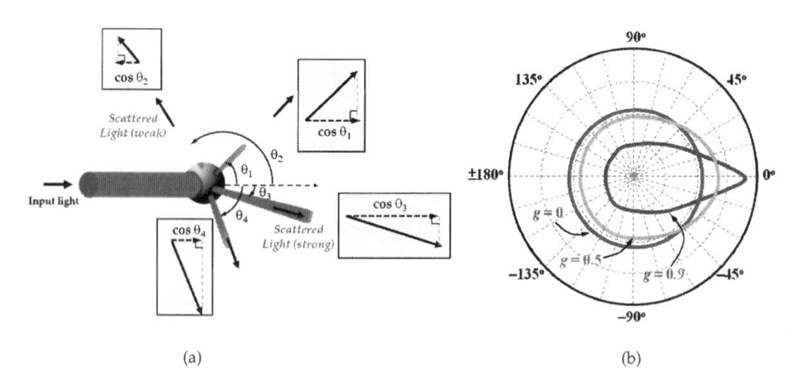

Figure 2.26 The anisotropy of scattering can be quantified by the g factor, which represents the average value of the cosine of the scattering angle in *all directions*. The cosine is essentially the component parallel to the incident direction (see the dash vectors shown in the 4 boxes in (a)). As illustrated in a simple example in (a), the light is scattered in four different directions (three in the forward direction and one in the backward direction), each of which has different scattered intensity. From this figure, we can observe that *on average* the light tends to be forward scattered. A more realistic picture is shown in (b), which is the scattered light intensity distribution (plotted in a polar coordinate system). When $g = 0$, the scattering is isotropic (the distribution is in a circular shape). In contrast, the majority of the light is forward scattered when $g = 0.9$ (the distribution is distorted toward the forward direction.

by means of geometrical optics, e.g., refraction or reflection at boundaries. Rayleigh and Mie scattering will be further discussed in the following sections as they represent the key scattering mechanisms in biological tissues and cells.

2.6.1 *Rayleigh Scattering*

Scattering by a very small particle, with the size much smaller than the wavelength, $<\sim\lambda/10$, is classified as Rayleigh scattering. The scattered intensity (or power) is inversely proportional to the fourth power of the wavelength of the incident light:

$$I \propto \frac{1}{\lambda^4} \tag{2.24}$$

It is this scattering that leads to the blue color of the sky during daytime but a reddish sky during sunset [1]. In a biological cell,

Rayleigh scattering is given by those ultrafine structures such as the lipid bilayer of the cell membranes (with a thickness of ~10 nm), cell subcompartments, and other extracellular components, e.g., the banded structure of collagen fibrils (with a periodicity of ~70 nm).

2.6.2 *Mie Scattering*

When the particles' sizes are comparable or larger than the wavelength, Mie scattering predominates. Mie scattering can be modeled reasonably well by Mie theory, which assumes that the scatterer has a spherical shape [9]. In this scattering regime, light scattering occurs preferably in the forward direction and is not strongly wavelength dependent. The scattered intensity is given by

$$I \propto \frac{1}{\lambda^x} \tag{2.25}$$

where $0.4 \leq x \leq 0.5$. Most of cellular structures fall in the Mie regime—the scattered light is highly anisotropic (forward)—for example, the mitochondria (0.5 to 2 μm depending upon the cell types), which are responsible for energy supply in cells, and smaller vesicle-type organelles (0.25 to 0.5 μm) such as lysosomes and perioxisomes.

From Eqs. 2.23 and 2.24, one can observe that the scattering strength generally follows a monotonically decreasing trend in both Rayleigh and Mie scattering—consistent with the wavelength dependence of the measured scattering coefficients shown in Figs. 2.24 and 2.25.

2.7 Light Transport in Tissue

In most biological tissues, multiple scattering is inevitable. Hence, the interaction of scattered light waves between neighboring scatterers cannot be neglected. Because of this effect (together with light absorption), most tissues are considered to be *turbid*. This makes studying and modeling of light transport in tissue not a trivial task, especially in deep tissue (> cm below the surface). For instance, it is in general challenging to image deep tissue structures with *cellular resolution* as the multiple scattering would disturb

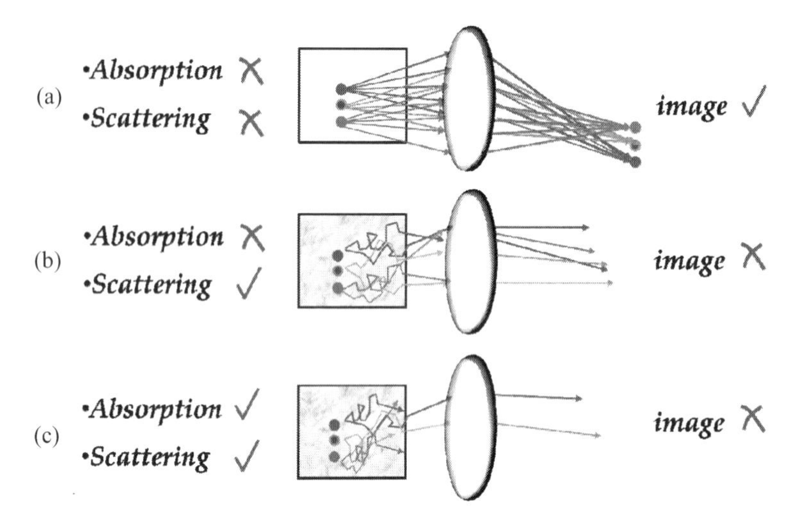

Figure 2.27 An illustration showing the impacts of the absorption and scattering on image formation of deep tissue features (i.e., three point sources embedded in the sample in this figure). (a) In a lossless (no absorption) and clear (no scattering) sample, the light rays emerging from the point sources can perfectly be routed to the image plane by the lens. An image can thus be formed. (b) In a lossless but turbid sample, the light scattering disturbs the trajectories of the light rays from the sources. Thus they cannot be routed to the image plane properly and a blurred image (or even no image) is formed. The situation can be even worse in a lossy and turbid sample, as shown in (c).

the light trajectories emerging from the object to be imaged and would make the light "diffuse" in the tissue. Therefore, it results in severe image blurring (Fig. 2.27). In this scenario, rather than directly capturing the image, one would need to reconstruct the image by a technique called *optical tomography (OT)* [23, 24]. This is essentially an inverse problem (it is thus computationally intensive) which relies on extracting the quantitative information of the light transport characteristics (e.g., absorption and scattering) in turbid tissues for image reconstruction. There are a number of models developed for studying the characteristics of light propagation. They include radiation transport theory, diffusion approximation theory, and the Monte Carlo approach.

2.7.1 *Radiative Transport Theory*

In principle, EM theory can serve as the fundamental approach to study light transport in tissue. Multiple scattering in the heterogeneous tissue structures, however, makes it very complicated to be directly applied to such problems. Instead, a model known as *radiative transport (RT) theory* can be employed. This theory deals with the light energy transport and neglects the wave properties of the light, namely coherence, diffraction, interference, and polarization.

Without going into the detailed derivation of the complete RT theory, we can follow a heuristic approach to comprehend the overall picture of the theory. The general idea of the RT theory is to consider changes in energy flow due to the incoming, outgoing, absorbed and emitted (e.g., fluroescent) photons in an infinitesimal volume dV within differential solid angle $d\Omega$ around the direction $\hat{\mathbf{s}}$ (Fig. 2.28). It is essentially a problem considering the conservation of energy in such volume. Here, it is useful to first define an energy flow quantity called *radiance* L (unit: $\mathrm{Wm^{-2}\,sr^{-1}}$),[a] which is defined as the energy flow per unit area per unit solid angle per unit time. Hence, radiance is a function of not only position and time but also the direction, i.e., $L(\mathbf{r}, \hat{\mathbf{s}}, t)$, where $\hat{\mathbf{s}}$ is the energy flow direction. Note that intensity I (unit: $\mathrm{W\ m^{-2}}$) is given by the total radiance integrated over the entire 4π solid angle. In the RT theory, there are in general four different mechanisms contributing to the overall energy change within the volume dV:

(i) *Divergence of the light from the volume.* This is the energy diverging out of the volume within the differential solid angle $d\Omega$ around the direction $\hat{\mathbf{s}}$. This is particularly relevant to the case when the incident light beam is not collimated. Mathematically, the fractional power loss δP_{div} (unit: W) within $d\Omega$ and dV due to this divergence is given by

$$\delta P_{\mathrm{div}} = [\hat{\mathbf{s}} \cdot \nabla L(\mathbf{r}, \hat{\mathbf{s}}, t)]\, d\Omega dV \qquad (2.26)$$

[a] sr, called steradian, is the unit of solid angle, i.e., to describe two-dimensional angular spread in a three-dimensional space. The solid angle of a sphere (i.e., including all directions) measured from a point is 4π sr.

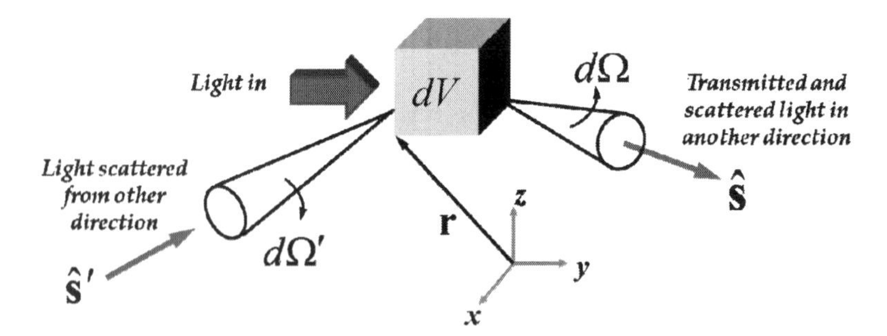

Figure 2.28 A schematic of describing the light propagation in the radiative transport (RT) theory. Note that the light along the direction $\hat{\mathbf{s}}$ include the scattered light, "residual" light from the absorption, and fluorescent light from the volume dV, as stated in Eqs. 2.26, 2.28, and 2.29.

(ii) *Scattering loss and absorption loss in the volume.* Specifically, it refers to the energy loss due to light scattered out of the volume and light absorption inside the volume. This power loss δP_l is given by

$$\delta P_{\mathrm{l}} = [(\mu_{\mathrm{a}} + \mu_{\mathrm{s}})\, L(\mathbf{r}, \hat{\mathbf{s}}, t)]\, d\Omega dV \qquad (2.27)$$

where the absorption coefficient μ_{a} (unit: m^{-1}) and the scattering coefficient μ_{s} (unit: m^{-1}) are used to characterize this effect.

(iii) *Scattering from other directions to the volume.* It refers to the "energy gain" due to the light scattered from all other directions to this volume. To evaluate this contribution, it is necessary to consider the probability of the light within the differential solid angle of $d\Omega'$ around its original direction $\hat{\mathbf{s}}'$ being scattered into the direction $\hat{\mathbf{s}}$ in Fig. 2.28. The functional form of this energy gain δP_{s} is given by

$$\delta P_{\mathrm{s}} = \left[\mu_{\mathrm{s}} \int_{4\pi} L(\mathbf{r}, \hat{\mathbf{s}}, t)\, p\left(\hat{\mathbf{s}}, \hat{\mathbf{s}}'\right) d\Omega'\right] d\Omega dV \qquad (2.28)$$

where the integral over the entire 4π solid angle means that the possible scattered light from all directions is included. The aforementioned probability is specified by the probability function $p\,(\hat{\mathbf{s}}, \hat{\mathbf{s}}')$.

(iv) *Internal light source.* This includes the light generated within the volume, e.g., fluorescence or other types of luminescence. This power gain is given by

$$\delta P_{\text{source}} = Q(\mathbf{r}, \hat{\mathbf{s}}, t)\, d\Omega dV \tag{2.29}$$

where $Q(\mathbf{r}, \hat{\mathbf{s}}, t)$ is the source power per unit volume per unit solid angle (unit: W m^{-3} sr^{-1}).

The total power change per unit volume per unit solid angle can be expressed as

$$\delta P_{\text{T}} = \frac{\partial \left(\frac{L(\mathbf{r}, \hat{\mathbf{s}}, t)}{c}\right)}{\partial t} d\Omega dV \tag{2.30}$$

where $L(\mathbf{r}, \hat{\mathbf{s}}, t)/c$ is the energy per unit volume per unit solid angle. Following energy conservation, the power change within this volume is given by

$$\delta P_{\text{T}} = -\delta P_{\text{div}} - \delta P_{\text{l}} + \delta P_{\text{s}} + \delta P_{\text{source}} \tag{2.31}$$

where the negative and positive signs specify energy loss and gain in this volume, respectively. Therefore, the complete formulism should be

$$\begin{aligned}\frac{1}{c}\frac{\partial L(\mathbf{r}, \hat{\mathbf{s}}, t)}{\partial t} &= -\hat{\mathbf{s}} \cdot \nabla L(\mathbf{r}, \hat{\mathbf{s}}, t) - (\mu_{\text{a}} + \mu_{\text{s}})\, L(\mathbf{r}, \hat{\mathbf{s}}, t) \\ &\quad + \mu_{\text{s}} \int_{4\pi} L(\mathbf{r}, \hat{\mathbf{s}}, t)\, p(\hat{\mathbf{s}}, \hat{\mathbf{s}}')\, d\Omega' + Q(\mathbf{r}, \hat{\mathbf{s}}, t) \qquad (2.32)\end{aligned}$$

which is the basic equation of the RT theory [25]. Table 2.3 summarizes the four main contributions in this theory. Note that if the incident light is a CW beam (i.e., $\partial L(\mathbf{r}, \hat{\mathbf{s}}, t)/\partial t = 0$), the medium is source-free (i.e., $Q(\mathbf{r}, \hat{\mathbf{s}}, t) = 0$), and the scattering effect is ignored (i.e., $\mu_s = 0$), Eq. 2.31 can be reduced to Beer's law (Eq. 2.8).

2.7.2 *Diffusion Approximation*

Direct analytical solution of the RT equation (Eq. 2.31) is complicated, and is available only for simple geometries. Therefore, in most occasions, brute-force numerical simulations of the complete RT equation would be implemented, or approximation of the RT

Table 2.3 Four main factors contribution to the total power change per unit volume per unit solid angle considered in the RT theory (see Eq. 2.31)

1.	Light loss due to divergence of the light from the volume	$\delta P_{div} = -[\hat{\mathbf{s}} \cdot \bar{\nabla} L(\mathbf{r}, \hat{\mathbf{s}}, t)] d\Omega dV$
2.	Light loss due to absorption and scattering	$\delta P_l = -[(\mu_a + \mu_s) L(\mathbf{r}, \hat{\mathbf{s}}, t)] d\Omega dV$
3.	Light loss gain to the scattered light from all directions to the direction $\hat{\mathbf{s}}$	$\delta P_s = +[\mu_s \int_{4\pi} P(\mathbf{s}, \hat{\mathbf{s}}, t) L(\mathbf{r}, \hat{\mathbf{s}}, t) d\Omega'] d\Omega dV$
4.	Light gain due to local light source (e.g., fluorescence)	$\delta P_{source} = +Q(\mathbf{r}, \hat{\mathbf{s}}, t) d\Omega dV$

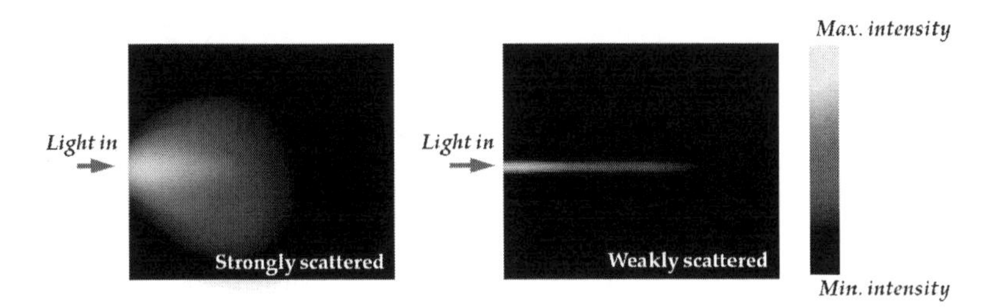

Figure 2.29 (a) When the scattering process is dominated, i.e., $\mu_s >> \mu_a$, the light scattering is almost isotropic, resembling a diffusion process. (b) When the absorption process is dominated, i.e., $\mu_s << \mu_a$, the light is weakly scattered and the propagation is primarily governed by Beer's law.

theory is employed. One commonly used approach is *diffusion approximation*, which assumes that the light transport is heavily dominated by the scattering process, i.e., $\mu_s >> \mu_a$. In this case, the significantly lower absorption allows the light to penetrate deeper into the tissue through scattering. The scattering is almost isotropic, resembling a diffusion process (Fig. 2.29). In this scattering-dominated regime, Eq. 2.31 can be simplified to a so-called *diffusion equation* [25]. The isotropic scattering is characterized by a so-called *reduced scattering coefficient* μ_s', which is given by

$$\mu_s' = (1 - g)\,\mu_s \tag{2.33}$$

μ_s' can be regarded as an effective isotropic scattering coefficient that represents the cumulative effect of several forward-scattering events (Fig. 2.30). For example, for $g = 0.9$, it takes an average of $1/(1 - g) = 10$ scattering events (each with a scattering mean free path $l_s = 1/\mu_s$) for the light to scatter in an isotropic manner (Fig. 2.30a). Recall that g in biological tissue is within $0.69 \leq g \leq 0.99$, which means the isotropic scattering effect can appear after less than 4 steps to 100 steps.

2.7.3 *Numerical Simulation: Monte Carlo (MC) Method*

When diffusion approximation fails to model the light transport, especially for the case of $\mu_s \approx \mu_a$, i.e., both the scattering and absorption coefficients have to be considered, full numerical simulation approaches are more preferable. One widely used

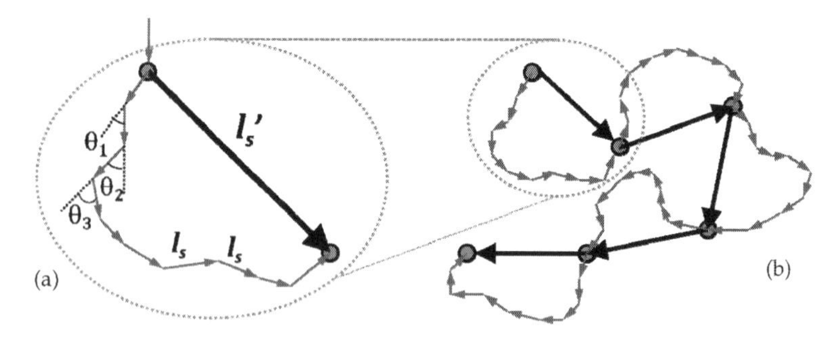

Figure 2.30 Schematics of explaining μ_s', which is regarded as an *effective* isotropic scattering coefficient that represents the cumulative effect of several forward-scattering events (each of the small arrows has a length of $l_s = 1/\mu_s$). In the other words, it is used to describe the "diffusion" of photons (isotropic scattering) in a random walk of step size, $l_s' = 1/\mu_s'$, which is the effective mean free path (i.e., length of large arrows). One *effective* scattering event is shown in (a) whereas five *effective* scattering events are shown in (b).

technique is called the Monte Carlo (MC) method. The propagation pathway of a photon is modeled as a random walk (Fig. 2.31a). The direction of each (scattering) step requires the knowledge of the prior propagation directions. The scattered direction and power for each step are calculated based on the probability distributions of scattering. By tracing a large enough number of photons (can be as large as >100,000), one can evaluate the overall light transport behavior in the medium—approaching to the exact solution of the RT theory or the diffusion approximation (Fig. 2.31a–c). Because of the large number of simulation runs involved for accurate prediction, MC method is often time-consuming and computational intensive. More details of the MC method can be found in Refs. [26, 27].

2.8 "Optical Window" for Biological Tissue

We have discussed that the major chromophores in tissue (water, oxyhaemoglobin, deoxyhaemoglobin, proteins, melanin, etc.) show their unique absorption spectra, which are all summarized in Fig. 2.32. Interestingly, there is a spectral window, between ~600

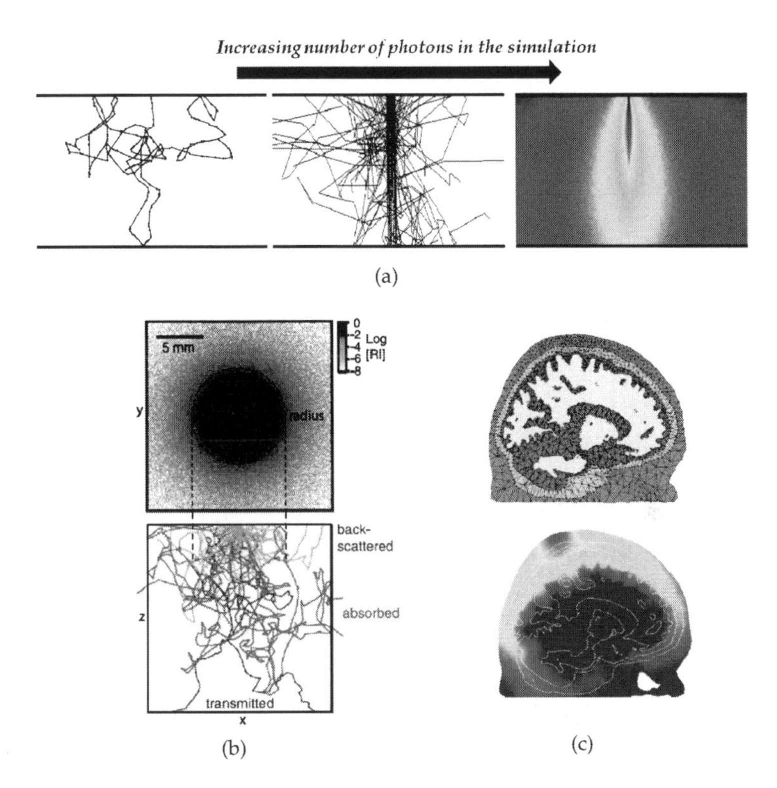

Figure 2.31 (a) Schematics of Monte Carlo simulation of the photon transport in a turbid medium. As number of photons increases (from 1 photon in the left to $\sim 10^5$ photons in the right), the light distribution can be accurately simulated and consistent with the RT theory or the diffusion approximation. (b) and (c) show two examples of using MC simulation to study the light transport in the biological tissues. (b) (Top) The top view of the backscattered light intensity distribution at the tissue surface from a focus at a depth of 100 μm into the tissue. (Bottom) Three sets of light paths (backscattered, absorbed, and transmitted) along the depth profile of tissue [28]. (c) (Top) A sagittal-cut view of the mesh of a human brain modeled in the MC simulation. The tissue layers, from exterior to interior, are scalp/skull, cerebrospinal fluid (CSF), gray matter, and white matter, respectively. (Bottom) The light distribution obtained from the MC simulation [29].

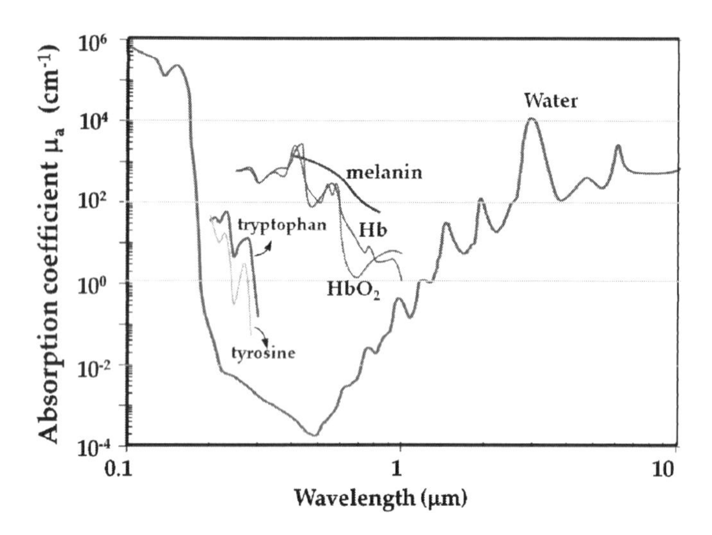

Figure 2.32 Absorption coefficients of major tissue chromophores in the UV–IR spectral region.

and ~1200 nm (i.e., NIR range), within which the absorption in tissue is relatively weak. This spectral window is commonly known as *diagnostic window* or *therapeutic window* for biophotonics. Within this window, the relatively low absorption implies that the light transport is mostly dominated by scattering, i.e., $\mu_s' >> \mu_a$ (Fig. 2.33). The light transport in tissue within this wavelength range can thus be modeled reasonably well under diffusion approximation. Outside the window, the absorption effect would be stronger than or comparable to the scattering effect depending upon the wavelengths, as shown in Fig. 2.33. For instance, light transport has to be accurately modeled by complete RT theory or MC simulation in ~1200–1400 nm and ~1600–1800 nm, where $\mu_s' \sim \mu_a$.

What makes this NIR window intriguingly special in biophotonics is its low absorption, permitting deep penetration into the tissue, which can reach up to several millimeters, and even to centimeters. The penetration depth can be quantified by defining an effective attenuation coefficient μ_{eff} (unit: cm^{-1}) [31]:

$$\mu_{\text{eff}} = \sqrt{3\mu_a\left(\mu_a + \mu_s'\right)} \tag{2.34}$$

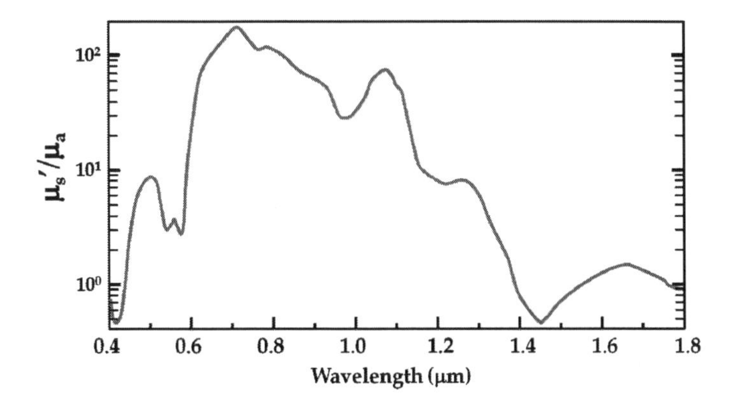

Figure 2.33 The ratio of reduced scattering coefficient to absorption coefficient of human skin [30].

The inverse of μ_{eff} is the effective penetration depth, d_{eff} (unit: cm), i.e.,

$$d_{\text{eff}} = \frac{1}{\mu_{\text{eff}}} \tag{2.35}$$

The wavelength dependence of the light penetration depth in the typical tissues (ranging from visible to IR) is depicted in Fig. 2.34.

Deep penetration is of great importance for deep tissue diagnostics of thick human tissues, such as the breast and the brain. A well-known example is pulse oximetry, which relies on the HbO_2 and Hb absorption measurements in the blood streams under the skin at ~600 nm (red light) and ~900 nm (NIR light) for *non-invasive* blood oxygenation evaluation. Serendipitously, there are a number of high-performance NIR light (laser) sources available for deep-tissue biological diagnostics, such as red and NIR laser diodes (e.g., InGaAs ~670 nm, and AlGaAs ~800 nm), superluminescent diodes (~800–1000 nm), solid-state Nd:YAG and Nd:YVO_4 lasers (1064 nm), Yb-doped fiber lasers (~1060 nm), and femtosecond Ti:sapphire lasers (~700–900 nm).

2.8.1 *Examples: NIR Spectroscopy and Tomography*

Another well-known example of making good use of this NIR window in biophotonics is to perform non-invasive cancer detection,

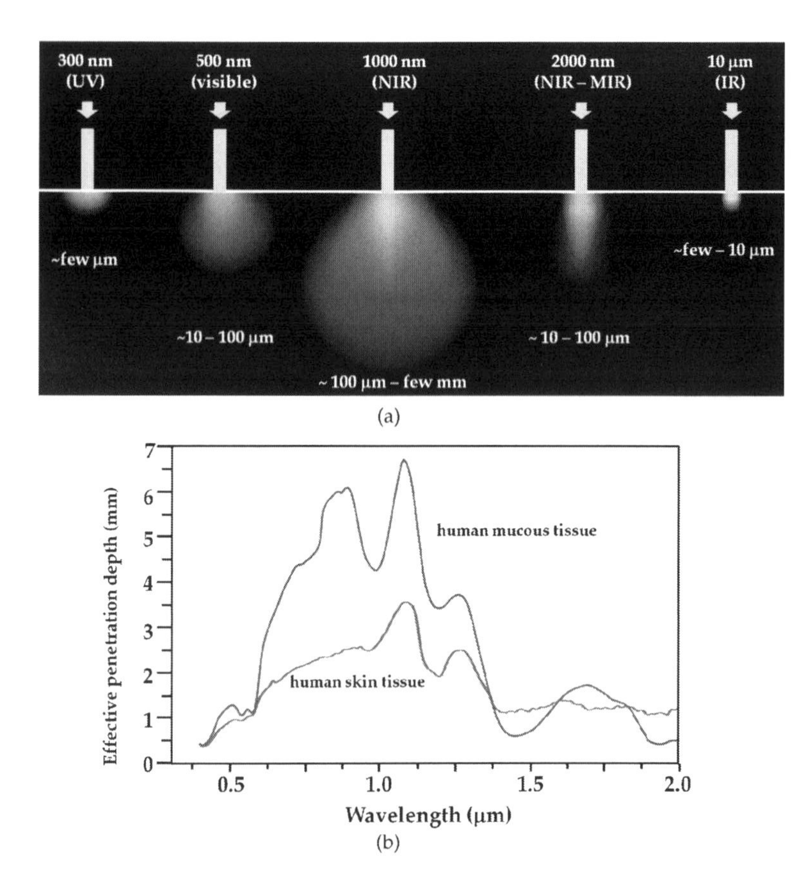

Figure 2.34 (a) An illustration of the light penetration into the biological tissue at different wavelengths from UV to IR. Note that the scattering strength decreases with wavelength. Therefore, compared to the case of the UV and visible light, the scattered light distributions of the longer NIR (~2000 nm) and IR (~10 μm) wavelengths are less "diffused." (b) The ratio of reduced scattering coefficient to absorption coefficient of human skin and human mucous tissues [32].

e.g., brain, breast, skin, and esophagus, based on *NIR spectroscopy* and *tomography*. It has been studied that tumor formation is typically associated with changes in blood oxygenation level, lipid content and water content, etc. Hence, spectroscopic measurements (namely absorption measurements) at NIR wavelengths can provide informative optical signatures of different physiological changes correlated to various pathologies, such as the growth of microvessels, fibrosis, and oxygen consumption [16, 34–37]. However, the fact that light transport in the NIR range is mainly dominated by scattering complicates the interpretation of the quantities of different chromophores, i.e., oxyhemoglobin, deoxyhemoglobin, lipids, and water. Accurate and quantitative measurements can only be made possible if the light transport characteristics in the tissues are known. In this case, modeling using the framework of the RT theory or a simpler version—diffusion approximation is necessary (see Sections 2.7.1 and 2.7.2). In many cases, they are inverse problems, which retrieve the optical parameters, namely scattering and absorption coefficients, from the measured data. Since inverse problems are ill-posed, a variety of signal processing techniques, such as optimization and statistical modeling are essential in order to obtain an accurate estimation [23–26, 38–40].

The technique of NIR spectroscopy can be further extended to imaging. However, direct imaging of deep tissue is challenging as the strong scattering of NIR light in tissues forbids high-resolution images. Hence, the most widely used approach is *diffuse optical tomography* (DOT) [23–26, 38, 41]. The central idea of DOT is to perform an image reconstruction of a spatial map of the optical properties (e.g., HbO_2 and Hb) inside the tissue based on non-invasive *in vivo* measurements of the back-scattered or/and transmitted light in the tissue. In practice, an array of light sources and photodetectors are required in order to obtain a complete spatial map (image). For each photodetector, one can measure the detected light contribution from each source. On the basis of the overall detected signal, we can reverse the RT model or the diffusion model to retrieve the unknown scattering and/or absorption coefficients at each location in the tissue, and thus to reconstruct the image (examples are shown in Figs. 2.35 and 2.36). Again, this is an inverse problem. DOT has successfully been employed to study the brain (or more precisely, cerebral) activity in real time, which is associated with the blood oxygenation level [42, 43]. DOT can also be used for early breast cancer detection based on, again, the changes in blood volume and blood oxygenation level [41].

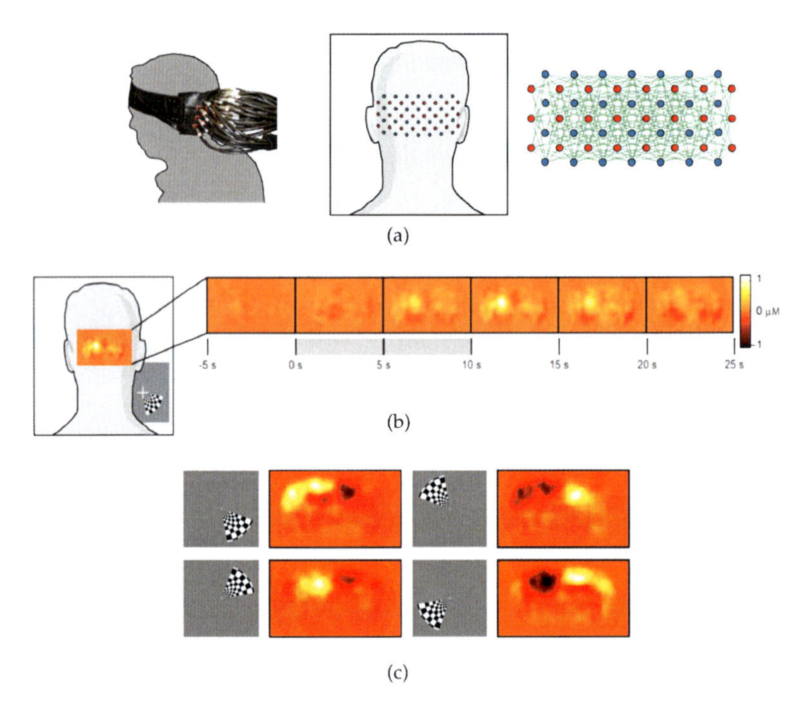

Figure 2.35 A DOT system for function hemodynamic brain imaging: retinotopic mapping of the visual cortex. (a) Schematic of the imaging grid over the visual cortex with 24 light sources (red) and 28 detectors (blue). Measurement pairs are represented by green lines. Fiber bundles relay light from the light sources to the subject's head. (b) Detection of visual cortex activations by the visual stimulus (shown in the inset: a reversing, radial grid). (c) Time course showing the temporal response of changes in oxygenated hemoglobin. The stimulus occurred during $t = 0$–10 s. (c) The response of changes in oxygenated hemoglobin to an angularly swept grid (with four different grid orientations). The results demonstrate that high-density DOT, as the portable and wearable technology, could be potentially useful for mapping function in the human cortex in the settings where the subjects are not able to access functional magnetic resonance imaging (fMRI) or positron emission tomography (PET), the current standard workhorses for functional brain imaging. Reproduced with permission from Ref. [33].

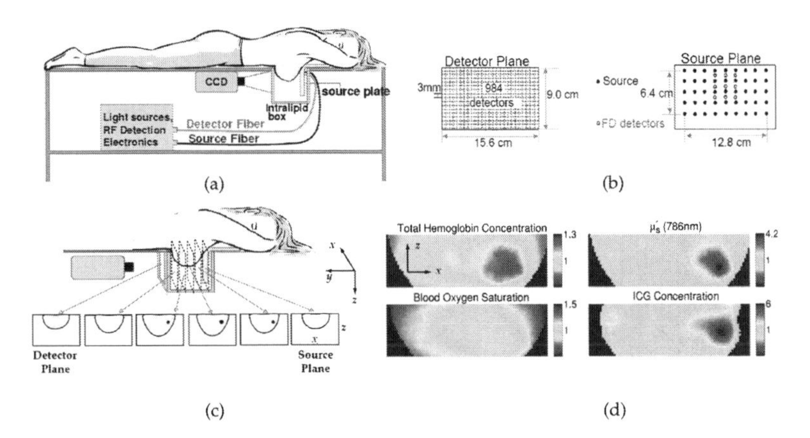

Figure 2.36 (a) Schematic of DOT instrument for breast cancer diagnosis. The subject lies down with breasts suspended in the breast box, which is designed to hold a matching fluid which has optical properties similar to human tissue. 45 sources and 9 detectors (frequency detector: FD) are positioned on the source plane as shown in (b). 984 detection points with 3 mm spacing on the detector plane are selected from CCD data for image reconstruction. The 9 FD are used for accurate quantification of bulk properties of human tissue and matching fluid, thus improving the initial guess for the image reconstruction [41]. (c) Orientation of 3-D reconstructed DOT image slices. (d) 3D diffused optical tomography (DOT) of breast with malignant cancer. Both endogenous (total hemoglobin concentration (THC), blood oxygen saturation and tissue scattering (μ_s')) and exogeneous fluorescence images (Indocyanine Green (ICG) concentration) sliced at 5 cm away from the source plane are shown. The results show that the regions with high values of, μ_s' and ICG concentrations are closely correlated with the tumor region confirmed by radiology reports – illustrating the potential of DOT in clinical in vivo cancer diagnosis. Reproduced with permission from Ref. [41].

2.9 Quantum Approach to Light–Matter Interaction

2.9.1 *General Concepts*

Light–matter interaction at the atomic/molecular level is best explained by using the quantum mechanical pictures in which we mostly deal with the *energy exchange* between the photon and the atom/molecule. The premises of quantum mechanics generally comprise of the following physical concepts:

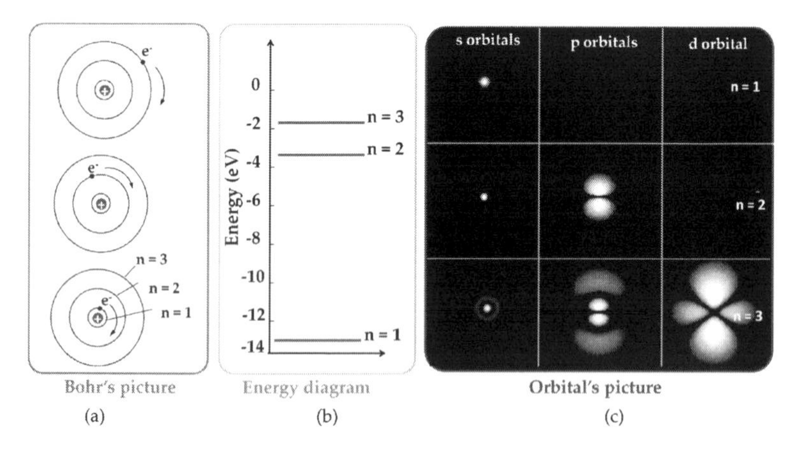

Figure 2.37 (a) Atomic structure of a hydrogen atom described by Bohr's model. The electron (e^-) can travel in a discrete set of circular orbits around the positively charged nucleus. The electron can gain or lose its energy to occupy different orbits ($n = 1, 2, 3 \ldots$) depending upon its energy. The more energetic the electron, the further away from the nucleus the orbit. (b) Energy diagram of the hydrogen atom. (c) The orbitals of the electron of the hydrogen atom. Note that energy level $n = 1$ has only one type of orbital (called *s orbital*). In contrast, in the higher energy levels, it is possible to have more than one orbital having the same energy. For examples, s and p orbitals for $n = 2$; s, p, and d orbitals for $n = 3$). It can be explained by the fact that there is more than one possible solution (wavefunction) to the Schrödinger equation for $n > 2$ [44].

(i) The fundamental (indivisible) unit of matter is *an atom*, which has a generic structure consisting of a positively charged nucleus surrounded by orbiting electrons. This is a model proposed by Niels Bohr in 1913 (Fig. 2.37a).

(ii) The energies of the orbiting electrons in an atom or a molecule can only exist in discrete, or called *quantized*, values. These allowed energy state can be depicted by an *energy diagrams* (Fig. 2.37b).

(iii) Matter can exhibit the characteristics of both particles and waves—an intriguing property called *wave-particle duality*, similar to light waves versus photons. This subtle picture becomes more apparent for microscopic or quantum particles, e.g., electrons, protons, and photons.

(iv) It is impossible to simultaneously determine the position and momentum of a particle with any great degree of accuracy. Similarly, there is an uncertainty of determining an absolute value of the energy of the particle at any time instant. A probabilistic description of the properties of the particle (e.g., energy, position) is essential. This is the infamous *Heisenberg uncertainty principle*—a concept at the heart of quantum mechanics.

The probabilistic interpretation in quantum mechanics echoes well to the concept of the wave nature of a particle. Let's take an atom as an example. An electron in an atom can exhibit different forms of wavefunction for different quantized electron energy levels. In fact, these wavefunctions are essentially the possible solutions of the *Schrödinger equation*—a central equation in quantum mechanics to describe the behaviors of any quantum systems (e.g., atoms, molecules) [44]. The physical meaning of the wavefunction, interpreted by Max Born, simply corresponds to a probability distribution of finding an electron in space. In chemistry, this distribution is mostly referred to as *electron orbital* (Fig. 2.37c). It means that we can know where it is highly probable to find an electron in a particular energy level (also named as *energy state*) in an atom by solving the Schrödinger equation. The electron can only change its energy to the permissible quantized levels, each of which is associated with one or more possible wavefunctions—meaning that the electron orbital would change accordingly. The relationships between the electron energy levels and the associated wavefunctions (or orbitals) can be seen in Fig. 2.37.

2.9.2 *Energy Diagrams*

The description of the energy states of a molecule is more complex. Let's take a diatomic molecule (e.g., carbon monoxide, CO) as an example. In this molecule, there are inevitable electron–electron (repulsive), nucleus–nucleus (repulsive), and electron–nucleus (attraction) interactions among the individual bonding atoms. Hence, the resultant electronic energy states and the *molecular orbital* (MO) are not simply the superposition of the individual atoms'

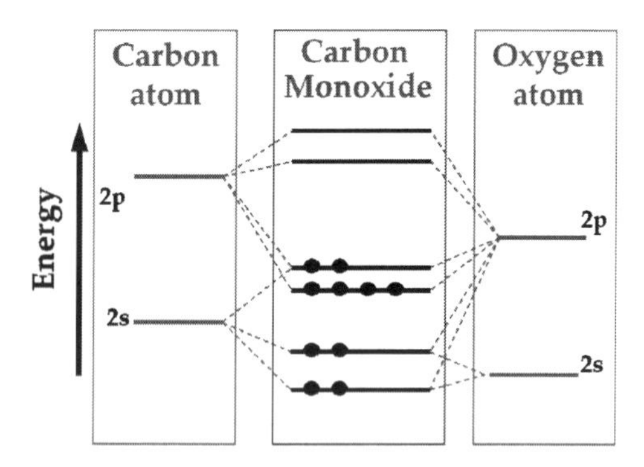

Figure 2.38 Energy states of a molecule (carbon monoxide in this example). The energy levels of the CO (middle) are the result of the redistribution and splitting of the energy levels of the carbon atom (left) and the oxygen atom (right). Only the orbitals 2s and 2p are shown. The dashed lines illustrate the splitting and redistribution of the energy levels.

energies and orbitals. Instead, the resultant MO is a result of the redistribution of the individual orbitals. Thus, the resultant MO energy levels are also "redistributed" and splitting in energy levels can even occur (Fig. 2.38). Such redistribution and splitting are governed by a set of quantum mechanical rules (namely Pauli's exclusion principle, Hund's rule, and the Aufbau principle [44]).

More precisely, each MO energy level (or sometime called bonding energy) is a function of the internuclear distance R. Such variation of the energy exhibits a potential-well-like characteristic as a function of R, as shown in Fig. 2.39a. The minimum point corresponds to the stable state at the *equilibrium internuclear separation* R_{eq0}, i.e., the bond length. The energy E_{D} refers to the *equilibrium dissociation energy*, which is required to dissociate the molecules into the individual atoms. If the electron gains enough energy to jump to another energy state, say to the next higher level, the overall profile of this higher energy state as a function of R follows the similar potential-well shape, except the equilibrium separation shifts to a longer R_{eq1}—indicating that the binding is weakened as the molecule is "more energetic" (Fig. 2.39a). In this case, the lowest energy level is called *ground state* S_0, whereas the higher energy level is called *excited state* S_1.

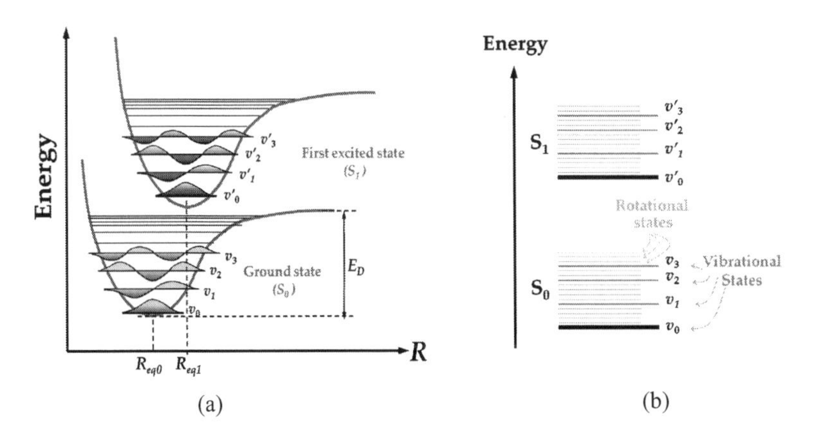

Figure 2.39 (a) Energy diagram of a diatomic molecule as a function of internuclear distance R. For each electronic state (E_0 or E_1), there is a set of vibrational energy levels, denoted as v_i and v_i' for E_0 and E_1 respectively ($i = 0, 1, 2, 3, \ldots$). (b) Vibrational states and rotational states of the molecule. Note that the rotational states are not shown in (a) for clarity.

Furthermore, both the electronic and nuclear motions should be taken into account when we consider the entire molecular structure as well as the complete energy diagram of the molecule. However, this would result in an intricate problem, which involves solving the *Schrödinger equation* considering both the spatial coordinates of the electrons and the nuclei. Fortunately, we could assume that the electronic motion and the nuclear motion in a molecule can be described separately, as postulated by Born and Oppenheimer in 1930. In the other words, it means that the wavefunction of a molecule can be decomposed into two components: electronic and nuclear. Therefore, on top of each electronic energy state (S_0, S_1...), we can define a set of quantized energy states which describe the *vibrational* and *rotational* motions of the nuclei (or essentially the bonding motions). These vibrational and rotational states are shown in Fig. 2.39a,b. Also shown in Fig. 2.39a are the vibrational wavefunctions at the individual states. Note that the energy separations of the electronic states are typically much larger than that of the vibrational and rotational states.

A simplified energy diagram, without considering the internuclear distance called *Jablonski diagram*, is rather commonly used to

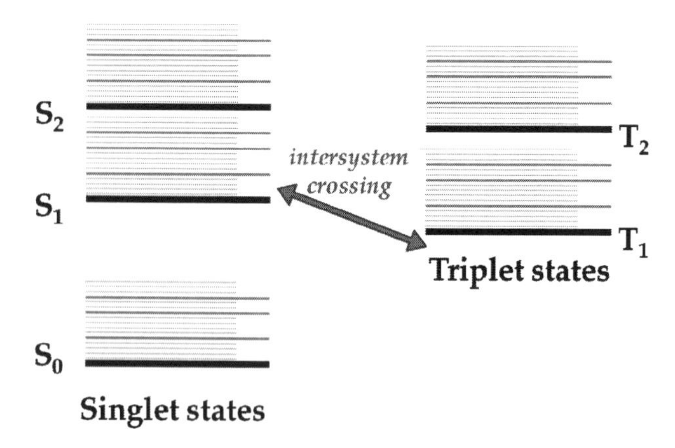

Figure 2.40 Jablonski energy diagram. The stack of lines on the right and left sides represent the energy levels of the singlet and triplet states, respectively. The vibrational ground states of each electronic state (S_0, S_1, S_2, T_1, T_2) are indicated with thick lines, whereas the higher vibrational and rotational states are indicated with thinner lines. The intersystem crossings between the singlet excited states and triplet states happen when the electron spin is "flipped." Such event is relatively rare compared to the transitions within the same systems (i.e., singlet state to singlet state, or triplet state to triplet state).

study how photons interact with the molecule (Fig. 2.40). In this diagram, an additional degree of freedom of describing electronic energy, called electron spin, is incorporated. Spin describes an electron's intrinsic angular momentum. From a classical point of view, we can think of an electron as a spinning sphere producing a magnetic field because of the electric current loops associated with its spinning charge. There are two different spin types: spin-up (magnetic field pointing up) or spindown (magnetic field pointing down) (Fig. 2.41). On the basis of the spin types, the Jablonski diagram consists of two different sub-systems: *singlet states* and *triplet states.*

In quantum mechanics, there is another key principle called *Pauli exclusion principle,* which tells us that the same energy level (or orbital) can accommodate at most two electrons provided that they have opposite spins [44]. In this case, the two electrons are said to be "paired." When an electron in a molecule in the ground state is

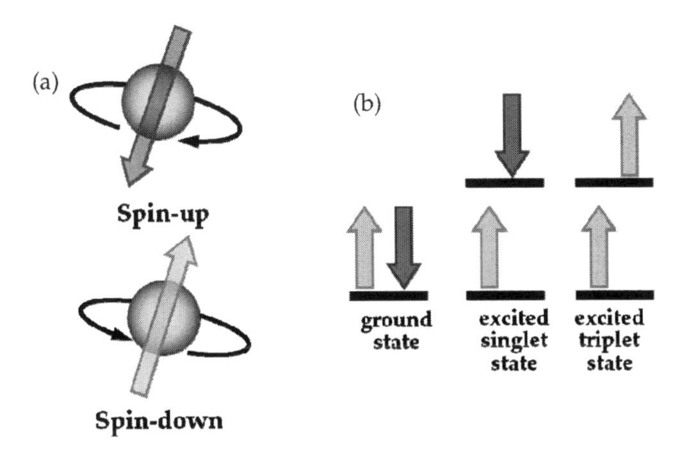

Figure 2.41 (a) A classical view of electron spin. (b) In the singlet state, the excited electron spin are "paired" with the electron's spin in the ground state (i.e., their spin directions are opposite). In the triplet state, the two spins are parallel, i.e., they are "unpaired."

excited to a higher energy level, it will keep the orientation of its spin, i.e., the two electrons are still paired. The electronic state in which electrons are paired is called a singlet state (Fig. 2.41b). In contrast, if the excited electron's spin is "flipped," it will have the same spin direction as the one in the ground state. The electron state in which the electrons are "unpaired" is called triplet state (Fig. 2.41b). From quantum theory, excitation to a triplet state involves an additional "forbidden" spin conversion; transition to the excited single state, instead of the excited triplet state, will much more likely form from the ground state. Nonetheless, electrons can undergo a so-called intersystem crossing between excited singlet and triplet states (Fig. 2.40). This relatively rare event is promoted by a process called spin–orbit interaction [44].

The final remark of the energy diagram is about the population distribution of the molecules as a function of energy. Under thermal equilibrium, it is reasonable to conceive that the majority of the molecules possess lower energies in the population. More precisely, the population distribution follows Maxwell–Boltzmann statistics, i.e., the number of molecules N_j at the energy level E_j and the number of molecules N_i at the energy level E_i is related by an

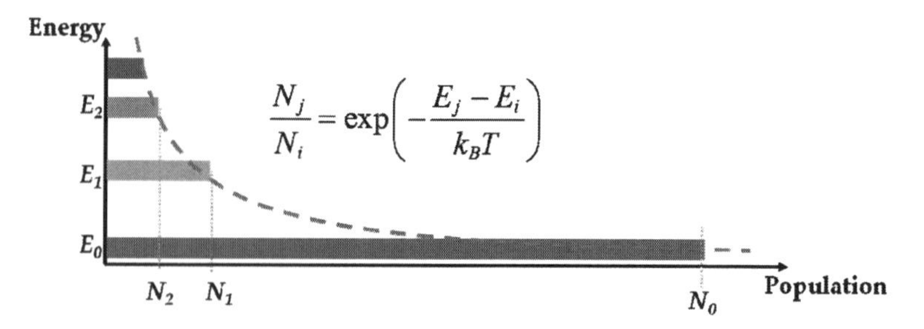

Figure 2.42 The population distribution of the molecules/atoms as a function of energy, which follows the Maxwell–Boltzmann statistics. N_0 is the population of the energy level E_0.

exponential law (see Fig. 2.42):

$$\frac{N_j}{N_i} = \exp\left(-\frac{E_j - E_i}{k_\mathrm{B} T}\right) \tag{2.36}$$

where $E_j > E_i$. k_B is called Boltzmann constant, which has a value of $\sim 1.38 \times 10^{-23}$ J/K[a]. T is the temperature. Equation 2.36 tells us that higher temperature results in the populated higher energy levels.

2.9.3 *Atomic Transitions through Radiative Processes*

Light–matter interaction can be described by how the energy of the molecule is changed from one atomic energy level to another by interacting with the incoming photon, which carries energy proportional to the light frequency (i.e., $E_2 - E_1 = h\nu_{12}$, see Fig. 2.43). This transition between the energy levels is called *atomic transition*. The atomic transitions involved in photons can primarily be categorized into three *radiative processes*—**absorption**, **spontaneous emission**, and **stimulated emission**, as proposed by Albert Einstein [45]. As any atomic transition involves only two energy levels, either going up or down, we will simplify the energy diagram as a "two-level system" to exemplify the three fundamental radiative processes (Fig. 2.43).

[a]Note that the quantity $k_\mathrm{B} T$ has a unit of energy. For example, at room temperature ($T = 273$ K): $k_\mathrm{B} T = 0.026$ eV.

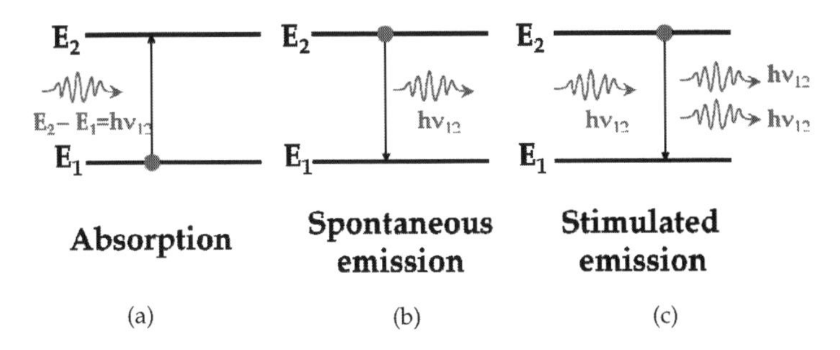

Figure 2.43 Three radiative atomic transitions: (a) absorption, (b) spontaneous emission, and (c) stimulated emission. The photon involved in these processes has the energy of $E_2 - E_1 = h\nu_{12}$, where ν_{12} is the frequency of the photon.

Absorption occurs when the atomic transition goes upward from a lower energy level to a higher one, with the energy difference (E_2–E_1) matching the photon energy, i.e., $E_2 - E_1 = h\nu_{12}$ (Fig. 2.43). In this case, the photon is annihilated in exchange with an upward atomic transition, i.e., the atom or molecule gains the energy from the photon. This can be related to the classical harmonic oscillator picture, in which the incoming photon drives the atom or molecule into a resonance vibration with the resonance frequency being equivalent to the frequency of the absorbed photon.

When the atom or molecule is initially in the higher energy level, it could spontaneously drop to the lower energy level and release its energy in a form of photon, with the photon energy of $h\nu_{12}$ (Fig. 2.43). This process is called **spontaneous emission**.

Similar to spontaneous emission, *stimulated emission* also refers to downward atomic transition, which results in the emission of a photon. However, this process is *stimulated* by an incoming photon with the energy matched with the energy difference (Fig. 2.44). The intriguing characteristic of stimulated emission is that the emitted photon has precisely the same properties as the original incoming photon—same frequency, same propagation direction, same polarization and same phase (in-phase). This process could continue in a cascaded manner which leads to *photon multiplication* (or amplification). This amplification effect is indeed the underlying operation principle of virtually all optical amplifiers and lasers.

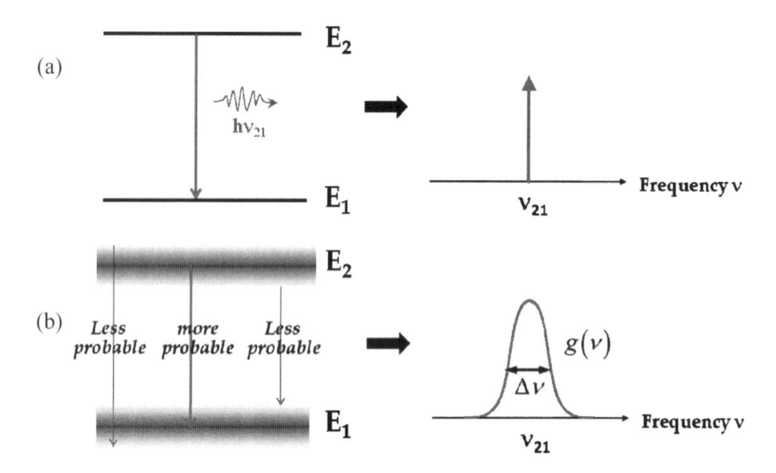

Figure 2.44 (a) (Left) An idealized energy diagram with "absolute" energy levels. The corresponding spontaneous emission spectrum (right) shows a delta function at the transition frequency ν_{21}. (b) In the real case, the fussiness in the energy levels results in an emission spectrum (called lineshape function $g(\nu)$) with a finite bandwidth $\Delta\nu$.

The energy diagrams are typically drawn with multiple discrete lines, each of which represents the individual energy level—implying that only a single-frequency (monochromatic) photon can be allowed to involve in transition between any two energy levels. However, this indeed violates the *Heisenberg uncertainty principle*, which precludes any absolute energy level. Instead, the energy level is "fuzzy." A heuristic explanation of the principle can be done by taking *spontaneous emission* as an example (Fig. 2.44): to obtain an absolute energy value of any energy level, the frequency of the emitted photon (as a result of the transition between two energy levels) has to perfectly be defined (i.e., a delta function in the emission frequency spectrum). In this case, the Heisenberg uncertainty principle requires an infinitely long photon measurement period, or the atom/molecule has to stay at the higher energy level forever. Indeed, this energy–time uncertainty echoes well the reciprocal relation between frequency and time (i.e., Fourier transform relation) as photon energy is proportional to frequency ($E = h\nu$).

Because of the fuzziness (i.e., probabilistic nature) in the atomic transition, photons at certain frequencies would more likely be emitted than the others (Fig. 2.44b). In the other words, there is a probability associated with the transition describing the strength/magnitude of the interaction with the photons at different frequencies. The probability distribution (as a function of frequency) is characterized by a ***lineshape function*** $g(\nu)$. The physical meaning of $g(\nu)$ can be understood in terms of $g(\nu)d\nu$, which is

- The probability that **a spontaneously emitted photon** appears between frequency ν and $\nu + d\nu$, or
- The probability that **a photon generated from stimulated emission** appears between frequency ν and $\nu + d\nu$, or
- The probability that **a photon is absorbed** between frequency ν and $\nu + d\nu$, and clearly, $\int_0^\infty g(\nu)d\nu = 1$.

Taking Fig. 2.44b as an example, the lineshape function $g(\nu)$ is center around the peak frequency ν_{21}, near which the transitions are most likely. The linewidth $\Delta\nu$ of $g(\nu)$ is defined as the full-width at half maximum (FWHM). The narrower the linewidth $\Delta\nu$, the less fussy the energy level, and vice versa. In the other words, the less fussy the energy level, the longer time the photon will be emitted for. This time value is so-called spontaneous decay lifetime τ. Note that taking the inverse Fourier transform of $g(\nu)$ would result in the temporal response of the transition, in which the lifetime τ can be evaluated directly. Typical lifetime values of the common laser materials and common fluorescent molecules are on the order of 1–10 ns. However, it can very over a wide range, from picoseconds to minutes.

Instead of using $g(\nu)$, another similar parameter, called **transition cross-section** $\sigma(\nu)$, is very often employed to describe the atomic transition strength of the atom/molecules (i.e., the absorption or emission strength). It represents an effective area $\sigma(\nu)$ (unit: m^2 or more commonly cm^2) over which radiative atomic transitions likely occur. Classically, this can be thought of an area surrounding the atom which can efficiently absorb a photon or capture it for stimulating another photon emission (similar to the idea shown in Fig. 2.14). It relates to the lineshape function and the

Table 2.4 Absorption cross-sections of the common laser media (HeNe, Ar, CO_2, ruby crystal, Nd:YAG, Ti:sapphire, semiconductors) and the selected fluorescent dyes

Medium	Wavelength	σ (cm^2)
HeNe gas	632 nm	1.0×10^{-13}
Ar ion gas	515 nm	2.5×10^{-12}
CO_2	10.6 μm	3.0×10^{-18}
Ruby (Cr^{3+}: sapphire)	690 nm	2.5×10^{-20}
Nd:YAG	1.064 μm	5.0×10^{-19}
Ti:sapphire	~500 nm	3.0×10^{-19}
Semiconductors	~visible–NIR	$1\text{-}5 \times 10^{-16}$
Fluorescent dye: Alexa Fluor 680	680 nm	5.4×10^{-16}
Fluorescent dye: Cy5.5	680 nm	7.3×10^{-16}
Fluorescent dye: IRDye 800CW	780 nm	9.2×10^{-16}

lifetime by

$$\sigma(\nu) = \frac{\lambda^2}{8\pi\tau} g(\nu) \tag{2.37}$$

where $\lambda = c/\nu$ is the wavelength of the photon. Table 2.4 shows the transition cross-sections of the common laser materials and the selected fluorescent dyes for optical bioimaging. Note that the absorption cross-section $\sigma(\nu)$ in Eq. 2.37 is equivalent to the classical definition σ_a shown in Eqs. 2.12 and 2.15.

As discussed in Section 2.2, light interactions (e.g., absorption) with a material can be classified into two main contributions: electronic and molecular. The same picture can also be explained by the quantum picture, which describes all radiative transitions, i.e., absorption, spontaneous emission, and stimulated emission. Fundamentally, light interaction entails atomic radiative transitions in a material. The photon energies released or absorbed in these transitions are thus entirely governed by the energy differences between any two energy states, which are between the electronic states, or between the much narrower molecular vibrational and rotational states. It is discussed in Section 2.9.2 that the energy separations of the electronic states are typically much wider than those of the molecular states. This difference results in a wide range

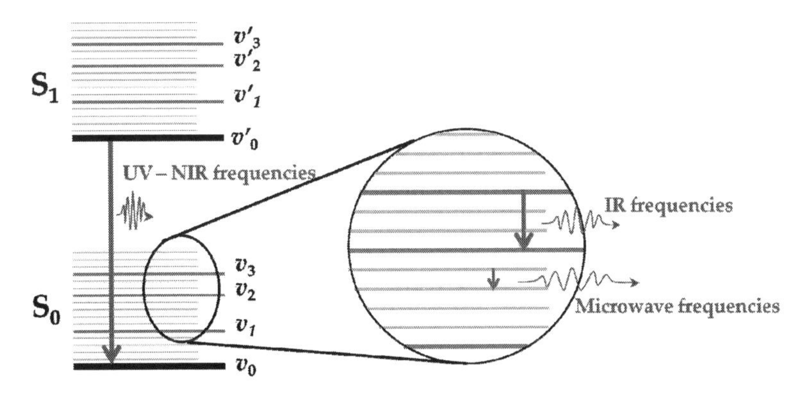

Figure 2.45 The frequency ranges of the photons (e.g., generated from spontaneous emissions) during transitions: electronic transition gives rises UV, visible, or NIR photons; vibrational transition can result in IR photons; rotational transition results in microwave emission. Note that at room temperature, radiative emission at IR and microwave frequencies are relatively more difficult to achieve in practice. It is because the thermal excitation or "de-excitation" can easily be achieved considering the energy differences between the vibrational and rotational is small ($<<0.1$eV).

of radiative transition energies: from $\sim 10^{-6}$–10^{-5} eV (microwave) to $\sim$1–10 eV (NIR–UV).

Generally speaking,

- long-wavelength photons [i.e., microwave ($\sim 10^{-6}$–10^{-5} eV), IR light ($\sim 10^{-2}$–10^{-1} eV)] involve in the *radiative* transitions which occur between the molecular states, either the vibrational or rotational state (Fig. 2.45). Note that such long-wavelength photon emission is relatively hard to observe in practice (at room temperature). It is because the transition can easily be made possible to thermal effect (i.e., heating or cooling). It can be readily understood by the Maxwell–Boltzmann statistics of the molecules (Eq. 2.36). As vibrational and rotational energy level differences are small, modest heating is already sufficient to excite the molecules to the next higher vibrational or rotational energy level without involving the radiative process.

- short-wavelength photons [i.e., UV (~1–10 eV), visible (~1 eV), and NIR light (~10^{-1}–1 eV)] involve electronic transitions (Fig. 2.45).

The likelihood (or the probability) of the electronic transition from one energy level to another is governed by the quantum mechanical rules [44]. Briefly, they essentially tell that the probability of a transition is proportional to the square of the overlap integral between the wavefunctions of the two states that are involved in the transition. In other words, it states that during an electronic transition, the molecule will more likely to change its energy from one vibrational energy level to another if the two vibrational wave functions overlap more significantly (or, are in similar shapes as indicated in Fig. 2.46). Interested readers can refer to Ref. [44] for more detailed explanation of this principle.

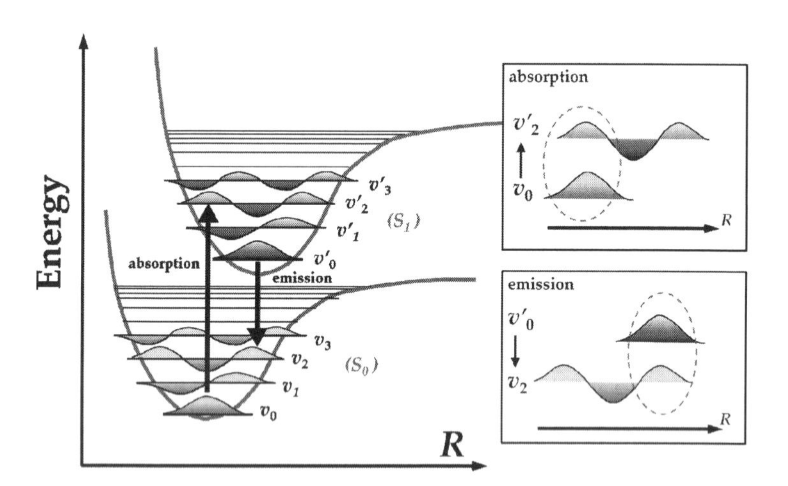

Figure 2.46 Schematic of explaining the probability of the electronic atomic transition. During an electronic transition, the molecule will more likely to change its energy from one vibrational energy level to another if the two vibrational wave functions overlap more significantly. For example: the absorption transition will be more likely (i.e., stronger absorption) to go from level ν_0 to ν'_2. This is because that the wavefunctions in these two states show the most substantial overlap (indicated by the dashed circles) among other possible pairs of states. Same argument applies to emission transition from level ν'_0 to ν_2. More rigorous mathematical treatment of the transition probability can be found in Ref. [44].

2.10 Fluorescence Basics

Fluorescence is a type of light emission phenomena (luminescence) from a substance that occurs when the substance relaxes back to the lower states from the electronically excited (singlet) state in a form of photons The use of fluorescence permeates in numerous applications in biological science and biotechnology, e.g., cellular and subcellular optical microscopy [46], DNA microarray [47], and flow cytometry [48], to name a few. In particular, fluorescence microscopy has been the workhorse for identifying the intracellular molecules (can even go down to the level of single-molecule detection) and the associated physiological processes—one of the greatest toolsets advancing the knowledge of life science and the development of biomedicine.

The processes involved in fluorescence can be best explained by the Jablonski diagram (Fig. 2.47). S_0 is the ground state representing

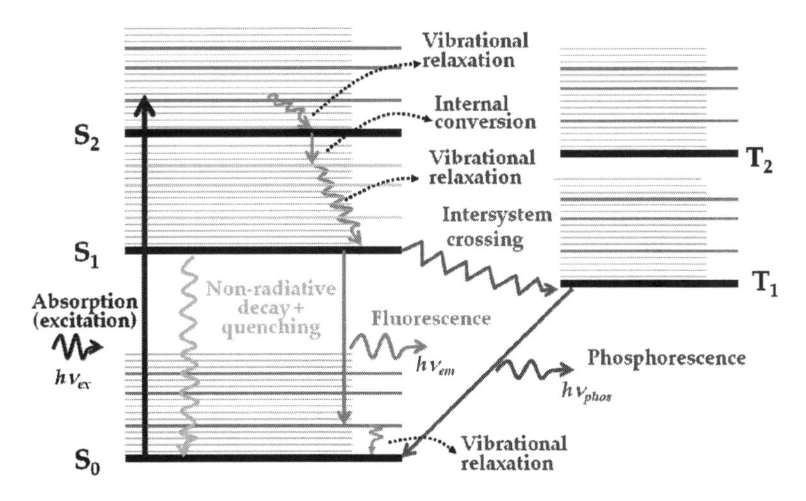

Figure 2.47 Jablonski energy diagram showing the key atomic transitions when the molecule is excited by an incoming photon: (1) absorption (excitation) by an incoming photon with a photon energy $h\nu_{ex}$, (2) vibrational relaxation, (3) internal conversion, (4) intersystem crossing between the singlet excited state and triplet state, (5) nonradiative decay from higher electronic state to the lower electronic state, (6) fluorescence emission with a photon energy $h\nu_{em}$, (7) long-lived phosphorescence emission with a photon energy $h\nu_{phos}$.

the energy of a molecule that is not excited. The first and second higher singlet energy states are denoted as S_1 and S_2, respectively. It should be reminded that these are the energy states in which an *outermost* electron is excited into the different orbitals. On the right side of Fig. 2.47 are the triplet states (T_1 and T_2) to which an outer electron undergoes an electron spin reversal.

Loosely speaking, it typically requires three main processes to give rise fluorescence: *light absorption (excitation)*, *nonradiative relaxation*, and *spontaneous emission*:

1. *Light absorption*: An incoming light (excitation) is first incident onto the molecule and the incoming photon energy is absorbed. If the photon energy is higher than is necessary for an electronic transition (from S_0 to S_1, or from S_0 to S_2), the excess energy is later converted into vibrational or rotational energy of the molecule. If the photon has insufficient energy to promote a transition, no absorption will occur. Thus, absorption is essentially an all-or-none phenomenon. This process happens in about femtoseconds (10^{-15} s), which is mostly regarded almost as an instantaneous process compared with the time scales of later transition processes.
2. *Nonradiative relaxation*: Once the molecule is excited, the vibrational energy of the molecule is transferred to the nearby molecules via (i) direct interaction (e.g., in aqueous medium, water is the likely energy recipient) or (ii) internal conversion (IC), in which a nonradiative transition takes place from the lowest vibrational state of a higher electronic state (S_2) to an excited vibrational state of a lower electronic state (S_1) (Fig. 2.47). Both processes last for about picoseconds (10^{-12} s) after absorption.
3. *Emission*: In this final process, a longer-wavelength photon is emitted during which the molecule relaxes back to the ground state S_0. This process typically lasts for about nanoseconds (10^{-9} s). This is the *fluorescence* photon. Note that the molecule does not necessarily relax back to the lowest vibrational state of S_0. Instead, it can relax to a certain higher vibrational state followed by a rapid nonradiative relaxation back to the lowest state of S_0 (in $\sim 10^{-12}$ s).

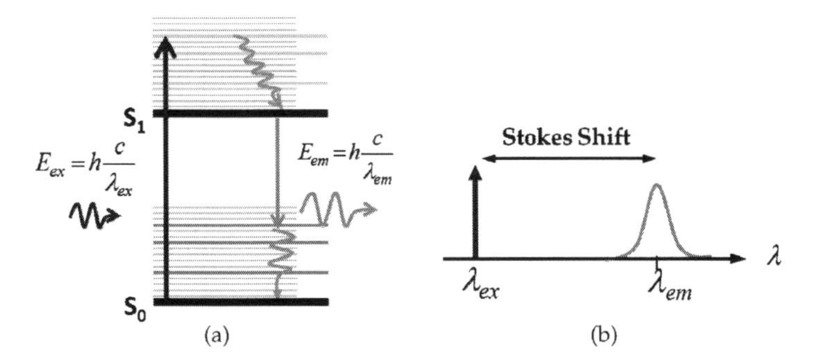

Figure 2.48 (a) Energy diagram showing the excitation and fluorescence emission. (b) The emission spectrum of fluorescence. Note that the finite bandwidth of the spectrum is a result of the uncertainty principle as shown in Fig. 2.44. The Stokes shift is defined as the wavelength (or frequency) difference between the excitation and the emission peak wavelengths.

Here below are the highlights regarding the key concepts of fluorescence:

Stokes shift. It is clear that the photon energy of the fluorescent emission is always less than that of the excitation light. In the other words, the fluorescent light has a longer wavelength than the excitation light. The wavelength (frequency) difference is called *Stokes shift* (Fig. 2.48). This shift stems from the energy loss to the nonradiative relaxation (decay) processes involved.

Emission spectra. The spectrum shown in Fig. 2.48 is incomplete. There are actually multiple possible pathways for fluorescent emission from the lowest state of S_1. Each of them is associated with a lineshape function. Hence, the overall emission spectrum is essentially the summation of all lineshape functions, each of which represents the possible transitions down to different vibrational and rotational states of S_0 (Fig. 2.49). In addition, as long as the excitation photon has enough energy to excite the molecule into its higher electronic states, the final emission spectra are usually independent of the excitation wavelength. It can be understood by that the time scale of the nonradiative decay of the excited molecule (10^{-12} s) is typically much shorter than that of the light emission (10^{-9} s), i.e., almost all the excited molecules quickly relax to the

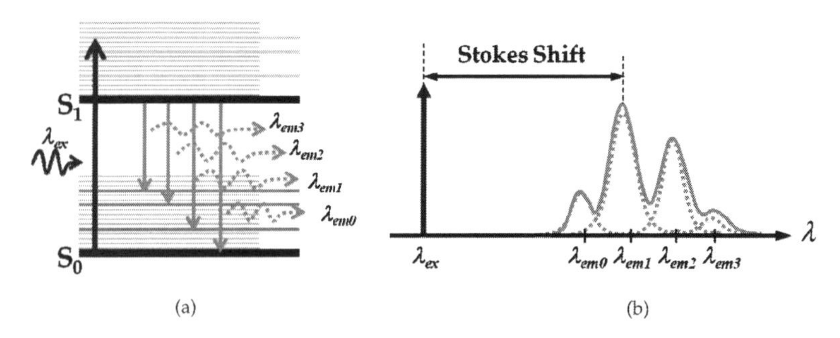

Figure 2.49 (a) Multiple possible pathways for fluorescence emission from S_1 to S_0 upon excitation. (b) The corresponding fluorescence emission spectrum which shows multiple peaks. The overall emission spectrum is the summation of all lineshape functions, each of which represents the possible transitions, peaked at the wavelengths λ_{em0}, λ_{em1}, λ_{em2}, and λ_{em3}. The magnitude of each lineshape function represents the strength (or the probability) of the corresponding transition. In this example, the transition giving rise to emission at λ_{em1} is the most likely, and thus shows the highest peak in the emission spectrum.

lowest vibrational state of S_1 well before the fluorescent emission is taken place.

Excitation spectra. Similar to the emission process, there are multiple pathways for excitation (i.e., absorption) from S_0 such that the excited molecule will eventually emit fluorescent light from the lowest vibrational state of S_1 (Fig. 2.50). The magnitudes of the excitation spectra generally scale with the likelihood of absorption.

Mirror symmetry between the emission and excitation spectra. Emission spectra usually show the mirror symmetry with respect to the excitation spectra (Fig. 2.50). This is a result of the equally probable transitions being involved in the similar vibrational energy states of S_0 and S_1. For example, vibrational state 0 of S_0 to vibrational state 3 of S_1 is comparable in likelihood to vibrational state 0 of S_1 to vibrational state 3 of S_0. Although this symmetry rule is often true, many exceptions occur. A detailed description can be found in Ref. [49].

Other competing processes. Several other relaxation pathways that have different degrees of probability compete with the fluorescence

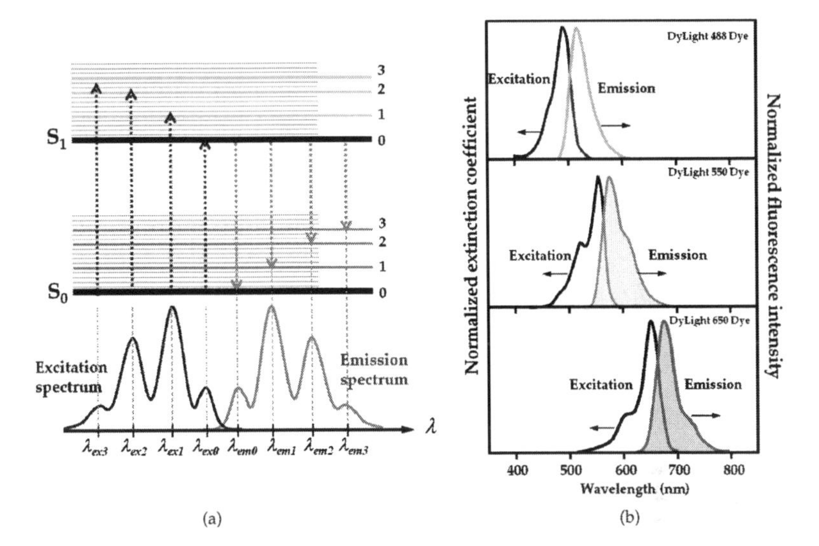

Figure 2.50 Mirror symmetry between the excitation spectrum and the emission spectrum. (b) Fluorescence spectra and excitation spectra of the selected fluorescent dyes, which in general follow the mirror symmetry. (*Source*: Thermo Fisher Scientific Inc, http://www.piercenet.com/.)

emission process. One common process is called *quenching*, in which the energy of the excited molecule is transferred nonradiatively to another molecule when they collide with each other or when they are in contact and form a new complex [49]. Another competing process is known as *intersystem crossing* to the excited triplet state. This event is relatively rare, but eventually results in the emission of a photon through a much slower process (10^{-6}–10^{1} s) called *phosphorescence*. A complete picture which describes all the essential processes involved in fluorescence is depicted in Fig. 2.47.

Fluorescence lifetimes. The fluorescence lifetime is defined by the average time the molecule stays in the excited state before relaxing back to the ground state. In its simplest form, the temporal variation of the fluorescence intensity right after the excitation can be expressed in an exponential decay function, i.e.,

$$I(t) = I_0 \exp\left(-\frac{t}{\tau}\right) \tag{2.38}$$

where I is the initial intensity at time $t = 0$. For a single exponential decay, the fluorescence lifetime τ is defined as the time taken for 63% (= 1 – 1/e) of initial intensity is attenuated (Fig. 2.51). Note that this lifetime is essentially as same as the spontaneous decay lifetime

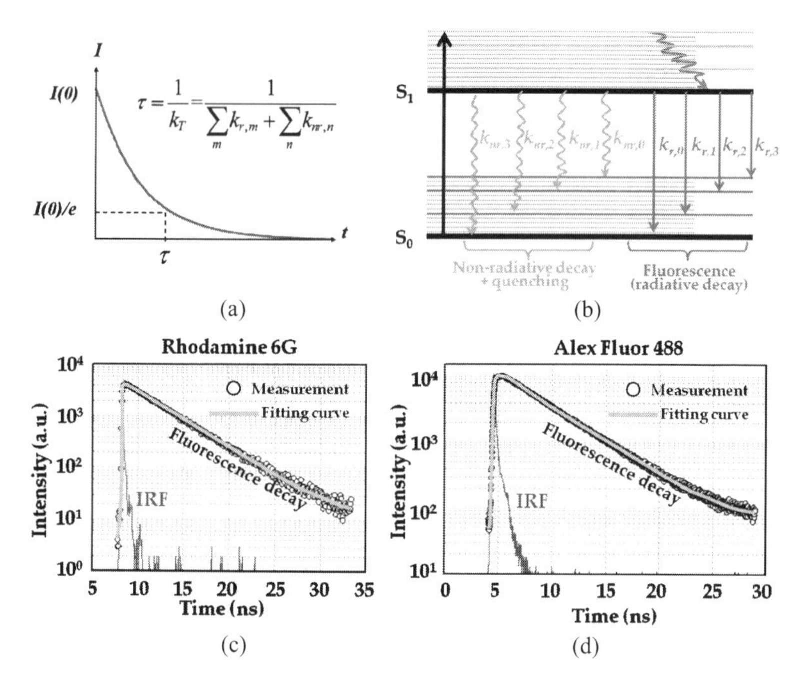

Figure 2.51 (a) Intensity decay of the fluorescence emission. The decay rate k_T is the summation of all the radiative decay transitions (i.e. $k_{r,0} + k_{r,1} + k_{r,2} + k_{r,3}$) plus all the nonradiative decay transitions ((i.e. $k_{nr,0} + k_{nr,1} + k_{nr,2} + k_{nr,3}$), as shown in (b). ((c)–(d)) The experimental lifetime measurements of two common fluorophores: (c) Rhodamine 6G and (d) Alex Fluor 488. Note that the decay curves are plotted in logarithmic scale in the y axis. It should also be emphasized that Eqs. 2.37 and 2.39 assume the delta function excitation in time, which is for sure not practical in reality. In practice, the measured fluorescence decay curve is a result of the convolution of the idealized exponential decay and the so-called instrument response function (IRF) (as shown in (c)). The IRF represents the temporal response of the measurement instrument, which takes into account the pulse shape of the excitation (usually the laser pulse) and the finite temporal response of the photodetector. With the known IRF, the lifetime is typically extracted by numerically fitting the measured decay curve (see the fitting curves in (c)). (*Source*: http://www.iss.com.)

τ discussed in Section 2.9.3. Typically the fluorescence lifetimes of the fluorescent molecules we deal with in the biological applications (e.g., fluorescence microscopy) are about 0.1–10 ns.

Another way of interpreting the meaning of lifetime is to consider the decay rate of the excited state population of the molecules (k_T, unit: s^{-1}). It turns out the rate of change of the excited state population, dN_1/dt, is directly proportional to the excited state population N_1, with k_T being the proportionality constant, i.e.,

$$\frac{dN_1}{dt} = -k_T N_1 = -\left(\sum_m k_{r,m} + \sum_n k_{nr,n}\right) N_1 \tag{2.39}$$

Note that the summations in the bracket term of Eq. 2.38 means that the decay rate k_T takes into account all the possible decay transitions, either the radiative, i.e., $k_{r,0} + k_{r,1} + k_{r,2} + \ldots$ or the nonradiative $k_{nr,0} + k_{nr,1} + k_{nr,2} + \ldots$pathway (see Fig. 2.51). The negative sign indicates that N_1 is decaying during this process. Equation 2.39 has a solution of

$$N_1(t) = N_1(0)\exp(-k_T t) \tag{2.40}$$

where $N_1(0)$ is the initial excited state population at $t = 0$. Comparing Eq. 2.40 with Eq. 2.38, it is easily conceivable that the decay rate of the excited state k_T equals to the inverse of the lifetime, i.e.,

$$\tau = \frac{1}{k_T} \tag{2.41}$$

As the fluorescence lifetime is the intrinsic parameter unique to the fluorescent molecule (see Table 2.5), lifetime measurement is one of the popular tools for quantitative studies in cell biology. The representative example is fluorescence lifetime imaging microscopy, or FLIM [46, 49, 50].

Quantum yield. From Fig. 2.47, we can see that not all the absorbed photons will be converted to the emitted (Stokes) photons because of the possible nonradiative decay process returning the molecule to the ground state. If we define the radiative rate of the emission to be k_r (unit: s^{-1}) and the rate of nonradiative decay to S to be k_{nr} (unit: s^{-1}), we can quantify the efficiency of converting the excitation

Table 2.5 Typical fluorescence lifetimes of the common fluorophores (for biological imaging, spectroscopy) [50]. It should be emphasized that the lifetime values are highly dependent on the local environment (e.g., pH value, temperature, ionic concentration). Therefore, the listed values may vary in different experimental conditions. DAPI: 4′,6-diamidino-2-phenylindole; PBS: phosphate buffered saline pH 7.4; DMSO: dimethyl sulfoxide; ICG: indocyanine green; CFP: cyan fluorescent protein; eGFP: enhanced green fluorescent protein; eYFP: enhanced yellow fluorescent protein; CdSe: cadmium selenide; CdTe: cadmium telluride; PbS: lead sulfide; PbSe: lead selenide

Fluorophores		Emission wavelength (nm)	Lifetime (ns)	Remarks
Intrinsic fluorophores	Phenylalanine	280	7.5	
	Tyrosine	300	2.5	
	Trytophan	350	3.03	
	NADH	450–500	0.3	
	Collagen	370–440	<5.3	
	Elastin	420–460	<2.3	
Organic dyes	DAPI	420–500	2.78	in water
(extrinsic fluorophores)	Fluorescein	450–500	4.0	in water
	Alexa488	500–560	4.0	in water
	Alexa532	500–560	2.53	in water
	Alexa546	520–560	4.06	in water
	Rhodamine 6G	520–600	4.0	in methanol
	Cy3	562	0.3	in PBS
	Alexa647	600–660	1.04	in water
	Bodipy 630/650	630–680	3.89	in water
		630–680	4.42	in ethanol
	Cy5	650–690	0.91	in water
	Alexa750	775	0.66	in water
	ICG	750–850	0.51	in methanol
		760–850	0.97	in DMSO
Fluorescent proteins	CFP	476	2.6	
	EGFP	509	2.71	
	EYFP	527	3.04	
	MCherry	610	1.4	
Quantum dots	CdSe	370–500	~10–30	
	CdTe	470–660		
	PbS/PbSe	>900		

Table 2.6 Quantum yield and extinction coefficient values of the common fluorophores. Note that the listed values are dependent on the local environment (pH value, temperature, ionic concentration, etc.). Therefore, the actual values may vary in different experimental conditions

Fluorophores	Absorption peak wavelength (nm)	ε (cm^{-1} M^{-1})	Q	Remarks	Ref.
DAPI	353	27,000	0.58	in DMSO	[51, 52]
Fluorescein	500	92,000	0.97	in ethanol	[53]
Alexa Fluor 488	495	73,000	0.92	in PBS	[54]
Alexa Fluor 532	531	81,000	0.61	in PBS	[54]
Alexa Fluor 546	556	112,000	0.79	in PBS	[54]
Rhodamine 6G	529	116,000	0.95	in ethanol	[52, 53]
Cy3	554	130,000	0.14	in PBS	[46]
Alexa Fluor 647	650	270,000	0.33	in PBS	[54]
Cy5	652	200,000	0.18	in PBS	[46]
Alexa Fluor 750	749	290,000	0.12	in PBS	[54]
ICG	781	210,000	0.2	in water/ methanol 75/25%	[55]
eCFP	434	26,000	0.4	in PBS	[56]
eGFP	489	55,000	0.6	in PBS	[56]
eYFP	514	84,000	0.61	in PBS	[56]
mRFP1	584	50,000	0.25	-	[57]
mCherry	587	72,000	0.22	-	[57]

photons to the emitted photons by a parameter called *quantum yield* Q, which is written as

$$Q = \frac{k_\mathrm{r}}{k_\mathrm{nr} + k_\mathrm{r}} \tag{2.42}$$

Essentially, Q is equivalent to the ratio of the number of emitted photons to the number of absorbed photons. Note that we lump all the possible nonradiative processes to the single rate constant k_nr. From Eq. 2.42, Q is always less than unity. In many applications, such as fluorescence spectroscopy and microscopy, it is desirable to have Q as close to unity as possible, which can be achieved by choosing the fluorescent molecule having the radiative process to be the dominant effect, i.e., $k_\mathrm{r} >> k_\mathrm{nr}$. Table 2.6 shows the Q values of some of the common fluorophores.

In Eqs. 2.12 and 2.15, we defined the molar extinction coefficient ε or the absorption cross-section σ_a to describe the "absorbing strength" of the molecule (chromophore). Both parameters tell us that not all the incident light intensity onto the molecule is absorbed. It can thus lead to the following estimation of the *fluorescence light budget*. Consider the intensity of the incident light I_{in}, the absorbed power **per molecule** P_{a} can be written as

$$P_{\text{a}} = \sigma_{\text{a}} I_{\text{in}} \tag{2.43}$$

Taking the quantum yield Q into account, we can write the fluorescence power **per molecule** P_{f} as

$$P_{\text{f}} = Q\sigma_{\text{a}} I_{\text{in}} \tag{2.44}$$

Because fluorescence is a spontaneous emission process, the Stokes (fluorescent) photons will randomly be emitted in all directions. In practice, we are not able to collect all the fluorescent light owing to the limitations imposed by the experimental setup, e.g., the finite NA of the objective lens, optical loss in the intermediate optical elements within the system (lenses, mirrors, filters, etc.). A collection efficiency ξ has thus to be taken into account when we need to estimate how much fluorescent light we can eventually collect. The collected fluorescent light power **per molecule** P_{c} is then

$$P_{\text{c}} = \xi Q\sigma_{\text{a}} I_{\text{in}} \tag{2.45}$$

If the excitation volume is V (unit: cm^3), and the concentration of the molecules is N_{a} (unit: cm^{-3}), the total collected fluorescent power $P_{N\text{c}}$ is written as

$$P_{N\text{c}} = \xi Q\sigma_{\text{a}} N_{\text{a}} V I_{\text{in}} = 3.823 \times 10^{-21} \xi Q\varepsilon N_{\text{a}} V I_{\text{in}} \tag{2.46}$$

which gives a quick estimation of the fluorescent light budget. Note that Eq. 2.15 is used to derive Eq. 2.46.

Photobleaching. It refers to the case in which a fluorescent molecule can no longer fluoresce due to light-induced chemical damage and modification. The fluorescence light fades away during continuous excitation (Fig. 2.52). Typically it is an irreversible process although in some special cases the fluorescence can be recovered in the molecule within a short period of time, which is called fluorescence recovery after photobleaching (FRAP) [46, 49]. While it is still a poorly understood effect, it is generally believed

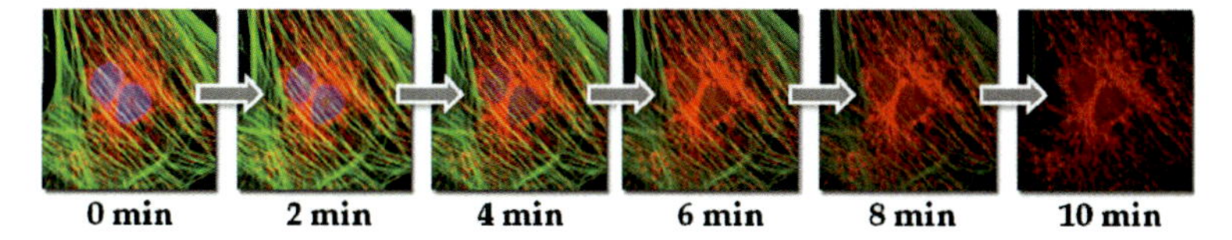

Figure 2.52 Photobleaching effect. Three fluorescent dyes are used to label the fibroblast cells. The nuclei, the mitochondria, and actin cytoskeleton were stained with a blue dye (DAPI), a red dye (MitoTracker Red), and a green dye (Alexa Fluor 488), respectively. Upon the continuous excitation over 10 minutes, the blue fluorescence intensity drops rapidly after two minutes and almost vanishes after 6 minutes. In contrast, the mitochondrial (red) and actin (green) stains are more resistant to photobleaching. (*Source*: http://www.olympusfluoview.com.)

to be caused by the interactions between the excited fluorescent molecule and the oxygen molecule in its triplet state [46, 49]. Some fluorescent molecules bleach quickly after emitting only a few photons, while others can undergo thousands or millions of excitation and emission cycles before bleaching (typically ~seconds to minutes).

2.10.1 *Fluorophores*

Fluorescence typically occurs from aromatic molecules. The fluorescent molecules are usually named as fluorochromes, or simply dyes. Fluorochromes that are conjugated to a larger macromolecule (e.g., nucleic acid, lipid, enzyme, or protein) are termed *fluorophores*. However, the meanings of the terms *fluorophores* and *fluorochromes* are nowadays becoming interchangeable. In general, fluorophores are divided into two broad classes: *intrinsic* (or endogenous) and *extrinsic* (or exogenous). Intrinsic fluorophores are those that occur naturally, e.g., aromatic amino acids, porphyrins, coenzymes (e.g., NADH), and fluorescent proteins[a]. Extrinsic fluorophores are the synthetic biochemicals that are tagged with other nonfluorescent

[a]Sometimes fluorescent proteins are regarded as extrinsic fluorophores because we are able to express a fluorescent protein gene (e.g., purified green fluorescent protein from jellyfish) in other living cells, tissues, or organisms and make them fluoresce. Enormous efforts are being made on such mutagenesis approach, which have led to a wide range of fluorescent protein variants—having the fluorescent color from blue to yellow and even red.

molecules or the biological cells/tissues to generate fluorescence. Examples are fluorescein, rhodamine, cyanine dyes, and quantum dots. Particularly, the "tagging" process using extrinsic fluorophores is highly specific, i.e., each fluorophore is designed to be only tagged with the specific targeted specimen. They are widely used for biomolecular sensing, fluorescence spectroscopy and microscopy [46, 49].

2.10.1.1 Intrinsic fluorophores

Many molecules inside the cells/tissues are able to give rise to fluorescence, called *autofluorescence*, without requiring any extrinsic fluorescent tagging/labeling. The autofluorescent emission wavelengths typically cover from the UV to visible regions. Autofluorescence can be typically employed to study the structural tissue matrices (e.g., elastin and collagen) or to monitor the different types of cellular metabolic pathways (e.g., NAD and NADH; lipopigments)—offering the possibilities for *label-free* diagnosis of diseases, such as cancer. Other examples of intrinsic fluorophores include aromatics amino acids, e.g., tryptophan, tyrosine, phenylalanine, and porphyrins. By monitoring the autofluorescent intensity, we are able to extract a good amount of useful information (either metabolic or structural) about the biological specimens. Autofluorescence correlates the concentration and spatial distribution of these biochemicals (which is the intrinsic fluorophores) with the local environment near the biochemicals (such as temperature and pH). One well-known example is NADH, a bright fluorophore in its reduced form emitting fluorescence at 460 nm but which is nonfluorescent in its oxidized form. Figure 2.53 shows the typical absorption and emission spectra of the intrinsic fluorophores [58].

2.10.1.2 Extrinsic fluorophores

Many biomolecules are not able to generate autofluorescence, or the autofluorescence is insufficient for practical measurements, e.g., spectroscopy and imaging. For instance, DNA and lipids generally do not exhibit any fluorescence. To monitor the cellular functions related to these molecules, or analyze their biochemical properties,

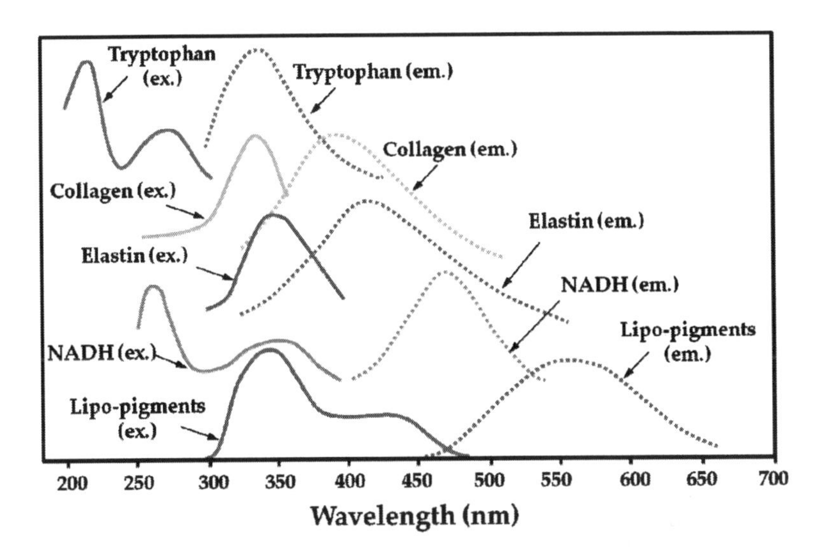

Figure 2.53 Fluorescence excitation (ex.) and emission (em.) spectra of selected intrinsic fluorophores. The spectra of each fluorophore are shifted in the vertical axis for clarity. Reproduced with permission from [58].

it is necessary to tag or label these molecules with the extrinsic fluorophores for the purpose of identification. These extrinsic fluorophores (or simply *dyes*) can be designed and engineered to specifically bind or be conjugated to a particular *target* of interest. *Targets* include specific groups of cells in a tissue, organelles, proteins, nucleic acids, lipids, ions (Ca^{2+}, Mg^{2+}, Na^{+}, etc.), and enzymes. In many applications, the molecular structure of the dye has to be modified, to form a *probe,* in order to detect specific biological targets. These probes are typically conjugated to a biomolecule (e.g., oligonucleotides) in order to obtain the specific (or selective) access to particular biological sites for biomolecular detection. For instance, fluorescein is a *dye* while fluorescein-labeled peptides are *probes.* Owing to this specificity as well as sensitivity, extrinsic fluorophores represent the indispensible tools which enable us to reveal the biochemical and the associated physiological information of the cells and tissues.

A number of key issues are relevant to choosing and designing the extrinsic fluorophores for most of the biomedical diagnoses:

1. **The excitation spectrum of the fluorophores.** It should well overlap with the wavelength of the light sources (e.g., gas discharge lamps and lasers) for optimum fluorescence excitation efficiency.
2. **The extinction coefficient ε (or equivalently the absorption cross-section σ) and the quantum yield Q.** They should be high in order to maximize fluorescence output (see Eq. 2.45).
3. **Photobleaching**. It is a particularly important for long-period imaging experiments, e.g., time-lapse imaging.
4. **Phototoxicity.** Closely related to photobleaching, it results from damage by photogenerated reactive oxygen species on biomolecules (e.g., proteins) and cellular components. The excitation source, e.g., excitation power and wavelengths, should be carefully considered in order to enhance the specimen viability.
5. **The sensitivity to the local environmental changes** (e.g., pH, temperature, interaction between different fluorophores). It can have strong influences on the fluorescence output. Such susceptibility to environmental change can be either detrimental or beneficial depending upon the applications. For instance, pH-sensitive fluorophores are utilized for studying the intracellular pH in bacteria and cells In contrast, pH-insensitive fluorophores are preferred when the pH variation is irrelevant or even disturbing in fluorescence imaging.
6. **The permeability and the solubility of the fluorophores**. They determine the strategies of fluorophore delivery and localization in the specimen. These factors can also dramatically affect the fluorescence output.
7. **Autofluorescence**. This background autofluorescence can deteriorate the final image contrast. In order to obtain plausible signal-to-background ratio, it is desirable to choose the extrinsic fluorophores with the excitation/emission wavelength ranges well separated from the autofluorescence range or/and with sufficient level of labeling (i.e., fluorophores concentration) in order to overwhelm the autofluorescence.
8. **Cytotoxicity.** It is particularly relevant to live cell imaging. Introduction of the extrinsic fluorophores could perturb the physiological behavior of the cells and could even induce

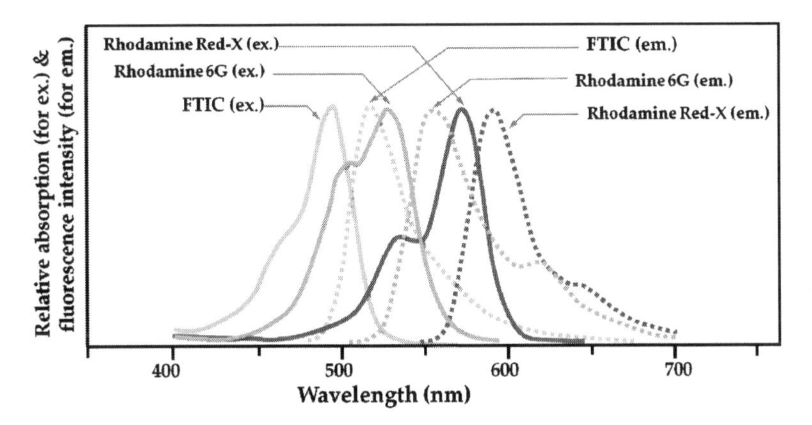

Figure 2.54 Excitation (ex.) and fluorescence emission (em.) spectra of fluorescein and rhodamines. From left to right: fluorescein isothiocyanate (FITC), Rhodamine 6G, and Rhodamine Red-X (*Source*: http://www.invitrogen.com.)

cytotoxicity and cell death. It is thus crucial to choose the optimal level of labeling which attains the best compromise between cytotoxicity and the image signal contrast.

Several commonly used extrinsic fluorophores are described below in order to further exemplify the above concerns. Fluoresceins and rhodamines are widely used as traditional extrinsic fluorophores (Fig. 2.54). Both fluorophores are frequently used for labeling antibodies (immunoglobulins) in fluorescence microscopy and in immunoassays.

Fluoresceins have absorption maxima near 480 nm, which coincides well with the lasing wavelengths of the argon-ion and krypton-argon lasers (488 nm) as well as mercury (436 nm) and xenon arc-discharge (467 nm) lamps. The absorption maxima of rhodamines (~600 nm) are close to the various laser lines of the helium-neon (He-Ne) laser, i.e., 543 nm, 594 nm, 612 nm, 633 nm. Both fluoresceins and rhodamines have been popular fluorophores in fluorescence imaging and spectroscopy because of their high quantum yield and high molar extinction coefficients (near 80,000 M^{-1} cm^{-1}). In addition, the excitation and emission wavelengths of these two fluorophores are relatively long—minimizing the background autofluorescence from biological samples (typically in the UV wavelength range).

However, fluoresceins are sensitive to pH variation and are prone to photobleaching. Rhodamines exhibit self-quenching of fluorescence as they tend to aggregate in aqueous solutions To this end, a vast number of new fluorophores have been developed for robust biomolecular detection. Popular examples are Alexa Fluor dyes, BODIPY dyes, and cyanine (Cy) dyes [46]. These dyes are all designed to match their excitation maxima to the wavelengths of the typical light sources and lasers (between 350 and 750 nm) (Fig. 2.55). Both the Alexa Fluor and Cy dyes have enhanced quantum yield and are more water-soluble and more photostable than their fluorescein and rhodamine counterparts. In contrast to the Alexa Fluor and Cy dye series, BODIPY dyes are nonpolar and relatively insoluble in water. They are thus mainly utilized in lipid trafficking and are also general-purpose membrane probes [46]. Further information on a wide range of fluorophores can be obtained from the Molecular Probes catalogue [54].

2.10.1.3 Near-infrared (NIR) fluorophores

While visible fluorophores are the mainstay in the field of fluorescence microscopy and spectroscopy, NIR fluorophores have in fact been gaining significant attention over the past decades. The main advantages are three-fold: (i) NIR fluorescence detection almost eliminates background autofluorescence from the biological specimens—enhancing the detected signal-to-background ratio. (ii) NIR fluorescence (from 700 to 900 nm) coincides well with the "optical window" of biological tissues where the best compromise between the light scattering and absorption in tissues can be obtained (see Section 2.8)—particularly favorable for deep-tissue in vivo fluorescence imaging in which highly sensitive detection is essential. (iii) NIR diode lasers for excitation and high-efficiency silicon photodiodes for fluorescence detection are easily available and of low cost. The most common NIR dyes are the Cy dyes, such as the Cy-5 and Cy-7. Other dyes such as IRDye 78 and IRDye 800 phosphoramidite (LI-COR Biosciences) can even push the

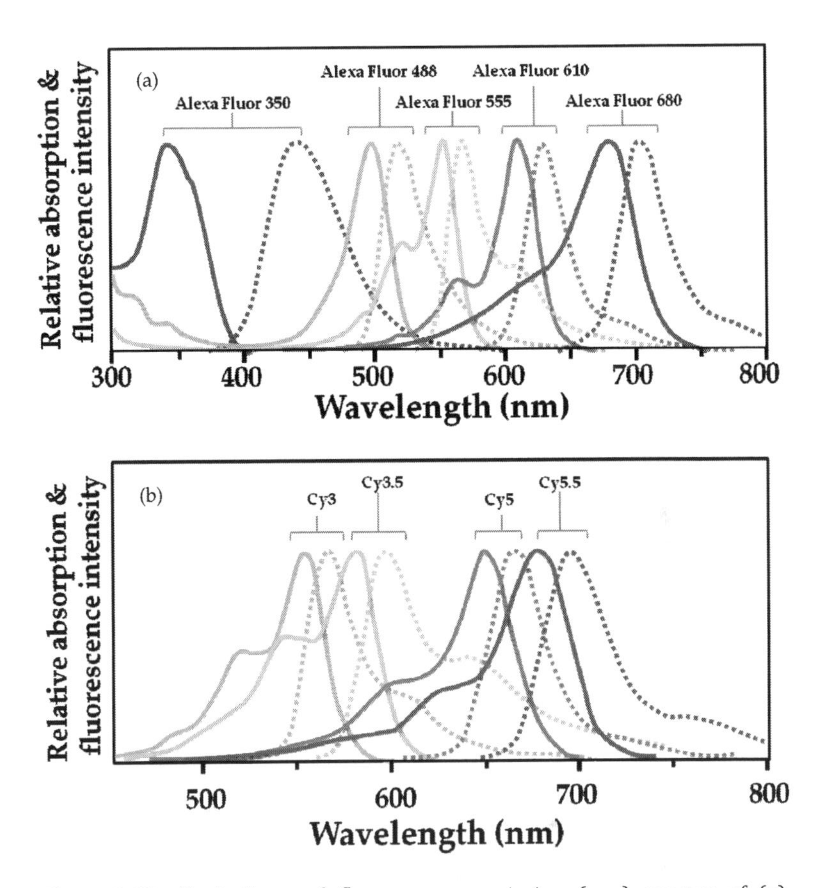

Figure 2.55 Excitation and fluorescence emission (em.) spectra of (a) the selected Alexa Fluor dyes and (b) cyanine (Cy) dyes. (*Source*: http://www.invitrogen.com/.)

fluorescence emission up to 800 nm where an optimal signal-to-background ratio for deep-tissue optical imaging can be achieved.

2.10.1.4 Quantum dots (QDs)

Quantum dots (QDs) are the nanometer-sized crystals typically made of inorganic semiconductors, such as cadmium selenide (CdSe), indium gallium phosphide (InGaP), and silicon. They are emerging as the versatile fluorescent labeling agents for fluorescence cellular imaging and in many immunocytochemical analyses [49, 59–62].

QDs are typically single crystals a few nanometers in diameter. In such a nano-scale, the energy levels of the QDs are *quantized*, with values related to their material compositions and sizes.[a] Hence, the excitation and emission wavelengths of the QDs can be flexibly engineered through precise control during synthesis (e.g., by controlling the processing temperature, duration, and other conditions [49, 59–62]). In general, the smaller the QD, the shorter the fluorescence emission wavelength (Fig. 2.56), e.g., CdSe QDs of 3 nm, emitted at 520 nm while one of 5.5 nm is emitted at 630 nm. In addition, the broad absorption spectra of QDs allow the simultaneous excitation of multiple QDs with a single light source, at any wavelength shorter than the emission peak wavelength (See Fig. 2.56). By far, CdSe is the most investigated QD material for biological applications. It consists of a CdSe core (~10 to 50 atoms in diameter) surrounded by a zinc sulfide (ZnS) shell, which is in turn coated by a hydrophilic polymer layer (Fig. 2.56). ZnS is used to improve the optical properties of QDs and to prevent the cytotoxic cadmium from leaking out of the core. The hydrophilic polymer shell is used to improve the water solubility of the QDs and to enhance bioconjugation to targeting molecules such as secondary antibodies, peptides, or streptavidin QDs based on other semiconductor materials are also possible. Recently, silicon quantum dots (or *nanoparticles*) have exhibited promising potential for biological applications owing to their biocompatibility and biodegradability [63].

These QD probes have significant benefits over the organic dyes and fluorescent proteins: They in general do not photobleach under light exposure—having long-term photostability. They exhibit high quantum yield (orders of magnitude higher than their organic dyes counterparts [61]), which provides higher detection sensitivity in live cell imaging compared with the traditional organic probes. They have a long lifetime, >10 ns [61], implying the narrow emission

[a]QDs made of the same semiconductor material can show many drastic differences in optical properties when compared with their bulk counterpart. This is a direct consequence of "quantum confinement" of the excited electron–hole pair in the semiconductor. More details can be found in Ref. [65].

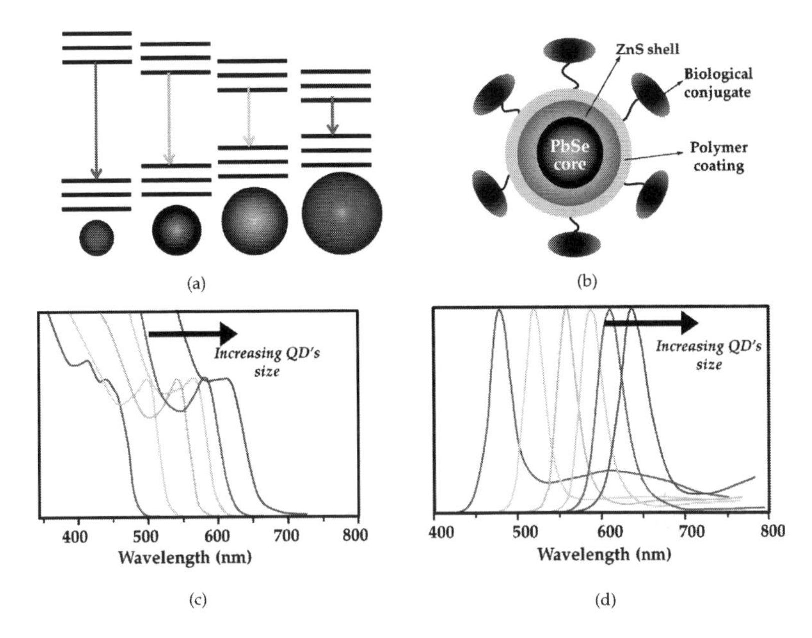

Figure 2.56 Schematic of the "size effect" of the QD on the corresponding energy diagram. This shows that the transition energy (thus the emission wavelength) can be engineered by varying the QD's size, as governed by the quantum theory. In general, the larger the size, the smaller the energy gap (and thus the shorter the emission wavelength). (b) Schematic of the typical QD structure for biological applications. (c) The absorption (excitation) spectra of lead selenide (PbSe) QDs with different sizes. (d) The fluorescence emission spectra of PbSe QDs. In both (b) and (c), the spectrum shifts to longer wavelengths as the QD size increases [60, 61].

spectral bandwidths. This facilitates distinguishing multiple QD probes simultaneously without severe spectral overlap. This makes QDs the ideal probes for multiple labeling in the same specimen, particularly useful for multicolor fluorescence imaging and high-throughput flow cytometry (in which multiple fluorescent probes are essential) [62, 64].

A remark worth noting is the cytotoxicity of QDs. Although numerous studies have been carried out to investigate whether QDs or nanoparticles induce toxic effects, such as the release of toxic metals (the core metals, e.g., cadmium, selenium) due to particle breakdown and the production of reactive oxygen species [66], these

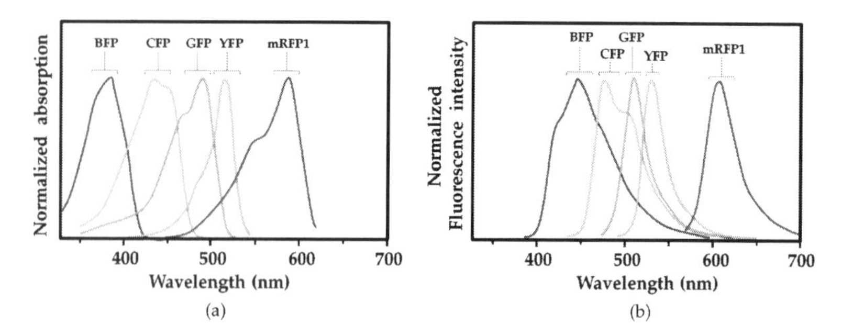

Figure 2.57 (a) Excitation and (b) fluorescence emission spectra of different fluorescent proteins. BFP: blue fluorescent protein; CFP: cyan fluorescent protein; GFP: green fluorescent protein; YFP: yellow fluorescent protein; mRFP1: monomeric red fluorescent protein. Reproduced with permission from [68].

results (whether from in vitro or in vivo studies) are yet to be proved conclusive. These studies thus made researchers, industry, and the general public cautious of translating this nanotechnology to clinical applications. Further research is definitely essential to verify its values and utilities in biomedicine.

2.10.1.5 Fluorescent proteins

The discovery of green fluorescent protein (GFP), isolated from the North Atlantic jellyfish, Aequorea victoria, in the early 1960s as well as the successful demonstrations of the cloning of GFP in the 1990s have undoubtedly created a new paradigm in the field of cell biology [67, 68]. Using the naturally occurring GFP and their derivatives as the fluorescent probes enables us to non-invasively visualize individual molecules and the associated intracellular processes in living organisms. Since the mid-1990s, a broad range of fluorescent protein variants, such as blue fluorescent protein (BFP), cyan fluorescent protein (CFP), and yellow fluorescent protein (YFP), have been developed with the fluorescence emission spectral range spanning the entire visible light spectrum (Fig. 2.57).

The most striking feature of fluorescent proteins is that by using recombinant complementary DNA cloning technology, the fluorescent protein gene can be expressed in a virtually unlimited range of cell types, tissues, and organisms, including mammals, almost without any detrimental biological effects. Hence, fluorescent proteins are particularly useful as reporters for gene expression studies in cultured cells, tissues, and living animals [67]. There are a wide range of different techniques developed for making fluorescent protein fusion products in order to express their genes in the cells/tissues/organisms. While these techniques are out of the scope of this book, interested readers could refer to many excellent review articles and books on fluorescent protein technology such as Refs. [49, 67–70].

The continual advances in exploring new fluorophores in conjunction with the technological innovation in optical microscopy make *fluorescence microscopy* the major workhorse in life science research as well as biomedical diagnostics. It has been proven to be a useful tool to image the structures and functions of biological tissues, cells, viruses, and bacteria, with impressively high chemical specificity and spatial resolution. Figure 2.58 shows some examples of the images taken by fluorescence microscopy. More details about the instrumentation and technological issues of different fluorescence microscopy modalities can be found in Ref. [46], such as laser scanning confocal microscopy, nonlinear optical microscopy, fluorescence lifetime imaging (FLIM), Förster resonance energy transfer (FRET) imaging, and so on.

2.11 Nonlinear Optics

The optical phenomena discussed so far are primarily "linear" in nature, i.e., the optical properties of the materials do not alter when the light intensity is varied. In the other words, they are intensity independent. In contrast, the term *nonlinear optical phenomena* connotes that the response of the material to the optical field (**E** field) depends *nonlinearly* on the strength of light field (or equivalently the optical intensity). Generally speaking, in nonlinear optics, an input light modifies the material properties, which in turn modify another light wave or the input light itself. Nonlinear optical

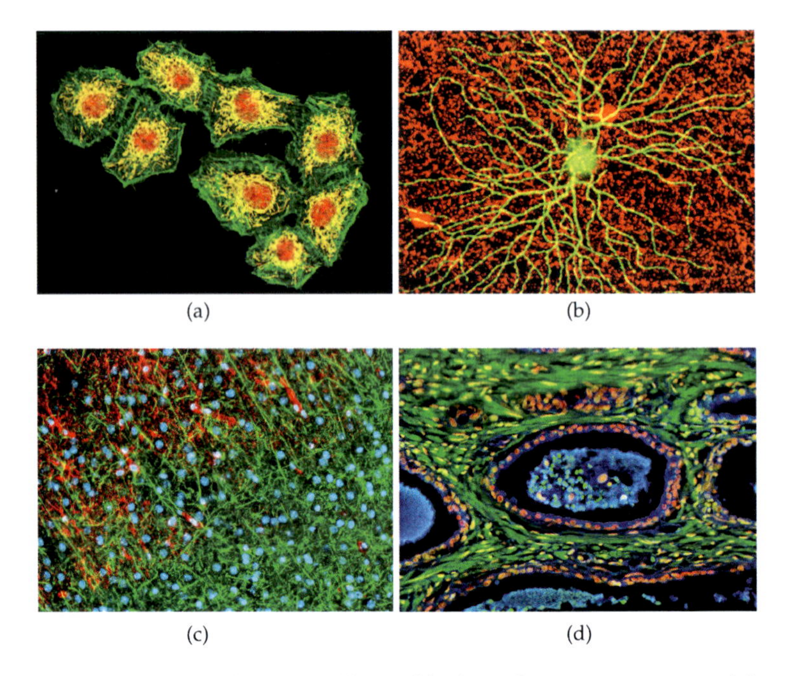

Figure 2.58 Selected image gallery of biological specimens captured by fluorescence microscopy. (a) Human brain glioma cells: mitochondria (yellow), F-actin network (green), and nuclear DNA (red). (b) Retina ganglion cell (yellow). (c) A coronal thin section of mouse brain: heavy chain neurofilament (green), glial fibrillary acidic protein (GFAP) expressed in astroglia and neural stem cells (red), and nuclear DNA (blue). (d) Human prostate gland tissue: filamentous actin (Green), Golgi complex (blue), and nucleus DNA (orange). (*Sources*: http://www.olympusfluoview.com, and http://www.microscopyu.com.)

effects can only be observed at high intensity, which can typically only be achieved by pulsed lasers. This explains why nonlinear optics, although perceived in the early 20th century, had not drawn significant attentions until the invention of the first laser (Ruby laser) in 1960, which is actually a pulsed laser [45]. Since then, along with the technological advancements in lasers, numerous nonlinear optical effects have been discovered and exploited in countless applications, especially nonlinear optical microscopy of biological tissues and cells [46].

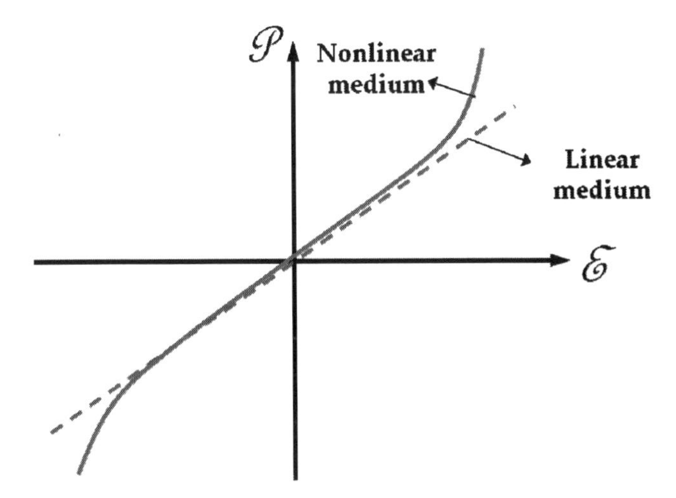

Figure 2.59 The relationship between the polarization ($\mathcal{P}$) and the input electric field ($\mathcal{E}$) in a linear medium (dashed straight line) and a nonlinear medium (solid curve). Note that for a small electric field (i.e., low intensity), the polarization more or less exhibits a linear behavior. In contrast, it deviates from such linearity as the electric field is increasingly stronger (i.e., high intensity), behaving as a nonlinear optical medium ($\mathcal{P}$ varies with $\mathcal{E}$ nonlinearly).

2.11.1 *General Descriptions of Optical Nonlinearity*

In Section 2.2, a classical interpretation was employed to explain general light–matter interaction, i.e., how the polarization of a medium (*not* the light polarization) depends on the strength of the **E** field of the light wave. In the case of linear optics (Eq. 2.1), the induced polarization varies linearly with the strength of the **E** field, with χ being the proportionality constant, known as linear electric susceptibility (Fig. 2.59). The material response can be modeled as a harmonic oscillator.

In the context of nonlinear optics, the polarization **P** (or the material response to light) generally does not depend on the **E**-field strength linearly, especially at a very high **E**-field amplitude comparable with inter-atomic electric field ($\sim 10^5$–10^8 V/m, which is equivalent to a high optical intensity ~ 1 kW/cm^2–1 GW/cm^2 in free space). In this case, Eq. 2.1 can be generalized by expressing the polarization as a power Taylor's series of the **E** field. For simplicity, we here consider only the scalar time-varying polarization P[a]:

$$\mathcal{P}(t) = \varepsilon_0 \left(\chi^{(1)}\mathcal{E}(t) + \chi^{(2)}\mathcal{E}^2(t) + \chi^{(3)}\mathcal{E}^3(t) + \cdots\right) \tag{2.47}$$

where $\mathcal{E}$ is the scalar electric field: $\mathcal{E}(t) = \frac{1}{2}\left(E_0 e^{+i\omega t} + \text{c.c.}\right)$. E_0 is the field amplitude. $\chi^{(1)}$, equivalent to χ in Eq. 2.1, is the linear susceptibility. The coefficients $\chi^{(n)}$ are the n-th-order susceptibilities of the medium which characterize the nonlinear optical properties of the medium. For examples, $\chi^{(2)}$ and $\chi^{(3)}$ are the second- and third-order nonlinear susceptibilities, respectively. And we can particularly consider the second-order polarization, which gives rise to second-order optical nonlinear effects, to be

$$\mathcal{P}^{(2)}(t) = \varepsilon_0 \chi^{(2)}\mathcal{E}^2(t) \tag{2.48}$$

and the third-order polarization, which gives rise to third-order optical nonlinear effects, to be

$$\mathcal{P}^{(3)}(t) = \varepsilon_0 \chi^{(3)}\mathcal{E}^3(t) \tag{2.49}$$

Because of the higher-order dependence on the **E** field, the nonlinear material response (i.e., nonlinear polarization) should be classically explained by an *anharmonic* oscillator model, instead of the harmonic one as in the case of linear optics (Section 2.2).

By far, the second- and third-order nonlinearities are the most commonly investigated nonlinear optical effects. This is because of the fact that $\chi^{(n)}$ becomes progressively smaller as we go to higher-order n. It means that extremely strong light intensity is required in order to achieve observable higher-order effects (see Eq. 2.1). For condensed matter, the magnitude of $\chi^{(2)}$ is typically on the order of $\sim 10^{-12}$ m/V whereas $\chi^{(3)}$ is about $\sim 10^{-24}$ m^2/V^2.

---

[a]The complete description of how to treat the vector nature of the induced polarization can be found in Ref. [71]. Other than the scalar assumption (i.e., no anisotropy), Eq. 2.43 has two other key assumptions: (i) the material response to optical field is *spatially local*, i.e., the response only occurs at the excitation location $\mathbf{r}$, not at other locations (that is why spatial dependence is dropped out in Eq. 2.46); (ii) the material response to optical field is *instantaneous*, i.e., the polarization will vanish as soon as the excitation is over. Clearly, in reality, the induced polarization can be spatially nonlocal and not instantaneous. The material exhibits loss and dispersion according to Kramers–Kronig relations. In this case, Eq. 2.46 has to be expressed in complete convolution integrals in both space and time [71].

Apart from the difference in optical field strength requirement, there is one more distinct difference between the second- and third-order nonlinear optical effects. Second-order effects occur only in noncentrosymmetric materials, whereas third-order effects can occur in both centrosymmetric and noncentrosymmetric materials. This difference makes second-order effects particularly useful for studying the biological specimens with specific molecular alignment or orientations, e.g., collagen, and microtubules (to be discussed later in more detail).

The nonlinear polarization $\mathcal{P}^{(n)}$ is a critical parameter describing the nonlinear optical phenomena because this polarization (which is time-varying) acts as a driving source of a new EM wave at the same frequency as $\mathcal{P}^{(n)}$. This can be rigorously proved by the wave equation, which is outside the scope of this book. Interested readers can refer to a number of excellent books on nonlinear optics, e.g., [5, 71]. In the following sections, the general concepts of nonlinear optics will be explained by some common nonlinear optical effects.

2.11.2 *Frequency Mixing*

Frequency mixing is one of the most prominent phenomena in nonlinear optics. It is a process in which a new EM wave (light) is generated at a frequency different from the input ones. More precisely, the newly generated frequency can be the algebraic sum or difference of all the frequencies of the input light waves. This results in a number of interesting effects, such as second-harmonic generation (SHG), third-harmonic generation (THG), sum-frequency generation (SFG), and difference-frequency generation (DFG). Let us consider the situation in which the input light wave consists of two different frequencies ω_1 and ω_2. The corresponding **E** field can be represented in the form

$$\mathcal{E}(t) = \frac{1}{2}\left(E_1 e^{-i\omega_1 t} + E_2 e^{-i\omega_2 t} + \text{c.c.}\right) \tag{2.50}$$

where E_1 and E_2 are the field amplitudes of the two components. Considering only the second-order effect, we can express $\mathcal{P}^{(2)}$

(Eq. 2.48) as

$$\mathcal{P}^{(2)}(t) = \frac{1}{2}\varepsilon_0\chi^{(2)}\left[\underbrace{E_1^2 e^{-i2\omega_1 t} + E_2^2 e^{-i2\omega_2 t}}_{\text{SHG}} + \underbrace{2E_1E_2e^{-i(\omega_1+\omega_2)t}}_{\text{SFG}} + \underbrace{2E_1E_2^* e^{-i(\omega_1-\omega_2)t}}_{\text{DFG}} + \text{c.c.}\right] + \underbrace{\varepsilon_0\chi^{(2)}\left(E_1E_1^* + E_2E_2^*\right)}_{\text{OR}} \quad (2.51)$$

Equation 2.51 shows that in second-order nonlinear effects, the two input frequency components "mix" with each other and can generate (i) the components with doubled frequencies $2\omega_1$ and $2\omega_2$ (the first two terms, i.e., SHG), (ii) the component with sum frequency $\omega_1+\omega_2$ (the third term, i.e., SFG), (iii) the component with difference frequency $\omega_1-\omega_2$ (the fourth term, i.e., DFG), and (iv) the DC component (the last term, called optical rectification, OR).

Equation 2.51 tells us that the output wave in this nonlinear process *in principle* consists of four different frequency components (except the DC component, which is the static electric field generated inside the medium). However, in practice no more than one of these components can be present with observable intensity (see the three examples shown in Fig. 2.60). One obvious reason is that the output intensity of a particular component depends on the input strengths (e.g., SHG intensity is $\propto|E_1|^2$ or $|E_2|^2$). Another reason, which is indeed the key determining factor, is that the new output component in this frequency mixing process can only be efficiently generated if a certain *phase-matching condition* is satisfied. Typically, this condition can only be satisfied for no more than one combination of all the frequencies involved. Generally speaking, phase matching is a condition under which a proper phase relationship among the interacting waves (including the newly generated wave) is maintained along the direction of propagation. In this way, the resultant field amplitudes of all waves will be in phase at the output—ensuring the optimal efficiency. However, the phase-matching condition in reality cannot be satisfied primarily because of *dispersion*. Figure 2.61 illustrates an example of the phase mismatch of SHG in the presence of dispersion. There are different

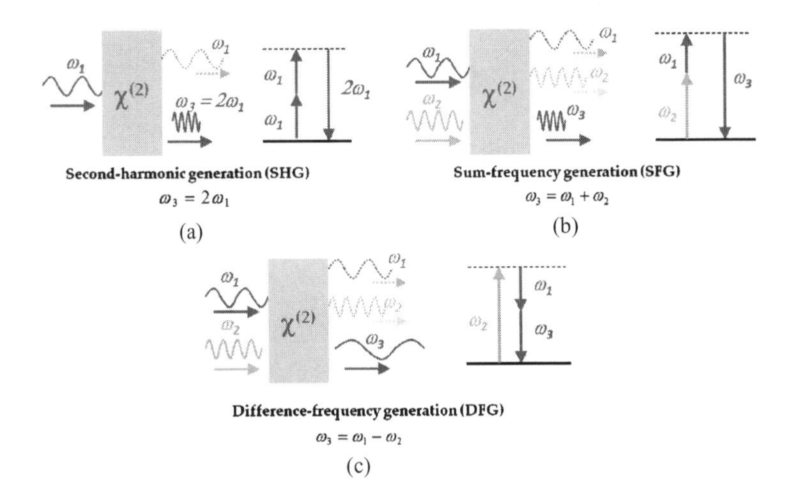

Figure 2.60 (a) Second harmonic generation (SHG), (b) sum-frequency generation (SFG), and (c) difference-frequency generation (DFG). Note that not all the incoming photons will be converted to the photons at the newly generated frequency (i.e., less-than-unity conversion efficiency). Therefore, the output light also consists of not only the newly converted light but also the transmitted, and attenuated, input light (see the dotted waves).

techniques and considerations regarding optimized phase-matching conditions. Interested readers can refer to Refs. [5, 71].

One common example of applying the frequency mixing effect in practice is to produce new light sources at different wavelengths, especially at the wavelength ranges in which no common laser source is readily available. For example, one can pass the 1064 nm output from Nd:YAG lasers or the 800 nm output from Ti:sapphire lasers to a nonlinear material which can convert these NIR outputs via SHG to the visible light, at 532 nm (green) or 400 nm (violet), respectively. The common SHG materials are monopotassium phosphate (KDP) and barium borate (BBO) (Fig. 2.62).

In the third-order nonlinearity, as many as four different frequency components can participate in the frequency mixing process. Consider a case of three different input waves at frequencies ω_1, ω_2, and ω_3. There are 19 new (nonnegative) frequency components generated in the third-order nonlinear polarization. Figure 2.63 shows some of the possible combinations. Two main categories can be identified in the third nonlinear frequency mixing (Fig. 2.63):

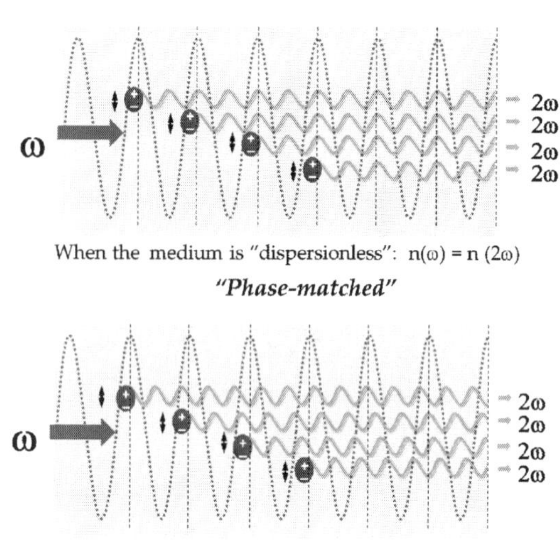

Figure 2.61 A crude picture illustrating the phase-matching condition in frequency mixing (SHG is shown in this example). Propagating in the medium, the input wave (also called fundamental wave) at frequency ω interacts with the atoms/molecules in the medium to continuously generate the frequency-doubled wave (at 2ω), i.e. SHG. When the medium has no dispersion ($n(\omega) = n(2\omega)$) as shown in (a), the maxima of the input wave coincide with the maxima of the second-harmonic (SH) waves. And all the newly generated SH waves are *in phase* and constructively interfered. Hence, the overall SHG output strength is strong—the phase-matching condition is satisfied. However, in the presence of dispersion $n(\omega) \neq n(2\omega)$, as shown in (b), the SH waves generated from different atoms/molecules are no longer in phase at the output. This frequency mixing (conversion) process is said to be phase mismatched. A more complete picture of the phase-matching condition can be understood by solving the wave equations of all the interacting waves, i.e., coupled wave equations. Detailed analysis can be found in Refs. [5, 71].

(i) *third harmonic generation (THG)*, i.e., the outputs at frequencies $3\omega_1$, $3\omega_2$, and $3\omega_3$, and

(ii) *four-wave mixing (FWM)*, i.e., the remaining outputs at frequencies of possible linear combinations of ω_1, ω_2, and ω_3. Examples are $\omega_1 + \omega_2 + \omega_3$, $2\omega_1 + \omega_2$, $2\omega_1 - \omega_2$, $2\omega_1 + \omega_3$, $\omega_1 + \omega_2 + \omega_3$, etc.

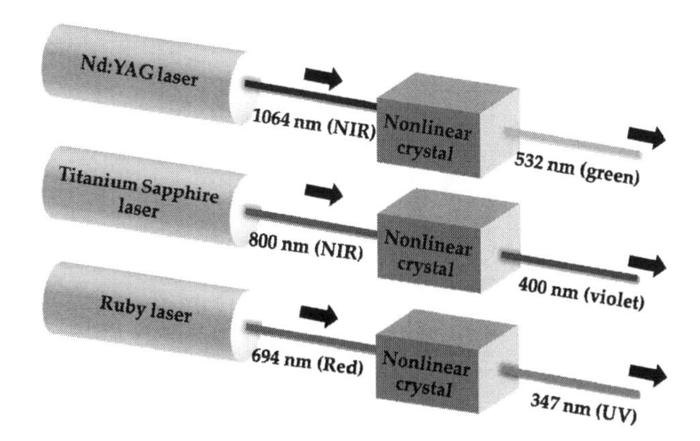

Figure 2.62 Examples of SHG for creating new laser sources at short wavelengths, such as green, violet, or even UV light, using the common red or NIR laser sources (such as Nd:YAG, titanium sapphire, and ruby lasers) together with the proper nonlinear crystals with strong $\chi^{(2)}$ as well as with the phase-matched configurations.

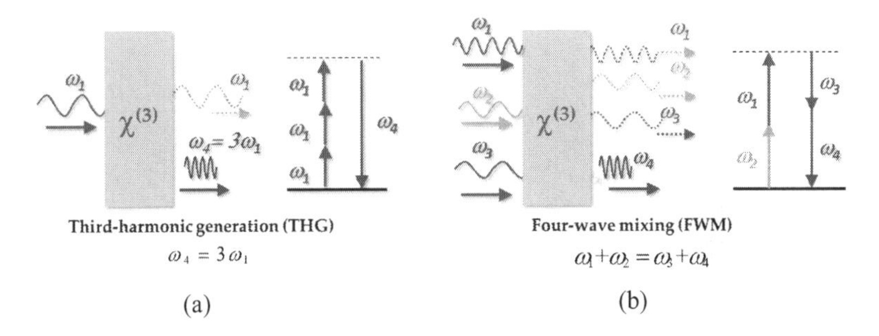

Figure 2.63 (a) Third harmonic generation (THG). (b) Four-wave mixing (FWM). Note that FWM can also be achieved by having $\omega_1 = \omega_2$. In this case, the FWM relation is written as $2\omega_1 = \omega_3 + \omega_4$. In practice, it can be achieved by using an intense input signal at ω_1 (called pump) and a signal at ω_3. The new signal at ω_4 (called idler) can be generated.

Again, the primary factor determining the frequency conversion efficiency is the phase-matching condition.

Frequency mixing processes can also be understood in terms of the energy conservation of all interacting photons, each of which corresponds to different frequency components of the light field.

This can be visualized in the picture of an energy-level diagram (Figs. 2.60 and 2.63). For example, SHG can be understood by considering that two photons of frequency ω are annihilated (upward transition) and a photon of frequency 2ω is simultaneously created (downward transition). The solid lines in these energy diagrams correspond to the atomic ground states, whereas the dashed lines are known as *virtual states*. Virtual states are not the atomic energy states, and these fictitious energy levels are adopted for representing the combined energy of the multiple photons of the interacting light fields plus the ground state energy.

2.11.3 *Intensity-Dependent Modification of Material Properties*

2.11.3.1 Optical Kerr effect

It is also possible to alter the refractive index of the materials through nonlinear polarization, particular for the third-order nonlinear polarization $\chi^{(3)}$. In this case, no frequency conversion is involved. Instead, the refractive index of the material depends linearly on the optical intensity of the input light field $I(\omega)$ at frequency ω, i.e.,

$$n = n_0 + n_2 I(\omega) \tag{2.52}$$

where n_0 is the intrinsic refractive index. n_2 (unit: m^2/W) is defined as *nonlinear refractive index*, which is directly proportional to $\chi^{(3)}$ [71]. Such intensity-dependent refractive index change, induced by the input field at ω, will in turn impose additional phase-shift to the input field itself at the same ω. That is why this effect is called self-phase modulation (SPM). In a more general terminology, the change in refractive index described by Eq. 2.52 is called the *optical Kerr effect*. The typical value of n_2 in a dielectric material is about $\sim 10^{-15}$ cm^2/W. Therefore, using a laser beam with an intensity of 10 MW/cm^2 (typical value for pulsed laser) can induce a refractive change of $\sim 10^{-8}$. Although this change is small, it is sufficient to lead to many nonlinear optical effects, including self-focusing, and spectral broadening, etc. [2, 5, 71] (Fig. 2.64).

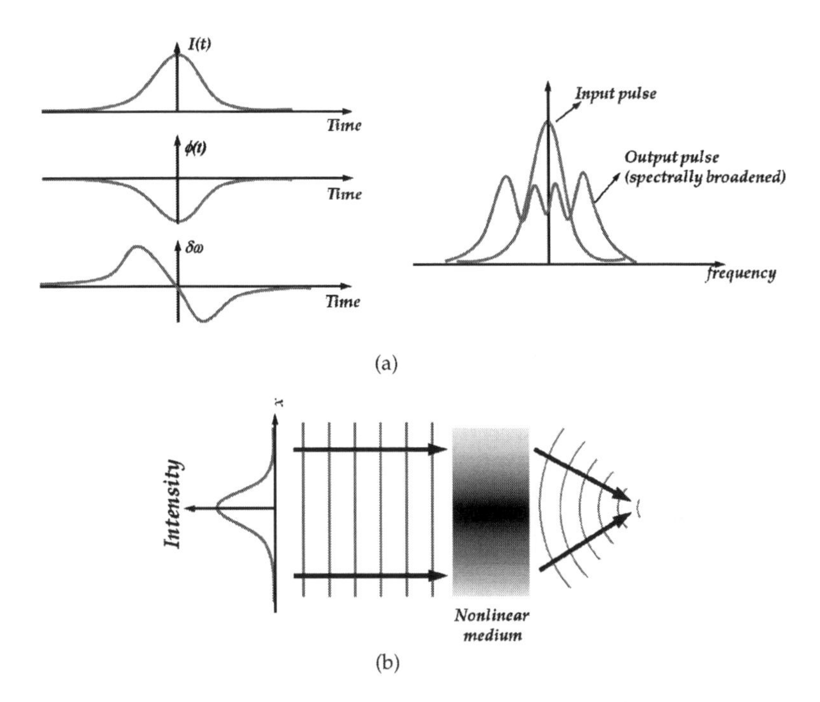

Figure 2.64 (a) Spectral broadening of an ultrashort pulse due to SPM. When an ultrashort pulse travels in a nonlinear medium, it will induce a time-varying refractive index of the medium due to the optical Kerr effect. Such refractive index variation introduces a time-varying phase shift in the pulse, i.e. $\phi(t) = kz - \omega_0 t = n(I)\omega z/c - \omega_0 t$, where ω_0 is the center frequency of the pulse. Note that the field of the pulse can be written as $A(z, t)exp(-j\phi(t))$, where $A(z, t)$ is the field envelope of the pulse. The time-varying phase shift leads to new frequency generation in the pulse as the instantaneous frequency $\omega(t)$ is defined as $\omega(t) = -d\varphi(t)/dt = \omega_0 - \delta\omega$, where $\delta\omega = (dn(I)/dt)z/c$ is the frequency shift of the pulse. $dn(I)/dt$ essentially follows the shape of the time derivative of the intensity profile of the pulse (see Eq. 2.52) (left). As a result, SPM effectively broadens the bandwidth of the pulse (right). Detailed theory of SPM can be found in Refs. [2, 71]. (b) Self-focusing due to SPM. The spatial dependence of the intensity of a light beam introduces the refractive index variation in the medium, which effectively "bends" and focuses the light as if it is a simple lens.

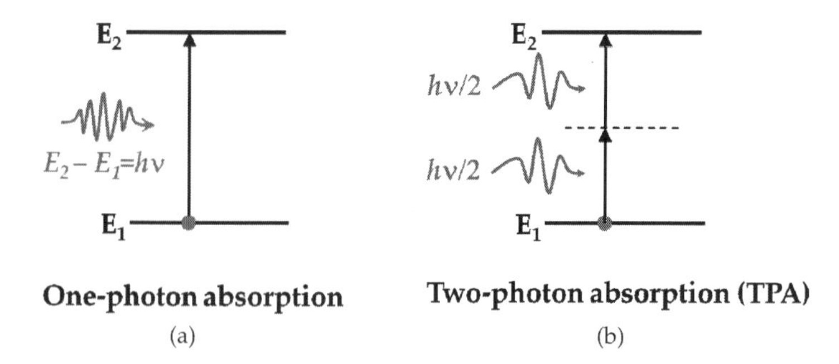

Figure 2.65 (a) One-photon absorption. (b) Two-photon absorption (TPA).

2.11.3.2 Multi-photon absorption

Similar to the linear susceptibility, the nonlinear susceptibility $\chi^{(n)}$ is in general a complex value. The imaginary part of the nonlinear susceptibility is strongly related to the atomic transition resonances of the material (generally absorption)—similar to the dependence of the imaginary part of the linear susceptibility on material's atomic transitions (see Section 2.2). Taking $\chi^{(3)}$ as an example, we can prove that the optical Kerr effects can be described by the real part of $\chi^{(3)}$ [2, 5, 71]. On the other hand, the imaginary part of $\chi^{(3)}$ leads to *two-photon absorption* (TPA). This is a third-order nonlinear optical effect in which an atomic transition occurs when two photons are simultaneously absorbed as the combined photon energy is sufficient to make such transition (Fig. 2.65). The overall absorption coefficient μ_a which takes the TPA into account can be expressed as

$$\mu_a = \mu_{a0} + \beta_2 I(\omega) \tag{2.53}$$

where μ_{a0} is the intrinsic material absorption coefficient. β_2 (unit: cm/W) is the TPA coefficient which characterizes the strength of the TPA effect. From Eq. 2.53, it can be seen that TPA is linearly dependent on the intensity of light. Another common way to define the strength of TPA is to use *two-photon cross-section* σ_2. Again, it is very much similar to how the "linear" absorption cross-section is defined (Eq. 2.37). Two-photon cross-section is usually quoted in the units of Goeppert-Mayer (GM) (named after the Nobel laureate Maria Goeppert-Mayer, who predicted multi-photon effects), where 1 GM is 10^{-50} cm^4 s photon^{-1}.

Following the relation described in Eq. 2.12, we can convert Eq. 2.53 into an overall *absorption cross-section* as

$$\sigma_a = \sigma_{a0} + \sigma_2 I(\omega) \quad (2.54)$$

where σ_{a0} is the linear absorption cross-section. Typically, the values of σ_{a0} of the common fluorophores are $\sim 10^{-15}$–10^{-17} cm^2, which are comparable to the size of fluorophores and can thus be understood intuitively. However, the unit of the *two-photon cross-section* σ_2 (cm^4 s photon^{-1}) requires further discussion, which can be understood from Eq. 2.43. We can first convert the absorbed power per molecule P_a in Eq. 2.43 into the number of absorbed photons per second N_{abs}.[a] Hence, this equation can be rewritten as

$$N_{abs} = \sigma_a I_N = \sigma_{a0} I_N + \sigma_2 I_N^2 \quad (2.55)$$

where I_N is the intensity with the unit of number of photons per unit area per unit time (i.e., photon/cm$^2\cdot$ s). Taking the typical values of $\sigma_2 \approx$ 10–100 GM (true for many fluorophores as shown in Fig. 2.66), we can immediately see that appreciable TPA (i.e., the second term in Eq. 2.55) can be observed only a very intense illumination light, i.e., $I_N \approx 10^{27}$–10^{29} photon/cm$^2\cdot$ s,[b] is employed. This can only be achievable by focusing the ultrashort pulsed laser (fs–ps) onto a miniature spot size ($\sim\mu$m^2).

The TPA effect can be generalized to multi-photon absorption. Of course, such effect corresponds to even a higher-order effect which requires much higher intensity (e.g., three-photon absorption corresponds to the $\chi^{(5)}$ effect)—making it less common to be observed in practice.

2.11.4 *Raman Effects*

2.11.4.1 Spontaneous Raman scattering

In Section 2.6, we briefly mentioned that light scattering can be inelastic in the sense that the photon energy of the scattered light is different than that of the incident light. In other words,

---

[a] $P_a = N_{abs} \times h\nu$, where $h\nu$ is the photon energy at frequency ν.

[b] For example, if a laser source at 800 nm is used, $I_N \approx 10^{27}$–10^{29} photon/cm$^2\cdot$ s corresponds to the intensity of $I_{in} \approx$ 100 MW/cm^2–10 GW/cm^2. Note that $I_{in} = I_N \times h\nu$. Such a high intensity can only be achieved by using high-power femtosecond laser sources, such as Ti:sapphire laser.

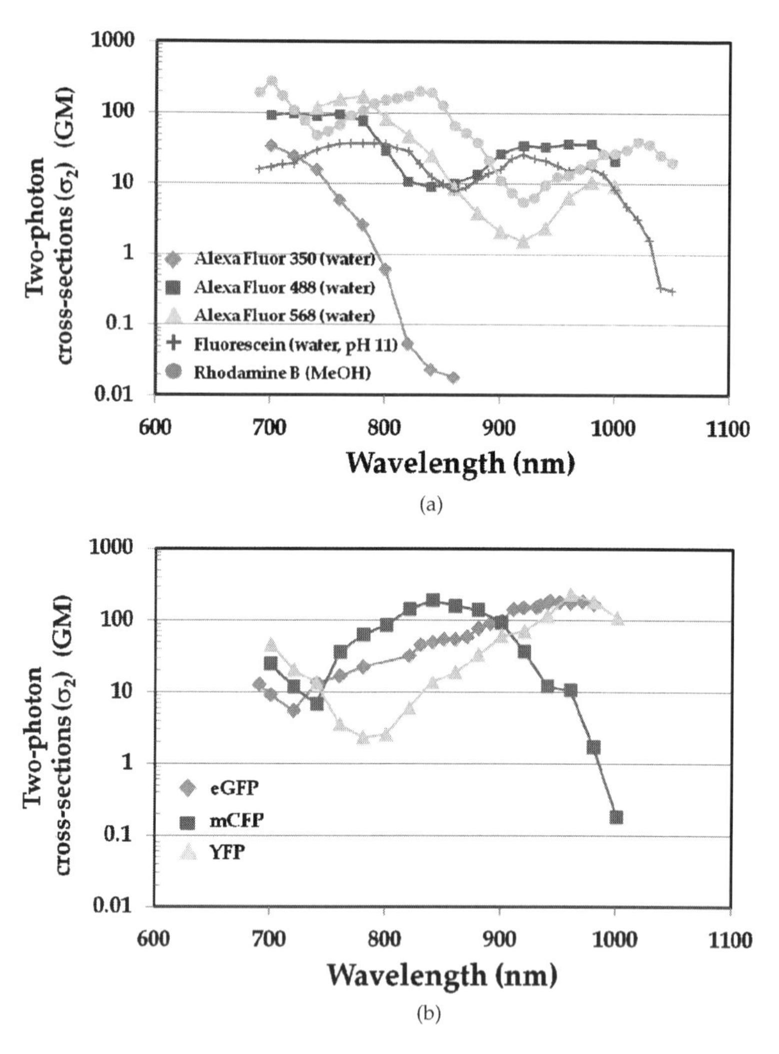

Figure 2.66 Two-photon cross-sections (σ_2) of (a) the common extrinsic fluorophores and (b) fluorescent proteins: enhanced GFP (eGFP), monomeric CFP (mGFP), and YFP. The unit of σ_2 is in Goppert-Mayer (GM): 1 GM $= 10^{-50}$ cm^{4}s/photon [72].

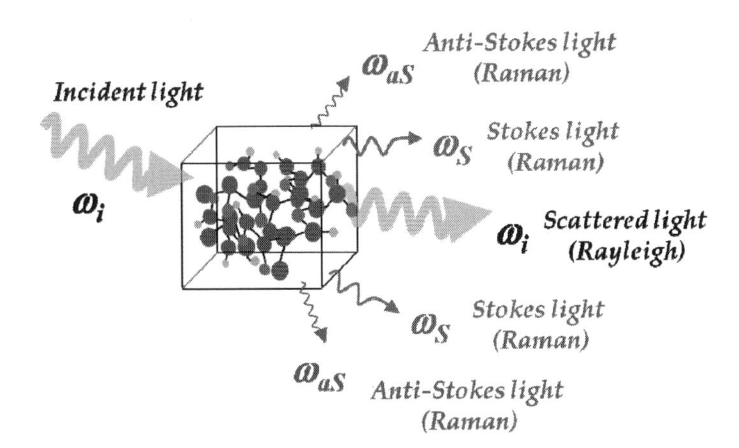

Figure 2.67 Spontaneous Raman scattering. Typically, it is an extremely weak scattering process. Only ~1 in 10^7 incident photons is Raman scattered. At room temperature, the scattered Stokes light intensity is stronger than the scattered anti-Stokes light intensity, as governed by the Maxwell–Boltzmann distribution.

the scattered light is frequency shifted. Perhaps the best-known inelastic scattering is *spontaneous Raman scattering* (Fig. 2.67), first discovered by C. V. Raman in 1928 (Raman was awarded the Nobel Prize in physics for this discovery in 1930). In spontaneous Raman scattering, a portion of the incident photon energy is absorbed by the molecule in the medium in order to drive the molecule into its vibrational motion. Then, the scattered light (i.e., the re-radiation) from the molecule has less photon energy and is thus frequency-down-shifted. The energy difference between the scattered and the input light equals the energy absorbed by the molecule. More specifically, the vibrational frequency of the excited molecule Ω equals the frequency difference between the scattered light ω_S and the input light ω_i, i.e., $\Omega = \omega_i - \omega_S$. Hence, the shift in frequency can be used to identify the particular vibrational modes in the molecules. Because of the unique molecular structure in each type of molecule, such frequency shifts can be regarded unique signatures or "fingerprints" for molecular identification. Thus, the Raman frequency shifts essentially represent the vibrational frequencies of the molecules, as shown in Table 2.1. This capability of fingerprint identification is indeed the primary motivation behind the development of molecular spectroscopy based on Raman effect, called *Raman spectroscopy* [74, 75].

The classical harmonic oscillator model can be used to explain the Raman effect. In brief, the molecular vibration at frequency Ω generates a sinusoidal modulation of the susceptibility. As the incident light wave induces polarization that is given by the product of the susceptibility and the incident wave (i.e., Eq. 2.1), the beating effect of incident wave oscillation (at frequency ω_i) and the susceptibility oscillation (at frequency Ω) will produce polarization at the frequencies $\omega_{aS} = \omega_i + \Omega$ and $\omega_S = \omega_i - \Omega$. The radiation (scattered light) generated by these two components is referred to as *anti-Stokes* and *Stokes* waves, respectively (Fig. 2.67). Different from Stokes generation, the anti-Stokes wave is generated by extracting the energy from the molecule. In the other words, the molecule in this case has to be initially excited. Quantitative description of the Raman effects based on the harmonic oscillator model can be found in Ref. [75].

The Raman effects can also be described using the energy diagrams as shown in Fig. 2.68. In Rayleigh scattering, the scattered

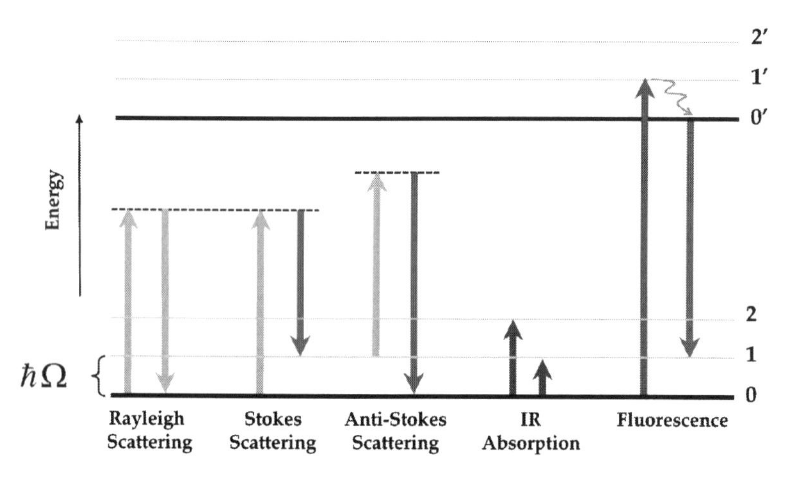

Figure 2.68 An energy diagram showing different atomic transition processes (left to right): Rayleigh scattering, Raman Stokes scattering, Raman anti-Stokes scattering, IR absorption, and fluorescence. 0, 1, 2 are molecular vibrational states of the electronic state S_0. 0′, 1′, 2′ are molecular vibrational states of the electronic state S_1.

photon results from a transition from the virtual state back to the original vibrational state (mostly the ground state). No photon energy change is involved during the process. Thus, it is an elastic scattering process (Fig. 2.68). In spontaneous Raman scattering, the incident photon (at frequency ω_i) interacts with the molecule such that the molecule (initially in the ground state (*0*)) is promoted to a higher energy excited vibrational state (*1*). The scattered photon at the frequency ω_S results from a transition from the virtual state to the excited vibrational state (n). The energy acquired by the molecule from the incident photon is $\hbar\Omega = \hbar\omega_i - \hbar\omega_S$, where $\hbar = h/2\pi$. This is called *Stokes scattering* (Fig. 2.68). In contrast, if the molecule initially is in the excited state (*1*), a photon with a higher energy of $\hbar\omega_{aS} = \hbar\omega_S + \hbar\Omega$ than that of the incident photon will be scattered. In this case, the molecule is relaxed from the excited state (*1*) back to the ground state (*0*). This process is called *anti-Stokes scattering* (Fig. 2.68). The atomic transitions of IR absorption and fluorescence are also shown in Fig. 2.68 for a more complete comparison. The light intensity ratio of the Stokes relative to the anti-Stokes is determined by the absolute temperature of the sample and the energy difference between the ground state and the excited state (following the Maxwell–Boltzmann statistics (see Eq. 2.36). At room temperature, the Stokes light is much stronger than the anti-Stokes as governed by the Maxwell–Boltzmann distribution (Fig. 2.69) [74, 75].

The intensity of the Raman Stokes scattered light is given by

$$I_{\text{Raman}} \propto \omega_i^4 I_i N \left(\frac{\partial \alpha}{\partial Q}\right)^2 \tag{2.56}$$

where I_i is the input "pump" intensity, N the number of molecules available for the process, α the polarizability of the molecule (a measure of how easy the molecule can be polarized), and Q the vibrational amplitude of the molecule. The quantity $\partial\alpha/\partial Q$ describes the change of polarizability with respect to a change in vibrational amplitude. The larger this change is, the stronger the Raman signal is. The expression in Eq. 2.56 also indicates a number of important considerations for Raman spectroscopy: (i) The Raman signal is linearly proportional to the concentration of the molecule. Hence, quantitative measurement is possible. (ii) Because of the

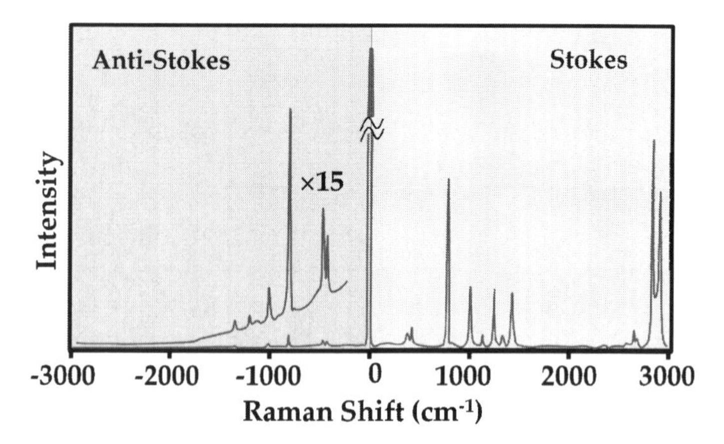

Figure 2.69 Raman spectrum of cyclohexane. Both the anti-Stokes and Stokes sides of the spectrum are shown. The anti-Stokes spectrum is magnified by 15 times for clarity. Also note that by convention, Stokes (anti-Stokes) frequencies are denoted as the positive (negative) Raman frequency shift, in the unit of cm^{-1}.

fourth-power dependence, using shorter wavelength excitation can significantly enhance the Raman signal strength. (iii) The Raman scattered light can also be increased by simply increasing the input power.

Raman scattering is inherently a weak process in that only one in every 10^6–10^8 photons which scatter is a Raman scattered photon. Hence, the scattered photons are primarily from Rayleigh scattering. Raman scattering is hardly observable with ambient light. Thanks to the advent of laser, which makes it feasible to easily achieve optical intensity beyond 100 MW/cm^2 to 1 GW/cm^2, the required input light intensity for obtaining a plausible Raman scattered signal. This opened up the entire avenue of Raman-based molecular spectroscopy used for chemical identification and biological imaging since the invention of laser in 1960s.

One appealing feature of Raman spectroscopy is its capability of obtaining a multitude of chemical information from the sample without the need for exogenous fluorescent labeling, which is widely adopted in fluorescence spectroscopy and bioimaging. Exogenous labeling, i.e., an extrinsic fluorophore, generally requires modification of the cellular structures. It is thus less favorable for

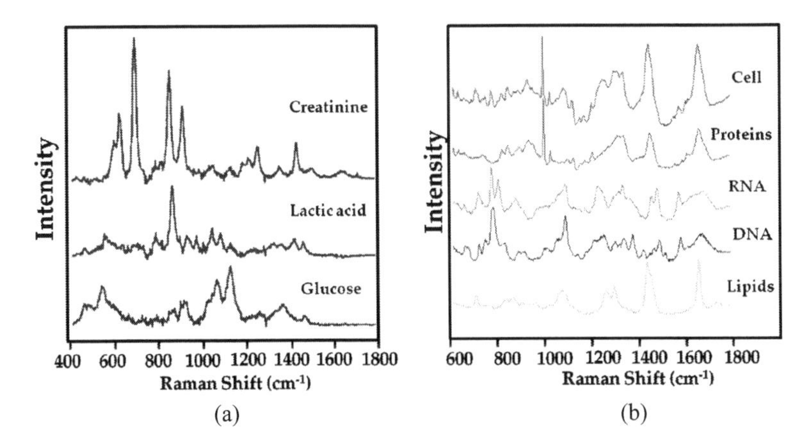

Figure 2.70 (a) Raman spectra of three biological analytes—glucose, lactic acid, and creatinine—in phosphate-buffered saline solution [75]. (b) Typical Raman spectra of main biomolecules found in the biological cells [76].

characterization of live cells in their native physiological states. The continuous loss of fluorescence intensity during measurement because of photobleaching and the potential for photodamage to the specimen that is due to the use of ultraviolet frequencies are also the fundamental problems of fluorescence microscopy [4]. In contrast, labeling-free Raman spectroscopy minimizes the sample preparation, and hence circumvents the problems related to the delivery, specificity, stability, and invasiveness of the exogenous labels. As a result, together with the ability of revealing multi-parametric "fingerprint" chemical information, Raman spectroscopy has been well recognized as a powerful tool for biomedical diagnostics, e.g., detecting the biological analytes found in blood (Fig. 2.70a) [75].

As another example, the unique Raman spectra of the fundamental set of single molecules in a single cell, e.g., DNA, RNA, protein, and lipids (Fig. 2.70b) [76], can be used to determine the biochemical composition of the cell (without the needs for extrinsic fluorophore labeling), which in turn can provide insight in the biochemical functions and the physiological processes of the cells. Raman spectroscopy has also been proven to be a useful tool for cancer diagnosis in vivo or ex vivo for the colon, esophagus, breast, skin, and brain [77]. It has also been gaining popularity in different areas of the pharmaceutical applications such as drug screening, including identification of active ingredients and contaminants in the drug mixtures [78].

2.11.4.2 Coherent Raman scattering

Because of its extremely weak scattering strength, spontaneous Raman spectroscopy suffers from prohibitively low signal-to-noise ratios [73]. Hence, typical spontaneous Raman measurements require long data-acquisition time (minutes to hours) in order to obtain a decent amount of signal. In addition, the spontaneous Raman signal is often masked by the auto-fluorescence from the biological cells/tissues. All these limitations impede the progress of further development of the Raman technique in clinical applications.

The weak spontaneous Raman signal can be greatly enhanced by the coherent Raman effects, including coherent anti-Stokes Raman scattering (CARS) [78–84] and stimulated Raman scattering (SRS) [85–92]. Briefly, these effects require two light beams (mostly laser), one of which is called pump beam (at frequency ω_i) and the other Stokes beam (at the frequency ω_S). When the frequency difference of the two beams (i.e., the beating frequency) matches with a particular molecular vibration frequency ($\Omega = \omega_i - \omega_S$), the two beams (pump and Stokes) create a force that stimulates molecular vibrations. This driving force will enhance molecular oscillations, which, in turn, will increase the amplitude of the Stokes. This positive feedback phenomenon is called SRS (Fig. 2.71a). On the other hand, it is also possible that the two beams generate a coherent anti-Stokes ($\omega_{aS} = \Omega + \omega_i$) beam through a resonant FWM process[a] between the pump and the Stokes under the phase-matching condition. This is called the CARS effect (Fig. 2.71b)[b]. In both SRS and CARS, the scattered

---

[a]The term "resonant" FWM describes that this is a wave-mixing process simultaneously involving molecular vibration at a frequency ($\Omega = \omega_i - \omega_S$) coinciding with the Raman molecular vibration frequency, i.e., in resonance. Such resonant oscillations are *coherently* driven by the incident wave and the Stokes waves, thereby generating an anti-Stokes signal at $\omega_{aS} = \Omega + \omega_i$.

[b]The energy diagram representation of the CARS process shown in Fig. 2.71b adopts the model suggested by Terhune [93], a widely adopted theory in CARS spectroscopy and microscopy. Nevertheless, this model suggests no net energy exchange with the medium during the process, especially not applicable for phase-matched CARS *at*

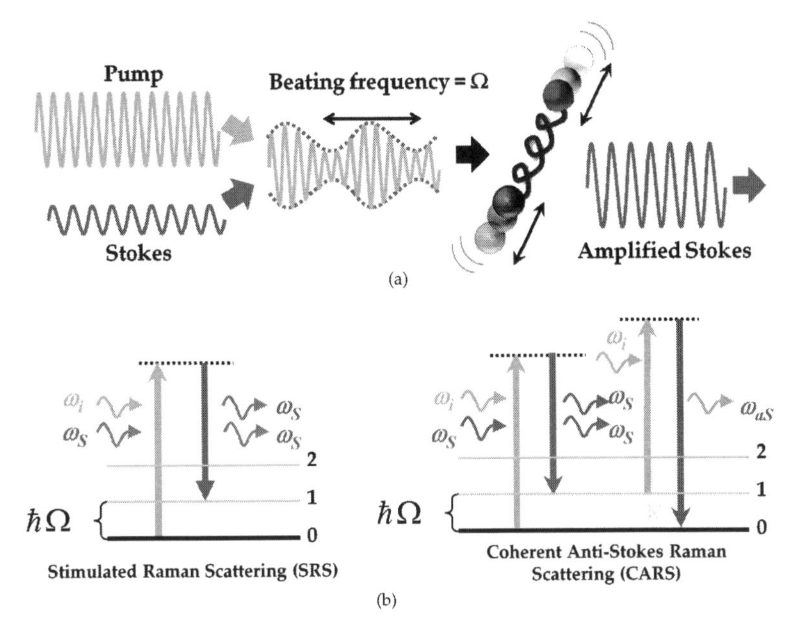

Figure 2.71 (a) Classical view of stimulated Raman scattering (SRS). The incoming Stokes will be amplified in this process. (b) Energy diagrams explaining (left) the SRS and (right) the CARS processes.

photons (Stokes and anti-Stokes) are all generated in phase and along a well-defined (preferred) scattered direction. This is in contrast to spontaneous Raman scattering, in which the scattered photons are in random phase and scattered in all directions. Hence SRS and CARS are said to be coherent. Such coherent nature makes the generated Raman signals (Stokes or anti-Stokes) significantly higher ($\sim 10^5$–10^7) than what is obtained by spontaneous Raman effect [84].

These two coherent Raman effects can be described by the third-order nonlinear polarization, which is induced by interaction among the pump, Stokes, and anti-Stokes waves in the material, i.e.,

$$\mathcal{P}^{(3)} = \varepsilon_0 \chi^{(3)} \mathcal{E}_{\mathrm{i}}^2 \mathcal{E}_{\mathrm{S}} \tag{2.57}$$

Raman resonance. In contrast, Vermeulen et al. [94] recently developed the model which states that the pump and Stokes photons are converted to an anti-Stokes and a pump photon with the molecular relaxation. This description is well consistent with many other well-established CARS theories for phase-matched conditions [95–98].

where $\mathcal{E}_i$ and $\mathcal{E}_S$ are the field amplitudes of the pump and Stokes waves, respectively. To be more precise, the third-order nonlinear effect, i.e., nonlinear susceptibility $\chi^{(3)}$, consists of two contributions:

(1) Raman (molecular) contribution $\chi_R^{(3)}$, which describes the molecular response to the input waves. This contribution gives rise to the Stokes amplification and/or anti-Stokes generation.
(2) Electronic contribution $\chi_E^{(3)}$, which describes the electronic response. It results in the Kerr effects (e.g., FWM) and TPA.

Hence, the overall third-order susceptibility can be generalized as [72, 79]

$$\chi^{(3)} = \chi_R^{(3)} + \chi_E^{(3)} = \frac{A}{\Omega - (\omega_i - \omega_S) - i\Gamma_R} + \chi_E^{(3)} \qquad (2.58)$$

where A is a constant representing the Raman scattering strength and Γ_R is the half-width at half-maximum of the Raman (i.e., molecular vibrational) resonance spectral line. Equation 2.58 tells us that the Raman contribution becomes significant when the frequency difference of the pump and Stokes coincides with the molecular vibrational frequency (i.e., $\Omega = \omega_i - \omega_S$). Otherwise, the electronic contribution, i.e., SPM and FWM, could be the dominant effects. Note that the Kerr effect is nonresonant in nature, whereas Raman is a resonance effect. This concept is particularly important for accurate CARS spectroscopic measurements, in which $\chi_R^{(3)}$ is of interest. These measurements are easily disturbed by the broadband background due to $\chi_E^{(3)}$. This problem has been extensively studied in CARS spectroscopy. A detailed discussion can be found elsewhere [72, 79–84].

2.11.5 *Applications of Nonlinear Optics in Biophotonics*

Perhaps one of the key applications of nonlinear optics in biophotonics is to use the nonlinear optical properties of the biological cells/tissues as the imaging contrast for optical microscopy. Advances in the laser technology along with other innovations now make it possible to reveal a multitude of molecular information based on different nonlinear effects of the biological specimens, such as multiphoton absorption, SHG, CARS, SRS, and so on—forming an impressive arsenal of *nonlinear optical microscopy*. The history, the theories and the development of nonlinear optical microscopy are well documented elsewhere [72]. Chapter 3 will also cover the key developments in multiphoton microscopy. Here we will highlight the key features of nonlinear optical microscopy.

A generic schematic of the typical nonlinear optical microscope is shown in Fig. 2.72. Because of the nonlinear optical effects, which can only be observable with sufficiently high optical intensities, the most viable imaging approach so far is to tightly focus the laser beam, by using a high-NA objectives lens (NA $>$ 0.8), into a small area of the specimen in order to obtain an appreciable signal-to-noise ratio in the captured image. The entire image is digitally reconstructed by raster-scanning the beam across the sample (Fig. 2.72), as in conventional laser-scanning confocal microscopy.

Tight focusing alone does not guarantee efficient nonlinear optical microscopy. A short-pulse laser (typical pulse widths of fs–ps) as the source for the microscope is also a necessary factor. This can be understood by the fact that the nonlinear optical effects depend on the instantaneous optical intensity rather than the time-average intensity. The instantaneous peak intensity of a laser pulse, I_{peak}, can be roughly related to the average power I_{ave} by the temporal FWHM of the pulse τ and the repetition rate of the laser R by

$$I_{\text{peak}} \approx \frac{I_{\text{ave}}}{R\tau} \tag{2.59}$$

A mode-locked titanium sapphire (Ti:S) pulsed laser, which is a popular laser used in nonlinear optical microscopy, can produce an ultrashort pulse with the pulse duration $\tau \approx 100$ fs at the repetition rate $R = 80$ MHz. Clearly, the advantage of using pulsed laser can be appreciated by the enhancement in the peak power compared to the averaged power, i.e., $1/R\tau \approx 10^5$!

In general, there are two main advantages of nonlinear optical microscopy compared to conventional "linear" microscopy (e.g., single-photon wide-field or confocal fluorescence microscopy):

(i) **Deep penetration depth in imaging.** Typical nonlinear optical microscopes employ NIR laser sources instead of using visible

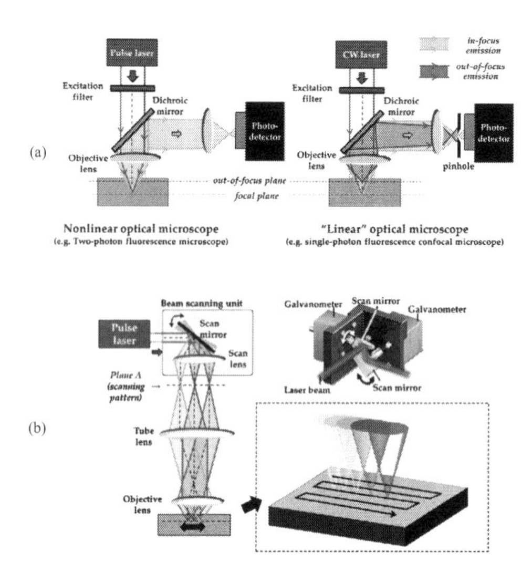

Figure 2.72 (a) Generic schematics of (left) a nonlinear optical microscope (e.g., two-photon fluorescence, SHG, or coherent Raman microscope) and (right) a linear optical microscope (e.g., single-photon fluorescence confocal microscope). One distinct difference between two microscopes is the requirement of a pinhole in single-photon confocal microscopy for rejecting the out-of-focus signal (scattered or fluorescence light) from the specimen—achieving optical sectioning. In contrast, optical sectioning can readily be attained without the pinhole in nonlinear optical microscopy. The absence of the pinhole within the focal volume of the imaging objective lens gives rise to a nonlinear optical signal (see (a)). Pulse laser, which is capable of delivering high peak power, is typically required for almost all nonlinear optical microscopes. In contrast, a continuous-wave (CW) laser source is mostly sufficient for single-photon microscopy. An excitation filter is used to eliminate any background light in the spectral range other than the excitation, such as the excitation light for fluorescence emission, or the pump and Stokes light for the generation of a coherent Raman signal (SRS or CARS). The dichroic mirror is a spectral filter which reflects and isolates the scattered light (CARS, SRS, or SHG) or fluorescence emission (two-photon fluorescence) from the excitation light. (b) Schematic of the common beam scanning strategy for a nonlinear optical microscope (also applicable to one-photon confocal microscope). The scanning pattern on *plane A* generated by the beam scanning unit is projected onto the focal plane of the specimen through the tube lens and the objective lens of the microscope. For 2D image acquisition, the focused beam on the specimen is raster-scanned by the two scan mirrors (one for x-axis motion and the other for y-axis motion). The scan mirrors are typically controlled by the galvanometers (see the inset in (b); *source*: www.microscopyu.com). The nonlinear optical signal generated from each scanned spot (diffraction-limited spot) is sequentially detected by the photodetector for subsequent image reconstruction.

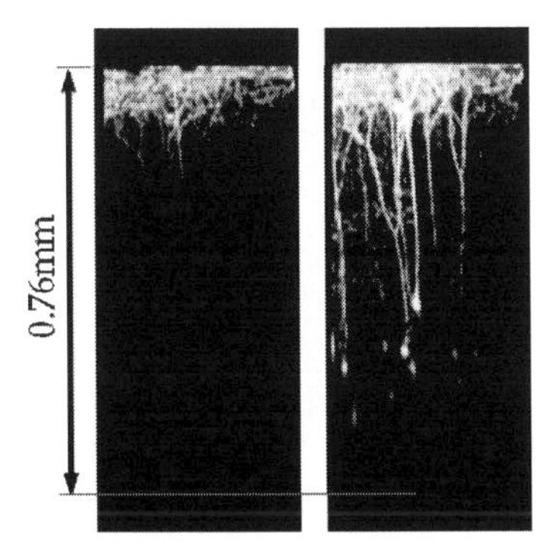

Figure 2.73 Comparison of the imaging depth between single-photon fluorescence (left) and two-photon fluorescence microscopy (right). The specimen is the cerebral cortex of a transgenic mouse (labeled with GFP). For single-photon fluorescence microscopy (excitation wavelength of 488 nm), the imaging depths can go down to only 400 μm. In contrast, two-photon microscopy (excitation wavelength of 920 nm) can achieve a penetration depth of ~750 μm. (*Source*: http://microscope.olympus-global.com.)

or UV light as in the case of conventional fluorescence microscopy. This is primarily because of the availability of high-power ultrashort laser sources in this wavelength regime, e.g., femtosecond Ti:S laser at 800 nm, picosecond Nd:YAG or Nd:YVO_4 laser at ~1064 nm. Remember that high power is required for generating observable nonlinear optical effects. On the other hand, in multiphoton fluorescence microscopy, longer-wavelength (NIR) photons are required for exciting fluorescence emission in the typical visible and UV fluorophores (see Fig. 2.74). Coincidently, such needs for NIR laser sources in nonlinear optical microscopy make it attractive for deep-tissue imaging. It is because the NIR wavelengths reside in the "optical window" of biological tissue (~600–1200 nm) (see Section 2.8), where the scattering loss is less prominent than that in the shorter wavelength, and the tissue absorption is low. These favorable conditions make nonlinear optical microscopy capable of achieving a deeper penetration depth through turbid tissue media (up to a few hundreds of micrometers to even millimeters), compared to its linear optical microscopy counterpart. As an illustration, Fig. 2.73 shows that the penetration depth for *two-photon fluorescence microscopy* (to be discussed further later) exceeds that for single-photon fluorescence microscopy. This is particularly of great importance for high-resolution in vivo imaging in different organs of living animals.

(ii) **3D optical sectioning.** The nonlinear optical signals depend on the high-order optical intensity, e.g., SHG signal strength scales with the squared excitation intensity. Two-photon excited fluorescence also shows quadratic dependence on the excitation intensity. This has an inherent *optical sectioning* property; i.e., only within the focal volume of the imaging objective lens can it give rise to a nonlinear optical signal. This is in contrast to the fact that single-photon fluorescence is excited at almost all regions within the focal volume (Fig. 2.74). Therefore it eliminates the need for a pinhole right in front of the detector to obtain the 3D imaging capability in confocal microscopy (Fig. 2.72a). In particular, two-photon excitation can render less photobleaching, phototoxicity, and photodamage because of the tight confinement of the excitation volume—essential for live cell/tissue imaging.

Among all the different nonlinear optical imaging techniques, three commonly used modalities will be highlighted below. Table 2.7 also summarizes the comparisons among them. The recent advances in nonlinear optical microscopy have been already been extended to exploit many other nonlinear optical effects as the image contrast, e.g., microscopy based on SPM [99], FWM [100], and third-harmonic generation (THG) [101–103]. In order to make the nonlinear optical imaging technology even more versatile, the current trend is to establish a single nonlinear microscope platform with a multimodal imaging capability, i.e., being able to simultaneously obtain different nonlinear optical image contrasts (e.g., two-photon fluorescence, SHG, THG, and CARS/SRS) from the complex biological specimens [104–107].

Two-photon fluorescence microscopy (TPFM). Two photons are absorbed "simultaneously" (typically < fs) such that the sum energy of the photons is sufficient to promote them to their excited state.

Table 2.7 Key features of three common nonlinear optical imaging modalities: two-photon fluorescence, SHG, and coherence Raman (CARS and SRS) imaging. Ti:S: titanium sapphire; OPO: optical parametric oscillator; Nd:YVO_4: neodymium vanadate

	Two-photon fluorescence	SHG	Coherent Raman
Energy diagram	Energy		SRS CARS
Optical nonlinearity	3rd-order	2nd-order	3rd-order
Extrinsic label	Required, except for intrinsic fluorophores	Label-free, except measuring membrane potential	Label-free
Common excitation wavelength	~700 nm – 1000 nm	~700 nm – 1000 nm	~700 nm – 1300 nm (Pump) ~1000 nm – 1500 nm (Stokes)
Common excitation pulse width	~100's of fs or less	~100's of fs or less	~1–10 ps
Laser sources	Ti: S laser	Ti: S laser	Ti: S laser, Nd: YVO_4 laser, OPO, or fiber lasers
Key applications	In-vivo brain imaging in small animals (e.g., calcium imaging of neuronal networks); imaging intrinsic fluorophores (e.g., NADH, hemoglobin)	Imaging structural protein arrays, collagen, actomyosin complexes (muscle structures); imaging membrane potential	Imaging cellular processes, such as lipid metabolism and storage; imaging drug delivery

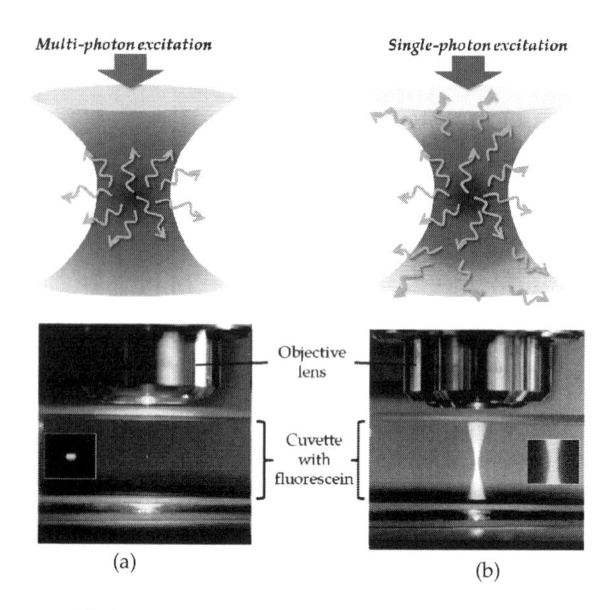

Figure 2.74 (a) (Top) In nonlinear optical microscopy, only the vicinity of the focal volume has optical intensity high enough to generate the nonlinear optical signal, e.g., two-photon fluorescence, SHG, SRS, or CARS. This automatically achieves *optical sectioning* for 3D imaging without the need of a confocal pinhole to reject the out-of-plane scattered light or fluorescence, as in single-photon fluorescence microscopy (see (b) and Fig. 2.72a). The two bottom images show the green fluorescence emission from a dilute fluorescein solution in a quartz cuvette, excited by (left) NIR laser (for two-photon fluorescence excitation) and (right) blue laser (for single-photon fluorescence excitation). Note that the green fluorescence is only observable within the focal volume (see the tiny spot) in two-photon fluorescence excitation. In contrast, the entire excitation pathway within the solution emits the fluorescence light in single-photon fluorescence excitation (*Source*: http://microscopy.berkeley.edu/.).

Then, it undergoes the normal fluorescence emission process, as described in Section 2.10 (see Table 2.7). A similar concept can be extended to three or more photons, which altogether result in a single fluorescence emission event. TPFM has been the most popular nonlinear optical imaging modality because it can not only enjoy essentially all the strengths of single-photon fluorescence microscopy—the mainstay of the optical imaging modality used in bioscience and biomedicine—but also reveal the biochemical information about the interaction between the fluorophores and the cells/tissues or the genetic expression of the probe in *deep tissue* while leaving the specimen intact [108–110]. It has been proven to be a robust tool for basic research in cell physiology, neurobiology, and tissue engineering [46, 72]. Examples include intracellular calcium ion dynamics imaging in single cells [111], blood flow imaging in the fine capillaries [112], and angiogenesis in cancer research [113].

Second harmonic generation (SHG) microscopy. In SHG, the two incident photons are simultaneously annihilated in exchange of the creation of a photon with doubled photon energy, i.e., the frequency of the converted (or scattered) photon is frequency-doubled (Fig. 2.60a). As opposed to TPFM, in which a portion of the incident photon energy is lost during the relaxation, SHG does not involve real atomic transition to the excited state. The overall photon energy is conserved during SHG and the scattered frequency-doubled signal preserves the coherence of the input. SHG can exist only in ordered or noncentrosymmetric media, e.g., structural protein, collagen, microtubules, and muscle myosin in cells/tissues (Fig. 2.75). In contrast to fluorescence imaging, which shows chemical specificity, SHG offers cellular imaging with information specific for structural configuration such as molecular symmetry, local morphology, orientation, and molecular alignment [46, 72, 114–118]. Note that this is achieved without the need for exogenous

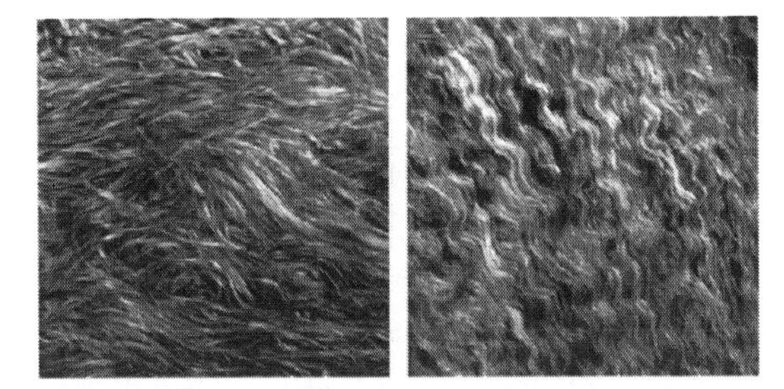

Figure 2.75 SHG images of the collagen fibers in (left) normal human ovary and (right) malignant ovary [118]. The clear morphological difference shows the utility of SHG microscopy in the analysis of diseased states of biological tissues/cells.

fluorescent labeling. To study membrane potential dynamics, extrinsic labeling to the cells could give rise to a considerable SHG signal correlated with the depolarization of the cells [116, 117].

The fact that SHG does not occur in centrosymmetric materials can be understood by Eq. 2.47. In a material system exhibiting inversion symmetry (centrosymmetry), the sign of the induced polarization $\mathcal{P}^{(2)}(t)$ has to be changed if the sign of the electric field $\mathcal{E}(t)$ is changed, i.e.,

$$-\mathcal{P}^{(2)}(t) = \varepsilon_0 \chi^{(2)} \left(-\mathcal{E}(t)\right)^2 = \varepsilon_0 \chi^{(2)} \mathcal{E}^2(t) \qquad (2.60)$$

Comparing Eq. 2.60 with Eq. 2.48, one can immediately realize that $\mathcal{P}^{(2)}(t)$ has to be equal to $-\mathcal{P}^{(2)}(t)$. It can be true *only if* $\chi^{(2)}$ vanishes and no SHG signal is expected in centrŏsymmetric media.

Coherent Raman microscopy. Although the conventional spontaneous Raman imaging techniques provide image contrast based on the intrinsic molecular vibrations of a sample without introducing extrinsic labels, they are extremely weak, and the strong autofluorescence background usually overwhelms the Raman signal. To this end, the coherent Raman effects (namely CARS and SRS), which provide orders of magnitude more sensitive and efficient detection than spontaneous Raman imaging, became increasingly eminent in the past decade. Because of the abundant CH_2-rich content in many biological specimens and its well-isolated Raman spectral feature ($\sim$2900–3000 cm^{-1}) from other molecular vibrational frequencies, which fall within the so-called *fingerprint region* ($\sim$800–2000 cm^{-1}), the vast majority of applications of CARS microscopy in biomedicine so far has been in imaging the structure and dynamics of lipids (lipids have long aliphatic chains full of $-CH_2$ groups). They are proven to be applicable at the levels ranging from the cell [119], the tissue [120, 121], to the organism [122, 123]. For instance, CARS imaging has been used as a sensitive tool to monitor the trafficking of lipid droplets in different cell types [119], and to visualize the distribution of the lipid deposits in atherosclerotic lesions [124]. CARS and SRS imaging can also be applied to monitor the dynamic processes during drug delivery—an important tool in pharmaceutical applications [79, 85].

In most of the coherent Raman imaging modalities, the pump and the Stokes laser beams (Fig. 2.71b) are derived from (i) two separate but synchronized pulsed lasers or (ii) a single laser source, which simultaneously consists of the pump and Stokes spectral regions. Typically, a supercontinuum (SC) source is chosen in the single laser configuration. To date, there are two approaches for CARS microscopy: single-frequency CARS and multiplex CARS. In single-frequency CARS, only the CARS signal from individual vibrational resonance, e.g., the stretching mode of lipids at $\sim$2845 cm^{-1} is detected at one laser focus point. The CARS response at that frequency is imaged by scanning the laser beams across the specimen, permitting real-time video-rate ($\sim$30 frames per second) CARS imaging with high chemical specificity.

Single-frequency CARS is by far the most viable approach if one only seeks qualitative contrast based on one particular vibrational frequency in the sample. However, it loses the original strength of spontaneous Raman measurements, i.e., providing the spectral information of the molecular vibrations and thus offering *quantitative* imaging capability. In this regard, multiplex CARS, in which a broad range of vibrational resonances can be probed simultaneously at one laser scan point, has been developed for truly vibrational hyperspectral imaging [125].

In CARS, the anti-Stokes is generated as a new but small signal out of the noise background in the anti-Stokes spectral range. In contrast, the stimulated Raman signal manifests as a minute intensity change of the incident beams, either as a *stimulated Raman gain* of the Stokes or as a *stimulated Raman loss* of the pump (Fig. 2.76). SRS microscopy represents a good alternative to CARS microscopy for vibrational imaging as it is free from the nonresonant background that appears in CARS microscopy [89]. It makes the spectral interpretation and analysis much more straightforward than CARS and thus is advantageous for quantitative hyperspectral imaging. However, the fact that such intensity change is extremely small, typically on the order of 10^{-6} for biological cell imaging, an extremely sensitive detection scheme, namely lock-in detection, has to be adopted. Such limitation together with other technological challenges explains why the SRS effect has been known since the 1960s [126] and yet practical SRS microscopy for bioimaging lay dormant for decades until 2008 [89].

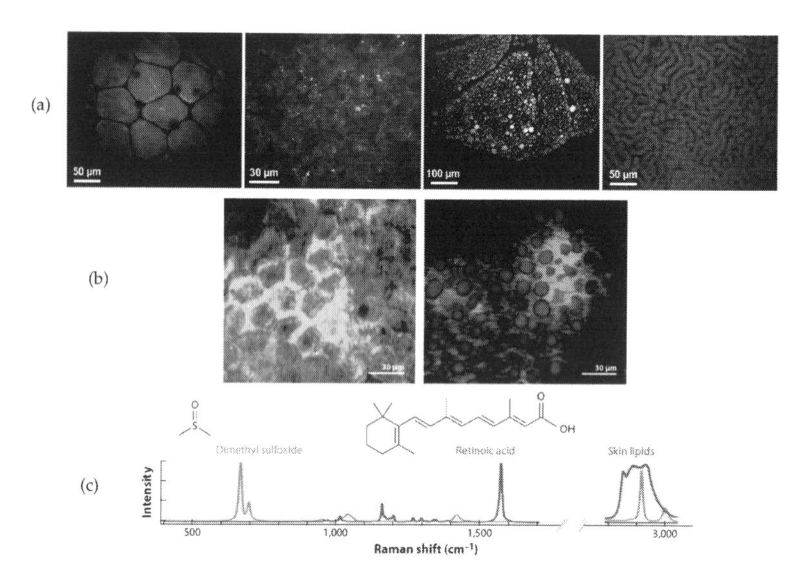

Figure 2.76 (a) Ex vivo CARS microscopy of different tissues with CH_2 contrast. (From left to right). White adipose tissue of mouse omentum majus, mouse lung tissue (showing the individual alveoli), the surface of the adipocyte-covered mouse kidney, and mouse kidney tissue at an imaging depth of 40 μm (revealing renal tubules) [79]. (b) SRS images showing the spatial distributions of (left) retinoic acid and (right) dimethyl sulfoxide (DMSO) applied in mouse ear skin. These SRS images were acquired at the Raman shifts of retinoic acid at 1570 cm^{-1}, DMSO at 670 cm^{-1}, and the CH_2 stretching vibration at 2845 cm^{-1}, as shown in (c) (corresponds to the skin lipid structure) [89].

2.12 More on Light–Tissue/Cell Interaction

We have discussed that the interactions of light with biological matter are highly determined by the irradiation wavelength, the absorption, and the scattering properties of the specimens. Apart from these parameters, the *light exposure time* (or equivalently the irradiation temporal pulse width) and the *pulse energy/power* are also important factors. Indeed, different combinations of the pulse width and the pulse energy result in drastically distinct photo-induced effects, which can roughly be classified into (i) photochemical, (ii) thermal (coagulation, vaporization, and ablation), and (iii) optical breakdown processes (Fig. 2.77). Hence, careful

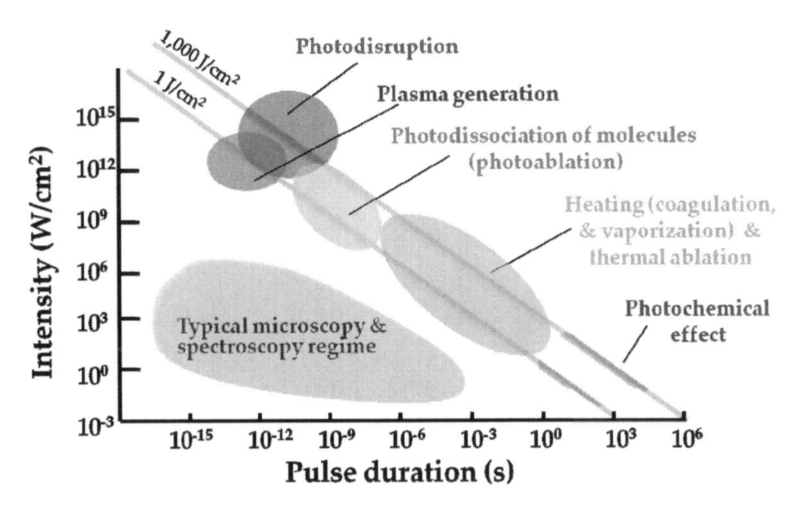

Figure 2.77 General classification of the key effects involved in laser–matter interaction in tissue in terms of the laser peak intensity (W/cm^2) and the laser pulse duration (s). Note that the shaded regions provide only a rough estimate of the laser parameters. The regimes of microscopy and spectroscopy (i.e., diagnostic applications) are also shown in the graph. Modified from [127].

choice of the laser pulse energy and pulse width, which leads to different photo-induced effects, can be adopted and tailored for very specific medical therapeutic applications, e.g., laser surgery and photodynamic therapy (PDT). It should be emphasized that we have discussed that the subsequent processes followed by light absorption by the cells/tissues can be either radiative or nonradiative (Section 2.10). *The photo-induced processes discussed in this section are primarily referred to as nonradiative processes*, i.e., no light emission (e.g., fluorescence) is involved.

Certainly, laser wavelength is the foremost parameter which we should consider for a specific therapeutic/diagnostic procedure. This is simply because (i) different chromophores in tissues, e.g., blood, water, fat, exhibit very distinct absorption spectra, and (ii) scattering is wavelength dependent—becoming weaker at longer wavelengths. These two factors together determine the tissue penetration (Section 2.8). For instance, high-energy Nd:YAG laser at 1064 nm can be used for recanalization of tumors inside the intestinal tract and coagulation of the blood vessels. This is because at this wavelength the penetration depth in tissue is almost the deepest among all wavelengths. Er:YAG laser emits light at 2.94 μm, which has a very high absorption in water and other hard tissues, e.g., bones, teeth, and thus a very short penetration depth. Thus it is useful for performing tissue ablation with minimal thermal damage. Er:YAG lasers have landed on applications in dentistry, dermatology, and cosmetic plastic surgery [128].

Given a certain laser wavelength (at which the tissue absorption is strong), the laser pulse width together with the pulse energy are, then, the key factors determining the particular photo-induced mechanism (Fig. 2.77). Here is a broad picture of the various photo-induced effects that are particularly relevant to therapeutic applications:

(i) At very low intensities ($\sim$1 W/cm^2) and long exposure times ranging from seconds to CW, the specific molecules can be excited by the incident photons, which in turn produce reactive oxygen species that can destroy the tumor. This is the basic mechanism of PDT. This is a nonthermal effect.

(ii) For CW or quasi-CW irradiation ($>$0.5 s) of energy fluence greater than 10 J/cm^2 (or the intensity of more than $\sim$10 W/cm^2), the tissue will efficiently convert the absorbed light energy to heat and thus undergo coagulation easily.

(iii) If the pulse width is shortened to 1 ms, the intensity will increase to $\sim$10^4 W/cm^2, given the same energy fluence of 10 J/cm^2. In this case, the heat cannot diffuse into the tissue within such a short period of time. Hence, local overheating occurs and the tissue can be vaporized (local temperature can go typically $>$300°C).

(iv) For the pulse width of $<\sim$1 μs (intensity of $\sim$10^7 W/cm^2), it does not have sufficient time even to vaporize the tissue. Instead, it will ablate explosively into fragments (photoablation). Although photoablation is generally regarded as an ablation effect driven by the *photochemical* mechanism, i.e., ionization or dissociation of the molecule upon electronic excitation, the complete picture of the underlying mechanisms of photoablation can also complicated by two other factors:

photothermal, i.e., overheating of the tissue after light absorption; and *photomechanical*, i.e., generation of mechanical stress, which is confined within the illuminated volume or propagates in the tissue.

(v) A shorter pulse with pulse width ranging from picoseconds to nanoseconds (intensity of $>10^{10}$–10^{13} W/cm^2) will cause optical breakdown, in which plasma (free electrons at very high density) will be created. The plasma will expand to produce a cavitation bubble, which later collapses. During this process, a shock wave will be generated (photodisruption).

2.12.1 *Photochemical Effect: Photodynamic Therapy*

When the molecules (chromophores) are excited by light, a variety of chemical reactions can be induced, such as photo-cross-linking, photofragmentation, photoxidation, and photohydration. Among all, photosensitized oxidation is the most popular process employed in clinical applications, such as PDT. While endogeneous chromophores, e.g., DNA, cholesterol, can undergo photochemical reactions, exogeneous chromophores (or called photosensitizers[a]) can also serve the same purpose. Photosensitization is the basic mechanism of cancer treatment by PDT. The general idea is to excite the photosensitizer which localizes to a target cell/tissue such that the excited photosensitizer transfers energy from light to ambient molecular oxygen. It results in the formation of reactive oxygen species (ROS), such as free radicals and singlet oxygen, which facilitate cellular toxicity (Fig. 2.78).

On the absorption of the photon, the photosensitizer is first promoted to an excited singlet state S_1, followed by an intersystem crossing to a long-lived triplet state T_1. Two alternative reaction mechanisms exist for the decay of the excited triplet state which are called Type I and Type II reactions (Fig. 2.79). In a Type I reaction, free or ionized radicals, peroxides, and superoxide anions

[a]A photosensitizer acts as a catalyst to sensitize photochemical processes. They are typically the "drugs" used for PDT. PDT drugs can generally be categorized into three classes: (i) the porphyrin platform (e.g., hematoporphyrin derivative, benzoporphyrin derivative, and 5-aminolevulinic acid [ALA], and texaphyrins), (iii) the chlorophyll platform (e.g., chlorins, purpurins, and bacteriochlorins), and (3) dyes (e.g., phtalocyanine and napthalocyanine) [129] (also see Table 2.8).

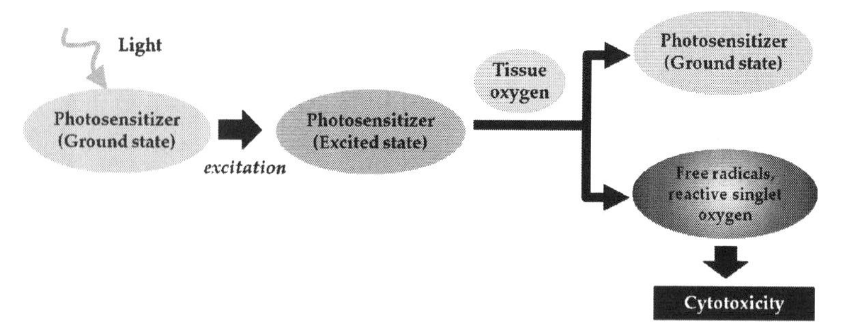

Figure 2.78 General mechanism of laser therapy based on photochemical reaction, called photodynamic therapy (PDT). PDT requires three key elements: light, a photosensitizer (also called PDT drug), and oxygen. When the photosensitizer is exposed to light at the wavelength such that it is excited from the ground state to an excited electronic state, the excited photosensitizer can reacts with the oxygen to generate reactive oxygen species (ROS), such as singlet oxygen and free radicals, which cause cellular toxicity (cytotoxicity). The photosensitizer is relaxed back to the ground state.

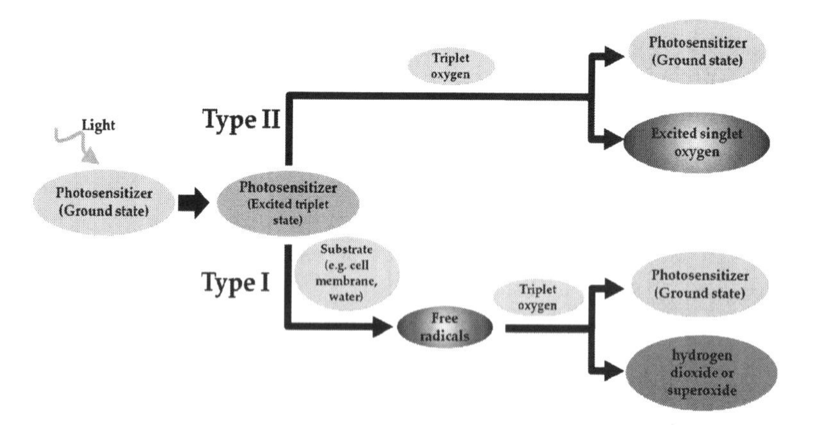

Figure 2.79 Type I and Type II reactions in PDT.

are generated by electron and hydrogen transfer with water or other biomolecules (or called "substrate") to induce the cytotoxic effect. The radicals further react with the triplet oxygen, which may lead to the formation of hydrogen dioxide or superoxide anions. A Type II reaction involves the conversion of an oxygen molecule from its normal triplet state to a highly reactive excited singlet state. The very reactive excited singlet oxygen easily leads to cellular oxidation and necrosis. That is why singlet oxygen is generally regarded as a toxic agent in tumor cells during photochemical reactions. Usually, both Type I and II reactions take place simultaneously. The respective contribution depends on the type of photosensitizer, the concentrations of biomolecules in the target cells (substrate) and oxygen. So far, the Type II reaction is widely recognized as the major mechanism for introducing the cytotoxic effects which can be used to destroy the malignant cells. A more detailed explanation of the photosensitization processes can be found elsewhere [129–131].

The major procedures involved in PDT are depicted in Fig. 2.80. In PDT, the photosensensitizer, or the PDT drug, is injected into

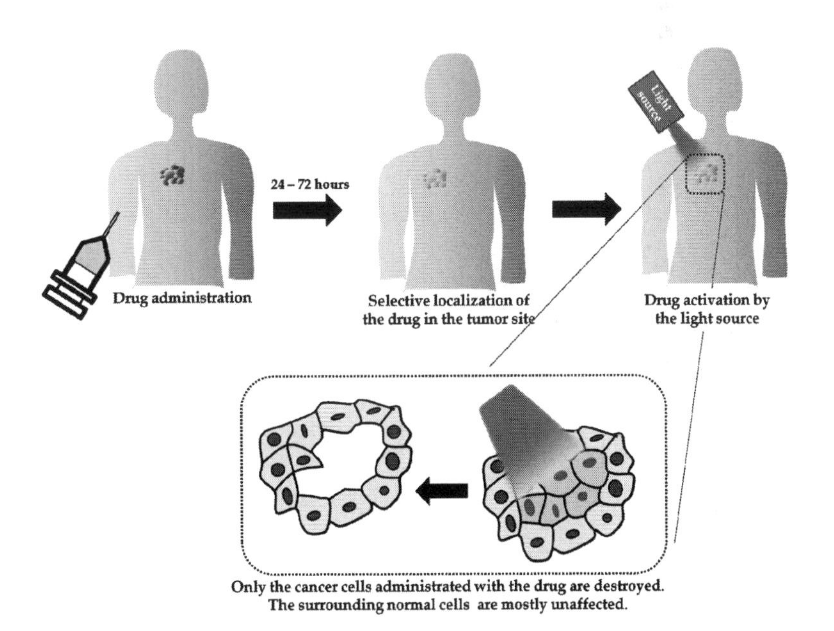

Figure 2.80 Typical procedures of PDT with the photosensitizer, simply called PDT drug. After injection, the drug slowly accumulates in the target tumor site. This process typically takes several hours to days depending on the type of drug. Irradiation from a laser source at a proper wavelength activates the drug and triggers a photochemical reaction to produce the cytotoxic effect in the tumor site. The surrounding healthy and normal cells/tissues are mostly unaffected.

the body and is selectively localized in the malignant tissue site after a certain time period ranging from hours to days. Then, a light source, at the wavelength range which matches with the PDT drugs' absorption spectrum, is used to illuminate the targeted site and produces the cytotoxic effect at the tumor site. The light can be either by either direct illumination for the case of superficial skin cancer or by miniaturized fiber-optic endoscopic systems for the case of cancers of internal organs, such as gastric cancer. Note that the light dose for PDT is low, typically 100–1000 mW/cm^2, in order to avoid any thermal effect or photodamage in the normal tissue induced by high intensity illumination.

PDT has been proven in clinical trials to be effective to treat patients with oesophageal cancer, gastric carcinoma, breast cancer, and pancreatic cancer. And it can also be combined with other well-established treatment options, such as chemotherapy or radiotherapy. Indeed, PDT is fairly (and slowly) being accepted in clinical practice for some types of skin cancer. Continual efforts are being made to further improve the effectiveness of PDT, e.g., the efficacy of selective localization of the PDT drugs (Table 2.8 shows the common PDT drugs for malignant diseases), with the aim of expanding PDT to other cancer treatments, such as cancers of the prostate, cervix, and brain [131–135].

2.12.2 *Thermal Effects*

The photo-induced thermal effects in tissues/cells generally originate from the conversion of the absorbed light energy to heat through nonradiative processes, e.g., vibrational relaxation, internal conversion, and intersystem crossing (see Section 2.10). Different thermal effects can primarily be classified according to the temperature raised in the specimen (because of absorption) and the specimen's exposure time. The major thermal events are (i) coagulation, (ii) carbonization, and (iii) vaporization.

The exact temperatures at which these effects will occur entirely depend on the tissue's thermal and mechanical properties. Roughly speaking, no irreversible structural change will be observed in the tissue below $\sim$40–50°C. If the temperature reaches $\sim$60°C and beyond, tissue coagulation begins to occur and the coagulated tissue

Table 2.8 Common photosensitizers for malignant diseases. 5-ALA, 5-aminolevulinic acid; BPD-MA, benzoporphyrin derivative-monoacid ring A; HPD, haematoporphyrin derivative; HPPH, 2-(1-hexyloxyethyl)-2-devinyl pyropheophorbide-alpha; mTHPC, metatetrahydroxyphenylchlorin; and SnET2, tin ethyl etiopurpurin [132]

Photosensitizer (PDT drug)	Trade name	Potential indications	Activation wavelength
HPD (partially purified), porfimer sodium	Photofrin	Cervical, endobronchial, oesophageal, bladder and gastric cancers, brain tumours	630 nm
BPD-MA	Verteporfin	Basal-cell carcinoma	689 nm
m-THPC	Foscan	Head and neck tumours, prostate and pancreatic tumours	652 nm
5-ALA	Levulan	Basal-cell carcinoma, head and neck, and gynaecological tumours	635 nm
5-ALA-methylesther	Metvix	Basal-cell carcinoma	635 nm
5-ALA benzylesther	Benzvix	Gastrointestinal cancer	635 nm
5-ALA hexylesther	Hexvix	Diagnosis of bladder tumours	375–400 nm
SnET2	Purlytin	Cutaneous metastatic breast cancer, basal-cell carcinoma, Kaposi's sarcoma, prostate cancer	664 nm
Boronated protoporphyrin	BOPP	Brain tumours	630 nm
HPPH	Photochlor	Basal-cell carcinoma	665 nm
Lutetium texaphyrin	Lutex	Cervical, prostate and brain tumours	732 nm

becomes necrotic. It is generally achieved by having CW or quasi-CW illumination (>0.5 s) of energy fluence greater than 10 J/cm^2 (or the intensity of more than ~10 W/cm^2). When the temperature rises up to 100–300°C, carbonization and vaporization will take place. Carbonization is a process that converts the tissue's organic components into carbon and makes the tissue char. At such a high temperature range, the tissue will easily vaporize and can thus be cut. Typically, pulsed laser sources with relatively high power should be used in order to attain these effects, e.g., 1 ms with an intensity greater than ~10^4 W/cm^2. Table 2.9 illustrate some of the key thermal effect as a function of temperature.

One important aspect—the spatial extent of the photo-induced thermal effects in the tissue—should also be taken into consideration, especially for laser cutting, welding, and surgery. Once the light

Table 2.9 Typical photothermal effects in biological tissues as different temperatures

Temperature (°C)	Tissue effect
37	No irreversible tissue damage
~45	Hyperthermia
60	Protein denaturation, coagulation and necrosis
80	Collagen denaturation
100	Desiccation
150	Carbonatization
>300	Evaporation, vaporization

enters the tissue, the absorbed heat energy leads to a rise in local temperature within the illumination volume. However, not only the illuminated volume is heated, but also its surrounding depending upon several factors. The primary factors include heat conduction and heat storage, which are both governed by the thermal properties of the tissue (e.g., thermal conductivity and specific heat capacity). In addition, heat dissipation by the blood flow in the vascular networks in the tissue should not be ignored. In many surgical applications, laser cutting or welding should be made as precise as possible to minimize the thermal damage in the peripheral area due to thermal diffusion. This can be resolved by shortening the illumination duration in order to promote significant heat diffusion. This explains why most of the laser surgical operations based on the photothermal effect employ pulsed laser (μs–ms) instead of CW illumination. The details of how these factors impact the thermal effects can be found elsewhere [127, 128].

As an illustration, we consider here a pulse of laser light illuminating a homogeneous tissue (Fig. 2.81a). First of all, because of absorption and scattering, the light intensity is attenuated as the light penetrates from the surface. As a result, a temperature gradient is established inside the tissue. We can roughly define several zones, each of which has a distinct characteristic, within this temperature gradient region. The tissue will be vaporized in the region where the temperature exceeds ~300°C (typically the region where the laser beam directly interacts with the tissue). The next outer zone is the region where tissue is carbonized (temperature >~150°C). Having

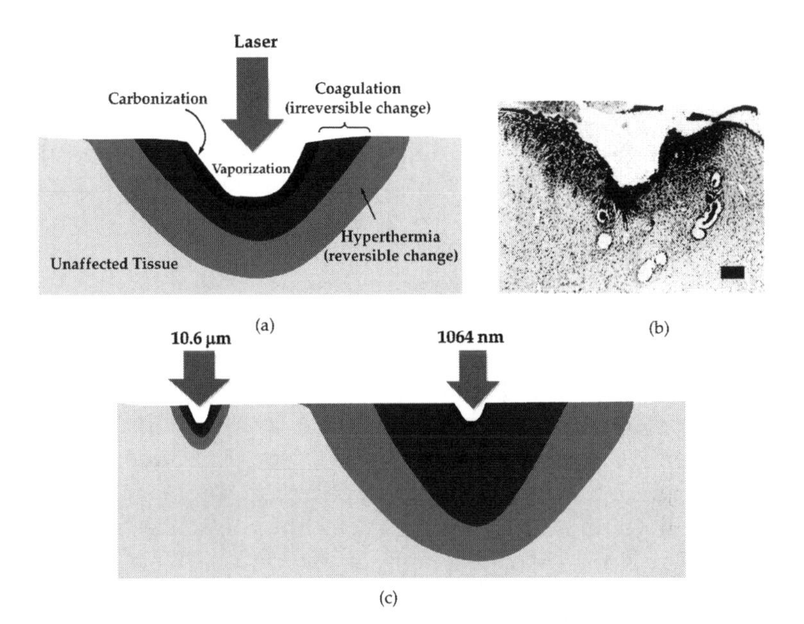

Figure 2.81 (a) Different photothermal effects following laser irradiation on the tissue. (b) Uterine tissue section of a wistar rat cut by a continuous-wave (CW) Nd:YAG laser (power: 10 W). The scale bar represents 80 μm) [127]. (c) Comparison of the photothermal action between the cases using NIR laser (1064 nm) and mid-IR laser (10.6 μm).

lower temperatures, the next zone is the coagulated tissue region, which is followed (surrounded) by the warmed region where the heating effect of the tissue is reversible, i.e., no severe tissue damage. The actual spatial extent of each zone highly depends on the tissue properties, laser parameters (e.g., pulse width and energy), and so on. Figure 2.81b shows the uterine tissue of a wistar rat cut with a CW Nd:YAG laser (power: 10 W). The coagulated tissue clearly appears darker than the other part of the tissue. Carbon dioxide (CO_2) pulsed laser, at a wavelength of 10.6 μm, could also be used for laser cutting of tissue. At this IR wavelength, the very high water absorption (see Fig. 2.20) renders an optical penetration depth as small as a few micrometers to 10 μm into the tissue. Hence, a small volume of tissue can be rapidly heated up to 100°C and vaporized. This is in contrast to the case of using an NIR laser source, which achieves deep penetration. The thermal diffusion effect is thus more widespread, degrading the cutting precision (Fig. 2.81c).

There are some other applications (both therapeutic and diagnostic) in biomedicine employing photothermal effects in tissue, e.g., selective photothermolysis and photoacoustic imaging. *Photothermolysis* is a process of selective thermal destruction of targeted chromophores in the tissue without injuring the surrounding parts. In this procedure, one has to carefully choose a laser source with an appropriate wavelength at which the absorption of the targeted chromophores is maximized while that of the surrounding tissue should be relatively small. Representative examples of selective photothermolysis are hair removal and tattoo removal [136–138]. In hair removal, melanin in the follicle is the targeted chromophore. In tattoo removal, laser at a specific wavelength heats up rapidly and fractures the tattoo ink/pigment granules. The cells stained with the pigment are also lysed (killed) in the operation.

Apart from therapeutic application, the photothermal effect can also be utilized for diagnosis. One of the most successful applications is *photoacoustic imaging* (tomography) [139–142]. This imaging technique has the potential to visualize an internal organ, such as the brain and breast, with high spatial resolution ($\sim$10 μm–1 mm) and high image contrast. It is particularly useful for visualizing vascular microstructure in deep tissue ($>$1–2 mm). Therefore one major potential clinical application of this technique is cancer diagnosis, i.e., identifying the tumors where the vasculature is unusual. The working principle of photoacoustic imaging is briefly described as follows. When a short-pulsed laser irradiates a tissue ($\sim$ns), it will be scattered throughout the tissue and strongly absorbed by the chromophores in the tissue (e.g., hemoglobin in the red blood cells). At the highly absorptive site, the absorbed energy is converted to heat rapidly and creates mechanical pressure, a process called thermoelastic expansion. The pressure propagates through the tissue as an ultrasonic wave, termed a photoacoustic wave. Hence, the chromophores' regions can be regarded as the initial acoustic source. The photoacoustic waves from these different acoustic sources reach the tissue surface with different time delays. An ultrasonic sensor (e.g., piezoelectric transducer [139], or Fabry-Perot interferometer based ultrasound sensor [143, 144]) is positioned on the tissue surface to detect these acoustic waves. Afterwards, the detected signal is further processed via a proper image reconstruction algorithm to determine the initial acoustic source distribution and thus to generate a spatial map (i.e., image) of the absorption (functional) properties of the tissue (Fig. 2.82a). The key advantage of photoacoustic imaging is its capability of delivering deep-tissue imaging (enabled by the weak tissue scattering of the ultrasonic wave) with high optical resolution and contrast (enabled by the chromophore absorption) (Fig. 2.82b). Photoacoustic imaging can quantify concentrations of multiple chromophores in vivo simultaneously, e.g., oxygenated and deoxygenated hemoglobin. It thus allows functional imaging of the concentration and oxygen saturation of hemoglobin—the important hallmarks for diagnosing cancer such as angiogenesis [140].

2.12.3 *Photoablation*[a]

When the laser pulse energy is above a certain threshold, explosive material removal can be achieved. This effect is laser-induced *ablation* (Fig. 2.83). Empirical studies have shown that such threshold, called ablation threshold varies according to laser parameters and tissue properties [127, 145]. Fundamentally, ablative material removal requires bond breaking which results in the removal of molecules or molecular fragments, or the formation of voids (i.e., bubble or crack) within the bulk tissue It is generally accepted that such molecular fragmentation, and void formation can be primarily governed by photochemical photothermal, and/or photomechanical mechanisms.

The photoablation effect cannot be initiated for laser illumination lower than the ablation threshold. Below this value, the tissue is heated upon absorbing the laser energy. Higher pulse energy (but still lower than the threshold) would lead to coagulation or

[a]The term *photoablation* is not well defined in literature. Literally, it simply describes an ablation phenomenon induced by light (photons), but it does not explicitly imply any other details. Although photoablation can in principle involve three main mechanisms—photochemical, photothermal, and photomechanical effects—this term is commonly used for indicating precise ablation caused by UV photons (i.e., predominantly photochemical effects). For the sake of generality, we refer here to photoablation as an effect governed by all three mechanisms.

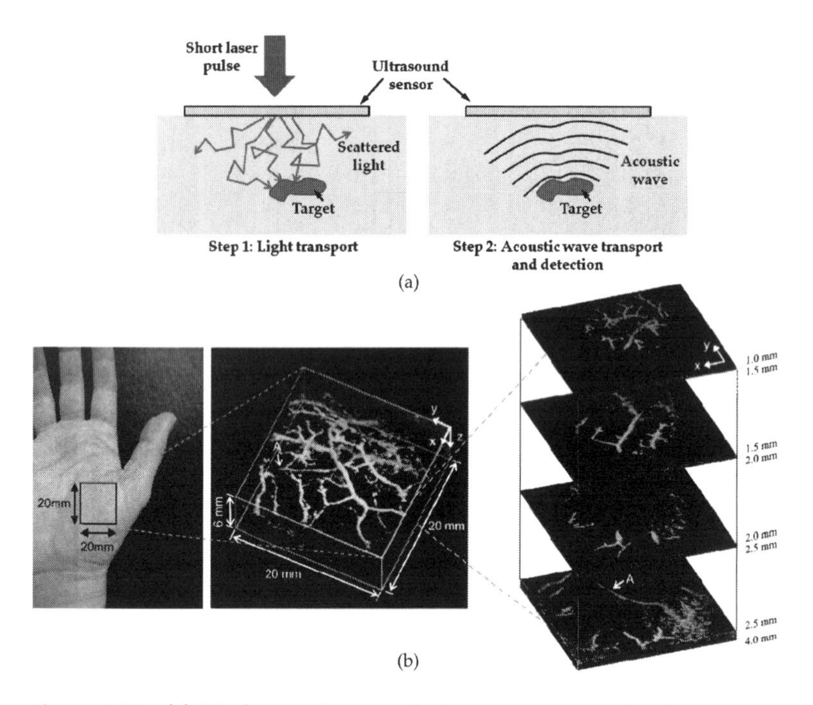

Figure 2.82 (a) Working principle of photoacoustic imaging (or tomography). In general, it is accomplished in two steps: (i) The pulse laser light is used to irradiate the tissue. The scattered and diffused light in the tissue eventually reaches the target region where the light absorption is high. (ii) The absorbed energy by the target region is then converted to heat. Pressure is also built up and it propagates away as stress wave at an ultrasonic frequency—an effect called thermoelastic expansion. Such ultrasound waves from the target region are detected by the ultrasound sensor placed on the surface of the tissue. The detected signal is further processed via a proper image reconstruction algorithm to determine the initial acoustic source distribution and thus to generate a spatial map (i.e., image) of the absorption (functional) properties of the tissue. (b) In vivo photoacoustic image of the vasculature in the palm using a laser wavelength at 670 nm. Photograph of the imaged region (left);rendered 3D image (middle); and image slices at different depths, down to 4 mm (right) [144].

vaporization of the tissue—photothermal effect, as discussed in the previous section. Above the threshold, ablation occurs and removes the tissue. The ablated depth (or the ablation rate) increases with the illumination energy. Further increasing the energy could make

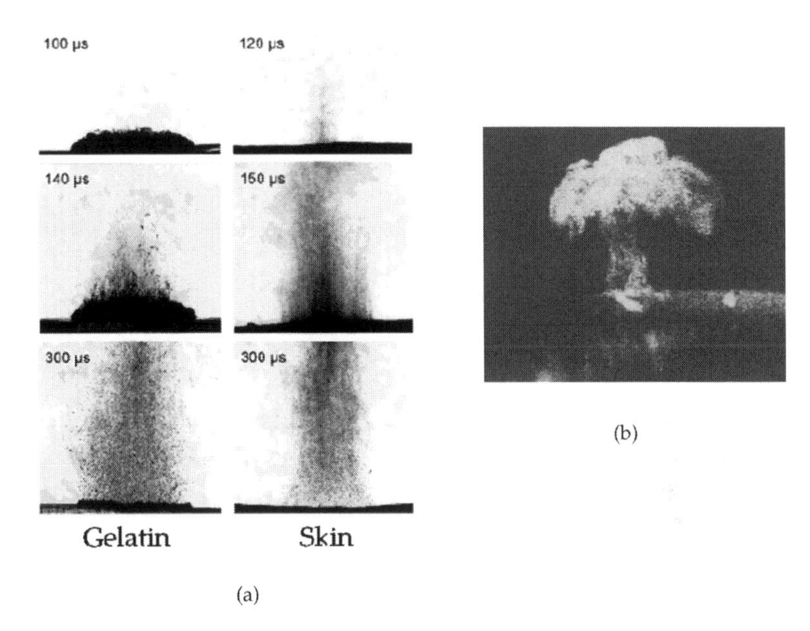

Figure 2.83 (a) High-speed imaging of the dynamic of Er:YAG laser (wavelength: 2.94 µm) ablation of gelatin with 70% water content, and skin (pulse energy of 4.6 J/cm^2, 5 mm spot size, and 200 μs pulse width) [146]. (b) Photograph of the excimer laser (UV) ablation of the cornea [147].

the ablation rate saturated when the plasma is formed—a situation in which the laser energy is strongly absorbed by the plasma instead of the tissue. Figure 2.84a shows the generic ablation curve of typical biological tissue. Of course, the threshold value depends on the tissue properties and the laser parameters. Anyhow, the general trend of the ablation curve applies to almost any type of tissue (e.g., Fig. 2.84b).

Being able to achieve very precise tissue ablation (cutting or removal) without significant or even no thermal damage to the surrounding tissue, photoablation has become one of the most popular effects utilized in laser therapy/surgery (Fig. 2.85), such as corneal tissue removal based on photorefractive keratectomy (PRK) and laser in situ keratomileusis (LASIK) [148]. Precise tissue ablation requires highly absorptive tissue at the proper laser wavelengths. High absorption generally implies a small laser

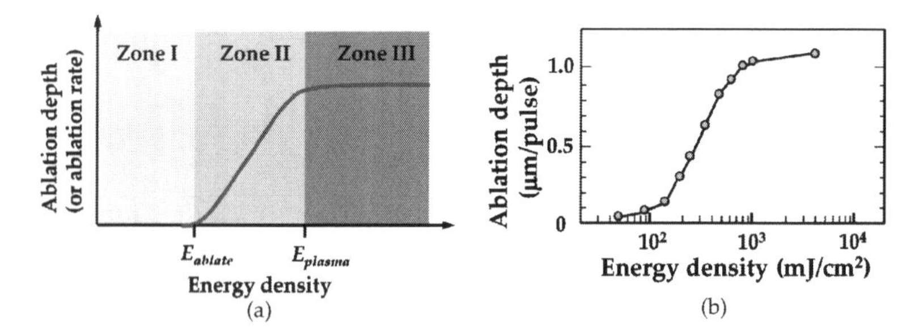

Figure 2.84 (a) Typical photoablation curve. Zone I: the tissue is heated by the laser beam only without material removal. Zone II: Starting from ablation threshold E_{ablate}, photoablation action takes place. The ablation depth and the ablation rate scale up with the input laser energy. Zone III: Beyond another threshold E_{plasma}, the ablation rate is saturated when the plasma is formed. In this case, the laser energy is strongly absorbed by the plasma instead of the tissue (b) Photoablation curve of rabbit cornea obtained with an ArF excimer laser (14 ns) [127].

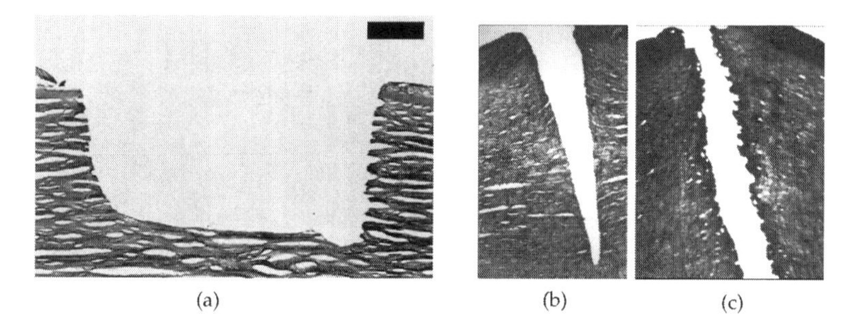

Figure 2.85 (a) Photoablation of corneal tissue by using an ArF excimer laser (14 ns, 180 mJ/cm^2). The scale bar represents 100 μm [127]. (b)–(c) Comparison of laser ablation of corneal tissue between using (b) 193 nm excimer laser and (c) 248 nm laser. A clear-cut boundary is seen in (b), whereas a darken border (due to heating effect) is observed in (c). This indicates that only deep UV (short enough) wavelength can minimize the tissue thermal damage [149].

penetration depth in tissue. Thus, the energy can be deposited and tightly confined within a small volume. Moreover, minimizing the spatial extent of thermal diffusion can ensure thermal confinement within the absorbed volume, and thus precise ablation. Therefore, "rapid heating" of tissue is essential. It can be achieved by pulsed laser radiation whose pulse duration (pulse width) is much shorter than the thermal diffusion time in the tissue, which is typically on the order of tens of nanoseconds to microseconds depending on the tissue's thermal properties. That is why most of the photoablation laser systems employ ns–μs laser pulses.

Another factor which influences the spatial extent of the photoablation effect is the thermoelastic stress induced by the rapid heating of tissue by pulsed laser illumination. The stress can propagate away from the illumination volume. The strength of the thermoelastic stress is determined by the speed of sound in the tissue, the temporal pulse width of the laser and the thermal properties of the tissue [145]. An interesting phenomenon called *stress confinement* can occur when the laser pulse width is smaller than or comparable to the characteristic time for a thermoelastic stress wave to propagate across the heated volume. Numerous studies have showed that such stress confinement effectively lowers the ablation threshold energy. This is because the thermoelastic stress facilitates bubble formation and vaporization, and even directly creates fracture of the tissue matrix—making the ablation easier to be initiated. It thus improves the ablation efficiency and minimizes thermal damage in the surrounding tissue.

2.12.3.1 UV versus IR photoablation

As discussed in Section 2.8, strong absorption in biological tissues occurs in both the UV and IR spectral ranges (Fig. 2.32). The primary chromophores in these ranges are water (both in UV and IR), the peptide bonds linking amino acids and proteins (in UV), and collagen (in IR). Therefore, the majority of the clinical applications of pulsed laser tissue ablation employ either the UV laser (e.g., ArF laser and KrF laser) or IR laser (e.g., Er:YAG laser, Er:YSGG laser) (Table 2.11). Because of the difference in wavelength, the mechanisms and thus the quality of the UV ablation are significantly different from those of IR ablation. It is generally accepted that tissue fragmentation and removal made by the UV laser ablation are chiefly mediated by photochemical processes, which directly introduce the breakage of the chemical bonds. In contrast, IR laser ablation is entirely governed by photothermal processes.

Table 2.10 Dissociation energies of selected chemical bonds. The corresponding photon wavelengths are also shown [145]

Chemical bond	Dissociation energy (eV)	Wavelength (nm)	Chemical bond	Dissociation energy (eV)	Wavelength (nm)
C=O	7.5	165	H–N	4.1	302
C=C	6.4	194	C–C	3.6	344
O=O	5.1	243	C–O	3.6	344
H–O	4.8	258	C–N	3	413
H–C	4.3	288	N–O	2.2	564
N=N	4.3	288	N–N	1.6	775

For photoblation *purely* mediated by the photochemical process, the illuminating photon should have high enough photon energy so that the molecule is excited to an electronic state which exceeds the bond energy. Typical bond (dissociation) energies of the chemical bonds range from a few electron-volts to 10 eV, as listed in Table 2.11. We can thus see that only UV photons have sufficient energy to dissociate these bonds and thus to decompose the tissue matrix. Therefore, a photochemical mechanism in photoablation can only be achieved by using UV laser (e.g., ArF laser [193 nm] and KrF laser [248 nm]) (compare Tables 2.10 and 2.11).

Owing to the heterogeneous structure of the biological tissue and its large water content, UV photoablation does not solely involve photochemical dissociation, but also the photothermal effect. The contribution of the photothermal effect becomes increasingly significant for longer ablation wavelength, i.e., IR photoablation (Fig. 2.85b,c). As mentioned earlier, the major tissue chromophores in UV are the proteins in the collagen fibrils. Studies have showed that these chromophores can be heated to high temperatures, and even photochemical decomposition of the proteins occurs [145]. Heat can be rapidly diffused from these proteins to the background tissue water. If the heating effect is strong enough so that a phase change occurs in tissue water (e.g., confined boiling and phase explosion), the heat exchange between tissue water and proteins offers a thermal pathway for material removal. This is the additional photothermal contribution in UV photoablation. But in this case, the ablation occurs at temperatures much lower than those required for a direct photothermal decomposition. The

Table 2.11 Wavelengths and the corresponding lasing photon energies of the typical lasers. Excimer lasers include argon fluoride (ArF), krypton fluoride (KrF), xenon chloride (XeCl), and xenon fluoride (XeF) lasers. Gas lasers include argon ion (Ar ion), helium-neon (HeNe) and carbon dioxide (CO_2) lasers. Solid-state lasers include ruby laser, neodymium-doped yttrium aluminum garnet $Nd{:}Y_3Al_5O_{12}$ (Nd:YAG), holmium-doped YAG (Ho:YAG), and erbium-dope YAG (Er:YAG) lasers

Laser	Wavelength (nm)	Photon energy (eV)
ArF	193	6.4
KrF	248	5
XeCl	308	4
XeF	351	3.5
Ar ion	514	2.4
He-Ne	633	2
Ruby	694	1.8
Nd:YAG	1064	1.2
Ho:YAG	2120	0.6
Er:YAG	2940	0.4
CO_2	10,600	0.1

reason is that the photochemical decomposition has already first weakened the tissue matrix and thus makes it relatively easier to be vaporized than in *pure* photothermal situations. In addition, the volatile products fragmented from photochemical decomposition of the tissue proteins serve as additional sites for nucleation and speed up the vaporization rate of tissue water. Therefore, UV photoablation is less explosive than IR photoablation. Of course, material ejection still occurs.

In contrast, IR photoablation is solely governed by the photothermal contribution. Under the thermally confined conditions (i.e., the laser pulse width is shorter than the thermal diffusion time), the ablation is achieved by a process of phase explosion or confined boiling which plays an important role in material fragmentation and ejection.

2.12.4 *Plasma-Induced Ablation and Photodisruption*

When the illumination intensity reaches certain threshold value, typically $>10^{11}$ W/cm^2 in solids and fluids, a phenomenon known as *optical breakdown* occurs. In this case, the corresponding **E** field of the light exceeds $\sim 10^7$ V/cm, which is comparable to the average intramolecular Coulomb electric fields. Thus, such an intense laser pulse is able to ionize the atoms or the molecules—leading to a *breakdown* of the tissue by creating a very high free electron density of $\sim 10^{18}/\text{cm}^3$ (i.e., plasma) within the illuminated volume in an ultrashort period of time ($\sim$100 fs–100 ps). In essence, the plasma generation results in another type of ablation, known as *plasma-induced ablation*.

To be more precise, there are two possible pathways of initial plasma generation: (i) Thermionic emission, in which the release of electrons due to thermal ionization. This process occurs when the absorptive tissues are severely heated via *linear* absorption of the incoming laser pulses. (ii) Multi-photon ionization (essentially a nonlinear absorption process) mediated by the high **E** field induced by the intense laser pulse. This could happen in weakly absorptive or even transparent tissues. In either case, the generated free electrons trigger *an avalanche effect*—free electrons and ions are accumulated in a multiplicative manner (Fig. 2.86).

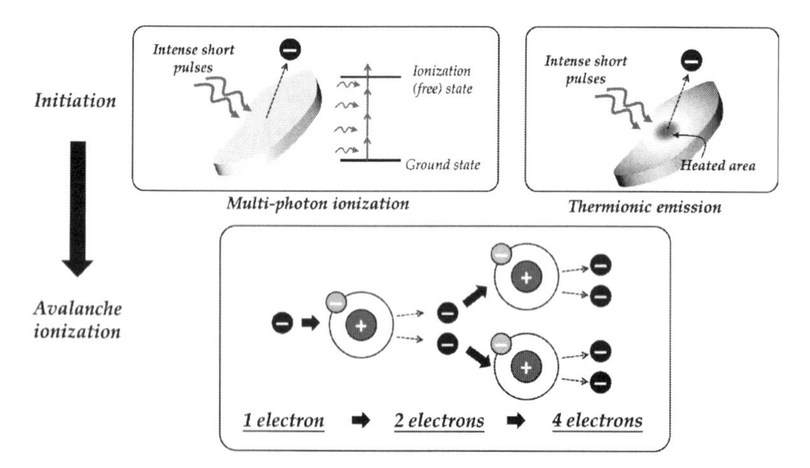

Figure 2.86 Schematic of plasma generation, which consists of two steps. Step 1 has two possible pathways: (i) multiphoton ionization and (ii) thermionic emission. This could initiate the generation of the first free electron. Step 2 is an avalanche ionization in which free electrons and ions are accumulated in a multiplicative manner.

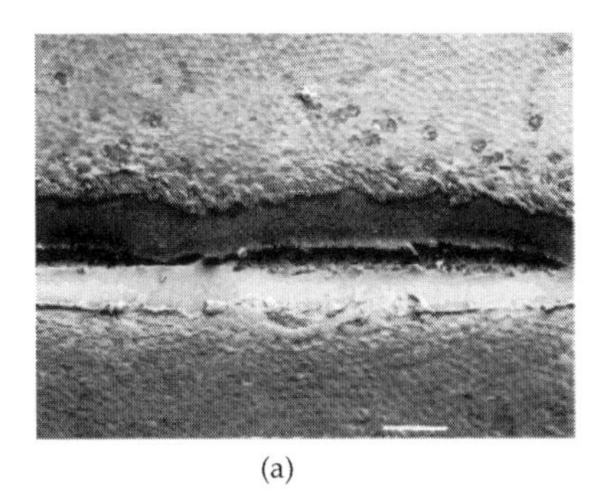

(a)

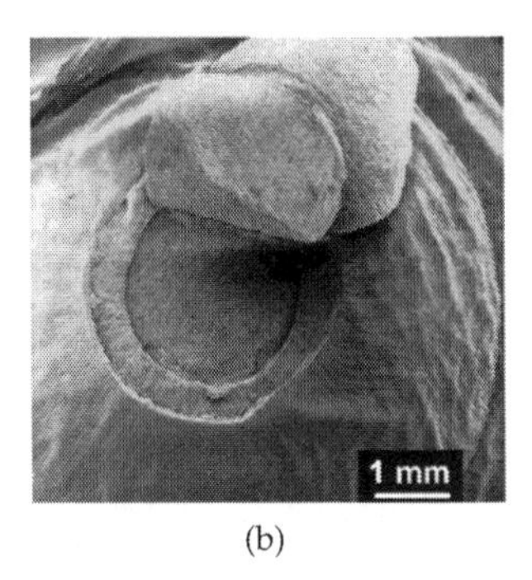

(b)

Figure 2.87 (a) Plasma-induced laser cut in Descemet's membrane in the cornea by 30 ps laser pulses at 1064 nm [156]. The scale bar represents 100 μm. (b) Lenticule dissected out of the corneal stroma using 110 fs pulses at 780 nm [153].

The high-energy plasma formed at the tissue surface is highly absorptive regardless of the tissue absorption strength. This effectively hinders further *linear* absorption of the irradiated tissue area—this area is said to be "shielded." As a result, with the appropriate laser parameters, plasma-induced ablation is able to achieve precise and well-defined tissue removal with negligible thermal or mechanical damage, irrespective of the tissue being absorptive or transparent. It is a particularly attractive tool for non-invasive surgery in ophthalmology, such as intraocular surgery [150] and intrastromal corneal refractive surgery [151–153]. Because of its exceptionally high ablation precision, it has also been applied for intracellular surgery [154, 155]. Figure 2.87 shows the representative images of plasma-induced ablation.

In fact, there are other secondary effects of the plasma that have to be taken into account. They are mostly the mechanical effects, e.g., shock wave generation, cavitation effect, and jet formation in fluid. They become more significant at even higher pulse energies than those required to initiate plasma-induced ablation. In generalized terms, such mechanical impact is usually called *photodisruption*.

Photodisruption can be generally regarded as a series of cascaded mechanical processes starting from optical breakdown. Once the plasma generation is initiated, the highly energetic plasma creates an abrupt change in the pressure gradient within the illuminated volume and makes the stress propagate outwards. Such abrupt effect is called *shock wave generation*—a stress wave moving from the boundary of the plasma at a hypersonic speed, slowing down gradually to the speed of sound in the tissue. For example, the speed of the laser-induced shock waves in water is typically ~5000 m/s whereas the speed of sound is 1483 m/s at 37°C. Because of the decreasing speed of the expanding shock wave, its internal pressure could eventually fall below the ambient pressure because of the increasing volume of the expanding "plasma bubble." As a result, the bubble collapses—an effect called *cavitation*. In fluid or near a solid boundary, a liquid jet could be developed as a result of the cavitation. This *jet formation* can cause drastic damage of solids. Major applications of photodisruption include posterior capsulotomy of the lens, and laser lithotripsy, which is used to remove impacted stones from the urinary tract [127].

On the basis of the appropriate laser parameters (pulse energy and pulse width), previous studies suggest the general working regimes for both plasma-induced ablation and photodisruption, as shown in Fig. 2.88. In general, an ultrashort laser pulse (fs–100 ps)

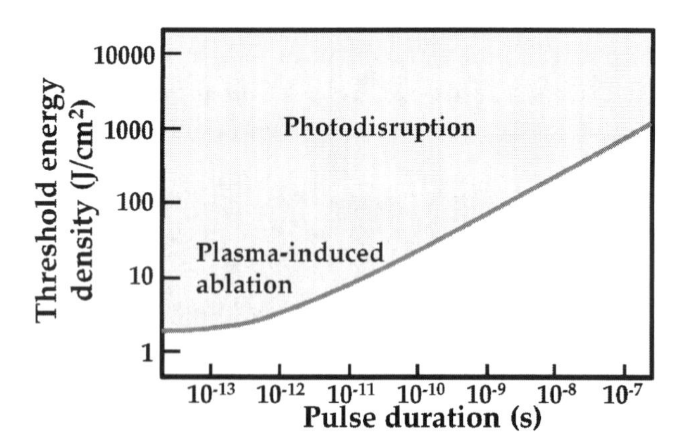

Figure 2.88 The regimes (in terms of laser pulse duration and the required threshold energy density) of plasma-induced ablation and photodisruption. In general, photodisruption requires higher threshold energy to be initiated. In addition, an ultrashort laser pulse (fs–100 ps) is favorable for both processes. The shorter the pulse, the lower the required threshold energy to initiate either process [127].

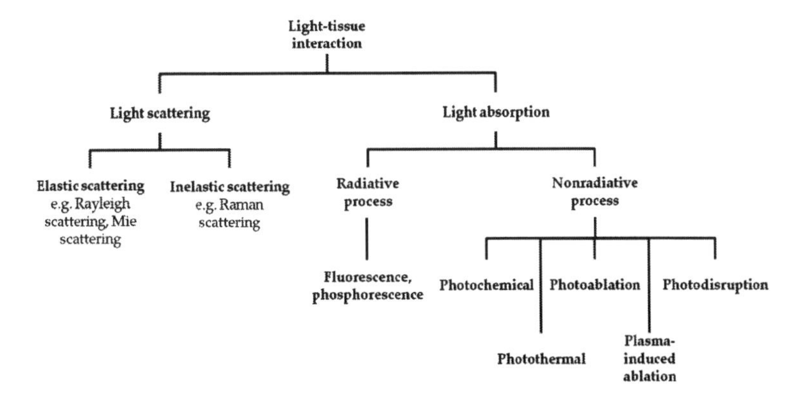

Figure 2.89 Common phenomena of light–tissue interaction in biophotonics.

is favorable for both processes. But photodisruption requires much higher pulse energies (>100 J/cm^2) than plasma-induced ablation (1–10 J/cm^2). Figure 2.88 also shows that the shorter the pulse, the lower the required threshold energy to initiate either process, i.e., the ablation efficiency is higher. That is why ultrashort laser has become increasingly popular for high-precision laser ablation. The fact that both plasma-induced ablation and photodisruption are initiated by plasma generation makes it challenging to clearly distinguish between these two processes in practice. Nevertheless, one can make distinction or determine which process is dominant roughly by assessing the ablation quality. In principle, plasma-induced ablation is spatially confined to the optical breakdown region. The ablated area should be highly localized and precisely defined by the laser beam geometry. In contrast, the ablation based on photodisruption is less localized as shock waves and cavitation effects propagate into the surrounding tissue.

As a summary, Fig. 2.89 shows the overall picture of all the common light–tissue interactions which would be encountered or applied in biophotonics, either in diagnostic or in therapeutic applications.

References

1. Hecht, E. (2001). *Optics* (4th ed.). Addison Wesley.
2. Saleh, B. E. A., and Teich, M. C. (2007). *Fundamentals of Photonics* (2nd ed.). Wiley-Interscience.
3. Griffiths, D. J. (1999). *Introduction to Electrodynamics* (3rd ed.). Addison Wesley.
4. Hale, G. M., and Querry, M. R. (1973). Optical constants of water in the 200-nm to 200-μm wavelength region. *Appl. Opt.* **12**:555–563.
5. Liu, J. M. (2005). *Photonic Devices.* Cambridge University Press.
6. Yariv, A. (1989). *Quantum Electronics* (3rd ed.). Wiley.
7. Chow, M. H., Yan, K. T. H., Bennett, M. J., and. Wong, J. T. Y. (2010). Birefringence and DNA condensation of liquid crystalline hromosomes. *Eukaryot. Cell* **9**:1577–1587.
8. Waterman-Storer, C. M., Sanger, J. W., and Sanger, J. M. (1993). Dynamics of organelles in the mitotic spindles of living cells: membrane and microtubule interactions. *Cell. Motil. Cytoskel.* **26**:19–39.
9. Dinh, T. V. (2010). *Biomedical Photonics Handbook* (2nd ed.). CRC Press.
10. Voet, D., Gratzer, W. B., Cox, R. A., and Doty, P. (1963). Absorption spectra of nucleotides, polynucleotides, and nucleic acids in the far ultraviolet. *Biopolymers* **1:**193–208.
11. Hanning, C. D., and Alexander-Williams, J. M. (1995). Pulse oximetry: a practical review. *BMJ.* **311**:367–370.
12. Sarna, T., and Sealy, R. C. (1984). Photoinduced oxygen consumption in melanin systems. Action spectra and quantum yields for eumelanin and synthetic melanin. *Photochem. Photobiol.* **39**:69–74.
13. Crippa, R. P., Cristofoletti, V., and Romeo, N. (1978). A band model for melanin deduced from optical absorption and photoconductivity experiments. *Biochim. Biophys. Acta* **538**:164–170.
14. Smith, B. C. (1995). *Fundamentals of Fourier Transform Infrared Spectroscopy.* CRC Press.
15. Burns, D. A., and Ciurczak, E. W. (eds.). (2007). *Handbook of Near-Infrared Analysis* (3rd ed.). CRC Press.
16. Siesler, H. W., Ozaki, Y., Kawata, S., and Heise, H. M. (eds.). (2002). *Near-Infrared Spectroscopy: Principles, Instruments, Applications.* Wiley-VCH.
17. Coates, J. (2000). Interpretation of Infrared Spectra, A Practical Approach. In *Encyclopedia of Analytical Chemistry* (R. A. Meyers, ed.), 10815–10837.

18. Goda, K., Solli, D. R., Tsia, K. K., and Jalali, B. (2009). Theory of amplified dispersive Fourier transformation. *Phys. Rev. A* **80**:043821.
19. Solli, D. R., Chou, J., and Jalali, B. (2008). Amplified wavelength–time transformation for real-time spectroscopy. *Nat. Photon.* **2**:48–51.
20. Yaroslavsky, A. N., Schulze, P. C., Yaroslavsky, I. V., Schober, R., Ulrich, F., and Schwarzmaier, H.-J. (2002). Optical properties of selected native and coagulated human brain tissues in vitro in the visible and near infrared spectral range. *Phys. Med. Biol.* **47**:2059–2073.
21. Tuchin, V. V. (2007). Tissue Optics—Light Scattering Methods and Instruments for Medical Diagnosis. SPIE Tutorial Texts in Optical Engineering Vol. TT38. SPIE Press.
22. Cheong, W. F., Prahl, S. A., and Welch, A. J. (1990). A review of the optical properties of biological tissues. *IEEE. J. Quantum Electron.* **26**:2166–2185.
23. Durduran, T., Choe, R., Baker, W. B., and Yodh, A. G. (2010). Diffuse optics for tissue monitoring and tomography. *Rep. Prog. Phys.* **73**:076701.
24. Jiang, H. (2010). *Diffuse Optical Tomography: Principles and Applications.* CRC Press.
25. Wang, L. V., and Wu, H. (2007). *Biomedical Optics: Principles and Imaging.* Wiley-Interscience.
26. Palmer, G. M., and Ramanujam, N. (2006). Monte Carlo-based inverse model for calculating tissue optical properties. Part I: theory and validation on synthetic phantoms. *Appl. Opt.* **45**:1062–1071.
27. Patterson, M. S., Chance, B., and Wilson, B. C. (1989). Time resolved reflectance and transmittance for the non-invasive measurement of tissue optical properties. *Appl. Opt.* **28**:2331–2336.
28. Saar1, B. G., Freudiger, C. W., Reichman, J., Stanley, C. M., Holtom, G. R., and Xie, X. S. (2010). Video-rate molecular imaging in vivo with stimulated Raman scattering. *Science* **330**:1368–1370.
29. Fang, Q. (2010). Mesh-based Monte Carlo method using fast ray-tracing in Plücker coordinates. *Biomed. Opt. Express* **1**:165–175.
30. Vogel, A., and Venugopalan, V. (2003). Mechanisms of pulsed laser ablation of biological tissues. *Chem. Rev.* **103**:577–644.
31. Ritz, J. P., Roggan, A., Isbert, C., Muller, G., Buhr, H., and Germer, C. T. (2001). Optical properties of native and coagulated porcine liver tissue between 400 and 2400 nm. *Lasers Surg. Med.* **29**:205–12.
32. Bashkatov, A. N., Genina, E. A., Kochubey, V. I., and Tuchin, V. V. (2005). Optical properties of human skin, subcutaneous and mucous tissues

in the wavelength range from 400 to 2000nm. *J. Phys. D: Appl. Phys.* **38:**2543–2555.

33. Zeff, B. W., White, B. R., Dehghani, H., Schlaggar, B. L., and Culver, J. P. (2007). Retinotopic mapping of adult human visual cortex with high-density diffuse optical tomography. *PNAS* **104:**12169–12174.
34. Rolfe, P. (2000). In vivo near-infrared spectroscopy. *Annu. Rev. Biomed. Eng.* **2:**715–754.
35. Ferrari, M., and Quaresima, V. (2012). A brief review on the history of human functional near-infrared spectroscopy (fNIRS) development and fields of application. *NeuroImage* 63:921–935.
36. Hoshi, Y. (2007). Functional near-infrared spectroscopy: current status and future prospects. *J. Biomed. Opt.* **12:**062106.
37. Kukreti, S., Cerussi, A., Tromberg, B., and Gratton, E. (2008). Intrinsic near-infrared spectroscopic markers of breast tumors. *Dis. Markers* **25:**281–290.
38. Gibson, A. P., Hebden, J. C., and Arridge, S. R. (2005). Recent advances in diffuse optical imaging. *Phys. Med. Biol.* **50:**R1–R43.
39. Arridge, S. R. (1999). Optical tomography in medical imaging. *Inverse Prob.* **15:**R41–R93.
40. Arridge S. R., and Hebden, J. C. (1997). Optical imaging in medicine: II. Modelling and reconstruction. *Phys. Med. Biol.* **42:**841–854.
41. Corlu, A., Choe, R., Durduran, T., et al. (2007). Three-dimensional in vivo fluorescence diffuse optical tomography of breast cancer in humans. *Opt. Express* **15:**6696–6716.
42. Yang, J., Zhang, T., Yang, H., and Jiang, H. (2012). Fast multispectral diffuse optical tomography system for in vivo three-dimensional imaging of seizure dynamics. *Appl. Opt.* **51:**3461–3469.
43. Culver, J. P., Durduran, T., Furuya, D., Cheung, C., Greenberg, J. H., and Yodh, A. G. (2003). Diffuse dptical tomography of cerebral blood flow, oxygenation, and metabolism in rat during focal ischemia. *J. Cereb. Blood Flow Metab.* **23:**911–924.
44. Eisberg, R., and Resnick, R. (1985). *Quantum Physics of Atoms, Molecules, Solids, Nuclei, and Particles.* (2nd ed.). Wiley.
45. Siegman, A. E. (1986). *Lasers*, University Science Books.
46. Pawley, J. (2006). *Handbook of Biological Confocal Microscopy* (3rd ed.). Springe.
47. Dufva, M. (2009). *DNA Microarrays for Biomedical Research: Methods and Protocols.* Humana Press.

48. Shapiro, H. M. (2003). *Practical Flow Cytometry* (4th ed.). Wiley-Liss.
49. Lakowicz, J. R. (2006). *Principles of Fluorescence Spectroscopy* (3rd ed.). Springe.
50. Berezin, M. Y., and Achilefu, S. (2010). Fluorescence lifetime measurements and biological imaging. *Chem. Rev.* **110**:2641–2684.
51. Du, H., Fuh, R. C. A., Li, J., Corkan, L. A., and Lindsey, J. S. (1998). PhotochemCAD: A Computer-Aided Design and Research Tool in Photochemistry and Photobiology. *Photochem. Photobiol.* **68**:141–142.
52. Dixon, J. M., Taniguchi, M., and Lindsey, J. S. (2005). PhotochemCAD 2. A refined program with accompanying spectral databases for photochemical calculations. *Photochem. Photobiol.* **81**:212–213.
53. Mujumdar, R. B., Ernst, L. A., Mujumdar, S. R., Lewis, C. J., and Waggoner, A. S. (1993). Cyanine dye labeling agents: sulfoindocyanine succidimidyl esters. *Bioconj. Chem.* **4**:105–111.
54. Molecular Probes®: Uhttp://www.invitrogen.com/site/us/en/home/brands/Molecular-Probes.htmlU
55. Wang, X., Qu, L., Zhang, J., Peng, X., and Xiao, M. (2003). Surface-related emission inhighly luminescent CdSe QDs. *Nano Lett.* **3**:1103–1106.
56. Patterson, G., Day, R. N., and Piston, D. (2001). Fluorescent protein spectra. *J. Cell Sci.* **114**:837–838.
57. Shaner, N. C., Campbell, R. E., Steinbach, P. A., Giepmans, B. N., Palmer, A. E., and Tsien, R. Y. (2004). Improved monomeric red, orange and yellow fluorescent proteins derived from Discosoma sp. red fluorescent protein. *Nat. Biotechnol.* **22**:1567–1572.
58. Wagnières, G. A., Star, W. M., and Wilson, B. C. (1998). In vivo fluorescence spectroscopy and imaging for oncological applications. *Photochem. Photobiol.* **68**:603–632.
59. Smith, A. M., and Nie, S. (2004). Chemical analysis and cellular imaging with quantum dots. *Analyst* **129**:672–677.
60. Mansur, H. S. (2010). Quantum dots and nanocomposites. *WIREs Nanomed. Nanobiotechnol.* **2**:113–129.
61. Resch-Genger, U., Grabolle, M., Cavaliere-Jaricot, S., Nitschke, R., and Nann, T. (2008). Quantum dots versus organic dyes as fluorescent labels. *Nat. Methods* **5**:763–775.
62. Hotz, C. Z., and Bruchez, M. (eds.). (2007). *Quantum Dots: Applications in Biology*. Humana Press.
63. Park, J. H., Gu, L., von Maltzahn, G., Ruoslahti, E., Bhatia, S. N., and Sailor, M. J. (2009). Biodegradable luminescent porous silicon nanoparticles for in vivo applications. *Nat. Mater.* **8**:331–336.

64. Chattopadhyay, P. K., Price, D. A., Harper, T. F., et al. (2006). Quantum dot semiconductor nanocrystals for immunophenotyping by polychromatic flow cytometry. *Nat. Med.* **12**:972–977.
65. Harrison, P. (2010). *Quantum Wells, Wires and Dots: Theoretical and Computational Physics of Semiconductor Nanostructures* (3rd ed.). Wiley.
66. Derfus, A. M., Chan, W. C. W., and Bhatia, S. N. (2004). Probing the Cytotoxicity of Semiconductor Quantum Dots. *Nano Lett.* **4:**11–18.
67. Hicks, B. W. (2002). *Green Fluorescent Protein: Applications & Protocols.* Humana Press.
68. Miyawaki, A., Sawano, A., and Kogure, T., (2003). Lighting up cells: labelling proteins with fluorophores. *Nat. Cell Biol.* **5**:S1–S7.
69. Shaner, N. C., Steinbach, P. A., and Tsien, R. Y. (2005). A guide to choosing fluorescent proteins. *Nat. Met.* **2**:905–909.
70. Day, R. N., and Davidson, M. W. (2009). The fluorescent protein palette: tools for cellular imaging. *Chem. Soc. Rev.* **38**:2887–2921.
71. Boyd, R. W. (2008). *Nonlinear Optics*, (3rd ed). Academic Press.
72. Masters, B. R., and So, P. (eds.). (2008). *Handbook of medical nonlinear optical microscopy.* Oxford University Press.
73. Smith, E., and Dent, G. (2005). *Modern Raman Spectroscopy: A Practical Approach*. Wiley.
74. Larkin, P. (2011). *Infrared and Raman Spectroscopy; Principles and Spectral Interpretation*, Elsevier.
75. Berger, A. J., Wang, Y., and Feld, M. S. (1996). Rapid, non-invasive concentration measurements of aqueous biological analytes by near-infrared Raman spectroscopy. *Appl. Opt.* **35**:209–212.
76. Notingher, I. (2007). Raman Spectroscopy Cell-based Biosensors. *Sensors* **7**:1343-1358.
77. Bégin, S., Bélanger, E., Laffray, S., Vallée, R., and Côte, D. (2009). In vivo optical monitoring of tissue pathologiesand diseases with vibrational contrast. *J. Biophoton.* **2**:632–642.
78. Sasic, S., and Ekins, S. (eds.). (2007). *Pharmaceutical Applications of Raman Spectroscopy*, Wiley-Interscience.
79. Evans, C. L, and Xie, X. S. (2008). Coherent anti-Stokes Raman scattering microscopy: chemical imaging for biology and medicine. *Annu. Rev. Anal. Chem.* **1**:883–909.
80. Begley, R. F., Harvey, A. B., and Byer, R. L. (1974). Coherent anti-Stokes Raman scattering. *Appl. Phys. Lett.* **25**:387–390.

81. Duncan, M. D., Reintjes, J., and Manuccia, T. J. (1982). Scanning coherent anti-Stokes Raman microscope. *Opt. Lett.* **7**:350–352.
82. Zumbusch, A., Holtom, G. R., and Xie, X. S. (1999). Three-dimensional vibrational imaging by coherent anti-Stokes Raman scattering. *Phys. Rev. Lett.* **82**:4142–4145.
83. Cheng, J. X., Jia, Y. K., Zheng, G., and Xie, X. S. (2002). Laser-scanning coherent anti-Stokes Raman scattering microscopy and applications to cell biology. *Biophys. J.* **83**:502–509.
84. Petrov, G. I., Arora, R., Yakovlev, V. V., Wang, X., Sokolov, A. V., and Scully, M. O. (2007). Comparison of coherent and spontaneous Raman microspectroscopies for non-invasive detection of single bacterial endospores. *Proc. Natl. Acad. Sci. USA* **104**:7776–7779.
85. Min, W., Freudiger, C. W. Lu, S., and Xie, X. S. (2011). Coherent Nonlinear Optical Imaging: Beyond Fluorescence Microscopy. *Annu. Rev. Phys. Chem.* **62**:507–530.
86. Kukura, P., McCamant, D. W., and Mathies, R. A. (2007). Femtosecond stimulated Raman spectroscopy. *Annu. Rev. Phys. Chem.* **58**:461–488.
87. Fang, C., Frontiera, R. R., Tran, R., and Mathies, R. A. (2009). Mapping GFP structure evolution during proton transfer with femtosecond Raman spectroscopy. *Nature* **462**:200–204.
88. Ploetz, E., Laimgruber, S., Berner, S., Zinth, W., and Gilch, P. (2007). Femtosecond stimulated Raman microscopy. *Appl. Phys. B* **87**:389–393.
89. Freudiger, C. W., Min, W., Saar, B. G., et al. (2008). Label-free biomedical imaging with high sensitivity by stimulated Raman scattering microscopy. *Science* **322**:1857–1861.
90. Nandakumar, P., Kovalev, A., and Volkmer, A. (2009). Vibrational imaging based on stimulated Raman scattering microscopy. *New J. Phys.* **11**:033026–35.
91. Ozeki, Y., Dake, F., Kajiyama, S., Fukui, K., and Itoh, K. (2009). Analysis and experimental assessment of the sensitivity of stimulated Raman scattering microscopy. *Opt. Express* **17**:3651–3658.
92. Saar, B. G., Freudiger, C. W., Reichman, J., Stanley, C. M., Holtom, G. R., and Xi, X. S. (2010). Video-rate molecular imaging in vivo with stimulated Raman scattering. *Science* **330**:1368–1371.
93. Terhune, R. W. (1963). Nonlinear Optics. *Bull. Am. Phys. Soc.* **8**:359.
94. Vermeulen, N., Debaes, C., and Thienpont, H. (2010). Models for coherent anti-Stokes Raman scattering in Raman devices and in spectroscopy *Proc. SPIE* **7728**:77281B-1-12.

95. Bobbs, B. and Warner, C. (1990). Raman-resonant four-wave mixing and energy transfer *J. Opt. Soc. Am. B* **7**:234–238.
96. Druet, S. A. J. and Taran, J. P. E. (1981). CARS spectroscopy *Prog. Quantum Electron.* **7**:1–72.
97. Bloembergen, N. (1996). *Nonlinear Optics* (4th ed.) World Scientific Pub Co Inc.
98. Druet, S. A. J. Attal, B. Gustafson, T. K. and Taran, J. P. (1978). Electronic resonance enhancement of coherent anti-Stokes Raman scattering *Phys. Rev. A* **18**:1529–1557.
99. Samineni, P., Li, B., Wilson, J. W., Warren, W. S., and Fischer, M. C. (2012). Cross-phase modulation imaging. *Opt. Lett.* **37**:800–802.
100. Wang, Y., Liu, X., Halpern, A. R., Cho, K., Corn, R. M., and Potma, E. O. (2012). Wide-field, surface-sensitive four-wave mixing microscopy of nanostructures. *Appl. Opt.* **51**:3305–3312.
101. Squier, J., Muller, M., Brakenhoff, G., and Wilson, K. R. (1998). Third harmonic generation microscopy. *Opt. Express* **3**:315–324.
102. Yelin, D., and Silberberg, Y. (1999). Laser scanning third-harmonic-generation microscopy in biology. *Opt. Express* **5**:169–175.
103. Débarre, D., Supatto, W., Pena1, A., et al. (2006). Imaging lipid bodies in cells and tissues using third-harmonic generation microscopy. *Nat. Methods* **3**:47–53.
104. Le, T. T., Langohr, I. M., Locker, M. J., Sturek, M., and Cheng, J. X. (2007). Label-free molecular imaging of atherosclerotic lesions using multimodal nonlinear optical microscopy. *J. Biomed. Opt.* **12**: 054007.
105. Chen, H., Wang, H., Slipchenko, M. N., Jet al. (2009). A multimodal platform for nonlinear optical microscopy and microspectroscopy. *Opt. Express* **17**:1282–1290.
106. Yue, S., Slipchenko, M. N., and Cheng, J. X. (2011). Multimodal nonlinear optical microscopy. *Laser Photonics Rev.* **5**:496–512.
107. Li, D., Zheng, W., Zeng, Y., and Qu, J. Y. (2010). In vivo and simultaneous multimodal imaging: Integrated multiplex coherent anti-Stokes Raman scattering and two-photon microscopy. *Appl. Phys. Lett.* **97**:223702.
108. Helmchen, F., and Denk, W. (2005). Deep tissue two-photon microscopy. *Nat. Methods* **2**:932–940.
109. Zipfel, W. R., Williams, R. M., and Webb, W. W. (2003). Nonlinear magic: multiphoton microscopy in the biosciences. *Nat. Biotech.* **21**:1369–1377.

110. Kobat, D., Horton, N. G., and Xu, C. (2011). In vivo two-photon microscopy to 1.6-mm depth in mouse cortex. *J. Biomed. Opt.* **16**:106014.

111. Helmchen, F. (2009). Two-Photon Functional Imaging of Neuronal Activity. in R. D. Frostig (ed.). *In Vivo Optical Imaging of Brain Function* (2nd ed.). CRC Press.

112. Chaigneau, E., Oheim, M., Audinat, E., and Charpak, S. (2003). Two photon imaging of capillary blood flow in olfactory bulb glomeruli. *Proc. Natl. Acad. Sci. USA* **10:**13081–13086.

113. Brown, E. B., Campbell, R. B., Tsuzuki, Y., Xu, L., Carmeliet, P., Fukumura, D., and Jain, R. K. (2001). In vivo measurement of gene expression, angiogenesis and physiological function in tumors using multiphoton laser scanning microscopy. *Nat. Med.* **7**:864–868.

114. Moreaux, L., Sandre, O., and Mertz, O. (2000). Membrane imaging by second-harmonic generation microscopy. *J. Opt. Soc. Am. B* **17:**1685–1694.

115. Zipfel, W. R., Williams, R. M., Christie, R., Yu Nikitin, A., Hyman, B. T., and Webb, W. W. (2003). Live tissue intrinsic emission microscopy using multiphoton-excited native fluorescence and second harmonic generation. *Proc. Natl. Acad. Sci. USA* **100:**7075–7080.

116. Campagnola, P. J., and Loew, L. M. (2003). Second-harmonic imaging microscopy for visualizing biomolecular arrays in cells, tissues and organisms. *Nat. Biotech.* **21:**1356–1360.

117. Campagnola, P. J., and Dong, C. Y. (2011). Second harmonic generation microscopy: principles and applications to disease diagnosis. *Laser Photonics Rev.* **5:**13–26.

118. Campagnola, P. J. (2011). Second harmonic generation imaging microscopy: applications to diseases diagnostics. *Anal. Chem.* **83:**3224–3231.

119. Nan, X., Cheng, J. X., and Xie, X. S. (2003). Vibrational imaging of lipid droplets in live fibroblast cells with coherent anti-Stokes Raman scattering microscopy. *J. Lipid Res.* **44:**2202–2208.

120. Wang, H. W., Fu, Y., Huff, T. B., Le, T. T., Wang, H., and Cheng, J. X. (2009). Chasing lipids in health and diseases by coherent anti-Stokes Raman scattering microscopy. *Vib. Spectrosc.* **50:**160–167.

121. Begin, S., Belanger, E., Laffray, S., Vallee, R., and Cote, D. (2009). In vivo optical monitoring of tissue pathologies and diseases with vibrational contrast. *J. Biophotonics* **2:**632–642.

122. Enejder, A., Brackmann, C., and Svedberg, F. (2010). Coherent anti-StokesRaman scattering microscopy of cellular lipid storage. *IEEE J. Sel. Top. Quantum Electron.* **16:**506–515.

123. Le, T. T., Duren, H. M., Slipchenko, M. N., Hu, C. D., and Cheng, J. X. (2010). Label-free quantitative analysis of lipid metabolism in living Caenorhabditis elegans. *J. Lipid Res.* **51:**672–677.

124. Le, T. T., Langohr, I. M., Locker, M. J., Sturek, M., and Cheng, J. X. (2007). Label-free molecular imaging of atherosclerotic lesions using multimodal nonlinear optical microscopy. *J. Biomed. Opt.* **12:**054007.

125. Pohling, C., Buckup, T., and Motzkus, M. (2011). Hyperspectral data processing for chemoselective multiplex coherent anti-Stokes Raman scattering microscopy of unknown samples. *J. Biomed. Opt.* **16:**021105.

126. Hellwarth, R. W. (1963). Theory of Stimulated Raman Scattering. *Phys. Rev.* **130:**1850–1852.

127. Niemz, M. H. (2007). *Laser-Tissue Interactions: Fundamentals and Applications* (3rd ed.). Springer.

128. Berlien, H., Müller, G. J., Breuer, H., Krasner, N., Okunata, T., and Sliney, D. (eds.) (2003). *Applied Laser Medicine*, Springer.

129. Allison, R., Downie, G. H., Cuenca, R., Hu, X. H., Childs, C. J., and Sibata, C. H. (2004). Photosensitisers in clinical photodynamic therapy. *Photodiagn. Photodyn. Ther.* **1**:27–42.

130. Kennedy, J. C., Pottier, R. H., and Pross, D. C. (1990). Photodynamic therapy with endogenous protoporphyrin: IX: Basic principles and present clinical experience. *J. Photochem. Photobiol. B* **6**: 143–148.

131. Hamblin, M. R., and Mroz, P. (eds.) (2008). *Advances in Photodynamic Therapy: Basic, Translational and Clinical.* Artech House Publishers.

132. Dolmans, D. E., Fukumura, D., and Jain, R. K. (2003). Photodynamic therapy for cancer. *Nat. Rev. Cancer* **3**:380–387.

133. Wilson, B. C., and Patterson, M. S., (2008). The physics, biophysics and technology of photodynamic therapy. *Phys. Med. Biol.* **53**:R61.

134. Agostinis, P., Berg, K., Cengel, K. A., et al. (2011). Photodynamic therapy of cancer: an update. *CA Cancer J. Clinic.* **61**:250–281.

135. Hamblina, M. R., and Hasan, T. (2004). Photodynamic therapy: a new antimicrobial approach to infectious disease? *Photochem. Photobiol. Sci.* **3**:436–450.

136. Dierickx, C. C. (2000). Hair removal by lasers and intense pulsed light sources. *Semin. Cutan. Med. Surg.* **19**:267–275.

137. Dierickx, C. C., Alora, M. B., and Dover, J. S. (1999). A clinical overview of hairremoval using lasers and light sources. *Dermatol. Clin.* **17**:357–366.

138. Choudhary, S., Elsaie, M. L., Leiva, A., and Nouri, K. (2010). Lasers for tattoo removal: a review. *Laser Med. Sci.* **25**:619–627.

139. Xu, M., and Wang, L. V. (2006). Photoacoustic imaging in biomedicine. *Rev. Sci. Instrum.* **77**:041101.

140. Hu, S., and Wang, L. V. (2010). Photoacoustic imaging and characterization of the microvasculature. *J. Biomed. Opt.* **15**:011101.

141. Wang, L. V. (ed.). (2009). *Photoacoustic Imaging and Spectroscopy*, CRC Press.

142. Wang, L. V., and Hu, S. (2012). Photoacoustic tomography: In vivo imaging from organelles to organs. *Science* **335**:1458–1462.

143. Laufer, J., Johnson, P., Zhang, E., et al. (2012). In vivo preclinical photoacoustic imaging of tumor vasculature development and therapy. *J. Biomed. Opt.* **17**:056016.

144. Zhang, E. Z., Laufer, J. G., Pedley R. B., and Beard, P. C. (2009). In vivo high-resolution 3D photoacoustic imaging of superficial vascular anatomy. *Phys. Med. Biol.* **54**:1035–1046.

145. Vogel, A., and Venugopalan, V. (2003). Mechanisms of pulsed laser ablation of biological tissues. *Chem. Rev.* **103**:577–644.

146. Nahen, K., and Vogel, A. (2002). Plume dynamics and shielding by the ablation plume during Er:YAG laser ablation. *J. Biomed. Opt.* **7**:165–178.

147. Stern, D., Krueger, R. R., and Mandel, E. R. (1987). High-speed photography of excimer laser ablation of the cornea. *Arch. Ophthalmol.* **105**:1255–1259.

148. Pallikaris, L. G., Papatzanaki, M. E., Stathi, E. Z., Frenschock, O., and Georgiadis, A. (1990). Laser in situ keratomileusis. *Lasers Surg. Med.* **10**:463–468.

149. Krueger, R. R., Trokel, S. L., and Schubert, H. D. (1985). Interaction of ultraviolet laser light with the cornea. *Invest. Ophthalmol. Vis. Sci.* **26**:1455–1464.

150. Steinert, R. F., Puliafito, C. A., and C. A., (1986). *The Nd:YAG Laser in Ophthalmology: Principles and Clinical Practice of Photodisruption*, Saunders.

151. Niemz, M. H., Hoppeler, T. P., Juhasz, T., and Bille, J. F. (1993). Intrastromal ablations for refractive corneal surgery using picoseconds infrared laser pulses. *Lasers Light Ophthalmol.* **5**:149.

152. Juhasz, T., Loesel, F. H., Kurtz, R. M., Horvath, C., Bille, J. F., and Mourou, G. (1999). Corneal refractive surgery with femtosecond lasers. *IEEE J. Sel. Top. Quantum Electron.* **5**:902–910.

153. Lubatschowski, H., Heisterkamp, A., Will, F., et al. (2002). Ultrafast laser pulses for medical applications. *Proc. SPIE* **4633**:38–49.

154. König, K., Riemann, I., Fischer, P., and Halbhuber, K. (1999). Intracellular nanosurgery with near infrared femtosecond laser pulses. *Cell. Mol. Biol.* **45**:195.

155. Venugopalan, V., Guerra, A., Nahen, K., and Vogel, A. (2002). Role of laser-induced plasma formation in pulsed cellular microsurgery and micromanipulation. *Phys. Rev. Lett.* **88**:078103.

156. Vogel, A., Capon, M. R. C., Asiyo, M. N., and Birngruber, R. (1994). Intraocular photodisruption with picosecond and nanosecond laser pulses: tissue effects in cornea, lens, and retina. *Invest. Ophthalmol. Vis. Sci.* **35**:3032–3044.

Chapter 3

Multiphoton Microscopy

Shuo Tang
Department of Electrical and Computer Engineering, University of British Columbia, Vancouver, BC V6T 1Z4, Canada
tang@ece.ubc.ca

3.1 Introduction

Optical imaging is capable of high-resolution imaging of cell morphology and metabolism—for example nuclear size, shape, and density, which are important for distinguishing cancer cells from normal cells. Optical imaging can provide molecular specificity because tissue autofluorescence can be used to distinguish metabolic cellular components and proteins, such as NADH, flavins, collagen, and elastin. Therefore, normal tissue and tumors can be distinguished on the basis of their physiological, biochemical, and metabolic properties by using optical imaging. Recent advancement in nanotechnology has developed new contrast agents such as quantum dots and gold nanoparticles. The advent of these molecular probes and contrast agents makes optical imaging a promising tool for diagnosing and treating cancer from the molecular level.

Multiphoton microscopy (MPM) is an emerging optical imaging technique which is based on exciting and detecting nonlinear optical

Understanding Biophotonics: Fundamentals, Advances, and Applications
Edited by Kevin K. Tsia

ISBN 978-981-4411-77-6 (Hardcover), 978-981-4411-78-3 (eBook)
www.panstanford.com

signals from tissues [1–6]. MPM has been shown to be a powerful tool for imaging cells, extracellular matrix, and vascular networks in turbid tissues with subcellular resolution. It uses femtosecond laser pulses for exciting nonlinear signals such as two-photon excited fluorescence (TPEF) and second harmonic generation (SHG) from tissues. MPM is capable of depth-resolved imaging because the excitation of nonlinear signals happens only at the focal plane of the laser beam by the imaging optics. MPM is a functional imaging technique where its contrast from collagen, elastin, NADH, and flavins is highly biochemically specific.

In 1931, Göppert-Mayer theoretically predicted the existence of two-photon absorption in her dissertation [7]. In 1990, Denk and colleagues developed the first two-photon fluorescence microscopy system and demonstrated the depth-resolved imaging on live cells [1]. Since then the development of MPM systems and applications has rapidly increased with commercial turn-key femtosecond Ti:sapphire lasers become available. To date, MPM has been widely used in cancer detection, brain imaging, and wound healing study [8–16]. Recently, new advances in MPM have demonstrated portable and miniaturized MPM endomicroscopy for clinical applications [17–23].

An important application area of MPM is skin imaging [8–11]. Lin et al. used MPM to detect basal cell carcinoma (BCC) from normal dermal stroma [8] and showed that BCC has masses of autofluorescent cells with relatively large nuclei in the dermis and reduced SHG signal in stroma. Paoli et al. characterized the morphological features of squamous cell carcinoma (SCC) and BCC using MPM [9]. Dimitrow et al. applied MPM on in vivo and ex vivo diagnosis of malignant melanoma [10]. They studied melanoma features such as cell morphology and presence of dendritic cells and showed that MPM can differentiate malignant melanoma from benign nevi with a sensitivity of up to 95% and specificity of up to 97%. Koehler et al. used MPM to assess human skin aging and found a negative relationship between the SHG to autofluorescence aging index of dermis (SAAID) and aging [11].

Brain imaging is another important application area of MPM [12–15]. Levene et al. used MPM and gradient refractive index (GRIN) lens probe to image intact mouse brain tissue for up to 1.5 mm deep below the cortex [12]. Barretto et al. used multiphoton endomicroscope composed of rod GRIN lens to study the progression of brain tumor in mouse models [13]. The endomicroscope were minimally invasively inserted into the mouse brain and the authors were able to monitor neuron morphology over multiple weeks.

In this chapter, we will describe the principles and instrumentation of multiphoton microscopy. We will also introduce the multimodal multiphoton microscopy and optical coherence tomography (MPM/OCT) system and MPM endomicroscopy system developed in our group.

3.2 Principles and Instrumentation

3.2.1 *One-Photon versus Two-Photon Fluorescence*

One-photon fluorescence is a contrast signal widely used in optical microscopy and spectroscopy, including fluorescence microscopy, confocal fluorescence microscopy, and fluorescence spectroscopy. Figure 3.1a shows the energy diagram of one-photon fluorescence. When a molecule absorbs the energy from a photon, the molecule can be excited from its ground energy state to an excited electronic energy state. The molecule is unstable at the excited state and will relax back to the ground state. When returning to the ground state, the reduced energy is given out as a new photon, which is called a fluorescence photon. In the excitation process, a high energy photon such as a blue or ultraviolet (UV) light photon is used in traditional one-photon fluorescence microscopy or spectroscopy.

In two-photon fluorescence, as shown in Fig. 3.1b, a molecule is excited to the excited state by absorbing two lower energy photons such as near infrared (NIR) photons simultaneously. The sum of the energy from the two NIR photons will be equivalent to a high-energy blue or UV light photon. The emission of the fluorescence photon happens when the molecule relaxes back to the ground state. This process is called the two-photon fluorescence. The difference between one- and two-photon fluorescence is in the excitation part. Two-photon fluorescence is a nonlinear process because the involvement of absorbing two photons simultaneously

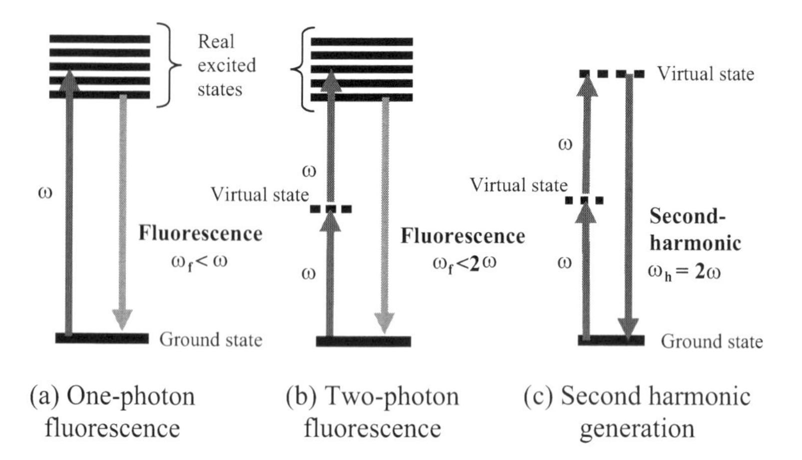

Figure 3.1 Energy diagrams of one- and two-photon fluorescence and second harmonic generation.

in the excitation. Two-photon fluorescence is a major contrast signal in MPM.

3.2.2 *MPM Contrast Signals*

The multiple contrasts obtained in MPM include two-photon excited fluorescence, second harmonic generation, three-photon excited fluorescence, and third harmonic generation. They are all related to the induced polarization of the material by the light wave. The response of a medium to an electromagnetic field generates induced polarization

$$P = \varepsilon_0 \left(\chi^{(1)} E + \chi^{(2)} E E^* + \chi^{(3)} E E^* E + \cdots\right) \tag{3.1}$$

where E is the applied electric field, ε_0 is the permittivity of free space, $\chi^{(1)}$ is the linear susceptibility, and $\chi^{(n)}$ is the nth-order nonlinear susceptibility (for $n > 1$) [3, 4]. Linear effects such as one-photon absorption and scattering are related to the linear susceptibility $\chi^{(1)}$, whereas nonlinear effects depend upon the higher-order susceptibilities $\chi^{(n)}$. For example, SHG is related to $\chi^{(2)}$ and TPEF to $\chi^{(3)}$, respectively. As a result of nonlinear effects, both the SHG and TPEF intensities depend quadratically on the incident laser power [1–6].

The energy diagrams of TPEF and SHG are illustrated in Fig. 3.1b,c. In TPEF, a susceptible molecule absorbs two photons simultaneously and is excited from ground state to a real excited state. When the molecule returns to the ground state, it emits a fluorescence photon which has less energy than the sum of the two excitation photons due to energy loss in the relaxation. Therefore, the TPEF signal has a frequency of $\omega_{\mathrm{f}}<2\omega$, where ω is the angular frequency of the incident light. In SHG, a molecule interacts with two photons simultaneously and is excited to a virtual excited state. When the molecule returns to the ground state, it emits an SHG photon, which has $\omega_{\mathrm{h}} = 2\omega$. There is no energy loss in the SHG process. Therefore, TPEF and SHG have different frequencies and can be separated using dichroic mirrors and filters.

A TPEF signal derives from intrinsic sources, such as elastin, NADH, and flavins, and exogenous fluorophores, such as various fluorescent dyes conjugated to molecular probes. An intrinsic TPEF signal has been observed from cells, collagen, and elastin fibers. Using exogenous probes, TPEF can image targeted subcellular structures and specific proteins. On the other hand, the generation of an SHG signal requires a noncentrosymmetric molecular structure. In tissue, collagen has been found to be the main tissue component which can generate SHG contrast. SHG is especially useful for imaging the extracellular matrix because collagen is the most abundant extracellular matrix protein in tissues.

The advantages of MPM come from nonlinear excitation. The probability of a molecule absorbing two photons simultaneously is very low and thus MPM requires high photon density for efficient excitation. With a high numerical aperture (NA) objective lens, photons are focused to a very high density at the focal volume. Therefore, efficient excitation happens only at the focal volume. Below or above the focal volume, there are not enough photons to excite multiphoton signals. Therefore, MPM has the inherent optical sectioning capability and is capable of depth-resolved imaging. Another advantage of MPM comes from the use of NIR light in the excitation. With NIR excitation, the photo damage is much reduced and cell viability is increased compared with UV or blue light excitation. NIR light can also penetrate deeper into tissue than visible or UV light because of the reduce scattering. Furthermore, the

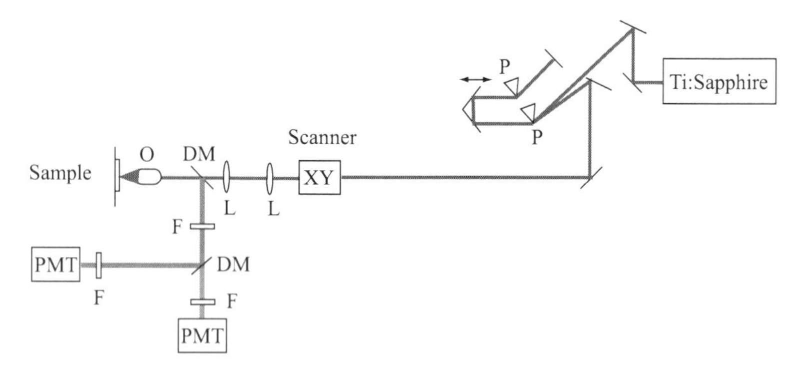

Figure 3.2 Schematics of an MPM system. DM denotes dichroic mirror; F, filter; L, lens; O, objective; P, prism; and PMT, photomultiplier tube.

excitation and emission photon wavelengths of MPM are separated far apart. Therefore, the emitted signal can be readily separated from the excitation light with dichroic mirrors and bandpass filters. The background signal due to residual excitation light is very low and thus its signal-to-noise ratio is very high. Therefore, MPM is especially suitable for high-resolution imaging of thick tissue and live cells.

3.2.3 *MPM Instrumentation*

A typical configuration of an MPM system is shown in Fig. 3.2. A femtosecond Ti:sapphire laser outputs femtosecond laser pulses. The laser beam passes through a dispersion precompensation unit which is composed of a pair of Brewster prisms. The prism pair precompensates the dispersion from the objective lens and other optics in the beam delivery path. The laser beam is raster scanned by two galvanometer mirrors in an *en-face* mode. The scanned laser beam is expanded by two lenses to fill the back aperture of the objective. A tightly focused laser beam is delivered to the sample by the objective. From the sample, the generated TPEF and SHG signals are collected by the same objective lens in a backward direction. The TPEF and SHG signals are separated from the excitation light by a dichroic mirror. TPEF and SHG are further separated by a second dichroic mirror and selected by suitable bandpass filters. The TPEF and SHG signals are detected by photomultiplier tubes (PMTs).

TPEF and SHG are nonlinear processes whose intensities depend quadratically on the incident laser power [1–6]. Because the quantum efficiency of the nonlinear excitation is very low, femtosecond pulsed lasers with high peak power are used to excited MPM signals. Femtosecond Ti:sapphire laser is the most frequently used laser source in MPM. The pulse width of a Ti:sapphire laser is typically 10–200 fs. The MPM excitation efficiency has been found to increase inversely as the pulse width decreases [24]. Therefore, shorter pulses tend to excited MPM signals more efficiently than longer pulses. The wavelength of Ti:sapphire lasers can be tuned from ~680 nm to ~1080 nm. For different fluorophores in tissue, the laser wavelength can be tuned to the respective optimum peak excitation wavelength.

Recently, femtosecond fiber lasers have also been used in MPM. Femtosecond fiber lasers are compact and portable and are especially suitable for developing MPM endomicroscopy for clinical applications [25–28]. There are two fundamental wavelength bands in fiber lasers, 1030 nm based on Yb-doped fiber and 1560 nm based on Er-doped fiber [29, 30]. The pulse width of a femtosecond fiber laser is typically 100–300 fs and the power is typically a few hundred milliwatts. An intrinsic SHG signal can be readily excited from collagen at both 1030 nm and 1560 nm. For TPEF imaging at 1030 nm and 1560 nm wavelengths, fluorescent staining is generally needed. For example, Alexa Fluro 750 can be used for TPEF imaging at 1560 nm. Using femtosecond fiber lasers can extend the MPM excitation wavelength from the Ti:sapphire band of ~680–1080 nm to a longer wavelength of 1560 nm. Exploring longer wavelength excitation for MPM can potentially interrogate deeper in turbid tissues and increase cell viability [31].

In MPM, femtosecond lasers with 10–200 fs pulse durations are often used. For femtosecond pulses, pulse broadening due to dispersion needs to be compensated. Dispersion happens because different wavelengths travel at different velocities. For example, red light travels faster than blue light in materials with normal dispersion [32]. Dispersion is introduced by thick optical components in the beam path, including the objective lens and beam expander lenses. Dispersion is more critical for shorter pulses with pulse width less than 100 fs. Femtosecond pulses can easily be broadened to picosecond pulses by dispersion if no compensation is applied. In order to excite MPM signals efficiently, we need to maintain ultrashort pulses at the sample location. In such a case, dispersion precompensation can be applied to compress the pulse duration to the shortest at the sample location. Dispersion precompensation can be achieved using prisms, gratings, or chirped mirrors. In compensating dispersion, for example, red light is allowed to propagate over a longer distance than the blue light so that all the wavelengths will arrive at the sample location at the same time. The difference in the propagation distance is created by using the prisms, gratings, or chirped mirrors to spatially separate the wavelengths. Prisms and gratings can compensate a larger amount of dispersion than chirped mirrors. Chirped mirrors use multilayered coating to reflect different wavelengths at different depths. Chirped mirrors can also be designed to compensate higher-order dispersions.

MPM is a laser scanning microscope. In MPM, the laser beam is focused to a diffraction-limited spot by a high-NA objective lens. MPM signal is excited from the focal spot. When the focal spot is raster scanned, a 2D image can be acquired. Furthermore, the focal depth can be scanned by moving the objective lens up and down. A stack of 2D images can be acquired and 3D view can be reconstructed. With laser scanning, 3D MPM imaging can be achieved with high spatial resolution. Raster scanning of the laser beam can be achieved by using galvanometer scanners, resonant scanners, microelectromechanical system (MEMS) scanners, etc. Galvanometer scanners are commonly used in MPM systems. Galvo scanners have very good linearity and large scanning angles. The speed of galvo scanners is up to a few hundred hertz. A resonant galvo scanner operates at its resonance frequency and thus its scanning speed is much faster than that of normal galvo scanners. The speed of a resonant scanner can be up to a few kilohertz. However, the motion of a resonant scanner is continuous and it cannot be stopped at a point of interest. For high-speed MPM imaging, multifocal scanning is also developed [33]. The laser beam is focused into multiple foci which can be scanned in parallel. Thus the imaging speed can be much improved because of the parallel acquisition. In multifocal MPM imaging, a detector array is needed to record the MPM signal from the multiple foci. MEMS scanner is a type of a miniaturized scanning device which has been used in MPM endomicroscopy [21, 22]. A MEMS mirror has a diameter on the millimeter scale and it can be packaged into a small probe.

PMTs are typically used in MPM for low light detection. There are two detection modes in MPM systems, analog mode and photon counting mode. In analog mode, the photocurrent detected by a PMT is converted to a voltage signal which is digitized by an analog-to-digital circuit board or a frame grabber. A frame grabber can digitize the voltage signal at very high speed and video rate MPM is possible. The photon counting mode is used when the signal level is very low. In photon counting mode, a discriminator/amplifier circuit is connected after the PMT. When photons strike on the PMT, photocurrent pulses are generated. The discriminator will only pass the pulses with signal level above a threshold and pulses below the threshold are considered as noise and are removed. The amplifier then amplifies the pulses to a standard TTL level. The TTL pulses are counted by counters to represent the number of arrived photons within a time window. Compared with analog mode, photon counting mode has higher sensitivity and less noise.

3.3 MPM Imaging of Cells and Tissues

MPM is capable of imaging intrinsic contrasts in cells and tissues with high resolution and deep penetration. It has been applied to study skin cancer, brain function, wound healing, etc. In tissue, cells have fluorescence and thus can be imaged with TPEF. The fluorescence contrast of cells is mainly from NADH. Some proteins also have fluorescence contrast, such as elastin fibers. Collagen has been found to have very strong SHG contrast. Collagen is the most abundant extracellular matrix in tissue. Therefore, MPM is capable of imaging the interaction of cells with extracellular matrix, which is important for studying caner progression, wound healing, and many other diseases.

Figure 3.3 shows the MPM image of fibroblasts embedded in collagen fibers. The TPEF signal shows the cells and the SHG contrast

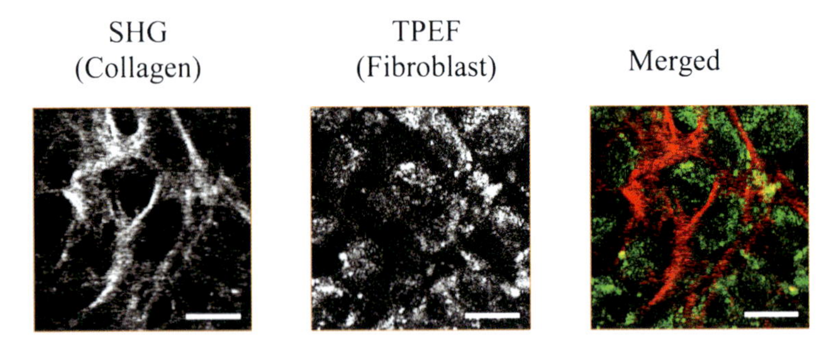

Figure 3.3 MPM imaging of cells and collagen fibers. The scale bar is 20 μm.

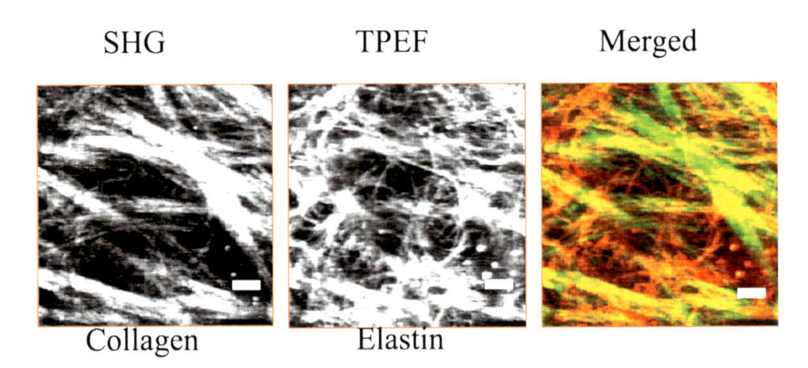

Figure 3.4 MPM imaging of collagen and elastin fibers in blood vessel wall. The scale bar is 10 μm.

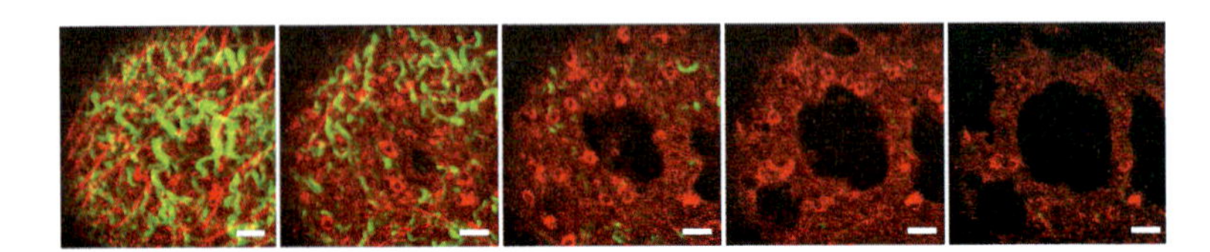

Figure 3.5 MPM imaging of rabbit lung tissue. Red is TPEF and green is SHG. The scale bar is 20 μm.

shows collagen fibers. Figure 3.4 shows the collagen and elastin fibers inside a blood vessel wall. Collagen fibers are shown in the SHG channel and elastin fibers in the TPEF channel. Figure 3.5 shows a series of MPM images at different depths of rabbit lung tissue. Both elastin and collagen fibers can be seen. Cells and alveoli of lung tissue are also observed in the TPEF channel.

Recently, there have been new advances in MPM system development for improving system capability and functionally. Some of those advances developed by the author's group are discussed below.

3.4 Combined MPM/OCT System

Optical coherence tomography (OCT) is another optical imaging technique which detects scattered light through interference [34–38]. MPM and OCT are both capable of non-invasive, high-resolution imaging of thick, turbid tissues. The primary signals detected in MPM are TPEF and SHG. TPEF derives from intrinsic sources (e.g., co-factors, proteins) and exogenous fluorophores, while strong SHG comes from noncentrosymmetric molecules, such as collagen, a common structural protein. OCT detects backscattered light from refractive index discontinuities that occur between tissues of different structures or compositions.

Combining MPM and OCT into a single platform creates a multimodality imaging system which can acquire structural and functional imaging of tissues simultaneously [39, 40]. The sensitivity of MPM to cells and extracellular matrix, and that of OCT to structural interfaces, enable direct monitoring of cell–cell and cell–matrix interactions during the development of neovasculature, cell migration, and extracellular matrix remodeling. These events are fundamentally important to nearly all biological processes, such as cancer growth, wound healing, aging, and diabetes. The combined system can thus provide a powerful imaging tool that has both the sensitivity and specificity necessary to detect precancerous and cancerous lesions on both tissue and cellular levels.

3.4.1 *MPM/OCT System Configuration*

A co-registered MPM/OCT system is developed using a Ti:sapphire laser with a 12 fs pulse width and 100 nm bandwidth [40]. The coherence length of the laser source is comparable to the focal depth

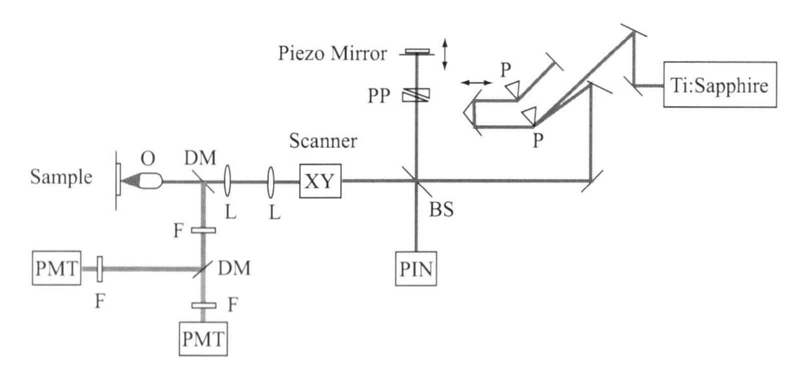

Figure 3.6 Schematics of a combined MPM/OCT system. BS denotes beam splitter; DM, dichroic mirror; F, filter; L, lens; O, objective; P, prism; PP, prism pair; and PMT, photomultiplier tube. Reproduced with permission from Ref. [40].

of a high-NA objective lens. Thus, the intrinsic OCT axial resolution from coherence gating matches with the axial resolution of the MPM. The system configuration is shown in Fig. 3.6. The center wavelength of the Ti:sapphire is 800 nm. Its output passes through a pair of fused silica Brewster prisms that precompensate the dispersion. Afterwards, the laser beam is split by a beam splitter into two arms, the sample arm and the OCT reference arm. In the sample arm, the laser beam is raster scanned by two galvanometer mirrors in an *en-face* mode. The laser beam is focused onto sample by a high-NA objective lens.

In the detection channel, TPEF and SHG signals are collected by the same objective lens in a backward direction. They are separated from the excitation source by a dichroic mirror (675DCSP, Chroma). The TPEF and SHG are separated by a second dichroic mirror (475DCLP, Chroma) and selected by bandpass filters. The TPEF and SHG signals are detected by two PMTs. For OCT imaging, the backscattered fundamental light is reflected by the beam splitter to a PIN detector where it is mixed with the reference beam. In the OCT reference arm, a scanning piezoelectric mirror generates a 5 kHz carrier frequency for OCT detection. In the *en-face* scanning mode, the scanning range of the piezoelectric mirror is set to be less than the coherence length of the light source so that it does not create depth scanning in the OCT imaging. OCT interference fringes are generated on the PIN detector when the path lengths of the reference and sample arms are matched. The envelope of the interference fringe is demodulated by a lock-in amplifier and digitized by an analog-to-digital board.

In the combined MPM/OCT system, three channels can simultaneously detect TPEF, SHG, and OCT signals, respectively. *En-face* imaging is achieved by raster scanning the two galvanometer mirrors. For Z scanning, the sample is vertically scanned by a linear translation stage so that there is minimal disruption of optical path length when taking stacks of images in three dimensions. The image size is 256 × 256 pixels.

The advantage of combining MPM and OCT is to acquire complementary structural and functional information about tissues. Such capability is tested on an organotypic RAFT tissue model [40]. The RAFT model consists of a basic polymerized collagen gel made up of type I rat-tail collagen and primary human dermal fibroblasts. Figure 3.7 shows images from the SHG (a), TPEF (b), and OCT (c)

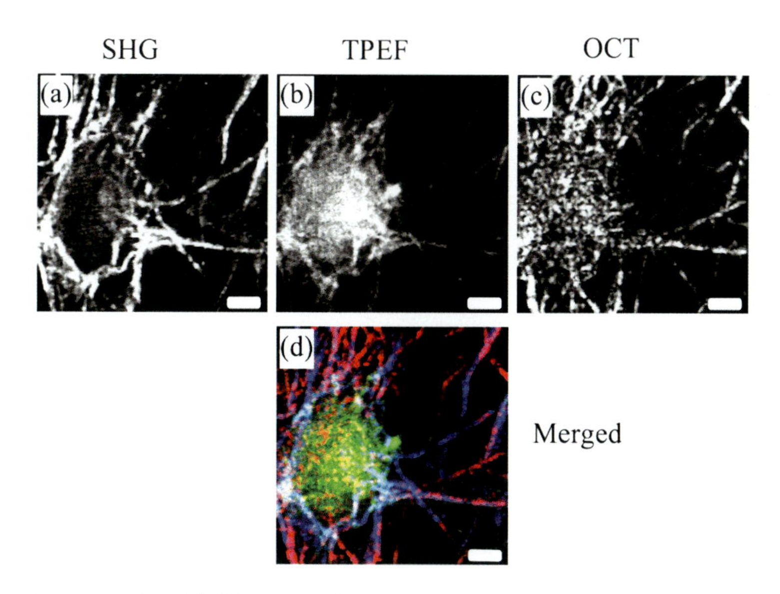

Figure 3.7 MPM/OCT images of an organotypic RAFT tissue model. (a) SHG from collagen matrix; (b) TPEF from fibroblasts; (c) OCT from scattering interfaces; (d) merged image with SHG, TPEF, and OCT signals in blue, green, and red. The scale bar is 5 μm. Reproduced with permission from Ref. [40].

channels, respectively. The SHG image shows the organization of the collagen matrix, the TPEF image shows the autofluorescence from a fibroblast, and OCT shows the morphology of the RAFT including the extracellular collagen matrix and the cell. A merged image of the three channels is displayed in Fig. 3.7d with SHG, TPEF, and OCT in blue, green, and red colors, respectively. The combination of the three channels provides a more complete picture of the tissue with both structural and compositional information that appear to be complementary.

3.4.2 *Imaging with the Combined MPM/OCT System*

The combined MPM/OCT system can acquire co-registered SHG, TPEF, and backscattered light simultaneously from the same sampling volume. This allows the use of TPEF imaging with fluorescence targeted probes to identify molecular components of subcellular scattering structures. The MPM/OCT system has been used to investigate the origin of scattering contrast in cells [41]. Images of single glioblastoma cells are shown in Fig. 3.8. The cells are embedded in 3D matrigel. Each row represents one cell with the TPEF (left), the OCT (middle), and the merged (right) images. Multiple fluorescent dyes have been used to distinguish the subcellular structures, including DAPI labeling for nuclei, rhodamine 123 for mitochondria, PKH67 for plasma membrane, and Alexa Fluor 488 conjugated to phalloidin for actin filaments. In Fig. 3.8, scattering contrast is observed from mitochondria, plasma membrane, actin filaments, and the boundary between cytoplasm and nucleus.

The MPM/OCT system has also been applied to study wound healing using engineered RAFT tissue model. An artificial wound is created by removing part of the RAFT tissue. MPM/OCT imaging is applied to monitor the migration of fibroblasts and remodeling of the collagen matrix. Figure 3.9 shows the MPM/OCT image of the RAFT tissue [38]. The OCT image indicates the boundary between the intact and the wounded areas of the tissue. The SHG shows the collagen matrix, and the TPEF shows a fibroblast which aligns in parallel with the boundary of the wounded area.

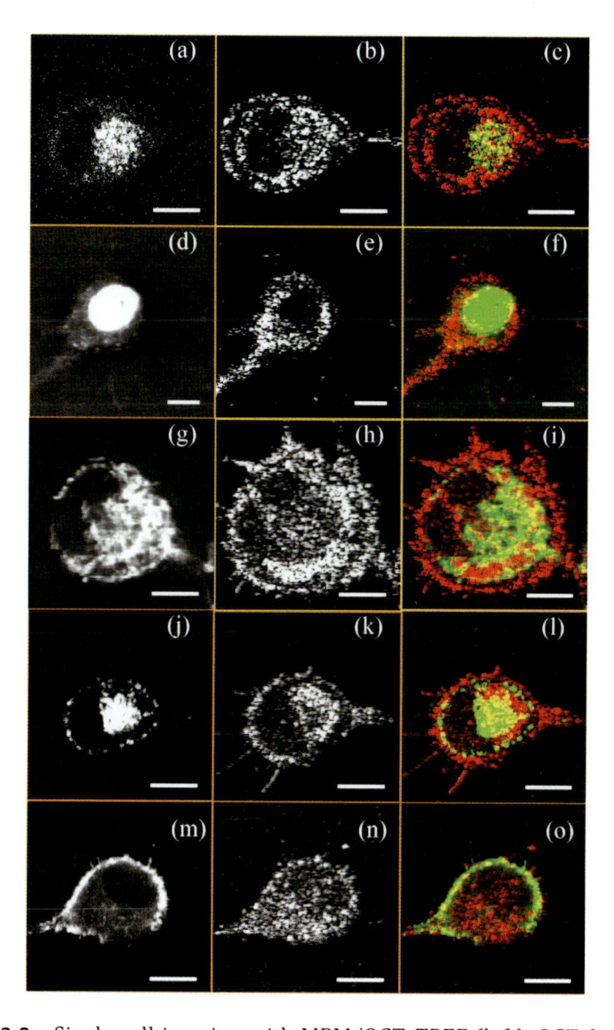

Figure 3.8 Single-cell imaging with MPM/OCT. TPEF (left), OCT (middle), and merged (right) images of single cells. (a)–(c) Single cell without labeling; (d)–(f) with DAPI labeling for nuclei; (g)–(i) with rhodamine 123 labeling for mitochondria; (j)–(l) with PKH67 labeling for plasma membrane; (m)–(o) with Alexa Fluor 488 conjugated to phalloidin for actin filaments. In the merged images, TPEF and OCT signals are in green and red, respectively. The scale bar is 10 μm. Reproduced with permission from Ref. [41].

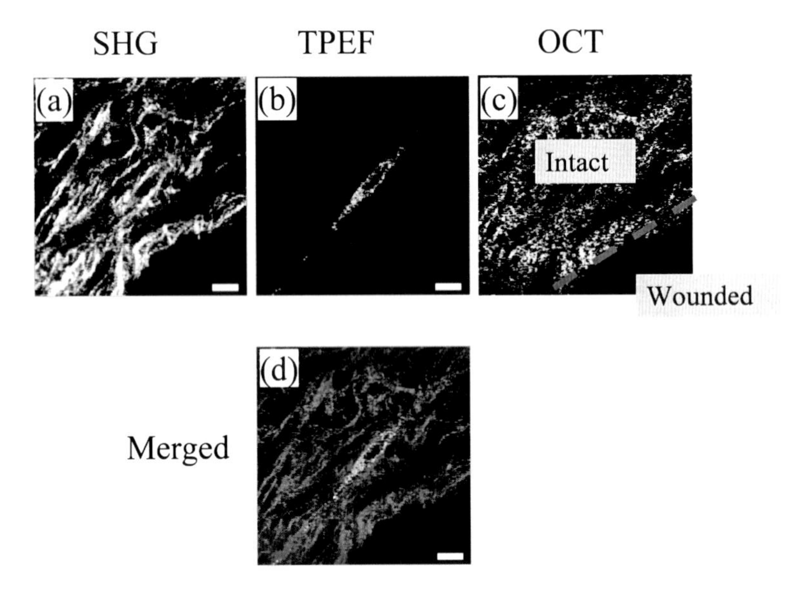

Figure 3.9 MPM/OCT images of an organotypic RAFT tissue model for studying wound healing. (a) SHG image shows collagen matrix. (b) TPEF image shows a fibroblast. (c) OCT image shows the boundary between intact and wounded areas. (d) Merged image of the above three channels. The scale bar is 10 μm. Reproduced with permission from Ref. [38].

3.5 Multiphoton Endomicroscopy

MPM systems have been mostly developed using free-space optics and microscope platforms. However, for in vivo imaging and clinical applications, a fiber-optic MPM endoscope is desirable where light can be delivered through a flexible fiber and images can be acquired using a miniature probe [17–23]. Delivering femtosecond pulses through a fiber and designing a miniature scanning probe are two challenges in MPM endomicroscopy.

A miniaturized third-harmonic-generation fiber microscope has been developed using a resonant MEMS scanner and large-mode-area photonic crystal fiber [23]. The fiber microscope has been applied for in vivo imaging of human skin. An MPM endomicroscope has also been developed using piezo tub scanner [18]. A fiber passes through the center of a piezo tub and is actuated as a cantilever. A double-cladding fiber (DCF) was used for single-mode delivery of

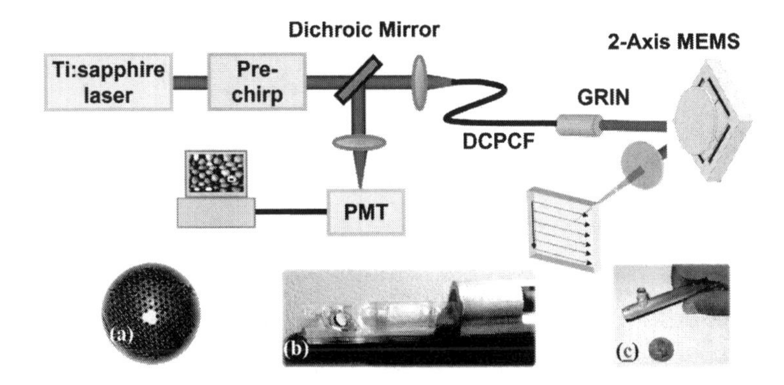

Figure 3.10 Schematic of the endoscopic MPM system using a two-axis MEMS scanner. (a) The cross-section of the DCPCF. (b) The MEMS mirror assembled with the DCPCF and GRIN lens. (c) Packaged MPM probe. Reproduced with permission from Ref. [22].

fs laser pulses and multimode collection of MPM signals. Photonic bandgap fiber (PBF) was used to compensate the dispersion of the DCF.

We have also developed a multiphoton endoscopy system using a two-axis MEMS mirror and double-cladding photonic crystal fiber (DCPCF) [21, 22]. Figure 3.10 shows the schematic of the MPM endoscopy system [22]. A femtosecond Ti:sapphire laser (Mira, Coherent) is used as the excitation light source. The wavelength of the Ti:sapphire is 790 nm and its bandwidth is ~10 nm. The pulse width of the laser output is ~170 fs. The laser beam first passes through a dispersion precompensation unit, which is composed of a pair of gratings. Afterwards, the excitation light transmits through a dichroic mirror (650 nm longpass, Chroma). A DCPCF is used for both light delivery and collection. The distal end of the endoscope is connected with an MPM probe, which is composed of a MEMS mirror, GRIN lens, and miniature lens.

Both the excitation light and the emitted MPM signal from the sample are delivered and collected by the same DCPCF. The emitted MPM signal is separated from the excitation light by the dichroic mirror, which transmits the excitation beam but reflects the emitted beam. Further elimination of the excitation light in front of the detector is achieved by passing through a bandpass filter (550 nm

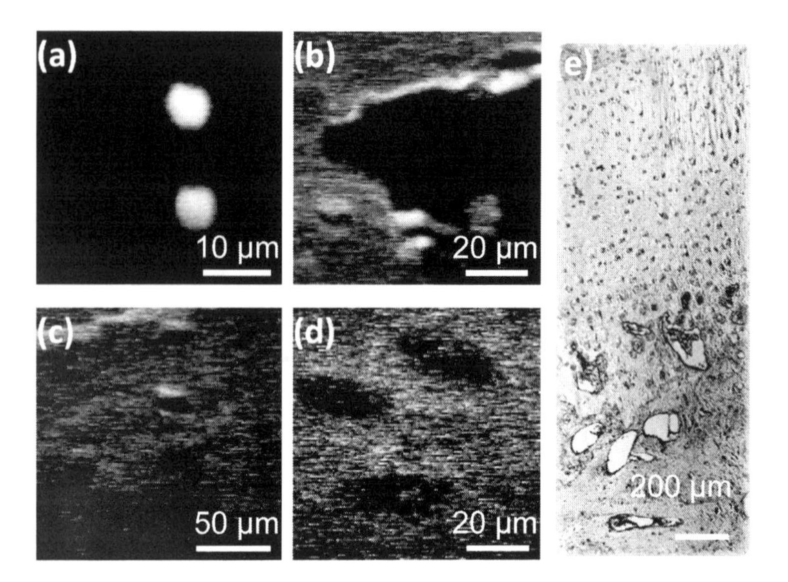

Figure 3.11 MPM images of fluorescent microspheres and bovine knee joint cartilage obtained with the endoscopic MPM system. (a) 6 μm beads, (b) bone structure, (c) (d) chondrocytes, and (e) white-light microscope photo showing the bovine knee joint cartilage sample. Reproduced with permission from Ref. [22].

bandpass, Chroma). The MPM signal is detected by a PMT with high detection sensitivity. Figure 3.10a shows the cross-section of the DCPCF, where the excitation light is coupled into the core of the fiber. Figure 3.10b shows the MEMS mirror assembled with the DCPCF and GRIN lens. Figure 3.10c shows the packaged MPM probe.

Typical images acquired with the MPM endoscope are shown in Fig. 3.11 [22]. The 6 μm diameter microspheres are shown in Fig. 3.11a. Bovine knee joint cartilage is also imaged with the endoscope. The structure is shown in the white-light microscope photo in Fig. 3.11e. The outer zone is cartilage with chondrocytes located in oblong spaces of lacunae. The inner zone is loose bone structure with large spaces. The sample is stained with fluorescein. Using the MPM endoscope, both the outer cartilage zone and the inner bone zone are imaged. The large spacing in the bone zone is observed in the MPM image as shown in Fig. 3.11b. The oblong lacunae and chondrocytes are clearly observed with the MPM endoscope as shown in Fig. 3.11c,d.

3.6 Summary

Multiphoton microscopy has become an important imaging tool in biosciences. With its inherent optical sectioning, deep penetration, and specific contrast from TPEF and SHG, MPM has found broad applications in studying cancer, brain function, wound healing, and other diseases. The next development of MPM will focus on more compact and portable systems for clinical applications. Fiber-based MPM endomicroscopy will be key for translating the MPM technique to clinical applications. New advancements in miniaturized scanners, photonic crystal fibers, femtosecond fiber lasers, and micro-optics will all be important for the development of MPM endomicroscopy. Other new developments will include multimodality systems which combine MPM with other imaging modalities to acquire multiple contrast signals and integrate structural imaging with functional imaging.

References

1. W. Denk, J. H. Strickler, and W. W. Webb (1990). Two-photon laser scanning fluorescence microscopy. *Science* 248:73–76.
2. W. R. Zipfel, R. M. Williams, and W. W. Webb (2003). Nonlinear magic: multiphoton microscopy in the biosciences. *Nat. Biotechnol.* 21:1368–1376.
3. P. J. Campagnola, M. D. Wei, A. Lewis, and L. M. Loew (1999). High-resolution nonlinear optical imaging of live cells by second harmonic generation. *Biophys. J.* 77:3341–3349.
4. P. J. Campagnola, and L. M. Loew (2003). Second-harmonic imaging microscopy for visualizing biomolecular arrays in cells, tissues and organisms. *Nat. Biotechnol.* 21:1356–1360.
5. A. Zoumi, A. Yeh, and B. J. Tromberg (2002). Imaging cells and extracellular matrix in vivo by using second-harmonic generation and two-photon excited fluorescence. *Proc. Natl. Acad. Sci. USA* 99:11014–11019.

6. B. G. Wang, K. Konig, and K. J. Halbhuber (2010). Two-photon microscopy of deep intravital tissues and its merits in clinical research. *J. Microsc.* 238:1–20.
7. M. Goppert-Mayer (1931). Elementary file with two quantum fissures *Annalen Der Physik* 9:273–294.
8. S. J. Lin, S. H. Jee, C. J. Kuo, R. J. Wu, W. C. Lin, J. S. Chen, et al. (2006). Discrimination of basal cell carcinoma from normal dermal stroma by quantitative multiphoton imaging. *Opt. Lett.* 31:2756–2758.
9. J. Paoli, M. Smedh, A. M. Wennberg, and M. B. Ericson (2008). Multiphoton laser scanning microscopy on non-melanoma skin cancer: morphologic features for future non-invasive diagnostics. *J. Invest. Dermatol.* 128:1248–1255.
10. E. Dimitrow, M. Ziemer, M. J. Koehler, J. Norgauer, K. Koenig, P. Elsner, et al. (2009). Sensitivity and specificity of multiphoton laser tomography for in vivo and ex vivo diagnosis of malignant melanoma. *J. Invest. Dermatol.* 129:1752–1758.
11. M. J. Koehler, K. Konig, P. Elsner, R. Buckle, and M. Kaatz (2006). In vivo assessment of human skin aging by multiphoton laser scanning tomography. *Opt. Lett.* 31:2879–2881.
12. M. J. Levene, D. A. Dombeck, K. A. Kasischke, R. P. Molloy, and W. W. Webb (2004). In vivo multiphoton microscopy of deep brain tissue. *J. Neurophysiol* 91:1908–1912.
13. R. P. J. Barretto, T. H. Ko, J. C. Jung, T. J. Wang, G. Capps, A. C. Waters, et al. (2011). Time-lapse imaging of disease progression in deep brain areas using fluorescence microendoscopy. *Nat. Med.* 17:223–229.
14. F. Helmchen, M. S. Fee, D. W. Tank, and W. Denk (2001). A miniature head-mounted two-photon microscope: High-resolution brain imaging in freely moving animals. *Neuron* 31:903–912.
15. J. C. Jung, A. D. Mehta, E. Aksay, R. Stepnoski, and M. J. Schnitzer (2004). In vivo mammalian brain Imaging using one- and two-photon fluorescence microendoscopy. *J. Neurophysiol.* 92:3121–3133.
16. A. T. Yeh, B. S. Kao, W. G. Jung, Z. P. Chen, J. S. Nelson, and B. J. Tromberg (2004). Imaging wound healing using optical coherence tomography and multiphoton microscopy in an in vitro skin-equivalent tissue model. *J. Biomed. Opt.* 9:248–253.
17. J. C. Jung and M. J. Schnitzer (2003). Multiphoton endoscopy. *Opt. Lett.* 28:902–904.
18. M. T. Myaing, D. J. MacDonald, and X. Li (2006). Fiber-optic scanning two-photon fluorescence endoscope. *Opt. Lett.* 31:1076–1078.

19. L. Fu, A. Jain, H. K. Xie, C. Cranfield, and M. Gu (2006). Nonlinear optical endoscopy based on a double-clad photonic crystal fiber and a MEMS mirror. *Opt. Express* 14:1027–1032.
20. C. J. Engelbrecht, R. S. Johnston, E. J. Seibel, and F. Helmchen (2008). Ultra-compact fiber-optic two-photon microscope for functional fluorescence imaging in vivo. *Opt. Express* 16:5556–5564.
21. W. Y. Jung, S. Tang, D. T. McCormic, T. Q. Xie, Y. C. Ahn, J. P. Su, et al. (2008). Miniaturized probe based on a microelectromechanical system mirror for multiphoton microscopy. *Opt. Letters* 33:1324–1326.
22. S. Tang, W. G. Jung, D. McCormick, T. Q. Xie, J. P. Su, Y. C. Ahn, et al. (2009). Design and implementation of fiber-based multiphoton endoscopy with microelectromechanical systems scanning. *J. Biomed. Opt.* 14: 034005.
23. S. H. Chia, C. H. Yu, C. H. Lin, N. C. Cheng, T. M. Liu, M. C. Chan, et al. (2010). Miniaturized video-rate epi-third-harmonic-generation fiber-microscope. *Opt. Express* 18:17382–17391.
24. S. Tang, T. B. Krasieva, Z. Chen, G. Tempea, and B. J. Tromberg (2006). Effect of pulse duration on two-photon excited fluorescence and second harmonic generation in nonlinear optical microscopy. *J. Biomed. Opt.* 11:020501.
25. A. C. Millard, P. W. Wiseman, D. N. Fittinghoff, K. R. Wilson, J. A. Squier, and M. Muller (1999). Third-harmonic generation microscopy by use of a compact, femtosecond fiber laser source. *Appl. Opt.* 38:7393–7397.
26. J. R. Unruh, E. S. Price, R. G. Molla, R. Q. Hui, and C. K. Johnson. (2006). Evaluation of a femtosecond fiber laser for two-photon fluorescence correlation spectroscopy. *Microsc. Res. Tech.* 69:891–893.
27. J. R. Unruh, E. S. Price, R. G. Molla, L. Stehno-Bittel, C. K. Johnson, and R. Q. Hui (2006). Two-photon microscopy with wavelength switchable fiber laser excitation. *Opt. Express* 14:9825–9831.
28. S. Tang, J. Liu, T. B. Krasieva, Z. P. Chen, and B. J. Tromberg (2009). Developing compact multiphoton systems using femtosecond fiber lasers. *J. Biomed. Opt.* 14:030508.
29. F. O. Ilday, J. Chen, and F. X. Kartner (2005). Generation of sub-100-fs pulses at up to 200 MHz repetition rate from a passively mode-locked Yb-doped fiber laser. *Opt. Express* 13:2716–2721.
30. J. R. Buckley, F. W. Wise, F. O. Ilday, and T. Sosnowski (2005). Femtosecond fiber lasers with pulse energies above 10 nJ. *Opt. Lett.* 30:1888–1890.

31. J. M. Squirrell, D. L. Wokosin, J. G. White, and B. D. Bavister (1999). Long-term two-photon fluorescence imaging of mammalian embryos without compromising viability. *Nat. Biotechnol.* 17:763–767.
32. B. E. A. Saleh and M. C. Teich (1991). *Fundamentals of Photonics*. John Wiley & Sons.
33. K. Bahlmann, P. T. C. So, M. Kirber, R. Reich, B. Kosicki, W. McGonagle, et al. (2007). Multifocal multiphoton microscopy (MMM) at a frame rate beyond 600 Hz. *Opt. Express* 15:10991–10998.
34. D. Huang, E. A. Swanson, C. P. Lin, J. S. Schuman, W. G. Stinson, W. Chang, et al. (1991). Optical coherence tomography. *Science* 254(5035):1178–1181.
35. J. G. Fujimoto, C. Pitris, S. A. Boppart, and M. E. Brezinski (2000) Optical coherence tomography: an emerging technology for biomedical imaging and optical biopsy. *Neoplasia* 2:9–25.
36. A. F. Fercher, W. Drexler, C. K. Hitzenberger, and T. Lasser (2003). Optical coherence tomography: principles and applications. *Rep. Prog. Phys.* 66:239–303.
37. W. Drexler (2004). Ultrahigh-resolution optical coherence tomography. *J. Biomed. Opt.* 9:47–74.
38. W. Drexler and J. G. Fujimoto (eds.) (2008). *Optical Coherence Tomography: Technology and Applications*. Springer.
39. E. Beaurepaire, L. Moreaux, F. Amblard, and J. Mertz (1999). Combined scanning optical coherence and two-photon-excited fluorescence microscopy. *Opt. Lett.* 24(14):969–971.
40. S. Tang, T. B. Krasieva, Z. P. Chen, and B. J. Tromberg (2006). Combined multiphoton microscopy and optical coherence tomography using a 12-fs broadband source. *J. Biomed. Opt.* 11:020502.
41. S. Tang, C. H. Sun, T. B. Krasieva, Z. P. Chen, and B. J. Tromberg (2007). Imaging subcellular scattering contrast by using combined optical coherence and multiphoton microscopy. *Opt. Lett.* 32:503–505.

Chapter 4

Biological Fluorescence Nanoscopy

David Williamson, Astrid Magenau, Dylan Owen, and Katharina Gaus

Centre for Vascular Research, University of New South Wales, Sydney, Australia

djw@unswalumni.com, k.gaus@unsw.edu.au

4.1 Introduction

Fluorescence microscopy is an invaluable tool, allowing the life sciences unprecedented access to the structural and biochemical processes within living tissue at scales covering several orders of magnitude—from an entire organism down to an individual cell. Fluorescence microscopy is also minimally invasive and permits fine tracking of dynamic processes over a wide range of time scales.

Decades of advances in optical microscopy, recording hardware, and image processing have pushed the resolving power of this technology to its limit, a literal limit imposed by the physical properties of light. It is at this resolution barrier that many of the most recent challenges to extend the reach of fluorescence microscopy have arisen and—through unique and surprising means—been met.

This chapter will outline some of the fundamental concepts to microscopy resolution and the diffraction barrier, the techniques which took microscopy to the edge of this limit and led to the

Understanding Biophotonics: Fundamentals, Advances, and Applications
Edited by Kevin K. Tsia

ISBN 978-981-4411-77-6 (Hardcover), 978-981-4411-78-3 (eBook)
www.panstanford.com

development of a diverse array of 'super resolution' imaging options for the biological scientist.

4.2 Resolution and Diffraction

Resolution, in imaging, is the smallest angular distance that two points can be apart for them to be distinguishable as two separate objects. The resolution of an imaging system is limited by many factors, from the nature of light itself to the sample properties, the components of the optical system, and the mechanism by which the image is acquired and recorded.

Microscopes provide a magnified image of an object and can observe very fine structural details in samples that are otherwise inaccessible to conventional observation. Optical microscopes use an arrangement of lenses to gather and converge light from a specimen upon an image plane. The quality and alignment of the optical components of a microscope will affect the quality of the images it produces. Resolution limits imposed by simple optical aberrations can be solved by improving the quality and geometric alignment of the optical components, often at high cost. However, there still remains a fundamental limit to the resolving power of a "perfectly constructed" microscope due to the diffraction of light.

The formation of an image in an optical fluorescence microscope, in the simplest example, begins with a single molecule radiating light in all directions. The emitted light travels as a spherical wavefront away from its source but is soon subjected to diffraction as it encounters changes in refractive index and various apertures as it travels within the sample and through the components of the microscope. The combined effects of diffraction within the microscope mean that a single point source of light arrives at the detector as a much larger, blurred spot. The three-dimensional intensity distribution of this spot is called the point spread function (PSF, Fig. 4.1a). For visible light, such as green light with a wavelength (λ) equal to 530 nm, the PSF in a conventional microscope is a fuzzy, ellipsoid volume, with the full width at half maximum (FWHM) approximately 250 nm in the lateral (xy) direction and 400–500 nm in the axial (z) direction. The lateral

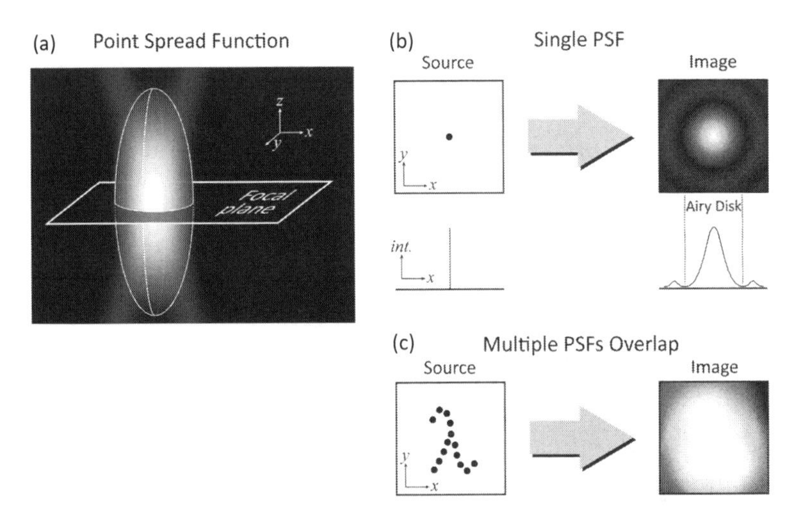

Figure 4.1 The point spread function and its effect on microscope imaging resolution. (a) The PSF is three-dimensional and is generally much longer in the axial (z) dimension than in the lateral (xy) dimensions. (b) A single point-source of photons arrives at the detector as a broad intensity distribution, the size and shape of which is determined by the microscope's point spread function (PSF). When imaged in the focal plane, the bright central part of the PSF is called the Airy disk. (c) When multiple photon sources are emitting simultaneously (e.g., fluorescent molecules in a specimen), their respective PSFs overlap at the detector, becoming indistinguishable from each other, and generating a diffraction-limited image.

image of a PSF (i.e., a section through the PSF, as viewed in the focal plane) is the *Airy disk*, which presents as a blurred spot surrounded by concentric rings of decreasing signal intensity (Fig. 4.1b). The size of the Airy disk governs the minimum distance at which two point sources can be distinguished as separate entities, i.e., the resolving power of the microscope. When the Airy disks of closely spaced single-point emitters overlap, it becomes impossible to distinguish the signal from each point (Fig. 4.1c).

When observing structures that are many times larger than the imaging wavelength of light, such as whole cells or tissues, the effects of diffraction and the resolving power of the microscope are largely irrelevant concerns. However, in order to resolve much finer structures, such as those found at the subcellular level, the limits imposed by diffraction are more apparent. When all fluorophores in a sample are emitting photons at the same time, their respective PSF intensity distributions overlap at the detector and individual fluorescent molecules soon become indistinguishable from their neighbors. This is further exacerbated by background noise, which contributes additional confounding PSFs to the detected signal. The observation of structures at smaller scales necessitates the construction of microscopes with ever higher resolving power.

Several ways to formally express the resolving power of optical microscopes have been described, of which the most widely recognized is that of Ernst Abbe in his 1873 "Theory of Image Formation," where he states that a microscope's resolving power is determined by the wavelength of light used for imaging (λ) and the numerical aperture (NA) of the objective lens:

$$\text{Resolution}_{\text{lateral}} = \frac{\lambda}{2\text{NA}}$$

$$\text{Resolution}_{\text{axial}} = \frac{2\lambda}{\text{NA}^2}$$

This relationship is also evident in the Rayleigh criterion, where the minimum resolvable distance (d, or when the maximum of one PSF overlaps with the first minimum of the other) is a function of wavelength and numerical aperture:

$$d = 1.22\frac{\lambda}{\text{NA}_{\text{condenser}} + \text{NA}_{\text{objective}}}$$

In most microscopes, the condenser and objectives are the same, so the expression is simplified to

$$d = 0.61\frac{\lambda}{\text{NA}}$$

The imaging resolution of a microscope can therefore be maximized by reducing the wavelength used to image a sample (i.e., using blue or UV light), or by increasing the numerical aperture of the objective. Numerical aperture can be increased by using higher refractive index media (e.g., oil immersion) or by increasing the size of the half angle of the lens, which is itself limited by the properties of optical glass.

In practice, most oil-immersion objectives have an NA less than 1.5, so that the diffraction-limited resolution of a microscope is approximated to a value equal to half the imaging wavelength. For example, if imaging a sample with UV laser ($\lambda = 405$ nm) through a 1.45 NA objective, then the best theoretical resolution will be 140 nm in the lateral dimensions and 385 nm in the axial dimension. These values increase, i.e., resolving power decreases, for more common fluorophores that operate at a longer wavelength, such as EGFP, and become even greater for fluorophores at the red and far-red end of the spectrum.

The diffraction barrier remained an accepted limitation of optical microscopy for hundreds of years, considered to be an impenetrable consequence of the physics of light. Developments in microscopy were able to improve and refine resolving power but were still working within the boundaries set by Abbe. However, a revolution in the approach to microscopic imaging came in the 1990s with the breaking of the diffraction barrier and the arrival of the first "super-resolution" microscopes. Since then, many techniques have come bearing the super-resolution moniker prompting the adoption of a new term, *nanoscopy*, to better represent techniques that deliver true nanoscale resolution.

4.3 Extending Resolution

Resolution improvement can be maximized using a detection aperture which is physically smaller than the diffraction limit (as in near-field techniques), reconfiguring the optical components to optimize resolving power in accordance with Abbe's expressions, or by reducing background fluorescence and therefore improving the detection of a specimen's fine structures.

Several biologically compatible imaging methodologies have evolved which boost imaging quality but still remain bound by the diffraction-limiting effects of light, which limits their resolving power.

Total internal reflection fluorescence microscopy (TIRFM) exploits the exponential decay of an evanescent field to selectively excite molecules within an extremely thin plane [1, 2]. Light refracts at the interface of two media with different refractive indexes (Fig. 4.2a). For light traveling from a high refractive index medium

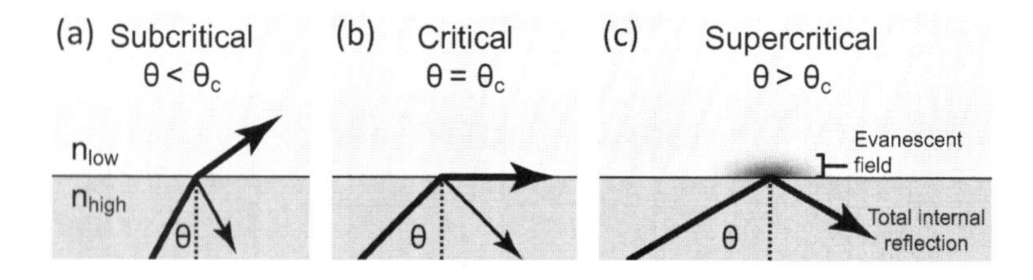

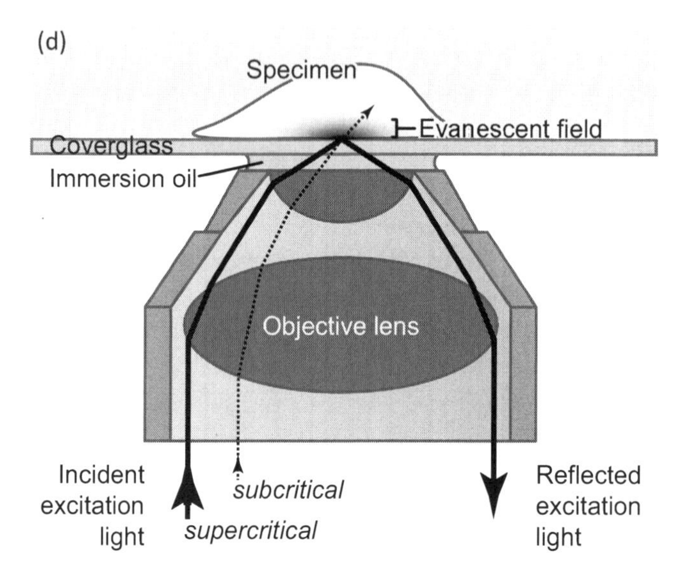

Figure 4.2 Principle of total internal reflection and its application in TIRF illumination. (a) Light traveling through two media of different refractive indexes is refracted at the interface. The extent of refraction is dependent on the incident angle θ. (b) At the critical angle, light is refracted along the media interface, traveling perpendicular to normal. (c) Beyond the critical angle, light is totally internally reflected back into the high-index medium, generating an evanescent field at the media interface which penetrates a few hundred nanometers into the low-index medium. (d) This evanescent field is exploited in biological fluorescence microscopy to illuminate a very thin plane of the basal cell membrane, yielding images with exceptionally low interference from out-of-focus fluorescent structures.

(e.g., a glass coverslip) to a lower refractive index medium (e.g., an aqueous sample buffer), the degree of refraction increases as the incident angle increases. At a certain angle, called the critical angle, the refraction of light becomes parallel to the interface (or 90° relative to normal, Fig. 4.2b). When the incident angle exceeds the critical angle, light is reflected back into the higher-index medium—a phenomenon called total internal reflection (Fig. 4.2c). A consequence of total internal reflection is the propagation of an evanescent wave across the interface, identical in wavelength to the incident light, and penetrating approximately 100 nm into the lower-index medium.

TIRFM exploits total internal reflection by directing excitation light at the sample coverglass (through the objective or a prism) at an angle greater than the critical angle, generating an evanescent field of excitation light in the aqueous sample medium at the glass–medium interface (Fig. 4.2d). This narrow excitation field significantly improves axial resolution by up to ten times compared to confocal microscopy, although as a widefield method TIRFM has considerably worse lateral resolution compared to confocal microscopy. The narrow excitation field also intrinsically restricts TIRFM to exploring features that are located very close to the coverglass, such as the basal membrane of cells and offers only limited access to cytoplasmic membrane-associated structures beyond.

For samples where membrane and associated structures are of interest, TIRFM offers another significant advantage as the selective illumination of fluorescent molecules located very close to the glass–water interface effectively eliminates the background fluorescence from other cellular components. This greatly increases the signal-to-noise ratio compared to conventional widefield epifluorescence microscopy. The improved signal-to-noise ratio affords TIRFM with extreme sensitivity and has made single-molecule fluorescence detection possible.

An altogether different approach to resolution improvement involves no optical manipulation at all, and can be performed "offline" after the images have been recorded. Deconvolution is a mathematical process to restore an image that has been degraded by blurring and noise [3]. In microscopy, this blurring is a function of the diffraction of light as well as aberrations within the optical components. With sufficient knowledge of how light from the specimen and the PSF are convolved within an optical system, it is possible to apply the inverse of these functions and deconvolve the recorded image into the "original" image.

This deconvolution process when applied to a two-dimensional image is referred to as deblurring. A three-dimensional "z-stack" image can be deblurred, with deconvolution applied discretely to each slice. Alternatively, since the PSF is a three-dimensional function, a z-stack can be deconvolved as a single entity, with information between slices used to derive higher-resolution image information in the z dimension. Deconvolution in this manner is termed *image reconstruction*.

Accurate deconvolution requires knowledge of the imaging conditions (specifically knowledge of the microscope's PSF, either measured or calculated) and appropriate selection of the deconvolution algorithm as improper processing parameters can introduce artifacts into the deconvolved image and different algorithms can yield different results.

Deconvolution is most frequently used to deblur widefield images, but it can also be applied to confocal images. Several deconvolution methods have shown considerable improvement in signal-to-noise ratio and resolution [4] with features resolvable to 30 nm laterally and 90 nm axially [5, 6].

Several far-field techniques have demonstrated improvements in resolving power through efforts to reduce the nonfocused background fluorescence signal. The most widely used is confocal laser scanning microscopy (CLSM), which places pinholes in front of the illuminating laser aperture and the detector. The laser pinhole increases focus of the excitation volume whereas the detector pinhole rejects most of the out of focus signal. When the size of the pinhole matches the size of the Airy disk, higher-order diffraction patterns are excluded and only the first harmonic is passed through. This increases the resolving power of the microscope by eliminating out-of-focus information, but this exclusion comes at the cost of decreased signal intensity.

Axial resolution, which is always worse than lateral resolution in a conventional microscope, was an early target for improvement. Interference microscopy uses dual opposing objectives to construct periodic interference patterns to improve the axial dimension of the PSF of either the illuminating laser (4Pi microscopy) or the emitted fluorescence (I^5 microscopy). The use of two objectives effectively increases the numerical aperture by enlarging the solid angle through which emitted light can be gathered.

4Pi microscopy [7] uses two opposing objective lenses in a confocal configuration to increase the system's effective numerical aperture. Further resolution enhancement is possible if the counter-propagating spherical excitation wavefronts from each objective are constructively interfered to sharpen the focal spot. Furthermore, if the emission wavefront collected by each objective can also be coherently summed at the detector, then there are three distinct 4Pi configurations implementing either coherent superposition at either the focal plane (type A), the detector (type B), or both (type C). This technique generates a much rounder PSF (the name "4Pi" is derived from the solid angle of a sphere, which is equal to 4π) although in practice the interference of spherical wavefronts introduces large side-lobes to the PSF. This effect is often minimized with mathematical deconvolution [8] after image acquisition or by using two-photon excitation [9].

Standing wave and interference microscopy (I^5M) [10, 11] also uses geometrically opposed objectives, but in a widefield configuration. A lamp supplies spatially incoherent light, which forms a flat standing wave parallel to the focal plane. Fluorescence emission is collected and constructively interfered at the detector as in 4Pi (type C) microscopy. The issue of interference-derived PSF side-lobes is also present and deconvolution is mandatory to remove "ghost image" artifacts.

Interference microscopy methods are able to improve axial resolution to around 100 nm. However, the technique offers no equivalent improvement in lateral resolution and remains diffraction-limited in these dimensions.

4.4 Super-Resolution Ensemble Nanoscopy: STED, GSD, SSIM

With the advent of nonlinear optics the relationship between very high imaging light intensity and emitted fluorescence signal could be exploited to improve resolution in far-field imaging systems. The introduction of saturable, sub-diffraction limited patterns into the excitation apparatus effectively modifies the spatial dimensions of the excitation PSF, thus revealing structures that are ordinarily inaccessible and beyond the limit of diffraction. Using a modified or structured excitation PSF, applied to a large ensemble of molecules at once, these techniques can precisely define the spatial and temporal coordinates at which a molecule can fluoresce.

REversible Saturable (or Switchable) Optically Linear Fluorescence Transitions (RESOLFT) [12] is a generalized scheme proposed by Stefan Hell in 1994 based on the controlled switching of fluorophores. The concept requires fluorophores exhibiting at least two reversibly switchable states, and that the transition can be controlled optically—for example, a fluorescent "on" state and a nonfluorescent "off" state. When switching a molecule from one state into another, the probability that the molecule remains stuck in the on state decreases exponentially with increasing intensity of the switching signal [13]. At the saturation switching intensity, half the molecules are switched into the off state and driving the intensity beyond saturation increases the likelihood that a fluorophore will be switched into the off state. In practice, a ring-shaped depletion beam is co-aligned with a standard excitation beam such that peripheral fluorophores are driven into the dark state while the remaining population of nonquenched molecules in the center of the ring is excited into fluorescence. The effective size of this central excitation spot is much smaller than the comparable confocal PSF leading to a significant increase in spatial resolution.

Super resolution images are obtained by raster scanning the saturated-depletion/excitation beam combination through the specimen, as in regular confocal microscopy. Raster scan imaging is much slower than taking a widefield image of the same area. However, single-molecule super-resolution techniques require multiple widefield acquisitions, whereas ensemble-based methods can build an image from a single scan through the sample. Furthermore, as the resolution enhancements are entirely derived from the configuration of the hardware, there is no need for additional processing post-acquisition to produce a super-resolution image although image resolution is often enhanced further by deconvolution algorithms.

The various specific techniques that derive from the RESOLFT principle are distinguished by the nature of the switchable states. Stimulated emission depletion (STED) employs ground (S_0) and excited singlet (S_1) states (Fig. 4.3b), whereas ground state depletion (GSD) and ground state depletion with individual molecule return (GSDIM) transition fluorophores between excited singlet and dark triplet states (Fig. 4.3c). Fluorescent proteins that exhibit stable, reversible *cis–trans* isomerization between bright and dark states, such as KFP1 or Dronpa, may also be used in RESOLFT.

STED was proposed alongside RESOLFT as a practical implementation of the concept [12] and then demonstrated experimentally [14] five years later. Here, an initial diffraction-limited pulse, tuned to the excitation wavelength, is used to excite fluorophores. This is immediately followed by a second pulse from the depletion laser (or STED laser) which is matched to the fluorophore's emission wavelength, and drives excited (S_1 state) molecules to the ground (S_0) state by stimulated emission.

Stimulated emission is a process by which an excited molecule interacts with a passing incident photon of a particular wavelength, forcing an excited electron to drop into a lower energy state. The energy released by this relaxation creates a second, new photon with identical characteristics to the first incident photon. Stimulated emission prevents an excited molecule from spontaneously relaxing and emitting a fluorescence photon (Fig. 4.3b).

The doughnut-shaped depletion pulse can be engineered by passing the depletion beam through a helical 2pi-phase plate [15]. The plate is a spiral ramp of gradually increasing thickness, constructed of optically transparent material. Circularly polarized light passing through the ramp is phase shifted by varying degrees along the spiral ramp creating a pocket of zero-intensity in the center of the exiting beam. It is through this "doughnut hole" that a small population of molecules can evade the depletion signal and remain fluorescent while all peripheral molecules are de-excited by stimulated emission (Fig. 4.3d).

As described by the RESOLFT concept, the doughnut configuration alone is not sufficient to obtain super-resolution information from a specimen. The intensity of the STED laser must be set sufficiently high (100–300 MW/cm^2) to achieve *saturable* depletion

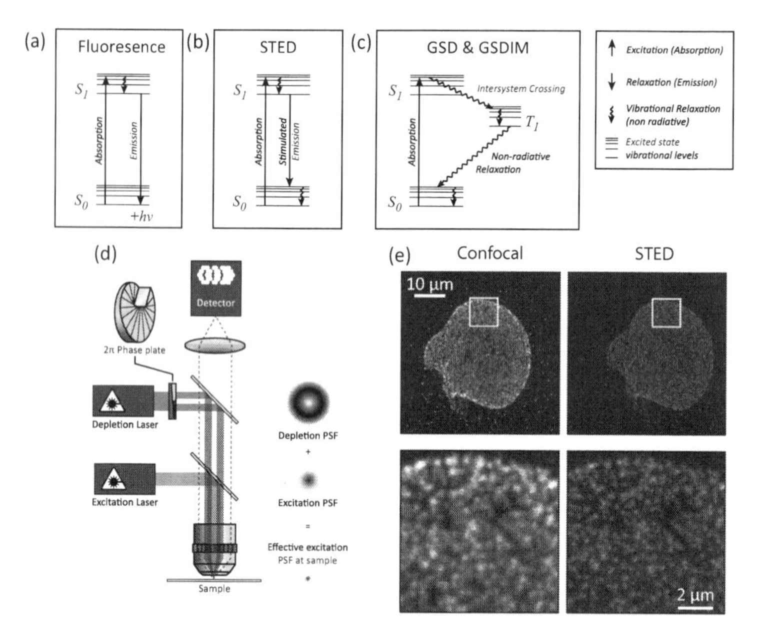

Figure 4.3 Principle of RESOLFT and its application in STED microscopy. (a) In standard fluorescence, a fluorophore in the ground state (S_0) absorbs a photon and is excited into a higher-energy state (S_1). Spontaneous relaxation from the excited state back to the ground state is accompanied by the emission of a fluorescence photon at a wavelength and energy is representative of the energy difference between the excited and ground states. (b) Stimulated emission, as used in STED, is the process where an excited-state molecule encounters a photon with wavelength comparable to the difference in energy between the molecule's excited and ground states. This interaction drives the molecule to release its energy as an emission photon identical to the incident stimulating photon, and then relax to the ground state. (c) In GSD, molecules are maintained in the higher-energy state through repeated excitation. This increases the probability of an intersystem crossing from S_1 into the dark triplet state T_1. Molecules in this state are "shelved" as relaxation from the triplet state occurs over a comparatively longer time period. (d) Schematic layout of a STED microscope. A phase mask creates the doughnut-shaped "STED PSF," which is co-aligned with the regular excitation PSF. Stimulated emission of molecules in the periphery of the excitation PSF yields a much smaller effective excitation PSF. A GSD microscope has a similar layout, except that the depletion and excitation beams are derived from the same laser source. (e) A HeLa cell immunostained with antinuclear protein antibodies conjugated to Alexa 488, observed with confocal microscopy and then with STED. The white box indicates the enlarged regions displayed beneath.

of peripheral molecules while maintaining the zero-depletion pocket in the center of the excitation beam. The size of the effective excitation PSF (which determines the resolution improvement) is inversely proportional to the STED laser intensity. Therefore, STED can theoretically generate an arbitrarily small excitation PSF. However, increasing depletion power to improve resolution comes at the cost of increasing the rate of bleaching and irreversible photodamage to the probes; in general, STED-suitable fluorophores exhibit high photostability [16].

The resolving power of STED microscopes can reach 30 nm [17] with high depletion intensity; however, at more biologically compatible levels of STED laser power, a resolving power of approximately 80 nm can be expected [18, 19].

STED has been employed across a variety of biological specimens either immunostained [20] or genetically tagged with fluorescent proteins [21], as well as in live specimens [22, 23]. STED-compatible synthetic dyes include most of the ATTO range [17, 24–26], Alexa Fluor 488 [27] and 594 [26], FITC [27], Chromeo 488 [27], Oregon Green [27], and DyLight 594 [26]. STED has been demonstrated with a more limited palette of fluorescent proteins including GFP [21], Citrine [22], and YFP [23].

STED has been implemented in a 4pi configuration—termed "isoSTED" for the nearly spherical PSF it generates—with improved axial resolution to within 30 nm [28]. Additional processing is required to compensate for the 4pi side-lobe artifacts, but the small and near-symmetrical PSF is highly desirable as an optical sectioning tool.

The recent introduction of STED implemented with continuous-wave lasers (CW-STED) [29] has addressed the high costs associated with pulsed lasers used to develop the technique and considerably simplified the setup as temporal synchronization of the depletion and excitation pulses is redundant. Supercontinuum laser sources have also been introduced to the STED family [25], and the broad emission spectrum of these lasers has also simplified the design of STED systems and provides the option of multicolor STED acquisition.

GSD

Ground state depletion (GSD) [30] is another technique based on the principles of RESOLFT, with the photoswitching mechanism involving controlled intersystem crossing of excited fluorophore electrons. GSD was proposed in 1995 following the introduction of STED and successfully demonstrated 12 years later [31].

As in STED, a doughnut-shaped depletion PSF with a central zero-intensity is superimposed over a regular excitation PSF and together these are raster-scanned through the sample. Unlike STED, the wavelength of both beams is set to match the excitation wavelength of the fluorophore. The population of molecules excited by the doughnut-PSF is driven from singlet ground state (S_0) into the singlet excited state (S_1). Maintaining these molecules in S_1 through repeated excitation with the higher energy of the "depletion beam" increases the probability of an electron spin-flip event to drive an intersystem crossing (a nonradiative process) into the dark triplet excited state (T_1). Excitation from S_0 directly to T_1 is very rare, as it requires the additional "forbidden" spin transition. Radiative decay from T_1 back to S_0 requires another electron spin reversal. The lifetime (τ) of intersystem crossing during relaxation is very slow—on the order of 10^{-8} to 10^{-3} seconds—so probes switched into a dark triplet state are effectively withdrawn from the population of fluorescent molecules (Fig. 4.3c). This leaves a much smaller population of molecules residing in the depletion doughnut "hole" which is accessible to the regular excitation beam. However, it is important for those peripheral fluorophores that are shelved in the dark triplet state to possess a suitably low τ such that they are able to relax to the ground state before the depletion beam has moved on.

Relaxation lifetimes are strongly dependent on molecular environment and this can be turned to an advantage by varying the buffer conditions in which the sample is imaged. The introduction of reducing agents, such as β-mercaptoethanol, quench triplet states (decreasing τ), whereas oxygen scavengers, such as glucose oxidase, preserve the triplet state and increase the relaxation lifetime [32]. Strategic application of these reagents can "tune" the behavior of fluorescent probes.

The depletion laser power required to populate the triplet state is on the order of 10–20 kW/cm^2, which is considerably less than the STED depletion beam power. This lower power is a consequence of the long-lived triplet-state shelving, which enables saturation to be reached much more efficiently compared to stimulated emission.

SIM

In the techniques presented above, fluorescence emission is controlled by manipulating the excited state. Fluorescence emission can also be controlled in a simple widefield configuration called structured illumination microscopy (SIM) [33], where a grid is placed in the path of the excitation beam to create alternate bands of bright and dark molecules at the sample.

High-frequency spatial features within the sample, which are otherwise not resolvable, interfere with the illumination pattern resulting in low-frequency moiré interference patterns (Fig. 4.4a). The moiré spatial frequency is a function of the difference between the uniform spatial frequency of the illumination pattern and the spatial frequency components of the fluorescent structures within the sample.

Multiple moiré patterns are recorded from a sample (usually by shifting and/or rotating the illumination pattern) and then processed by Fourier Transformation using the known spatial frequency of the illumination pattern to extract the sample's unknown spatial frequencies and construct a higher-resolution image (Fig. 4.4b–d).

SIM, as a linear excitation tool, can only increase resolution by a factor of 2 because the illumination pattern cannot focus beyond half the wavelength of the excitation light. Nonlinear excitation combined with structured illumination can extend the resolution improvement further, with two such techniques, including saturated structural illumination microscopy (SSIM) [34] and saturated pattern excitation microscopy (SPEM) [35]. The nonlinear effect of saturated illumination yields a distorted fluorescence intensity pattern containing higher-order harmonics, which can generate images with enhanced lateral resolution down to 130 nm and axial resolution to 350 nm. Saturated excitation light increases the

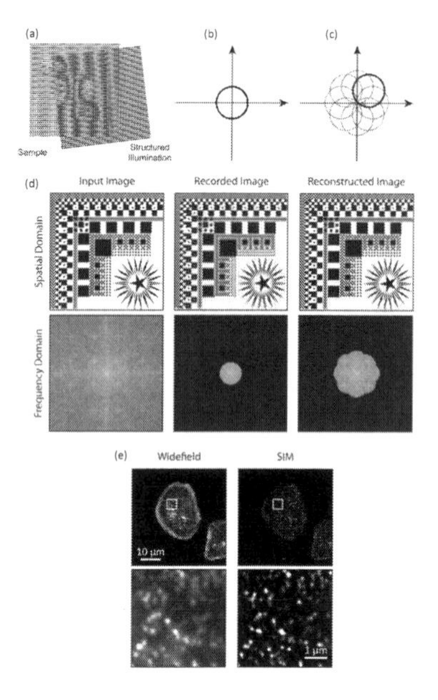

Figure 4.4 Principle of structured illumination microscopy. (a) When a regular, repeating illumination pattern (usually a grid) is applied to a sample, a moiré pattern is generated. This pattern contains lower spatial frequency information that is dependent upon the spatial frequencies contained within the sample. Importantly, this lower frequency information is accessible and detectable by the microscope, where it can be processed against the known spatial frequency of the illumination grid to derive the unknown spatial information contained within the sample. (b) When viewed in frequency space, the amount of spatial information that is accessible to the microscope is limited, as represented by the area enclosed by the circle. (c) Structured illumination does not change the range of accessible frequencies, but relocates them by changing the angle and phase of the grid. By combining spatial frequency information gathered from several grid rotations and phase shifts, the overall "aperture" of the microscope is increased, allowing higher frequency information, such as fine cellular structures, to be resolved. (d) This is demonstrated a sample image, above a Fourier transform of the image, which shows the full range of spatial frequencies contained within the input image. Reducing the range of accessible spatial frequencies—as occurs in a regular microscope—results in a blurry image lacking in fine detail. By moving the aperture, and combining the spatial information acquired, much of the image detail can be recovered. (e) Jurkat T cells immunostained with anti-LAT antibodies and Chromeo 488, observed with widefield epifluorescent microscopy and then with SIM. The white box indicates the enlarged regions which are shown beneath.

risk of fluorophore photodamage, although this may be avoided by changing to a different nonlinear process such as stimulated emission or reversible photoactivation. SIM has also been married to an I^5M configuration, termed I^5S, with improved resolution of 100 nm in both lateral and axial dimensions [36].

Structured illumination methods rely on specific hardware configurations and extensive post-acquisition software processing. However, as a wide-field technique the rate of raw image acquisition can be high compared to point-scanning and single-molecule techniques, although the rate is still affected by the need to acquire around a dozen images at different grid angles and phases. Imaging illumination intensities are generally lower compared to other super-resolution imaging techniques and structured illumination can be performed with any existing fluorophore, making it considerably more accessible to researchers than many other super-resolution techniques that require specific fluorescent markers.

4.5 Super-Resolution Single Molecule Nanoscopy: (F)PALM and STORM

RESOLFT microscopy is complemented by an alternative approach to breaking the diffraction barrier based upon widefield imaging of stochastically switched fluorescent probes. While the spatial patterning techniques based on RESOLFT subvert the diffraction limit through the manipulation of the PSF, this alternative strategy resolves closely spaced molecules by observing individual molecules sequentially.

In most cases the density of fluorophores in a sample is high enough that many share the same diffraction-limited volume. Ensemble activation of these probes makes it impossible to localize individual molecules. Introducing the additional property of stochastic photoactivation allows a small subset of molecules to become visible and thus distinguishable from each other.

Single molecule imaging encompasses a family of closely related techniques. Eric Betzig and Harald Hess developed the concepts of photoactivated localization microscopy (PALM) [37] while Samuel Hess simultaneously developed the functionally equivalent technique of fluorescence photoactivation localization microscopy (FPALM) [38], both of which use photoswitchable or photoactivatable fluorescent proteins. Xiaowei Zhuang developed another equivalent technique called stochastic optical reconstruction microscopy (STORM) [39] but distinct from (F)PALM in that STORM uses a pair of closely spaced fluorescent dyes, rather than proteins, to generate a photoswitchable signal [40]. An extension of STORM, called direct STORM (or dSTORM) [41] improves on the technique by using conventional carbocyanine dyes and a reducing buffer to provide suitable conditions for reversible photoswitching.

Ground state depletion with individual molecule return (GSDIM) [42] is another stochastic, single-molecule technique. As in GSD, molecules are optically shelved into a T_1 dark state. However, unlike regular GSD, the molecules are allowed to spontaneously relax back to S_0 ground state, where they can be repeatedly excited to produce fluorescence. Sequential widefield acquisition and localization of individual molecules as they fluoresce allows a super-resolution image to be constructed. Furthermore, repeated excitation of molecules into the "bright" excited S_1 state eventually drives the molecule into the dark excited triplet T_1 state, effectively switching the molecule off and preventing a gradual buildup of "returned" molecules from obscuring newly returning molecules. GSDIM is a considerably simpler system compared to GSD and STED as there is no requirement for complicated doughnut optics, or activation lasers as GSDIM is theoretically compatible with any fluorescent species having appropriate triplet-state properties. GSDIM has been demonstrated imaging four conventional fluorophores using only two detection channels and a single laser line [43]. An important distinction between GSDIM and dSTORM is that GSDIM assumes the population of shelved molecules reside in the dark excited triplet state, whereas dSTORM applies no specification to the nature of the long-lived dark-states that molecules may occupy.

These techniques all use probes that are photoactivatable (moving between a dark and a bright state) or photoswitchable (moving from one fluorescent emission spectrum to another). Unlike RESOLFT-based methods, it is desirable for the activation or switching to be non-reversible so that each molecule is only sampled once.

A key component of PALM and STORM is that the switching signal is at a sufficient intensity so as to only activate or switch a small proportion of molecules. The stochastic nature of this process means that molecules that switch will generally be separated from each other by a distance greater than the size of their PSFs. When these switched molecules are excited to fluoresce, they yield discrete image spots at the detector. Once the PSFs have been recorded the molecules are photobleached and another set of molecules can be switched, imaged, and bleached (Fig. 4.5a). This process is repeated many times (for up to 20,000 image frames) or until the supply of photoswitchable molecules is exhausted. The resulting raw data is an image stack of thousands of frames, each one containing the diffraction-limited fluorescence PSFs of a small subset of activated molecules.

The center of each PSF in each frame is then determined, such as by statistical curve fitting of the PSF intensity distribution to a Gaussian function. The center of the fitted curve localizes the lateral position of the original molecule to within tens of nanometers. Processing the entire stack builds a coordinate map of for the locations of every sampled molecule, from which a super-resolution image of the specimen can be composed (Fig. 4.5b).

The localization precision, or how accurately a molecule's position can be determined from its recorded PSF, is determined by several factors but primarily by the number of photons collected from each molecule. The more photons that are emitted by a molecule, the more precisely that molecule's location can be calculated. Each recorded photon is equivalent to making one measurement of the molecule's location. It is important therefore to select a probe with a high quantum yield, and hence high photon emission rates.

Other factors affecting localization precision (σ_μ), in addition to the number of photons collected (N), include the pixel size (a) of the CDD detector and the level of background noise (b, as standard deviation from background) and standard deviation (width) of the fitted Gaussian distribution (s). Localization precision can be

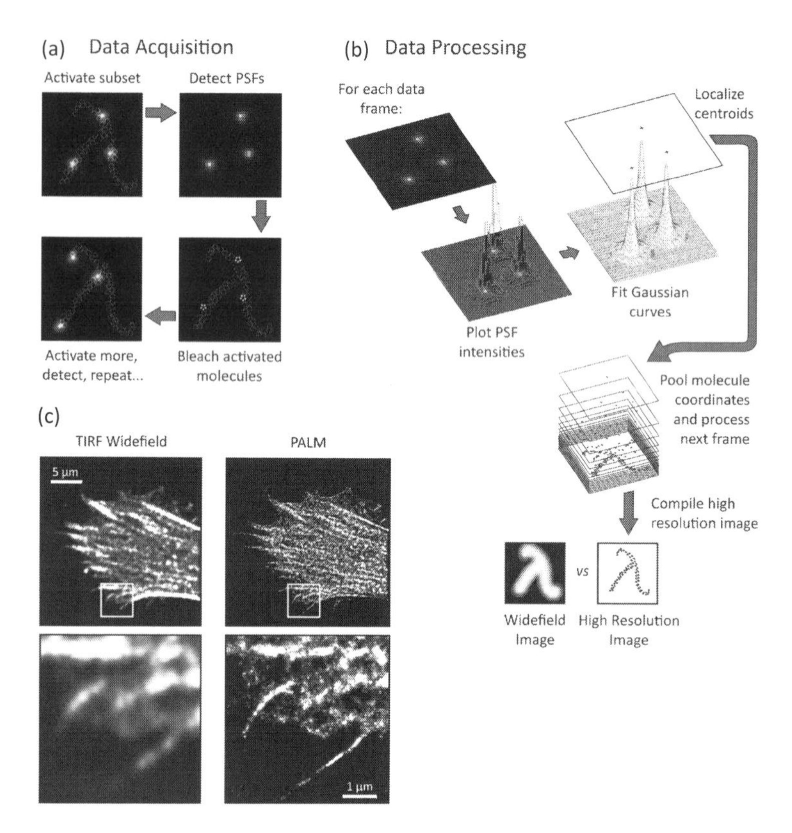

Figure 4.5 Principle of single molecule localization microscopy. (a) For data acquisition, a small subset of the total population of fluorescent molecules is activated and their PSFs are recorded. The molecules are bleached and the next subset of molecules is activated and recorded. This process is repeated many thousands of times or until all the molecules in the sample have been recorded. (b) For data processing, the intensity profile of each recorded PSF is fitted to a three-dimensional Gaussian curve. The center of this curve (at the FWHM) is located, and the PSF coordinates are stored. The molecular coordinates from all recorded PSFs are flattened to generate a super-resolution image. The discrete phases of data acquisition and localization processing can be performed sequentially, as in this example, or simultaneously. (c) A HeLa cell expressing Lifeact (an F-actin binding protein) fused to the photoswitchable protein tdEos, observed with TIRF widefield microscopy and then with TIRF-PALM. The white box indicates the regions which are enlarged beneath, to demonstrate that the substantial increase in resolution can be seen with PALM.

calculated according to the following formula [44]:

$$\sigma_\mu = \sqrt{\left(\frac{s^2}{N}\right) + \left(\frac{a^2/12}{N}\right) + \left(\frac{8\pi s^4 b^2}{a^2 N^2}\right)}$$

Background signal—a combination of background fluorescence and camera noise—adversely affects PSF detection and localization. The use of EMCCD cameras is critical to reducing camera noise. Background fluorescence can be minimized using PALM in a TIRF configuration. Photon output can be enhanced and fluorophore recycling minimized by using oxygen-rich buffers (or cysteamine) to quench excursions into dark triplet states.

A distinct advantage of these methods is that no specialized optical equipment is required. The technique can be performed on many existing inverted fluorescence microscopes capable of single-molecule imaging.

The resolution of PALM/STORM is often expressed simply as the average of the molecular localization precisions for all localized molecules in an image. However, PALM/STORM resolution is also determined by the overall labeling density within the sample, which has a direct consequence on the resolvability of structural features within a specimen. The Nyquist sampling theorem dictates that the distance between adjacent, localized molecules must be at least half the distance of the resolution for a structure to be optimally resolved. If the overall labeling density in a specimen is lower than the Nyquist sampling interval (i.e., the labeled molecules are spaced too far apart), then the underlying structures will be under-sampled and poorly resolved; thus it is pointless to have an image with very small localization precisions if there are not enough molecules present to reveal the shape or distribution of the protein are attached to. Achieving this level of sampling density is usually only problematic for target proteins which are expressed at low levels (either endogenously such as membrane proteins or, in the case of PALM, as poorly expressed fusion constructs) or where STORM antibodies exhibit low specificity. Even with high labeling density, it is difficult to detect continuous structures with PALM or STORM.

The need to repeatedly acquire subsets of data necessitates much longer experimental timescales when compared to ensemble super-resolution imaging methods, or even traditional confocal and widefield microscopy. For this reason single molecule imaging of biological specimens is often conducted on fixed samples. The extended acquisition periods also introduces a sensitivity to stage and sample drift during the course of an experiment. Even simple thermal expansion of a sample during acquisition, which may cause a feature to drift only one pixel across the detector, can introduce obvious and unwanted artifacts. The use of high-quality stage drives and close observance of equilibration practices can minimize sample drift. Residual drift can be corrected by including fiducial markers (such as gold beads) that can be identified by software and used to realign localization coordinates.

Long image acquisition times of 2 to 12 hours are implicit due to the successive nature of switching, excitation, imaging, and bleaching of fluorophores. PALMIRA (PALM with independently running acquisition [45, 46]) significantly improves on this time by continuously recording the spontaneous activation, excitation, and bleaching of fluorophores. A delicate balance is required between the power of the lasers used for switching and excitation, and the rate at which data is recorded from the camera to obtain a sufficient density of discrete, "blinking" molecules. By combining all the imaging modes (switching, excitation, bleaching) of regular PALM with continuous data acquisition, PALMIRA can record 10,000 frames of data within 5 minutes.

4.5.1 *Probe Selection for PALM/STORM*

Probes used in super-resolution microscopy fall into two distinct categories: the synthetic fluorophores, which include cyanine dyes and quantum dots, and fluorescent proteins (FPs). Not all dyes are suitable for every super-resolution imaging technique, as each method dictates its own requirements for spectra, photoconversion, photostability, quantum yield, etc.

Fluorescent proteins can be further classified according to the different ways their fluorescence properties can be changed. Photo*activatable* fluorescent proteins (PA-FPs) change from a non-fluorescent dark state to a bright fluorescent state. Photo*switchable* fluorescent proteins (PS-FPs) change their absorption and emission spectra, usually toward longer-wavelength excitation and emission peaks. To further complicate matters, these changes can be reversible in some instances, although this is usually more common in PA-FPs (which can be flipped between dark and bright states) than in PS-FPs (which are often irreversibly locked into their switched state). Synthetic fluorophores are presently only available as photoactivatable dyes. There are currently no reports of synthetic dyes being able to change spectra in the manner of photoswitchable fluorescent proteins.

A suitable probe must have high molar extinction coefficient (ε_{abs}, a measure of how strongly the probe absorbs excitation light) and also a high quantum yields (φ, the efficiency with which absorption yields a fluorescence photon). The product of these two properties is the "brightness" of the molecule's fluorescence, and brighter molecules have higher contrast against the background signal. Synthetic fluorophores have very high molar extinction coefficients and also efficient quantum yields (e.g., ATTO 532 $\varepsilon_{532\ nm} = 115{,}000\ M^{-1}\ cm^{-1}$, and $\varphi = 0.90$), conferring a significant advantage over the less bright fluorescent proteins (e.g., PS-CFP2, in its switched state, $\varepsilon_{490\ nm} = 47{,}000\ M^{-1}\ cm^{-1}$, and $\varphi = 0.23$).

Photon emission rate is also a primary attribute to consider when selecting a probe for super-resolution single molecule imaging as this will dictate whether a molecule is imaged as brief but very bright spot (and hence highly localizable) or whether the same "photon budget" is spent over a much longer time frame (perhaps with dark triplet-state excursions), imaged as a dimmer molecule with poorer localization precision, especially if the image acquisition rate is running much faster than the on time of the fluorophore. This becomes especially important where high temporal resolution is required—for example, in live cell imaging—and the detector acquisition frame rate is high. Here it is necessary to have a fluorescent protein that will generate the maximum number of photons per molecule per unit time, but balanced against added risk of a faster photobleaching rate. In the case of PALM, a rapid emission of photons followed by photobleaching is desirable as it allows good localization while preventing unwanted resampling of the same molecule.

Photoactivatable fluorescent protein reactivation is possible [47], perhaps through repeated excursions into an excited dark state, as occurs for reversibly photoactivatable proteins such as Dronpa. Unwanted reactivation is comparatively rare compared to the rate of bleaching events and any repeated presentation of a molecule at identical coordinates can be identified and dealt with during post-processing.

Photoswitchable proteins must have well separated spectra between their various personalities. Probes must maintain a low rate of spontaneous conversion in the absence of the directed optical switching signal. For super-resolution methods that rely on repeated cycling, such as STED, it is important for the probe to have a robust switching kinetics to survive multiple switching cycles.

Among the options for PA/PS-FPs are monomers, dimers, and tetramers. Obligate multimers are often the configuration in which the original fluorescent protein is isolated and characterized (for example native avGFP forms a dimer) and much subsequent work is performed to generate a monomer while preserving the original protein's spectral characteristics. Monomeric fluorescent proteins have a distinct stoichiometric advantage in that a monomeric fluorescent protein represents only the protein to which it has been fused. Multimeric fluorescent proteins are noted for often disrupting (sometimes quite severely) the trafficking, localization, and even function of the host protein. Monomeric PA- and PS-FPs have been engineered either by selective mutagenesis to disrupt multimeric interfaces or to confer photoswitchable properties on existing monomeric fluorescent proteins.

Target specificity can be conferred upon small dye molecules by attaching them to a carrier molecule, such as a protein (e.g., antibody IgG), or coupling them to a substrate molecule which is recognized by the target protein or an enzyme genetically fused to the target protein. The potential for nonspecific interactions of these conjugated dye molecules means that they cannot presently match the degree of target specificity that is inherent in using fluorescent proteins which are genetically fused to their host protein, although this problem is greatest for antibody-conjugated dyes.

For immunofluorescence methods, the additional bulk of the IgG molecule introduces localization uncertainty by pushing the fluorescent moiety some distance away from the true position of the target molecule. This effect can be somewhat ameliorated by conjugating to smaller F'ab fragments rather than whole IgG. If the target is intracellular, then cells must be fixed and permeabilized as conjugated antibodies are too large to diffuse across the cell membrane. However, it is not always desirable to introduce exogenous fused proteins into cells and immuno-staining with antibody conjugated probes is the best way to access endogenous proteins. For delicate cells, of limited number, or poor transfectability (e.g., primary cells) immuno-staining may also present a more feasible and reproducible approach than fused fluorescent proteins.

Advances in protein-tagging technologies have introduced new possibilities to labeling proteins in cells, promising to unify the high labeling specificity and efficiency of genetically encoded tags with the excellent optical properties of synthetic dyes. Cells expressing the SNAP-tag, CLIP-tag, or HaloTag sequences recognize a diverse range of tag-specific substrates coupled to small synthetic dyes. The biarsenical compounds FlAsH, ReAsH, and CoAsH recognize a specific 6-amino-acid tetracysteine tag motif each, although this system currently suffers from high background signal generated by unbound probes. A 16-amino-acid tag confers specificity to biotinylation by the bacterial enzyme biotin ligase [48] (BirA). Fluorophores conjugated to streptavidin can then be applied. Another option uses fluorophore-conjugated NTA probes recognizing His6-tagged proteins.

4.5.2 *Dual-Color PALM/STORM*

Multicolor imaging requires fluorophores with non-overlapping emission spectra. If using photoswitchable probes, the native (nonswitched) emission spectra must also be sufficiently separated from the other probes so as not to introduce background signal. In this situation, photoactivatable fluorescent proteins are especially useful as they emit no fluorescence in their non-activated state (although nonfluorescent samples can be difficult to locate when preparing for acquisition!). Ideally, the wavelengths of the switching signal should also be distinct, however, this is rarely possible

Table 4.1 Photoswitchable proteins (color changing)

Name	Native fluorescence				Switching wavelength (nm)	Converted fluorescence				
	Ex (nm)	Em (nm)	EC ($\times 10^{-3}$)	QY		Ex (nm)	Em (nm)	EC ($\times 10^{-3}$)	QY	Photons
PS-CFP2 [49]	400	468	43.0	0.2	405	490	511	47.0	0.23	260
Dendra2 [50]	490	507	45.0	0.5	405–488	553	573	35.0	0.55	ND
EosFP [51]	506	516	72.0	0.70	405	571	581	41.0	0.55	ND
mEos2 [52]	506	519	56.0	0.74	405	573	584	46.0	0.66	500
tdEos [53]	506	516	34.0	0.66	405	569	581	33.0	0.60	750
mKikGR [54]	505	515	49.0	0.69	405	580	591	28.0	0.63	970

Table 4.2 Photoactivatable proteins (dark to bright)

Name	Native fluorescence					Activating wavelength (nm)	Converted fluorescence				
	Ex (nm)	Em (nm)	EC ($\times 10^{-3}$)	QY	Photons		Ex (nm)	Em (nm)	EC ($\times 10^{-3}$)	QY	Photons
PA-GFP [55]	400	515	20.7	0.13	70	405	504	517	17.4	0.79	300
PA-mCherry [56]	NA	NA	NA	NA	NA	405	564	595	18.0	0.46	ND

Table 4.3 Reversible photoactivatable proteins (bright to dark, with recovery)

Name	Ex (nm)	Em (nm)	EC ($\times 10^{-3}$)	QY	Dark-state switching wavelength (nm)	Bright-state recovery wavelength (nm)
Dronpa [57]	503	517	95.0	0.85	503	405
Dronpa2 [58]	486	513	ND	ND	486	405
Dronpa-3 [58]	487	514	58.0	0.33	487	405
Padron [59]	503	522	43.0	0.64	405	488
bsDronpa [59]	460	504	45.0	0.5	488	405
rsFastLime [60]	496	518	39.1	0.77	488	405
rsCherry [61]	572	610	80.0	0.02	450	550
rsCherryRev [61]	572	608	84.0	0.005	550	450

Table 4.4 Synthetic dyes for STORM and dSTORM

Name	Ex (nm)	Em (nm)	EC ($\times 10^{-3}$)	QY	Photons
Cy5	649	664	250.0	0.28	6000
Cy5.5	675	694	190.0	0.23	6000
Cy7	743	767	250.0	ND	1000
Alexa Fluor 647	649	666	240.0	0.33	6000
DyLight 649	652	670	250.0	ND	ND

as most probes are activated or converted by short-wavelength violet/ultraviolet light.

These restrictions significantly limit the options for performing two-color PALM or STORM imaging, and strategic selection of probes is necessary. For example, even though the green emission spectra of Dronpa overlaps with the non-switched green emission spectra of mEos2, Dronpa absorbs light at 488 nm and emits at 503 nm (much like GFP) but rapidly "bleaches" into a dark state. Absorption of UV light enables these dark-state Dronpa molecules to recover their fluorescence. UV light switches mEos2 into a red-fluorescence state. For dual-color imaging both molecules are exposed to UV, but only mEos2 is imaged in the red channel. Then the sample is exposed to intense 488 nm light, driving Dronpa into its dark state, while also bleaching any residual mEos2 molecules. A second application of UV stochastically recovers Dronpa molecules from the dark state, whereupon they can be imaged in the green channel. An alternative strategy might involve two photoactivatable probes which are switched by the same UV light, then sequentially imaged and bleached at their respective emission wavelengths.

In the case of STORM, dual-color imaging is achieved through careful selection of spectrally separated activator molecules paired to the existing palette of carbocyanine dyes. Sequential dual-color PALM/STORM imaging has proven useful in tracking total protein distribution by PALM alongside STORM using immunostaining with antibodies that specifically recognize activated proteins (e.g., phosphorylated residues or an activated conformation) within the same sample.

4.5.3 *Live-Cell PALM*

Imaging biological processes in living cells often demands a high temporal resolution, a feature not implicit in PALM's prescription of repeated photoactivation—imaging—bleaching cycles. PALM acquisition rate can be accelerated by increasing the power of the activation and fluorescence lasers, which hastens the respective rates of photoswitching and bleaching. However, the additional energy burden can generate toxic oxygen radicals and live specimens must be closely observed for morphological changes, such as membrane blebbing, and other indicators of phototoxicity. PALMIRA, with its asynchronous imaging approach, offers the best solution for live-cell imaging and here the rate of acquisition is generally dictated by the sensitivity and readout times of the EMCCD camera.

Furthermore, temporal resolution in live-cell imaging is degraded by the need to compile enough localized molecules from multiple PALM data frames into a time-point "image frame" to accurately discern features of the target protein's structure or distribution. There is an inherent tradeoff between temporal resolution and obtaining enough molecules to generate a super-resolution image that is representative of the specimen. However, live-cell PALM is still a nascent field and many avenues remain open to improving probes and acquisition schemes to improve temporal resolution.

Immunofluorescence-based techniques such as STORM and dSTORM are generally restricted to examining fixed specimens due to necessary preparative techniques such as cell permeabilization or harsh sample buffers. The advent of direct labeling of proteins by coupling dyes to either endogenous substrate for the target protein or via SNAP-tag substrates has permitted live cell imaging with STORM [62]. In dSTORM, the thiol additives and oxygen scavengers necessary to induce blinking of small synthetic dyes are also toxic to living cells. However, at least one study has demonstrated dSTORM imaging in living cells [63] (also with SNAP-tag fused proteins) by using endogenous glutathione to render the substrate-conjugated dyes photoswitchable.

4.5.4 *3D PALM/STORM*

With the advent and maturation of PALM and STORM as two-dimensional imaging techniques (as a consequence of using a TIRF excitation geometry to minimize background), considerable effort has been invested in recent years to extend single molecule localization into the axial dimension, yielding some unique approaches with promising results.

Biplane FPALM [64], from Sam Hess's laboratory, simultaneously images two focal planes and compares how a single molecule appears in each plane to deduce its axial location. With only a small modification to the standard PALM beam path, the two focal planes can be spatially separated and imaged on the same CCD detector. Molecules that are in focus in one plane will be defocused and blurry in the other plane. A point lying directly between the two planes will appear blurred to an equivalent degree in both of the imaged planes. By analyzing spots in both planes simultaneously, and with knowledge of the point spread function, a fitting algorithm can determine the axial position of the emitting molecule relative to the two planes. In a proof-of-principle study, biplane FPALM can resolve fluorescent proteins to 75 nm in the axial direction, 30 nm radially, with a depth of field approaching 1 μm [64].

3D STORM has been demonstrated using a similar defocusing technique which introduces optical astigmatism [65]. A weak cylindrical lens is inserted into the beam path to give slightly different focal planes in the *x* and *y* dimensions. The shape of a molecule's PSF therefore changes with its axial position—a molecule located at the focal plane presents a normal circular-shaped PSF to the detector. A molecule below this focal plane becomes sharper in the *x* direction but blurred in the *y* direction, generating an "upright" elliptical-shaped PSF. A molecule above the focal plane presents a horizontally oriented elliptical PSF, as the molecule is now more in focus in the *y* direction but blurred in the *x* direction. Fitting of a 2D elliptical Gaussian function allows the relative axial position of the molecule to be determined, with resolution to 50 nm in a field approximately 600 nm deep.

Yet another technique to access the third dimension is interference PALM (iPALM) [66]. In an arrangement similar to 4pi, opposing objectives are used to gather fluorescent light from a specimen from two sides. Fluorescent emission from a molecule can then be recovered along two paths. The signal from one objective is then able to serve as the reference for the other and the signals are recombined in a three-way beam-splitter where they undergo self-interference. The difference in path length will increase for molecules located further from the focal plane. Information about the depth of a fluorescent molecule can be determined from the phase difference in the recombined waves. The axial resolution of iPALM has been measured to 15 nm with fluorescent proteins and is the only technique to demonstrate finer axial resolution than lateral resolution. However, iPALM is best suited to thin samples as interferometry is less tolerant of the optical distortions introduced by thicker specimens.

A different approach again involves "temporal focusing" to generate a stack of several super-resolution images from samples up to 10 μm thick [67]. Femtosecond laser pulses are first broadened through a diffraction grating and then focused onto the sample using a telescope. This results in a dispersion of the two-photon excitation light everywhere except at the focal plane, where there remains a narrow plane of intense excitation light. Temporal focusing enables selective activation of molecules within a thin plane, approximately 1.6 μm thick. A super-resolution image can be acquired from the molecules within this illumination plane, effectively collapsing the illumination volume into a single 2D image. Optical sectioning can then be performed as a series of super-resolution images acquired at different axial depths, and used to build a 3D representation of the sample. When deployed in a PALM configuration two-photon temporal focusing can be applied to either the photoswitching laser or the fluorescence excitation laser, or both.

3D PALM has also been demonstrated in a technique where a fluorescent protein's PSF is modified to include two prominent lobes that rotate with the molecule's axial position, creating a double-helix shape as the PSF propagates [68]. Additional lenses and a liquid crystal spatial light modulator are employed to generate the double-helix PSF (DH-PSF) each image point. Lateral position is determined from the midpoint of the lobes and axial position is inferred from the angle of the lobes, derived from a depth-calibration plot. This technique suffers from a significant degree of signal loss (>75%) during the DH-PSF convolution and is still undergoing refinement, but nevertheless has demonstrated 20 nm axial resolution for synthetic dyes over a range of 2 μm.

4.6 Conclusions and Future Directions

This chapter describes some of the unique approaches taken to defeating the diffraction barrier and producing high-resolution images of biological specimens. Many biological structures and processes operate within nanoscale dimensions, and super-resolution fluorescence microscopy is already beginning to yield surprising and splendid results as the technology progresses from development to application.

Driving the field forward are the emergences of novel labeling methods, new fluorescent probes, and refinements to microscopy hardware. Such advances deliver more accurate representation and permit better interpretation of the underlying physiology of a specimen. Faster hardware is especially relevant for live-cell applications, where the goal is to obtain usable data (which often comes down to being able to collect a sufficient number of photons) as quickly as possible.

It is unlikely that there will ever be an ideal, unified imaging system that delivers the supreme spatial and temporal resolution required to access multiple targets, simultaneously in living cells. However, with developments in biological fluorescence nanoscopy evolving at a rapid pace, recent investigations have demonstrated imaging of biological samples with high resolution in four dimensions [62, 69, 70]. An expanding catalogue of commercially available super-resolution microscopes will ensure that biologists will soon have easy access to this technology and confidence of even greater resolution improvement in the future.

References

1. Axelrod, D. (1981). Cell-substrate contacts illuminated by total internal reflection fluorescence. *J. Cell Biol.* **89**:141.
2. Axelrod, D. (2003). Total internal reflection fluorescence microscopy in cell biology. *Methods Enzymol.* **361**:1–33.
3. Agard, D. A. (1984). Optical sectioning microscopy: cellular architecture in three dimensions. *Annu. Rev. Biophys. Bioeng.* **13**:191–219.
4. Biggs, D. S. C. (2010). 3D deconvolution microscopy. *Curr. Protoc. Cytom.* **Chapter 12**, Unit 12.19.1–20.
5. Vicidomini, G., Mondal, P. P., and Diaspro, A. (2006). Fuzzy logic and maximum a posteriori-based image restoration for confocal microscopy. *Opt. Lett.* **31**:3582.
6. Mondal, P. P., Vicidomini, G., and Diaspro, A. (2007). Markov random field aided Bayesian approach for image reconstruction in confocal microscopy. *J. Appl. Phys.* **102**:044701.
7. Hell, S., and Stelzer, E. H. K. (1992). Properties of a 4Pi confocal fluorescence microscope. *J. Opt. Soc. Am. A* **9**:2159.
8. Schrader, M., Hell, S. W., and van der Voort, H. T. M. (1998). Three-dimensional super-resolution with a 4Pi-confocal microscope using image restoration. *J. Appl. Phys.* **84**:4033.
9. Hell, S., Lindek, S., and Stelzer, E. (1994). Enhancing the Axial Resolution in Far-field Light Microscopy: Two-photon 4Pi Confocal Fluorescence Microscopy. *J. Mod. Opt.* **41**:675–681.
10. Gustafsson, M. G. L. (1995). *Sevenfold improvement of axial resolution in 3D wide-field microscopy using two objective lenses. Proc. SPIE* **2412**:147–156.

11. Gustafsson, M. G. L., Agard, D. A., and Sedat, J. W. (1999). I5M: 3D widefield light microscopy with better than 100 nm axial resolution. *J. Micros.* **195**:10–16.
12. Hell, S. W., and Wichmann, J. (1994). Breaking the diffraction resolution limit by stimulated emission: stimulated-emission-depletion fluorescence microscopy. *Opt. Lett.* **19**:780.
13. Harke, B., Keller, J., Ullal, C. K., et al. (2008). Resolution scaling in STED microscopy. *Opt. Express* **16**:4154.
14. Klar, T. A., and Hell, S. W. (1999). Subdiffraction resolution in far-field fluorescence microscopy. *Opt. Lett.* **24**:954.
15. Xia, Y., and Yin, J. (2005). Generation of a focused hollow beam by an 2pi-phase plate and its application in atom or molecule optics. *J. Opt. Soc. Am. B* **22**:529.
16. Hotta, J.-I., Fron, E., Dedecker, P., et al. (2010). Spectroscopic rationale for efficient stimulated-emission depletion microscopy fluorophores. *J. Am. Chem. Soc.* **132**:5021–5023.
17. Meyer, L., Wildanger, D., Medda, R., et al. (2008). Dual-color STED microscopy at 30-nm focal-plane resolution. *Small* **4**:1095–100.
18. Willig, K. I., Rizzoli, S. O., Westphal, V., Jahn, R., and Hell, S. W. (2006). STED microscopy reveals that synaptotagmin remains clustered after synaptic vesicle exocytosis. *Nature* **440**:935–939.
19. Fitzpatrick, J. A. J., Yan, Q., Sieber, J. J., et al. (2009). STED nanoscopy in living cells using Fluorogen Activating Proteins. *Bioconjug. Chem.* **20**:1843–1847.
20. Donnert, G., Keller, J., Medda, R., et al. (2006). Macromolecular-scale resolution in biological fluorescence microscopy. *Proc. Natl. Acad. Sci. USA* **103**:11440–11445.
21. Willig, K. I., Kellner, R. R., Medda, R., et al. (2006). Nanoscale resolution in GFP-based microscopy. *Nat. Methods* **3**:721–723.
22. Hein, B., Willig, K. I., and Hell, S. W. (2008). Stimulated emission depletion (STED) nanoscopy of a fluorescent protein-labeled organelle inside a living cell. *Proc. Natl. Acad. Sci. USA* **105**:14271–14276.
23. Nägerla, U. V., Willigb, K. I., Heinb, B., Hell, S. W., and Bonhoeffer, T. (2008). Live-cell imaging of dendritic spines by STED microscopy. *Proc. Natl. Acad. Sci. USA* **105**:18982–18987.
24. Rankin, B. R., Kellner, R. R., and Hell, S. W. (2008). Stimulated-emission-depletion microscopy with a multicolor stimulated-Raman-scattering light source. *Opt. Lett.* **33:**2491–2493.

25. Wildanger, D., Rittweger, E., Kastrup, L., and Hell, S. W. (2008). STED microscopy with a supercontinuum laser source. *Opt. Express* **16**:9614.
26. Wildanger, D., Medda, R., Kastrup, L., and Hell, S. W. (2009). A compact STED microscope providing 3D nanoscale resolution. *J. Microsc.* **236**:35–43.
27. Moneron, G., Medda, R., Hein, B., et al. (2010). Fast STED microscopy with continuous wave fiber lasers. *Opt. Express* **18**:1302.
28. Dyba, M., and Hell, S. W. (2002). Focal spots of size lambda/23 open up far-field fluorescence microscopy at 33 nm axial resolution. *Phys. Rev. Lett.* **88**:163901.
29. Willig, K. I., Harke, B., Medda, R., and Hell, S. W. (2007). STED microscopy with continuous wave beams. *Nat. Methods* **4**:915–918.
30. Hell, S. W., and Kroug, M. (1995). Ground-state-depletion fluorscence microscopy: A concept for breaking the diffraction resolution limit. *Appl. Phys. B* **60**:495–497.
31. Bretschneider, S., Eggeling, C., and Hell, S. W. (2007). Breaking the diffraction barrier in fluorescence microscopy by optical shelving. *Phys. Rev. Lett.* **98**:218103.
32. Aitken, C. E., Marshall, R. A., and Puglisi, J. D. (2008). An oxygen scavenging system for improvement of dye stability in single-molecule fluorescence experiments. *Biophys. J.* **94**:1826–1835.
33. Gustafsson, M. G. L. (2000). Surpassing the lateral resolution limit by a factor of two using structured illumination microscopy. *J. Microsc.* **198**:82–87.
34. Gustafsson, M. G. L. (2005). Nonlinear structured-illumination microscopy: wide-field fluorescence imaging with theoretically unlimited resolution. *Proc. Natl Acad. Sci. USA* **102**:13081–13086.
35. Heintzmann, R., Jovin, T. M., and Cremer, C. (2002). Saturated patterned excitation microscopy—a concept for optical resolution improvement. *J. Opt. Soc. Am. A* **19**:1599.
36. Shao, L., Isaac, B., Uzawa, S., et al. (2008). I5S: wide-field light microscopy with 100-nm-scale resolution in three dimensions. *Biophys. J.* **94**:4971–4983.
37. Betzig, E., Patterson, G. H., Sougrat, R., et al. (2006). Imaging intracellular fluorescent proteins at nanometer resolution. *Science* **313**:1642.
38. Hess, S. T., Girirajan, T. P. K., and Mason, M. D. (2006). Ultra-high resolution imaging by fluorescence photoactivation localization microscopy. *Biophys. J.* **91**:4258–4272.

39. Rust, M. J., Bates, M., and Zhuang, X. (2006). Sub-diffraction-limit imaging by stochastic optical reconstruction microscopy (STORM). *Nat. Methods* **3**:793–795.

40. Bates, M., Blosser, T., and Zhuang, X. (2005). Short-Range Spectroscopic Ruler Based on a Single-Molecule Optical Switch. *Phys. Rev. Lett.* **94**:108101.

41. Heilemann, M., van de Linde, S., Schüttpelz, M., et al. (2008). Subdiffraction-resolution fluorescence imaging with conventional fluorescent probes. *Angew. Chem. Int. Ed. Engl.* **47**:6172–6176.

42. Fölling, J., Bossi, M., Bock, H., et al. (2008). Fluorescence nanoscopy by ground-state depletion and single-molecule return. *Nat. Methods* **5**:943–945.

43. Testa, I., Wurm, C. A., Medda, R., et al. (2010). Multicolor fluorescence nanoscopy in fixed and living cells by exciting conventional fluorophores with a single wavelength. *Biophys. J.* **99**:2686–2694.

44. Thompson, R. E., Larson, D. R. and Webb, W. W. (2002). Precise Nanometer Localization Analysis for Individual Fluorescent Probes. *Biophys. J.* **82**:2775–2783.

45. Egner, A., Geisler, C., von Middendorff, C., et al. (2007). Fluorescence nanoscopy in whole cells by asynchronous localization of photoswitching emitters. *Biophys. J.* **93**:3285–3290.

46. Geisler, C., Schönle, A., von Middendorff, C., et al. (2007). Resolution of λ /10 in fluorescence microscopy using fast single molecule photoswitching. *Appl. Phys. A* **88**:223–226.

47. Annibale, P., Scarselli, M., Kodiyan, A., and Radenovic, A. (2010). Photoactivatable Fluorescent Protein mEos2 Displays Repeated Photoactivation after a Long-Lived Dark State in the Red Photoconverted Form. *J. Phys. Chem. Lett.* **1**:1506–1510.

48. Chen, I., Howarth, M., Lin, W., and Ting, A. Y. (2005). Site-specific labeling of cell surface proteins with biophysical probes using biotin ligase. *Nat. Methods* **2**:99–104.

49. Chudakov, D. M., Verkhusha, V. V., Staroverov, D. B., et al. (2004). Photoswitchable cyan fluorescent protein for protein tracking. *Nat. Biotechnol.* **22**:1435–1439.

50. Gurskaya, N. G., Verkhusha, V. V., Shcheglov, A. S., et al. (2006). Engineering of a monomeric green-to-red photoactivatable fluorescent protein induced by blue light. *Nat. Biotechnol.* **24**:461–465.

51. Wiedenmann, J., Ivanchenko, S., Oswald, F., et al. (2004). EosFP, a fluorescent marker protein with UV-inducible green-to-red fluorescence conversion. *Proc. Natl Acad. Sci. USA* **101**:15905–15910.

52. McKinney, S. A., Murphy, C. S., Hazelwood, K. L., Davidson, M. W., and Looger, L. L. (2009). A bright and photostable photoconvertible fluorescent protein. *Nat. Methods* **6**:131–133.

53. Nienhaus, G. U., Nienhaus, K., Hölzle, A., et al. (2006). Photoconvertible fluorescent protein EosFP: biophysical properties and cell biology applications. *Photochem. Photobiol.* **82**:351–358.

54. Habuchi, S., Tsutsui, H., Kochaniak, A. B., Miyawaki, A., and van Oijen, A. M. (2008). mKikGR, a monomeric photoswitchable fluorescent protein. *PLoS ONE* **3**:e3944.

55. Patterson, G. H., and Lippincott-Schwartz, J. (2002). A photoactivatable GFP for selective photolabeling of proteins and cells. *Science* **297**:1873.

56. Subach, F. V., Patterson, G. H., Manley, S., et al. (2009). Photoactivatable mCherry for high-resolution two-color fluorescence microscopy. *Nat. Methods* **6**:153–159.

57. Ando, R., Mizuno, H., and Miyawaki, A. (2004). Regulated fast nucleocytoplasmic shuttling observed by reversible protein highlighting. *Science* **306**:1370.

58. Flors, C., Hotta, J., Uji-i, H., et al. (2007). A stroboscopic approach for fast photoactivation-localization microscopy with Dronpa mutants. *J. Am. Chem. Soc.* **129**:13970–13977.

59. Andresen, M., Stiel, A. C., Fölling, J., et al. (2008). Photoswitchable fluorescent proteins enable monochromatic multilabel imaging and dual color fluorescence nanoscopy. *Nat. Biotechnol.* **26**:1035–1040.

60. Stiel, A. C., Trowitzsch, S., Weber, G., et al. (2007). 1.8 A bright-state structure of the reversibly switchable fluorescent protein Dronpa guides the generation of fast switching variants. *Biochem. J.* **402**:35–42.

61. Stiel, A. C., Andresen, M., Bock, H., et al. (2008). Generation of monomeric reversibly switchable red fluorescent proteins for far-field fluorescence nanoscopy. *Biophys. J.* **95**:2989–2997.

62. Jones, S. A., Shim, S.-H., He, J., and Zhuang, X. (2011). Fast, three-dimensional super-resolution imaging of live cells. *Nat. Methods* **8**:499–505.

63. Klein, T., Löschberger, A., Proppert, S., et al. (2011). Live-cell dSTORM with SNAP-tag fusion proteins. *Nat. Methods* **8**:7–9.

64. Juette, M. F., Gould, T. J., Lessard, M. D., et al. (2008). Three-dimensional sub-100 nm resolution fluorescence microscopy of thick samples. *Nat. Methods* **5**:527–529.

65. Huang, B., Wang, W., Bates, M., and Zhuang, X. (2008). Three-dimensional super-resolution imaging by stochastic optical reconstruction microscopy. *Science* **319**:810–813.
66. Shtengel, G., Galbraith, J. A., Galbraith, C. G., et al. (2009). Interferometric fluorescent super-resolution microscopy resolves 3D cellular ultrastructure. *Proc. Natl. Acad. Sci. USA* **106**:3125–3130.
67. Vaziri, A., Tang, J., Shroff, H., and Shank, C. V. (2008). Multilayer three-dimensional super resolution imaging of thick biological samples. *Proc. Natl. Acad. Sci. USA* **105**:20221–20226.
68. Pavani, S. R. P., Thompson, M. A., Biteen, J. S., et al. (2009). Three-dimensional, single-molecule fluorescence imaging beyond the diffraction limit by using a double-helix point spread function. *Proc. Natl. Acad. Sci. USA* **106**:2995–2999.
69. Baddeley, D., Crossman, D., Rossberger, S., et al. (2011). 4D super-resolution microscopy with conventional fluorophores and single wavelength excitation in optically thick cells and tissues. *PLoS ONE* **6**:e20645.
70. Williamson, D. J., Owen, D. M., Rossy, J., et al. (2011). Pre-existing clusters of the adaptor Lat do not participate in early T cell signaling events. *Nat. Immunol.* **12**:665–662.

Chapter 5

Raman Spectroscopy of Single Cells

Thomas Huser[a,b,d] and James Chan[a,c]

[a]*NSF Center for Biophotonics Science and Technology,*
[b]*Department of Internal Medicine,*
[c]*Department of Pathology and Laboratory Medicine,University of California, Davis, 2700 Stockton Blvd., Suite 1400, Sacramento, CA 95817, USA*
[d]*Department of Physics, University of Bielefeld, Universitätsstrasse 25, 33501 Bielefeld, Germany*
thomas.huser@physik.uni-bielefeld.de

Raman spectroscopy is a powerful technique to dynamically characterize and image living biological cells without the need for fluorescent staining. In this chapter we explain in simple terms the fundamental physics necessary to understand the phenomenon of inelastic light scattering by molecular bond vibrations. We provide the reader with the basic background to enable him or her to delve into Raman spectroscopy of single cells by providing an overview of the strengths and weaknesses of the different optical spectroscopy systems that have been developed over the last decade. Different strategies utilizing confocal detection optics, multi-spot and line illumination, extending all the way to wide-field and even light sheet illumination have been used to further improve the speed and sensitivity to analyze single cells by Raman spectroscopy. To analyze and visualize the large data sets obtained during such experiments, powerful multivariate statistical analysis tools are

Understanding Biophotonics: Fundamentals, Advances, and Applications
Edited by Kevin K. Tsia

ISBN 978-981-4411-77-6 (Hardcover), 978-981-4411-78-3 (eBook)
www.panstanford.com

required to reduce the data and extract key parameters. We discuss the state of the art in single-cell analysis by Raman spectroscopy to bring the reader new to this subject up to speed and conclude the chapter with a brief outlook into the future of this rapidly evolving and expanding research area.

5.1 Introduction

Over the last few decades, Raman spectroscopy has emerged as a powerful alternative to fluorescent probes in biomedical research. Much of this can be attributed to its ability to provide chemical information about a sample in the absence of exogenous fluorophores. This analytical power of Raman spectroscopy is a result of the inelastic scattering of photons by molecular bonds within a sample. Molecular bonds can be excited to undergo vibrations when a photon with an energy that is higher than the vibrational energy of the bond is scattered by the bond. Similarly, photons can also de-excite bonds that are already undergoing vibrations. Raman spectroscopy analyzes the loss or gain in energy of the scattered photons to permit the identification of specific molecular groups of which a molecule is comprised. Because of its non-invasive analytical power, Raman spectroscopy allows for live cell chemical analysis and imaging without destroying or altering the biological material under investigation. In particular, this technique works really well at the cellular and subcellular level, because it can be easily integrated with optical microscopy. In this chapter we will provide a brief introduction to Raman spectroscopy, explain the principles behind Raman spectroscopy, and discuss applications and recent advancements in the chemical analysis of living cells.

The Raman effect was initially discovered and described in 1928 by C. V. Raman and K. S. Krishnan [1], who used sunlight, narrow spectral filters, and their bare eyes to characterize the results of inelastic light scattering. C. V. Raman was subsequently awarded the Nobel Price in Physics in 1930 "for his work on the scattering of light and for the discovery of the effect named after him." In the following years mercury arc lamps became the main source for Raman spectroscopy. The effect of inelastic light scattering is, however, rather small—especially when compared to fluorescence spectroscopy. It was not until the mid to late 1960s, after the invention of the laser as an intense, coherent source of light, that Raman spectroscopy underwent a renaissance and began to be adopted widely by physicists, chemists, and biologists. Since then, this area of research has undergone continuous development. During the last two decades, many of the tools utilized for Raman spectroscopy have been refined leading to new, inexpensive steep-edge filters and more sensitive detectors. These technological advancements have enabled the development of compact (even hand-held) commercial devices, which are now used by law enforcement, as well as in the pharmaceutical industry for process control. In its combination with microscopy, Raman spectroscopy permits the nondestructive chemical analysis of volumes as small as 1 femtoliter, or approximately one bacterial cell, which, of course, made it a highly interesting tool for the life sciences.

The main interest in Raman spectroscopy by life scientists, however, is based on the fact that Raman spectroscopy enables the dynamic chemical analysis of living cells with subcellular spatial resolution. In essence, no other technique can claim a similar combination of high sensitivity, high spatial resolution, and the ability for nondestructive chemical analysis. Standard techniques for cellular imaging and analysis, such as immunoassays, blotting, chromatography, and optical and fluorescence microscopy, provide high sensitivity but suffer from a lack of molecular specificity, perturb the sample by the introduction of exogenous fluorescent probes and antibodies, or require destructive processes such as cell fixation or lysis. Other molecular analysis techniques, such as mass spectrometry, in particular matrix-assisted laser desorption/ionization time-of-flight mass spectrometry (MALDI-TOF), can achieve similar spatial resolution but typically require the use of special substrates and destroy cells in the process of the analysis. Mass spectrometry is, however, an interesting technique that can be used in combination with Raman spectroscopy to obtain elemental maps of the sample, since Raman spectroscopy provides complementary information about molecular bonds. Secondary ion mass spectrometry, especially when used with tightly focused gallium ion beams to achieve a resolution on the nanometer scale (termed "NanoSIMS"), can obtain even higher spatial resolution and sensitivity, but is also not compatible with the analysis of living cells since it destroys them in the process. Nuclear magnetic resonance (NMR) spectroscopy and its imaging analog magnetic resonance imaging (MRI) cannot compete with the superior spatial resolution and sensitivity provided by Raman spectroscopy. Infrared (IR) absorption spectroscopy provides similar information, but it cannot achieve subcellular spatial resolution and is generally difficult to perform on living cells because IR wavelengths are highly absorbed by water. Synchrotron sources are typically required to achieve sufficient brightness and signal strength in order to perform cellular-level analyses. As mentioned before, the only major caveat of Raman spectroscopy is that it is a process with a fairly low efficiency requiring high sample concentrations (albeit minute sample volumes). This issue is currently being addressed by a number of research groups around the world, leading to novel concepts such as the parallelized detection of spectral peaks from the same sources, or coherent Raman spectroscopy. The use of Raman scattering is destined to continue its rapid growth in the biochemical sciences, and we believe that it will become a major research tool in the life sciences in the foreseeable future.

5.2 The Principles of Raman Scattering: Inelastic Light Scattering

Raman spectroscopy is the detection and dispersion of photons that are inelastically scattered by molecular vibrations. When a monochromatic laser source with frequency ω_i (i.e., photon energy E_i) is used to probe a sample, most of the interacting photons encounter elastic scattering where the frequency of the photons remains unchanged after the scattering event. This process is called Rayleigh scattering. A small fraction of photons, however, that scatters inelastically off molecular bonds exchanges energy with the vibrational energy levels of the molecules in the sample. These photons are said to be *Raman* scattered (see the schematic depiction in Fig. 5.1). These scattered photons, with a new frequency ω_s, will show up as discrete peaks in a Raman spectrum, where each peak corresponds to a distinct type of molecular vibration. Typically these peaks are represented in terms of a wavelength-independent relative energy unit called "wavenumber (cm^{-1})," which reflects the amount of energy a photon has exchanged with a molecular vibration during the scattering event. Since molecular vibrations are strongly dependent on the molecular conformation and environment, spectral analysis of the scattered photons can be used not only to identify a molecular bond but also to assess the chemical micro-environment in which it is found. For example, the schematics in Fig. 5.1a shows a C=C double bond that can only undergo a stretching vibration along the axis connecting both carbon atoms. If this vibration is momentarily in the ground state, the incident photon (shown in green) can excite the bond to the first excited vibrational state, i.e., the stretching vibration with a fixed frequency. The scattered photon will then have transferred parts of its energy to this vibration and has a lower energy than the incident photon, which is represented by its red color. Molecular bond vibrations are quantized, so only certain well-defined amounts of energy can be exchanged between photons and molecular bonds, leading to a Raman spectrum with discrete peaks, each of which can be assigned to a specific vibrational mode.

At this point it is useful to remind the reader that the energy of a photon depends on the frequency (wavelength) of the electromagnetic wave packet that it represents and it can be written as $E = h\nu = \hbar\omega$. For consistency, we will use either the energy E or frequency ω to describe photons for the remainder of this text. The expression ν will be reserved for the representation of different vibrational energy levels of the molecules (starting from Fig. 5.2 on).

Let us briefly consider another analogy to the process of Raman scattering. In very simplistic terms, a specific molecular bond can be represented in terms of a harmonic oscillator, such as the one shown in Fig. 5.1b. Here, the two atoms that form the bond have specific masses M_1 (e.g., a carbon atom) and M_2 (e.g., a hydrogen atom). The bond itself is represented by a spring with a spring constant k. Since the energy of a stretching vibration along the axis of the spring is proportional to its frequency, i.e., the inverse of the wavelength, in spectroscopy it is common to use a simplified notation and describe molecular vibrations in terms of an inverse wavelength with the

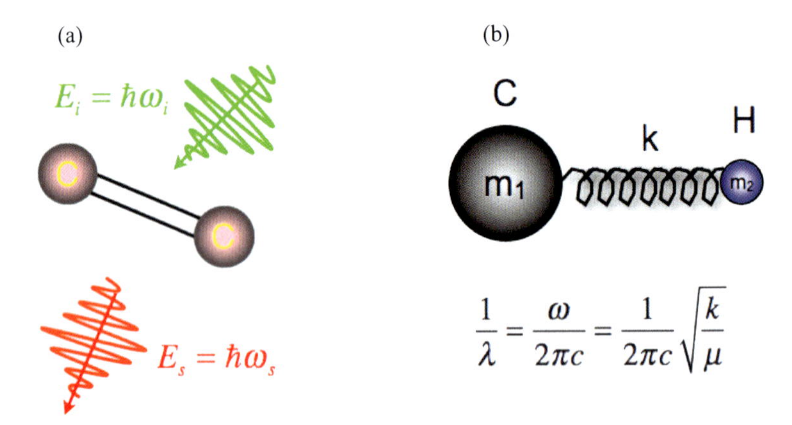

Figure 5.1 Schematics of Raman scattering. (a) An incoming photon with an energy higher than the energy needed to excite a molecular bond vibration is scattered by a C=C bond. The outgoing photon has lost the exact amount of energy that now excites the bond vibration. (b) Depiction of a molecular bond as spheres connected by a spring.

unit "inverse centimeter" (cm^{-1}). With this convention the energy of the stretching vibration can be calculated quite easily by the reduced mass μ of the two atoms forming the bond, and the spring constant between them. This simple mechanical model is usually sufficient for the description of diatomic molecules. It is, however, no longer valid for complex molecules and we have to resort to a more complicated description, which, for large molecules, can only be solved numerically. A number of excellent software packages exist that can calculate the normal-mode vibrations of complex molecules within reasonable limits (typically up to the size of amino acids).

In inelastic light scattering there are two possible modes of interaction between photons and molecular vibrations. If a molecule is in the vibrational ground state ν_0 during its interaction with the photon, it can pick up energy from the photon and is then excited to the vibrational state ν_1 after the scattering event. In this case the photon has transferred part of its energy to the molecular vibration, leading to a shift to longer wavelengths for the scattered photon. This event is called Stokes-shifted Raman scattering (see Fig. 5.2a). Alternatively, the molecule could also be in the excited vibrational state ν_1 just before the interaction. Now, the incident photon can

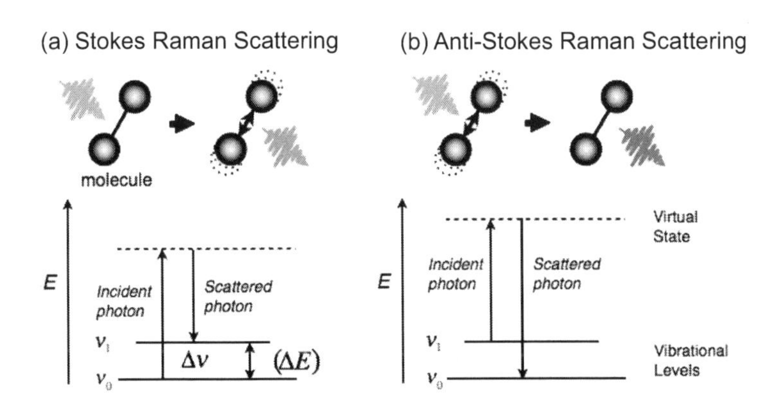

Figure 5.2 Stokes and anti-Stokes Raman scattering depicted as scattering by a molecular bond that is (a) momentarily at rest or (b) undergoing a stretching vibration.

pick up energy from the vibration, returning the molecule to the vibrational ground state, and the energy of the scattered photon is then shifted to shorter wavelengths (termed anti-Stokes Raman scattering). This is depicted in Fig. 5.2b. During the scattering event the molecular vibration and the photon are indistinguishable because of Heisenberg's uncertainty principle. This short-lived state is what we call a "virtual state," and it is depicted by the combined energy levels of the photon and the molecule in Fig. 5.2.

In order to better understand the inelastic scattering process it is important to revisit how light interacts with molecules. Since simple molecules are much smaller than the wavelength of light at optical frequencies a photon can be approximated as a uniform oscillating electric field. This oscillating electric field can affect the size and the shape of a molecule. The result of this interaction is often described on two different scales: the microscopic scale, which describes the response of a single molecule to the interaction, and the macroscopic scale, which describes the average response of bulk material. On the microscopic scale the vector of the induced electric dipole moment, **p**, is a measure of how much the electric field distorts the electron distribution within a molecule. At low field strengths **p** is simply proportional to the electric field, **E**:

$$\vec{p} = \alpha \vec{E} \tag{5.1}$$

Here, **p** is linearly dependent on **E** and is related to **E** by a second-rank tensor α, the polarizability. The elements of the polarizability tensor correspond to the amount a molecule is deformed along the x, y, and z directions by light that is polarized in some arbitrary direction.

The polarizability, α, as defined above, can be thought of as a measure of how difficult it is for an electric field to stretch a particular chemical bond. Just like a rubber band, it is reasonable to imagine that the more the bond is stretched the more difficult it becomes to stretch it, i.e., α may change as a function of nuclear displacement. Nuclear displacement here means the amount by which the nuclei of the atoms that are partners in a chemical bond are displaced from their normal position due to an incident electric field. To better describe the relationship between α and the nuclear displacement we can expand the polarizability tensor α in a Taylor series relative to the extent the bond has stretched. If we consider only a single vibration this expansion can be written as

$$\alpha = (\alpha)_0 + \left(\frac{\partial \alpha}{\partial r}\right)_0 r + \frac{1}{2}\left(\frac{\partial^2 \alpha}{\partial r^2}\right)_0 r^2 + \ldots \tag{5.2}$$

where $(\,)_0$ indicates the value at equilibrium and r is the coordinate along which the molecule is being deformed. We can further make another simplification by using the mechanical harmonic approximation which only retains terms that are linearly dependent on the nuclear displacement, in this case just the first two terms. Dealing with the anharmonic, or higher-order terms, is outside the scope of this chapter. This approximation ensures that we stay within the limits of the simple mechanical model that we introduced in the previous section and the molecule acts like a system of balls and springs (see Figs. 5.1 and 5.2). Since we are assuming that this vibrating molecule will act just like a harmonic oscillator we can also assume that the displacement along r will vary sinusoidally. Thus we can define an expression for the polarizability as the molecule vibrates:

$$\alpha = (\alpha)_0 + \frac{1}{2}\left(\frac{\partial \alpha}{\partial r}\right)_0 r_0 \sin(\omega_k t + \delta) \tag{5.3}$$

Here, r_0 is the equilibrium displacement, ω_k the frequency of a normal mode of vibration of the molecule, and δ the phase shift that accounts for the fact that in a large sample each molecule can vibrate with its own, discrete but random phase. Assuming that the **E** field also oscillates sinusoidally, $\mathbf{E} = E_0 sin(\omega_l t)$, we can use Eqs. 5.1 and 5.3 to obtain an expression for the induced dipole moment:

$$p = (\alpha)_0 E_0 \sin(\omega_l t) + \frac{1}{2}\left(\frac{\partial\alpha}{\partial r}\right)_0 r_0 E_0 \sin(\omega_k t + \delta)\sin(\omega_l t) \quad (5.4)$$

The first term in this equation describes elastic or "Rayleigh" scattering since the frequency of the induced dipole moment will oscillate with frequency ω_l, which is the same as that of the incident **E** field. The only selection rule for Rayleigh scattering is that the polarizability, $(\alpha)_0$, be non-zero at equilibrium but this is virtually always the case so this effect is typically observed. By using a simple trigonometric identity we find that the second term corresponds to inelastic scattering:

$$\sin(A)\sin(B) = \frac{1}{2}[\cos(A - B) - \cos(A + B)] \quad (5.5)$$

By using this identify, the second term in Eq. 5.4 becomes

$$p_{\text{Raman}} = \frac{1}{4}\left(\frac{\partial\alpha}{\partial r}\right)_0 r_0 E_0[\cos((\omega_l - \omega_k)t + \delta) - \cos(\omega_l + \omega_k)t + \delta] \quad (5.6)$$

Equation 5.6 shows that the interaction of the oscillating molecule with the sinusoidally varying electric field results in two inelastically scattered contributions: $(\omega_l\text{–}\omega_k)$, the Stokes-shifted contribution where the incident laser photons are red-shifted by ω_k, and $(\omega_l + \omega_k)$, the anti-Stokes-shifted photons, which are blue-shifted (shifted to higher energies) by ω_k. Unique molecular modes of vibration will appear as distinct lines in a Raman spectrum and appear at $\omega_l \pm \omega_k$. These can then be used to infer structural and environmental information. The connection between vibrational energy levels and the Raman spectrum is shown schematically in Fig. 5.3. Traditionally, however, a Raman spectrum is shown as the intensity of the scattered photons plotted against their Raman shift (here: ω_k). This is shown in Fig. 5.4 on the example of the fingerprint spectrum obtained from a fatty acid, oleic acid. Here, both the anti-Stokes scattered spectrum as well as the Stokes-shifted spectrum are shown. While these spectra exhibit a wealth of peaks that can all be attributed to specific vibrational modes within the molecules, it has

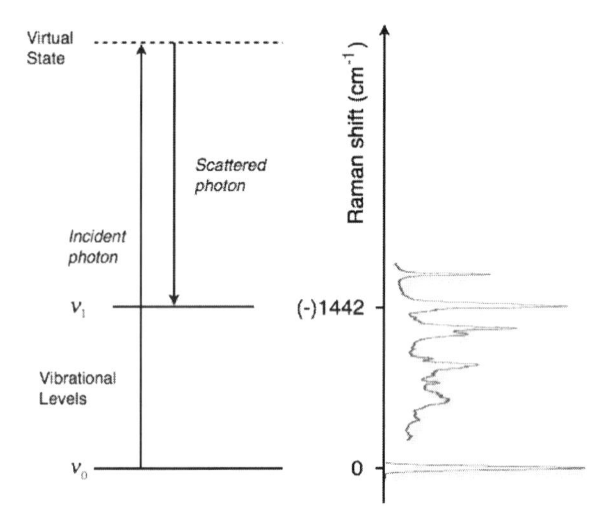

Figure 5.3 The energy diagram describing Raman scattering as related to the actual Raman spectrum of a molecule. This is highlighted on the example of a lipid CH deformation mode at 1442 cm^{-1}.

to be noted that in most cases one would obtain even more peaks if all vibrational modes were Raman active. Raman activity, however, as briefly mentioned before is defined by certain selection rules. Unlike infrared (IR) spectroscopy, which requires a changing dipole moment, the selection rule for Raman scattering requires only that at least one of the tensor terms, $(\partial\alpha/\partial r)_0$, be non-zero.

Although Raman spectroscopy can provide substantial chemical information, the signals are very weak with typical scattering cross-section of $\sim 10^{-30}$ cm^2. By comparison, fluorescence typically has an absorption/excitation cross-section of $\sim 10^{-16}$ cm^2. Several techniques have been developed to "boost" this signal, including resonance Raman scattering (RRS), surface-enhanced Raman scattering (SERS), and coherent Raman scattering (CRS).

As mentioned, Raman spectroscopy provides rich chemical information. Most of this information is extracted on the basis of peak assignments and relative intensities but can also involve peak shapes and dependence on the polarization of the laser light. The strength of the Raman signal is related to the polarizability by the following expression:

$$I_{\text{Raman}} = NL\Omega \frac{\hbar}{2m\omega_k} \left(\frac{\partial \alpha}{\partial r}\right)^2 \frac{\omega_i^4}{c^4} I_l \tag{5.7}$$

Here, I_{Raman} and I_{l} are the intensities of the Raman line and the excitation laser respectively, N is the number of bonds, L the length of the focal volume, Ω the solid angle over which the signal is collected, and m the reduced mass of the vibrating oscillator (Raman-active molecular group). Most notably, Raman scattering scales with a ω^4 dependence of the scattering signal on the excitation wavelength. The result of this dependence is that Raman spectra generated with green (e.g., 532 nm) or blue (488 nm) laser light are significantly more intense than the ones generated in the red part of the optical spectrum. This potential advantage can, however, be offset by the excitation of autofluorescence at these wavelengths, which can easily overwhelm the Raman scattered signal. This is particularly problematic for biological samples, where some amino acids and certain proteins exhibit particularly strong autofluorescence. Therefore, although shorter excitation wavelengths provide a stronger Raman signal, when analyzing biological samples, red (632.8 nm, 647 nm) or near-infrared (785 nm, 830 nm) laser light is often used to minimize contributions from autofluorescence and also maximize tissue penetration. It should also be noted that the Raman signal intensity is inversely proportional to the molecular vibrational frequency, ω_{k}, so that the greater the Raman shift the less efficient the scattering will be. This dependence is typically not limiting since most biologically relevant Raman lines are found at wavenumbers $<3400\ \text{cm}^{-1}$ but must be considered when attempting quantitative analysis. Another important effect that occurs with different excitation wavelengths is that the *absolute* Raman shift obtained from a molecule scales with the laser wavelength. This is due to the fact that photons at the blue end of the optical spectrum carry more energy than red photons. Since the energy required to excite a molecular bond vibration is fixed, the ratio of energy transferred from the photon relative to the photon energy is smaller for blue photons than it is for red photons. Thus, if a Raman spectrum is plotted on a wavelength scale it will appear more compressed when obtained with a blue light source

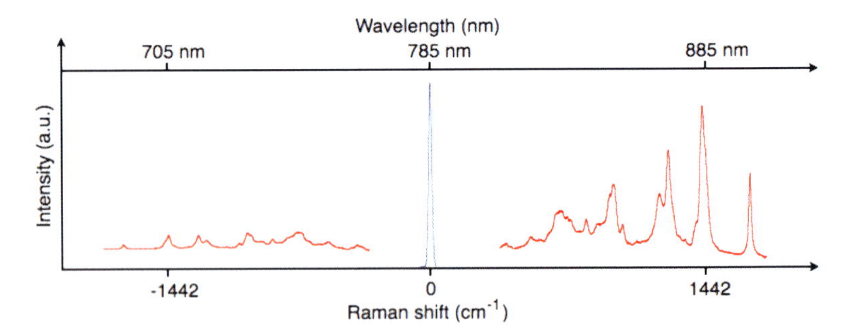

Figure 5.4 Stokes (right) and anti-Stokes (left) Raman spectra of oleic acid. The scattered photons are shifted to the red (blue) depending on the bond vibration and the scattering process. At room temperature anti-Stokes Raman scattering is always less efficient than Raman scattering, because only a very small fraction of molecules will be in a vibrationally excited state.

rather than a red light source. Because of this effect, Raman spectra are typically shown plotted against a *relative* Raman shift, which removes the dependence on excitation laser wavelength. This is also the appropriate place to mention that the polarity of the sign in front of the relative Raman shift is rather randomly assigned. Different authors will plot the Stokes-shifted Raman line to the left or to the right of the laser line and also change signage rather randomly. Strictly speaking, the sign of the Stokes-shifted Raman peaks should be negative since these represent the amount of energy transferred from the photon to the molecule. Most authors will, however, plot the Stokes-shifted Raman peaks with positive labels because the majority of spectroscopists only collect Stokes-shifted Raman lines, and plotting negative Raman shifts might appear ambiguous and difficult to understand to the average reader.

Equation 5.7 can be simplified by equating the change in polarizability, $\partial\alpha/\partial r$, to the differential scattering cross-section, $d\sigma/d\Omega$:

$$\frac{\partial\sigma}{\partial\Omega} = \frac{\hbar}{2m\omega_v}\left(\frac{\partial\alpha}{\partial q}\right)^2 \frac{\omega^4}{c^4} \tag{5.8}$$

The intensity of the Raman line can then be expressed as a function of the scattering cross-section:

$$I_{\text{Raman}} = NL\Omega\left(\frac{\partial\sigma}{\partial\Omega}\right) I_l \tag{5.9}$$

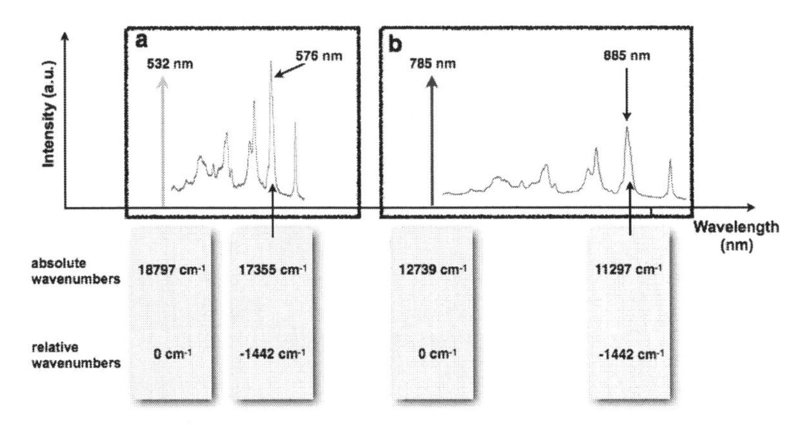

Figure 5.5 The strength of the Raman scattering process depends on the wavelength of the source: the intensity of the inelastically scattered light scales with λ^{-4}. Raman spectra are more intense when probed with a green laser (box a) rather than a red laser (box b). In addition, because green light has a higher photon energy than red light, Raman scattering by the same molecular bond can lead to more "condensed" or "stretched" spectra depending on the laser wavelength. This is the reason why Raman spectra are typically plotted in terms of the energy difference between the source wavelength and the signal wavelength. These spectra are independent of the wavelength at which they were obtained.

Given accurate estimates for the length of the focal volume, the collection angle, the laser intensity, and the bond density, the cross section can be straightforward to determine. Alternatively, given an a priori knowledge of the cross section (often found in look-up tables or published literature) the bond density can be determined with a high level of accuracy. Although bond densities and cross sections are useful to determine they may not be sufficient for correctly assigning Raman lines or determining environmental effects on the Raman spectrum. For such determinations the symmetry of the vibration that results in the measured peak must often be determined.

By returning to Eq. 5.7 we can see the dependence of the scattering intensity on the tensor, $(\partial\alpha/\partial r)_0$. The elements of this tensor can be calculated for a molecule of interest using group theory, which can then be used to determine peak assignments. Usually these calculations are only useful when analyzing the spectrum of a neat compound or determining how that compound contributes to the spectrum of a non-interacting mixture. Often, unknown chemical products or interactions can contribute lines to a Raman spectrum and in order to determine the origin of those lines the symmetry of the vibration must be determined without a priori knowledge.

While fully characterized spectra are very useful and relatively straightforward to obtain in neat or bulk chemical samples, it is typically not possible to obtain them in complex chemical environments, such as the cytoplasm or nucleus of a living cell. In these cases, a useful alternative approach to take is difference spectroscopy, where spectra of a cell in an unaltered state and in an altered state are obtained and subtracted from each other to determine shifts in peak locations and changes in peak intensities due to controlled stimuli. The spectra from cells and tissue represent superimposed spectra from the many contributing constituents, the spectra of which can be obtained from neat compounds. In this case, cytometric measurements can also be made using spectral deconvolution algorithms, e.g., utilizing neural networks and multivariate statistical methods, such as principal component analysis.

To conclude this introductory section, let us consider the spatial resolution afforded by confocal micro-Raman spectroscopy for the analysis of single cells. An example is shown in Fig. 5.6. Here, the sample consists of frozen bull sperm cells that were thawed, diluted in salt solution, and then an aliquot was allowed to adhere to a calcium fluoride glass slide. Calcium fluoride and magnesium fluoride are excellent sample carriers for Raman analysis because they exhibit essentially no Raman-active peaks and thus add no background contributions throughout the entire fingerprint spectral range from $\sim$500 cm^{-1} to 1800 cm^{-1}. These materials, are, however, relatively expensive especially as polished substrates with high optical quality and they are rather brittle. Instead, quartz substrates work well, too, and can even be obtained in coverslip thickness ($\sim$170 μm), but add some background signals to the Raman spectrum that have to be carefully characterized and removed from the final sample spectra. For upright microscope systems, simple metal mirrors (e.g., silver mirrors) also work well as substrates

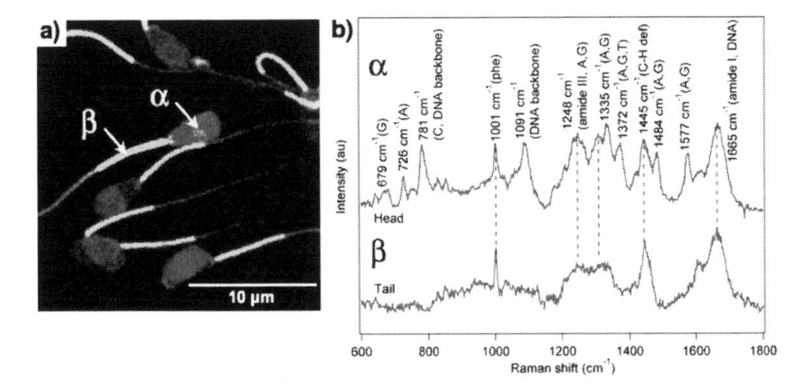

Figure 5.6 Raman spectra obtained from bovine sperm cells. (a) Intact, whole sperm cells are imaged in a confocal Raman microscope based on their autofluorescence. Specific areas, i.e., the sperm head (α) or sperm tail (β), can then be addressed by point spectroscopy. The resulting spectra and their assigned bond vibrations (see (b)) show that the sperm head is mostly composed of DNA and proteins (α), while the major constituent of the tail is mostly protein (β).

with the added benefit that the collection efficiency for scattered photons is also increased. For the sample shown in Fig. 5.6, an autofluorescence image of the sperm cells was first obtained by raster-scanning the CaF_2 substrate with a 488 nm laser beam focused to a diffraction-limited spot using a 100$\times$ long working distance air objective. The resulting image clearly displays the cells by their strong autofluorescence, as can be seen in Fig. 5.6a. Sperm heads and tails are clearly visible by their different contrast and structure. The confocal Raman detection system is then used to probe different areas of the sperm cell (head and tail) to obtain Raman spectra from different parts of the cell. The spectra of the head region reveals distinct peaks that can be assigned to DNA that is tightly packaged by protamine, an arginine-rich protein, whereas the tail region only exhibits Raman peaks characteristic for lipids and proteins and contains no evidence for DNA, as expected (see Fig. 5.6b). In this image, typical Raman acquisition times were approximately 5 minutes at 1 mW laser power. Although a blue laser was used to obtain these spectra, it should be noted, that these data were first taken about 10 years ago, when powerful near-infrared lasers were still scarce and expensive. Today, these spectra would ideally be obtained with lasers in the near-infrared part of the optical spectrum, which reduces the risk for sample damage because biological samples have significantly less absorption in the near-IR. Contributions from autofluorescence excitation are also much lower or absent altogether in this wavelength range. This also results in much faster spectral acquisition times because the sample can tolerate higher laser powers.

5.3 Single-Cell Analysis by Raman Scattering

5.3.1 *Micro-Raman Spectroscopy Systems*

Prior to 1990, studying single cells with Raman spectroscopy had been extremely difficult due to the low Raman scattering cross-sections of biological macromolecules. Although resonance Raman spectroscopy using ultraviolet excitation light for acquiring single-cell spectra had been demonstrated [2], this approach was far from ideal because of the likelihood of inducing sample damage with such highly energetic photons. Puppels et al. [3] were the first to demonstrate the acquisition of nonresonant Raman spectra from subcellular regions within individual living cells using a highly sensitive confocal Raman microspectrometer. Since this seminal work, considerable advancements in the design and configuration of single-cell micro-Raman spectroscopy systems have been made to improve their speed and efficiency for acquiring Raman spectra and images from biological cells. This section summarizes the development of key components in a single-cell Raman spectroscopy instrument. The different optical geometries are summarized in Fig. 5.7.

5.3.1.1 Confocal detection

The confocal geometry is a critical component of any micro-Raman spectroscopy system because it enables the acquisition of Raman spectra from individual cells with high spatial resolution. In a typical configuration, point illumination is achieved by focusing a laser beam with a Gaussian mode profile through a microscope objective

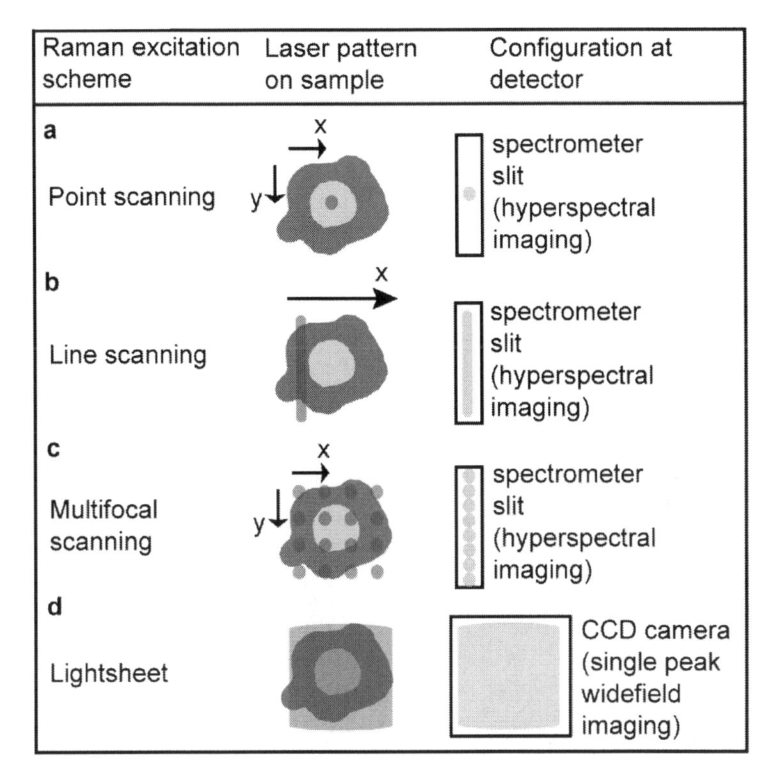

Figure 5.7 Illustration of the different laser excitation schemes most commonly used for Raman spectroscopy and imaging of cells. (a) Point scanning involves raster scanning a tightly focused laser spot in two dimensions (x, y) to collect spectra from each Raman pixel. At the detector, signals are delivered through the spectrometer slit in a single spot. (b) Line scanning involves reshaping the laser into a linear pattern and scanning the pattern in one dimension (x) across the cell. Signals from the line are projected through the entire slit and individual spectra are collected simultaneously along the vertical axis of the CCD chip. (c) In multifocal scanning, multiple laser spots are arranged in an array at the sample plane. Spectra at each spot are collected by individual fibers, which are aligned onto the spectrometer split. (d) Light sheet microscopy generates Raman images at single Raman peaks. A thin light sheet illuminates the entire sample, and the Raman image is collected orthogonally with a large area CCD camera.

with high numerical aperture onto a biological cell. A spatial pinhole is placed in an optical plane conjugate to the sample plane and in front of the detector, which serves to eliminate out-of-focus signal while allowing only Raman signals generated close to the focal plane to be detected. The spatial filter is very important because the out-of-focus signal (e.g., fluorescence, Raman signal from substrates) can often have a much higher intensity than the Raman signal from the biological sample. This background signal would otherwise overwhelm the weaker Raman signal if confocal detection is not used, preventing the detection of Raman spectra from individual cells. Another advantage is improved optical resolution, particularly in the axial direction, which allows for three-dimensional sectioning of different regions within a single cell at diffraction-limited spatial resolution. While confocal detection offers improved rejection of out-of-focus signals and increased spatial resolution, a trade-off is reduced Raman signal intensity from the sample since more of the useful scattered light from the sample is blocked by the pinhole. Longer exposure times are therefore required. Uzunbajakava et al. [4] illustrated the effect of different confocal pinhole sizes on the Raman spectra of nuclei in live peripheral blood lymphocytes. A reduction of the pinhole diameter by a factor of 2 led to a decrease in the contribution of the background contribution from water by a factor of 2, and a slight decrease in the Raman signal intensity. The smaller pinhole size also enabled much better axial sectioning of the nucleus, enabling the axial distribution of macromolecules to be profiled based on the Raman signal intensities.

5.3.1.2 Laser wavelength

The Raman effect involves the absorption and emission of a photon through an intermediate virtual energy level. Therefore, unlike an electronic excitation process, Raman spectra can be acquired using any frequency of the excitation light because it is not a resonant effect. In practice, wavelengths from the visible through the near infrared have been used for the Raman analysis of biological cells. Shorter excitation wavelengths are desirable because of improved spatial resolution and Raman scattering efficiency due to the increase in the Raman scattered radiation with the fourth power of the excitation frequency. However, short wavelength excitation light may generate sufficient autofluorescence in certain biological samples that will obscure the Raman spectra of the sample. Moreover, another important consideration to make when choosing the proper excitation wavelength is laser-induced sample photodamage. Already early on, researchers have investigated the effect of different laser wavelengths on cell viability. Puppels et al. [5] showed that several milliwatts of 514.5 nm radiation focused to a 0.5 μm diameter spot resulted in the degradation of single cells (human lymphocytes), as evidenced by visible paling of the samples accompanied by a decrease in the intensity of the Raman signals over time. However, no cell degradation was observed for longer wavelength excitation at 660 nm. Photochemical reactions between the higher energy photons and macromolecules within cells and chromosomes were postulated as being responsible for the sample degradation. Other laser damage studies have been performed on *E. coli* bacteria [6] and Chinese Hamster Ovary (CHO) cells [7]. *E. coli* bacteria cells tethered to a glass coverslip were exposed to different wavelengths of light while their rotation rates were monitored. The LD_{50} time, defined as the time at which the rotation rate decreased to 50% of its initial value, was recorded. 870 nm and 930 nm were the most damaging wavelengths, while 830 and 970 nm were the least damaging wavelengths for bacteria. These results were consistent with prior studies that studied the effect of different wavelengths on the cell cloning efficiency of CHO cells, suggesting similar mechanisms of photodamage. Largely as a result of these studies, it is generally accepted that the near infrared excitation range around 785–830 nm should be used for micro-Raman spectroscopy of biological samples to avoid photodamage. These wavelengths also fall within the spectral range of minimum absorbance for most biological materials, which reduces the probability of inducing photothermal damage. In fact, many Raman spectrometers and detectors are now designed for this wavelength region specifically for biological applications. However, it should be noted that, in practice, the decision of which wavelength to use should really be evaluated on a case-by-case basis depending on the specific biological system being investigated. Choosing an arbitrary wavelength in the near-infrared to avoid photodamage and photothermal damage may not necessarily work, as discussed in the studies mentioned above that showed that the most and least favorable wavelengths can be separated by as little as a few tens of nanometers. Also, near-infrared light does not necessarily ensure that autofluorescence will be avoided. For example, a deeper NIR Raman system operating at 1064 nm had to be developed [8] to probe single cyanobacteria, a well-known model organism for photosynthesis research, because of its strong fluorescence emission even when excited with 800 nm light. Remarkably, a commonly used glass substrate, borosilicate glass (BK7), also exhibits fluorescence due to contaminants when excited at 785 nm.

5.3.1.3 Optical excitation and detection geometries

Point geometry

The most common micro-Raman spectroscopy systems use point illumination with a tightly focused monochromatic laser source, a single spatial pinhole (as discussed earlier) for confocal detection, and a spectrometer equipped with a grating and a cooled charge-coupled device (CCD) camera to disperse the spectral signals onto the camera for spectral detection. Depending on the grating that is chosen, the spectral range that is covered lies typically between 400 cm^{-1} to 2500 cm^{-1}. While the camera has a large active area (e.g., 100 vertical × 1340 horizontal pixels), only a fraction of these pixels are actively used. The Raman signals generated from within the small focal volume of the excitation beam are dispersed across all of the horizontal pixels but are projected onto only a small fraction of the vertical pixels (e.g., 15–20 pixels). Spatially resolved Raman spectroscopy can be performed by scanning the cell, which is attached to a substrate surface, relative to the point illumination source using an x–y–z scan stage. With this configuration, Raman images can also be generated to obtain information of the spatial distribution of the different chemicals present in the sample. In Raman hyperspectral imaging, the sample is divided into small squares (i.e., Raman pixels) in the x–y sample plane, whose minimum size is limited by the probe beam diameter and smallest step size of the scan stage. Raman spectra are acquired from each of these defined Raman pixel regions, and each pixel can be assigned an intensity value based on the strength of a particular Raman marker. By performing this methodology sequentially for all pixels, multiple chemical images can be generated based on different Raman "bands" that depict the intracellular distribution of different chemical species within a single cell.

The number of studies that have utilized a point illumination Raman system for spectroscopy and imaging of cells is too numerous to describe in detail. Here, a few applications are highlighted to illustrate the type of information that can be obtained from this spectral analysis and the potential applications of this technology. In 2009, Huser et al. [9] showed that Raman spectra could be obtained from individual sperm cells, and that the vibrational signatures could be used to assess DNA packaging efficiency in the sperm heads. Sperm cells attached to a substrate were imaged first by autofluorescence, and the sperm heads were probed by the point illumination beam. A correlation was found between the packing efficiency and the shape of the sperm head. More importantly, it was found that the DNA packing efficiency in sperm cells with normal head shape morphology varies greatly, suggesting that selecting viable sperms for in vitro fertilization solely by morphology, which is the currently accepted practice, may not be sufficient. Single-cell Raman spectroscopy could therefore be a powerful tool to aid in the selection of quality sperm cells for in vitro fertilization. Wood et al. [10, 11] used single-cell Raman spectroscopy to monitor the hemoglobin deoxygenation and oxygenation process of human erythrocytes over the span of 30 minutes. Moreover, photoinduced and thermal degradation of the cells were observed after prolonged exposure to the laser beam, and Raman markers indicative of heme aggregation were identified. These results suggest the potential use of single-cell Raman spectroscopy for the analysis of erythrocyte disorders characterized by heme aggregation, such as sickle cell disease and malaria. Huang et al. [12] reported the discovery of a specific Raman spectroscopic signature reflecting the metabolic activity of mitochondria in a yeast cell. Raman images were generated of yeast cells with GFP labeled mitochondria, which were directly compared to fluorescence images. Overlay of the GFP image and the Raman image of the same cell generated using the intensity of the 1602 cm^{-1} Raman band indicated excellent signal colocalization. The addition of KCN, a respiration inhibitor, was found to induce a drop in intensity of the 1602 cm^{-1} band, indicating that the marker reflects the metabolic activity of the mitochondria. This Raman marker was termed the Raman spectroscopic signature of life. Recently, Zoladek et al. [13] used confocal Raman microspectroscopy to obtain time-stepped spectral images of live human breast cancer cells undergoing apoptosis after being exposed to the drug etoposide. Raman images of DNA (788 cm^{-1}) and lipids (1659 cm^{-1}) were acquired from the same cell every 2 hours over a 6 hour period. It was observed that, relative to control cells, drug treated cells exhibited an increase in DNA band intensities due to DNA condensation and lipid band intensities reflecting the high accumulation of membrane phospholipids and unsaturated non-membrane lipids. These results suggest the potential of using Raman spectroscopy for in vitro toxicological studies, drug testing and screening, and further applications that require the real-time continuous monitoring of single-cell dynamics.

Linescan geometry

Many biological cells are much larger in diameter than the spot size of a focused laser beam. Therefore, a limitation of point illumination for Raman spectroscopy is that it can take a relatively long time to sample the *entire* cell volume to obtain a spectrum representative of the cell. Similarly, for Raman microscopy, it can take a long time to generate a Raman image of an entire cell purely by spontaneous Raman scattering. The large number of Raman pixels that can comprise the Raman image of a cell, coupled with the long integration times needed to generate a Raman spectrum for each pixel due to the low Raman scattering cross sections, results in overall imaging times that can be many tens of minutes to hours. This often prohibits live cell imaging, and most samples therefore need to be chemically fixed before they can be imaged. An alternative optical excitation and detection geometry based on line scan illumination has been developed to reduce the spectral and imaging acquisition time. A laser beam with a circular Gaussian mode profile is reshaped into a line pattern using cylindrical lenses and focused onto the cell with a microscope objective. The line pattern allows a larger fraction of the cell to be illuminated by the laser beam than point illumination, and therefore allows Raman signals from different regions within a cell to be generated simultaneously. The Raman signals from this illuminated region are collected by projecting the line through the linear slit of a spectrometer and onto the CCD camera. The line pattern is spatially dispersed onto the vertical pixels of the camera, and the Raman spectra at each position of the line pattern are dispersed in the horizontal direction. Therefore, with this geometry, the majority of the CCD pixel area is actively being used. For Raman imaging, scanning of the line in only one direction (x) at the sample plane is needed, which significantly reduces the acquisition time compared to point illumination, which requires two-dimensional point scanning.

With this configuration, live-cell imaging with spontaneous Raman spectroscopy within a few minutes of image acquisition time has been demonstrated. Hamada et al. [14] demonstrated dynamic Raman imaging of the molecular distribution inside living HeLa cells based on this scheme. Here, the slit scanning detection was combined with an excitation wavelength at 532 nm optimized for detecting cytochrome c, protein beta sheets, and lipids while still generating minimal autofluorescence (which is sample-specific, though). Time-resolved images were obtained of these cells during cytokinesis by acquiring images with a 185 second total acquisition time, with an interval between images of 115 seconds. Similarly, Notingher et al. [15] used a Raman microspectroscopy system with a line shape configuration to detect Raman biochemical markers during the differentiation of murine embryonic stem cells that could be used to assess the differentiation status of the entire cell (rather than line scanning). With a dimension of roughly 5 μm by 10 μm, the line-shaped spot enabled a large fraction of each stem cell, which had typical dimensions of less than 10 um in diameter, to be probed in a single spectral acquisition.

Multifocal geometry

Another optical configuration based on a multifocal laser array has recently been demonstrated [16] for Raman spectroscopy/

microscopy of living cells. This design enables Raman spectra to be acquired simultaneously along both, the x and y axes of the sample plane. This configuration uses a microlens array to split a laser beam into many individual beamlets, which are then projected through a confocal pinhole array and focused with an objective. A pattern of 8×8 independent foci was generated at the sample plane. Raman signals from the foci were delivered through the confocal pinhole array and collimated into one end of a fiber bundle consisting of 64 fibers arranged into an 8×8 rectangular pattern. This fiber bundle directed the Raman signals to the slit of the spectrograph. At the entrance slit, the fiber bundle was arranged into a 1×64 linear stack geometry. This linear array of Raman spectra from the fiber bundle analyzed on the CCD camera in much the same way as in the line scan arrangement. Because spacing between adjacent foci at the sample plane was still present, sample scanning in the x and y direction was still necessary to acquire the full Raman image of the cell, but over much shorter distances compared to the single-point illumination geometry. Imaging of a 16×12 μm area in less than 1 minute (20 seconds) was reported for the monitoring of budding yeast cells.

Light sheet geometry

Spectral and imaging acquisition speeds can be further increased if the entire cell can be illuminated at the same time with wide-field illumination. However, the challenge with implementing wide-field illumination for micro-Raman spectroscopy is the loss of optical sectioning, primarily in the axial direction. Consequently, the generation of strong background signals in optical planes above and below the sample plane would interfere with and overwhelm the Raman signals from the sample. For instance, a low numerical aperture objective lens can be used to increase the field of investigation, but such a system typically suffers from poor axial resolution. To address this issue, light sheet single-plane illumination Raman microscopy has recently been developed to enable wide-field Raman imaging with high optical sectioning capability. A cylindrical lens is used to reshape a Gaussian mode laser beam into a thin light sheet that illuminates a single plane of the sample from the side. Raman signals generated from the sample plane are detected by a microscope objective and CCD camera placed orthogonal to the illumination direction. The objective is positioned such that its focal plane overlaps with the illumination plane. Narrow bandpass filters are placed in front of the CCD camera to allow for chemical imaging of specific Raman bands. Oshima et al. [17] used this system to image solvents. In this case, Raman scattering was achieved by a tunable laser, which eliminated the need to use a different bandpass filter to generate images at each Raman band. Only one bandpass filter was needed because the laser could be tuned such that the Raman peak for every molecular vibration was fixed at the same wavelength. More recently, this scheme was used to demonstrate wide-field Raman imaging with optical sectioning in the axial direction. Wide-field Raman imaging of 20 μm diameter solid polystyrene spheres dried on a glass surface was demonstrated [18]. Each frame was 50 × 150 μm with an axial step size of 3 μm, with 10 second acquisition time per frame. Application of this technique for imaging of single biological cells remains to be demonstrated.

5.4 Data Processing and Analysis of Raman Spectral Data

The raw Raman spectral data obtained from any of the Raman systems described above will typically need to be processed before any meaningful, interpretable information can be obtained. Processing methods such as smoothing, background subtraction, and normalization are typically performed before the intensities of individual specific peaks from different spectra can be quantified and compared to each other. Alternatively, multivariate data reduction methods can also be employed to enable the comparison of spectra while taking into account the full spectral information of the data. This section briefly summarizes these key data processing methods. Readers are referred to the publications cited in this section for more detailed information and descriptions of other techniques not discussed herein.

5.4.1 *Smoothing*

Removal of noise from the Raman spectra is an important operation in data processing. Noise can mask relevant peaks in the Raman spectra, which are often quite weak in the case of biological samples, and can complicate the identification of Raman features that can be used to separate cell types. Cosmic rays impinging on the CCD detector array can also generate sharp spikes that contaminate the Raman spectra. A simple, straightforward algorithm to remove noise is a "moving average" filter that takes an average of the intensity of a fixed subset of adjacent wavelengths in the Raman spectrum. The subset is moved forward and the new subset of intensity values is averaged. After this is done over the entire spectrum, the averaged values are used to create a smoothed Raman spectrum that removes the noise contributions. A drawback of this simple approach is that this averaging technique may flatten features leading to loss of spectral information. Therefore, the Savitzky–Golay smoothing filter is more routinely used to smooth Raman spectra, as it can better preserve the height and width of the spectral features. It calculates a new value for each point in the spectrum by performing a local polynomial fit of order n on a series of values around that point. At least $2n+1$ equally spaced, neighboring points (including the point of interest) are needed for this fit.

5.4.2 *Background Subtraction*

Autofluorescence from organic molecules in the biological cell and from the substrate in close proximity to the cell, both of which can be several orders of magnitude greater than the weak Raman signal, can contribute significantly to a Raman spectrum. The strength and variability of this background signal can make it difficult to analyze, interpret, and compare Raman spectra. Pre-processing of the spectra to remove this background signal is therefore necessary in order to extract relevant information from the Raman spectrum. Because the background signal typically has a broad low-frequency spectral profile, polynomial curve fitting is typically used. This technique relies on the user selecting several points in the Raman spectrum within regions that contain no known Raman peaks and then fitting a polynomial curve to those points. Subtraction of the polynomial curve fit from the Raman spectrum results in a background-corrected spectrum. This technique, however, can be subjective and time-consuming since the user needs to process each spectrum individually. Furthermore, it is not always straightforward to identify with certainty, those regions that don't contain any Raman features. The result is that manual polynomial curve fitting can lead to variability in the background subtraction from spectrum to spectrum. To improve upon these problems, Lieber et al. [19] developed a modified polynomial fit method that removes the subjectivity and enables systematic, automated subtraction of the background. A polynomial fit is first performed on the initial Raman spectrum. This fit is modified such that all data points in the generated curve that have a higher intensity value than their original value in the input spectrum are reassigned their original value, while the fit result is retained for all the other points. Curve fitting is performed on this new modified fit curve and the intensity values are reassigned again. This process is repeated for several hundred iterations until the high frequency Raman peaks are removed, and what is left is the broad underlying background curve. This curve is subtracted from the initial raw spectrum to yield the final corrected Raman spectrum.

As an extension to this polynomial fitting method, Beier and Berger [20] developed a routine that also takes into account the spectral contributions from a known contaminant that is present either in the sample or in the optical system. If this contaminant has a varying concentration, the conventional polynomial fitting method would leave residual contaminant-related spectral artifacts in the Raman spectrum of the biological sample. In this new scheme, a reference spectrum is first obtained from the known contaminant. The initial Raman spectrum is least squares fit to a simple model using the reference spectrum and an initial constant value that approximates the signal strength or concentration of this contaminant. This constant is adjusted and optimized such that when the fluorescence-only component (obtained by subtracting the fitted contaminant spectrum from the Raman spectrum) is fitted to a polynomial, the sum of squares of the residual will be minimized. This background, which is now modeled with two polynomial curves (one accounting for the contaminant and the other for the fluorescence) is next processed using the modified polynomial fit by Lieber et al. to remove the Raman peaks, which yields the final background curve. This background serves as the initial spectrum and is processed through the entire algorithm again. This is repeated until a termination criterion is met, such as when the root mean square of the difference between the background spectra reach an empirically determined threshold.

It should be noted for completeness that there are other methods that have been developed to remove background signals. These techniques seek to eliminate the background signals during the experimental acquisition of the spectra, not in post-processing of the data, by developing novel optical detection schemes. For example, shifted excitation Raman difference spectroscopy (SERDS) [21, 22] was developed where two laser frequencies that are slightly shifted to acquire two Raman spectra from the sample are utilized to obtain Raman spectra. The broad backgrounds in the two spectra remain relatively unchanged with regard to the excitation laser wavelength, but the sharp Raman bands will be shifted with respect to each other in the two spectra. Subtraction of the spectra results in a background-free derivative-like spectrum with clearly identifiable Raman features. De Luca et al. [23] developed a modulated Raman spectroscopy technique for online fluorescence suppression based on the SERDS principle, but incorporating two new components: (i) a continuously modulated excitation wavelength (modulation frequency is 0.4 Hz, $dv = 60$ GHz) and (ii) multichannel lock-in detection. While the Raman peaks will also be modulated at this frequency, the background remains essentially constant. Synchronization of the detection with the modulated excitation wavelength enables the modulated Raman signals to be detected from the unmodulated background signal. This technique has the added benefit of a slightly improved detection limit for the spontaneous Raman scattered signal.

5.4.3 *Principal Component Analysis (Multivariate Analysis Methods)*

Conventional analysis of the pre-processed Raman spectra usually involves the calculation of the intensity or area of individual peaks, calculation of mean and standard deviation values, and comparison of these values for different spectra to identify unique Raman features that can be used to discriminate cell types or monitor cellular processes. These intensity values can also be used to generate Raman images of the cell. Because a Raman data cube can be highly multidimensional (a large number of individual Raman spectra consisting of >1000 individual data points), it can often be laborious to implement these univariate methods to analyze the full data set simultaneously and determine the degree of correlation between variables. Multivariate techniques can be powerful tools for Raman spectral analysis because they can be used to analyze the entire data stack to identify combinations of highly correlated variables that best explains the data variance. A widely used representative of multivariate analysis tools in Raman spectroscopy is principal component analysis (PCA). Its main purpose is to explain the major trends within the data using a combination of the original variables to form new principal component (PC) variables. By finding combinations of the original dimensions that describe the largest variance between the data sets, the dimensionality of the data is reduced and the data can be represented by a much smaller number of factors, or PCs. PCA can reduce Raman spectra significantly to just two or three primary PCs that can describe the most significant variance between the spectra, as shown in Fig. 5.8. The less significant PCs describe mostly random noise and can be discarded. Therefore, PCA can represent a highly multidimensional Raman spectral data set within a 2D scatter plot plotting PC 1 and 2 on the axes, and each data point on the scatter plot represents an individual Raman spectrum of an independent measurement of the sample. Such scatter plots are ideal for visualizing group/sample separation based on Raman differences.

PC values can also be used to create PCA images of a cell [24]. This is achieved by analyzing the spatially resolved spectra from the cell with PCA to generate PCs for each spectrum. For each PC, its values for all spectra are scaled between 0 (black) and 255 (white) and used to create monochrome intensity maps depicting the abundance of the PC in the imaged cell. For an RGB image, three monochrome maps reflecting three different PCs can be used to generate a pseudo-colored map with mixed colors describing the

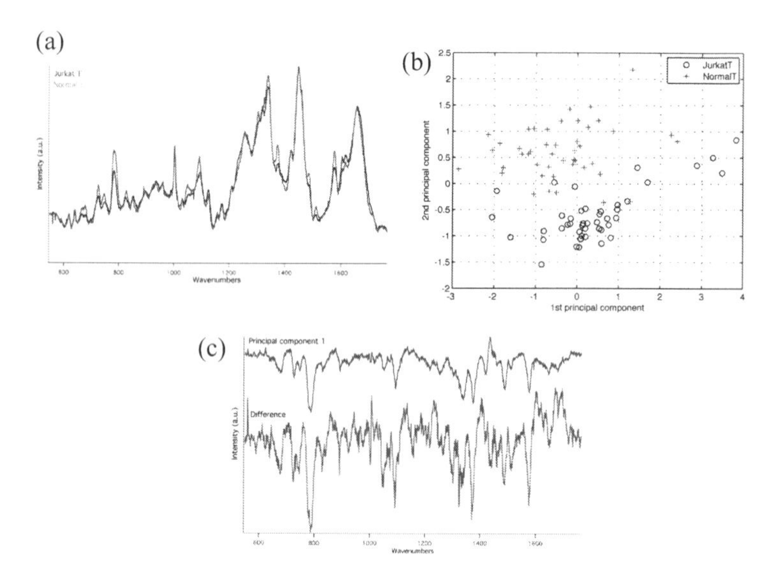

Figure 5.8 (a) Mean Raman spectra of normal and Jurkat (leukemia) T cells. (b) PCA scatter plot of the Raman data shows two distinct groups for the two cell populations. (c) Loadings from the first principal component are similar to the difference between the mean spectra of the two cell types, indicating that the first component captures most of the differences.

relative intensity values of each of the three PCs. These maps show the relative distribution of the molecules in the cell as determined by each of the three PCs chosen.

A limitation of PCA is that the maximum variance is determined in the measured spectral data set without taking into account the different sample groupings. Therefore, PCA may be good for data reduction and visualizing data variance, but is not optimal for the discrimination and classification of groups. For this, linear discriminant analysis (LDA) is more appropriate, since it finds linear combinations of variables to define directions in spectral space that maximize the between group variance while minimizing the within group variance according to Fisher's criterion. However, LDA requires that the number of input variables into the LDA algorithm must be smaller than the number of observations (i.e., number of independent spectra of different cells). Therefore, Raman spectra cannot be directly analyzed by LDA, since a Raman spectral data set typically has more variables (wavelengths) than observations (cells). Frequently, a combined PCA-LDA approach [25] is used in Raman analysis where PCA first reduces the dimensionality of the data, and the significant PCs are used as input variables for LDA.

An n-fold cross validation method is often used to test the accuracy of the LDA model to predict an unknown sample type. In this procedure, Raman data are randomly divided into n different groups, and one group is treated as the blind sample and the remaining $n - 1$ groups are implemented into LDA to define the discriminant functions. This process is repeated for each group until all samples are classified and classification sensitivity and specificity values can be calculated. Performing the classification in this manner is ideal for data sets with a limited number of samples and ensures that the data to be classified does not bias the algorithm.

5.5 Laser Tweezers Raman Spectroscopy

The development of the technique of laser tweezers Raman spectroscopy (LTRS) by our group and others [26–29] was a significant advancement in the field of single-cell micro-Raman spectroscopy that has helped to further its use in biological and biomedical research. While all single-cell Raman spectroscopy systems discussed above require single cells to be adhered to a surface for Raman interrogation, LTRS-enabled single non-adherent cells to be probed while suspended in solution. In the simplest implementation of this method, LTRS uses a *single* tightly focused laser beam to optically trap a suspended cell within the laser focus and to simultaneously acquire its Raman spectrum using the same laser beam for excitation. The tight focusing condition is key, since it generates a high electric field gradient in the transverse direction that draws the particle to the center of the beam, which is the equilibrium position where there is no net lateral force due to the symmetry of the Gaussian-shaped beam. The axial gradient force negates the scattering force, resulting in stable trapping in the axial direction. There are many benefits of using this geometry: It significantly improves the signal to noise ratio of the Raman spectrum via a combination of optimal overlapping of the laser foci with the particle and lifting the cell away from the glass surface to avoid undesired background signals from the glass substrate. LTRS also facilitates the analysis of naturally non-adherent cells and small micrometer-sized (e.g., bacteria) and sub-micrometer-sized biological particles (e.g., isolated organelles, such as mitochondria). Another advantage of this configuration is its ability to be integrated with microfluidic systems, which enables unique Raman experiments to be performed and allows for the development of novel Raman-based optofluidic instrumentation. More recently, studies have also demonstrated the ability to measure both the mechanical properties and Raman chemical fingerprint at the same time using optical tweezers Raman spectroscopy. This section highlights key developments of this method throughout the past 15 years.

5.5.1 *Different Optical Trapping Geometries*

While the single-beam geometry is the simplest optical excitation design for LTRS, other configurations have been implemented. In another scheme two separate collinear laser beams at different wavelengths are used for optical trapping and Raman excitation [28]. This is desirable for situations in which higher laser powers are needed to stably trap an object, but where potential photodamage at visible and near-infared wavelengths limits the maximum powers that can be used. In this case, a long wavelength (e.g., 1064 nm) laser is used for optical trapping while a shorter wavelength (e.g., 633, 785 nm) is used for the Raman excitation. Decoupling of the optical trapping and excitation beam is also advantageous because it allows the Raman excitation beam to be scanned across different parts of the cell, or the focus of the excitation beam to be relaxed and reshaped in order to probe larger fractions of the cell. Creely et al. [30] developed a more advanced two color LTRS system in which the trapping beam at 1064 nm was split into multiple (3 or 4) laser foci using a spatial light modulator, which was used to trap different regions on the periphery of a cell to ensure the cell is immobilized and not allowed to rotate. By moving the foci pattern, the cell is scanned relative to the fixed 785 nm Raman excitation beam to enable different regions of the cell to be probed and to generate Raman images of the floating cell. Another configuration that has been demonstrated uses two divergent counterpropagating beams [31] emanating from two optical fibers opposite one another to trap a cell. A third focused beam is used as the Raman probe beam. The advantage of this geometry is that large cells can be trapped within the opposing divergent beams while minimizing the prospect of laser damage. Furthermore, the probe beam can scan across the cell to probe different regions of the large cell.

5.5.2 *Biological and Biomedical Applications of LTRS*

LTRS has shown promise as a single-cell tool for cancer detection at the single-cell level. Our group was the first to apply LTRS to the detection of cancer, specifically leukemia—a cancer of non-adherent blood cells [32, 33]. Using normal and cancer cells from both cell lines and isolated from patient samples, unique Raman markers were identified that enabled accurate detection and classification of the cell populations. Several Raman peaks that were noticeably different between normal and cancer cells included 785, 1093, 1126, 1575, and 1615 cm^{-1}, which are related to DNA and protein vibrations. Cancer cells exhibited a lower DNA/protein ratio compared to normal cells. Using a combination of PCA, LDA, and leave-n-out cross-validation methods, the sensitivity and specificity for discriminating cell populations were determined to be routinely greater than 90% for both cancer cell lines and clinical samples.

The ability to monitor cancer cell response to chemical agents and drugs in real time has important implications for applications such as in vitro cancer treatment monitoring, drug discovery/screening, or toxicological testing. Using a leukemia cell-chemotherapy drug system, we have demonstrated that Raman spectroscopy can detect the time-resolved intracellular molecular changes of drug-exposed cells [34] (Fig. 5.9). Raman spectra were obtained from cancer cells at different times (e.g., 24, 36, 72 h) following exposure to the chemotherapeutic drug doxorubicin at two different concentrations. Three distinct Raman signatures of drug exposed cells were observed, and PCA performed on the data

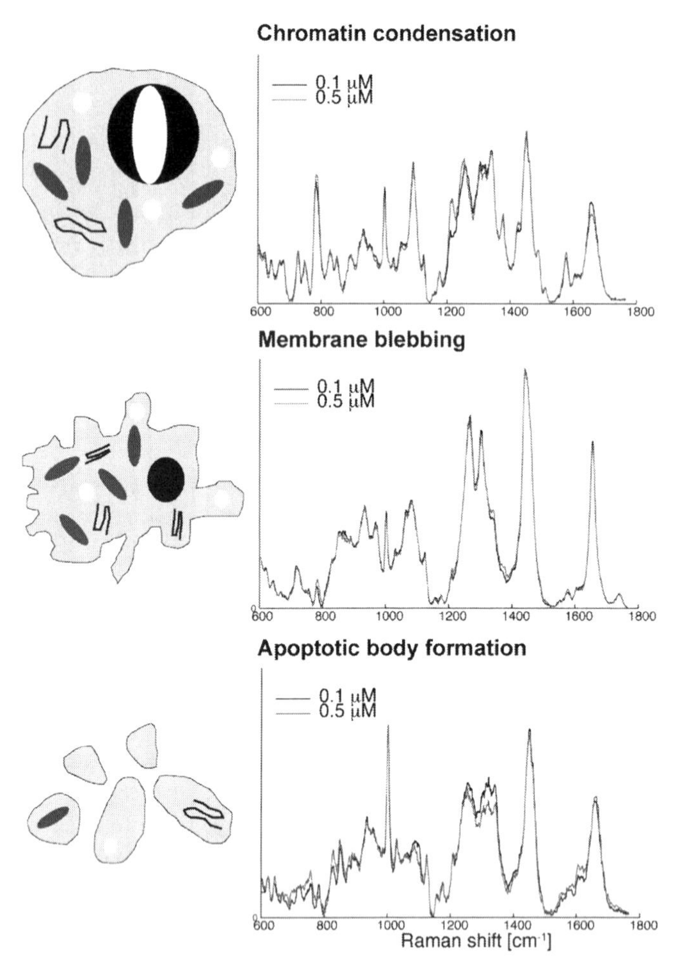

Figure 5.9 Raman spectra of doxorubicin-treated leukemia T cells show a time and drug concentration-dependent evolution of the cell spectra. The spectral profiles are consistent with known processes during cellular apoptosis (i.e., chromatin condensation, membrane blebbing, apoptotic body formation).

confirmed that these signatures group into distinct regions (Group 1, 2, 3) separate from the control spectra in the PCA scatter plot. Group 1 spectra were observed early in the drug interaction process and represent an increase in intracellular lipids. Group 2 spectra were observed next, which showed an intensity increase in DNA related Raman peaks. Group 3 spectra were observed last, showing a drop in overall signatures and increase in the phenyalanine peak. These time-resolved changes are consistent with membrane blebbing, chromatin condensation, and cytoplasmic components that are known to occur in that sequence during cellular apoptosis induced by doxorubicin. It was observed that cells at specific time points span multiple groups in the PCA plot, indicating that each cell, due to its stochastic nature, did not have the same temporal response time to the drug. However, despite the differences in the time-dependent behavior from cell to cell, the evolution of the spectral profiles remains consistent for all cells.

Other groups have demonstrated the detection and discrimination of other cancer types as well, including colorectal [35], prostate, and bladder cancer [36]. Aside from cancer, LTRS has a much broader range of potential applications. For example, our group has demonstrated that LTRS could be used to identify different blood cell phenotypes, characterize the composition of intranuclear proteins [37] relevant in neurodegenerative disease, determine the degree of saturation of lipid particles in the blood [38], monitor protein overexpression in bacteria cells [39], and detect the antibiotic drug susceptibility of bacteria cells [40, 41]. Other groups have demonstrated additional applications. Xie et al. [42] showed that Raman signatures of different bacterial species (*Bacillus cereus, Enterbacter aerogenes, Escheria coli, Streptococcus Pyrogenes, Enterococcus faecalis*, and *Streptococcus salivarius*) were sufficiently different for each species to be accurately identified. LTRS may therefore be a very attractive tool for rapid bacterial identification in hospital or point-of-care settings, especially for samples with a low cell concentration where it might compete with more costly and time-consuming techniques, such as polymerase chain reaction (PCR). It has also been shown that the germination kinetics [43] of individual bacterial spores can be studied in real time by LTRS by monitoring the evolution of Raman markers specific to calcium dipicolinic acid (CaDPA). A significant decrease in their signal intensity signified the release of CaDPA from the spore, which is a hallmark signature of the initiation of the Stage I germination process. Ajito et al. [44] demonstrated the analysis of single synaptosomes and the quantification of the concentration of glutamate that was released upon addition of a K+ channel blocker. This information is potentially important for understanding the role of glutamate in neural cells. Tang et al. [45] applied LTRS to optically trap individual mitochondria isolated from liver tissue, heart muscle, and kidney and monitored chemically induced changes in the physiological state of individual isolated mitochondria. Changes in the 1602 cm^{-1} band related to changing bioactivity of mitochondria were observed. Ojeda et al. [46] showed that individual human chromosomes could be captured and manipulated by optical tweezers, and Raman spectra acquired from chromosomes 1, 2, and 3 exhibited unique spectral profiles that enabled positive identification after the use of multivariate statistical analysis. The spectra reflected the different order and ratio of bases for the different chromosomes as well as the different associated proteins. Zachariah et al. [47] showed that Raman tweezers could characterize the oxidative stress of single red blood cells chemically induced by exposure to OH radicals. A sampling of these different LTRS studies in this section serves to reinforce several important points. (i) LTRS can be broadly applied to many different biological systems primarily because the technique is not limited by the availability and use of specific reagents or other molecular labels. (ii) An important aspect of LTRS research is the application of the technology to different biological systems and biomedical applications for the purpose of discovering new, useful Raman markers. (iii) LTRS can be used both as a basic research tool to investigate fundamental cellular properties and dynamics or as a clinical instrument for improving the detection, diagnosis, and treatment of diseases.

5.5.3 *Microdevices with Raman Tweezers Systems*

The manual operation of LTRS makes cell sampling an arduous and time-consuming procedure when large populations of cells need to be sampled. Raman analysis and sorting of cells can be automated by integrating LTRS with microfluidic devices. Such Raman-based optofluidic systems would significantly improve the performance efficiency of Raman cytometry, which is important for potential clinical applications of this technology. LTRS is highly amenable for integration with microfluidic systems given the ability of optical tweezers to capture and move cells suspended or flowing in solution. Xie et al. [48] first demonstrated Raman sorting of small cell populations using a simple system consisting of two chambers connected by a microchannel. A small population of cells were loaded into the sample chamber. After individual cells were trapped and analyzed by a single-beam LTRS system, they were transported through the channel using the laser trap and placed in the collection chamber at specific array patterns. After loading the collection chamber with all cells of interest, additional cellular analyses can be performed. Lau et al. [49] developed a more sophisticated approach for sampling larger numbers of cells by integrating a single-beam LTRS with multichannel microfluidic devices. The Raman activated cell sorter (RACS) is analogous in principle to commercial fluorescence activated cell sorters (FACS) that analyze and sort cells based on fluorescence signals. In the RACS scheme, the microfluidic device hydrodynamically focuses a stream of cells to the laser trap. A single trapped cell is pulled out of the cell stream into an adjacent channel, where its Raman spectrum is acquired. Manipulation of the cell out of the cell stream into the adjacent channel during the Raman spectral acquisition is needed to avoid the possibility of cells upstream displacing the trapped cell during the Raman acquisition, which can take several tens of seconds. After the analysis, cells can be sorted into channels downstream. Sorting of two different leukemia cell lines with the RACS system was demonstrated. Dochow et al. [50] reported the development of a similar Raman sorting system for the identification and discrimination of erythrocytes, leukocytes, acute myeloid leukemia cells, and breast tumor cells. Their system utilized counter propagating fiber lasers to trap a cell flowing in a microchannel orthogonal to the fiber lasers. A separate laser beam focused by a microscope objective was used to obtain Raman spectra from the trapped cells. Two different systems based on capillary and microchip designs were developed. The microchip was fabricated of glass due to its favorable optical properties. Isotropic wet etching was used to create the microchannels, and two mirror substrates were bonded face to face with thermal diffusion. A separate port allowed for insertion of the optical fibers. A cavity was implemented into the trapping region of the device to reduce the flow speed such that lower forces would be needed to trap the cell in the flowing media.

Microfluidic LTRS systems are also important in biological research because it facilitates the study of single-cell dynamics. The microfluidic devices enable fine control of the local chemical environment around the trapped cell as time resolved Raman spectra are acquired. Ramser et al. [51] measured the oxygenation cycle of red blood cells by using a microfluidic system to introduce different buffers to the trapped cell. The microfluidic system consisted of three adjacent reservoir chambers connected by two microchannels, and was molded in PDMS and sealed onto a glass coverslip. Cells were placed in the center reservoir, and the end reservoirs contained buffer solution. A voltage applied between electrodes placed in the two end reservoirs induced electro-osmotic flow, which allowed for fresh buffer or buffer containing sodium dithionite (which induces deoxygenation of red blood cells) to be flushed through the channels and into the center chamber containing the cells. A microfluidic-based LTRS system was also used to study *E. coli* cells overexpressing wild type neuroglobin (NGB) and its E7Leu mutant [52]. A flow cell was constructed from two cover glasses tightly sealed with a Plexiglas ring, which was connected to a continuous pump that can flush the cell with two buffer solutions purged with air or N_2. Aerobic conditions in the flow cell could be finely controlled during the spectral monitoring. The system was used to study the influence of environmental changes on the oxidation/oxygenation state of the NGB variants in *E. coli* cells. By switching between the two buffer solutions, the ratio of two Raman bands at 1361 and 1374 cm^{-1}, which provides a measure of the rate at which oxygen binds to wtNGB, could be monitored continuously, showing that these states could be switched back and forth.

5.5.4 *Multifocal Raman Tweezers*

A drawback of current LTRS systems is its low analytical throughput, making it slow and impractical for many single-cell studies. Single cells are sampled one at a time sequentially, and it typically takes several tens of seconds to minutes to acquire a Raman spectrum of the cell. Consequently, these systems are not designed to sample a significant number of cells or to monitor the response of an individual cell throughout its interaction with the environment and to do this for many cells simultaneously. To address this issue, a novel multifocal LTRS (M-LTRS) method has been developed that combines LTRS with multifocal optical tweezers, which expands the confocal detection principle to a parallel optical system that enables multiple individual suspension cells to be analyzed simultaneously. A 1D multifocal LTRS (M-LTRS) configuration was developed using a time-sharing trapping scheme [53], in which repeated rapid scanning of a single laser beam generates multiple foci where each spot can trap an individual particle. The Raman spectra of these particles are simultaneously detected by a single spectrometer and charge-coupled device (CCD) detector. A galvanometric scanning mirror system scans the 785 nm CW excitation laser beam such that the axial position of the laser beam does not change as the mirror is scanned. 1D scanning of the galvo-mirrors in a staircase wave pattern generates multiple time-sharing optical traps aligned along one direction in the focal plane of the objective. The full Raman spectra from the trapped objects can then be simultaneously collected with the objective, focused through the slit of the spectrometer, and imaged onto a thermoelectrically cooled (TE) 100 $\times$ 1340 pixel CCD camera. Instead of using a physical pinhole, confocal detection of the Raman signals from each trapped object is achieved through the combination of the narrow spectrometer slit width ($\sim$100 μm), the tight excitation laser focus, and selective software binning of the vertical pixels on the CCD camera. The application of M-LTRS for the Raman analysis of biological cells was demonstrated using red blood cells (RBCs). Here, the laser focus dwells on each cell for 0.1 second during each scanning period and the total laser illumination time is 15 seconds per cell. Up to 10 cells are analyzed at a time, resulting in a 10-

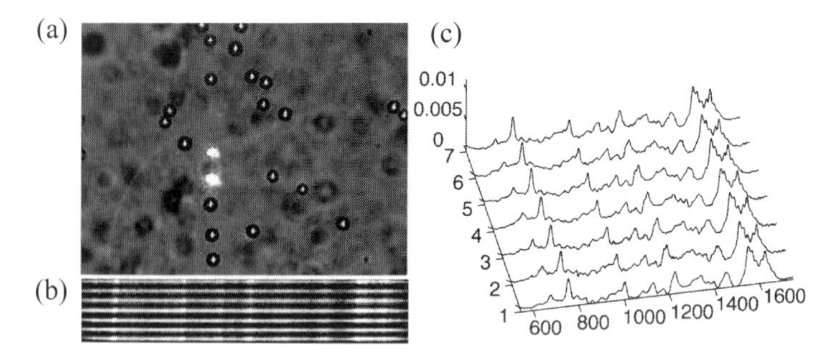

Figure 5.10 (a) Brightfield image of polymer particles trapped in a row using a time-shared optical trapping scheme (b). Image of individual Raman spectra positioned at discrete positions along the vertical axis of the CCD camera behind a monochromator. The spectra are due to single red blood cells trapped in a row. (c) Raman spectra of seven red blood cells trapped simultaneously.

fold improvement over previous single-beam LTRS systems. The schematics of multifocal optical trapping and Raman spectroscopy are shown in Fig. 5.10. A similar system was developed by Zhang et al. [54] for monitoring the dynamics of bacterial spore germination. Their system had a slightly different design in that it integrated a second scanning mirror directly in front of the spectrometer slit that was synchronized with the mirrors that scanned the beam at the sample plane. This allowed for decoupling of the excitation and detection components of the system allowing for the Raman spectra to be independently positioned at specific, defined regions of the spectrometer's CCD detector along the vertical axis.

5.5.5 *Multimodal LTRS System for Measuring Mechanical and Chemical Properties of Single Cells*

Optical tweezers have become an established tool for applying forces to biological cells. This capability can be exploited in an LTRS system to induce or measure mechanical properties of a single cell while simultaneously acquiring its Raman spectra. Acquiring both mechanical and biochemical information of a single cell can be advantageous for studying mechanically induced biochemical changes of a cell, or for using this multimodal information to improve the identification and diagnosis of cellular diseases. Rao et al. [55] reported the first direct measurement of the mechanically induced transition between deoxygenated and oxygenated states of hemoglobin in individual red blood cells using laser tweezers Raman spectroscopy. A dual-spot optical trap was created by splitting a 1064 nm beam using an interferometric arrangement. By adjusting the mirrors of the interferometer, the relative distance of the two spots was adjusted. This dual spot trap stretched single RBCs from its equilibrium position with forces on the order of tens of picoNewtons. A separate 785 nm Raman excitation beam measured the Raman spectrum of the cell while it was being stretched. Changes in key oxygenation Raman markers could be observed as the cell was stretched, indicating that mechanical stretching induces a transition to a deoxygenized state when an oxygenated RBC undergoes mechanical stretching. It was proposed that this transition is due to enhanced neighbor–neighbor hemoglobin (Hb) interactions and Hb-membrane rates of interaction as strain was applied to the cell and the Hb molecules. A similar study [56] investigating mechanically induced oxygenation changes of RBCs with LTRS was reported. One, two, and four time-sharing optical traps were generated by scanning a 1064 nm with a galvanometric mirror. A second laser at 532 nm was used to probe the Raman spectrum. Of note in this study was the demonstration that the mechanically induced deoxygenation transitions were reversible as individual cells were stretched and unstretched back to their equilibrium position. More recently, we have also shown that RBCs in a single-beam trap can be forced to deoxygenate simply by increasing the laser trap power. The result suggests that the higher laser power increases the force imposed on the cell, causing it to fold or squeeze in the trap and become more deoxygenated [57] (Fig. 5.11). The potential for this technology to characterize blood diseases has also been reported. Thalassemia, an RBC disease, is characterized by reduced rate of synthesis of either the α or β globin chains, leading to an imbalance of the chains and an excess of the normal globin chains. Free globin chains can bind and damage the cell membrane, leaving the cell more prone to mechanical injury. LTRS has been used to measure the oxygenation capability of β-thalassemic erythrocytes [58]. The results showed that the

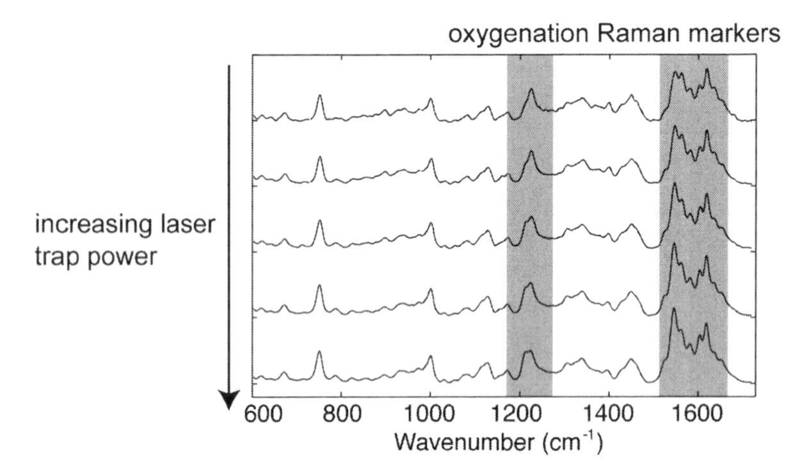

Figure 5.11 Mean Raman spectra of red blood cells trapped at different laser powers. An increase in laser trap power induces changes in the oxygenation related Raman bands, which suggests that increasing optical forces induce deoxygenation.

oxygenation capability of Hb was reduced in β-thalassemic RBCS. Also, after prolonged laser trapping times, Raman signals were reduced by 50% for normal cells compared to 80% for thalassemic cells, indicating that thalassemic cells were more sensitive to photo-oxidation. It is speculated that the increased sensitivity to oxidant stress of thalassemic cells is due to the excessive oxidation of the unstable free globin chains and the subsequent release of oxygen radicals that damage the cell membrane. Mechanical properties of the cells were also obtained by capturing white light images of cells in their relaxed and optically stretched positions and recording the maximum induced cell stretching length. From this data, diseased cells exhibited a 40% higher membrane shear modulus compared to healthy cells, indicative of the rigidity of the cell membranes.

5.6 Outlook

As we have seen, Raman spectroscopy has rapidly gained acceptance as a powerful companion to or even replacement of optical fluorescence microscopy in the life sciences. It is currently still limited by its relatively long signal acquisition times, which are mostly attributable to the significantly lower scattering cross section when compared with fluorescence excitation. This problem can be dealt with either by utilizing coherent Raman scattering techniques, such as stimulated Raman scattering (SRS) or coherent anti-Stokes Raman scattering (CARS), which greatly increase the signal obtained from Raman scattering, but at the cost of significantly higher system complexity. This problem can, on the other hand, also be dealt with for example by utilizing spectral shaping. Here, the Raman spectrum is dispersed onto a multichannel optically active device, such as a spatial light modulator (SLM) or a micromirror device to shape the spectrum in such a way that the entire spectroscopic signature of a substance can be combined into a single broadband signal, which can then be imaged by a point detector [59, 60]. This increases the signal-to-noise ratio significantly to the point where imaging can be conducted by up to a full order of magnitude faster compared to regular spontaneous Raman spectroscopy. Additional improvements in optical components, mostly to improve the efficiency with which spectrometers disperse and detect the light will further improve upon these limitations. We expect scientific CMOS cameras, especially once back-illuminated devices become available, to eventually replace CCD detectors, which will dramatically decrease the read-rates of Raman spectra. This development will be most significant for line or full field-of-view illumination modes, where hundreds of spectra are obtained and need to be read simultaneously.

Additionally, significant work on alternative optical geometries, and their combination with advanced sample handling schemes will help to unleash the full power of Raman spectroscopy. We envision the development of fully integrated compact optofluidic systems that combine optical trapping, optical stretching, and Raman spectroscopy for the mechanical and biochemical analysis of single cells. Such a system could be attractive for many potential applications in clinical diagnostics, basic and applied research, or point-of-care testing. One of the keys to achieving a high level of compactness with this technology, which is particularly important for point-of-care applications, will be the direct integration of optical components with microfluidic systems. It will not suffice to simply develop microfluidic devices for handling biological samples and coupling these devices to a large Raman microscope and detectors. Several studies have already begun exploring this integration of optical components into microfluidic systems for Raman spectroscopy. Ashok et al. [61] demonstrated the first fiber-based microfluidic Raman spectroscopy scheme, termed waveguide confined Raman spectroscopy (WCRS), in which optical fibers were directly embedded into a PDMS microfluidic chip. The fibers were placed orthogonal to each other in a fixed position and in direct contact with the fluid in the microchannel. In addition to the improved compactness of the system, other advantages of this design include the elimination of the background signals from the PDMS that would interfere with the Raman spectrum and the alignment free operation of the overall system. Solution phase detection of reaction products and chemicals in microdroplets was demonstrated. In another study [62], an optofluidic monolithic chip was fabricated in a fused silica glass substrate using femtosecond laser pulses. Both 3D optical waveguides and microfluidic channels were fabricated within a single glass substrate using femtosecond laser writing by using different laser power thresholds. Two counterpropagating waveguides were fabricated that were orthogonally adjacent to a center microchannel. 1070 nm laser light was coupled into these waveguides and the divergent beams emanating from the waveguide into the microchannel were able to trap and stretch a cell floating through the microchannel. This monolithic geometry is a considerable advancement compared to prior geometries [63, 64] that integrated optical fibers with glass capillaries and polymer microdevices, which require very careful alignment of the optical and microfluidic components. We anticipate that the integration of Raman spectroscopy into such a system should be relatively straightforward, especially given that fused silica has optimal optical properties (low intrinsic fluorescence and Raman signals) as a substrate for single-cell Raman spectroscopy.

Overall, Raman spectroscopy has evolved from fringe status, where only a handful of select research groups were able to perform single-cell spectroscopy to an entire research topic by itself, enriching and complementing other optical tools. Such multimodal imaging and analysis tools are currently in the early phase of making their way into the market place, but based on ever-increasing demands on minimally invasive experiments and diagnostics in the life sciences, laser Raman microspectroscopy is one of the most powerful tools available to analyze and characterize living cells and their components.

References

1. Raman, C. V., and Krishnan, K. S. (1928). A new type of secondary radiation. *Nature* **121**:501–502.
2. Baek, M., Nelson, W. H., Britt, D., and Sperry, J. F. (1988). UV-Excited Resonance Raman Spectra of Heat Denatured Lysozyme and *Staphylococcus epidermidis*. *Appl. Spectrosc.* **42**:1312–1314.
3. Puppels, G. J., de Mul, F. F. M., Otto, C., et al. (1990). Studying single living cells and chromosomes by confocal Raman microspectroscopy. *Nature* **347**:301–303.
4. Uzunbajakava, N., Lenferink, A., Kraan, Y., et al. (2003). NonresonantRaman imaging of protein distribution in single human cells. *Biopolymers* **72:**1–9.
5. Puppels, G. J., Olminkhof, J. H., Segers-Nolten, G. M., et al. (1991). Laser irradiation and Raman-spectroscopy of single living cells and chromosomes: sample degradation occurs with 514.5 nm but not with 660 nm laser light. *Exp. Cell Res.* **195**:361–367.
6. Neuman, K. C., Chadd, E. H., Liou, G. F., Bergman, K., and Block, S. M. (1999).Characterization of photodamage to *Escherichia coli* in optical traps. *Biophys. J.* **77**:2856–2863.
7. Liang, H., Vu, K. T., Krishnan, P., et al. (1996). Wavelength dependence of cell cloning efficiency after optical trapping. *Biophys. J.* **70**:1529–1533.
8. Ando, M., Sugiura, M., Hayashi, H., and Hamaguchi, H. O. (2011). 1064 nm Deep near-infrared (NIR) excited raman microspectroscopy for studying photolabile organisms. *Appl. Spectrosc.* **65:**488–492.
9. Huser, T., Orme, C. A., Hollars, C. W., Corzett, M. H., and Balhorn, R. (2009). Raman spectroscopy of DNA packaging in individual human sperm cells distinguishes normal from abnormal cells. *J. Biophotonics* **2:**322–332.
10. Wood, B. R., Caspers, P., Puppels, G. J., Pandiancherri, S., and Mc-Naughton, D. (2007). Resonance Raman spectroscopy of red blood cells

using near-infrared laser excitation. *Anal. Bioanal. Chem.* **387**:1691–1703.

11. Wood, B. R., Hammer, L., Davis, L., and McNaughton, D. (2005). Raman microspectroscopy and imaging provides insights into heme aggregation and denaturation within human erythrocytes. *J. Biomed. Opt.* **10**:014005.
12. Huang, Y. S., Karashima, T., Yamamoto, M., Ogura, T., and Hamaguchi, H. (2004). Raman spectroscopic signature of life in a living yeast cell. *J. Raman Spectrosc.* **35:**525–526.
13. Zoladek, A., Pascut, F. C., Patel, P., and Notingher, I. (2011). Non-invasive time-course imaging of apoptotic cells by confocal Raman micro-spectroscopy. *J. Raman Spectrosc.* **42**:251–258.
14. Hamada, K., Fujita, K., and Smith, N. I., et al. (2008). Raman microscopy for dynamic molecular imaging of living cells. *J. Biomed. Opt.* **13**: 044027.
15. Notingher, I., Bisson, I., Bishop A. E., et al. (2004). In situ spectral monitoring of mRNA translation in embryonic stem cells during differentiation in vitro. *Anal. Chem.* **76**:3185–3193.
16. Okuno, M., and Hamaguchi, H. O. (2010). Multifocus confocal Raman microspectroscopy for fast multimode vibrational imaging of living cells. *Opt. Lett.* **35:**4096–4098.
17. Oshima, Y., Furihata, C., and Sato, H. (2009). Light sheet direct Raman imaging technique for observation of mixing of solvents. *Appl. Spectrosc.* **63**:1115–1120.
18. Barman, I., Tan, K. M., and Singh, G. P. (2010). Optical sectioning using single-plane-illumination Raman imaging. *J. Raman Spectrosc.* **41**:1099–1101.
19. Lieber, C. A., and Mahadevan-Jansen, A. (2003). Automated method for subtraction of fluorescence from biological Raman spectra. *Appl. Spectrosc.* **57:**1363–1367.
20. Beier, B. D., and Berger, A. J. (2009). Method for automated background subtraction from Raman spectra containing known contaminants. *Analyst* **134:**1198–1202.
21. Shreve, A. P., Cherepy, N. J. and Mathies, R. A. (1992). Effective Rejection of Fluorescence Interference in Raman-Spectroscopy Using a Shifted Excitation Difference Technique. *Appl. Spectrosc.* **46**:707–711.
22. Mosier-Boss, P. A., Lieberman, S. H., and Newbery, R. (1995). Fluorescence Rejection in Raman Spectroscopy by Shifted-Spectra, Edge-Detection, and FFT Filtering Techniques. *Appl. Spectrosc.* **49**:630–638.

23. De Luca, A. C., Mazilu, M., Riches, A., Herrington, C. S., and Dholakia, K. (2010). Online fluorescence suppression in modulated Raman spectroscopy. *Anal. Chem.* **82**:738–745.

24. Miljkovic, M. Chernenko, T., Romeo, M. J., et al. (2010). Label-free imaging of human cells: algorithms for image reconstruction of Raman hyperspectral datasets. *Analyst* **135**:2002–2013.

25. Notingher, L., Jell, G., Notingher, P. L., et al. (2005). Multivariate analysis of Raman spectra for in vitro non-invasive studies of living cells. *J. Mol. Struct.* **744**:179–185.

26. Xie, C. G., Dinno, M. A., and Li, Y. Q. (2002). Near-infrared Raman spectroscopy of single optically trapped biological cells. *Opt. Lett.* **27**:249–251.

27. Chan, J. W., Esposito, A. P., Talley, C. E., et al. (2002). Reagentless identification of single bacterial spores in aqueous solution by confocal laser tweezers Raman spectroscopy. *Anal. Chemi.* **76**:599–603.

28. Creely, C. M., Singh, G. P., and Petrov, D. (2005). Dual wavelength optical tweezers for confocal Raman spectroscopy. *Opt. Commun.* **245:**465–470.

29. Ramser, K., Logg, K., Goksör, M., et al. (2004). Resonance Raman spectroscopy of optically trapped functional erythrocytes. *J. Biomed. Opt.* **9**:593–600.

30. Creely, C. M., Volpe, G., Singh, G. P., Soler, M., and Petrov, D. V. (2005). Raman imaging of floating cells. *Opt. Express* **13**:6105–6110.

31. Jess, P. R. T., Garcés-Chávez, V., Smith, D., et al. (2006). Dual beam fibre trap for Raman microspectroscopy of single cells. *Opt. Express* **14**:5779–5791.

32. Chan, J. W., Taylor, D. S., Lane, S. M., et al. (2008). Nondestructive identification of individual leukemia cells by laser trapping Raman spectroscopy. *Anal. Chem.* **80**:2180–2187.

33. Chan, J. W., Taylor, D. S., Zwerdling, T., et al. (2006). Micro-Raman spectroscopy detects individual neoplastic and normal hematopoietic cells. *Biophys. J.* **90**:648–656.

34. Moritz, T. J., Taylor, D. S., Krol, D. M., Fritch, J., and Chan, J. W. (2010). Detection of doxorubicin-induced apoptosis of leukemic T-lymphocytes by laser tweezers Raman spectroscopy. *Biomed. Opt. Express* **1**:1138–1147.

35. Chen, K., Qin, Y. J., Zheng, F., Sun, M. H., and Shi, D. R. (2006). Diagnosis of colorectal cancer using Raman spectroscopy of laser-trapped single living epithelial cells. *Opt. Lett.* **31:**2015–2017.

36. Harvey, T. J., Faria, E. C., Henderson, A., et al. (2008). Spectral discrimination of live prostate and bladder cancer cell lines using Raman optical tweezers. *J. Biomed. Opt.* **13**:064004.
37. Moritz, T. J., Brunberg, J. A., Krol, D. M., et al. (2010). Characterisation of FXTAS related isolated intranuclear protein inclusions using laser tweezers Raman spectroscopy. *J. Raman Spectrosc.* **41**:33–39.
38. Chan, J. W., Motton, D., Rutledge, J. C., Keim, N. L., and Huser, T. (2005). Raman spectroscopic analysis of biochemical changes in individual triglyceride-rich lipoproteins in the pre- and postprandial state. *Anal. Chem.* **77**:5870–5876.
39. Chan, J. W., Winhold, H., Corzett, M. H., et al. (2007). Monitoring dynamic protein expression in living E. coli. Bacterial cells by laser tweezers Raman spectroscopy. *Cytometry A* **71**:468–474.
40. Moritz, T. J., Polage, C. R., Taylor, D. S., et al. (2010). Evaluation of Escherichia coli cell response to antibiotic treatment by use of Raman spectroscopy with laser tweezers. *J. Clin. Microbiol.* **48**:4287–4290.
41. Moritz, T. J., Taylor, D. S., Polage, C. R., et al. (2010). Effect of cefazolin treatment on the nonresonant Raman signatures of the metabolic state of individual Escherichia coli cells. *Anal. Chem.* **82**:2703–2710.
42. Xie, C., Mace, J., Dinno, M. A., et al. (2005). Identification of single bacterial cells in aqueous solution using conflocal laser tweezers Raman spectroscopy. *Anal. Chem.* **77**:4390–4397.
43. Kong, L. B., Zhang, P. F., Setlow, P., and Li, Y. Q. (2010). Characterization of Bacterial Spore Germination Using Integrated Phase Contrast Microscopy, Raman Spectroscopy, and Optical Tweezers. *Anal. Chem.* **82**:3840–3847.
44. Ajito, K., Han, C. X., and Torimitsu, K. (2004). Detection of glutamate in optically trapped single nerve terminals by Raman spectroscopy. *Anal. Chem.* **76**:2506–2510.
45. Tang, H., Yao, H., Wang, G., et al. (2004). NIR Raman spectroscopic investigation of single mitochondria trapped by optical tweezers. *Opt. Express* **15**:12708–12716.
46. Ojeda, J. F., Xie, C., Li, Y. Q., et al. (2006). Chromosomal analysis and identification based on optical tweezers and Raman spectroscopy. *Opt. Express* **14**:5385–5393.
47. Zachariah, E., Bankapur, A., Santhosh, C., Valiathan, M., and Mathur, D. (2010). Probing oxidative stress in single erythrocytes with Raman Tweezers. *J Photoch. Photobio. B* **100**:113–116.

48. Xie, C. G., Chen, D., and Li, Y. Q. (2005). Raman sorting and identification of single living micro-organisms with optical tweezers. *Opt. Lett.* **30**:1800–1802.
49. Lau, A. Y., Lee, L. P., and Chan, J. W. (2008). An integrated optofluidic platform for Raman-activated cell sorting. *Lab Chip* **8**:1116–1120.
50. Dochow, S., Krafft, C., Neugebauer, U., et al. (2011). Tumour cell identification by means of Raman spectroscopy in combination with optical traps and microfluidic environments. *Lab Chip* **11:**1484–1490.
51. Ramser, K., Enger, J., Goksör, M., et al. (2005). A microfluidic system enabling Raman measurements of the oxygenation cycle in single optically trapped red blood cells. *Lab Chip* **5:**431–436.
52. Ramser, K., Wenseleers, W., Dewilde, S., et al (2007). Micro-resonance Raman study of optically trapped Escherichia coli cells overexpressing human neuroglobin. *J. Biomed. Opt.* **12**:044009.
53. Liu, R., Taylor, D. S., Matthews, D. L., and Chan, J. W. (2010). Parallel analysis of individual biological cells using multifocal laser tweezers Raman spectroscopy. *Appl. Spectrosc.* **64:**1308–1310.
54. Zhang, P. F., Kong, L. B., Setlow, P., and Li, Y. Q. (2010). Multiple-trap laser tweezers Raman spectroscopy for simultaneous monitoring of the biological dynamics of multiple individual cells. *Opt. Lett.* **35:**3321–3323.
55. Rao, S., Balint, S., Cossins, B., Guallar, V., and Petrov, D. (2009). Raman study of mechanically indued oxygenation state transition of red blood cells using optical tweezers. *Biophys. J.* **96**:2043–2043.
56. Rusciano, G. (2010). Experimental analysis of Hb oxy-deoxy transition in single optically stretched red blood cells. *Phys. Medica.* **26**:233–239.
57. Liu, R., Zheng, L., Matthews, D. L., Satake, N., and Chan, J. W. (2011). Power dependent oxygenation state transition of red blood cells in a single beam optical trap. *Appl. Phys. Lett.* **99**:043702.
58. De Luca, A. C., Rusciano, G., Ciancia, R., et al. (2008). Spectroscopical and mechanical characterization of normal and thalassemic red blood cells by Raman Tweezers. *Opt. Express* **16**:7943–7957.
59. Smith, Z. J., Strombom, S., and Wachsmann-Hogiu, S. (2011). Multivariate optical computing using a digital micromirror device for fluorescence and Raman spectroscopy. *Opt. Express* **19**:16950–16962.
60. Davis, B. M., Hemphill, A. J., Maltaş, D. C, et al. (2011). Multivariate hyperspectral Raman imaging using compressive detection. *Anal. Chem.* **83**:5086–5092.

61. Ashok, P. C., Singh, G. P., Rendall, H. A., Krauss, T. F., and Dholakia, K. (2011). Waveguide confined Raman spectroscopy for microfluidic interrogation. *Lab Chip* **11**:1262–1270.

62. Bragheri, F., Ferrara, L., Bellini, N., et al. (2010). Optofluidic chip for single cell trapping and stretching fabricated by a femtosecond laser. *J. Biophotonics* **3**:234–243.

63. Guck, J., Ananthakrishnan, R., Mahmood, H., et al. (2001). The optical stretcher - A novel laser tool to micromanipulate cells. *Biophys. J.* **80**:277A–277A.

64. Lincoln, B., Erickson, H. M., Schinkinger, S., et al. (2004). Deformability-based flow cytometry. *Cytometry. A* **59**:203–209.

Chapter 6

High-Speed Calcium Imaging of Neuronal Activity Using Acousto-Optic Deflectors

Benjamin F. Grewe
James H. Clark Center for Biomedical Engineering and Sciences, Stanford University, Stanford, California, USA
grewe@stanford.edu

One of the most challenging questions in neuroscience is how neuronal circuits in different brain areas process incoming stimuli such as tones (auditory cortex), contrasts (visual cortex), or touch (somatosensory cortex). While working principles of such neuronal networks can be investigated to some extent by imaging cultured cell networks and acute brain slice preparations, gaining a more complete real-world picture requires monitoring of tens or hundreds of neurons in the intact brains of living animals. Because most neurons are located relatively deep below the brain surface (>100 μm), optical measurement techniques used have to provide a sufficient imaging depth, combined with the ability to resolve the activity of individual cells within a neuronal network. Two-photon laser scanning microscopy meets these criteria and has become the standard technique for functional in vivo measurements of neuronal populations. In addition, because communication between neuronal cells usually happens on a time scale of tens of milliseconds,

Understanding Biophotonics: Fundamentals, Advances, and Applications
Edited by Kevin K. Tsia

ISBN 978-981-4411-77-6 (Hardcover), 978-981-4411-78-3 (eBook)
www.panstanford.com

it is also desirable to record neuronal activity at similar or even higher sampling rates. This chapter therefore focuses on a recent technological development that uses acousto-optic deflectors (AODs) as laser scanners to improve the scanning speed when imaging large areas of neuronal cell networks. Beginning with the underlying technical basics of two-photon microscopy, acousto-optics, and laser beam dispersion issues, important advantages and disadvantages of the novel high-speed scanning method are discussed, followed by a description of possible cell-scanning methods. Finally, an example of an in vivo imaging experiment from mouse visual cortex is presented to demonstrate the advantages of the AOD system for high-speed scanning for in vivo two-photon microscopy.

6.1 Introduction

One of the most powerful advantages of using two-photon microscopy is the ability to image deep inside light-scattering tissue up to a depth of 1 mm [1]. Two-photon imaging is thus often used in live animal preparations to image neuronal cell networks that are located deep below the cortical surface. In combination with functional fluorescence indicators that report neuronal spiking activity, two-photon imaging allows the characterization of spatiotemporal network activity pattern in the intact brain (for rev. see [2]).

Neuronal spiking activity, or action potentials (APs), although primarily carried by sodium currents, can be linked to a substantial increase in the intracellular calcium concentration [3]. The calcium concentration in turn can be transformed into a fluorescence signal using a calcium sensitive indicator. This link makes it possible to reconstruct the underlying electrical firing pattern of neurons by analyzing their fluorescence signals. Today, calcium-sensitive fluorescent dyes that transform the electrical activity of cells into a fluorescence signal, are sufficiently advanced to provide signal-to-noise ratios (SNR) great enough to extract AP firing. In addition, various labeling methods that range from single-cell labeling to bulk labeling of large populations in the order of hundreds of cells are currently available for use in live animal (in vivo) preparations. For bulk loading, new forms of synthetic acetomethyl-esther (AM) dyes allow staining of large populations of neuronal cell populations with fluorescent calcium indicators, e.g., Oregon Green BAPTA-1 (OGB-1), Fluo-4, or Rhod-2. Such AM indicators can be bolus injected directly into neural tissue using glass micropipettes [4]. After a short incubation time, fluorescent calcium traces of neuronal cell populations can be recorded and spiking activity can be extracted to a resolution of even single APs [5, 6]. More recently, novel genetically encoded Ca^{2+} indicators (GECIs) that employ fluorescent Ca^{2+}-sensitive proteins have been used to image neuronal population activity repeatedly over days and months [7]. Thus, despite some general limitations of calcium imaging, it is improvements in microscopy technology rather than the development of better fluorescent reporters that has become the bottleneck for optimizing fast optical measurements of neural population activity [8–12]. In order to improve two-photon imaging techniques, two key directions have emerged in recent years, which include (i) increasing the size of the sampled area or volume, and the thus the proportion of sampled neurons within the whole network, and (ii) increasing the imaging speed, i.e., increasing the effective sampling rate per cell. In nearly every laser scanning method, area or volume of the network sampled and sampling speed relate inversely to each other. This is because a minimum fluorescence signal integration time is needed per pixel or per cell in order to capture enough fluorescence to achieve an adequate SNR. For this reason, most recent technological developments have been aimed at improving either one or the other parameter. Improving imaging speed is an important goal for accurately recording neuronal information processing that usually happens on a time scale of tens of milliseconds. Following the network activity precisely in time thus requires fast imaging methods, which enable measurements of neuronal activity patterns with a sufficient temporal resolution. Inadequate imaging speed may be the reason why, to date, most in vivo two-photon imaging studies of cortical circuit dynamics have focused on mapping the spatial organization of neuronal tuning properties in response to sensory stimuli [8–12] rather than investigating dynamic responses of such networks.

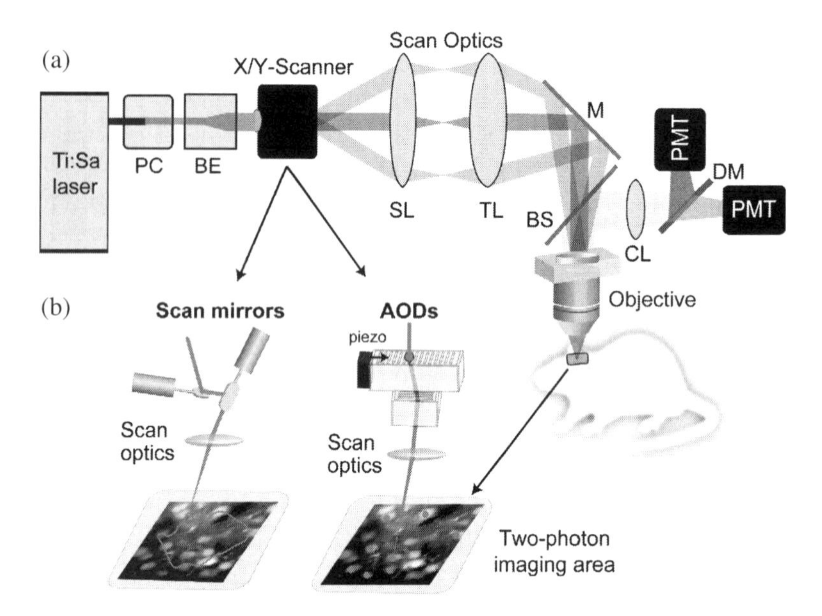

Figure 6.1 (a) Schematized setup of a two-photon microscope system, which includes the titanium:sapphire laser oscillator (Ti:Sa), a Pockels cell (PC), a beam expander (BE), an x/y-laser scanner, the scan optics (scan lens SL and tube lens TL), a fluorescence collection condenser lens (CL), a broadband near infrared mirror (M), a two-photon beam splitter (BS), a dichroic mirror (DM), a fluorescence photon detector such as a photomultiplier (PMT), and the microscope objective (OBJ). (b) Simplified schemes of different laser scanning devices which can be galvanometric or resonant scanning mirrors for fast frame- or line-scanning (left) or AODs for high-speed random-access scanning (right), (b) was adapted with permission from [2].

Figure 6.1 shows a simplified diagram of the basic setup of a standard two-photon microscope. This includes a titanium:sapphire (Ti:Sa) laser source, a Pockels cell (PC) for laser intensity regulation (although combinations of neutral density filters may also fulfill this purpose), x/y-laser scanners to direct the laser beam, standard scan optics (a scan lens and a tube lens), a dichroic beam splitter to reflect the fluorescence light into the detection pathway, photo multipliers (PMTs) as fluorescence detectors, and finally the microscope objective.

To improve speeds of laser beam scanning, AODs can be used to replace the standard voltage-driven galvanometric scan mirrors (Fig. 6.1b). Following this approach, we recently demonstrated dynamic high-speed population calcium imaging using an AOD microscope, which allowed action potentials to be detected with near-millisecond precision and increased scanning speed severalfold [5].

6.2 The AOD Two-Photon Microscope Design

The main body of the AOD-imaging system is designed as a "moving-stage" microscope that allows the entire microscope to be moved while keeping the sample fixed (Fig. 6.2a). An advantage of this design is that it provides a large space below the objective lens, allowing access for other equipment such as table-mounted manipulators for electrophysiological recordings. The entire microscope is built around an aluminum column mounted onto a motorized x/y stage (Fig. 6.2b).

Several important issues need to be considered when designing this type of microscope. In the set up shown in Fig. 6.2, the AODs were selected to allow adequate scan-resolution over a large field of view (FOV) and optimal laser beam transmission, featuring an active aperture of 10 mm, 47 mrad maximum scan-angle, and about 55% transmission (x/y-AOD) specified for 850 nm wavelength. Although AODs can still function over a small range of wavelengths (e.g., ± 20 nm for an 850 nm AOD), they are designed and optimized for a single wavelength and should be selected with respect to the calcium indicator excitation wavelength. AOD specifications must also be considered with respect to the objective and the scan- and tube-lenses as these determine the image resolution and the FOV size. As the maximal AOD scan angle Θ is limited (47 mrad), the number of resolvable spots N is also limited due to the intrinsic beam divergence and depends on the laser beam diameter at the AOD active aperture (see also Eq. 6.4).

If best-possible resolution is the main interest, the laser beam should fill the AOD aperture (to maximize N) while FOV size should be bigger than $S * N$ (where S is the diffraction-limited spatial resolution determined by the effective numerical aperture [NA] of

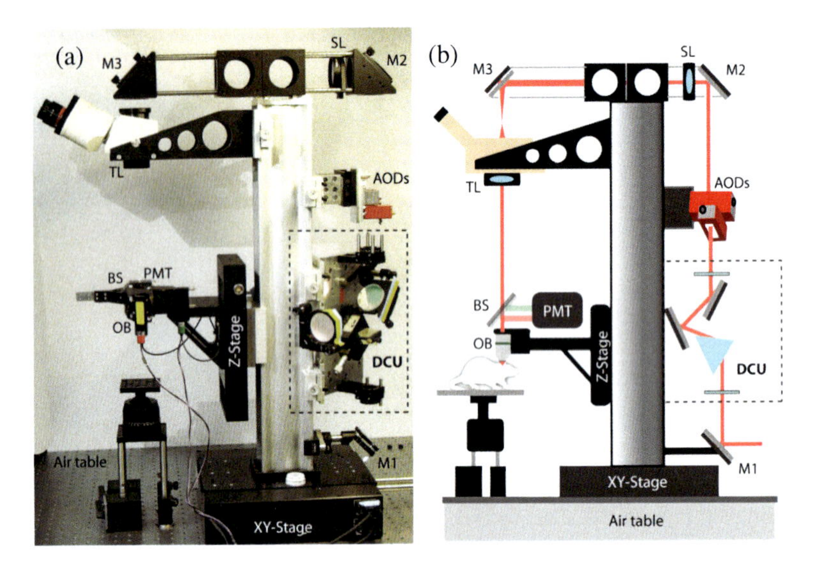

Figure 6.2 AOD-based microscope setup for high-speed calcium imaging. (a) Photo of the two-photon microscope setup mounted on an air-table. A motorized x/y-stage supports movements of the entire microscope over the sample. Above the dispersion compensation unit (DCU) a pair of orthogonal AODs was used for x/y-scanning. Scan lens (SL) and tube lens (TL) were chosen to provide a large FOV with a 40x objective (OB). Fluorescence is collected in two channels (red and green) using two photomultipliers (PMTs). M1, M2 and M3 are near-infrared dielectric broadband mirrors. BS is two-photon beam splitter. (b) Simplified schematic drawing of the same two-photon setup.

the microscope objective). If maximizing the FOV for functional imaging of large neuronal cell populations is the main goal, it is also possible to adapt the combination of scan and tube lens with respect to a particular microscope objective. The final FOV size or scan area under the objective can be calculated according to simple lens equations using the focal lengths of the objective, scan and tube lenses, and the scan angle of the AODs.

Finally, every two-photon microscope "ends" with the most essential parts, the microscope objective and the fluorescence detection unit. In the example above, these were directly mounted on a motorized z-stage. For separate detection of fluorescence photons at different wavelengths (here green and red) the fluorescence detection pathway includes a dichroic beam splitter, allowing green and red fluorescence photons to be detected separately on either of the two photomultipliers (PMTs). Depending on the fluorescence detection wavelength, suitable chromatic filters are installed before the PMTs.

6.3 Laser Scanning with Acousto-Optic Devices

Acousto-optic devices are very interesting for two-photon microscopy because they enable fast scanning of random-positions within the FOV at kHz rates. Such random-access scanning allows adaptation of the scanning path to match regions of interest in biological samples and thus combines fast scanning of these regions with an optimal SNR. Nevertheless, using AODs for laser beam scanning also creates some technical challenges as AODs usually employ highly dispersive optic media and have a relatively poor laser beam diffraction efficiency of about 75% per AOD.

6.3.1 *Operation Modes of Acousto-Optic Deflectors (AODs)*

In a standard AOD, a sinusoidal radio-frequency (RF) signal (typically 76–114 MHz, slow shear oscillation mode) is applied to a piezoelectric transducer that is fused to an acousto-optic crystal. The piezoelectric transducer generates an RF acoustic wave of compression and relaxation acting like a "phase grating" that travels through the crystal with an acoustic velocity depending on the optic material and an acoustic wavelength that matches the RF signal. Due to this dynamic grating within the crystal, any incident laser beam will be diffracted in an angle proportional to the sound-wave frequency, which allows operating the AOD as a simple 1D laser scanner.

By varying the AOD construction properties such as the length of the piezoelectric transducer (L) or the refractive index of the optically active material (n), AODs can be operated in two different basic interaction modes, the Bragg and the Raman–Nath mode. To determine in which mode the AOD is operating, the parameter Q has been defined as the "quality factor" of the interaction mode that can

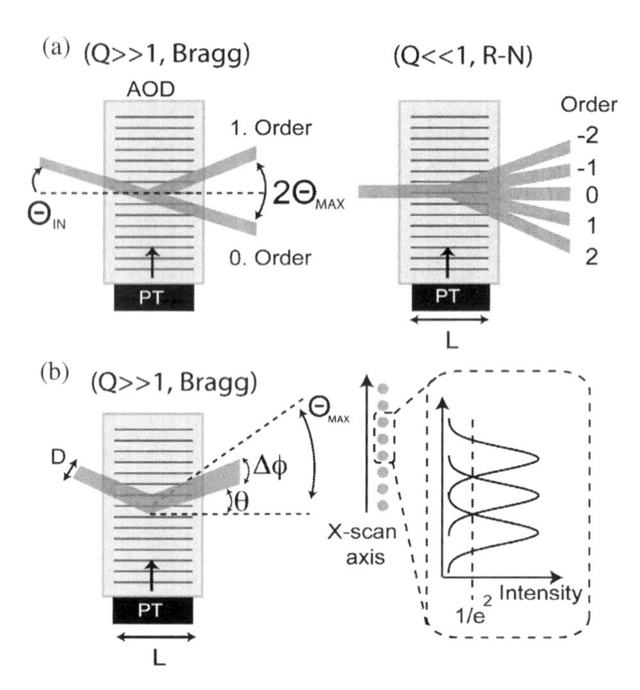

Figure 6.3 (a) Schematic drawing of an acousto-optic deflector that can be used in the "Bragg" (left) or in the "Raman–Nath" (right) operation mode depending on its construction properties. (b) Left: Schematic of laser beam scanning using an acousto-optic crystal in the Bragg mode. A standing sound wave is produced by a piezoelectric transducer and acts as a diffraction grating, deflecting a major part of the laser beam into the first diffraction order. The angle of laser beam deflection is thereby proportional to the frequency of the acoustic wave, which in turn is determined by the oscillation frequency of the attached piezo-transducer (PT). Right: Schematic illustration of the resolution properties of an acousto-optic scanner. In this scheme D is the laser beam diameter, Θ_{IN} the laser beam incident angle, Θ_{MAX} the maximum possible scan angle, θ the dynamic scan angle of the first diffraction order, $\Delta\varphi$ the divergence of the diffracted laser beam, and L the length of the piezo-transducer.

be either Bragg ($Q >>1$) or Raman–Nath ($Q <<1$) (Fig. 6.3a). This quality factor Q is given by

$$Q = \frac{2\,\pi}{n}\,\frac{\lambda\,L}{\Lambda^2} \tag{6.1}$$

where λ is the wavelength of the incident laser beam, n is the refractive index of the crystal, L is the distance the laser beam travels throughout the active acoustic area, and Λ is the acoustic wave length. In the Bragg regime ($Q >> 1$), a certain fixed incident angle Θ_{IN} produces mainly one order of diffraction with more than 50% of the incident beam intensity, while the other orders are mostly annihilated by destructive interference. When operating AODs in the Bragg mode (Fig. 6.3b, left), the dynamic deflection angle of the first order (θ) is given by

$$\theta = \frac{\lambda f}{V_{\text{acoustic}}} \tag{6.2}$$

where λ is the laser light wavelength, f is the sound wave frequency and V_{acoustic} is the sound velocity within the acousto-optic material. Most acousto-optic deflectors are operated in the Bragg-regime, as only this mode enables laser beam diffraction into the first order of up to 75% per AOD (slow shear mode).

6.3.2 *Scanning Resolution of Acousto-Optic Devices*

The static scan resolution of an AOD can be defined as the maximum number of resolvable scanning points or directions of the deflected laser beam over the total scanning angle (Fig. 6.3b, right). For a collimated laser beam with the diameter d(typically $1/e^2$, TEM00-type), the natural beam divergence $\Delta\varphi$ is given by

$$\Delta\varphi = \frac{\lambda}{d} \tag{6.3}$$

Knowing the maximum scan angle of the AODs ($\Theta_{\max}$) and the maximum laser beam diameter ($D_{\max}$), the number of resolvable spots can be calculated as

$$N = \frac{\pi}{4}\,\frac{\Theta_{\max}}{\Delta\phi} = \frac{\pi}{4}\,\frac{\Theta_{\max} * D_{\max}}{\lambda} \tag{6.4}$$

where $D_{\max}$, the maximum laser beam diameter, is usually limited by the active aperture of the AOD scanners. While Eq. 6.4 holds in for all types of scanners, for AODs the maximum scanning angle ($\Theta_{\max}$) needs to be calculated as

$$\Theta_{\max} = \lambda\,\frac{\Delta F_{\Theta}}{V_{\text{acoustic}}} \tag{6.5}$$

where λ is again the wavelength of the incident laser beam, ΔF_{Θ} is the acousto-optic frequency bandwidth, and V_{acoustic} is the acoustic velocity within the crystal. Combining the Eqs. 6.4 and 6.5 results in

$$N = \frac{\pi}{4}\frac{\Delta F_{\Theta} D_{\max}}{V_{\text{acoustic}}} = \frac{\pi}{4}\Delta F_{\Theta} \cdot \Delta T_{\text{access}} \qquad (6.6)$$

In this equation, the number of resolvable spots (N) is equal to the acousto-optic bandwidth (ΔF_{Θ}) multiplied by the access or transition time (ΔT_{access}) of the AODs. This transition time corresponds to the time that the acoustic wave needs to travel through the active aperture of the AODs and thus to move the laser beam from one scanning position to another. Often an AOD is thus characterized by the time–bandwidth product (TBP), which is given by TBP $= \Delta F_{\Theta} * \Delta T_{\text{access}}$. When using a pair of crossed AODs in series for 2D scanning the total transition time will not change when the driving signals of the AODs are synchronized.

6.3.3 *Specifications of Acousto-Optic Deflectors*

For most imaging applications a high scanning precision (high N) in combination with good diffraction efficiency is necessary to scan precisely with the laser beam. Therefore large-aperture tellurium dioxide (TeO_2) AODs (10–30 mm aperture size, slow shear mode) are usually used to deflect large-diameter laser beams. This minimizes the relative intrinsic optical divergence and thus increases the scanning precision or scan resolution for imaging applications (see Eq. 6.4). The main parameters that characterize the performance of an acousto-optic deflector are listed in the Table 6.1.

Figure 6.4 illustrates how a commercially available acousto-optic deflector (AA optoelectronic) is integrated into the electric circuit. The acousto-optic crystal, the piezoelectric transducer, and the housing are also pictured. Using two of such deflector types that are orthogonally crossed in series, a fast laser beam scanning in two dimensions can be realized (Fig. 6.4b).

Table 6.1 Typical parameters of AODs that are optimized for the slow shear operation mode (76–114 MHz), largely used for laser scanning in microscopy applications, where high diffraction efficiency is needed. All values shown are for TeO_2 AODs

Characteristic	Parameter	Typical Values
Center frequency	f_c	70–100 MHz
Acoustic frequency bandwidth	ΔF_{Θ}	30–40 MHz
Total scan angle	Θ	30–50 mrd
Active aperture	D	1–30 mm
Max. diffraction efficiency	η_D	1–75%
Sound velocity (slow shear)	$V_{acoustic}$	~650 m/s
Access time	ΔT_{access}	1–50 μs
Time-bandwidth product	TBP	30–2000 MHz*μs
RF input power	P	1–3 W

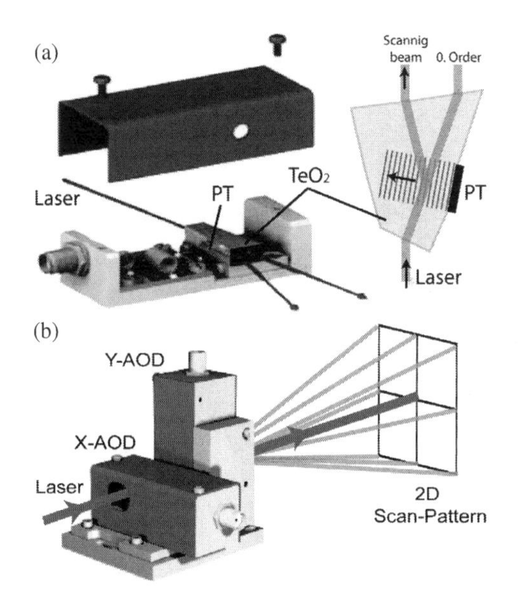

Figure 6.4 (a) Left: Technical (CAD) drawing of a commercially available acousto-optic deflector. Right: Improved design of the acousto-optic crystal allows easy collinear alignment while maintaining the optical axes within the optical pathway of the microscope. (b) Technical (CAD) drawing of two orthogonally crossed AODs for fast two-dimensional laser scanning. AOD mages in (a) and (b) were adapted with permission from http://opto.braggcell.com.

6.4 Dispersive Effects on Femtosecond Laser Pulses

Most acousto-optic crystals that are used in AODs are birefractive, such that their dispersive nature causes spatial and temporal dispersion, especially when using femtosecond laser pulses that are not monochromatic. This usually results in temporal pulse broadening (temporal dispersion) and distortions of the laser beam profile, because different light wavelengths are refracted at different angles (spatial dispersion, $\sigma \approx 0.0085^{\circ}/\text{nm}$) (Fig. 6.5a). In microscopy, these effects can lead to a strongly attenuated fluorescence signal and a decrease in image resolution. For our AOD microscope, if not compensated, the image resolution is reduced by a factor of a 3–4 times along the diagonal axes (Fig. 6.5b). These effects are relatively large when using AODs for two-photon imaging, and need to be compensated for. Because both spatial and temporal dispersion are caused by the acousto-optic grating both effects always accompany each other, which allows simultaneous compensation for both kinds of dispersion [13, 14].

6.4.1 *Spatial Dispersion Compensation*

As shown in Fig. 6.3 an active acousto-optic crystal can be described by a grating that deflects the incident laser beam according to its wavelength. Using the dynamic deflection angle (θ, Eq. 6.2), the spatial dispersion constant σ_{AOD} can be derived with regard to the wavelength (λ) as

$$\sigma_{\text{AOD}} = \frac{d\theta}{d\lambda} = \frac{f}{V_{\text{acoustic}}} \tag{6.7}$$

For a single AOD (1D scanning) with a center frequency of 95 MHz and an acoustic sound velocity of 650 m/s (slow shear mode), the spatial dispersion constant calculates as follows: $\sigma_{\text{AOD-1D}} = 360^{\circ}/2\pi * (f/V_{\text{acoustic}})/1000 = 0.0085^{\circ}/\text{nm}$. When switching from one AOD to a pair of orthogonally arranged AODs, the same compensation principles hold. In 2D AOD scanning, however (i), the direction of spatial dispersion of the 2D AOD is rotated by 45° with respect to the 1D AOD scanner; (ii) the amount of dispersion that needs to be compensated is larger according to the vector addition when compared to a single AOD. The compensation scheme can be

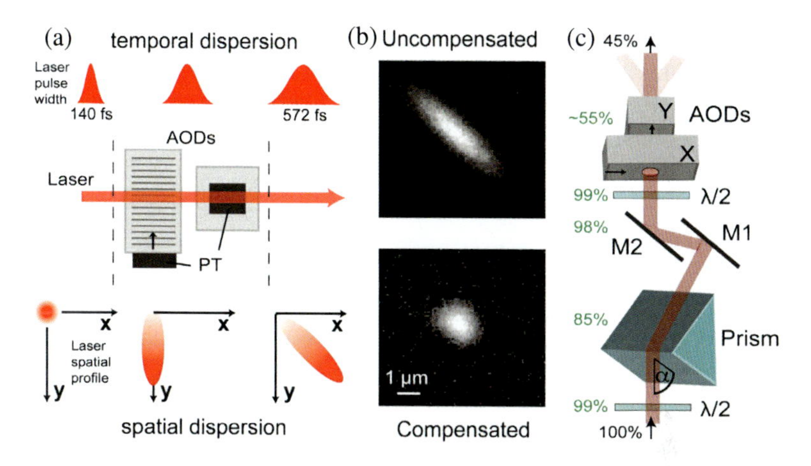

Figure 6.5 (a) Ultrashort laser pulses that are deflected by an AOD experience temporal (above) and spatial dispersion (below), which leads to laser pulse broadening and distortions of the laser beam profile, respectively. PT is the piezo-transducer. Adapted from [13]. (b) Two-photon images of 500 nm fluorescent beads with (below) and without (above) compensation of the spatial dispersion. (c) A single prism can be used to compensate for both spatial and temporal dispersion simultaneously. The optical pathway through the single-prism dispersion compensation unit (DCU) consists of two $\lambda/2$ waveplates and a custom prism oriented at a 45° angle to the x/y-AOD scanner. The single-prism DCU almost completely restores the original pulse width under the objective and the microscope resolution in the center of the FOV. Laser beam transmission numbers (green) of each individual component finally result in a total transmission of about 45% for the complete compensation and AOD-scanning unit. M1 and M2 are near-infrared (NIR) mirrors, (b) and (c) were modified with permission from [5].

adapted by tilting the dispersive compensation element by 45° to adjust the altered orientation of the spatial dispersion caused by the 2D AODs. Additionally, the angle of incidence at the prism has to be increased to compensate for the larger spatial dispersion. For the combined 2D AOD scanner, $\sigma_{\text{AOD-2D}}$ can be calculated as a vector combination (angle of 90°) of two separate but equal dispersions: $\sigma_{\text{AOD-2D}} = \sigma_{\text{AOD-1D}} * 0.012°/\text{nm}$. Assuming a spectral pulse width of ~10 nm (FWHM at 140 fs pulse length), the total dispersion angle would be 0.12°, which is about 4% of the total AOD-scan angle (47 mrad or 2.7°). The final goal when compensating for spatial dispersion is that $\sigma_{\text{AOD-2D}}$ and spatial dispersion constant from the compensation unit σ_{COMP} cancel each other out. This can be achieved by tuning the laser beam's incident angle on the prism to the theoretical value ($\sigma_{\text{AOD-2D}}$). Using this method spatial dispersion effects can be compensated nearly perfectly for the center of the FOV, which restores the imaging resolution.

6.4.2 *Temporal Dispersion Compensation*

In addition to the spatial dispersive effects, the length of ultra short laser pulses can be substantially broadened, because different spectral components travel at different speeds through the acousto-optic crystal. This effect is known as temporal dispersion (Fig. 6.5a). Using nonmonochromatic femtosecond laser pulses from a Ti:Sa laser, this effect leads to a strong reduction in fluorescence excitation efficiency und thus to a strongly attenuated fluorescence signal. Compensating for the effects of temporal dispersion is thus crucial when designing an AOD two-photon microscope.

Standard pulse compressors normally consist of two dispersive devices (usually two prisms) that form a pair configuration [15]. As a basic principle the first device introduces an angular dispersion while the second re-collimates the light beam. As a result the light beam propagates with its angular dispersion between the two dispersive elements while the different wavelength components travel on different optical pathways (longer or shorter), thereby introducing a group-delay dispersion (GDD). The negative GDD provided by this configuration can compensate for the positive GDD introduced by the material dispersion of AOD crystals and other optical elements in the system, such as the objective. For our AOD setup the GDD produced by the two orthogonally crossed AODs was around 14.000 fs^2, which led to a four- to fivefold broadening of the laser pulse. For the crossed 2D AOD scanners the negative GDD can be fully compensated, in principle by any two dispersive elements that provide enough angular dispersion such as prisms, gratings, or additional AODs arranged in a pair configuration. If the second dispersive element within this paired configuration is replaced with an AOD scanner, the principle of dispersion compensation still holds, because the AOD acts as the second dispersive element. Optimizing the distance of the prism in front of the AODs (in our case about 35 cm) nearly restored the original laser pulse length behind the AODs, which was measured using an autocorrelator device.

In summary, both the spatial and temporal dispersion caused by the two crossed AODs can be compensated for simultaneously while the original scanning function of the AOD is not disturbed. As shown in Fig. 6.5c, a simple prism approach (custom prism, tilted by 45°) can be adapted and optimized to provide sufficient compensation of spatial and temporal dispersion caused by a 2D AOD scanner. Two $\lambda/2$ waveplates in front of the prism and the AODs turn the polarization of the laser beam, thereby maximizing the transmission through the prism (85%) and the AOD pair (55%), which results in a final transmission of the compensation and the AOD-scanning unit of about 45% of the incoming laser light.

6.5 Example of Application: High-Speed Population Imaging in Mouse Barrel Cortex

Before functional in vivo two-photon imaging can be performed, some essential steps are necessary to prepare the animal and stain the neuronal cell populations for imaging. Before starting, please ensure that all animal handling and surgical procedures have been approved by the veterinary department in accordance with local guidelines of animal welfare. After the mouse has been anesthetized a small craniotomy of around 1 mm × 1 mm is made in the skull and the dura is removed (for a detailed discussion of surgical protocols please see [16]). While standard protocols use wild-type mice, most staining methods also work in rats or other rodents [6, 17]. The surgery and staining procedures are illustrated by a simplified drawing in Fig. 6.6a, showing a small craniotomy above the barrel cortex. The membrane-permeable (AM-ester) form of a calcium indicator dye, e.g., Oregon Green BAPTA-1 (OGB-1), can then directly be injected into the neural tissue. This staining technique is also known as multicell bolus loading [4]. To dampen heartbeat- and breathing-induced motion during imaging, the cranial window is usually filled with agarose and covered with an immobilized glass cover slip. Adding dissolved Sulforhodamine 101 to the dye solution in the injection pipette allows counterstaining of cortical astrocytes [18].

6.5.1 *The Random-Access Pattern Scanning (RAPS) Approach*

The main advantage of using AOD laser scanners is that the laser focus can be shifted between any positions in the scanning area within a few microseconds (about 5–15 μs). While our active AOD aperture of 10 mm theoretically limits the minimal point-to-point transition time to 15.4 μs (sound velocity 650 m/s), this transition time can be shortened without any measurable effects down to about 10 μs [19]. To improve the SNR of fluorescence recordings from neuronal somata, the signal integration time per neuron can be increased. Fast AOD random-access scanning can then be combined with a new scanning mode, in which a pre-defined spatial pattern of a few points is scanned on each neuron (Fig. 6.6b). This scanning mode avoids stationary parking of the laser focus on one location in the cell body while minimizing the effects of both photobleaching and cell damage. After completing all points within the pattern on each neuron, the laser scanner jumps to the next neuron where the next "point pattern scan" is started (Fig. 6.6b, upper panel).

The collection of fluorescence photons thus continues as long as the laser is scanning individual neurons, and only stops during the transition period between neurons (Fig. 6.6b, lower panel). We termed this scanning mode "random-access pattern scanning" (RAPS). Figure 6.6c illustrates a typical scanning path using a 5-point RAPS pattern to sequentially scan a subset of neurons. The new scanning method allows high-speed scanning of sets of arbitrarily pre-selected neurons in vivo with sampling rates of up to 500–1.000 Hz [5]. In addition, the scan times for individual cells can be adjusted to maintain a sufficient SNR for detection of single action potential evoked fluorescence changes. For the standard AOD system, 5-point RAPS can record from 16.700 locations per second (83.500 single points) while the number of sampled cells (N_{RAPS}) always trades off against the effective sampling rate per cell (f_{RAPS}) such that

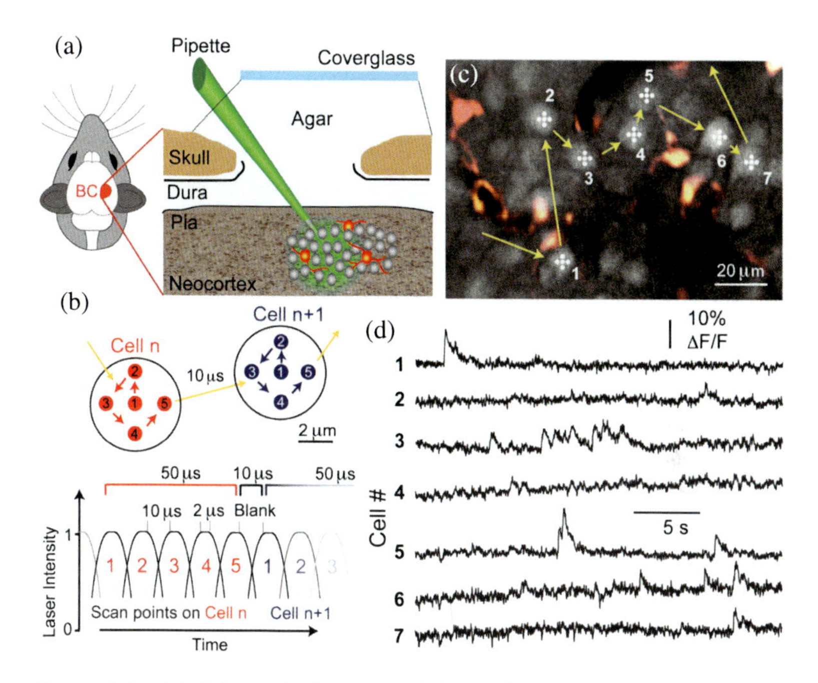

Figure 6.6 (a) Schematic drawing of the rodent barrel cortex position including a close up of the bolus injection technique used to stain neuronal cell populations with a fluorescent calcium indicator (gray cells), Sulforhodamine 101 stained astrocytes shown in red. (b) Upper panel: Scheme of the RAPS imaging mode, exemplifying a 5-point pattern scanned onto two cell somata sequentially. Lower panel: Signal integration protocol for 5-point RAPS. Cellular fluorescence signals are integrated over the entire 5-point-scan period while signal integration for the 10 μs transition between cells is stopped. (c) Neuronal cell population (grey) in layer 2/3 of mouse barrel cortex after labeling with the calcium indicator OGB-1. Astrocytes were labeled by co-injecting Sulforhodamine 101 and overlayed in red. Subsequently, high-speed random access pattern scanning was performed using 5-point patterns targeting 55 cells in total (only seven are shown). (d) Spontaneous calcium transients from 7 neurons indicated in (C), recorded with high speed RAPS scanning at 298 Hz sampling rate. Panels (b), (c), and (d) were modified with permission from [5].

$N_{\text{RAPS}}{}^{*} f_{\text{RAPS}} = 83.500/P_{\text{RAPS}}$ (where P_{RAPS} is the number of points in the RAPS pattern). Figure 6.6d shows spontaneous spiking activity of several neurons measured in the barrel cortex of mice using high-speed in vivo calcium imaging. In total, populations of about

50–60 neurons were scanned with a five-point pattern scaled to match the size of a typical cell soma. This resulted in a final signal integration time of 50 μs per cell. For automated extraction of visually evoked spiking activity from fluorescent calcium traces, a "spike-peeling algorithm" was used, which is described below.

6.5.2 *Automated Extraction of Spiking Activity from Fluorescent Data Traces*

Defining a direct relation between the exact intracellular calcium concentration during an AP and fluorescence evoked is difficult because every fluorescent indicator that binds to calcium also acts as an exogenous calcium buffer. However, during neuronal spiking activity, a relatively large change in intracellular calcium concentration causes a substantial increase in of the fluorescence signal. Standard transient peak values for a single action potential for cortical pyramidal neurons range between 7–10% $\Delta F/F$, with decay time constants of about 400–1000 ms using the standard OGB-1 fluorescence indicators [5, 6]. The function over time (f_{Ca}) of the fluorescent transient decay can be approximated by a double exponential function:

$$f_{\text{Ca}}(t) = (A_l e^{-(t-\tau_0)/\tau_l}) + (A_2 e^{-(t-\tau_0)/\tau_2}) \quad \text{for } t > t0 \tag{6.8}$$

$$f_{\text{Ca}}(t) = 0 \quad \text{for } t \leq t0 \tag{6.9}$$

where $A_{0,1}$ are the transient peak amplitudes and $\tau_{0,1}$ are the decay time constants. Although higher frequency spiking leads to overlapping fluorescent calcium transients, these superimposed transients can be resolved by approximating a linear summation of single action potential (1AP) calcium transients, which allows the extraction of short, higher frequent AP bursts.

While spike extraction approaches based on deconvolution allow neuronal firing patterns to be extracted from calcium fluorescence traces in principle [20], it is—for certain investigations—highly desirable to determine the precise spike times of individual APs. Precise reconstruction of spike times can be achieved by employing a combination of thresholding and template matching algorithms that are integrated within an iterative "spike-peeling" strategy,

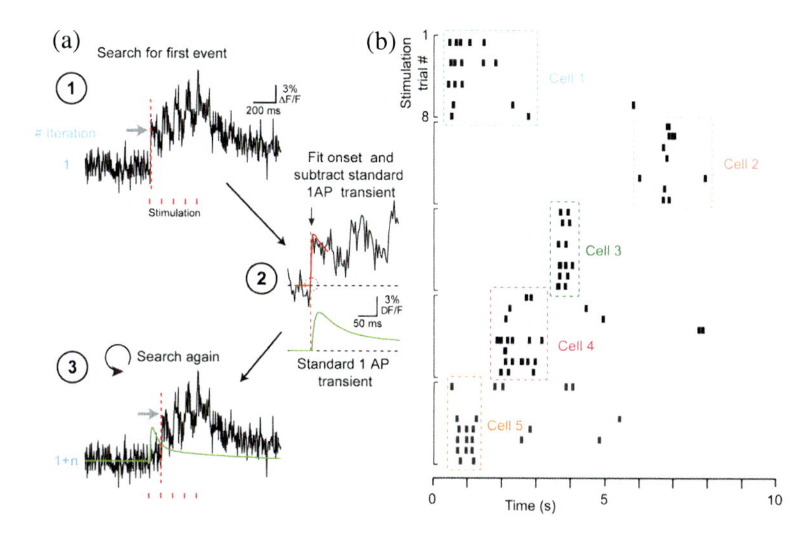

Figure 6.7 (a) Illustration of the step-by-step protocol of the automated "spike-peeling" procedure. In the initial step the first "event" is detected using a customized Schmitt-trigger threshold routine. In the second step, the onset of the detected calcium transient is fitted with a rising exponential function within a short time window (red curve) in order to obtain a good estimate of the starting point of the event (red circle). Subsequently in the third step a stereotypic single action potential (1AP) evoked calcium transient is placed with its start at this time point and is subtracted from the original fluorescence trace. Then the algorithm starts the next iteration on the remaining trace until only the residual trace without any transient signals remains. (b) Final results show chronological raster-plots of the reconstructed spiking activity of five neurons in mouse visual cortex that responded to eight repetitions of a short visual movie presentation. Neuronal activity is indicated by ticks for neurons that showed reliable activation during specific epochs during movie presentation (marked by dashed boxes). (a) and (b) have been modified with permission from [5].

which allows the reconstruction of neuronal spike times from high-speed calcium imaging data [5].

A modified Schmitt-trigger thresholding algorithm allows precise detection of short trains of up to five high-frequency action potentials. The Schmitt-trigger searches the complete trace for events that first pass a preset high threshold and then stays high until a second lower threshold is reached for at least a minimum duration. After the first AP has been detected and the onset phase fitted with an exponential rising function, the procedure subtracts a standard AP waveform at the position of the detected spike and starts searching from the beginning of the trace again (Fig. 6.7a). This iterative procedure can then be repeated until only the residual noise trace without any calcium transients remains.

The standard single AP waveform used for subtraction was previously derived by averaging and fitting numerous 1AP calcium transients, which had been verified by electrical recordings. Figure 6.7b illustrates the final results of the spike extraction procedure for five neurons in mouse visual cortex that reliably responded to a short movie representation. When combining this spike extraction algorithm with acousto-optic high-speed RAPS imaging, it becomes possible to extract individual neuronal spike times with a temporal precision of some milliseconds (5–15 ms) while short trains of overlapping APs of up to 20–30 Hz can be resolved [5].

6.6 Discussion

In comparison to previous in vivo imaging studies that use standard galvanometric scan mirrors (1–30 Hz frame rate [6, 11, 21, 22]), the AOD-based imaging approach can provide up to 1 kHz cell sampling, thereby gaining a factor of 30–1000 in temporal resolution. Although clever user-defined line-scanning approaches can be used to increase sampling speed [12, 23, 24], it is unclear whether they can reach an SNR comparable to the AOD-based RAPS measurements. By scanning a small point pattern on each individual neuron, the AOD scanning approach achieves a high SNR while minimizing effects of photobleaching and photodamage. However, with the current AOD system it will be difficult to make functional recordings at depths much greater than 300 μm due to the slightly reduced spatial resolution of the AOD scanners and laser beam transmission losses (diffraction efficiency of 2D AOD scanner and DCU is about 45%). While the AOD microscope discussed here is only capable of scanning neuronal networks in two dimensions, the design should allow adaptation to record in three dimensions using a mechanical *z*-dimension scanning device such as a fast piezo *z*-focusing device [23, 25]. Fast three-dimensional measurements would be especially beneficial for analyzing spike sequences in specific neuronal subsets after screening of large populations for relevant sub ensembles. Nevertheless only 2D AOD two-photon calcium imaging currently offers the advantage of high-speed capture of neuronal activity patterns from large cell populations within a local neuronal network.

6.7 Future Directions

The technical and methodological advances that are summarized above create new opportunities to investigate neuronal ensembles in the working brain (in vivo). In particular the combination of spike train reconstruction and high-speed calcium imaging now opens the door for various investigations of neural coding. The new technique would, for example, allow better characterization of bursting activity in neurons, which could be particularly interesting when comparing awake and inactive animals, where neuronal bursting seems to be enhanced. Another interesting aspect might be to investigate phase relationships of AP firing in rat hippocampus with regard to neuronal theta oscillations (4–10 Hz) [26]. Moreover, we are positive that the technique will enable a detailed investigation of neuronal network plasticity, which seems to be governed by small differences in relative spike timing in the order of a few milliseconds [27, 28].

In addition two-photon imaging can be combined with genetic markers or post-hoc immunohistochemistry, which allows to visualize and to further identify subpopulations and subtypes of neurons. This has been recently demonstrated in the visual cortex of mice, by revealing the orientation tuning responses of different subtypes of neurons such as interneurons [29, 30]. Furthermore cell-type specific expression of genetically expressed calcium indicators (GECIs) makes it possible to target individual neuronal subtypes for investigating their function within the brain. Future studies of network reconfiguration will be facilitated when using the new high-speed technique in combination with long-term expression of GECIs, which allows repeated in vivo imaging of the same neuronal network over days and weeks [7, 31]. Another important direction is to use high-speed in vivo population calcium imaging to investigate the network dysfunctions in mouse models of brain diseases. For example a recent two-photon calcium imaging study revealed a redistribution of spontaneous neuronal activity in mice models of Alzheimer's disease, with hyperactive neurons that seem to appear exclusively in the vicinity of amyloid plaques [32]. Using quite similar imaging tools, pathological issues involving interactions between neurons and glial cells have been addressed [33, 34].

However, while standard two-photon population imaging is often used for mapping of neuronal tuning properties [11, 12], the fast scanning approach now allows real-time investigation of ongoing network dynamics. Bringing together novel scanning techniques with chronic, long-term imaging of cell-type specific GECIs in head-fixed animals performing real-world tasks will create an exciting future for studying the dynamics of cortical circuits in great detail and most importantly in realistic behavioral scenarios. In addition, improved algorithms for extracting neuronal spiking from noisy fluorescence traces [5, 6, 35] should greatly increase the reliability and precision of spike train reconstruction. With regard to these future applications, novel and improved two-photon imaging techniques seem very promising for providing novel insights of important aspects of neural network dynamics in various brain diseases.

6.8 Final Conclusions

In summary, emerging optical techniques such as the AOD two-photon imaging as well novel genetic staining methods are revolutionizing the investigation of neural dynamics while bridging the gap between our understanding of circuits at the neuronal network level and at the level of interconnected brain areas. This book chapter thereby mainly focused on novel high-speed two-photon imaging techniques within the context of the rapidly advancing field of in vivo population imaging. Taking all these developments together, the direct observation of neuronal network dynamics in awake, behaving animals no longer seems unrealistic. This will help to uncover fundamental principles of network dynamics in the brain, hopefully in the near future.

Acknowledgments

I thank Edward Bracey and Ninja Grewe, Jerome Lecoq, and Jeff Chi-Tat Law for proofreading the manuscript and for their comments on it. This work was supported by a postdoctoral fellowship from the Swiss National Science Foundation (B.F.G.).

References

1. Helmchen, F., and Denk, W. (2005). Deep tissue two-photon microscopy. *Nat. Methods*, 2(12):932–940.
2. Grewe, B. F., and Helmchen, F. (2009). Optical probing of neuronal ensemble activity. *Curr. Opin. Neurobiol.*, 19(5):520–529.
3. Helmchen, F., Imoto, K., and Sakmann, B. (1996). Ca^{2+} buffering and action potential-evoked Ca^{2+} signaling in dendrites of pyramidal neurons. *Biophys. J.*, 70(2):1069–1081.
4. Stosiek, C., Garaschuk, O., Holthoff, K., and Konnerth, A. (2003). In vivo two-photon calcium imaging of neuronal networks. *Proc. Natl. Acad. Sci. USA*, 100(12):7319–7324.
5. Grewe, B. F., Langer, D., Kasper, H., Kampa, B. M., and Helmchen, F. (2010). High-speed in vivo calcium imaging reveals neuronal network activity with near-millisecond precision. *Nat. Methods*, 7(5):399–405.
6. Kerr, J. N. D., Greenberg, D., and Helmchen, F. (2005). Imaging input and output of neocortical networks in vivo. *Proc. Natl. Acad. Sci. USA*, 102(39):14063–14068.
7. Lütcke, H., Murayama, M., Hahn, T., et al. (2010). Optical recording of neuronal activity with a genetically-encoded calcium indicator in anesthetized and freely moving mice. *Front. Neural Circuits*, 4:9.
8. Bandyopadhyay, S., Shamma, S. A., and Kanold, P. O. (2010). Dichotomy of functional organization in the mouse auditory cortex. *Nat. Neurosci.*, 13(3):361–368.
9. Kara, P., and Boyd, J. D. (2009). A micro-architecture for binocular disparity and ocular dominance in visual cortex. *Nature*, 458(7238):627–631.
10. Kerr, J. N., de Kock, C. P., Greenberg, D. S., et al. (2007). Spatial organization of neuronal population responses in layer 2/3 of rat barrel cortex. *J. Neurosci.*, 27(48):13316–13328.

11. Ohki, K., Chung, S., Ch'ng, Y. H., Kara, P., and Reid, R. C. (2005). Functional imaging with cellular resolution reveals precise micro-architecture in visual cortex. *Nature*, 433(7026):597–603.
12. Rothschild, G., Nelken, I., and Mizrahi, A. (2010). Functional organization and population dynamics in the mouse primary auditory cortex. *Nat. Neurosci.*, 13(3):353–360.
13. Zeng, S., Lv, X., Bi, K., et al. (2007). Analysis of the dispersion compensation of acousto-optic deflectors used for multiphoton imaging. *J. Biomed. Opt.*, 12(2):024015.
14. Zeng, S., Lv, X., Zhan, C., et al. (2006). Simultaneous compensation for spatial and temporal dispersion of acousto-optical deflectors for two-dimensional scanning with a single prism. *Opt. Lett.*, 31(8):1091–1093.
15. Fork, R. L., Martinez, O. E., and Gordon, J. P. (1984). Negative dispersion using pairs of prisms. *Opt. Lett.*, 9(5):150–152.
16. Garaschuk, O., Milos, R. I., and Konnerth, A. (2006). Targeted bulk-loading of fluorescent indicators for two-photon brain imaging in vivo. *Nat. Protoc.*, 1(1):380–386.
17. Li, Y., Van Hooser, S. D., Mazurek, M., White, L. E., and Fitzpatrick, D. (2008). Experience with moving visual stimuli drives the early development of cortical direction selectivity. *Nature*, 456(7224):952–956.
18. Nimmerjahn, A., Kirchhoff, F., Kerr, J. N., and Helmchen, F. (2004). Sulforhodamine 101 as a specific marker of astroglia in the neocortex in vivo. *Nat. Methods*, 1(1):31–37.
19. Otsu, Y., Bormuth, V., Wong, J., et al. (2008). Optical monitoring of neuronal activity at high frame rate with a digital random-access multiphoton (RAMP) microscope. *J. Neurosci. Methods*, 173(2):259–270.
20. Sasaki, T., Takahashi, N., Matsuki, N., and Ikegaya, Y. (2008). Fast and accurate detection of action potentials from somatic calcium fluctuations. *J. Neurophysiol.*, 100(3):1668–1676.
21. Rochefort, N. L., Garaschuk, O., Milos, R. I., et al. (2009). Sparsification of neuronal activity in the visual cortex at eye-opening. *Proc. Natl. Acad. Sci. USA*, 106(35):15049–15054.
22. Sato, T. R., Gray, N. W., Mainen, Z. F., and Svoboda, K. (2007). The Functional Microarchitecture of the Mouse Barrel Cortex. *PLoS Biol.*, 5(7):e189.
23. Gobel, W., Kampa, B. M., and Helmchen, F. (2007). Imaging cellular network dynamics in three dimensions using fast 3D laser scanning. *Nat. Methods*, 4(1):73–79.

24. Lillis, K. P., Eng, A., White, J. A., and Mertz, J. (2008). Two-photon imaging of spatially extended neuronal network dynamics with high temporal resolution. *J. Neurosci. Methods*, 172(2):178–184.
25. Gobel, W., and Helmchen, F. (2007). New angles on neuronal dendrites in vivo. *J. Neurophysiol.*, 98(6):3770–3779.
26. Hartwich, K., Pollak, T., and Klausberger, T. (2009). Distinct firing patterns of identified basket and dendrite-targeting interneurons in the prefrontal cortex during hippocampal theta and local spindle oscillations. *J. Neurosci.*, 29(30):9563–9574.
27. Caporale, N., and Dan, Y. (2008). Spike timing-dependent plasticity: a Hebbian learning rule. *Annu. Rev. Neurosci.*, 31:25–46.
28. Kampa, B. M., Letzkus, J. J., and Stuart, G. J. (2007). Dendritic mechanisms controlling spike-timing-dependent synaptic plasticity. *Trends Neurosci.*, 30(9):456–463.
29. Kerlin, A. M., Andermann, M. L., Berezovskii, V. K., and Reid, R. C. (2010). Broadly tuned response properties of diverse inhibitory neuron subtypes in mouse visual cortex. *Neuron*, 67(5):858–871.
30. Runyan, C. A., Schummers, J., Van Wart, A., et al. (2010). Response features of parvalbumin-expressing interneurons suggest precise roles for subtypes of inhibition in visual cortex. *Neuron*, 67(5):847–857.
31. Mank, M., Santos, A. F., Direnberger, S., et al. (2008). A genetically encoded calcium indicator for chronic in vivo two-photon imaging. *Nat. Methods*, 5(9):805–811.
32. Busche, M. A., Eichhoff, G., Adelsberger, H., et al. (2008). Clusters of hyperactive neurons near amyloid plaques in a mouse model of Alzheimer's disease. *Science*, 321(5896):1686–1689.
33. Kuchibhotla, K. V., Lattarulo, C. R., Hyman, B. T., and Bacskai, B. J. (2009). Synchronous hyperactivity and intercellular calcium waves in astrocytes in Alzheimer mice. *Science*, 323(5918):1211–1215.
34. Takano, T., Han, X., Deane, R., Zlokovic, B., and Nedergaard, M. (2007). Two-photon imaging of astrocytic Ca^{2+} signaling and the microvasculature in experimental mice models of Alzheimer's disease. *Ann. N. Y. Acad. Sci.*, 1097:40–50.
35. Vogelstein, J. T., Watson, B. O., Packer, A. M., Yuste, R., Jedynak, B., and Paninski, L. (2009). Spike inference from calcium imaging using sequential Monte Carlo methods. *Biophys. J.*, 97(2):636–655.

Chapter 7

Intravital Endomicroscopy

Gangjun Liu and Zhongping Chen
Department of Biomedical Engineering, Beckman Laser Institute,
University of California, Irvine, CA 92612, USA
z2chen@uci.edu

Although medicine has evolved rapidly with advances in biotechnology, many therapeutic procedures still require diagnosis of disease at an early stage to enable effective treatment and prevent irreversible damage. Direct visualization of cross-sectional tissue anatomy and physiology provides important information for the diagnosis, staging, and management of disease. Recent advances in fiber optics, compact laser, miniature optics, and microelectromechanical systems (MEMS) technologies have accelerated the development of high-resolution optical imaging technologies for in vivo endomicroscopic imaging of internal organs. Optical imaging techniques, such as optical coherence tomography (OCT), two-photon excited fluorescence (TPEF), second-harmonic generation (SHG) and coherence anti-Stokes Raman scattering (CARS), offer high resolution, unique contrast for 3D visualization of tissue structure and functions. Miniaturization of these imaging systems is essential for translating these technologies from benchtop microscopy to clinical imaging devices. This chapter reviews the

Understanding Biophotonics: Fundamentals, Advances, and Applications
Edited by Kevin K. Tsia

ISBN 978-981-4411-77-6 (Hardcover), 978-981-4411-78-3 (eBook)
www.panstanford.com

emerging endomicroscopy technologies, describes component and system design consideration, and discusses recent advances in this exciting field.

7.1 Introduction

Endoscopy is a diagnostic modality used to identify lesions or other pathologic disorders at various sites within the body. Most common-surface and near-surface visceral organ abnormalities are accessed endoscopically. Conventional light-based endoscopes use fiber-optical devices, such as fiber bundles or a rod lens, to transmit the illumination light into the body and to collect the reflected light. These white-light endoscopes provide forward and angled surface visualization for still imaging or video but do not offer depth penetration. Thus, endoscopy usually guides the clinician through the lumen of internal organs or spaces and aids in site selection and performance of excisional biopsies. Other image-guided diagnostic biopsy approaches, which include MRI and ultrasound guided biopsies, have been available for more than two decades. However, these techniques have their limitations, such as resolution, size, or accessibility to regions that are difficult to reach.

Optical tomography techniques, such as OCT, TPEF, SHG, and CARS, use visible and near-infrared photons for 3D imaging. Such techniques have the advantage of high contrast without contact requirement. OCT is an interferometric technique based on optical coherent gating [1]. In OCT, imaging contrast originates from the inhomogeneities of sample scattering properties that are linearly dependent on the sample's refractive indexes. OCT offers an axial resolution of 2–15 μm and a penetration depth of around 2–3 mm. OCT was first used clinically in ophthalmology for the imaging and diagnosis of retinal disease [2]. Recently, it has been applied to image subsurface structure in skin, vessels, oral cavities as well as respiratory, urogenital, and gastrointestinal tracts.

For TPEF, SHG, and CARS, the interaction between light and the sample is nonlinear and more than one photon is involved in the process. They are also called multiphoton microscopes (MPMs) or nonlinear microscopes [3]. Fluorescence signals, for example, give exceptional insights into the biophysical and biochemical interactions of biological tissue with molecular specificity and sensitivity. In TPEF microscopy, a molecule absorbs energy from two photons to reach an excited state, and then emits a single photon through a normal fluorescence-emission pathway [4–6]. Compared to a single-photon fluorescence confocal microscope, a TPEF microscope has the advantage of reduced photodamage and photobleaching [6]. SHG is a sum frequency process where two photons with an identical frequency combine into one with twice the energy when interacting with the molecule [3]. Only noncentrosymmetric structures can produce SHG. SHG signals reveal orientation, organization, and local symmetry of biomolecules. An SHG microscope can be used to image collagen, myosin heads in muscle, and microtubules [3]. TPEF and SHG microscopes typically use an infrared mode-lock ultrafast pulse laser. CARS is a third-order nonlinear process in which a pump beam with frequency ω_p and a Stokes beam with frequency ω_S are mixed in the sample via a four-wave mixing process [3, 7]. When the frequency difference between the pump and Stokes matches the frequency of a Raman-active molecular vibration ($\omega_R = \omega_p - \omega_S$), the signal at anti-Stokes with frequency ($\omega_{aS} = 2\omega_p - \omega_S$) is generated. CARS is a label-free, highly sensitive technique that provides a high measure of chemical selectivity. However, it requires two synchronized pulse lasers.

Since the first demonstration of MPM in 1990 [4], MPM has been widely used to image morphology and functions of various cells and tissues [3, 5]. For the past few years, deep-tissue and in vivo MPM imaging has been reported [8, 9]. MPM has found wide applications in neurobiology and developmental biology to monitor calcium dynamics and image neuronal plasticity and to evaluate neurodegenerative disease in animal models. MPM has also been shown to be a valuable tool to study angiogenesis and metastasis and to characterize cell/extracellular matrix interactions in cancer research [5]. The impact of CARS microscopy technique for clinical applications in diagnosis of lipid-related diseases [10–12] and myelin-related neurodegenerative diseases has also been demonstrated [13–15].

There are a number of advantages using MPM for in vivo tissue imaging. First, nonlinear interaction provides intrinsic sectioning capability. Second, the use of near-infrared light reduces scattering and allows for greater penetration depth. Third, the excitation volume for MPM is limited to the focal region, which minimizes photodamage. Finally, large separation between excitation, fluorescence, and second-harmonic spectra allows highly sensitive detection.

Optical imaging modalities such as OCT, TPEF, SHG, and CARS offer high resolution, unique contrast for 3D visualization of tissue structure and function. Miniaturization of these imaging systems is essential for translating these technologies from benchtop microscopy to clinical imaging devices. In many clinical applications, a miniature probe or catheter is essential to access the region of interest, especially the internal organs. The challenge for designing an optical imaging system compatible for clinical use is to fulfill the criteria of robustness, flexibility, and patients' comfort without significantly compromising the quality of the images. In this book chapter, we will describe the development of intravital endomicroscopy capable of in vivo imaging for animal model studies and clinical use. In particular, we will focus on a number of technical challenges, including fiber-based light delivery of femtosecond (fs) pulse for MPM microendoscopy, signal collection, and miniaturizing the focusing optics and scanning probes for both OCT and MPM microendoscopy. Different embodiments of these microendoscopies will be discussed.

7.2 Development of Intravital Endomicroscopy

There are a number of technical challenges to translate from a benchtop microscope to a clinical endomicroscope. First, different from traditional white light medical endoscopes, optical tomography endoscopes usually require point-to-point beam scanning. The beam scanning mechanism becomes one of the most challenging parts of the endoscope. Second, for MPM endomicroscopy, a high intensity short pulse laser is required. Laser beam delivery media, signal collection efficiency, and miniature focusing optics are also very important components that need to be addressed, especially for MPM endoscopes. OCT and MPM endoscopes can generally share similar scanning mechanisms. However, the considerations for choosing a proper light guiding media are different for OCT and MPM. OCT uses broadband continuous wave (CW) and quasi-CW laser sources [16] and MPM usually adopts high-energy, ultrashort pulse lasers as light sources [3]. For OCT, single-mode operation is the most important parameter to be considered. Linear effects such as dispersion may affect the performance of OCT, but usually dispersion can be compensated for with either hardware or software methods. For MPM, both temporal and spectral widths of the light interacting with the sample are very important for efficient MPM signal generation. These parameters are related to both linear and nonlinear optical effects in the delivery media. In addition to the single mode consideration, nonlinear optical effects are the most important issues that need to be considered for delivery of an ultrashort pulse in MPM applications, because the nonlinear effects will change both temporal and spectral profiles of the ultrashort pulse, and such effects cannot be compensated in most cases. Third, because MPM requires high intensity for increasing efficiency, it is essential to generate a very tiny focusing spot for high-speed imaging. Therefore, a high numerical aperture focusing component is necessary for MPM applications. On the other hand, OCT requires a large depth of view, and therefore, a focusing optics with moderate numeric aperture is a better choice. In the following section, we will discuss in detail the general considerations of designing and developing miniature intravital endomicroscopy, including light-guiding media, scanning mechanisms, and focusing optics.

7.2.1 *Light-Guiding Media*

The GRIN rod lens has been widely used commercially in fiber-optic rigid medical endoscopes. In a GRIN rod lens, the refractive index is not uniform in the radial direction of the glass and the ray path inside is not straight but curved. In general, a parabolic index distribution along the radial direction follows the equation below [17]:

$$n(r) = n_0 \left(1 - \frac{k}{2} r^2\right) \qquad (7.1)$$

where n_0 is the refractive index at the center of the rod lens, r is the distance from the center of the lens, and $\sqrt{k}$ is the gradient constant.

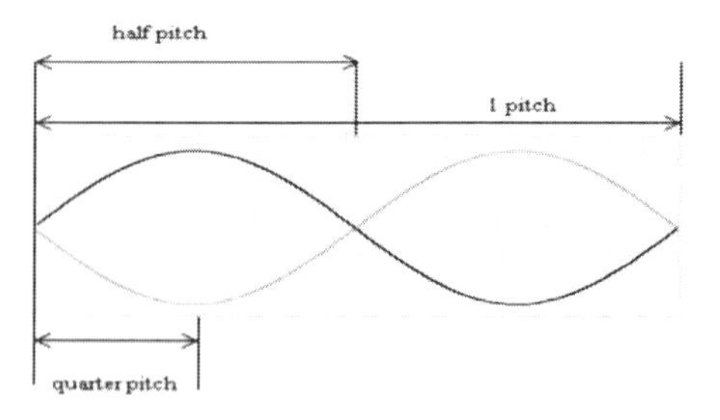

Figure 7.1 A GRIN rod lens with 1 pitch length.

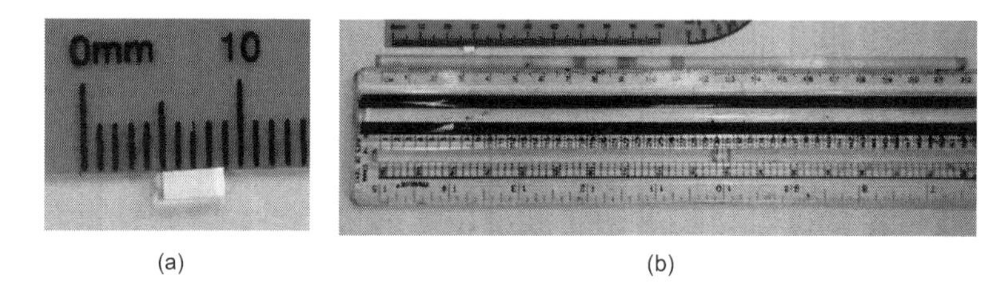

Figure 7.2 (a) 0.23 pitch miniature GRIN rod lens. (b) A long GRIN rod lens together with the miniature GRIN lens shown in (a).

The ray traces inside the GRIN lens will follow a sinusoidal curve and the light will be focused periodically if the GRIN lens is long enough. A length equal to one full cycle of the sinusoidal trace inside the GRIN lens is called a pitch, as shown in Fig. 7.1. One pitch length equals $2\pi/\sqrt{k}$. The diameter of a GRIN rod lens varies from a few hundred micrometers to a few millimeters. Figure 7.2a shows the photo of a 0.23 pitch miniature GRIN lens. Figure 7.2b shows the photo of a large-diameter long GRIN rod lens together with the miniature GRIN lens shown in Fig. 7.2a. The pitch length of the GRIN lens is usually proportional to the diameter of the GRIN rod.

The limitation with having a GRIN rod lens as a light-guiding medium is that it is rigid and can only be used for specific applications. The development of optical fibers in telecommunication offers a much better and economic way of light guiding and delivery. A traditional optical fiber for optical communications has a high refractive index material (n_1) as its core, surrounded by a lower-index cladding material (n_c). Optical fibers enable flexible delivery of light from a laser light source to the distal tip of a miniature probe. The mechanical flexibility offered by optical fibers has accelerated the integration of optical fibers in imaging and sensing systems that align well with clinical requirements. In optical fibers, light is guided in the core through total internal reflection, which puts restrictions on the angular variability of the incident rays. Incident light that falls within the angular acceptance cone of the fiber, which is usually referred to the numerical aperture (NA) of the fiber, is supported for propagation. The NA is calculated by [18]

$$\text{NA} = \sqrt{n_1^2 - n_c^2} \tag{7.2}$$

and the acceptance angle can be obtained according to [18]

$$\text{Angle} = \arcsin(\text{NA}) \tag{7.3}$$

Propagating light is confined to transverse modes. The number of modes that a fiber can support is dependent on the size of the core and core-clad index difference [19]:

$$V = \kappa_0 a \sqrt{n_1^2 - n_c^2} \tag{7.4}$$

where V is called normalized frequency, $\kappa_0 = 2\pi/\lambda$, λ is the light wavelength, and a is the radius of the core. Fibers that support only one transverse mode ($V <= 2.405$) are called single-mode fibers (SMF), whereas fibers that support multiple transverse modes ($V > 2.405$) are referred to as multimode fibers (MMF). The transverse modes a fiber supports are also related to the wavelength. The wavelength above which the fiber propagates only the fundamental mode is called the cutoff wavelength. Single-mode fibers operating in the visible and near-infrared range typically have core diameters on the order of a few micrometers (2–10 μm) with a clad of up to a few hundred micrometers. Light propagating in a fiber is governed by the following equation [19]:

$$\frac{\partial A}{\partial z} = (\hat{D} + \hat{N})A \tag{7.5}$$

where A is the slow-varying pulse envelope, z is the light propagation length along the fiber, $\hat{D}$ is a differential operator that accounts for linear effects, and $\hat{N}$ is a nonlinear operator that accounts for nonlinear effects of the fiber. These linear and nonlinear effects originate from the fiber material. For the linear effects, the response of the material depends linearly on the applied optical field. Linear effects include dispersion and losses (absorption and scattering). When a laser light is sufficiently intense to alter the optical properties of a material, the response of the material depends, in a nonlinear manner, on the strength of the optical field. Nonlinear effects require high laser intensity and usually high-energy pulse lasers are used to produce such nonlinear effects. For endoscopic MPM using fibers, the pulse propagating inside the fibers will be affected by both linear and nonlinear effects. The length of the fiber used is typically a few meters and the transmission loss for traditional SMF is usually very small and negligible. Therefore, the dispersion effect will be the only linear effect considered for traditional SMF. Dispersion occurs when different wavelengths (or frequencies) of light travel at different speeds, a phenomenon known as group velocity dispersion (GVD). There are two kinds of dispersion that need to be considered in fibers: material and waveguide dispersion. Material dispersion results from the wavelength dependent refractive index of the material. Waveguide dispersion arises from waveguide effects: the dispersive phase shift is different for a wave in a waveguide and for a wave in a homogeneous medium. The nonlinear effects include second-order nonlinear effects, third-order nonlinear effects, and high-order nonlinear effects. For typical fibers, the material is fused silica (SiO_2), and because fused silica is a symmetric molecule, optical fibers do not normally exhibit second-order nonlinear effects [19]. The lowest-order nonlinear effects in optical fibers to be considered are third-order nonlinear effects. These effects include self-phase modulation (SPM) and cross-phase modulation (XPM), third-harmonic generation (THG), fourwave mixing (FWM), stimulated Raman scattering (SRS) and stimulated Brillouin scattering (SBS) [19].

Typical OCT systems use either a CW broadband SLD or quasi-CW broadband swept source laser [16]. The typical power delivered by these sources are in the range of tens of milliwatts (mW). Therefore, nonlinear effects can be neglected and only linear effects should be considered. Typical endoscopic OCT systems use light sources with a central wavelength of around 1.3 μm and are usually fiber based. The traditional communication SMF, the Corning SMF-28, is one of the best options for OCT applications. The transmission loss is very low because usually only a few meters of fiber is used. The linear chromatic dispersion is the only effect needed to be considered. The mismatched dispersion between the reference arm and sample arm in an OCT system will decrease the depth resolution [16, 20, 21]. The dispersion introduced can be compensated for by using a matched fiber length in the reference arm or by numerical compensation methods [16, 20, 21].

For TPEF and SHG imaging, broadband fs pulse lasers are usually preferred and used. The most popular light source for TPEF or SHG is the mode-lock Ti:sapphire laser with a pulse width around 100 fs. The strength of the produced TPEF or SHG signal is inversely proportional to the pulse width of the incident light pulse [3, 4]. A short pulse is essential and preferred for these methods. When a light pulse propagates in a fiber, both the linear and nonlinear effects will affect the pulse. GVD will increase the pulse duration and broaden the pulse width. Temporal broadening lowers the peak energy of the pulses, which reduces the magnitude of the induced nonlinear signal. This broadening of the pulse is dependent on the fiber length and pulse width. The effects of temporal broadening become more significant for shorter pulses. For instance, after propagation in a 1.0 m long fused silica fiber, GVD stretches a transform-limited pulse of 100 fs at 800 nm to 375 fs, whereas a 10 fs pulse will be stretched to 3.6 ps. However, the GVD induced pulse broadening can be compensated for by pre-conditioning the pulses with prisms, grating pairs, or chirp mirrors [22, 23].

Besides linear broadening effects, the temporal and spectral profiles of the pulses are affected by nonlinear optical effects, such as SPM and XPM within the fiber. In addition, for CARS imaging, nonlinear generation of radiation at the anti-Stokes frequency ($\omega_{aS} = 2\omega_p - \omega_s$) through four-wave mixing in the fiber can substantially interfere with CARS imaging experiments [24]. In fibers, these nonlinear effects are related to a nonlinear parameter γ, which can be expressed as [19]

$$\gamma = \frac{n_2 \cdot \omega}{c \cdot A_{\text{eff}}} \tag{7.6}$$

where n_2 is the nonlinear index coefficient, ω is the light frequency, c is the speed of light in the vacuum, and A_{eff} is the effective mode area. The n_2 is related to one component of third-order susceptibility $\chi^{(3)}$ (a fourth-rank tensor) of the fiber material.

In silica fibers, the nonlinear index coefficient n_2 is 2.2–3.4 $\times$ $10^{-20}\text{m}^2/\text{W}$ [19]. The effective mode area A_{eff} is related to the radius of the fiber core and also the core-clad index difference (n_1– n_c). Typical step index SMF fibers with a cutoff wavelength around 920 nm will have a mode diameter of 6.2 um for a wavelength at 1060 nm (e.g., Corning Hi-1060 specialty fiber). For typical step index SMF with cutoff wavelength at around 720 nm, the mode diameter is around 4.6 mm for a wavelength at 780 nm (e.g., Corning PureMode Hi-780). It is difficult to overcome these nonlinear effects produced in fibers. To minimize the nonlinear effects, the nonlinear parameter γ must be reduced. One way is to utilize a large core size fiber so that the effective mode area increases while still maintaining the total transmission power. However, as shown in Eq. 7.4, increasing the core size may increase the transverse modes, and the laser light cannot be focused to a single diffraction-limited point if multiple modes propagate in the fiber [25].

One possible way to deliver short pulses with MMF can be done by excitation of only one fundamental mode in a MMF. For example, a small-diameter fiber taper before the MMF may be used as a spatial filter to strip off higher modes [25, 26]. An alternative solution is to use photonic crystal fibers (PCFs). A photonic crystal structure, formed from air holes in a silica matrix, constitutes the mechanism for confining and guiding the light. The major advantage of using such a guiding mechanism is that the optical properties of the fiber can be tailored by changing the pattern of the photonic crystal structure. Figure 7.3 shows the cross-section diagrams for three kinds of PCFs. Single-mode, large core size fibers based on PCF are commercially available now. These fibers are usually called large mode area (LMA) fibers, and Fig. 7.3a shows the cross-section diagram of a LMA PCF. MPMs delivered with these LMA fibers have been demonstrated by several groups [27–33]. Another way to reduce nonlinear parameter γ is to reduce the nonlinear index coefficient n_2. This could be realized with another kind of PCF- photonic bandgap (PBG) fiber. A special class of PBG fibers

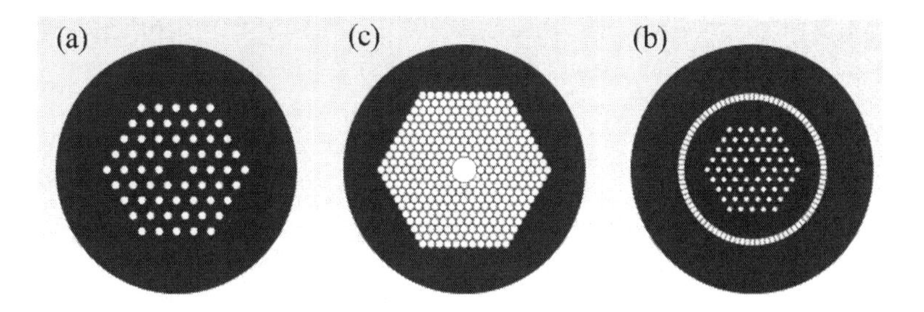

Figure 7.3 Cross-section schematic diagrams of three PCFs: (a) LMA fiber, (b) hollow-core PCF, and (c) LMA double-clad PCF (from the website of NKT Photonics).

is the hollow core fiber (Fig. 7.3b) [34–36]. In the hollow-core fiber, the light is confined to the air core surrounded by a photonic bandgap material. Since the light is guided in air, the nonlinear index coefficient n_2 is zero and there is no optical nonlinear effects in the fiber. In addition, the GVD properties of the fiber can be tuned through the design of the photonic bandgap structure, allowing dispersion free propagation in the near-infrared window.

Besides delivering the light to the sample at the distal end, optical fibers are also used to collect the signal generated from the sample at the distal end. For OCT, a single SM fiber is used for both delivery and signal collection. For MPM, the signal is usually very weak and increasing collection efficiency is critical. In addition, the signal wavelength is different from the excitation wavelength. The SM or LMA fibers used to deliver the excitation light for MPMs are not efficient for the MPM signal collection due to chromatic dispersion of the focusing optics, limited NA and core diameter. For example, the LMA PCF fiber LMA-20 from NKT Photonics (NKT Photonics, Birkerød, Denmark) has an NA of only 0.04 at a wavelength of 780 nm. The hollow-core fibers usually have a limited bandwidth. The transmission loss increases sharply if the wavelength is out of this range. Typically this transmission wavelength range is about $\pm 10\%$ of the designed pass wavelength. Low-cost MMFs offer large cores and high NA and are very efficient for MPM signal collection. However, a separate fiber for signal collection will increase the system complexity. Recently, several groups have demonstrated the application of double-clad PCFs for endoscopic MPM imaging

[28–32]. Double-clad PCFs consist of an LMA core surrounded by a microstructured inner-clad and a silica outer-clad. Between the inner- and outer-clad are air gaps and microscopic silica bridges, which transform the inner-clad into a multimode waveguide. The NA of the inner-clad is high (up to 0.65), and the diameter of the inner-clad is large (>600 μm). These features ensure efficient collection of weak signals. Double-clad PCF is unique because it can support single–mode propagation through a relatively large area core while simultaneously supporting the multimode propagation through the inner-clad. The relatively large core area can significantly reduce the power density of the short-pulse light in the core area and, therefore, minimize the nonlinear effects. Double-clad PCFs thus enable undistorted delivery of a short pulse laser through the LMA core as well as excellent signal collection by the inner-clad.

Finally, fiber bundle is another kind of medium that can be used for light delivery and signal collection. A fiber bundle consists of thousands of step index fibers and may have a diameter from a few hundred micrometers to a few millimeters [37].

7.2.2 *Beam Scanning Mechanisms*

Depending on the location of the scanning device relative to the two ends of a delivery media, scanning mechanisms can be classified as proximal scanning or distal scanning mechanisms. In a proximal scanning setup, the scanning device is located at the proximal end of the beam delivery media, e.g., fiber, fiber bundle, or gradient lens. For endoscopic application, the proximal scanning device is usually located outside the body of the imaging subject so that the size of the scanning device is not limited by the size of the region of interest to be imaged. In addition, there are scanning mechanisms that are based on moving the fiber (push-pull or rotate) with external equipment. Proximal scanning mechanisms are referred to as the external scanning mechanism in this book chapter because the scanning devices are located outside of the subject body. In a distal scanning setup, the scanning device is located at the distal end of the beam delivery media, and usually this scanning device is inside the body of the subject. Because the size of a distal end scanning device must be small enough to access the imaging subject, the scanning component is generally small and can be packaged into a miniature probe head. We briefly review the main features of proximal and distal scanning mechanisms in the following section.

7.2.2.1 External scanning mechanism

For proximal endoscopic scanning, the choice of a beam scanner is flexible and virtually any benchtop scanning module can be used. The most common device used is a scanning unit that employs galvanometer-based scanning mirrors for controlling the angular deviation of the laser beam [37, 38]. This kind of scanner can provide large deflection angles, fast scanning speeds, and flexible scanning patterns. The scanning pattern can be controlled by the voltage waveforms applied to each of the two axes. The most common pattern is the raster scan pattern, which can be achieved by applying sawtooth waveforms to both axes. For fast scanning speeds, the galvanometer mirrors must be driven by a sinusoidal waveform with a frequency equal to the galvanometer mirror's resonant frequency. Because a sinusoidal waveform is used instead of a linear waveform for resonant scanning, correction is needed to reconstruct the image. Commercially available resonant scanning mirrors can achieve a resonant frequency of up to tens of kilohertz and deflection angles greater than 40°.

In proximal scanning endoscopic systems, a beam delivery medium is needed that relays the scanning pattern made at the proximal site to the distal site of the beam delivery medium, as illustrated in Fig. 7.4. The two most common types of relay optics are fiber bundles (Fig. 7.4a) and the GRIN rod lens (Fig. 7.4b) relay [37, 38]. Fiber bundles relay the scanning pattern projected onto the proximal end to the distal end of the probe. Hence, the image quality depends on the number of fibers in the bundle and the suppression of cross-talks between the fibers. This form of proximal scanning has been successfully used in OCT, TPEF, and CARS microscopy [37, 39]. Similarly, a compact GRIN lens relay can be used to transfer the scanning pattern and scanning spot from the focusing objective side to the sample side [38]. Unlike fiber bundles, GRIN lens relay is suitable for use in rigid millimeter-sized probes, which can be inserted directly into tissues or mounted onto a freely moving

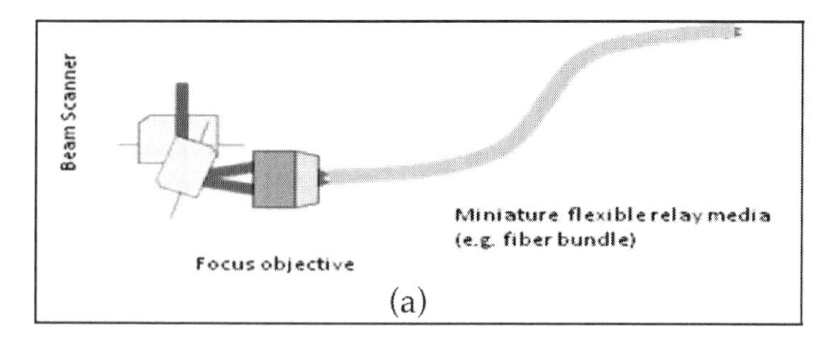

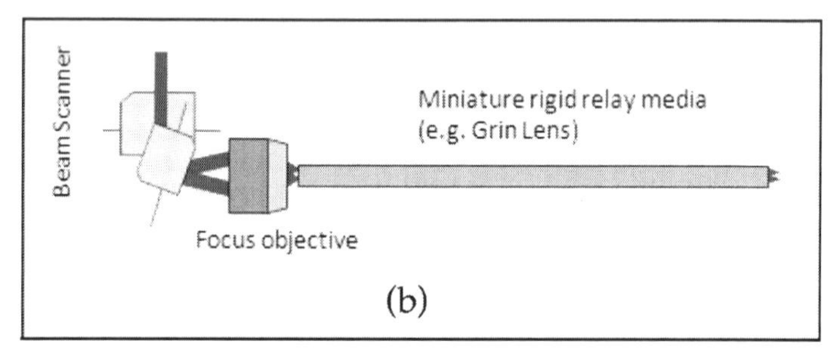

Figure 7.4 Proximal scanning. (a) Proximal scanning scheme with a flexible relay medium. (b) Proximal scanning scheme with a rigid relay medium.

animal. A GRIN lens-based relay probe has been demonstrated in OCT, TPEF, and CARS endoscopic applications [38, 40–42].

Another form of an external scanning scheme is to move the fiber with an external stage or motor as shown in Fig. 7.5. The movement of the fiber will be transmitted from the proximal end to the distal end of the fiber. One simple scanning scheme is using a linear stage or galvanometer to pull and/or push the fiber so that the probe at the distal end of the fiber will have a linear movement along the fiber (Fig. 7.5a). Endoscopic systems based on this design have been demonstrated for OCT applications [43–46]. Another design uses a rotation motor and a fiber rotary joint (FRJ) to rotate the fiber, as shown in Fig. 7.5b. The FRJ allows uninterrupted transmission of an optical signal while rotating along the fiber axis. The fiber probe attached to the rotary joint will rotate with the rotation motor. Rotational torque is transferred from the fiber probe's proximal end to the distal tip, and a circumferential scanning pattern is achieved. Catheters based on this scanning

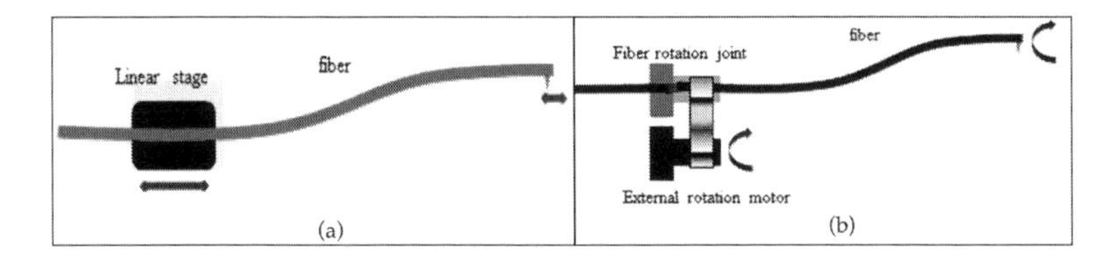

Figure 7.5 Proximal scanning. (a) Proximal scanning scheme with a linear scanner. (b) Proximal scanning scheme with a rotational scanner.

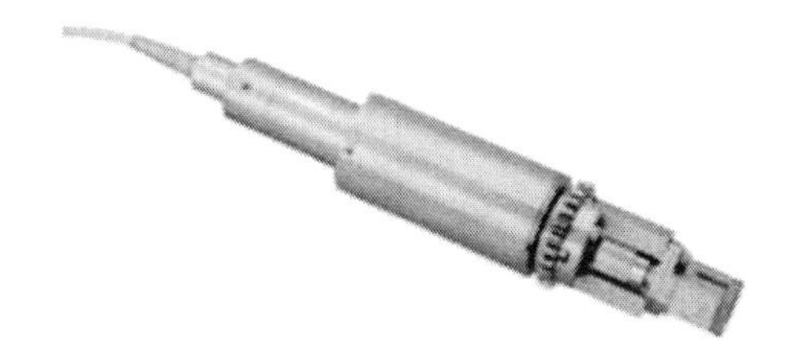

Figure 7.6 Photograph of a fiber rotation joint (from the website of Princetel Inc.).

mechanism have been demonstrated for intravascular OCT imaging applications [47–49]. The probes that adopt this scanning method can be made to have a diameter of less than 1 mm and have been used for imaging internal structure of tubular organs, such as blood vessels and gastrointestinal and respiratory tracts. For OCT imaging, the rotational scan combined with OCT A-scans can generate a 2D image. If the rotational scan is combined with the pull-back scanning mechanism, 3D OCT images can be generated [50].

7.2.2.2 Distal-end scanning mechanism

In the distal-end scanning scheme, a single fiber is typically used to deliver the excitation beam to the probe head. In this configuration, both the scanning device and the focusing components are an integral part of the probe head. Naturally, the components should be small enough to minimize the total mass of the probe to facilitate the insertion of the probe into tissues and body cavities. Below we will discuss several scanning mechanisms that are compatible with the strict requirements for size in endoscopy.

A scanning mechanism suitable for miniaturized probes is the fiber vibration actuator [47, 51–53]. In this scheme, the distal side of the delivery fiber is attached to a miniature actuator in such a way that a fiber cantilever is formed. By oscillating the actuator, the fiber tip is physically shaken into motion in a controlled manner. The vibrating fiber tip is placed in the image plane of the miniature focusing optics. The position of the fiber tip is mapped onto the conjugate sample plane, enabling rapid scanning of the focal spot through the oscillatory motion of the fiber cantilever. The actuators can be driven by piezoelectric, electrostatic, and electromagnetic effects, while piezoelectric driven actuators are the ones mostly used [47, 51–53]. A piezoelectric material, such as lead zirconate titanate (PZT), can produce a mechanical deformation when an electric field is applied. PZT-based tube actuators are widely used in scanning tunneling microscopy, scanning probe microscopy and atomic force microscopy [53]. The use of PZT tubes for driving the fiber cantilever has been successfully demonstrated in the field of TPEF microscopy and OCT [54–60].

When used for fiber optical endoscopes, PZT tubes usually have outer diameters from a few hundred micrometers to a few millimeters. The outer metal coating of the PZT tube is sectioned into four quadrants and two pairs of electrodes are formed. Figure 7.7 shows a photo of a 3 mm diameter PZT tube. When voltage is applied to the electrodes, the tube will bend. The final deflection of the PZT tube is governed by the applied voltage, the piezoelectric coefficient, the outer diameter of the tube, the wall thickness, and the length of the tube [61]. In addition, the scanning range of the fiber tip is also affected by the frequency of the applied voltage. When the driving voltage frequency matches the mechanical resonant frequency of the fiber cantilever, its scanning range reaches maximum due to resonance enhancement. When both axes of the PZT tube are driven with constant amplitude and resonant frequency voltages, the 2D scanning pattern is a Lissajous pattern and algorithms or look-up tables can be used to reconstruct the image [60]. Furthermore, a spiral scanning pattern is also possible by modulating the resonant frequency sinusoidally driven waveform with a triangular envelope [56]. Similarly, electromagnetically driven actuators have been used for fiber vibration scanning in endoscopic applications [62,

Figure 7.7 A PZT tube with a diameter of 3 mm and with quadrant electrodes on the outer surface.

63]. In this latter configuration, a fiber is coated or attached to magnetic material. The fiber is forced to vibrate at specific resonance frequencies by electromagnetic forces. Most endoscopic scanning devices based on fiber vibration provide a field of view (FOV) of a few hundred micrometers.

An alternative scheme for distal-end beam scanning is the use of miniature scanning mirrors with MEMS technology [30, 64–68]. Modern photolithography and micromachining technologies allow the realization of structures and devices that are capable of manipulating light with subwavelength precision. The small physical size of MEMS optical components is highly advantageous when building miniaturized, portable handheld scanners and minimally invasive endoscopic probes. In addition, due to the small size and low inertia of microdevices, MEMS technology allows the realization of miniature optical systems exhibiting both high speed and low power operation. Today, MEMS mirrors with diameters ranging from several hundred micrometers to a few millimeters are commercially available. A MEMS-based scanner has the advantage of small size, low cost, and high speed.

There are various MEMS mirror actuation mechanisms, such as electrothermal, electrostatic, magnetic, and pneumatic actuation. Among these mechanisms, an electrostatic actuator enables the fabrication of a low-mass device with low power consumption, making it a suitable candidate for integration into a miniature probe head [65–67, 69]. An attractive feature of MEMS mirrors for endoscopic applications is that 2D scanning can be achieved with a

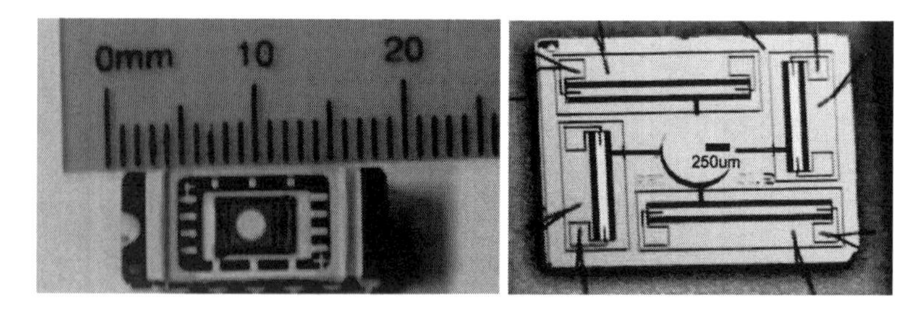

Figure 7.8 Left: A packaged two-axis MEMS mirror with a 2 mm diameter. Right: A zoomed in view of a MEMS scanner.

single micro-mirror. Figure 7.8 shows a photo of a packaged two-axis MEMS mirror. A single MEMS mirror can achieve 2D raster scanning with a high degree of accuracy and reproducibility, which avoids the need for image reconstruction. When driven at frequencies that correspond to mechanical resonances, MEMS mirrors can achieve video rate imaging speeds. The resonant frequency of a MEMS mirror may be as high as 40 kHz, and the maximum optical deflection angle can be more than 80° [68]. MEMS mirror-based scanners have been successfully implemented in confocal microscopy, OCT and MPM imaging modalities [30, 64–69].

The MEMS rotational motor is another scanning technology that is promising for intravascular and endoscopic applications [70–72]. Unlike the scanning mechanisms discussed above, the MEMS rotational motor provides an extended circular FOV. In this scheme, a mirror or reflecting prism is attached to the shaft of the rotational motor. Figure 7.9 shows a photo of such MEMS motor with a diameter of 1.9 mm. Upon rotating the reflector, a 360° circular scanning pattern is produced. If the rotational motor is combined with an external linear scanning mechanism, a cylindrical FOV of the tissue can be achieved [70–72]. The rotational speed of such MEMS motors can be as fast as several thousand revolutions per minute without applying a load on the shaft. The speed will be reduced when the shaft is loaded. With diameters as small as 1 mm, the rotational MEMS scanning mechanism has the potential to be used for imaging circular organs such as vessels and airways. Traditional intravascular probes use a mechanical cable to transfer the rotational torque from the endoscope proximal end to the distal

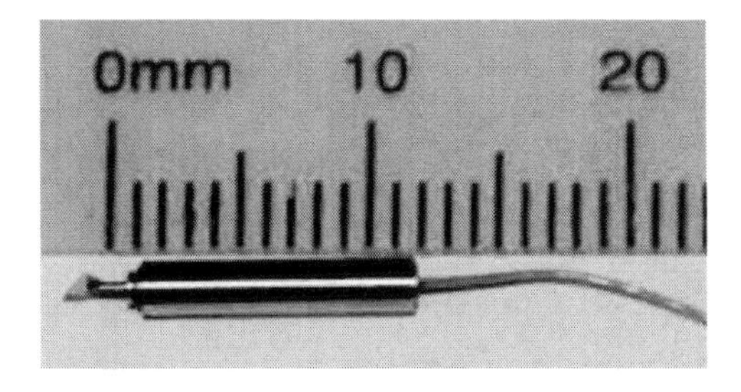

Figure 7.9 A MEMS motor with a prism glued to the shaft.

tip. In such design, the rotation of the whole endoscope, which is usually curved in the body of a subject, can cause unstable vibration and uneven rotational speed. This effect is more severe when the length of the endoscope increases. The MEMS rotational motor overcomes this limitation. This scheme has been integrated in OCT and MPM fiber coupled probes [70–72].

7.2.3 *Focusing Optics*

Conventional microscope lenses offer superior diffraction limited resolution. For endoscopy, miniaturization of the lenses is essential. There are commercially available microscope lenses with diameters as small as 1.3 mm (Ultra-Slim MicroProbe 'Stick' Objectives, Olympus) that could be used for endoscopic applications. A benchtop MPM microscope can be assembled using these lenses. In vivo TPEF and CARS imaging of living animals with a self-assembled miniature probe has been demonstrated [33, 73].

An alternative miniature focusing element is the GRIN lens. The GRIN technology offers a flexible route for designing compact miniature lenses with a wide range of focusing properties and NA values of up to 0.6. GRIN lenses with diameters as small as 0.35 mm have been successfully used in miniature nonlinear optical probes [30, 48, 58, 67, 68, 74, 75]. However, GRIN lenses suffer from chromatic dispersion and limited NA. By combining a GRIN lens with a miniature plano-convex lens, the NA of the focusing system could be further raised to 0.85, which significantly improves the imaging resolution [42].

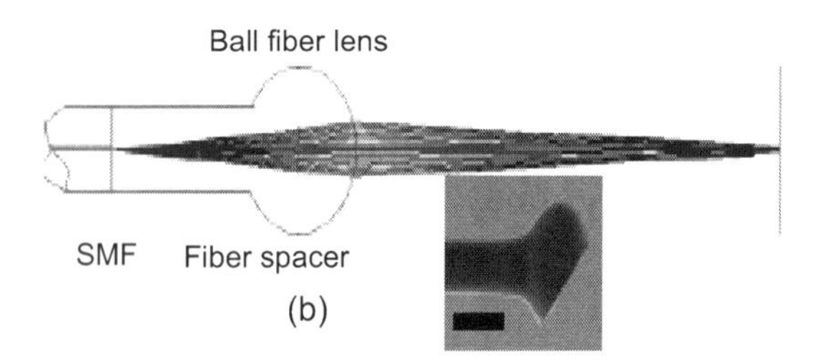

Figure 7.10 Schematic diagram of a fiber lens. Inset: scanning electron micrograph of the ball fiber lens tips fused with angle-polished beam director. Scale bar is 0.2 mm (from [76]).

Another interesting miniature objective is the fiber lens [52, 76]. In a fiber lens, a fiber spacer is fused with the delivery fiber, and the fiber fusing technique is used to form a ball shape structure at the fiber tip as shown in Fig. 7.10. By tailoring the length of the spacer and the diameter of the ball, the beam feature and the focusing characters of such ball lens can also been controlled. The diameter of the lens can be as small as 140 μm. The lens may be polished to provide a side view instead of a forward view. The fiber ball lens is very compact and robust because it is actually part of the fiber. This fiber ball lens based probe has been demonstrated for endoscopic OCT and MPM applications by several groups [52, 76–78].

7.3 Embodiment of Introvital Endoscopic System

Endoscopic systems are mainly designed for in vivo imaging of internal organs of human beings or living animals. The embodiments of the systems are dependent on the organs to be imaged and the way these organs are accessed. Generally, a flexible or catheter-based probe is necessary for internal lumens inside the body, and rigid or handheld devices are suitable for access to regions during open-field surgical procedure. Although the OCT and MPM endoscope share some common elements, such as the scanning mechanisms, other considerations may vary. For MPM endoscopic systems, additional issues need to be considered, such as undistorted beam delivery, signal collection efficiency, and high NA miniature focusing lens. There are a few excellent reviews about the MPM and OCT endoscopic systems [75, 79]. In this chapter, we will focus our discussion on a few recent advances from our group.

7.3.1 *OCT Endoscopic Imaging*

For endoscopic OCT applications, the single mode fiber is used for the delivery of light and the collection of back-scattered or back-reflected signals. The main issues to be considered for designing endoscopic OCT probes are the scanning mechanism and the focusing optics.

A proximal scanning scheme for OCT endoscopic probes was demonstrated as early as 1997 [80], and it has been used widely for in vivo imaging of human subject and animal models. T. Xie et al. demonstrated a proximal scanning probe based on a fiber imaging bundle [37]. A GRIN rod lens-based handheld probe with a diameter of 2.7 mm that can adjust focus dynamically has been demonstrated [38]. By adjusting the distance between the scan head lens system and the entrance surface of a GRIN rod lens, a focusing range of 0–7 mm was achieved without moving the reference mirror and the probe. The probe can provide a forward FOV, or a side FOV by gluing a prism at the distal end [Fig. 7.11]. A probe based on this design has been used for minimum invasive surgery and for office-based laryngoscopic examination [81–83]. In vivo noninvasive and

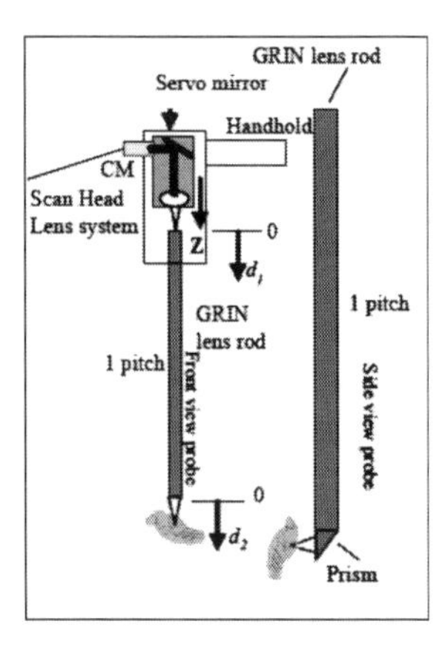

Figure 7.11 Schematic diagram of a GRIN lens rod-based dynamic focusing probe (from [38]).

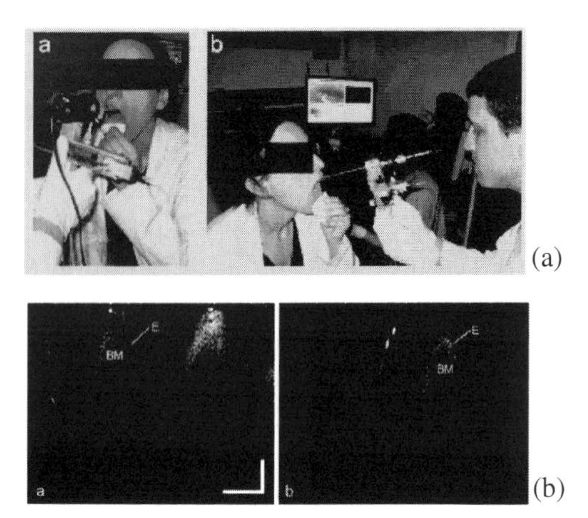

Figure 7.12 Top: OCT probe attached to the laryngoscope for office-based laryngoscopic examination. Dual-channel endoscope and OCT signals shown in a monitor. Bottom: OCT images of vibrating vocal cord captured during office-based laryngoscopic examination: (a) 120Hz and (b) 200 Hz vibration frequency; scale bar: 500 μm (from [84]).

noncontact vocal-fold vibration imaging of awake patients without the use of anesthesia has also been demonstrated [84]. Figure 7.12 shows photos of an office-based laryngoscopic examination with a GRIN rod lens based handheld probe, and OCT images of vibrating vocal folds. The vibration frequencies were also obtained by analyzing these images [84].

MEMS mirror-based OCT probes have also been demonstrated by several groups [64–66, 85, 86]. A two-axis MEMS mirror for OCT applications is shown in Fig. 7.13 [85]. The MEMS scanning actuator is a monolithic, single crystal silicon, two-dimensional, gimbal-less, vertical comb-driven structure. The mirrors are fabricated in a separate SOI process and later bonded to the actuator. This design approach allows the mirror and the actuator to be independently optimized. The MEMS mirror exhibits x- and y-axis resonant frequencies of 1.8 kHz and 2.4 kHz, respectively. The optical scanning angle of each axis can be as high as 20°. Figure 7.13 shows the schematic diagram and the picture of the packaged endoscopic probe based on two-axis MEMS scanning mirror. The representative

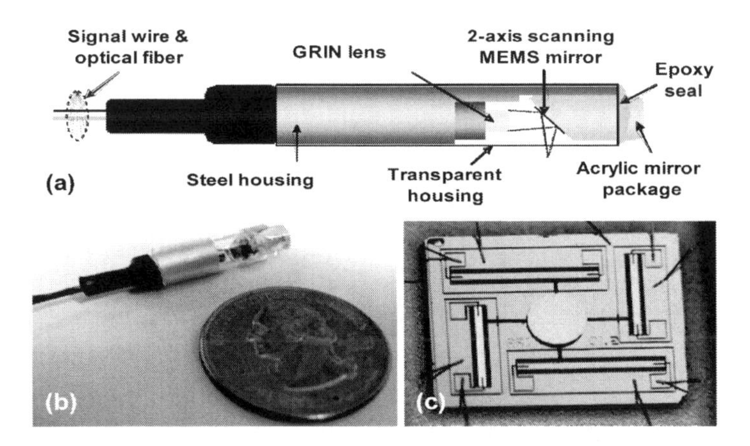

Figure 7.13 An OCT probe based on an MEMS mirror and scanning mirror: (a) schematic of mirror-based OCT probe; (b) packaged probe compared for size to a U.S. quarter coin; (c) a stereo microscope image of a two-axis MEMS mirror (from [85]).

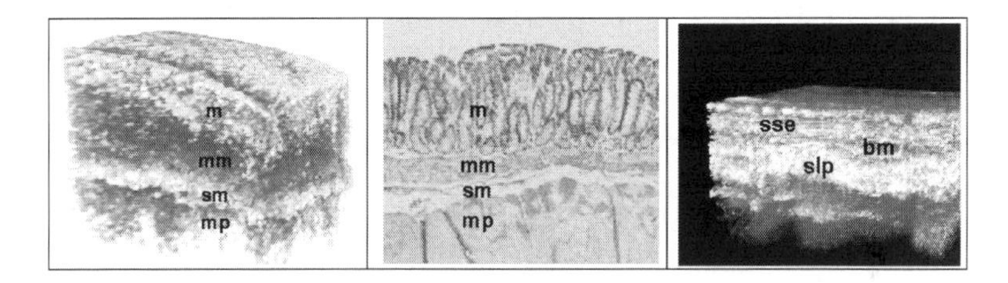

Figure 7.14 In vivo 3D OCT images obtained using the 2D MEMS probe. Image size is 1 mm × 1 mm × 1.4 mm. (a) In vivo 3D OCT images of a rabbit rectum; (b) histology image at the same site; (c) in vivo 3D imaging of true human vocal cord. 3D OCT image in (a) correlated very well with histology in (b); mucosa (m), muscularis mucosa (mm), submucosa (sm), muscularis propria (mp) (from [85]).

in vivo 3D images from rabbit and human tissues are shown in Fig. 7.14.

The cantilever type scanning probes have been demonstrated by several groups [51, 52, 60, 80, 87, 88]. The scanning probes are driven by either PZT, electrostatic, or magnetic force [51, 52, 60, 80, 87, 88]. For a 2D cantilever scanner, the nonresonant actuation has the advantage of wider freedom of operation, especially for the capability of the simple but nearly ideal 2D raster scan, which requires a far different scan frequency for each axis. However, the scanning amplitude of a nonresonant scanner is usually too small for OCT applications. The first OCT demonstration of the fiber-cantilever piezotube scanner was performed with a time-domain OCT system at a kilohertz resonance frequency; the scanner scans a sample spirally with the *lateral-priority* 3D scan strategy [87]. This approach is not applicable to Fourier-domain OCT, which must first acquire depth-resolved A-lines. The resonance frequency gets lower as the length of the fiber cantilever increases. For Fourier-domain OCT with an A-line rate of a few hundred kilohertz, the fast scan axis needs to be below 100 Hz. The resonance frequency obtainable with a compact cantilever is several kilohertz, which is far higher than the desired range of the scan frequency. For example, a resonance frequency of 60 Hz is obtained when the length of the cantilever fiber with a standard diameter of 125 μm exceeds 40 mm. This is unacceptably long to make a compact endoscope catheter. Alternatively, the frequency can be reduced by increasing the mass of the oscillating body or by putting a weight at the end of the cantilever [60, 88]. However, the scan characteristic of the weighted fiber cantilever is vulnerable to environmental changes when driven at resonance frequency. In addition, the scan characteristic is more complex compared with other types of resonant scanners due to its nonlinearity [89]. The possible instability of scan patterns introduces distortions to the acquired OCT images, with deformed image morphology particularly in the *en face* views. Moreover, the spiral scan strategy is not well suited for OCT images because the central scan area tends to be *oversampled* while the area at the outer boundary is *undersampled*. Recently, a new scan strategy of the semiresonant Lissajous scan was proposed to overcome these limitations [60]. A forward-view OCT scanning catheter has been developed on the basis of a fiber-cantilever piezotube scanner by using a semiresonant scan strategy for a better scan performance. A compact endoscope catheter was fabricated by using a tubular piezoelectric actuator with quartered electrodes in combination with a resonant fiber cantilever (Fig. 7.15a) [60]. A cantilever weight was attached to the fiber cantilever to reduce the resonance frequency down to 63 Hz, well in the desirable range for

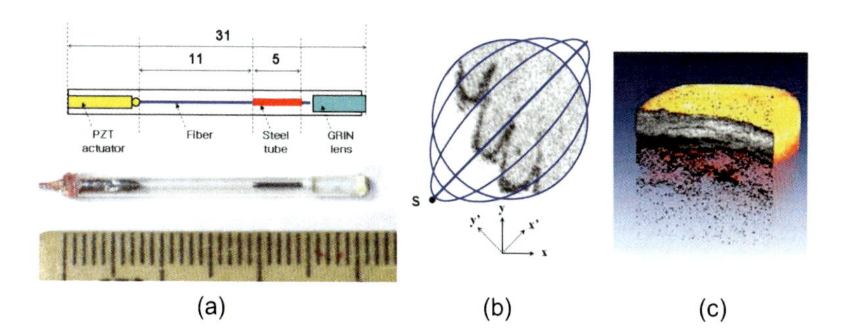

Figure 7.15 (a) Schematic design and picture of the fabricated OCT scanning catheter. The dimensions are given in millimeters. (b) Schematic Lissajous scan pattern (blue lines) in the *XY* plane laid over an OCT *en face* image. (c) The 3D-rendered tomogram (from [60]).

Fourier-domain OCT. The resonant-cantilever scanner was driven at semiresonance frequencies that were well out of the resonance peak but within a range of partial resonance. This driving strategy has been found to minimize the phase difference between the two scan axes, for better scan stability against environmental perturbations, as well as for a driving simplicity [60]. By driving the two axes at slightly different frequencies, a low-order Lissajous pattern has been obtained for a 2D area scan (Fig. 7.15b). 3D OCT images have been successfully acquired in an acquisition time of 1.56 seconds for a tomogram volume of $2.2 \times 2.2 \times 2.1$ mm^3 (Fig. 7.15c).

For imaging circular organs such as vessels and airways, the rotational scanning probe has many advantages over the linear scanning probe. The first OCT endoscope uses a mechanical cable to transfer the rotational torque from the endoscope proximal end to the distal tip [90]. This technique has been extended for intravascular imaging where a more stringent limitation on probe size is required [91, 92]. Because the scanning devices are located at the proximal end, the probe/catheter which will be inserted into the body of the imaged subject can be as small as several hundred micrometers, and be used for neuroendovascular imaging [93]. An alternative design uses a miniature MEMS motor at the distal end (Fig. 7.16) [70–72]. The major advantage of the MEMS motor-based endoscope is that coupling the rotation torque from the proximal end of the traditional endoscope is not necessary. The aluminum-

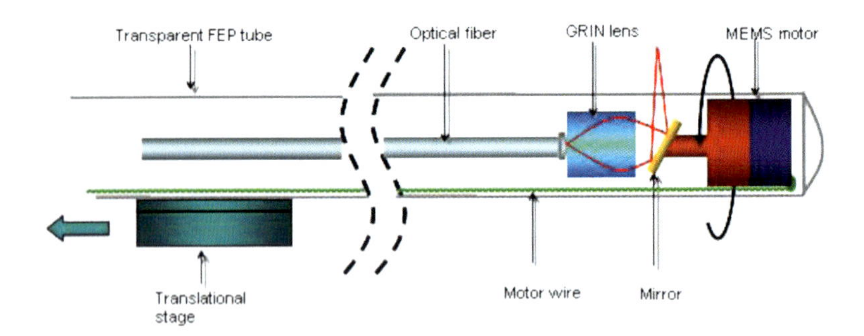

Figure 7.16 Schematic of MEMS rotation motor-based endoscope.

coated prism is the sole moving part in achieving a 360° full circular view. Because either a lightweight micro-prism or a 45° cut in the shaft can be used to deflect the light, a real-time frame rate (>50 frames/second) could be achieved easily by the MEMS motor's output torque. Since rotation of the entire endoscope is eliminated, a metal reinforcement sheath in the previous endoscope design is no longer needed. Hence, endoscope flexibility is increased. In addition, the FRJ between the traditional rotational endoscope and static sample arm fiber is unnecessary and, thus, decreases coupling power fluctuation.

Figure 7.17a shows sequential photos of the OCT probe in 3D working mode. Figures 7.17b shows a 2D in vivo OCT image of human left lower lobe bronchus. The motor wire blocked part of the signal, thereby creating a gap in the image. Figure 7.17c shows the

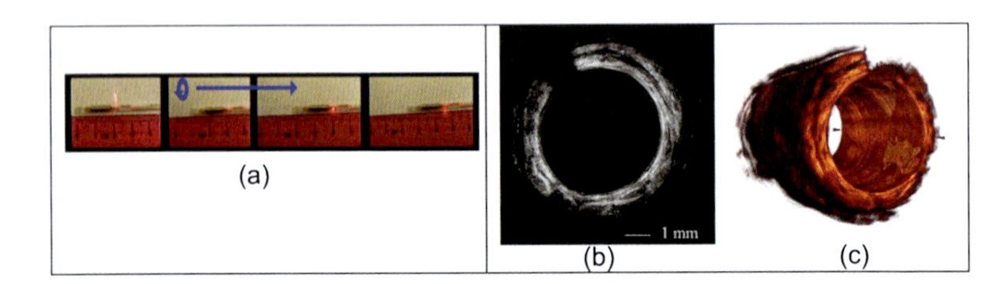

Figure 7.17 (a) Endoscope in 3D working mode. The whole endoscope moves from left to right. The red dots indicate the output beam orientation reflected by the micro-prism. (b) One slice of the 3D OCT images of human left lower lobe bronchus. (c) 3D reconstruction of 400 slices from part (b). The longitudinal length is 8.2 mm (from [71]).

3D reconstruction image volume, which was rebuilt from 400 slices of 2D frame. The lesion at the inside wall can be clearly identified.

7.3.2 *MPM Endoscopic Imaging*

Although MPM is a relatively mature technique and has been established in many laboratories for in vitro or in vivo small animal model studies, most are benchtop microscopes that are not suitable for clinical applications. In the past, several groups have tried to develop an MPM endoscope with limited success because of the difficulty in the delivery of a fs laser through the fiber and the lack of a miniaturized probe. The development of new photonic crystal fibers has enabled the delivery of fs pulses through optical fibers with more than adequate power for MPM endoscopy [28–32, 94, 95].

The design of MPM probes may vary with the applications. For some applications, fiber delivery/collection may not be necessary. A benchtop MPM microscope can be converted to a MPM endoscope with the help of miniature relay lenses [96, 97]. The proximal end scanning mechanism is used for this rigid probe. MPM probes that use proximal end scanning mechanisms are also possible with the help of fiber bundle relay [35]. However, fiber bundle degrades image quality due to pixilation and the ultrashort pulse may be distorted by the nonlinear effects.

To access internal organs, flexible probes are necessary. Early multiphoton endoscopy used a flexible probe with a SMF for excitation pulse delivery [25, 74]. However, typical excitation pulse wavelength for MPM excitation is around 800 nm and the traditional SMF at this wavelength has only a modal field diameter of less than 6 μm. Intense short laser pulses in such a small core diameter will generate significant nonlinear effects which will broaden the excitation pulse.

MPM probes with a more compact and simpler design can be realized by using double-clad PCF [28–32, 94, 95]. We have developed a fiber-based system that integrates double-clad PCF and a 2D MEMS probe [31, 32, 95]. Efficient delivery of short pulses and collection of multiphoton signals are essential for an endoscopic MPM system. To maintain sufficient probability for multiphoton excitation at the sample, an optical fiber is expected to deliver high power and fs optical pulses without significant attenuation or broadening of the pulse. On the other hand, the receiving optical fiber should have maximized efficiency to receive the blue-shifted fluorescence signal. In addition, single-mode propagation of the excitation beam is preferred so that the beam can be focused to a diffraction limited spot on the sample. A conventional single-mode fiber can deliver a single-mode beam over a long distance. However, because of the small core size and the high peak intensity of the short-pulse laser, self-phase modulation and other nonlinear effects will significantly broaden the pulse in both time and spectral domains. In addition, dispersion from the fiber will also broaden the fs pulse. However, if the nonlinear effects are negligible, the dispersive effect of the fiber can be pre-compensated with a pre-chirping optical element such as a grating pair. Therefore, the best approach is to minimize the nonlinear effects while keeping single-mode propagation. PCF offers a large core area while still supporting single-mode propagation.

There are two novel PCFs that have been used to deliver fs pulses. The first one is the hollow-core PCF. Hollow-core PCF has the advantage of minimal nonlinearity because light is guided through air. Thus, it can support high-energy fs pulses. In addition, the hollow-core PCF has relatively large waveguide dispersion. One can design hollow-core PCF with zero dispersion at a specific wavelength. However, because the MPM signal is significantly blue-shifted, the transmission bandgap that supports an excitation laser may not cover the fluorescence and harmonic spectra.

The second one is the double-clad PCF. Double-clad PCF for two-photon fluorescence measurement was first reported by Myaing et al. in 2003 [28]. Since then, several groups have demonstrated the applications of double-clad PCF for endoscopic two-photon fluorescence imaging [28–32, 94, 95]. Double-clad PCF is unique because it can support single–mode propagation through a relatively large area core while simultaneously supporting multimode propagation through the inner-clad. The relatively larger core helps significantly reduce the power density of the short-pulse light in the core area and minimize nonlinear effects. Figure 7.18a shows a commercial double-clad fiber that we have acquired and tested for MPM [31, 32]. The double-clad PCF has a core diameter of 16 μm and an

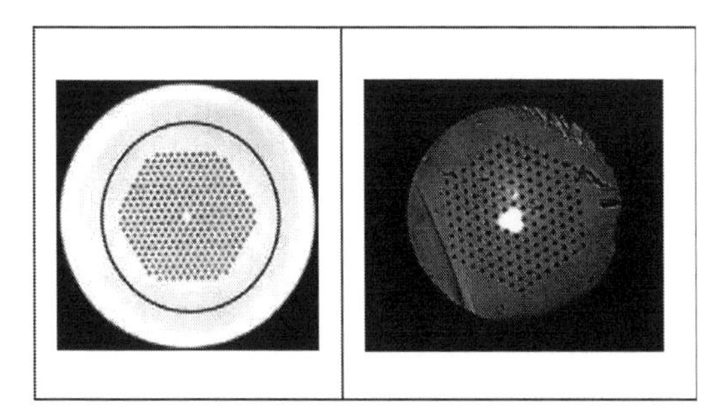

Figure 7.18 (a) Image of a double-clad PCF; (b) output image of double-clad PCF when a femtosecond laser is coupled into the core and propagated through a 2 m long fiber.

inner-clad diameter of 163 μm. Figure 7.18b shows an image of output surface from a double-clad PCF (2 m long) when a fs pulse from a Ti:sapphire laser is coupled into the core. Although there is some leakage from the core to the inner-clad (the inner-clad would be totally dark without leakage), most light is concentrated in the core region and the beam profile of the central core has a near-Gaussian profile. This result indicates that despite the large core for minimizing nonlinear effects, the double-clad fiber can support single-mode propagation of short pulses in the core region. We have also performed an experiment to measure pulse broadening when a 2 nJ, 170 fs pulse propagates through the core of a 2 m long PCF (Fig. 7.19). Without pre-chirping to compensate for the dispersion, the pulse width of the output light is broadened to 2.5 ps. However, when a grating pair is inserted to pre-chirp the fs pulse, the pulse width of the output light is 200 fs. This result confirms that PCF can be used for fs pulse delivery.

In addition to supporting single-mode propagation of the short pulse with minimal nonlinear effect, the double-clad fiber also supports collection of the fluorescence in the inner-clad cladding. Because of the double-clad PCF's large-diameter inner-clad and multimode propagation with a large NA, much higher collection efficiency can be achieved in comparison to using a conventional SMF. Our calculation indicates that when a double-clad fiber is used for excitation (through the core) and collection of excited fluorescence through the core and inner cladding, the collection

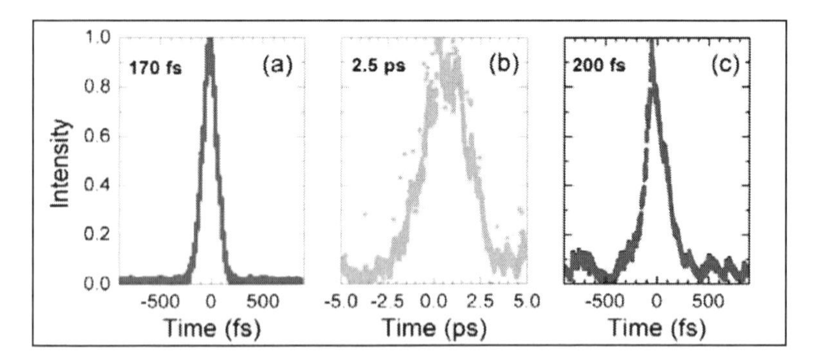

Figure 7.19 Pulse broadening in a double-clad PCF and the compression of the pulse width with dispersion pre-compensation. (a) Pulse from the laser; (b) pulse broadened after propagating in fiber; (c) pulse compressed back with dispersion pre-compensation.

efficiency of the double-clad fiber is up to 1,000 times larger than that of a conventional SMF. These unique features make double-clad PCF ideal for endoscopic MPM.

Figure 7.20 shows a schematic diagram for the fiber-based MPM microendoscope. A short-pulse beam generated from the fs fiber laser source first goes through the pre-chirp unit, which consists of a grating pair. The use of a large-core double-clad PCF minimizes the pulse broadening due to nonlinear effects. However, dispersion from the double-clad PCF could still significantly broaden the fs pulses. Fs pulses from the laser source will be negatively pre-chirped by a

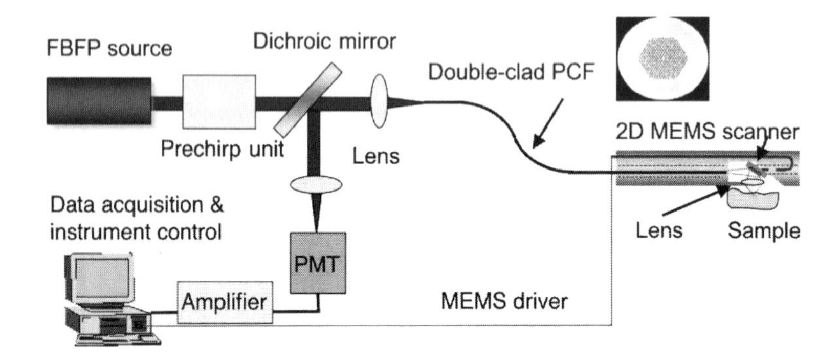

Figure 7.20 Schematic diagram of the fiber-based MPM endoscope.

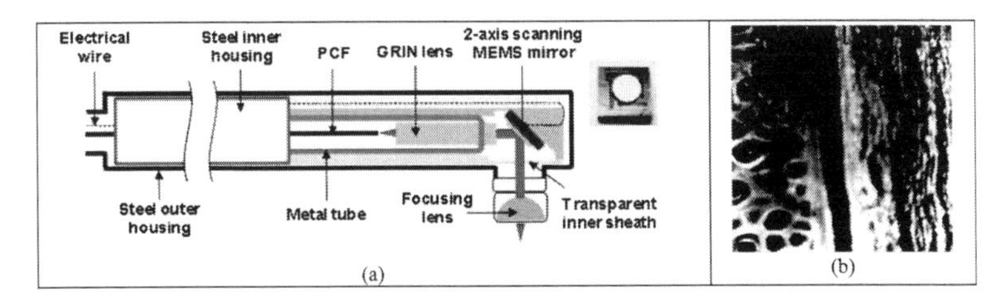

Figure 7.21 (a) Schematic of a 2D MEMS probe for MPM imaging. The inset shows the photograph of a 2 mm two-axis MEMS mirror. (b) MPM image of a rabbit ear (from [67]).

grating-based pulse stretcher to compensate for pulse broadening caused by positive dispersion in the double-clad PCF core. The pre-chirped pulses will pass through a dichroic mirror and be coupled into the double-clad PCF. Short-pulse light exiting from the double-clad PCF will be focused on the sample tissue through a miniature probe made by a 2D rotational MEMS scanner. The fluorescence or harmonic generation signal will be collected through the lens in the endoscopic probe and coupled into both the core and inner cladding of the fiber. The collected MPM signal propagates through the double-clad fiber and passes through the dichroic mirror that filters the excitation beam. And then the MPM signal detected by the photomultiplier tube (PMT) is amplified and digitized. The MEMS scanner and analog-to-digital converter (ADC) board are controlled and synchronized by a personal computer.

We have demonstrated in vitro imaging of a biological tissue using a MEMS scanner (Fig. 7.21) [31, 32]. An MPM image of a rabbit ear obtained with the fiber-based MPM system with a MEMS probe is shown in Fig. 7.21b.

In addition to the linear scaning, a rotational scanning probe-based MPM system has also been demonstrated (Fig. 7.22) [95]. Figure 7.22 shows the diagram of the probe distal end. The probe adopted a composite scanning mechanism that used a fiber distal end rotational scanning by a MEMS motor and a linear stage pullback. A GRIN lens is used to focus the excitation light, and a microprism glued to the shaft of the MEMS motor is used to redirect the light. SHG images of unstained rat-tail tendon and fish scale

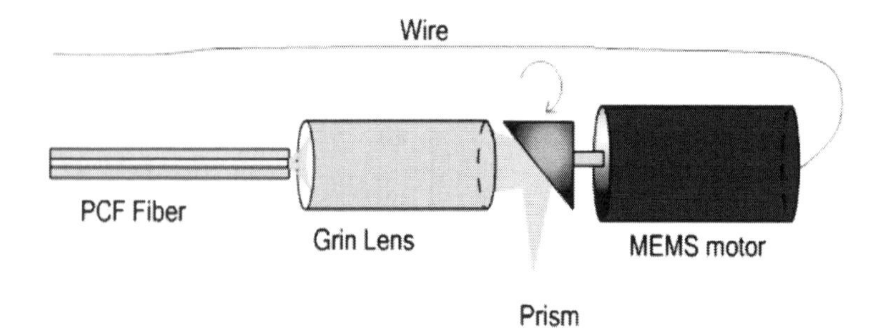

Figure 7.22 Diagram of the micro-rotational probe, which consists of a DCPCF fiber, a GRIN lens, a prism, and a rotational MEMS motor.

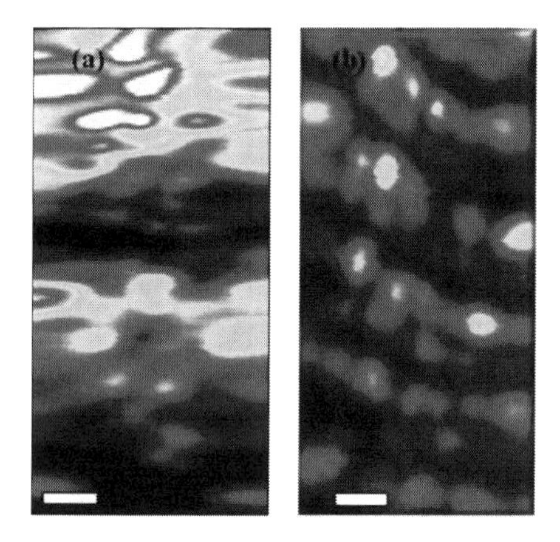

Figure 7.23 SHG images of (a) rat tail tendon and (b) fish scale. Scale bar is 25 μm (from [95]).

were obtained as shown in Fig. 7.23. A fiber fs laser is used for the excitation and the whole system is compact and potentially portable.

Compared to OCT and SHG/TPEF endoscopy, the development of fiber-based CARS endoscopy is still in its infant stage [24, 33, 98–100]. There are a number of challenges for fiber-based CARS, including efficient propagation of two ultrafast pulse trains of different color through an optical fiber for delivery of the excitation light to the sample, and efficient collection of the backward scattered signal into the fiber for detection. An important design consideration is the suppression of optical nonlinearities in the delivery fiber,

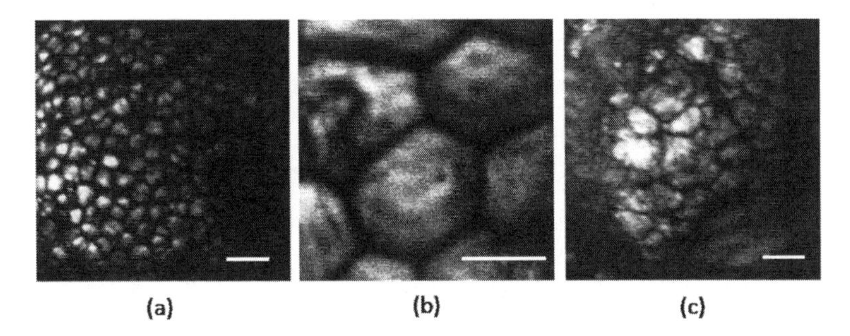

Figure 7.24 CARS images of tissue samples ex vivo at 2845 cm^{-1}, obtained with a fiber-coupled probe. (a) Small adipocytes in mouse skin. (b) Adipocytes in the subcutaneous layer of rabbit skin. (c) Meibomian gland in a mouse eyelid. Images were acquired in 2 s. Scale bar is 50 μm (from [24]).

such as SPM and SRS, which generally lead to unwanted spectral broadening of the pulses. In principle, the picosecond laser pulses that are often used in CARS microscopy have pulse energies that are low enough to minimize the effects of SPM and SRS. This permits the use of SMF, which has been successfully used in initial designs of CARS fiber probes [98]. Compared to SMF, the use of PCF is more attractive because they can be optimized for the delivery of both picosecond and fs pulse trains [24]. A major problem in fiber-based CARS probe design is the presence of a very strong anti-Stokes component in silica delivery fibers generated through a FWM process [24]. Without proper spectral filtering, this component affects the CARS image from the tissue sample [24]. We have developed a scheme that efficiently suppresses this spurious anti-Stokes component through the use of a separate fiber for excitation delivery and for signal detection which allows the incorporation of dichroic optics for anti-Stokes rejection [24]. CARS images from an ex vivo sample have been obtained with a miniature fiber probe (Fig. 7.24) [24].

7.3.3 *Multimodal Endoscopic Imaging*

There has been great interest in combining high-resolution imaging techniques derived from complementary signals to provide simultaneous structural and functional imaging of tissues. Combining MPM and OCT into a single system creates a novel multimodality imaging technique that provides structural and functional imaging of tissues simultaneously [101–105]. A multimodal system that combines photoacoustic microscopy and OCT to provide both the microanatomy and microvasculature of a sample has been reported [106]. An integrated OCT and fluorescence laminar optical tomography system to provide tissue structural and molecular information in 3D and in millimeter imaging scale has been demonstrated [106]. Multimodal nonlinear imaging that combines TPEF, SHG, and CARS has been used to analyze the impact of diet on macrophage infiltration and lipid accumulation on plaque formation in ApoE-deficient mice [107]. However, most of the current integrated systems are in microscopy platform. Further miniaturization of these combined systems into a single miniature endoscopic probe is essential for translation of the technology from laboratory benchtop to clinical applications.

Combing nonlinear optical contrast with OCT will greatly enhance clinical applications of these imaging modalities. It has been demonstrated in studies of wound healing and oral cancer that MPM and OCT can provide complementary structural and functional information about tissues [108, 109]. OCT imaging was able to distinguish an injured area from its surroundings, SHG revealed disruption and reconstruction of collagen fibrils, and TPEF imaging showed fibroblast migration in response to injury. With information obtained from OCT and MPM together, the sensitivity and specificity of cancer detection in oral tissues were shown to be substantially increased [109]. We recently developed a fiber-based OCT/MPM combined endomicroscopic system [110]. The combined OCT/MPM system is schematically illustrated in Fig. 7.25. A fiber-based fs laser (FBFL) was used as the light source for both OCT and MPM. The laser source has a central wavelength of 1.04 μm and delivers a sub-100 fs pulse with an average power of more than 1 W. The system used a 2 × 2 fiber coupler and specifically, a DCF-based multimode device: a (2+1):1 pump/signal combiner was introduced to enhance the MPM signal collection and to separate the OCT and MPM signals Co-registered OCT/MPM images of fluorescence beads and ex vivo tissue are shown in Fig. 7.26.

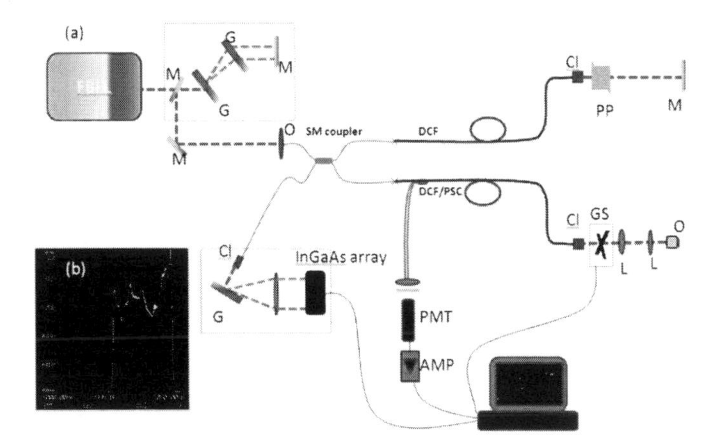

Figure 7.25 Schematic of the fiber-based OCT/MPM system. FBFL: fiber-based femtosecond laser; M: mirror; G: grating; O: objective; SM coupler: single-mode coupler; DCF: double-clad fiber; PSC: pump/signal combiner; CL: collimator; PP: prism pair; GS: galvo mirror scanner; L: lens; PMT: photomultiplier tube; AMP: low-noise pre-amplifier.

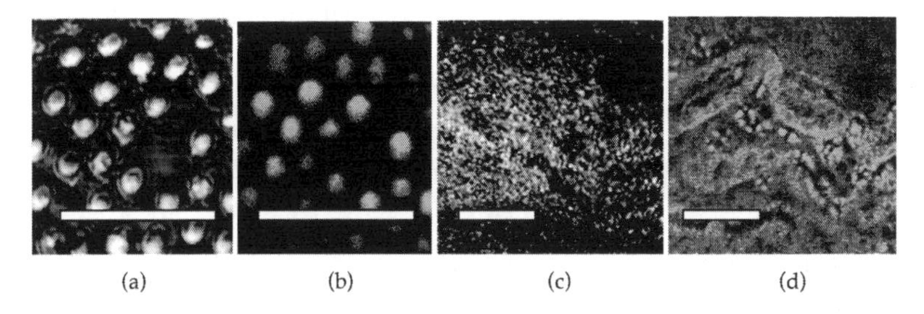

Figure 7.26 (a) OCT and (b) TPEF images of red fluorescence beads. (c) OCT and (d) SHG images of a thin slice of fixed rabbit heart stained with hematoxylin and eosin. Scale bar: 50 μm (from [110]).

Atherosclerosis is the leading cause of morbidity and mortality in the United States and is becoming the preeminent health problem worldwide. Detection and diagnosis of atherosclerosis relies on medical imaging techniques [111]. Intravascular ultrasound (IVUS) imaging has become a standard imaging modality for atherosclerosis diagnosis since it provides direct visualization of vessel walls. Recently, OCT has been applied to intravascular imaging because it offers high resolution imaging to gain new insights into the microstructure of atherosclerosis and tissue responses to stent implantation [112–114]. Research has been conducted to compare diagnostic accuracy of OCT and IVUS during which separate OCT and IVUS systems were used to image the same sites of interest [115, 116]. It has been pointed out that OCT and IVUS are complementary in the application of intravascular imaging, and the combination of the two can offer advantages which cannot be achieved by using either modality alone [117].

An integrated OCT-US system will be capable of offering high resolution, which is essential for visualizing microstructure of plaque, and also large penetration depth which is necessary for visualizing structures deep within the vessel wall. Moreover, improved safety for the patient can be achieved since only a minimum amount of the flushing agent will be needed for OCT under the guidance of US. An integrated OCT-US probe can provide both OCT and US imaging simultaneously so that both cost and the physician's time will be reduced significantly compared to using separate probes. We have developed an integrated intravascular OCT/US system (Fig. 7.27) [49, 118, 119]. The scanning mechanism is based on the external rotational motor (Fig. 7.27b).

Several different dual-modality intravascular probes that combine OCT/US imaging function have been demonstrated [49, 118, 119]. Figure 7.28 shows an integrated probe based on a high-frequency ultrasound ring transducer [118]. The homemade 50 MHz focused ring transducer has an effective aperture of 2 mm outer diameter with a 0.8 mm hole at the center to make room for the OCT probe which has an outer diameter of 0.7 mm. The coaxial US and light beams have a common focal length of 4 mm, and both are steered into tissue by a 45° mirror along their pathway. The glass mirror, coated with aluminum, is fixed close to the anterior surface of the hybrid probe to ensure both beams are focused at the tissue target. The mirror and hybrid probe are properly aligned and packaged in a brass tube housing on which a window is made to allow US and light beams to exit. This design allows a focused US beam and a light beam to be launched coaxially toward a common imaging spot, enabling automatically co-registered US and OCT images simultaneously [118]. An integrated intravascular OCT/US probe with a diameter of less than 1 mm has been developed [119].

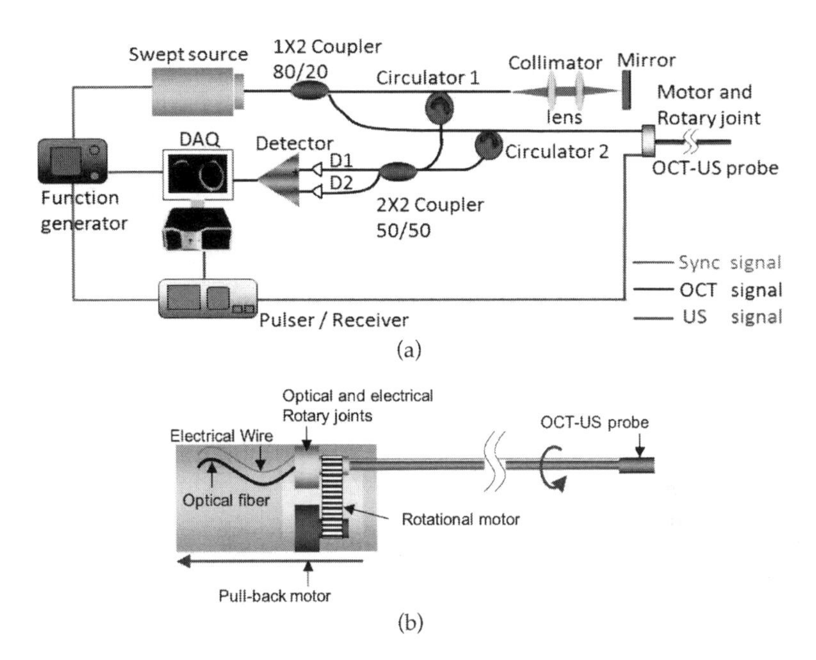

Figure 7.27 (a) Schematic of integrated US-OCT system. (b) Schematic of probe.

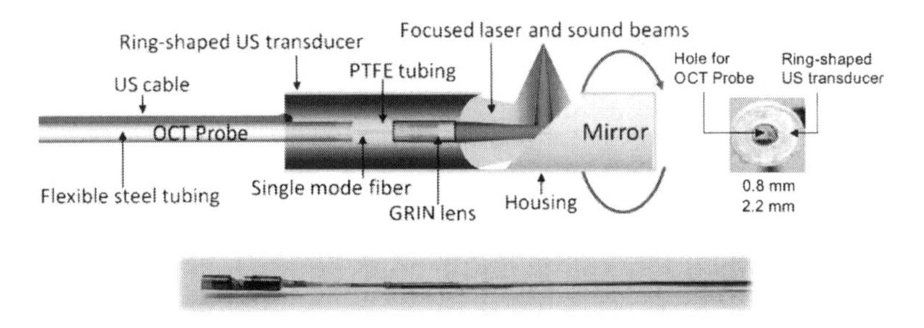

Figure 7.28 Schematic and picture of integrated OCT/US probe based on focused ultrasound ring transducer (from [118]).

The co-registered in vitro OCT/US images of normal rabbit aorta are shown in Fig. 7.29. The results clearly demonstrate the advantage of high resolution of OCT and deep penetration of US, which can offer complementary information for atherosclerosis diagnosis that

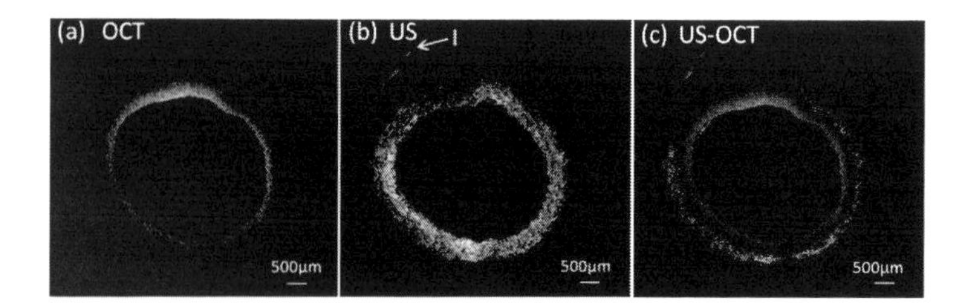

Figure 7.29 In vitro OCT/US images of a rabbit aorta from an integrated OCT/US system. (a) OCT image, (b) US image, (c) over layer of co-registered OCT/US images (from [118]).

cannot be obtained by either modality alone. In vivo imaging in animal models has also been demonstrated recently [119].

7.4 Summary

Optical imaging techniques, such as OCT, TPEF, SHG, and CARS, offer high resolution, unique contrast for 3D visualization of tissue structure and functions. Recent advances in fiber optics, compact lasers, miniature optics, and MEMS technologies have made it possible to translate these imaging technologies from benchtop microscopy to clinical endomicroscopy. The emerging endomicroscopy technology will enable real-time in situ imaging of tissue structure and physiology with molecular contrast. A multimodality imaging probe that combines different imaging technologies with complementary contrast is still in the early stage of development, and may have a significant impact in biomedical science and clinical medicine.

Acknowledgments

We would like to thank many of our colleagues who have contributed to the endomicroscopy project at the Beckman Laser Institute and the Department of Biomedical Engineering at UCI, particularly the students and postdoctoral fellows whose hard work made it possible for us to review many of the exciting results in this chapter.

Dr. Chen would like to acknowledge the research grants awarded from the National Institutes of Health (EB-10090, EB015890, HL-125084, HL-105215). Institutional support from the Air Force Office of Scientific Research (FA9550-04-0101) and the Beckman Laser Institute Endowment is also gratefully acknowledged. Dr. Chen has a financial interest in OCT Medical Imaging Inc., which, however, did not support this work.

References

1. Huang, D., et al. (1991). Optical coherence tomography. *Science*, 254:1178–1181.
2. Lee, M. R., Izatt, J. A., Swanson, E. A., Huang, D., Schumun, J. S., Lin, C. P., Puliafito, C. A., and Fujimoto, J. G. (1995). Optical coherence tomography for ophthalmic imaging: new technique delivers micron-scale resolution. *IEEE Eng. Med. Biol. Mag.*, 14:67–76.
3. Masters, B. R., and So, P. T. C., eds. (2008). Handbook of biomedical nonlinear optical microscopy. (Oxford University Press, USA).
4. Denk, W., Strickler, J. H., and Webb, W. W. (1990). Two-photon laser scanning fluorescence microscopy. *Science*, 248:73–79.
5. Zipfel, W. R., Williams, R. M., and Webb, W. W. (2003). Nonlinear magic: multiphoton microscopy in the biosciences. *Nat. Biotechnol.*, 21:1369–1377.
6. Helmchen, F., and Denk, W. (2005). Deep tissue two-photon microscopy. *Nat. Methods*, 2:932–940.
7. Evans, C. L., and Xie, X. S. (2008). Coherent anti-stokes Raman scattering microscopy: chemical imaging for biology and medicine. *Annu. Rev. Anal. Chem.*, 1:883–909.
8. Zipfel, W. R., Williams, R. M., Christie, R. A., Nikitin, Y., and Hyman, B. T. (2003). Live tissue intrinsic emission microscopy using multiphoton-excited native fluorescence and second harmonic generation. *Proc. Nat. Acad. Sci. USA*, 100:7075.
9. Sun, C. K., Chen, C. C., Chu, S. W., Tsai, T. H., Chen, Y. C., and Lin, B. L. (2003). Multiharmonic-generation biopsy of skin. *Opt. Lett.*, 28:2488–2490.
10. Le, T. T., Yue, S., and Cheng, J. X. (2010). Shedding new light on lipid biology with coherent anti-Stokes Raman scattering microscopy. *J. Lipid Res.*, 51:3091–3102.

11. Zhu, J., Lee, B., Buhman, K. K., and Cheng, J. X. (2009). A dynamic cytoplasmic triacylglycerol pool in enterocytes revealed by ex vivo and in vivo coherent anti-Stokes Raman scattering imaging. *J. Lipid Res.*, 50:1080–1089.
12. Lee, B., Fast, A. M., Zhu, J., Cheng, J. X., and Buhman, K. K. (2010). Intestine specific expression of acyl coa:diacylglycerol acyltransferase 1 (DGAT1) reverses resistance to diet-induced hepatic steatosis and obesity in Dgat1-/- mice. *J. Lipid Res.*, 51:1770–1780.
13. Henry, F., Côté, D., Randolph, M. A., Rust, E. A. Z., Redmond, R. W., Kochevar, I. E., Lin, C. P., and Winograd, J. M. (2009). Real-time in vivo assessment of the nerve microenvironment with coherent anti-Stokes Raman scattering microscopy. *Plast. Reconstr. Surg.*, 123:123S–130S.
14. Huff, T. B., and Cheng, J. X. (2007). In vivo coherent anti-Stokes Raman scattering imaging of sciatic nerve tissues. *J. Microsc.*, 225:175–182.
15. Huff, T. B., Shi, Y., Yan, Y., Wang, H., and Cheng, J. X. (2008). Multimodel nonlinear optical microscopy and applications to central nervous system. *IEEE J. Sel. Top. Quantum Electron.*, 14:4–9.
16. Drexler, W., and Fujimoto, J. G., eds. (2008). *Optical coherence tomography: technology and applications.* (Springer-Verlag, Berlin, Heidelberg, New York).
17. Iniewski, K., ed. (2009). *Medical imaging: principles, detectors, and electronics.* (John Wiley & Sons New York).
18. Crisp, J., and Elliott, B. (2005). *Introduction to Fiber Optics.* (Newnes, Burlington, MA).
19. Agrawa, G. (2006). *Nonlinear Fiber Optics.* 4th ed. (Academic Press, New York).
20. Wojtkowski, M., Srinivasan, V. J., Ko, T., Fujimoto, J. G., Kowalczyk, A., and Duker, J. S. (2004). Ultrahigh-resolution high speed Fourier domain optical coherence tomography and methods for dispersion compensation. *Opt. Express*, 12:2404–2422.
21. Fercher, A. F., and Hizenberger, C. K. (2002). Optical Coherence Tomography, *in Progress in Optics*, Wolf, E., Editor. (Elsevier, North-Holland). p. 215.
22. Xi, P., Andegeko, Y., Pestov, D., Lovozoy, V. V., and Dantus, M. (2009). Two-photon imaging using adaptive phase compensated ultrashort laser pulses. *J. Biomed. Opt.*, 14:014002.
23. Tang, S., Liu, J., Krasieva, T. B., Chen, Z., and Tromberg, B. J. (2009). Developing compact multiphoton systems using femtosecond fiber lasers. *J. Biomed. Opt.*, 14:030508.

24. Balu, M., Liu, G., Chen, Z., Tromberg, B. J., and Potma, E. O. (2010). Fiber delivered probe for efficient CARS imaging of tissues. *Opt. Express*, 18:2380–2388.

25. Helmchen, F., Tank, D. W., and Denk, W. (2002). Enhanced two-photon excitation through optical fiber by single-mode propagation in a large core. *Appl. Opt.*, 41:2930–2934.

26. Mirgorodskii, V. I., Gerasimov, V. V., and Peshin, S. V. (1996). The possibility of investigating the spatial distribution of sources of incoherent radiation by correlation processing. *Zh. Tekh. Fiz.*, 66:196–202.

27. Ouzounov, D. G., Moll, K. D., Foster, M. A., Zipfel, W. R., Webb, W. W., and Gaeta, A. L. (2002). Delivery of nanojoule femtosecond pulses through large-core microstructured fibers. *Opt. Lett.*, 27:1513–1515.

28. Myaing, M. T., Ye, J. Y., Norris, T. B., Thomas, T., Baker, J. R. J., Wadsworth, W. J., Bouwmans, G., Knigh, J. C., and Russell, P. S. J. (2003). Enhanced two-photon biosensing with double-clad photonic crystal fibers. *Opt. Lett.*, 28:1224–1226.

29. Myaing, M. T., macdonald, D. G., and Xingde, L. (2006). Fiber-optic scanning two-photon fluorescence endoscope. *Opt. Lett.*, 31:1076–1078.

30. Fu, L., Jain, A., Xie, H., Cranfield, C., and Gu, M. (2006). Nonlinear optical endoscopy based on a double clad photonic crystal fiber and a MEMS mirror. *Opt. Express*, 14:1027–1032.

31. Jung, W. G., Tang, S., mccormick, D. T., Xie, T., Ahn, Y.-C., Su, J., Tomov, I., Krasieva, T. B., Tromberg, B., and Chen, Z. (2008). Miniaturized probe based on a microelectromechanical system mirror for multiphoton microscopy. *Opt. Lett.* 12:1324–1326.

32. Tang, S., Jung, W., mccormick, D., Xie, T., Su, J., Ahn, Y. C., Tromberg, B. J., and Chen, Z. (2009). Design and implementation of fiber-based multiphoton endoscopy with microelectromechanical systems scanning. *J. Biomed. Opt.*, 14:034005.

33. Wang, H., Huff, T. B., and Cheng, J. X. (2006). Coherent anti-Stokes Raman scattering imaging with a laser source delivered by a photonic crystal fiber. *Opt. Lett.*, 31:1417–1419.

34. Bouwmans, G., Luan, F., Knight, J., St, J. R. P., Farr, L., Mangan, B., and Sabert, H. (2003). Properties of a hollow-core photonic bandgap fiber at 850 nm wavelength. *Opt. Express*, 11:1613–1620.

35. Gobel, W., Nimmerjahn, A., and Helmchen, F. (2004). Distortion-free delivery of nanojoule femtosecond pulses from a Ti:sapphire laser

through a hollow-core photonic crystal fiber. *Opt. Lett.*, 29:1285–1287.

36. Tai, S. P., Chan, M. C., Tsai, T. H., Guol, S. H., Chen, L. J., and Sun, C. K. (2004). Two-photon fluorescence microscope with a hollow-core photonic crystal fiber. *Opt. Express*, 12:6122–6128.
37. Xie, T., Mukai, D., Guo, S., Brenner, M., and Chen, Z. (2005). Fiber-optic-bundle-based optical coherence tomography. *Opt. Lett.*, 30:1803–1805.
38. Xie, T., Guo, S., Chen, Z., Mukai, D., and Brenner, M. (2006). GRIN lens rod based probe for endoscopic spectral domain optical coherence tomography with fast dynamic focus tracking. *Opt. Express*, 14:3238–3246.
39. Göbel, W., Kerr, J. N. D., Nimmerjahn, A., and Helmchen, F. (2004). Miniaturized two-photon microscope based on a flexible coherent fiber bundle and gradient-index lens objective. *Opt. Lett.*, 29:2521–2523.
40. Jung, J. C., and Schnitzer, M. J. (2003). Multiphoton endoscopy. *Opt. Lett.*, 28:902–904.
41. Jung, J. C., Mehta, A. D., Aksay, E., Stepnoski, R., and Schnitzer, M. J. (2004). In vivo mammalian brain imaging using one- and two-photon fluorescence microendoscopy. *J. Neurophysiol.*, 92:3121–3133.
42. Barretto, R. P. J., Messerschmidt, B., and Schnitzer, M. J. (2009). In vivo fluorescence imaging with high-resolution microlenses. *Nat. Methods*, 6:511–512.
43. Brenner, M., et al. (2007). Detection of acute smoke-induced airway injury in a New Zealand white rabbit model using optical coherence tomography. *J Biomed. Opt.*, 12:051701.
44. Brenner, M., et al. (2008). In vivo optical coherence tomography detection of differences in regional large airway smoke inhalation induced injury in a rabbit model. *J. Biomed. Opt.*, 13:034001.
45. Ridgway, J. M., et al. (2008). Optical coherence tomography of the newborn airway. *Ann. Otol. Rhinol. Laryngol.*, 117:327–334.
46. Yang, V. X., et al. (2003). High speed, wide velocity dyhamic range Doppler optical coherence tomography (part III): in vivo endoscopic imaging of blood flow in the rat and human gastrointestinal tracts. *Opt. Express*, 11:2416–2424.
47. Tearney, G. J., Boppart, S. A., Bouma, B. E., Brezinski, M. E., Weissman, N. J., Southern, J. F., and Fujimoto, J. G. (1996). Scanning single-mode fiber optic catheter-endoscope for optical coherence tomography. *Opt. Lett.*, 21:543–545.

48. Tearney, G. J., Brezinski, M. E., Bouma, B. E., Boppart, S. A., Pitris, C., Southern, J. F., and Fujimoto, J. G. (1997). In vivo endoscopic optical biopsy with optical coherence tomography. *Science*, 276:2037–2039.
49. Yin, J., Yang, H. C., Li, X., Zhang, J., Zhou, Q., Hu, C., Shung, K. K., and Chen, Z. (2010). Integrated intravascular optical coherence tomography ultrasound imaging system. *J. Biomed. Opt.*, 15:010512.
50. Yun, S. H., et al. (2006). Comprehensive volumetric optical microscopy in vivo. *Nat. Med.*, 12:1429–1433.
51. Munce, N. R., et al. (2008). Electrostatic forward-viewing scanning probe for Doppler optical coherence tomography using a dissipative polymer catheter. *Opt. Lett.*, 33:657–659.
52. Min, E. J., Na, J., Ryu, S. Y., and Lee, B. H. (2009). Single-body lensed-fiber scanning probe actuated by magnetic force for optical imaging. *Opt. Lett.*, 34:1897–1899.
53. Binnig, G., and Smith, D. P. E. (Single-tube three-dimensional scanner for scanning tunneling microscopy). 1986. *Rev. Sci. Inst.*, 57:1688–1689.
54. Helmchen, F., Fee, M. S., Tank, D. W., and Denk, W. (2001). A miniature head-mounted two-photon microscope: high-resolution brain imaging in freely moving animals. *Neuron*, 31:903–912.
55. Flusberg, B. A., Jung, J. C., Cocker, E. D., Anderson, E. P., and Schnitzer, M. J. (2005). In vivo brain imaging using a portable 3.9 gram two-photon fluorescence microendoscope. *Opt. Lett.*, 30:2272–2274.
56. Liu, X., Cobb, M. J., Chen, Y., Kimmey, M. B., and Li, X. (2004). Rapid-scanning forward-imaging miniature endoscope for real-time optical coherence tomography. *Opt. Lett.*, 29:1763–1765.
57. Myaing, M. T., macdonald, D. J., and Li, X. (2006). Fiber-optic scanning two-photon fluorescence endoscope. *Opt. Lett.*, 31:1076–1078.
58. Engelbrecht, C. J., Johnston, R. S., Seibel, E. J., and Helmchen, F. (2008). Ultra-compact fiber-optic two-photon microscope for functional fluorescence imaging in vivo. *Opt. Express*, 16:5556–5564.
59. Wu, Y., Xi, J., Cobb, M. J., and Li, X. (2009). Scanning fiber-optic nonlinear endomicroscopy with miniature aspherical compound lens and multimode fiber collector. *Opt. Lett.*, 34:953–955.
60. Moon, S., Lee, S. W., Rubinstein, M., Wong, B. J., and Chen, Z. (2010). Semi-resonant operation of a fiber-cantilever piezotube scanner for stable optical coherence tomography endoscope imaging. *Opt. Express*, 18:21183–21197.

61. Chen, C. J. (1992). Electromechanical deflections of piezoelectric tubes with quartered electrodes. *Appl. Phys. Lett.*, 60:132.
62. Dhaubanjar, N., Hu, H., Dave, D., Phuyal, P., Sin, J., Stephanou, H., and Chiao, J. C. (2006). A compact optical fiber scanner for medical imaging. *Proc. SPIE*, 6414:64141Z.
63. Min, E. J., Na, J., Tyu, S. Y., and Lee, B. H. (2009). Single-body lensed-fiber scanning probe actuated by magnetic force for optical imaging. *Opt. Lett.*, 34:1897–1899.
64. Pan, Y., Xie, H., and Fedder, G. K. (2001). Endoscopic optical coherence tomography based on a microelectromechnic mirror. *Opt. Lett.*, 26:1966–1968.
65. Jung, W. G., Zhang, J., Wang, L., Chen, Z., mccormick, D. T., and Tien, N. C. (2005). Three-dimensional optical coherence tomography employing a 2-axis microelectromechanical scanning mirror. *IEEE J. Sel. Top. Quantum Electron.*, 11:806–8010.
66. Jung, W. G., Zhang, J., Wang, L., Chen, Z., mccormick, D. T., and Tien, N. C. (2006). Three-dimensional endoscopic optical coherence tomography by use of a two-axis microelectromechanical scanning mirror. *Appl. Phys. Lett.*, 88:163901–163903.
67. Jung, W., Tang, S., mccormic, D. T., Xie, T., Ahn, Y. C., Su, J., Tomov, I. V., Krasieva, T. B., Tromberg, B. J., and Chen, Z. (2008). Miniaturized probe based on a microelectromechanical system mirror for multiphoton microscopy. *Opt. Lett.*, 33:1324–1326.
68. Piyawattanametha, W., Barretto, R. P., Ko, T. H., Flusberg, B. A., Cocker, E. D., Ra, H., Lee, D., Solgaard, O., and Schnitzer, M. J. (2006). Fast-scanning two-photon fluorescence imaging based on a microelectromechanical systems two- dimensional scanning mirror. *Opt. Lett.*, 31:2018–2020.
69. Chen, Z., Jung, J. C., Ahn, Y.-C., Sepehr, A., Armstrong, W. B., Brenner, M., mccormick, D. T., and Tien, N. C. (2007). High speed three-dimensional endoscopic OCT using MEMS technology. *Proc. SPIE*, 6466:64660H1–64660H8.
70. Tran, P. H., Mukai, D. S., Brenner, M., and Chen, Z. (2004). In vivo endoscopic optical coherence tomography by use of a rotational microelectromechanical system probe. *Opt. Lett.*, 29:1236–1238.
71. Su, J., Zhang, J., Yu, L., and Chen, Z. (2007). In vivo three-dimensional microelectromechanical endoscopic swept source optical coherence tomography. *Opt. Express*, 15:10390–10396.
72. Liu, G., Xie, T., Tomov, I. V., Su, J., Yu, L., Zhang, J., Tromberg, B. J., and Chen, Z. (2009). Rotational multiphoton endoscopy with a 1 μm fiber laser system. *Opt. Lett.*, 34:2249–2251.

73. Williams, R. M., Flesken-Nikitin, A., Ellenson, L. H., Connolly, D. C., Hamilton, T. C., Nikitin, A. Y., and Zipfel, W. R. Strategies for high-resolution imaging of epithelial ovarian cancer by laparoscopic nonlinear microscopy. *Transl. Oncol.*, 3:181–194.
74. Bird, D., and Gu, M. (2003). Two-photon fluorescence endoscopy with a micro-optic scanning head. *Opt. Lett.*, 28:1552–1554.
75. Flusberg, B. A., Cocker, E. D., Piyawattanametha, W., Jung, J. C., Cheung, E. L., and Schnitzer, M. J. (2005). Fiber-optic fluorescence imaging. *Nat. Methods*, 2:941–950.
76. Mao, Y., Chang, S., and Flueraru, C. (2010). Fiber lenses for ultra-small probes used in optical coherent tomography. *J. Biomed. Sci. Eng.*, 3:27–34.
77. Bao, H., and Gu, M. (2009). A 0.4-mm-diameter probe for nonlinear optical imaging. *Opt. Express*, 17:10098–10104.
78. Yang, V. X., Mao, Y. X., Munce, N., Standish, B., Kucharczyk, W., Marcon, N. E., Wilson, B. C., and Vitkin, I. A. (2005). Interstitial Doppler optical coherence tomography. *Opt. Lett.*, 30:1791–1793.
79. Yaqoob, Z., Wu, J., mcdowell, E. J., Heng, X., and Yang, C. (2006). Methods and application areas of endoscopic optical coherence tomography. *J. Biomed. Opt.*, 11:063001.
80. Boppart, S. A., Bouma, B. E., Pitris, C., Tearney, G. J., Fujimoto, J. G., and Brezinski, M. E. (1997). Forward-imaging instruments for optical coherence tomography. *Opt. Lett.*, 22:1618–1620.
81. Xie, T., Liu, G., Kreuter, K., Mahon, S., Colt, H., Mukai, D., Peavy, G. M., Chen, Z., and Brenner, M. (2009). In vivo three-dimensional imaging of normal tissue and tumors in the rabbit pleural cavity using endoscopic swept source optical coherence tomography with thoracoscopic guidance. *J. Biomed. Opt.*, 14:064045.
82. Guo, S., Yu, L., Sepehr, A., Perez, J., Su, J., Ridgway, J. M., Vokes, D., Wong, B. J., and Chen, Z. (2009). Gradient-index lens rod based probe for office-based optical coherence tomography of the human larynx. *J. Biomed. Opt.*, 14:014017.
83. Guo, S., Hutchison, R., Jackson, R. P., Kohli, A., Sharp, T., Orwin, E., Haskell, R., Chen, Z., and Wong, B. J. (2006). Office-based optical coherence tomographic imaging of human vocal cords. *J. Biomed. Opt.*, 11:30501.
84. Yu, L., Liu, G., Rubinstein, M., Saidi, A., Wong, B. J., and Chen, Z. (2009). Office-based dynamic imaging of vocal cords in awake patients with swept-source optical coherence tomography. *J. Biomed. Opt.*, 14:064020.

85. Jung, W., mccormick, D. T., Ahn, Y. C., Sepehr, A., Brenner, M., Wong, B., Tien, N. C., and Chen, Z. (2007). In vivo three-dimensional spectral domain endoscopic optical coherence tomography using a microelectromechanical system mirror. *Opt. Lett.*, 32:3239–3241.
86. Zara, J. M., Yazdanfar, S., Rao, K. D., Izatt, J. A., and Smith, S. M. (2003). Electrostatic micromachine scanning mirror for optical coherence tomography. *Opt. Lett.*, 28:828–830.
87. Liu, X., Cobb, M. J., Chen, Y., Kimmey, M. B., and Li, X. (2004). Rapid-scanning forward-imaging miniature endoscope for real-time optical coherence tomography. *Opt. Lett.*, 29:1763–1765.
88. Huo, L., Xi, J., Wu, Y., and Li, X. (2010). Forward-viewing resonant fiber-optic scanning endoscope of appropriate scanning speed for 3D OCT imaging. *Opt. Express*, 18:14375–14384.
89. Quinn, Y., Smithwick, J., Reinhall, P. G., Vagners, J., and Seibel, E. J. (2004). A nonlinear state-space model of a resonating single fiber scanner for tracking control: theory and experiment. *J. Dyn. Sys. Meas.*, 126:88–101.
90. Tearney, G. J., Brezinski, M. E., Bouma, B. E., Boppart, S. A., Pitvis, C., Southern, J. F., and Fujimoto, J. G. (1997). In vivo endoscopic optical biopsy with optical coherence tomography. *Science*, 276:2037–2039.
91. Tearney, G. J., Brezinski, M. E., Boppart, S. A., Bouma, B. E., Weissman, N., Southern, J. F., Swanson, E. A., and Fujimoto, J. G. (1996). Catheter-Based Optical Imaging of a Human Coronary Artery. *Circulation*, 94:3013–3013.
92. Jang, I. K., et al. (2002). Visualization of coronary atherosclerotic plaques in patients using optical coherence tomography: comparison with intravascular ultrasound. *J. Am. Coll. Cardiol.*, 39:604–609.
93. Mathews, M. S., Su, J., Heidari, E., Levy, E. I., Linskey, M. E., and Chen, Z. (2011). Neuroendovascular optical coherence tomography imaging and histological analysis. *Neurosurgery*, 69:430–439.
94. Ouzounov, D. G., Moll, K. D., Foster, M. A., Zipfel, W. R., Webb, W. W., and Gaeta, A. L. (2002). Delivery of nanojoule femtosecond pulses through large-core microstructured fibers. *Opt. Lett.*, 27:1513–1515.
95. Liu, G., Xie, T., Tomov, I. V., Su, J., Yu, L., Zhang, J., Tromberg, B. J., and Chen, Z. (2009). Rotational multiphoton endoscopy with a 1 microm fiber laser system. *Opt. Lett.*, 34:2249–2251.
96. Jung, J. C., and Schnitzer, M. J. (2003). Multiphoton endoscopy. *Opt. Lett.*, 28:902–904.

97. Konig, K., Ehlers, A., Riemann, I., Schenkl, S., Buckle, R., and Kaatz, M. (2007). Clinical two-photon microendoscopy. *Microsc. Res. Tech.*, 70:398–402.

98. Legare, F., Evans, C. L., Ganikhanov, F., and Xie, X. S. (2006). Towards CARS Endoscopy. *Opt. Express*, 14:4427–4432.

99. Murugkar, S., Smith, B., Srivastava, P., Moica, A., Naji, M., Brideau, C., Stys, P. K., and Anis, H. (2010). Miniaturized multimodal CARS microscope based on MEMS scanning and a single laser source. *Opt. Express*, 18:23796–23804.

100. Saar, B. G., Johnston, R. S., Freudiger, C. W., Xie, X. S., and Seibel, E. J. (2011). Coherent Raman scanning fiber endoscopy. *Opt. Lett.*, 36:2396–2398.

101. Beaurepaire, E., Moreaux, L., Amblard, F., and Mertz, J. (1999). Combined scanning optical coherence and two-photon-excited fluorescence microscopy. *Opt. Lett.*, 24:969–971.

102. Tang, S., Krasieva, T. B., Chen, Z., and Tromberg, B. (2006). Combined multiphoton microscopy and optical coherence tomography using a 12 femtosecond, broadband source. *J. Biomed. Opt.*, 11:020502.

103. Vinegoni, C., Ralston, T. S., Tan, W., Luo, W., Marks, D. L., and Boppart, S. A. (2006). Integrated structural and functional optical imaging combining spectral-domain optical coherence and multiphoton microscopy. *Appl. Phys. Lett.*, 88:053901.

104. Tang, S., Sun, C. H., Krasieva, T. B., Chen, Z., and Tromberg, B. (2007). Imaging sub-cellular scattering contrast using combined optical coherence and multiphoton microscopy. *Opt. Lett.*, 32:503–505.

105. Yazdanfar, S., Chen, Y. Y., So, P. T. C., and Laiho, L. H. (2007). Multifunctional imaging of endogenous contrast by simultaneous nonlinear and optical coherence microscopy of thick tissues. *Microsc. Res. Tech.*, 70:628–633.

106. Jiao, S., Xie, Z., Zhang, H. F., and Puliafito, C. A. (2009). Simultaneous multimodal imaging with integrated photoacoustic microscopy and optical coherence tomography. *Opt. Lett.*, 34:2961–2963.

107. Lim, R. S., Kratzer, A., Barry, N. P., Miyazaki-Anzai, S., Miyazaki, M., Mantulin, W. W., Levi, M., Potma, E. O., and Tromberg, B. J. (2010). Multimodal CARS microscopy determination of the impact of diet on macrophage infiltration and lipid accumulation on plaque formation in apoe-deficient mice. *J. Lipid Res.*, 51:1729–1737.

108. Yeh, A. T., Kao, S., Jung, W. G., Chen, Z., Nelson, J. S., and Tromberg, B. (2004). Imaging wound healing using optical coherence tomography and multiphoton microscopy in an in vitro skin-equivalent tissue model. *J. Biomed. Opt.*, 9:248–253.

109. Wilder-Smith, P., Krasieva, T., Jung, W. G., Zhang, J., Chen, Z., Osann, K., and Tromberg, B. (2005). Non-invasive imaging of oral premalignancy and maliganancy. *J. Biomed. Opt.*, 10:051601.

110. Liu, G., and Chen, Z. (2011). Fiber-based combined optical coherence and multiphoton endomicroscopy. *J. Biomed. Opt.*, 16:036010.

111. Pasterkamp, G., Falk, E., Woutman, H., and Borst, C. (2000). Techniques characterizing the coronary atherosclerotic plaque: Influence on clinical decision making. *J. Am. Coll. Cardiol.*, 36:13–21.

112. Farooq, M. U., Khasnis, A., Majid, A., and Kassab, M. Y. (2009). The role of optical coherence tomography in vascular medicine. *Vasc. Med.*, 14:63–71.

113. Tearney, G. J., Jang, I. K., and Bouma, B. E. (2006). Optical coherence tomography for imaging the vulnerable plaque. *J. Biomed. Opt.*, 11:021002.

114. Liu, Y., et al. Assessment by optical coherence tomography of stent struts across side branch. -Comparison of bare-metal stents and drug-eluting stents. *Circ. J.*, 75:106–112.

115. Kawasaki, M., Bouma, B. E., Bressner, J., Houser, S. L., Nadkarni, S. K., macneill, B. D., Jang, I. K., Fujiwara, H., and Tearney, G. J. (2006). Diagnostic accuracy of optical coherence tomography and integrated backscatter intravascular ultrasound images for tissue characterization of human coronary plaques. *J. Am. Coll. Cardiol.*, 48:81–88.

116. Rieber, J., et al. (2006). Diagnostic accuracy of optical coherence tomography and intravascular ultrasound for the detection and characterization of atherosclerotic plaque composition in ex-vivo coronary specimens: a comparison with histology. *Coron. Artery Dis.*, 17:425–430.

117. Sawada, T., et al. (2008). Feasibility of combined use of intravascular ultrasound radiofrequency data analysis and optical coherence tomography for detecting thin-cap fibroatheroma. *Eur. Heart J.*, 29:1136–1146.

118. Li, X., Yin, J., Hu, C., Zhou, Q., Shung, K. K., and Chen, Z. (2010). High-resolution coregistered intravascular imaging with integrated

ultrasound and optical coherence tomography probe. *Appl. Phys. Lett.*, 97:133702.

119. Yin, J., et al. (2011). Novel combined miniature optical coherence tomography ultrasound probe for in vivo intravascular imaging. *J. Biomed. Opt.*, 16:060505.

Chapter 8

Nanoplasmonic Biophotonics

Luke Lee
Department of Bioengineering, University of California, Berkeley, CA, USA
lplee@berkeley.edu

8.1 Conventional Biophotonics versus Nanoplasmonic Biophotonics

In a broad sense, the term *biophotonics* refers to the use of light (photons) to investigate biological processes (Fig. 8.1.1a). Culminating in a research area that intersects biology, chemistry, physics, and engineering, biophotonics has allowed many optical science and technological developments to bring new insights into life sciences for both fundamental and clinical interests. For instance, transmission light microscopy combined with various contrast enhancement methods (e.g., phase contrast or differential interference contrast) is widely adopted to observe tissues, cells and organelles. Spectroscopy (e.g., absorption, luminescence, infrared, or Raman spectroscopies) permits measurements of sample concentration and identification of material contents [1]. Fluorescent probes (e.g., organic dyes,

Understanding Biophotonics: Fundamentals, Advances, and Applications
Edited by Kevin K. Tsia

ISBN 978-981-4411-77-6 (Hardcover), 978-981-4411-78-3 (eBook)
www.panstanford.com

quantum dots and fluorescent nanodiamonds) have enabled molecular imaging and have gained considerable attention because of their molecular-level sensitivities and multiplexing capabilities [2–6]. Together with fluorescent protein reporters, fluorescence microscopy and high-throughput fluorescence-activated cell sorting techniques are routinely used in real-time monitoring of gene expression and regulation in single cells [2, 7]. In addition to the characterization of biological materials, the term *biophotonics* can also refer to optical techniques that physically or genetically control specimens. In fact, since the first demonstration of the laser in the 1960s, progress in laser development has not only impacted our lives through improved microscopy or spectroscopy techniques but has also enabled unprecedented manipulation and actuation of materials of a diverse range For example, photodynamic therapy is widely used to guide and activate chemical compounds for drug release [8]. Additionally, optical tweezers that employ highly focused laser beams can individually trap micrometer-sized objects [9]. Such actuators also enable measurements of the force generated by biomolecular interactions at the piconewton scale to investigate the biophysics of single molecules More recently, optogenetic control was demonstrated to allow manipulation through guided light [10, 11]. This report opened up new venues for researching biological events in live cells One common theme for both light-enabled imaging and actuation is that the photonic system in question must allow for light collection and manipulation within the length and time scale of biological interest.

However, in researching the essential constituents of life, a problem inevitably arises when seeking to probe smaller biological structures within the wavelength scale of the illumination light. A fundamental resolution limit, which is approximately one-half of the wavelength ($\lambda/2 \approx 300$ nm) of visible light, establishes a barrier for how optical energy can be focused Although conventional biophotonic techniques often satisfy the requirements of cell biologists, this diffraction limit of light fundamentally restricts the applicability of biophotonic tools to the nanometer scale in which active key biological ingredients such as proteins, binding molecules and genetic materials (DNA and RNA), reside (Fig. 8.1a). Without a means to reach the subwavelength scale, progress in biological

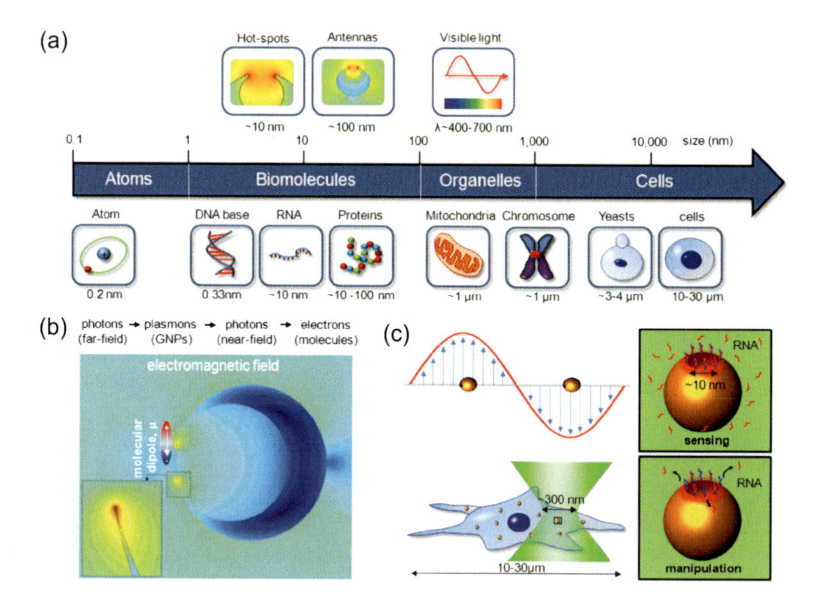

Figure 8.1 Nanoplasmonic biophotonics for biomolecular sensing and manipulation. (a) Relative scale of cells, biomolecules, wavelength of light and plasmonic hot spots. Note that the wavelength of visible light is 1–2 orders of magnitude greater than the size of the biomolecules and hot spots. (b) Schematic illustration of localized surface plasmon resonance on a gold nanocrescent, which enhances the electromagnetic field of the incident light. The bottom-left inset shows the hot spot on the tip of the nanocrescent [12, 13]. (c) The top-left inset shows the relative scale of nanoantennas versus the wavelength of light. In the bottom inset, light-absorbing plasmonic nanoantennas (e.g., gold nanocrescents, yellow dots) are promising probes and vectors for sensing and manipulating biomolecules (e.g., RNA) with nm-scaled resolution both *in vivo* and *in vitro* [14–16].

science and technology is impeded by the traditional biophotonic techniques.

There have been a few attempts to address this issue with the development of nano-optic measurement techniques, such as scanning near-field microscopy and super-resolution microscopy [17–19]. While these novel nanotools have broken the diffraction limit in terms of imaging capability, their use of sophisticated bulky optical systems might not always be suitable for molecular or cellular biological investigations, especially *in vivo*. For examination and manipulation at the molecular level, a more suitable alternative for biological applications is *nanoplasmonics* through techniques that rely on subwavelength optical fields enhanced by collective charge oscillation (or plasmons) of a metal nanostructure (or a nanoantenna, Fig. 8.1a) [14–16]. Nanoplasmonic devices based on localized surface plasmon resonance (LSPR) offer a promising route to overcome the diffraction limit of light while simultaneously generating novel *in vitro* and *in vivo* biological applications [14, 20–23]. Therefore, *nanoplasmonic biophotonics* is the emerging biophotonic area that seeks to develop optical techniques to probe and manipulate biological matter based on a new paradigm, which creatively uses tightly focused subwavelength optical energy found at the vicinity of metallic nanostructures [24–29].

Shaped through bottom-up chemical synthesis or top-down lithographic fabrication methods, plasmonic antennas generate highly confined electromagnetic near-field patterns (hot spots) [14]. These hot spots have a characteristic size on the order of tens of nanometers, which are useful for biological molecules that have a comparable size (e.g., DNA, RNA, and proteins). In fundamental terms, the synergy between a light harvesting plasmonic photonic device and a target biomolecule of interest can be described as follows (Fig. 8.1b): (i) far-field photons scatter around nanostructures and result in near-field electromagnetic patterns; (ii) the collective oscillation of electrons from metals (plasmons) are a means to increase the local density of photons; and (iii) the increased number of near-field photons, in turn, permeate biomolecules and interact with them through their electronic polarization (i.e., the strongest being the molecular dipole) [30, 31]. As reviewed in this text, utilizing plasmon-tailored optical energy while interacting with biomolecules has led to new insights in both biology and optics because of the ability to probe, image, and manipulate individual biomolecules, both *in vitro* and *in vivo* (Fig. 8.1c) [14, 20, 21, 23–29, 32, 33].

The goal of this chapter is to describe the fundamentals and applications of nanoplasmonics as related to the fields of biology and medicine. First, we will review the fundamentals of nanoplasmonics by examining the optical properties of metallic nanoparticles. An explanation of how to tune the LSPR of nanoparticles through their size, shape and coupling strategies to match the absorption of target molecules will be provided. A solid understanding of the foundations will better equip the reader with the knowledge required to understand the emerging nanofabrication techniques and computer simulation methods, which are useful for producing novel structures that leverage the unique photophysical properties of nanoscale metallic structures.

The applications will be organized by how the plasmon resonance is used. First, LSPR-based detections can be sensitive to the local environment. For example, based on the refractive index-induced LSPR shift, nanoplasmonic molecular rulers have been used to investigate the hydrolysis of DNA by monitoring the length of conjugated DNA molecules around single nanoantennas [34, 35]. In the plasmon resonance energy transfer (PRET) method, a transfer of plasmonic energy from nanoantennas to the surrounding biomolecular acceptors allows scientists to probe molecular absorption spectra at the nanometer scale both *in vitro* and *in vivo* [36–39].

Second, nanoplasmonic applications can be based on the local enhancement of physicochemical properties. Hot-spots-mediated surface enhancements have gained considerable attention recently because they are able to lower the detection thresholds and also improve the temporal resolution. For instance, metal-enhanced fluorescence (MEF), which is based on strong coupling between antennas and fluorophores, has been applied to reduce the lifetimes of fluorescence excited states (i.e., increase the emission rate) and simultaneously increase the fluorescence emission intensities [40–44]. Another popular surface-enhanced technique is based on surface-enhanced Raman scattering (SERS). This method offers enhancements of more than 10 orders of magnitude, which allows the detection of Raman scattering from single molecules [32, 45–53]. Both MEF and SERS are widely applicable to biomedical diagnoses and imaging both *in vitro* and *in vivo*.

Finally, nanoplasmonic antennas have also recently been shown to function as subwavelength energy harvesters through light-to-heat (photothermal effect) and light-to-force (nanoplasmonic trapping) conversion, which has promise to facilitate many biological manipulations by precisely delivering molecules through the guide of light. As a possible application, DNA and RNA molecules conjugated with plasmonic nanoantennas can be released to regulate gene expression in live cells via the localized photothermal effect [22, 31, 54–56]. Similarly, localized strong hot spots enable the trapping of subwavelength materials. The diameter of confinement in nanoplasmonic trapping is approximately 10 times smaller than that in conventional optical trapping. The trapping of single biological particles and molecules at the nanometer scale will be reviewed [57–62].

8.2 Fundamentals of Nanoplasmonics

8.2.1 *History of Nanoplasmonic Materials*

The use of gold colloids in medicinal formulations dates back to ancient China, Egypt, and India (few millennia BC). Likewise, colloidal suspensions of gold–silver alloys in decorative pieces, such as ceramics or stained glass, were also observed in the works of ancient Chinese, Roman, or Medieval artisans [63, 64]. Figure 8.2 presents a few examples of the use of metal colloids in ancient ceramics, medieval stained glass, and Faraday's microscope slides. Whether for their potential as a therapeutic agent or vivid optical properties discernible to the naked eye, gold nanoparticles have long been an object of fascination. Obviously, the ancient doctors and glassmakers did not possess considerable knowledge of electrodynamics, surface chemistry, or nanofabrication infrastructure. With the advent of nano-optics and nanofabrication, we are better able to systematically examine the unique photophysical properties of metallic nanostructures, which are named nanoplasmonics.

8.2.2 *Overview of the Electrodynamics of Nanoplasmonics*

A *plasmon* is the basic unit of a collective charge oscillation (plasma) in a noble metal. The optical response of a metal results from the field contribution of a considerable amount of conduction electrons in oscillation under the external driving force of illumination. The applications of *plasmonics* are inspired by the unique optical properties of metal structures that occur near a *plasmon resonance,*

Figure 8.2 Examples of nanoplasmonic materials used by our ancestors [63, 64]. (a), Gold nanoparticles were used in the paint of some ancient Chinese ceramics to bring out bright and sharp reddish colors. (b) The beautifully crafted Roman Lycurgus Cup, which depicts the death of the king of Thracians from Homer's *Illiad*, displays striking shades of bright red and dark green [64]. (c) The stained glass in the windows of medieval cathedrals, such as la Sainte-Chapelle or Notre-Dame de Paris, contains gold and silver colloids [65]. (d) The collection of Michael Faraday's microscope slides is often referenced as one of the first studies of gold colloids. The inset shows one of Faraday's slides that is partially coated with metal nanoparticles [66].

in which free and bound electrons oscillate well with the incident light. Surface plasmon resonance occurs in bulk and dot-like metallic structures. In *nanoplasmonics*, focus is placed on the ability of subwavelength metallic nanostructures to collect and focus optical energy beyond the diffraction limit of light via the *localized surface plasmon resonance* (LSPR) [23]. These metallic structures have a characteristic size, a, that is smaller than the wavelength, λ, of the incident light. With expanding fabrication toolsets and substantially improved computational powers, we expect the emergence of creative nanoplasmonic biophotonic devices based on their predicted and prescribed electrodynamic response.

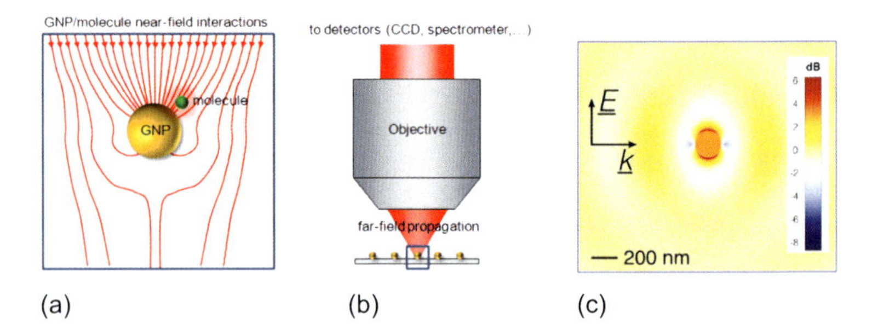

Figure 8.3 Fundamentals of nanoplasmonics [68]. (a) Schematic illustration of localized surface plasmon resonance on gold nanoparticles (GNP, yellow dots), focusing an electromagnetic field (red line) and inducing nanolens effects. Probing biomolecular interactions is localized by hot spots of nanoplasmonic antennas. Optical properties of molecules (e.g., scattering and fluorescence emission) can be enhanced by near-field interactions with plasmonic nanoparticles. (b) Schematic illustration of coupling near-field plasmonic nanoparticles/molecules interactions to far-field detection. (c) Electric field amplitude surrounding a gold nanoparticle with an incident light based on Mie theory [67].

Conceptually, a nanoplasmonic structure is antenna-like; it converts the received optical energy into its near-field and far-field counterparts [67, 68]. Therefore, a nanoplasmonic structure has large absorption and scattering cross-sections (Fig. 8.3a,b). The near-field represents the evanescent portion of the optical energy that tends to stay near the nanostructure (exponentially decays away from object boundary interfaces at the nanoscale lengths), whereas the far-field is the propagating part capable of traveling to a longer distance (propagation wave). A quantitative understanding of the relationship between light and metallic nanostructures is essential. The optical response of a nanoplasmonic probe is determined by the size, shape and frequency dependent dielectric constant, $\varepsilon(\omega)$, of a material [14, 23, 31, 32]. The optical response translates the collective charge oscillation into polarizable electromagnetic field patterns. The Drude-Lorentz model is often introduced to describe the optical response of plasmonic materials to provide a first-order picture. In terms of applications, stress is placed on the importance of being able to rationally design electromagnetic fields near a given nanoplasmonic structure. The starting point to quantitatively describe the electromagnetic field patterns is through the pedagogical Mie theory of light scattering of an arbitrary spherical particle. The rigorous Mie scattering gives a full vectorial analytical description of the multipolar electromagnetic field patterns dressing a sphere under illumination (Fig. 8.3c) [68]. However, because the multipolar description of scattering is quite involved (some description in the Appendix), a more intuitive quasi-static dipole approximation is often used in a quick analysis, which is a good estimate for particles that are less than 10–30 nm, i.e., in the *Rayleigh limit*. For an arbitrary shape that deviates from a Mie sphere, or a complex assembly of shapes that involve strong interparticle coupling, a combination of numerical techniques are available and can be used to obtain a rigorous quantitative prescription for nanoplasmonic designs.

8.2.3 *Dielectric Constant of a Noble Metal*

The frequency-dependent dielectric constant, $\varepsilon(\omega)$, (or relative permittivity) describes the ability of a material to displace its electromagnetic field,

$$\mathbf{D}(\omega) = \varepsilon_0 \mathbf{E}(\omega) + \mathbf{P}(\omega) = \varepsilon_0 \varepsilon(\omega) \mathbf{E}(\omega)$$

where ε_0 is the vacuum permittivity, **E** and **D** denote the electric and the displacement fields, respectively, and **P** is the polarization vector. The dielectric constant, $\varepsilon(\omega) = \varepsilon_1(\omega) + i\varepsilon_2(\omega)$, is generally a complex function that contains a real part and an imaginary part related to the amplitude and phase difference between the electric field and its displaced fields [31]. The dielectric constant is used to characterize the propagation and attenuation of electromagnetic energy inside a medium (e.g., meal and water). Note that the well-known refractive index, m, of a material is related to the relative permittivity, ε, and the relative permeability, μ through $m = \sqrt{\varepsilon\mu}$.

The optical response of a noble metal is determined by the dynamics between the incident field and the polarization of charged constituents inside the material [31, 69, 70]. Consider an incident electromagnetic wave that penetrates a metal with a skin depth δ. As a conductor, metals contain overlapping energy bands in which a considerable number of free valence and bound interband electrons can easily move. This combined optical response of the free and bound electrons of a metal is described by a dielectric function, $\varepsilon_{\mathrm{M}}(\omega)$. As the incident electric plane wave, $\mathbf{E} = \mathbf{E}_0 e^{-i\omega t}$, drives the electrons inside the metal to oscillate at a frequency ω, a time-varying electron displacement, $\mathbf{r} = \mathbf{r}_0 e^{-i\omega t}$, causes local polarization via $\mathbf{P} = -ne\mathbf{r}$ with n and e denoting the electron density and the fundamental charge unit, respectively. To a first-order approximation in which the bound electrons are neglected, the optical response of a bulk metal is described through a damped harmonic oscillation of the free electrons,

$$m_{\mathrm{e}} \frac{\partial^2 \mathbf{r}}{\partial t^2} + m_{\mathrm{e}} \Gamma \frac{\partial \mathbf{r}}{\partial t} = -e\mathbf{E}_0 e^{-i\omega t}$$

where m_{e} denotes the effective mass of an electron and Γ is the damping constant related to the rate of collisions of electrons. With oscillation occurring at the plasma frequency ω_{P}, the solution to the oscillator gives the Drude–Lorentz dielectric constant of metal,

$$\varepsilon_{\mathrm{M}}(\omega) = 1 - \frac{\omega_{\mathrm{P}}^2}{\omega^2 + i\Gamma\omega} = 1 - \frac{\omega_{\mathrm{P}}^2}{\omega^2 + \Gamma^2} + i\frac{\Gamma\omega_{\mathrm{P}}^2}{\omega\left(\omega^2 + \Gamma^2\right)} = \varepsilon_1 + i\varepsilon_2$$

Permittivity data (or dielectric function) obtained from direct measurements, such as those by Johnson and Christy, are often used in real calculations and simulations (Fig. 8.4a) [71–73]. In terms of applications, gold (Au) and silver (Ag) are popular materials of choice due to their strong plasmon resonance in the visible and near-infrared regions. Gold is sometimes preferred for biological applications because it is well characterized, chemically stable, biologically safe, and can be easily conjugated with DNAs and other molecules.

8.2.4 *The Rigorous Field Distributions of a Metallic Sphere from Mie Theory*

In 1908, Gustav Mie rigorously solved the light scattering and absorption problem of a sphere using a full vector electrodynamics approach [68, 77–79]. Consider a sphere of arbitrary radius a with a material permittivity ε_{M} and permeability μ_{M} buried in a surrounding medium characterized by ε_{D} and μ_{D}. According to Mie theory, the incident, scattered, and internal electromagnetic

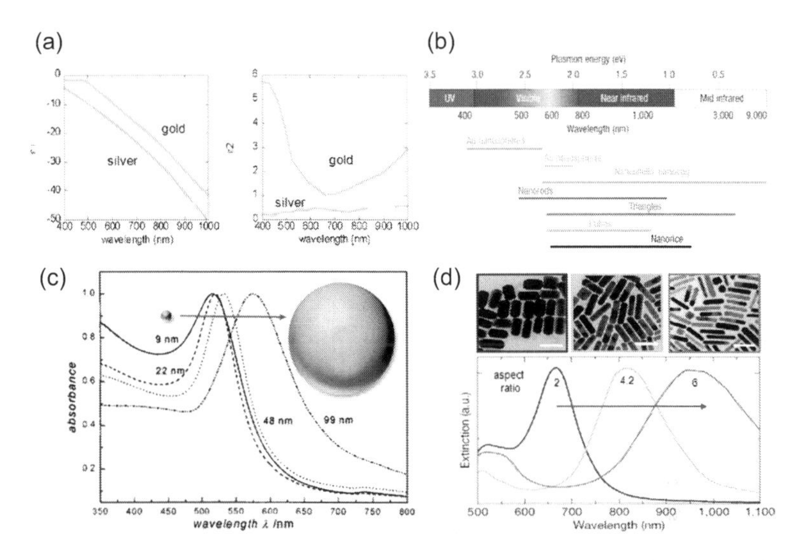

Figure 8.4 (a) Real (left) and imaginary (right) parts of the dielectric functions of silver and gold [71, 72]. (b) A wide range of surface plasmon resonance peaks are tuned by the material, size, and shape of nanoantennas [14, 20]. (c) Normalized absorption spectra of GNPs with various diameters [74, 75]. (d) SEM images of gold nanorods with various aspect ratios (top) along with their corresponding extinction spectra (normalized by peaks, bottom) [76].

fields (both inside and outside the sphere) are expanded as a linear superposition of weighted multipoles (monopole, dipole, quadrupole, etc.), as further described in the Appendix. Therefore, the scattered electromagnetic fields around a sphere can be solved, as shown in Fig. 8.3c. From a functional point of view, the plasmonic sphere can act as an optical nanolens because it distorts the flow of incident electromagnetic energy at specific frequencies (Fig. 8.3a).

8.2.5 *The Simplified Field Distributions of a Small Metallic Sphere*

For a small subwavelength particle ($a \approx 10 - 30$ nm), the wave retardation of the incident electromagnetic wave can be neglected [30, 31, 71]. The metallic nanosphere roughly experiences a uniform incident plane wave $\mathbf{E} \approx \mathbf{E}_0$ when its size is greater than the mean path length and smaller than the skin depth A sphere in this regime is known as a Rayleigh sphere. The electrostatic (or quasi-static) approximation can be used to simplify the electromagnetic problem. Thus, the electric field pattern of a dipole inside, $\mathbf{E}_{\mathrm{M}}$, and outside, $\mathbf{E}_{\mathrm{D}}$, of the metal sphere are respectively given by [69, 70, 80]

$$\mathbf{E}_{\mathrm{M}} = \frac{3\varepsilon_{\mathrm{D}}}{\varepsilon_{\mathrm{M}}(\omega) + 2\varepsilon_{\mathrm{D}}}\mathbf{E}_0$$

$$\mathbf{E}_{\mathrm{D}} = \mathbf{E}_0 + \frac{3\mathbf{n}(\mathbf{n}\cdot\mathbf{p}) - \mathbf{p}}{4\pi\varepsilon_0\varepsilon_{\mathrm{D}}}\frac{1}{r^3}$$

where $\mathbf{n}$ is a unit direction vector and r is the radius of the particle. The dipole moment is expressed as $\mathbf{p} = \alpha\varepsilon_0\varepsilon_D\mathbf{E_0}$, with an induced polarizability α in the form of the well-known Clausius-Mossotti expression,

$$\alpha = 4\pi a^3 \frac{\varepsilon_{\mathrm{M}}(\omega) - \varepsilon_{\mathrm{D}}}{\varepsilon_{\mathrm{M}}(\omega) + 2\varepsilon_{\mathrm{D}}}$$

The field strength of this plasmonic dipole radiation pattern is affected by the size of the sphere and the material permittivity of the metal at different frequencies. In the Fröhlich condition (the frequency satisfying $\mathrm{Re}\{\varepsilon_{\mathrm{M}}(\omega)\} = -2\varepsilon_{\mathrm{D}}$), nanoplasmonic particles have the strongest dipole.

8.2.6 *Extinction, Scattering, and Absorption Cross Sections*

The extinction cross section represents the energy removed from the incident beam of light due to absorption and scattering events [30, 31, 71, 77, 81]. The light-scattering power of a particle is expressed in terms of the scattering cross-section, σ_{sca}, which represents a hypothetical area for which a photon is scattered in an outward direction. The mutual relationship between the scattering, extinction and absorption cross-sections is $\sigma_{\mathrm{ext}} = \sigma_{\mathrm{abs}} + \sigma_{\mathrm{sca}}$. For a small sphere in the Rayleigh regime with polarizability α, the scattering and absorption cross sections can be simplified as

$$\sigma_{\mathrm{sca}} = \frac{k^4}{6\pi\varepsilon_0^2}|\alpha(\omega)|^2 = \frac{8\pi k^4 a^6}{3}\left|\frac{\varepsilon_{\mathrm{M}}(\omega) - \varepsilon_{\mathrm{D}}}{\varepsilon_{\mathrm{M}}(\omega) + 2\varepsilon_{\mathrm{D}}}\right|^2$$

$$\sigma_{\mathrm{abs}} = \frac{k}{\varepsilon_0}\mathrm{Im}\{\alpha(\omega)\} = 4\pi k a^3 \mathrm{Im}\left\{\frac{\varepsilon_{\mathrm{M}}(\omega) - \varepsilon_{\mathrm{D}}}{\varepsilon_{\mathrm{M}}(\omega) + 2\varepsilon_{\mathrm{D}}}\right\}$$

where $k = 2\pi/\lambda$. The scattering and absorption cross sections exhibit a maximum due to plasmon resonance when $\mathrm{Re}\{\varepsilon_\mathrm{M}(\omega)\} = -2\varepsilon_\mathrm{D}$. For the Fröhlich condition with a water permittivity $\varepsilon_\mathrm{D} \approx 1.7$, according to the dielectric function of silver and gold (Fig. 8.4a), the resonant peak for spherical gold nanoparticles in water is approximately 520 nm (visible spectrum) and that of silver particles is in the UV region at approximately 400 nm (Fig. 8.4b) [31, 71, 82–84]. These equations also indicate that the cross-sections and resonance peaks of scattering and absorption scale to the particle size, a. A redshift occurs as the particle size increases (Fig. 8.4c) [85]. Furthermore, increasing ε_D also results in a redshift of the plasmon resonance wavelength, which implies that polarization is sensitive to the permittivity, ε_D, of the local environment and can be used for local sensing.

With extremely large absorption and scattering cross sections, metallic nanoparticles can be observed using simple and inexpensive illumination setups, such as a dark-field microscope [83]. Under illumination with white light, gold nanoparticles appear as very bright spots, similar to fluorescent dyes but free of quenching and photobleaching. The light-scattering signal from a 60 nm gold nanoparticle is roughly equal to fluorescent emission from 3×10^5 fluorescein dyes. In addition, silver nanoparticles can scatter light 2000 times more efficiently than polystyrene particles of the same size [26, 27].

8.2.7 *Complex Nanoplasmonic Geometries*

For a single metallic sphere, the parameters describing its optical performance can be described by Mie theory using the dielectric function of the metal. However, in addition to the size and material composition of a nanoplasmonic particle, its shape can also affect its photophysical properties (Fig. 8.4) [31, 71]. For example, a nanoshell can be viewed as the result of hybridization between a coupling spherical void plasmonic structure and a plasmonic sphere (with plasmons on the inner and outer surfaces of the metallic shells) [86]. As a result of hybridization, a metal nanoshell contains symmetric (bonding) and antisymmetric (antibonding) plasmon modes with dissimilar energy levels. As progress in the fabrication of metal nanostructures leads to a proliferation of plasmonic geometries, such as nanospheres, nanorods, nanotriangles, and nanocrescents, just to name a few, plasmonic particles cover different portions of the scattering spectrum and can be used in diverse applications (Fig. 8.4b). Many theories and simulations are being developed to design complex nanoplasmonic antennas. For example, nanorods can be modeled analytically using Gans Theory to explain their field distribution and redshift [31, 87, 88]. In this description, spheres are elongated into ellipsoids through an aspect ratio length over diameter (b/a) to explain their redshift (Fig. 8.4d) [88, 89].

Moreover, to quantitatively describe the optical response of arbitrary complex metallic structures, various electromagnetic numerical methods can be employed, including the discrete-dipole approximation, finite-difference time-domain, T-matrix, or transformation optics [90–95]. Each of these methods has its own strengths and weakness in terms of accuracy, intuitiveness, ease of application and computational power. In FDTD, the general Maxwell's equations describing electromagnetic waves are solved numerically within a discretized space. Boundary conditions, material dielectric constants and incident fields are established at the initial time. A recurrence approach is taken to solve the time evolution of the EM fields in small temporal incremental steps. In DDA, the complex structure is approximated as a system of interacting lowest energy dipole arrays in the quasi-static regime. This approach yields simple and intuitive solutions and is applicable when spheres are small and weakly coupled because higher radiation modes are ignored. The accuracy might suffer otherwise. In the T-matrix approach, a generalized Mie theory is devised in which a multiple multipole approach is necessary to account for successive light scattering among strong coupling subwavelength particles. This method applies to strong coupling spheres; however, the number of equations to solve increases with increasing numbers of spheres. Finally, the transformation optics approach takes advantage of the covariant property of Maxwell's equations to distort the spatial coordinates of the structure of a known field solution into a more complicated shape that may not exhibit the same simplifying symmetry. This method can be useful for the analysis of metamaterials and plasmonic designs.

8.2.8 *Strongly Coupled Nanoplasmonic Structures*

The plasmonic resonance properties (i.e., resonance peak and the intensity of hot spots) can be changed by strongly interacting plasmonic particles [31, 34, 71, 96]. Analogous to molecular orbital theory, the different near-field plasmonic modes mutually couple together to yield a complex electromagnetic response. In linear structures with strong interparticle coupling, scattering fields are obtained not only through the superposition of individual responses but also from the mutual interaction between scatterers along a highly polarizable axis. In coupled structures where the interparticle near-fields significantly overlap, additional resonances are also possible due to the strong interactions between localized plasmon modes (Fig. 8.5a). For example, spherical plasmon dimer pairs strongly couple depending on their gap size. Such dimer redshifts and enhances its near-fields monotonically with decreasing gap size [97]. Quantum effects might also need to be considered when the coupling gap size becomes so small that electron transfers may occur between two adjacent conductive nanostructures [98–101]. For nonclassical quantum plasmonic particles with narrow gaps, electron couplings through a short metal–dielectric–metal junction and nonlocal screening effects of the induced fields also occur to alter the near-field and far-field electromagnetic responses. Another pairing effect caused by coupled plasmonic structures is the plasmonic Fano resonance, which results from the constructive or destructive interference between a spectrally overlapping broad resonance and a narrow discrete one. A plasmonic nanoparticle exhibits a tunable Fano resonance with a varying degree of interparticle distance (Fig. 8.5b). A large and unique induced spectral shift can be applied in chemical and biological sensing [102].

Strongly coupled plasmonic nanostructures can interact with molecules through their electronic dipole to enhance their emission signals (for instance, fluorescence and Raman signals) [32, 40–44, 47–49]. In addition, the sensing volume (defined by the size of hot spots) is greatly reduced in this scale. Various nanogap metallic structures have achieved large magnitude enhancements in fluorescence and Raman spectra. However, the majority of nanogap

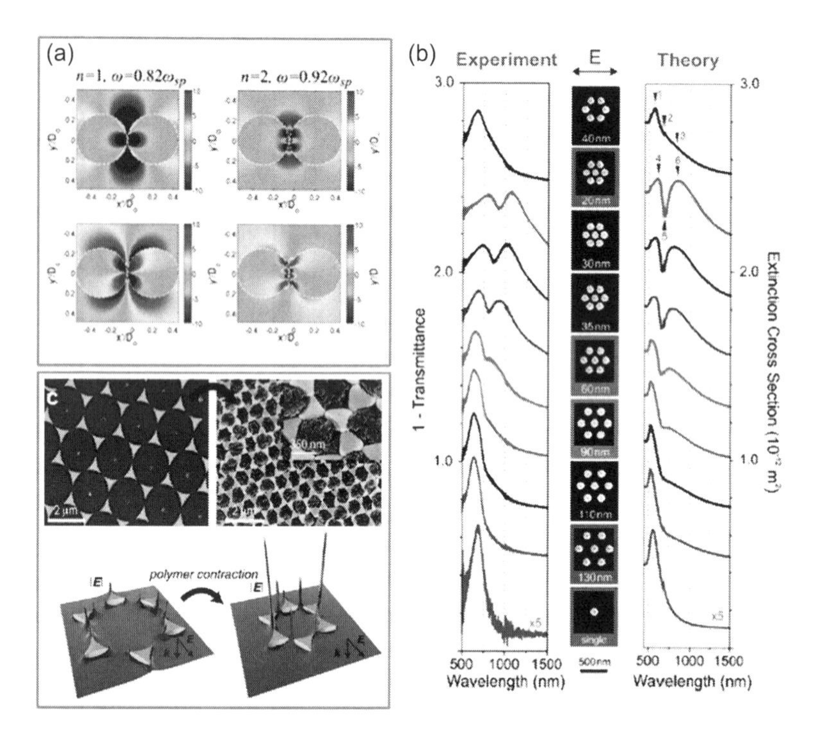

Figure 8.5 Simulation and design of plasmonic coupling and hot spots. (a) Near-field enhancement of touching sphere as calculated semi-analytically through the use of Transformation Optics [32]. (b) Demonstration of Fano resonances in a variety of plasmonic disk oligomers patterned lithographically. The measured and theoretical transmission spectra show the tuning of the asymmetric Fano resonance with different interparticle separations [106]. (c) Top: SEM images of gold nanoprisms on polymer substrates before (left) and after (right) contraction. Bottom: A finite element simulation of electric field amplitude around nanoprisms and nanogaps [107].

structures have been produced using expensive and sophisticated nanofabrication techniques, such as electron beam lithography and focused ion beam techniques. Lowering the fabrication threshold and developing microfluidic integratable enhancement substrates would widen their applicability and accessibility for biomedical applications. For example, researchers have recently developed simple methods for fabricating large-scale, high-density nanogap structures on inexpensive polymer substrates [103, 104]. By simply leveraging the stiffness mismatch of materials, metal deposition and subsequent heating above the glass transition temperature of polymer sheets cause the substrates to retract and nonshrinkable metallic materials to buckle. The enhancement of the intensity of hot spot is due to the small nanogaps. Nanometer-sized gaps that are effective in surface-enhanced sensing have been demonstrated [105]. In addition, these metallic structures can be easily integrated into polymer-based microfluidic devices or deformed into complex 3D geometries.

8.3 Nanoplasmonic Molecular Ruler

One of the key avenues to understanding how biomolecules interact at the single molecule level is through measuring their intramolecular distance. Fluorescence resonance energy transfer (FRET) is a technique that is widely adopted for this purpose and is popular due to its single molecule sensitivity and ease of specific targeting. However, FRET is unsuitable for measuring large conformational changes because its performance is often hampered by its short detection range (~5 nm). To improve the detection range, long-range molecular rulers based on shifts in single-antenna LSPR are used to overcome these limitations [34, 35, 96, 108, 109]. In nanoplasmonic molecular rulers, the LSPR extinction peaks are a function of the local refractive index change, Δn, of the environment in the vicinity of the nanoantennas. Therefore, molecular interactions on the surface of a metal can be locally probed by monitoring the scattering spectrum shift of nanoantennas, $\Delta\lambda_{\text{max}}$, as given by

$$\Delta\lambda_{\text{max}} = m\Delta n \left[1 - \exp\left(\frac{-2d}{l_{\text{d}}}\right)\right]$$

where m is the bulk refrative index of the environment, Δn is the change in the refractive index due to molecular interactions around the antennas, d is the effective thickness of the absorbate layer, and l_{d} is the characteristic evanescent field length into the sample (normally on the order of 10 nm) [34, 35, 108, 109].

One typical experimental setup of a nanoplasmonic molecular ruler is shown in Fig. 8.6a [34, 35, 38, 108, 109]. A glass plate coated

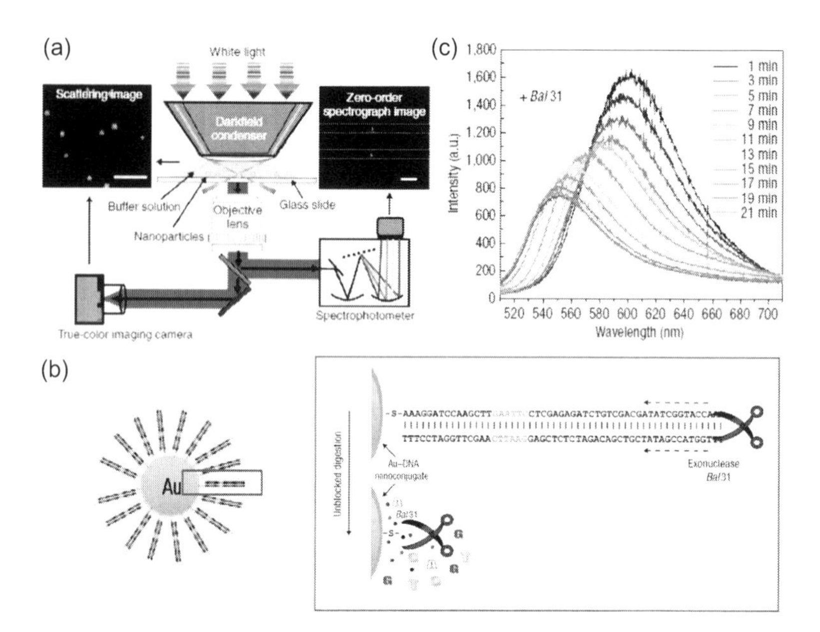

Figure 8.6 Nanoplasmonic molecular rulers for real-time monitoring of biomolecular interactions [35, 38]. (a) Configuration of the experimental system for dark-field imaging and recording the scattering spectrum of single nanoantennas. (b) Design of Au–DNA nanoplasmonic molecular rulers for the real-time monitoring dsDNA hydrolysis by Bal31 exonuclease. (c) Time-dependent scattering spectra from a single plasmonic ruler during DNA hydrolysis.

with nanoantennas was mounted on a dark-field optical microscope for the experiment. Excitation of the sample was performed using a dark-field condenser lens (NA = 1.2–1.4) with a broadband white light lamp. The scattered intensity of a single nanoantenna was collected by a microscope objective lens (NA $\approx$ 0.65, smaller than that of the dark-field condenser lens) and detected by a true color charge-coupled device (CCD) camera. After obtaining the images, the corresponding scattering spectrum can also be measured using a spectrometer.

To investigate the shift in the scattering spectra induced by molecular interactions, 20 nm Au nanoparticles were conjugated with 54 bp (length $\approx$ 18 nm) double-stranded DNA (dsDNA) molecules (Fig. 86b) [35]. DNA hydrolysis caused by Bal31 exonuclease was monitored in real-time by tracking the shift in the plasmon resonance wavelength of individual probes, which corresponds to the length of dsDNA. An average wavelength shift of ~1.24 nm/bp was observed, and the range of detection was close to 20 nm, which is greater than the effective distance of FRET (Fig. 8.6c). In addition to long-range detection, nanoplasmonic rulers are photostable, free of photobleaching and blinking over organic dyes, thereby allowing long-term monitoring of single biomolecular interactions with high temporal resolution. It has been demonstrated that photostable gold nanoparticles ensure long-term (and short-term) detection of DNA foot-printing and nuclease activity. In addition coupling-type plasmonic rulers have been applied to various biological applications, such as protein-induced DNA bending and DNA cleavage by restriction enzymes [34, 35, 108, 109].

8.4 Plasmon Resonance Energy Transfer

Absorption spectroscopy that utilizes the interaction of light and molecules has been used for a considerable amount of time. However, conventional absorption spectrometers have a major limitation for use as a molecular biophotonic tool. The spatial resolution of optical systems cannot break the diffraction limit of light. Therefore, it is difficult to obtain absorption spectra for single (or few) molecules. As recently suggested, the plasmon energy from nanoplasmonic particles can be transferred to surrounding biomolecular acceptors, which allows nanometer-scaled measurements of molecular absorption spectra. When the plasmon resonance peak overlaps well with the biomolecule absorption peaks, the PRET effect contributes to specific quenching dips in the Rayleigh scattering spectra of biomolecule-conjugated nanoantennas (Fig. 8.7a) [36–39].

For example, when nanoantennas were conjugated with reduced cytochrome c, dramatic quenching was observed in their scattering spectra and the positions of the dips matched well with the absorbance peaks of the adsorbed molecules (Fig. 1.7b,c) [38]. This label-free nanoplasmonic method provides an ultrasensitive biomolecular absorption spectroscopy technique for probing a small

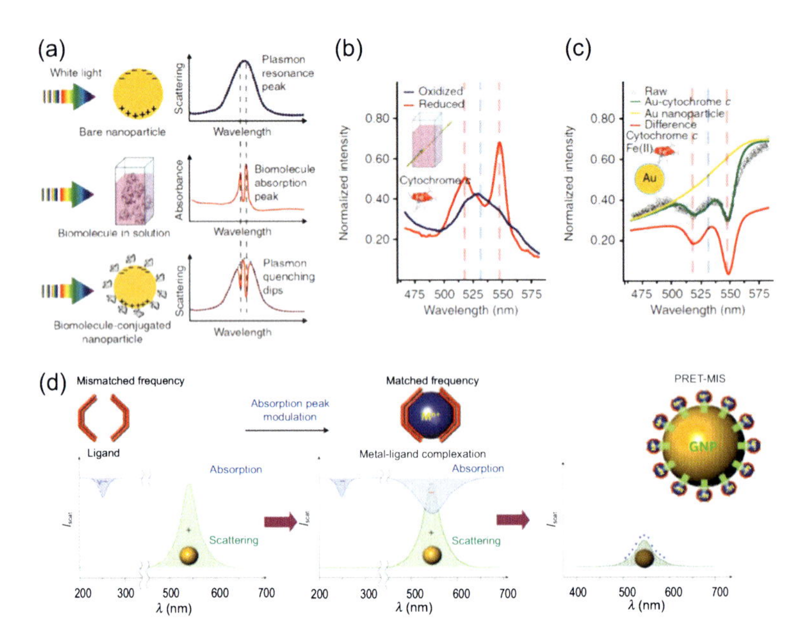

Figure 8.7 Plasmon resonance energy transfer [36–39]. (a) Plasmon quenching dips (bottom) in the scattering spectrum of a GNP (top) match well with the absorption speaks of biomolecules in bulk solution (middle). (b) The absorption spectra of oxidized (blue solid line) and reduced (red solid line) cytochrome c in bulk solution. (c) Plasmon quenching dips (red line) of GNPs/ reduced cytochrome c complex, which was obtained by subtracting the raw data (green line) from the scattering spectrum of bare GNPs (yellow line). (d) PRET-based sensing of a metal ion. The electronic absorption peak of the ligand/metal ion complex can be tuned to ~540 nm, which matched the SPR spectrum of GNPs (left and middle). The local concentration of metal ions can be measured from the resonant quenching in the Rayleigh scattering spectrum (right).

number of biomolecules. In addition to *in vitro* measurements, PRET-based spectroscopy techniques have been applied in the real-time monitoring of the expression of cytochrome c in living cells [36]. During ethanol-induced apoptosis, HepG2 cells released cytochrome c molecules from the mitochondria to the cytoplasm, which caused specific dips in the scattering spectra of the inter-cellular antennas. More recently, by conjugating with metal specific ligands, PRET probes were used to detect metal ions (e.g. Cu), which play important roles in molecular biology, cell biology and environmental research (Fig. 8.7d) [39]. In addition to a several orders of magnitude increase in sensitivity, this method exhibited high selectivity depending on the conjugated ligands.

8.5 Metal-Enhanced Fluorescence

Fluorescence is one of the most widely used light-matter interactions for probing the functions of biological systems because of its high sensitivity and multiplexing capabilities. Organic dyes and fluorescent proteins are two types of probes that are widely adopted for this purpose. However, the low quantum yield, photobleaching and blinking problems of these probes inevitably restrict their applications for long-term observations *in vitro* or *in vivo*. In recent years, fluorescent nanoparticles (such as quantum dots and fluorescent nanodiamonds) have gained considerable attention because of their high photobleaching thresholds. To further improve the signal-to-noise ratio of fluorescence signals, the MEF technique utilized metallic nanostructures in which the surface plasmons resonate with the molecular dipole of the fluorophores to reduce their fluorescence excited state lifetimes and simultaneously increase their emission intensities [40–44]. In fluorescence emission, the quantum yield (Q_0) and lifetime (τ_0) of fluorophores in free space are defined by the radiative rate (Γ) and the nonradiative rate (k_{nr}), [110]

$$Q_0 = \frac{\Gamma}{\Gamma + k_{nr}}$$

$$\tau_0 = \frac{1}{\Gamma + k_{nr}}$$

In the presence of a nearby metallic nanostructure, an additional radiative rate factor (Γ_m) is introduced that consequently increases the quantum yield (Q_m) while reducing the excited state lifetime (τ_m), [110]

$$Q_m = \frac{\Gamma + \Gamma_m}{\Gamma + \Gamma_m + k_{nr}}$$

$$\tau_m = \frac{1}{\Gamma + \Gamma_m + k_{nr}}$$

The effect of fluorophore–metal interactions is distance dependent [111]. Experiments using single plasmonic particles/single dyes revealed that the maximum fluorescence enhancement occurred at a separation distance of 5–10 nm between the surface of the metal and the fluorophores (Fig. 8.8a). Furthermore, short gap distances (0–5 nm) can induce fluorescence quenching. It has been suggested that covering the metallic nanostructures with nanometer-thick buffer layers is essential for achieving the largest MEF effect [41, 42, 111, 112]. Typically, a 2- to 10-fold enhancement ($\gamma_{em}/\gamma^0_{em}$) in

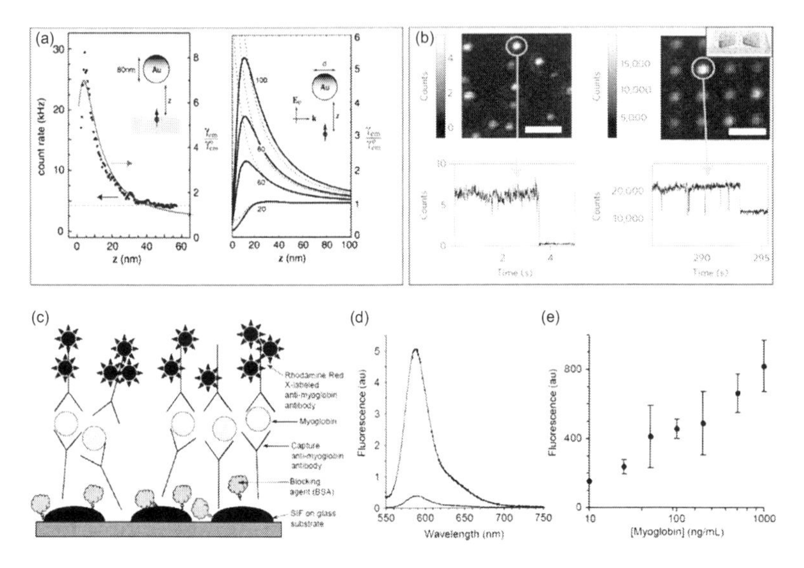

Figure 8.8 Metal-enhanced fluorescence (MEF). (a) Interaction of a metallic nanoparticle and a fluorophore in the near-field. The metal–fluorophore coupling increases both the absorption and emission rates of the fluorophores. Fluorescent emission rate (left axis) and factors of enhancement (right axis) as a function of the gap between the GNPs and fluorophores [111]. (b) Fluorescent images (top), together with time traces of intensities (bottom), of single dye molecules without (~7 counts, left) and with enhancement (~20,000 counts, right) by bowtie nanostructures (top-right inset) [44]. (c) Schematic illustration of a metal-enhanced immunoassay by silver island film (SIF) [41]. (d) Fluorescence spectra of an anti-myoglobin antibody (Rhodamine Red-X-labeled)/myoglobin sandwich on SIF (with MEF, strong intensity) and on a glass slide (without MEF, weak intensity). (e) The detection limit of myoglobin can reach 10 ng/ml with the use of MEF techniques.

the fluorescence emission can be obtained. To overcome such low efficiency of enhancement due to weak hot spots, nanotechnologies have been utilized. For instance, coupled nanoparticles, metal tips, nanopockets, optical antennas, and feed gaps have achieved 10- to 1,000-fold enhancements in the fluorescence emission and reduced the excitation volumes to atto- to zeptoliter [40–43, 111, 113–120].

In single-molecule experiments, bowtie nanoantennas have been used to facilitate the detection of single dye molecules and exhibited a 1340-fold increase in fluorescence intensity (Fig. 8.8b) [117]. The spatial extent of the enhancement hot spots, as small as 15 nm, has also been measured by single molecule imaging [43]. Furthermore, the MEF technique has been used in biomedical applications to improve the detection limit of immunoassays and DNA hybridization (Fig. 8.8c) [41]. Silver island films are also general enhancement substrates [41, 42]. For example, the cardiac marker myoglobin was captured by antibody-coated plasmonic substrates and then further labeled by dye-labeled antimyoglobin antibodies. Compared to glass substrates without the MEF effect, an approximate 10-fold enhancement in the fluorescence signal was observed, which improved the detection limit to 10 ng/mL and is less than the cut-off concentrations for myoglobin in clinical studies (Fig. 8.8d,e). To lower the fabrication threshold and to develop microfluidic integratable metallic nanostructures, our group recently presented a simple, ultra-rapid and robust method to create sharp nanopetal structures in a shape memory polymer substrate over large surface areas [113, 114]. Extremely strong surface plasmon resonance due to the high-density multifaceted petal structures increases the fluorescence emission by thousands-fold. This microfluidic-integratable material enables us to apply MEF techniques in point-of-care biomedical diagnoses.

8.6 Surface-Enhanced Raman Spectroscopy

Raman scattering is an interaction between light and matter that occurs when the incident photons are inelastically scattered from molecules [47]. The resulting photons may gain or lose energy after scattering (Fig. 8.9a). Under irradiation ($h\nu_L$), a redshift in scattering is referred to as a *Stokes shift* if the interaction causes the vibrating molecule with energy $h\nu_M$ to re-emit a photon of signal $h\nu_S$, which corresponds to an energy difference between a higher vibrational energy and the incident excitation levels ($h\nu_S = h\nu_L - h\nu_M$). In contrast, the light is blue-shifted, and its shift is known as *Anti-Stokes* if the signal $h\nu_{AS}$ occurs with a difference between a lower vibration energy and the incident excitation levels ($h\nu_{AS} = h\nu_L + h\nu_M$). The vibrational energy $h\nu_M$ is characteristic of the molecule in question and can serve as a unique spectroscopic signature. However, unlike the elastic Rayleigh scattering process, the Raman one occurs less frequently. The weak effect of Raman is evaluated at a small absorption cross-section on the order of $\sigma^R \approx 10^{-30}\text{cm}^2$ per molecules, whereas that of the fluorescence is at $\sigma^R \approx 10^{-16}$ cm^2 [47–49].

The weak Raman signal can be enhanced by several orders of magnitude in SERS [47, 121]. This enhancement is achieved chemically or electromagnetically by the adsorption of molecules near the surface of metals. The electromagnetic enhancement plays the primary role because SERS scales roughly with the "fourth" power of the electric field enhancement (hot spots). Such strong enhancement makes Raman scattering from single molecules sufficiently strong to be detected (Fig. 8.9b) [122, 123]. Polarized confocal microscopy with spectroscopy was used to probe ultra-low concentrations of Rhodamine 6G molecules. The $\sim 10^{15}$-fold enhancement in Raman scattering from single silver particle measurements was considerably greater than the ensemble-averaged values obtained using traditional methods. Such SERS effect provides single-molecule Raman signals, which are stronger and more stable than single-molecule fluorescence signals [47, 122–124].

Although single-molecule SERS has opened a new route in label-free ultra-sensitive detection, there are some additional concerns with respect to SERS-based molecular detection. For instance, because the enhancement occurs a few nanometers away from the surface of the SERS substrate, it requires the analytes to be extremely close to the surface. Especially in bulk single-molecule detection, the possibility for the analyte reaching hot spots becomes very low. In the bulk solution, the mean free path of molecules

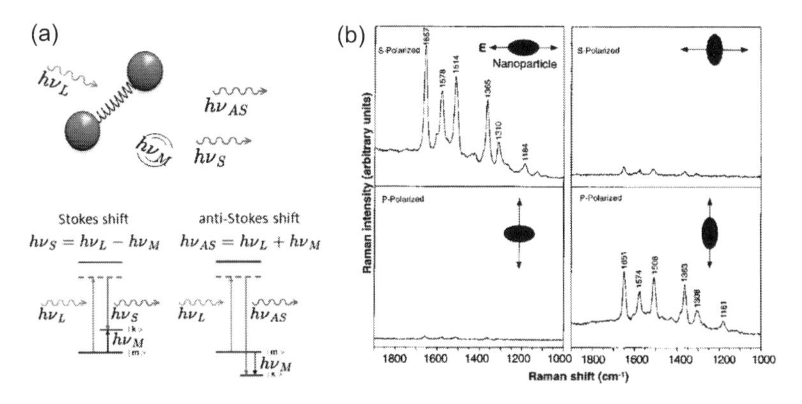

Figure 8.9 Fundamentals of surface enhanced Raman spectroscopy (SERS) [48, 49]. (a) Interplay between the incident photon and vibrational energies of a molecule undergoing nonresonant Raman scattering. The Stokes shift corresponds to a reduction in photon energy (redshift with a longer scattering wavelength), whereas the anti-Stokes corresponds to an increase in the photon energy (blueshift with a shorter scattering wavelength). The upper purple dotted lines represent virtual states for the fact that, contrary to fluorescence, Raman scattering is a nonresonant light absorption process and can occur without resonant absorption that corresponds to electronic states. (b) The hot spots around metallic nanoparticles can enhance the Raman signal of molecules over 10^{15}-fold, which resulted in a finger-print-like SERS spectrum. Single-molecular SERS was demonstrated by a specific SERS signal and controlled by the direction of the laser polarization and the orientation of the particle [122].

(depending on the diffusion coefficient of the molecules) is on the scale of micrometers, which is considerably smaller than the scale of the sample chambers. This mismatch results in dramatic decreases in sensitivity and makes the detection time extremely long. To address this problem, microfluidic chips basically lower the channel height to approach the mean free path of analytes and consequently increase the accessibility of the surface hot spots to molecules. SERS substrates have been integrated into micro- and nanofluidic chips [125–130]. For example, a silver nanowell array was fabricated in microfluidic devices composed of polydimethylsiloxane (PDMS). Compared to the smooth silver surface, the nanowell-based SERS substrates exhibit a 10^7-fold of enhancement in the SERS signal [130]. It is possible to detect molecules with concentrations as low

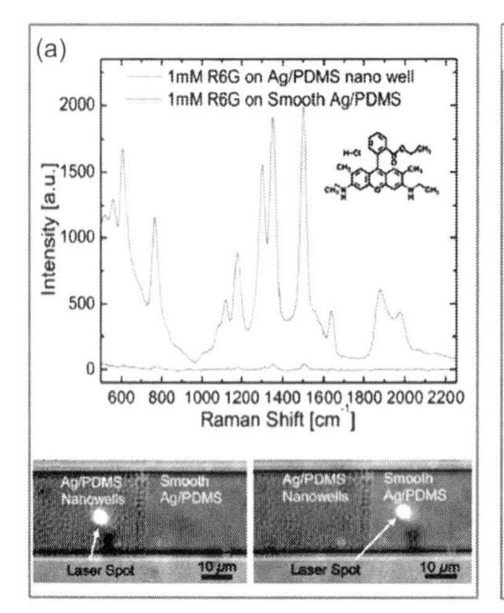

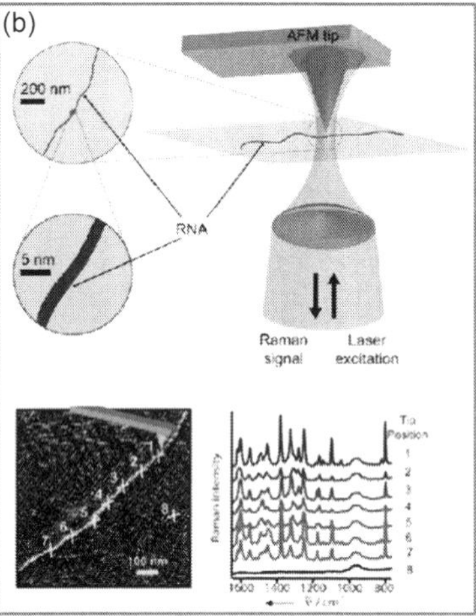

Figure 8.10 Applications of SERS *in vitro*. (a) Images of nanowell arrays fabricated using soft lithography for biomolecular SERS detections in integrated microfluidic channels (bottom) [130]. Top: SERS spectra of R6G molecules collected from Ag/PDMS nanowell arrays (red) and from a smooth Ag/PDMS surface (blue). (b) Top: Configuration of the experimental system for scanning single RNA stands using tip-enhanced Raman spectroscopy [46]. Bottom: A topography image (left) and corresponding tip-enhanced Raman scattering spectra collected from various positions along a single strand of RNA (right), which demonstrates the detection of a single RNA strand.

as 10 fM (Fig. 8.10a). In addition to improvements in sensitivity, SERS substrates/microfluidic systems also make high-throughput analysis possible, which is consistent with their multiplexing capabilities.

Rather than locating samples on SERS substrates, an alternative method is to move the hot spots close to the samples. For instance, tip-enhanced Raman scattering (TERS) utilizes the hot spot at a metal-coated atomic-force microscope (AFM) tip to locally enhance molecules with a spatial resolution of a few nanometers (Fig. 8.10b) [46, 131–133]. This method has been applied to obtain submolecular Raman scattering spectra from various biomolecules, such as DNA, RNA, and peptides. Specific Raman spectra were obtained along single RNA strands and demonstrated the potential of being a method for direct sequencing.

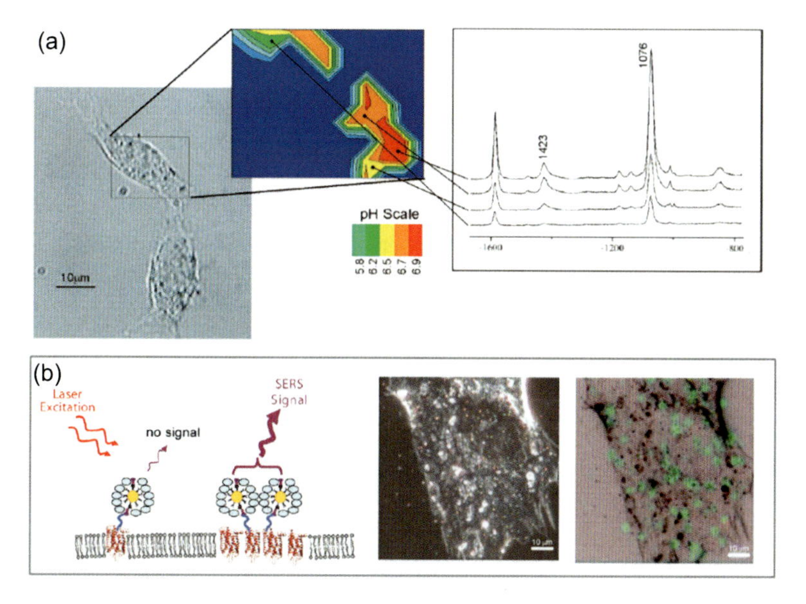

Figure 8.11 Applications of SERS for investigating the dynamics of single cells. (a) Subcellular mapping of the pH in living cells by measuring the SERS spectra of 4-mercaptobenzoic acid (pH-sensitive probes) [50]. (b) Cell membrane receptor-specific binding antibodies were conjugated with GNPs (left) for detection of receptor dynamics on the cell membrane. A dark-field image shows a cell binding with GNPs (middle). Strong coupling of GNPs enhanced the SERS signal (green dots) and indicated the aggregation of receptors on the cell membrane [45].

Recently, SERS has been utilized in cellular experiments for mapping the pH inside living mouse fibroblasts by monitoring the SERS spectra of pH-sensitive molecules (e.g., 4-mercaptobenzoic acid and pMBA). GNP-pMBA nanoantennas were introduced into the cells, and the obtained SERS spectra from individual probes provided a map of pH values based on the ratio of the spectral lines at 1423 and 1076 cm^{-1} (Fig. 8.11a) [50]. Dynamic monitoring of pH changes provides a new method to study cell cycles and metabolic processes with subcellular resolution. In addition to the applications of single SERS particles, dimer-enhanced SERS was used to study the dynamics of the aggregation of receptors on the cell membrane (Fig. 8.11b). GNPs were conjugated with receptor-specific binding antibodies and targeted on corresponding receptors. The aggregation of receptors induced the formation of GNP-GNP dimers. Strong coupling between GNPs further enhanced the intensity of the hot spots and consequently generated a stronger SERS signal. Studying the dynamics of cell membrane proteins might provide more detailed information regarding cellular interactions and communication [45].

In addition to *in vitro* detection, SERS-based diagnosis has been investigated for *in vivo* applications [53, 134–137]. For example, tumor targeting and detection has been demonstrated using SERS nanoparticles. Pegylated colloidal gold nanoparticles were conjugated with Raman reporters and single-chain variable fragment (ScFv) antibodies for specific targeting of epidermal growth factor (EGF) receptors on cancer cells (Fig. 8.12a) [53]. The intensity of the SERS nanoparticles was considerably brighter

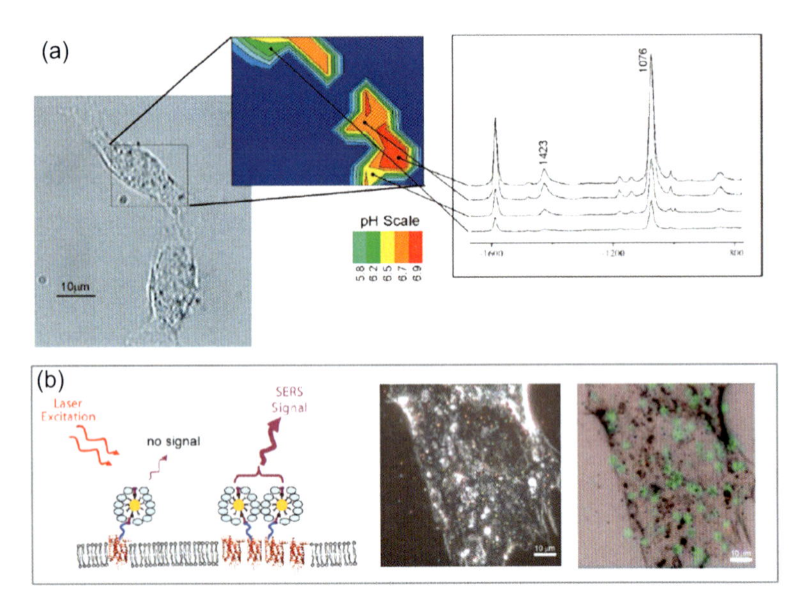

Figure 8.12 Applications of SERS *in vivo*. (a) Preparation of antibody-conjugated SERS nanoparticles for targeting cancer cells with spectroscopic detection. (b) SERS spectra recorded from EGFR-positive cancer cells (red line) and from EGFR-negative cancer cells (green line), along with other controls. (c) Tumor targeting and SERS detection by antibody-conjugated SERS nanoparticles in a small animal. (d) SERS spectra obtained from the tumor (red line) and the liver (blue) *in vivo*.

than semiconductor quantum dots in the near-infrared biological window. After incubation with cancer cells (expressing EGF surface markers), ScFv-SERS nanoparticles can specifically target those cells and provide a strong SERS signal (Fig. 8.12b). In contrast, in negative control experiments (without ScFv conjugation in SERS particles or without EGF expression in cells), the SERS signal was significantly weak. To demonstrate the capability of *in vivo* cancer targeting, nanoparticles were injected into the tails of nude mice, and then, the spectra were recorded 5 hours after injection (Fig. 8.12c). A NIR laser with a wavelength of 785 nm was equipped for penetrating the skin and exciting the SERS particles. The detected SERS signal from tumors was stronger than nontumor spots, such as the liver (Fig. 8.12d).

8.7 Plasmonic Enhanced Photothermal Effect

Upon laser excitation using a pulsed or continuous wave, the photothermal response of metallic nanoparticles involves many energy conversions that occur at different time scales [31, 138]. The response can be *internal* to the plasmonic particle in which the particle becomes a source of heat due to inner dynamic of particles or *external* in which the core and the electromagnetic energy determine how the heat is transferred to the environment. Following the absorption of photons by metallic nanoparticles, plasmons are excited via LSPR. Inside the particles, the interplay between photons, plasmons and phonons inside the conductor—quanta of light, charge oscillations or lattice vibrations—determine the internal conversion of energy to heat and the time scales of each step. Toward the outside, heat is transferred across the boundary of the metal-surrounding interface and is dissipated to the immediate surroundings.

Following the absorption of a fs pulse laser, LSPR occurs inside the subwavelength metal nanoparticle from the oscillation of the collective conduction electrons and rapidly dephases (Fig. 8.13a) [31, 138–140]. The broadening of the resonance line width is due to surface free electron scattering surfaces for small particles in the quasi-static regime and to radiation damping for larger ones.

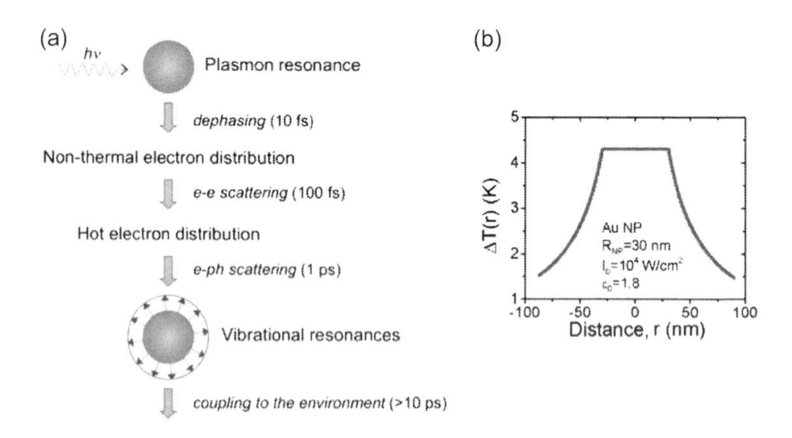

Figure 8.13 Localized photothermal conversion based on plasmonic nanoparticles. (a) Sequence of photothermal conversion and heat diffusion following electromagnetic radiation [31]. (b) Calculated temperature profile (cross-sectional) for a single GNP under irradiation [138].

The broadening can be quantitatively explained by examining the complex dielectric function. The real part determines the position of the resonance and the imaginary part provides the dephasing. The line width is defined by [31]

$$\Gamma = \gamma + \frac{Av_{\mathrm{F}}}{l_{\mathrm{eff}}} + 2\hbar\kappa V$$

where γ is the bulk damping constant, A is a constant that characterizes the electron scattering, v_{F} is the Fermi velocity, l_{eff} is the effective path length of electrons, $\hbar = h/(2\pi)$ is the reduced Planck's constant, κ is a phenomenological damping efficiency constant, and V is the polarization volume. Note that the third term is volume dependant and is negligible for small nanospheres. Usually, the LSPR line width is on the order of 20 nm. The dephasing time is very rapid and is estimated to be approximately 10 fs.

The energy from dephasing due to surface electron scattering is then partially converted nonthermally over the conduction bands through interbands or intraband transitions in which electrons are excited (Fig. 8.13a) [31, 138]. This electron distribution reaches equilibrium through electron–electron scattering on a few 100 fs time scale before they create a hot electron distribution with a new equilibrium temperature that is greater than the lattice phonon temperature, which causes energy transfer from the electrons to the lattice phonons. This process is referred to as electron–phonon relaxation and occurs within a few picoseconds, and this is manifested as vibrations of the lattice and increases the overall temperature of the nanoparticle that is in contact with its surrounding medium [141].

The EM contribution arising from the field inside of the nanoplasmonic structures can be described through a heat transfer equation, which provides an evaluation of the amount of temperature change the hot plasmonic heat source provides the surrounding environment (Fig. 8.13b). Without a phase transformation, the outward heat transfer at the metal–water interface at R is governed by

$$\rho_M c_M \frac{\partial T(\mathbf{r}, t)}{\partial t} = \begin{Bmatrix} \kappa_M \nabla^2 T(\boldsymbol{r}, t) + Q(\boldsymbol{r}, t) & \text{inside} \\ \kappa_W \nabla^2 T(\boldsymbol{r}, t) & \text{outside} \end{Bmatrix}$$

where ρ_M is mass density of a metal, c_M is its specific heat, $T(\boldsymbol{r}, t)$ is the temperature field, and k_M and k_W are the thermal conductivities of metal and water, respectively. $Q(\boldsymbol{r}, t)$ is the local heat intensity as a function of the average joule heating. $Q(\boldsymbol{r}, t) = \langle \boldsymbol{J}(\boldsymbol{r}, t) \cdot \boldsymbol{E}(\boldsymbol{r}, t) \rangle$ given by the time-averaged value of Joules heating (energy lost resistively) and can be calculated with the knowledge of the fields inside the sphere. The above heat transfer equation is usually solved numerically, except in the case of a perfectly spherical nanoparticle, which can yield an analytical solution. The analysis can be performed for a continuous wave or pulse [31, 138]. In the steady state, the expression for a metallic nanosphere of radius r with a volume V_{NP} is

$$\Delta T(\boldsymbol{r}) = \frac{V_{NP} Q}{4\pi k_0 r}$$

where k_0 is the thermal conductivity of the surrounding medium. An example of the difference in the temperature of a GNP sphere as a function of its radial distance is plotted in Fig. 8.13b.

To remotely manipulate gene delivery, a new method based on the photothermal effect of plasmonic nanoparticles has been demonstrated for precisely controlling gene expression in live cells using a near-infrared (NIR) laser [22, 55, 56, 142, 143]. The NIR laser has a long tissue penetration depth, which is suitable for

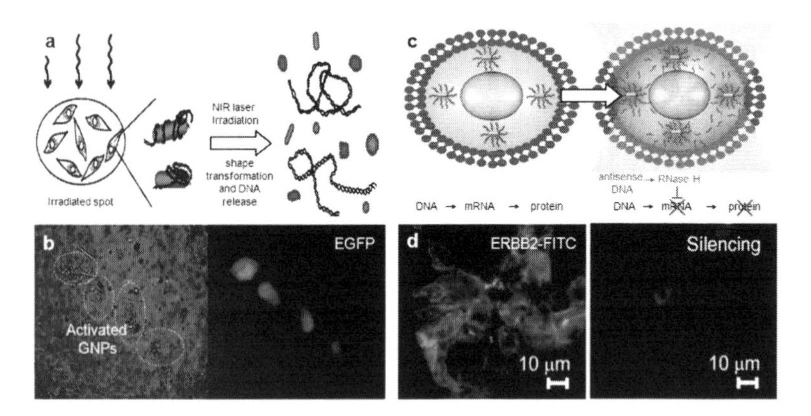

Figure 8.14 Light control over the release of genetic materials in live cells. (a) Schematic illustration for the remote controlled release of DNA (with green fluorescence protein, GFP, gene) in live cells by laser-triggered shape transformation of GNRs [142]. (b) Bright-field (left) and fluorescence (right) images of cells with high GFP expression (inside white circles) after laser irradiation. (c) Schematic illustration for the remote controlled release of antisense DNA in live cells by locally heating GNRs. (d) Fluorescence images of cells with (right) and without (left) irradiation [22, 56, 143].

in vivo experiments. Gene delivery and interference are important steps in controlling cell behaviors and reprogramming cells by introducing biomaterials, such as DNA, RNA, small molecules, and proteins. To perform remote light-induced molecular delivery, NIR-absorbed gold nanorods (GNRs) are often used to convert light into thermal energy. After absorbing a high power of pulsed NIR laser, rod-like gold nanoparticles can transform into a spherical shape, which results in the rearrangement of surface atoms and the release of molecules. For example, the shape transformation of GNRs conjugated with enhanced green fluorescent protein (EGFP) plasmid DNA broke the chemical bonds between the gold and DNA and consequently triggered the release of the gene (Fig. 8.14a) [142]. HeLa cells that had uptaken EGFR-gold nanorods exhibited strong GFP expression after irradiation (Fig. 8.14b). It has been demonstrated that plasmonic nanoparticles are promising materials for use as vectors for remotely controlling gene delivery without causing serious damage to the cell.

In contrast, under low power irradiation, nanoplasmonic particles can be slowly heated without dramatic changes in their shape [22, 56, 143]. Heating above the melting temperature of conjugated molecules can also induce local molecular delivery. For example, downstream gene interference can be achieved through nanoplasmonic particle-mediated small DNA or RNA deliveries. After antisense DNA conjugated GNRs are introduced into BT474 breast carcinoma cells, small DNA molecules can be released into the cytosol by heating the GNRs to greater than the melting temperature of the DNA complex (Fig. 8.14c). After hybridization of the antisense DNA/target ERBB2 mRNA in cytosol, ubiquitous RNase H-mediate gene silencing down regulates the expression of ERBB2 proteins, which was demonstrated by fluorescence imaging (Fig. 8.14d).

An additional advantage with respect to plasmonic GNRs is that their light absorption properties are subject to change depending on their aspect ratio. For investigating gene circuits in which multiple gene perturbations are required, it is possible to employ different colors (i.e., wavelengths) of light to address specific GNRs which absorb light at different wavelengths (Fig. 8.15a) [54, 56]. This possibility expands the scope of directly probing signaling pathways in live cells, which might provide critical insights in both cell biology and translational medicine. To test this concept, two different sequences of siRNA molecules were conjugated to two types of GNRs with separate absorption peaks at 780 nm and 650 nm, respectively. Therefore, by simply utilizing two different laser wavelengths (λ_1 and λ_2 in Fig. 8.15a), a light-mediated method for dynamically reconfiguring gene circuits has been suggested. For instance, gene expressions of p65 can be turn on or off, depending on the wavelength of the laser irradiation (Fig. 8.15b,c). Moreover, the feasibility of time-dependent modulations of photonic gene circuits with corresponding downstream protein expression has been investigated by immunostaining. In the future, as inspired by optogenetics, nanoplasmonic genetics might also be able to more precisely regulate multiple genes in living cells and small animals.

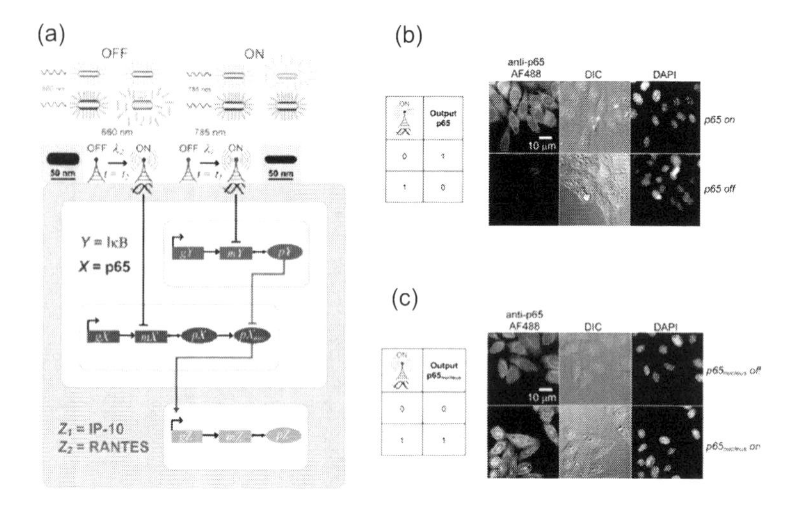

Figure 8.15 (a) Controlling the release of siRNA for manipulating gene circuits using two optically addressable siRNA-Au nanoantennas (absorption at 660 and 785 nm, respectively). Genes (gX, gY, and gZ), mRNAs (mX, mY, and mZ) and proteins (pX, pY, and pZ) can be regulated by siRNA-Au nanoantennas with different colors of incident light. (b) Logic table for OFF-switch photonic gene circuit (left) and fluorescence cell imaging of OFF-switch photonic gene circuit (right). Lowering the expression of p65 proteins was demonstrated by anti-p65-AF488 immunostaining. (c) Logic table for ON-switch photonic gene circuit (left) and fluorescence cell imaging of ON-switched cells (right). The expression of p65 proteins was greater after the ON switch [54, 56].

8.8 Nanoplasmonic Trapping

Conventional optical tweezers, which are based on diffraction-limited laser spots, are not suitable for subwavelength objects (Fig. 8.16a) [61, 62]. Two undesirable effects against the trapping operation occur when the particle size decreases, as follows: 1) the trapping potential well becomes shallower due to a decrease in the magnitude of the restoring force, and 2) the viscous drag is reduced, which causes a decrease in the damping of the trapped object. Because the random thermal energy becomes more significant for smaller particles, a rule of thumb given by Ashkin et al. stipulates that the potential depth should be at least 10 $k_B T$ to compensate for the specimen from escaping from the trap due to random stochastic Brownian motion [61, 62, 144]. In general, it is difficult to trap particles that are smaller than the wavelength of light using conventional optical tweezers.

The hot spots that result from a tight plasmonic focus can overcome the limit of conventional optical trapping. The intense electric and magnetic field tailored by subwavelength focusing through LSPR can contribute to a force that is capable of stable nanoscale trapping with increased confinement and a deep trapping potential depth (Fig. 8.16b) [62, 69]. The optical force and trapping potential can be calculated from focused fields through the Maxwell stress-tensor. Examples of trapping potentials for a spherical particle are illustrated in Fig. 8.16a and 16b. Although the trapping force arises purely from an electromagnetic contribution of the field around the plasmonic structure, it is difficult to quantify the respective amount that will actually result in optical trapping rather than in photothermal heating. LSPR can also cause an internal heating and dissipation of the particle surface energy through the medium, which results in convection and thermophoresis. It is therefore important to mention the components of both optical forces and thermally induced forces.

Figure 8.16c illustrates a nanoplasmonic substrate for the trapping of nanoparticles. Two-dimensional arrays of gold nanodots dimers with a diameter of 134 nm were fabricated on a glass substrate using electron beam lithography (left, Fig. 8.16c) [61]. The gap in the middle of the dimers supported a localized and intense hot spot, which was confirmed by numerical simulations (right, Fig. 8.16c). In conventional optical trapping, a polystyrene bead (200 nm in diameter, refractive index of 1.6) was trapped by a focused laser beam and its trajectory demonstrated that the confinement was c.a. 800 nm in diameter (left, Fig. 8.16d). In contrast, in the presence of a hot spots supported by a dimer of gold nanodots, the movement of the polystyrene bead in solution was confined to the scale of 100 nm, which is smaller than the diffraction limit of light (right, Fig. 8.16d). More recently, nanoplasmonic trapping of ~10 nm dielectric particles, bacteria, and even single proteins has been demonstrated. For example, plasmonic enhanced optical trapping of single bovine

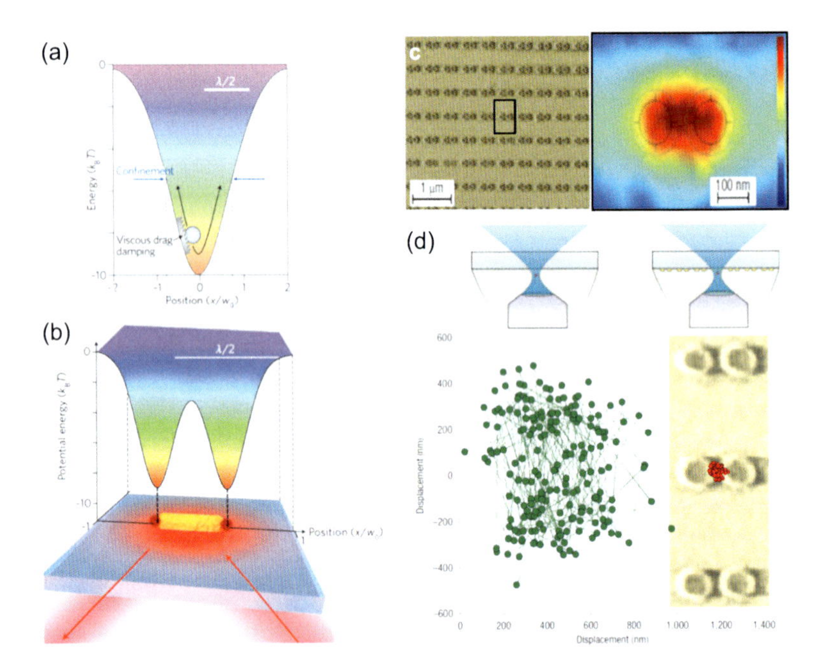

Figure 8.16 Conventional optical trapping versus nanoplasmonic trapping [61, 62]. (a) Conventional optical trapping potential for a polystyrene sphere. (b) Nanoplasmonic trapping potential landscape resulting from a plasmonic gold nanorod. Note that the scale of confinement in nanoplasmonic trapping (i.e., $\sim\lambda/4$) is smaller than that in conventional optical trapping (i.e., $\sim\lambda$). (c) Micrograph of nanodot pairs (left) and the corresponding simulation of the electromagnetic field intensity in the near-fields. (d) Trapping of single polystyrene spheres by conventional optical trapping (left, green line) and by a pair of plasmonic nanodots (right, red line).

serum albumin (BSA) molecules has been demonstrated using a double nanohole in a Au film (Fig. 8.17a) [57]. The hot spot in the gap strongly trapped a single protein and caused an increase in the reflective index (Fig. 8.17b). Because of the strong optical force in the gap, unfolding of the protein was also observed. High concentration, together with localization, makes nanosized hot spots a promising tool for trapping single particles and single molecules [57, 58, 61, 62].

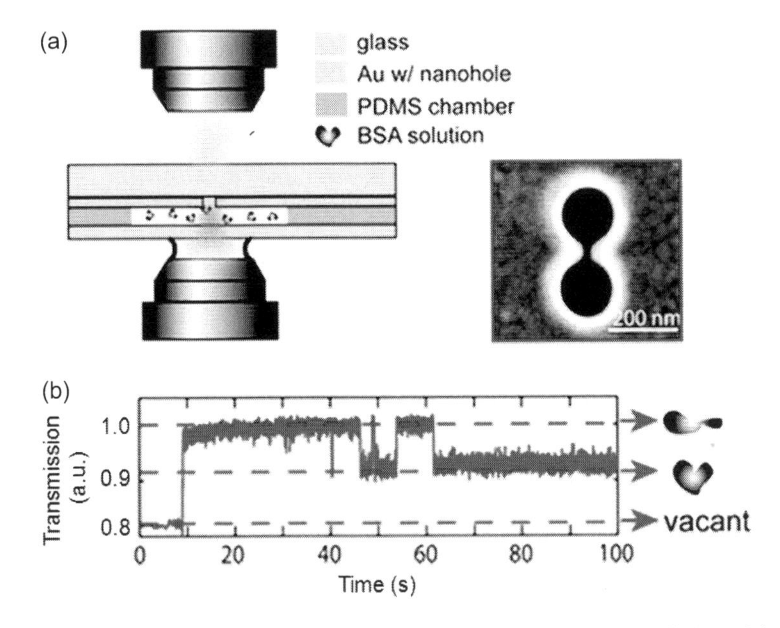

Figure 8.17 Trapping of single proteins by plasmonic nanoholes. (a) Schematic illustration of the single bovine serum albumin trapping experiment (left) using double-nanohole structures (SEM image, right) [57]. (b) Time traces of light transmission demonstrated that plasmonic nanoholes can perform trapping of a single protein. The nanoplasmonic force not only trapped a single protein (indicated by the middle arrow) but also induced unfolding of the trapped protein (indicated by the top arrow).

8.9 Future Outlook

We conclude with an outlook on the use of nanoplasmonic-integrated nanofluidics for medical diagnostics and *in vivo* nanoplasmonic manipulation of genetic circuits. On the one hand, considering point-of-care applications and the science of miniaturization, micro- and nanofluidic chips that creatively utilize the unique optical properties of metals could bring new optofluidic applications with refined spatiotemporal and environmental control, which is promising for creating simple and robust conditions for multiplexing [15, 145]. On the other hand, applications of multiple plasmonic nanomaterial-controlled deliveries are promising for reconstruction of gene circuits in live cells [56, 146]. Through inspirations from optogenetics, [55, 147–153]. Nanoplasmonic gene regulation mediated by photothermal effects might provide multiple manipulations with high spatial and temporal resolution. Investigating nanoplasmonic-mediated sensing and gene regulation could provide new avenues for understanding how biological systems function at the molecular level.

8.9.1 *Nanoplasmonics and Fluidic Integration for Point-of-Care Diagnosis*

In point-of-care diagnosis, one of the requirements for the miniaturization of biophotonic features on a microfluidic device is to overcome the bulkiness in light sources and detection components that are currently used in conventional microscopy; integrating them onto a chip would require new technical paradigms that mold light beyond the diffraction limit. At the same time, the emerging field of optofluidics offers new ways to bring optical capabilities down to the biochip scale and shed light on controlled biological flow.

Although the fields of plasmonics and microfluidics have developed independently, some forms of convergence have recently been observed [145, 154–156]. Indeed, in plasmonic fluidic systems, metallic nanostructures are promising candidates for achieving seamless integration of subwavelength miniaturization of biophotonic manipulation and detection functions, whereas fluidic control at the small scale seeks to automate and optimize biochemical and biomedical analyses (Fig. 8.18a). For example, in the work by Eftekahri et al. (Fig. 8.18b), the plasmonic nanohole functions as a biosensor based on the extraordinary transmission of subwavelength apertures and also as a flow-through fluidic device [157]. Another example is observed in the work of Oh et al. (Fig. 8.18c) in which a nanocap plays a triple role as a hot spot, a plasmonic resonator and a nanofluidic trapping substrate for target molecules [124].

The two fields of plasmonics and nanofluidics meet at the small scales because they can mutually benefit from the advantages of miniaturization that utilize the interplay between the short-scale evanescent near-field, convection, diffusion and surface-bound

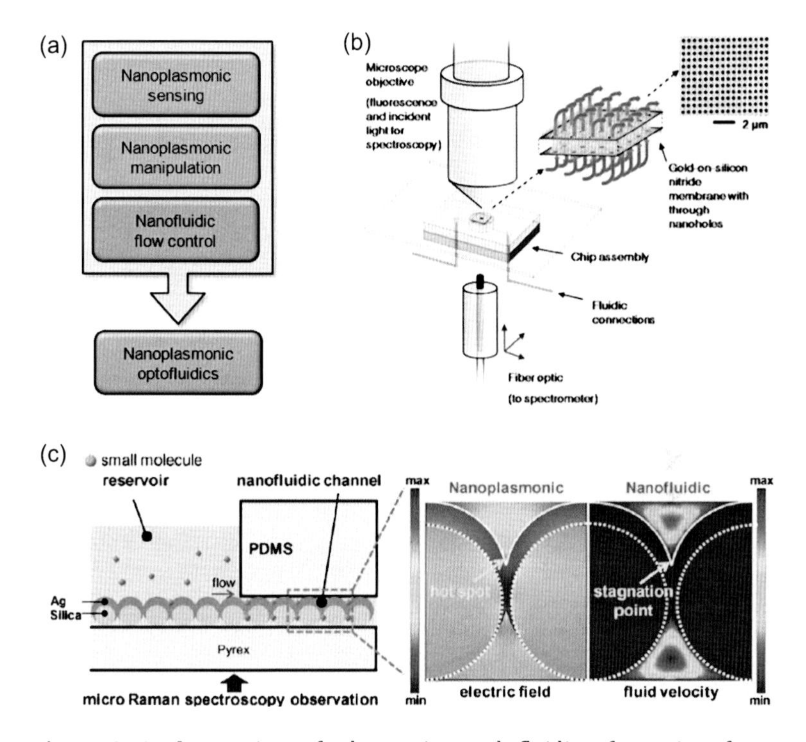

Figure 8.18 Integration of plasmonics and fluidics for point-of-care diagnosis. (a) At the nanoscopic scale, optical control for sensing and manipulation using nanoplasmonics is integrated with nanofluidic flow control. (b) A plasmonic nanohole-based biosensor is also a nanopore array for the nanofluidic flow of molecules [157]. (c) A plasmonic fluidic SERS substrate that utilizes nanocaps for its triple functions in localizing hotspots, LSPR resonator, and fluid trapping [124].

reactions [158, 159]. In this regard, the marriage of nanoplasmonic and microfluidic technologies is expected to render more quantitative molecular control In conjunction, microfluidic platforms developed in-house can provide precise spatial and temporal control of the local extracellular environment of the cell under study while plasmonic nanoparticles systematically perturb its intracellular interior. Through the creative combination of both plasmonic and fluidic devices at the micro- and nanometer scales, numerous exciting opportunities exist for molecular diagnostics and single-cell characterization.

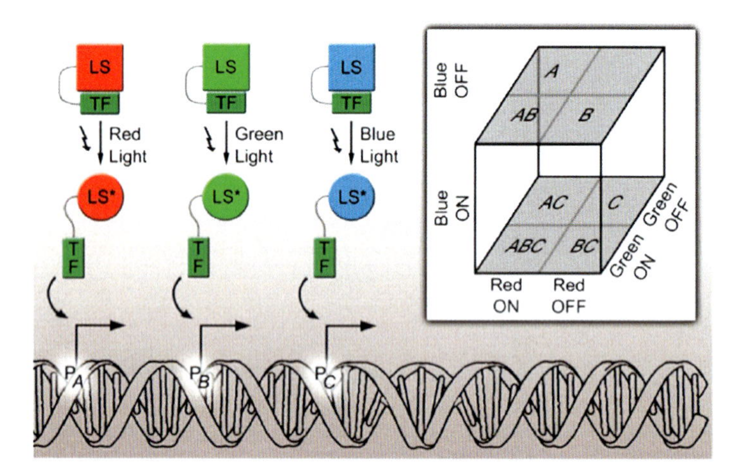

Figure 8.19 The concept of multichromatic control of gene circuits [150]. Inactivated light-sensor (LS)/transcription factor (TF) complexes can be turned on (or activated) individually by irradiating with different colors of light (i.e., blue, green and red). After activation (LS*), TFs facilitate the cells to express corresponding genes and proteins (i.e., P_A, P_B, and P_C). The inset shows different combinations of irradiation results in different patterns of gene expression.

8.9.2 *Multiplex Gene Manipulation Using Tunable Optical Antennae*

Researchers have long been fascinated by remotely and non-invasively switching genes on and off [147–153]. The flexible manipulation of genetic circuits has been studied using *Escherichia coli* and bacteria based on genetically encoded light sensors (Fig. 8.19) [147, 150]. Recently, there has been a resurgence of interest in leveraging light for gene regulation from bacteria to animals [149, 152]. Genetic modifications are finding increased utility for multichromatic control of gene expression. By irradiating with different colors of light, it is possible to force cells to express certain combinations of proteins with high temporal resolution.

However, one concern with respect to optogenetic methods is that making genetically modified animals or cells is sophisticated and might cause the formation of cancer. This concern limits the scope of their applications. It has been demonstrated that nanoplasmonic-mediated DNA/RNA release is capable of controlling gene expression in cells. Such safety and flexibility holds a number of advantages over traditional optogenetic methods [56, 146]. Without gene modification, it is possible to perform the deconstruction of genetic circuits using the spatial and temporal control of multiple deliveries of genetic materials. For further studies on genetic circuits, nanoplasmonic-mediated gene regulation makes it possible to expend target genes or pathways to more than three by addressing different vectors (i.e., gold nanorods and nanospheres with various absorption peaks) [55, 56, 146, 160, 161]. In addition, gold nanoparticles can serve as multi-function materials (e.g., simultaneously acting as vectors and probes). The long-range photothermal triggering, together with noncytotoxicity and ease of surface functionalization, makes nanosized plasmonic antennae promising vectors for tissue-specific gene regulation and also as probes for real-time monitoring of the responses.

8.10 Appendix

8.10.1 *Mie Theory*

In Mie theory, the incident, scattered, and internal electromagnetic fields $(\mathbf{E}, \mathbf{H})$ inside and outside of the sphere are expanded as a linear superposition of weighted multipoles (monopole, dipole, quadrupole, etc.), which are numbered by indexes (mn) and expressed in terms of vector spherical harmonics $(\mathbf{N}_{mn}^{(s)}\mathbf{M}_{mn}^{(s)})$ [77–79, 102, 162]. Consequently, the scattered fields $(\mathbf{E}_{\text{sca}},\mathbf{H}_{\text{sca}})$ around a Mie sphere take the form

$$\mathbf{E}_{\text{sca}} = \sum_{n=1}^{\infty}\sum_{m=-n}^{n} \mathrm{i}E_{\text{mn}} \left[a_{mn}\mathbf{N}_{mn}^{(3)}+b_{mn}\mathbf{M}_{mn}^{(3)}\right]$$

$$\mathbf{H}_{\text{sca}} = \frac{1}{\eta}\sum_{n=1}^{\infty}\sum_{m=-n}^{n} E_{\text{mn}} \left[b_{mn}\mathbf{N}_{mn}^{(3)}+a_{mn}\mathbf{M}_{mn}^{(3)}\right]$$

where E_{mn} is a multipole constant, $\eta = (\mu_0\mu_D/(\varepsilon_0\varepsilon_D))^{1/2}$ is the wave impedance of the ambient medium and (a_{mn},b_{mn}) are scattering expansion coefficients. With $\mathbf{E} \propto \mathbf{M}_{mn}^{(s)}$, transverse electric

(TE) modes are described, whereas with $\mathbf{E} \propto \mathbf{N}_{mn}^{(s)}$ transverse magnetic (TM) modes are represented. The scattering expansion coefficients (a_{mn}, b_{mn}) are obtained through the establishment of boundary conditions between the sphere and the surrounding medium and depend on a size parameter, $x = 2\pi m_{\mathrm{D}} a/\lambda$ (defining a ratio between the particle size, a, and the illumination wavelength inside the medium, λ/m_{D}) and a relative refractive index contrast, $m_{\mathrm{r}} = m/m_{\mathrm{D}} = (\varepsilon\mu/\varepsilon_{\mathrm{D}}\mu_{\mathrm{D}})^{1/2}$. Upon illumination, the different electromagnetic modes constitute the antenna field pattern of the sphere and describe the flow of incident optical energy around it. For a metal, the plasmonic enhancement of optical modes enters Mie theory through the relative refractive index, m_{r}, where the optical response of the collective oscillation of the metal electrons is included in the dielectric function $\varepsilon = \varepsilon_{\mathrm{M}}(\omega)$. From a functional point of view, the plasmonic sphere can act as an optical nanolens because it distorts the flow of incident electromagnetic energy and is an efficient light collector at specific frequencies.

8.10.2 *Far-Field Scattering of Metallic Spheres*

When making a far-field measurement (e.g., transmission of light through a sample in a cuvette), the attenuation of light is given by the Beer-Lambert law,

$$I = I_0 e^{-n\sigma_{\mathrm{ext}} l}$$

where I_0 represents the incident intensity, n is the number density of weakly interacting scatterers, σ_{ext} is the extinction cross-section, and l is the optical path length. The extinction cross-section represents the energy removed from the incident beam of light due to absorption and scattering events. The light-scattering power of a particle is expressed in terms of the scattering cross-section σ_{sca}, which represents a hypothetical area for which a photon is scattered in an outward direction. The far-field cross-section σ_{sca} calculates the projection of the scattering electromagnetic energy (through a Poynting vector) on the surface of the hypothetical "large" sphere that encloses the scatterer,

$$\mathbf{S}_{\mathrm{sca}} = \frac{1}{2}\mathrm{Re}\left\{\mathbf{E}_{\mathrm{sca}} \times \mathbf{H}_{\mathrm{sca}}^{*}\right\}$$

$$\sigma_{\text{sca}} = -\frac{1}{I_0} \oiint \mathbf{S}_{\text{sca}} \cdot d\mathbf{A}$$

where I_0 is the incident intensity and the surface integral is evaluated for a large volume that includes the particle. The scattering and extinction cross-sections as calculated by Mie theory are,

$$\sigma_{\text{sca}} = \frac{2\pi}{k^2} \sum_{n=1}^{\infty} (2n+1) \left[|\alpha_n|^2 + |\beta_n|^2 \right]$$

$$\sigma_{\text{ext}} = \frac{2\pi}{k^2} \sum_{n=1}^{\infty} (2n+1) \, \text{Re} \left\{ \alpha_n + \beta_n \right\}$$

where k is the incoming wavenumber and n is an integer that represents the multipole order (i.e., dipole, quadrupole, etc.). The nth Lorenz–Mie coefficients (α_n, β_n) that affect the strength of the scattered fields are

$$\alpha_n = \frac{m\psi_n{}'(x)\,\psi_n(mx) - \psi_n(x)\,\psi_n{}'(mx)}{m\xi_n{}'(x)\,\psi_n(mx) - \xi_n(x)\,\psi_n{}'(mx)}$$

$$\beta_n = \frac{\psi_n{}'(x)\,\psi_n(mx) - m\psi_n(x)\,\psi_n{}'(mx)}{\xi_n{}'(x)\,\psi_n(mx) - m\xi_n(x)\,\psi_n{}'(mx)}$$

where $\psi_n(\rho) = \rho j_n(\rho)$ and $\xi_n(\rho) = \rho h_n(\rho)$ with $j_n(\rho)$ and $h_n(\rho)$ are the Bessel and spherical Hankel of the first order, respectively. The primes indicate differentiation with respect to the argument in parentheses. For a small sphere in the Rayleigh regime, the expression for cross sections simplifies and is related to the polarizability $\alpha(\omega)$.

8.10.3 *Theory of Plasmonic Enhanced Photothermal Effects*

The hot spots that result from a tight plasmonic focus go around the limit of conventional optical trapping. The intense electric and magnetic field tailored by subwavelength focusing through LSPR can contribute to a force capable of stable nanometric subwavelength trapping, with increased confinement and a deep depth of trapping potential. The optical force can be calculated from focused fields (**EH**) through the Maxwell stress tensor, [69]

$$\mathbf{T} = \varepsilon_0 \varepsilon \mathbf{EE} + \mu_0 \mu \mathbf{HH} - \frac{1}{2}\left(\varepsilon_0 \varepsilon E^2 + \mu_0 \mu H^2\right)$$

$$\mathbf{F}_{\text{trap}}\left(\boldsymbol{r}, t\right) = \iint_{\partial V} \mathbf{T} \cdot \mathbf{n} dS$$

The trapping potential $U\ (\boldsymbol{r})$ of a particle located at a point **r** is then given by

$$U\ (\boldsymbol{r}) = -\int_{\infty}^{\mathbf{r}'} \mathbf{F}_{\text{trap}}\left(\boldsymbol{r}'\right) d\boldsymbol{r}'$$

Although the trapping force arises purely from an electromagnetic contribution of the field around the plasmonic structure, it is difficult to quantify the respective amount that will actually result in optical trapping rather than in photothermal heating. LSPR can also cause an internal heating and dissipation of the particle surface energy through the medium, which results in convection and thermophoresis. It is therefore important to mention that both components of optical forces and thermally induced forces are present.

In a liquid medium, the optical force in solution also needs to balance other counter interacting ones, such as the viscous forces. According to Stoke's law for hydrodynamics, the time-average viscous force field exerted on a spherical particle is given by

$$\langle \mathbf{F}_{\text{drag}}(\boldsymbol{r}t) \rangle = 6\pi \eta r \mathbf{v}$$

where η is the viscosity of the liquid medium, r is the submersion particle size (assumed spherical here), and **v** is the velocity vector of the particle.

References

1. Weiss, S. (1999). Fluorescence spectroscopy of single biomolecules. *Science*, **283**:1676–1683.
2. Michalet, X., et al. (2005). Quantum dots for live cells, *in vivo* imaging, and diagnostics. *Science*, **307**:538–544.

3. Lippincott-Schwartz, J., and Patterson, G. H. (2003). Development and use of fluorescent protein markers in living cells. *Science*, **300**:87–91.
4. Lacoste, T. D., et al. (2000). Ultrahigh-resolution multicolor colocalization of single fluorescent probes. *Proc. Natl. Acad. Sci. USA*, **97**:9461–9466.
5. Fu, C. C., et al. (2007). Characterization and application of single fluorescent nanodiamonds as cellular biomarkers. *Proc. Natl. Acad. Sci. USA*, **104**:727–732.
6. Gao, X. H., Cui, Y. Y., Levenson, R. M., Chung, L. W. K., and Nie, S. M. (2004). *In vivo* cancer targeting and imaging with semiconductor quantum dots. *Nat. Biotechnol.*, **22**:969–976.
7. Yu, J., Xiao, J., Ren, X. J., Lao, K. Q., and Xie, X. S. (2006). Probing gene expression in live cells, one protein molecule at a time. *Science*, **311**:1600–1603.
8. El-Sayed, I. H., Huang, X. H., and El-Sayed, M. A. (2006). Selective laser photo-thermal therapy of epithelial carcinoma using anti-EGFR antibody conjugated gold nanoparticles. *Cancer Lett.*, **239**:129–135.
9. Butt, H. J., Cappella, B., and Kappl, M. (2005). Force measurements with the atomic force microscope: Technique, interpretation and applications. *Surf. Sci. Rep.*, **59**:1–152.
10. Gradinaru, V., et al. (2010). Molecular and cellular approaches for diversifying and extending optogenetics. *Cell*, **141**:154–165.
11. Deisseroth, K. (2011). Optogenetics. *Nat. Methods*, **8**:26–29.
12. Lu, Y., Liu, G. L., Kim, J., Mejia, Y. X., and Lee, L. P. (2005). Nanophotonic crescent moon structures with sharp edge for ultrasensitive biomolecular detection by local electromagnetic field enhancement effect. *Nano Lett.*, **5**:119–124.
13. Ross, B. M., and Lee, L. P. (2008). Plasmon tuning and local field enhancement maximization of the nanocrescent. *Nanotechnology*, **19:**275201.
14. Anker, J. N., et al. (2008). Biosensing with plasmonic nanosensors. *Nat. Mater.*, **7**:442–453.
15. Lee, S. E., and Lee, L. P. (2010). Biomolecular plasmonics for quantitative biology and nanomedicine. *Curr. Opin. Biotechnol.*, **21**:489–497.
16. Li, Y., Jing, C., Zhang, L., and Long, Y.-T. (2012). Resonance scattering particles as biological nanosensors *in vitro* and *in vivo*. *Chem. Soc. Rev.*, **41**:632–642.
17. Hell, S. W. (2007). Far-field optical nanoscopy. *Science*, **316**:1153–1158.

18. Fernandez-Suarez, M., and Ting, A. Y. (2008). Fluorescent probes for super-resolution imaging in living cells. *Nat. Rev. Mol. Cell Biol.*, **9**:929–943.
19. Huang, B., Bates, M., and Zhuang, X. W. (2009). Super-resolution fluorescence microscopy. *Annu. Rev. Biochem.*, 78:993–1016 (Annual Reviews, Palo Alto).
20. Lal, S., Link, S., and Halas, N. J. (2007). Nano-optics from sensing to waveguiding. *Nat. Photonics*, **1**:641–648.
21. Zheng, Y. B., Kiraly, B., Weiss, P., and Huang, T. (2012). Molecular plasmonics for biology and nanomedicine. *Nanomedicine(Lond)*, **7**:751–770.
22. Lee, S. E., and Lee, L. P. (2010). Nanoplasmonic gene regulation. *Curr. Opin. Chem. Biol.*, **14**:623–633.
23. Willets, K. A., and Van Duyne, R. P. (2007). Localized surface plasmon resonance spectroscopy and sensing. *Annu. Rev. Phys. Chem.*, **58**:267–297.
24. Khlebtsov, N. G., and Dykman, L. A. (2010). Optical properties and biomedical applications of plasmonic nanoparticles. *J. Quant. Spectrosc. Radiat.*, **111**:1–35.
25. Rosi, N. L., and Mirkin, C. A. (2005). Nanostructures in biodiagnostics. *Chem. Rev.*, **105**:1547–1562.
26. Yguerabide, J., and Yguerabide, E. E. (1998). Light-scattering submicroscopic particles as highly fluorescent analogs and their use as tracer labels in clinical and biological applications - II. Experimental characterization. *Anal. Biochem.*, **262**:157–176.
27. Yguerabide, J., and Yguerabide, E. E. (1998). Light-scattering submicroscopic particles as highly fluorescent analogs and their use as tracer labels in clinical and biological applications - I. Theory. *Anal. Biochem.*, **262**:137–156.
28. Chen, K., Liu, Y., Ameer, G., and Backman, V. (2005). Optimal design of structured nanospheres for ultrasharp light-scattering resonances as molecular imaging multilabels. *J. Biomed. Opt.*, **10:**024005.
29. Zamborini, F. P., Bao, L. L., and Dasari, R. (2012). Nanoparticles in measurement science. *Anal. Chem.*, **84**:541–576.
30. Willets, K. A., and Van Duyne, R. P. (2007). Localized surface plasmon resonance spectroscopy and sensing. *Annu. Rev. Phys. Chem.*, 58:267–297 (Annual Reviews, Palo Alto).
31. Hartland, G. V. (2011). Optical studies of dynamics in noble metal nanostructures. *Chem. Rev.*, **111**:3858–3887.

32. Halas, N. J., Lal, S., Chang, W. S., Link, S., and Nordlander, P. (2011). Plasmons in strongly coupled metallic nanostructures. *Chem. Rev.*, **111**:3913–3961.

33. Kumar, A., Boruah, B. M., and Liang, X. J. (2011). Gold Nanoparticles: Promising Nanomaterials for the Diagnosis of Cancer and HIV/AIDS. *J. Nanomater.*, 2011 Article ID 202187, 17 pages.

34. Sonnichsen, C., Reinhard, B. M., Liphardt, J., and Alivisatos, A. P. (2005). A molecular ruler based on plasmon coupling of single gold and silver nanoparticles. *Nat. Biotechnol.*, **23**:741–745.

35. Liu, G. L., et al. (2006). A nanoplasmonic molecular ruler for measuring nuclease activity and DNA footprinting. *Nat. Nanotechnol.*, **1**:47–52.

36. Choi, Y., Kang, T., and Lee, L. P. (2009). Plasmon resonance energy transfer (PRET)-based molecular imaging of cytochrome c in living cells. *Nano Lett.*, **9**:85–90.

37. Zheng, Y. B., et al. (2009). Active molecular plasmonics: controlling plasmon resonances with molecular switches. *Nano Lett.*, **9**:819–825.

38. Liu, G. L., Long, Y.-T., Choi, Y., Kang, T., and Lee, L. P. (2007). Quantized plasmon quenching dips nanospectroscopy via plasmon resonance energy transfer. *Nat. Methods*, **4**:1015–1017.

39. Choi, Y., Park, Y., Kang, T., and Lee, L. P. (2009). Selective and sensitive detection of metal ions by plasmonic resonance energy transfer-based nanospectroscopy. *Nat. Nanotechnol.*, **4**:742–746.

40. Lakowicz, J. R., et al. (2008). Plasmon-controlled fluorescence: a new paradigm in fluorescence spectroscopy. *Analyst*, **133**:1308–1346.

41. Aslan, K., et al. (2005). Metal-enhanced fluorescence: an emerging tool in biotechnology. *Curr. Opin. Biotechnol.*, **16**:55–62.

42. Fort, E., and Gresillon, S. (2008). Surface enhanced fluorescence. *J. Phys. D Appl. Phys.*, **41:**013001.

43. Cang, H., et al. (2011). Probing the electromagnetic field of a 15-nanometre hotspot by single molecule imaging. *Nature*, **469**:385–388.

44. Kinkhabwala, A., et al. (2009). Large single-molecule fluorescence enhancements produced by a bowtie nanoantenna. *Nat. Photonics*, **3**:654–657.

45. Kennedy, D. C., et al. (2009). Nanoscale aggregation of cellular beta2-adrenergic receptors measured by plasmonic interactions of functionalized nanoparticles. *Acs Nano*, **3**:2329–2339.

46. Bailo, E., and Deckert, V. (2008). Tip-enhanced Raman spectroscopy of single RNA strands: Towards a novel direct-sequencing method. *Angew. Chem. Int. Ed. Engl.*, **47**:1658–1661.

47. Drescher, D., and Kneipp, J. (2012). Nanomaterials in complex biological systems: insights from Raman spectroscopy. *Chem. Soc. Rev.*, **41**:5780–5799.
48. Kneipp, J., Kneipp, H., and Kneipp, K. (2008). SERS - a single-molecule and nanoscale tool for bioanalytics. *Chem. Soc. Rev.*, **37**:1052–1060.
49. Porter, M. D., Lipert, R. J., Siperko, L. M., Wang, G., and Narayanana, R. (2008). SERS as a bioassay platform: fundamentals, design, and applications. *Chem. Soc. Rev.*, **37**:1001–1011.
50. Kneipp, J., Kneipp, H., Wittig, B., and Kneipp, K. (2010). Following the Dynamics of pH in Endosomes of Live Cells with SERS Nanosensors. *J. Phys. Chem. C*, **114**:7421–7426.
51. Cho, H. S., Lee, B., Liu, G. L., Agarwal, A., and Lee, L. P. (2009). Label-free and highly sensitive biomolecular detection using SERS and electrokinetic preconcentration. *Lab Chip*, **9**:3360–3363.
52. Ando, J., Fujita, K., Smith, N. I., and Kawata, S. (2011). Dynamic SERS Imaging of Cellular Transport Pathways with Endocytosed Gold Nanoparticles. *Nano Lett.*, **11**:5344–5348.
53. Qian, X. M., et al. (2008). *In vivo* tumor targeting and spectroscopic detection with surface-enhanced Raman nanoparticle tags. *Nat. Biotechnol.*, **26**:83–90.
54. Anikeeva, P., and Deisseroth, K. (2012). Photothermal genetic engineering. *Acs Nano*, **6**:7548–7552.
55. Braun, G. B., et al. (2009). Laser-Activated Gene Silencing via Gold Nanoshell-siRNA Conjugates. *ACS Nano*, **3**:2007–2015.
56. Lee, S. E., et al. (2012). Photonic gene circuits by optically addressable siRNA-Au nanoantennas. *ACS Nano*, **6**:7770–7780.
57. Pang, Y. J., and Gordon, R. (2012). Optical trapping of a single protein. *Nano Lett.*, **12**:402–406.
58. Righini, M., et al. (2009). Nano-optical trapping of rayleigh particles and escherichia coli bacteria with resonant optical antennas. *Nano Lett.*, **9**:3387–3391.
59. Branton, D., et al. (2008). The potential and challenges of nanopore sequencing. *Nat. Biotechnol.*, **26**:1146–1153.
60. Venkatesan, B. M., and Bashir, R. (2011). Nanopore sensors for nucleic acid analysis. *Nat. Nanotechnol.*, **6**:615–624.
61. Grigorenko, A. N., Roberts, N. W., Dickinson, M. R., and Zhang, Y. (2008). Nanometric optical tweezers based on nanostructured substrates. *Nat. Photonics*, **2**:365–370.

62. Juan, M. L., Righini, M., and Quidant, R. (2011). Plasmon nano-optical tweezers. *Nat. Photonics*, **5**:349–356.
63. Thakor, A. S., Jokerst, J., Zavaleta, C., Massoud, T. F., and Gambhir, S. S. (2011). Gold nanoparticles: a revival in precious metal administration to patients. *Nano Lett.*, **11**:4029–4036.
64. Wagner, F. E., et al. (2000). Before striking gold in gold-ruby glass. *Nature*, **407**:691–692.
65. Stockman, M. I. (2010). Nanoscience: Dark-hot resonances. *Nature*, **467**:541–542.
66. Tweney, R. D. (2006). Discovering discovery: How Faraday found the first metallic colloid. *Perspect. Sci.*, **14**:97–121.
67. Ross, B. M., and Lee, L. P. (2009). Comparison of near- and far-field measures for plasmon resonance of metallic nanoparticles. *Opt. Lett.*, **34**:896–898.
68. Kam, Z. (1983). Absorption and scattering of light by small particles - Bohren, C. F., Huffman, D. R. *Nature*, **306**:625–625.
69. L. Novotny, B. H. (2006). *Principles of nano-optics* (Cambridge University Press, Cambridge, MA).
70. Maier, S. A. (2007). *Plasmonics-fundamentals and applications* (Springer, Bath).
71. Mayer, K. M., and Hafner, J. H. (2011). Localized surface plasmon resonance sensors. *Chem. Rev.*, **111**:3828–3857.
72. Johnson, P. B., and Christy, R. W. (1972). Optical constants of noble metals. *Phys. Rev. B*, **6**:4370–4379.
73. Hao, F., and Nordlander, P. (2007). Efficient dielectric function for FDTD simulation of the optical properties of silver and gold nanoparticles. *Chem. Phys. Lett.*, **446**:115–118.
74. Link, S., and El-Sayed, M. A. (2000). Shape and size dependence of radiative, non-radiative and photothermal properties of gold nanocrystals. *Int. Rev. Phys. Chem.*, **19**:409–453.
75. Eustis, S., and El-Sayed, M. A. (2006). Why gold nanoparticles are more precious than pretty gold: Noble metal surface plasmon resonance and its enhancement of the radiative and nonradiative properties of nanocrystals of different shapes. *Chem. Soc. Rev.*, **35**:209–217.
76. Zijlstra, P., Chon, J. W. M., and Gu, M. (2009). Five-dimensional optical recording mediated by surface plasmons in gold nanorods. *Nature*, **459**:410–413.
77. Bohren, C. F., and Huffman., D. R. (1983). *Absorption and Scattering of Light by Small Particles* (Wiley, New York).

78. Mie, G. (1908). Beiträge zur Optik trüber Medien, speziell kolloidaler Metallösungen. *Annalen der Physik, Vierte Folge* **25**, 377–445.
79. Born, W. (1999). *Principles of Optics* 7th ed. (Cambridge University Press).
80. Jackson, J. D. (1999). *Classical Electrodynamics*, 3rd ed. (John Wiley & Sons, Inc. New York).
81. Kreibig, U., and Michael, V. (1995). Optical properties of metal clusters (Springer Series in Materials Science).
82. Link, S., Wang, Z. L., and El-Sayed, M. A. (1999). Alloy formation of gold-silver nanoparticles and the dependence of the plasmon absorption on their composition. *J. Phys. Chem. B*, **103**:3529–3533.
83. Link, S., and El-Sayed, M. A. (1999). Spectral properties and relaxation dynamics of surface plasmon electronic oscillations in gold and silver nanodots and nanorods. *J. Phys. Chem. B*, **103**:8410–8426.
84. Zou, S. L., Janel, N., and Schatz, G. C. (2004). Silver nanoparticle array structures that produce remarkably narrow plasmon lineshapes. *J. Chem. Phys.*, **120**:10871–10875.
85. Link, S., and El-Sayed, M. A. (1999). Size and temperature dependence of the plasmon absorption of colloidal gold nanoparticles. *J. Phys. Chem. B*, **103**:4212–4217.
86. Prodan, E., Radloff, C., Halas, N. J., and Nordlander, P. (2003). A hybridization model for the plasmon response of complex nanostructures. *Science*, **302**:419–422.
87. Gans, R. (1915). Form of ultramicroscopic particles of silver. *Annalen der Physik (Berlin, Germany)* **47**:270–284.
88. Link, S., Mohamed, M. B., and El-Sayed, M. A. (1999). Simulation of the optical absorption spectra of gold nanorods as a function of their aspect ratio and the effect of the medium dielectric constant. *J. Phys. Chem. B*, **103**:3073–3077.
89. Prescott, S. W., and Mulvaney, P. (2006). Gold nanorod extinction spectra. *J. Appl. Phys.*, **99**:12.
90. Wriedt, T. (2009). Light scattering theories and computer codes. *J. Quant. Spectrosc. Radiat.*, **110**:833–843.
91. Mackowski, D. W. (1991). Analysis of radiative scattering for multiple sphere configurations. *Proc. R. Soc. London Ser. A-Math. Phys. Eng. Sci.*, **433**:599–614.
92. Xu, Y. L. (1995). Electromagnetic scattering by an aggregate of spheres. *Appl. Opt.*, **34**:4573–4588.

93. Pendry, J. B., Aubry, A., Smith, D. R., and Maier, S. A. (2012). Transformation optics and subwavelength control of light. *Science*, **337**:549–552.

94. Fernandez-Dominguez, A. I., Maier, S. A., and Pendry, J. B. (2010). Collection and concentration of light by touching spheres: a transformation optics approach. *Phys. Rev. Lett.*, **105**:266807.

95. Fernandez-Dominguez, A. I., Luo, Y., Wiener, A., Pendry, J. B., and Maier, S. A. (2012). Theory of three-dimensional nanocrescent light harvesters. *Nano Lett.*, **12**:5946–5953.

96. Reinhard, B. M., Sheikholeslami, S., Mastroianni, A., Alivisatos, A. P., and Liphardt, J. (2007). Use of plasmon coupling to reveal the dynamics of DNA bending and cleavage by single EcoRV restriction enzymes. *Proc. Natl. Acad. Sci. USA*, **104**:2667–2672.

97. Khlebtsov, B., Melnikov, A., Zharov, V., and Khlebtsov, N. (2006). Absorption and scattering of light by a dimer of metal nanospheres: comparison of dipole and multipole approaches. *Nanotechnology*, **17**:1437–1445.

98. Scholl, J. A., Koh, A. L., and Dionne, J. A. (2012). Quantum plasmon resonances of individual metallic nanoparticles. *Nature*, **483**:421–427.

99. Zuloaga, J., Prodan, E., and Nordlander, P. (2010). Quantum plasmonics: optical properties and tunability of metallic nanorods. *Acs Nano*, **4**:5269–5276.

100. Marinica, D. C., Kazansky, A. K., Nordlander, P., Aizpurua, J., and Borisov, A. G. (2012). Quantum plasmonics: nonlinear effects in the field enhancement of a plasmonic nanoparticle dimer. *Nano Lett.*, **12**:1333–1339.

101. Esteban, R., Borisov, A. G., Nordlander, P., and Aizpurua, J. (2012). Bridging quantum and classical plasmonics with a quantum-corrected model. *Nat. Commun.*, **3**.

102. Luk'yanchuk, B., et al. (2010). The Fano resonance in plasmonic nanostructures and metamaterials. *Nat. Mater.*, **9**:707–715.

103. Ross, B. M., Wu, L. Y., and Lee, L. P. (2011). Omnidirectional 3D nanoplasmonic optical antenna array via soft-matter transformation. *Nano Lett.*, **11**:2590–2595.

104. Chi-Cheng, F., et al. (2010). Bimetallic nanopetals for thousand-fold fluorescence enhancements. *Appl. Phys. Lett.*, **97**:203101–203103.

105. Wang, H. H., et al. (2006). Highly Raman-enhancing substrates based on silver nanoparticle arrays with tunable sub-10 nm gaps. *Adv. Mater.*, **18**:491–495.

106. Aubry, A., et al. (2010). Plasmonic light-harvesting devices over the whole visible spectrum. *Nano Lett.*, **10**:2574–2579.
107. Hentschel, M., et al. (2010). Transition from isolated to collective modes in plasmonic oligomers. *Nano Lett.*, **10**:2721–2726.
108. Rong, G., Wang, H., and Reinhard, B. M. (2010). Insights from a nanoparticle minuet: two-dimensional membrane profiling through silver plasmon ruler tracking. *Nano Lett.*, **10**:230–238.
109. Jun, Y. -W., et al. (2009). Continuous imaging of plasmon rulers in live cells reveals early-stage caspase-3 activation at the single-molecule level. *Proc. Natl. Acad. Sci. USA*, **106**:17735–17740.
110. Lakowicz, J. R. (2006). Plasmonics in biology and plasmon-controlled fluorescence. *Plasmonics*, **1**:5–33.
111. Anger, P., Bharadwaj, P., and Novotny, L. (2006). Enhancement and quenching of single-molecule fluorescence. *Phys. Rev. Lett.*, **96**: 113002.
112. Szmacinski, H., and Lakowicz, J. R. (2008). Depolarization of surface-enhanced fluorescence: An approach to fluorescence polarization assays. *Anal. Chem.*, **80**:6260–6266.
113. Fu, C. -C., et al. (2009). Tunable nanowrinkles on shape memory polymer sheets. *Adv. Mater.*, **21**:4472–4476.
114. Fu, C. C., et al. (2010). Bimetallic nanopetals for thousand-fold fluorescence enhancements. *Appl. Phys. Lett.*, **97:**203101.
115. Garcia-Parajo, M. F. (2008). Optical antennas focus in on biology. *Nat. Photonics*, **2**:201–203.
116. Kawata, S., Inouye, Y., and Verma, P. (2009). Plasmonics for near-field nano-imaging and superlensing. *Nat. Photonics*, **3**:388–394.
117. Kinkhabwala, A., et al. (2009). Large single-molecule fluorescence enhancements produced by a bowtie nanoantenna. *Nat. Photonics*, **3**:654–657.
118. Sanchez, E. J., Novotny, L., and Xie, X. S. (1999). Near-field fluorescence microscopy based on two-photon excitation with metal tips. *Phys. Rev. Lett.*, **82**:4014–4017.
119. Levene, M. J., et al. (2003). Zero-mode waveguides for single-molecule analysis at high concentrations. *Science*, **299**:682–686.
120. Muhlschlegel, P., Eisler, H. J., Martin, O. J. F., Hecht, B., and Pohl, D. W. (2005). Resonant optical antennas. *Science*, **308**:1607–1609.
121. Kneipp, K. Moskovits, M., and Kneipp, H. (2006). *Surface-Enhanced Raman Scattering: Physics and Applications* (Springer, Berlin).

122. Nie, S. M., and Emery, S. R. (1997). Probing single molecules and single nanoparticles by surface-enhanced Raman scattering. *Science*, **275**:1102–1106.

123. Kneipp, K., et al. (1997). Single molecule detection using surface-enhanced Raman scattering (SERS). *Phys. Rev. Lett.*, **78**:1667–1670.

124. Oh, Y. J., et al. (2011). Beyond the SERS: Raman enhancement of small molecules using nanofluidic channels with localized surface plasmon resonance. *Small*, **7**:184–188.

125. Huber, D. L., Manginell, R. P., Samara, M. A., Kim, B. I., and Bunker, B. C. (2003). Programmed adsorption and release of proteins in a microfluidic device. *Science*, **301**:352–354.

126. Lu, Y., Liu, G. L., and Lee, L. P. (2005). High-density silver nanoparticle film with temperature-controllable interparticle spacing for a tunable surface enhanced Raman scattering substrate. *Nano Lett.*, **5**:5–9.

127. Wang, M., Jing, N., Chou, I. H., Cote, G. L., and Kameoka, J. (2007). An optofluidic device for surface enhanced Raman spectroscopy. *Lab Chip*, **7**:630–632.

128. Tong, L., Righini, M., Ujue Gonzalez, M., Quidant, R., and Kall, M. (2009). Optical aggregation of metal nanoparticles in a microfluidic channel for surface-enhanced Raman scattering analysis. *Lab Chip*, **9**:193–195.

129. Choi, D., Kang, T., Cho, H., Choi, Y., and Lee, L. P. (2009). Additional amplifications of SERS via an optofluidic CD-based platform. *Lab Chip*, **9**:239–243.

130. Liu, G. L., and Lee, L. P. (2005). Nanowell surface enhanced Raman scattering arrays fabricated by soft-lithography for label-free biomolecular detections in integrated microfluidics. *Appl. Phys. Lett.*, **87:**074101.

131. Hayazawa, N., Saito, Y., and Kawata, S. (2004). Detection and characterization of longitudinal field for tip-enhanced Raman spectroscopy. *Appl. Phys. Lett.*, **85**:6239–6241.

132. Knoll, B., and Keilmann, F. (1999). Near-field probing of vibrational absorption for chemical microscopy. *Nature*, **399**:134–137.

133. Hartschuh, A., Sanchez, E. J., Xie, X. S., and Novotny, L. (2003). High-resolution near-field Raman microscopy of single-walled carbon nanotubes. *Phys. Rev. Lett.*, **90**.

134. Giljohann, D. A., et al. (2010). Gold nanoparticles for biology and medicine. *Angew. Chem. Int Ed. Engl.*, **49**:3280–3294.

135. Boisselier, E., and Astruc, D. (2009). Gold nanoparticles in nanomedicine: preparations, imaging, diagnostics, therapies and toxicity. *Chem. Soc. Rev.*, **38**:1759–1782.

136. Sperling, R. A., Rivera gil, P., Zhang, F., Zanella, M., and Parak, W. J. (2008). Biological applications of gold nanoparticles. *Chem. Soc. Rev.*, **37**:1896–1908.
137. Davis, M. E., Chen, Z., and Shin, D. M. (2008). Nanoparticle therapeutics: an emerging treatment modality for cancer. *Nat. Rev. Drug Discovery.* **7**:771–782.
138. Govorov, A. O., and Richardson, H. H. (2007). Generating heat with metal nanoparticles. *Nano Today*, **2**:30–38.
139. Qin, Z. P., and Bischof, J. C. (2012). Thermophysical and biological responses of gold nanoparticle laser heating. *Chem. Soc. Rev.*, **41**:1191–1217.
140. Baffou, G., and Rigneault, H. (2011). Femtosecond-pulsed optical heating of gold nanoparticles. *Phys. Rev. B*, **84**.
141. Baffou, G., Quidant, R., and de Abajo, F. J. G. (2010). Nanoscale control of optical heating in complex plasmonic systems. *Acs Nano*, **4**:709–716.
142. Chen, C. C., et al. (2006). DNA-gold nanorod conjugates for remote control of localized gene expression by near infrared irradiation. *J. Am. Chem. Soc.*, **128**:3709–3715.
143. Lee, S. E., Liu, G. L., Kim, F., and Lee, L. P. (2009). Remote optical switch for localized and selective control of gene interference. *Nano Lett.*, **9**:562–570.
144. Grier, D. G. (2003). A revolution in optical manipulation. *Nature*, **424**:810–816.
145. Kim, J. (2012). Joining plasmonics with microfluidics: from convenience to inevitability. *Lab Chip*, **12**:3611–3623.
146. Anikeeva, P., and Deisseroth, K. (2012). Photothermal genetic engineering. *ACS nano*, **6**:7548–7552.
147. Levskaya, A., et al. (2005). Engineering Escherichia coli to see light - These smart bacteria 'photograph' a light pattern as a high-definition chemical image. *Nature*, **438**:441–442.
148. Levskaya, A., Weiner, O. D., Lim, W. A., and Voigt, C. A. (2009). Spatiotemporal control of cell signalling using a light-switchable protein interaction. *Nature*, **461**:997–1001.
149. Jenkins, M. W., et al. (2010). Optical pacing of the embryonic heart. *Nat. Photonics*, **4**:623–626.
150. Camsund, D., Lindblad, P., and Jaramillo, A. (2011). Genetically engineered light sensors for control of bacterial gene expression. *Biotechnol. J.*, **6**:826–836.

151. Tabor, J. J., Levskaya, A., and Voigt, C. A. (2011). Multichromatic control of gene expression in escherichia coli. *J. Mol. Biol.*, **405**:315–324.
152. Yanik, M. F., Rohde, C. B., and Pardo-Martin, C. (2011). Technologies for micromanipulating, imaging, and phenotyping small invertebrates and vertebrates. *Annu. Rev. Biomed. Eng.*, Vol. 13 (eds. Yarmush, M. L., Duncan, J. S., and Gray, M. L.) 185–217.
153. Yizhar, O., Fenno, L. E., Davidson, T. J., Mogri, M., and Deisseroth, K. (2011). Optogenetics in Neural Systems. *Neuron*, **71**:9–34.
154. Psaltis, D., Quake, S. R., and Yang, C. H. (2006). Developing optofluidic technology through the fusion of microfluidics and optics. *Nature*, **442**:381–386.
155. Monat, C., Domachuk, P., and Eggleton, B. J. (2007). Integrated optofluidics: A new river of light. *Nat. Photonics*, **1**:106–114.
156. Fainman, Y., Lee, L., Psaltis, D., and Yang, C. (2009). Optofluidics: Fundamentals, Devices, and Applications (McGraw-Hill Professional, 1st edition).
157. Eftekhari, F., et al. (2009). Nanoholes as nanochannels: flow-through plasmonic sensing. *Anal. Chem.*, **81**:4308–4311.
158. Kang, T., Hong, S., Choi, Y., and Lee, L. P. (2010). The effect of thermal gradients in SERS spectroscopy. *Small*, **6**:2649–2652.
159. van den Berg, A., Craighead, H. G., and Yang, P. D. (2010). From microfluidic applications to nanofluidic phenomena. *Chem. Soc. Rev.*, **39**:899–900.
160. Svoboda, K., and Yasuda, R. (2006). Principles of two-photon excitation microscopy and its applications to neuroscience. *Neuron*, **50**:823–839.
161. Melancon, M. P., Zhou, M., and Li, C. (2011). Cancer theranostics with near-infrared light-activatable multimodal nanoparticles. *Acc. Chem. Res.*, **44**:947–956.
162. Arfken, G., and Weber, H. J. (2005). *Mathematical Methods for Physicists*, 6th ed., (Academic, Orlando).

Chapter 9

Label-Free Detection and Measurement of Nanoscale Objects Using Resonance Phenomena

Ş. K. Özdemir, L. He, W. Kim, J. Zhu, F. Monifi, and L. Yang
Department of Electrical and Systems Engineering, Washington University, St. Louis, Missouri 63130, USA
ozdemir@ese.wustl.edu

Detection and characterization of natural and synthesized nano-objects are crucial for a variety of applications, including environmental monitoring and diagnostics, detection of individual pathogens, airborne viruses and bio/chemical warfare agents, and engineering the composition, size and shape of nano-objects for use in optoelectronics, medicine and biology. Due to their small size and low refractive index contrasts with the surrounding medium, most of these nanoscale objects have very low polarizabilities, which lead to weak light–matter interactions, hindering their label-free detection using optical methods. Micro- and nanophotonic devices with their small mode volumes and high quality factors provide a platform to significantly enhance the light–matter interactions, enabling the detection and measurement of nano-objects with unprecedented resolution and accuracy. In this chapter, we will first briefly review the resonance phenomena used for sensing

Understanding Biophotonics: Fundamentals, Advances, and Applications
Edited by Kevin K. Tsia

ISBN 978-981-4411-77-6 (Hardcover), 978-981-4411-78-3 (eBook)
www.panstanford.com

applications and then focus on the current status of microresonator-based detection of synthetic and biological nanoparticles, and discuss our results obtained using whispering gallery mode passive and active toroidal shaped microresonators.

9.1 Introduction and Overview

Interest in the development of new sensing technologies has been largely driven by recent developments in three different application areas. The first of these areas grew out of a number of advancement in nanotechnology that led to an ever-increasing interest in synthesizing nanoscale objects with various shapes, structures and functionalities. Over the last decade, newly synthesized nanoparticles (1–100 nm in diameter) have found many applications in areas ranging from biomedical diagnostics and therapeutics to optoelectronics, actuators, and biosensors [1–10]. For example, gold nanoparticle based biosensors have been developed, and silver nanoparticles have been used for solar cell applications. At the same time, massive quantities of nanoparticles are produced as end/byproducts of industrial processes such as particulate emission from combustion of fossil fuels. With the increasing presence of nanoparticles in our daily lives, there is a growing interest in assessing the many benefits of nanomaterials and their potential risks and toxicological effects [11–13]. It has been shown that nanoparticles with sizes in the range of 1–100 nm can enter cells altering signaling processes essential for basic cell functions, 1–40 nm can enter the cell nucleus, and 1–35 nm can pass through the blood–brain barrier and enter the brain. Sizes of particles can vary depending on the relative humidity (RH), that is, the particles can grow in size by absorbing water with the increase in RH or can shrink in size when water evaporates with the decrease of RH [14]. Size dependency of these changes becomes significantly more pronounced for particles with dry sizes below 20–30 nm. Such humidity dependent changes modify deposition pattern of inhaled particles in the humid human respiratory system leading to direct effects on human health. Thus, accurate size measurement of nanoparticles prior to exposure to living cells or animals will contribute to understanding their size dependent toxicity, to tailoring their size for target cells or tissues, or to designing systems for exposure prevention. Similarly, atmospheric aerosol particles affect the radiation balance by modifying the reflection, absorption, and emission of electromagnetic radiation via alterations in particle concentrations and optical properties, and by aerosol effects on cloud properties due to changes in the concentration of cloud condensation nuclei [15]. All of these effects depend on the shape, size, chemical composition, and water content of individual particles [14–22]. Therefore, accurate sizing of nanoparticles and aerosols will also help to identify their benefits and risks on technology, climate, environment, and cloud formation.

The second area grew out of advancements in medicine, molecular biology, and biochemistry which elucidated the pathways for complex diseases leading to the discovery of associated biomarkers for diagnostic and prognostic purposes. Identifying these biomarkers in body fluids can provide a powerful tool for early detection of disease, monitoring the effects of prescribed treatments, and developing targeted medicines [23–27]. Detection of biomarkers requires sensing platforms with high sensitivity, very low limits of detection, low false alarm rates, and high specificity.

Finally, the third area relates to the need for rapid detection, measurement, and identification of pathogenic threats from air-borne contaminants, pandemic viruses and bacteria, hazardous and volatile chemicals, and bio/chemical warfare agents. The ability to detect, assess, and identify such threats is vital to protect the public, as this will help design response strategies for handling the risks and searching for appropriate medical treatment in a timely manner [28–31]. Especially, detection and identification of viruses are very challenging due to their very small sizes which lie in the range of 1–200 nm.

A necessary first step in these endeavors is to establish reliable techniques for detecting a single nanoparticle, regardless of whether it is synthetic, biological, organic or inorganic, and measuring its size in various media including air, serum, buffer solution, and various aquatic environments. Microscopy and spectroscopy techniques such as confocal, near-field optical microscopy, Raman spectroscopy, fluorescence microscopy, and photothermal imaging have played central roles in single-nanoparticle/molecule detection [32–37]; however, their widespread use is limited by bulky and expensive instrumentation, long processing times, or specific labeling of the particles. In the last decade, a variety of label-free techniques based on electrical conductance measurements [38, 39], light scattering and interferometric methods [40–45], surface plasmon resonance [46–48], nanomechanical [49–55] and optical resonators [56–63] with single particle, virus, or molecule resolution have been demonstrated. Among these techniques micro-/nanosized photonic and electromechanical resonators are emerging as front-runners due to their immense susceptibility to perturbations in their environment which enhances sensitivity and resolution.

In this chapter, we will give a brief overview of resonance phenomena-based sensing schemes with the ability to detect and measure single particles, viruses, and molecules, focusing primarily on optical whispering gallery mode resonators but also inclusive of nanoelectromechanical resonators and surface Plasmon techniques. The goal is to introduce various techniques and approaches and discuss how well they address the challenges of detection and measurement with single-particle resolution. We will give a survey of recent developments with the metrics that are used to characterize their sensing performances.

9.2 Quality Factor and Mode Volume of a Resonator

Resonant structures, in particular optical and mechanical resonators, have been under intensive investigation for label-free single-particle/molecule detection. An important figure-of-merit for any resonant structure is the quality factor Q. The Q factor of a resonator is expressed as the ratio of the energy stored in the resonator to the amount of energy dissipated during each cycle of oscillation:

$$Q = \omega \frac{\text{Stored Energy}}{\text{Dissipated Power}} = \omega\tau \tag{9.1}$$

where ω and τ are the resonance angular frequency and lifetime of the oscillation until it is completely decayed. A resonator with a high Q has low loss and long decay time, and consequently it has narrower spectral linewidth, i.e., $Q = \omega/\Delta\omega = \lambda/\Delta\lambda$ where λ is the resonance wavelength, and $\Delta\omega$ and $\Delta\lambda$ are the linewidths of the resonance in the frequency and wavelength domains, respectively. For optical resonators, Q and hence τ describes the photon lifetime of the mode with frequency ω. A resonator mode with $Q = 10^{10}$ at wavelength $\lambda = 1.55$ μm allows a photon lifetime of $\tau \sim 82$ ns, thus providing a candidate medium for light storage. Such a long photon lifetime effectively increases the interaction of the cavity field with the surrounding medium making the resonators attractive sensor platforms. Moreover, longer photon life time enables the build-up of significant circulating power in the cavity with low pump powers. In optical resonators, material absorption in the resonator and the surrounding, scattering, tunneling, and coupling losses are among the main contributions to loss mechanism [64].

Various high-Q optical resonators have been fabricated, including whispering-gallery-mode (WGM) cavities, micro-post cavities, and one-, two-, and three-dimensional photonic-crystal (PhC) cavities. For WGM silica microsphere and microtoroid resonators, $Q > 10^9$ [65, 66] and $Q > 10^8$ [59, 67–69] at the $\lambda = 1.5$ μm spectral band have been reported, respectively. Quality factors of microring and microdisk WGM resonators are usually of the order of 10^6 [70–72]. Savchenkov et al. have reported a record high Q value of $Q > 2 \times 10^{10}$ at $\lambda = 1.319$ μm for calcium fluoride [73]. Same authors also reported $Q > 2 \times 10^8$ and $Q > 10^9$ for lithium niobate and lithium tantalite resonators, respectively [73]. Photonic crystal cavities with Q values as high as 10^6 have been reported. Takahashi et al. achieved $Q = 2.5 \times 10^6$ in a two-dimensional PhC cavity [74]. An array of PhC resonators having $Q \approx 10^6$ has been reported for a width-modulated line-defect cavity in a two-dimensional triangular-lattice air-hole photonic crystal [75].

In mechanical resonators, on the other hand, clamping losses and damping induced by the surrounding environment in which the resonator is placed strongly affect the their Q [50, 76]. These resonators perform best in a vacuum, because energy dissipated by a resonant mechanical resonator is very small in a vacuum; thus, the vibration response is sharply peaked at the resonant frequency, leading to higher Q. When these resonators are operated in air or in fluidic environment, their performance is strongly degraded.

Damping becomes more severe when the resonators are placed in fluidic environments, leading to significant decrease in the Q factor due to viscous damping (i.e., Q factors are usually of the order of 200 in fluidic environment). Burg et al. report that the Q of a vacuum-sealed cantilever resonator increases to 400 from its value of 85 in air [50].

Another important parameter for a resonator is the mode volume V_{mode} which quantifies how much the resonator field is confined. Mode volume V_{mode} of a resonator mode is dependent on the distribution of the resonant field and hence on the particular geometry of the resonator. In general it is defined as [64]

$$V_{\text{mode}} = \frac{\int_V \epsilon\left(\vec{r}\right) \left|E_{\text{mode}}\left(\vec{r}\right)\right|^2 d^3\vec{r}}{\max\left\{\epsilon\left(\vec{r}\right) \left|E_{\text{mode}}\left(\vec{r}\right)\right|^2\right\}} \tag{9.2}$$

where $\epsilon\left(\vec{r}\right) = n^2\left(\vec{r}\right)$ denotes the variation of the refractive index in space, $E_{\text{mode}}\left(\vec{r}\right)$ represents the electric field of the cavity mode, and the integration is carried out over all space including the regions where the field is evanescent. A smaller V_{mode} implies that the optical energy is tightly confined in a very small physical volume. Smaller V_{mode} with a high Q leads very high energy density in the cavity which is crucial for nonlinear optics and quantum optics where strong interactions between matter and light are required. Thus, in applications where strong light–matter interactions are required, one should design resonators with maximal Q/V_{mode}, that is a Q as high as possible and a V_{mode} as small as possible. However, mode volume cannot be made arbitrarily small without sacrificing Q as the shrinking device size leads to increase in the radiation losses which in turn decreases Q. For WGM mode resonators, mode volumes are of the order of tens of cubic micrometer whereas for PhC resonators, mode volumes are of the order of cubic nanometer. The PhC resonators have achieved very small mode volumes because light confinement in the photonic bandgap is effective at the wavelength scale.

9.3 Surface Plasmon Resonance Sensors

A surface plasmon (SP) wave is a charge density oscillation that takes place at a metal–dielectric interface leading to a resonance, surface plasmon resonance (SPR), when the energy of the p-polarized light incident to the interface is absorbed by the surface electrons of the metal [77–79]. Excited SP polaritons (SPP) propagate along the interface, and decay evanescently in directions normal to the interface. While the propagation distance along the interface is of the order of a few tens of micrometers, the propagation in the normal direction reaches a few hundreds of nanometers enabling subwavelength evanescent confinement of electric field in directions normal to the interface. This, in turn, leads to enhancement of the field at the interface. SPR phenomenon relies on the collective oscillation of free electrons at the interface of a metal and a dielectric, and thus any physical factor that perturbs the plasmon dispersion relations significantly affects the resonance condition, making SPs excellent probes for surface sensing applications. Consequently, variations in the refractive indexes and thicknesses of materials (metal and dielectric layers, recognition layers, liquid or gas environment, etc.) forming the SPR configuration generate significant changes in the SPR spectrum. This has enabled the construction of highly sensitive platforms for characterizing and quantifying biomolecular interactions, detecting chemical and biological species, monitoring analytes related to food and environmental safety, as well as for measuring humidity, pressure, and temperature [80–85].

SPR can be excited if the wave vector $\mathbf{k_{inc}}$ of the incident light has an interface-parallel component $\mathbf{k_x}$ which matches the surface plasmon wavevector $\mathbf{k_{sp}}$, i.e., $\mathbf{k_x} = \mathbf{k_{sp}}$ This, however, cannot be satisfied with light directly incident from air onto the smooth metal surface, because energy and momentum conservation cannot be simultaneously satisfied. In order to overcome this problem, the wavevector of the incident light can be increased to match that of the SP by using couplers employing attenuated total reflection or diffraction. Prism coupling in Otto and Kretschmann configurations, waveguide coupling and grating coupling are the most commonly used techniques to excite SP waves (Fig. 9.1) [77–80]. The SP wavevector at metal–dielectric interface is given as

$$\mathbf{k_{sp}} = \frac{2\pi}{\lambda}\sqrt{\frac{\epsilon_d \epsilon_m}{\epsilon_d + \epsilon_m}} \tag{9.3}$$

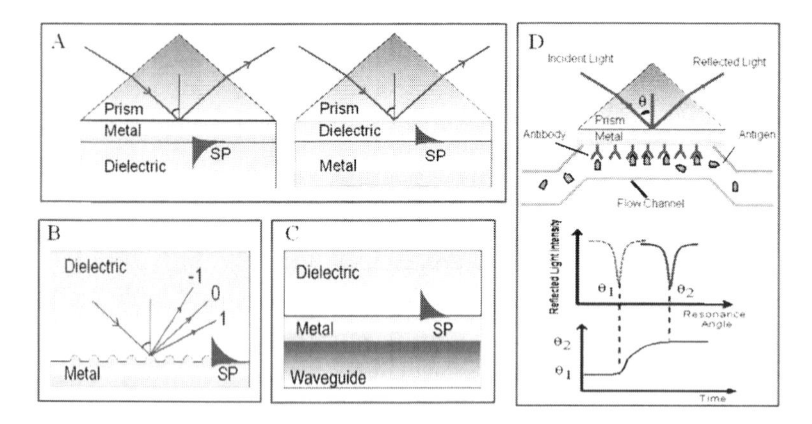

Figure 9.1 Illustration of surface Plasmon coupling schemes and biosensing methodology. (A) Prism coupling with Kretschmann (left) and Otto (right) configuration, (B) grating coupler, and (C) waveguide coupler for surface Plasmon wave excitation. (D) Biosensing using SPR. Metal surface is functionalized by specific Antibodies. Binding of antigens to antibodies shifts the SPR angle.

where ϵ_d and ϵ_m are the dielectric permittivities of the dielectric and metal layers, respectively, and λ is the wavelength in vacuum. It should be noted that dielectric permittivity of metals have a significant imaginary component leading to non-zero imaginary component for $\mathbf{k_{sp}}$ and hence attenuation in the direction of propagation. It is clear that any change in ϵ_d and ϵ_m will affect the $\mathbf{k_{sp}}$ A light field with wave vector $\mathbf{k_{inc}}$ incident at the interface with an angle θ through a prism coupler with refractive index n_p has the parallel component $\mathbf{k}_x = \mathbf{k}_0 n_p \sin\theta = \frac{2\pi}{\lambda} n_p \sin\theta$. Thus in order to excite the SPR, real part of $\mathbf{k_{sp}}$, $\Re(\mathbf{k_{sp}})$, should be equal to $\mathbf{k_x}$, that is

$$\Re\left(\mathbf{k_{sp}}\right) = \frac{2\pi}{\lambda} n_p \sin\theta \tag{9.4}$$

should be satisfied. This, in turn, shows that one can satisfy phase matching conditions by changing either the incidence angle θ (i.e, angular interrogation, AI) while keeping the wavelength λ of the incident light constant or by changing the λ of the incident light (spectral interrogation, SI) while keeping the θ constant. The angle or the wavelength at which a resonance dip is observed in the reflected light intensity is referred to as the resonance angle θ_{spr}

or the resonance wavelength λ_{spr}. In the case of a grating coupler, the phase matching condition can be written as $\mp\Re\left(\mathbf{k_{sp}}\right) = \frac{2\pi}{\lambda} n_p \sin\theta + m\frac{2\pi}{\Lambda}$ where Λ is the grating period and m is an integer denoting the diffraction order.

An SPR sensor is usually composed of three interfaces: a coupler (e.g., prism, grating, waveguide), a metal layer, a recognition layer to enable specific response, and the surrounding medium (e.g., fluid, air, or gas). Binding of a specific analyte to the recognition layer leads to change in the resonance angle or resonance wavelength, which is observed using AI or SI methods. SPR sensor platforms capable of simultaneous detection of multiple analytes have been developed using either prism couplers or waveguide schemes. Since the pioneering work of Rothenhausler and Knoll on SP microscopy [86], detection and characterization of biomolecules, particles, and viruses using SP imaging techniques, which rely on the scattering of plasmon waves by the nanoscale objects, have been the focus of increasing interest and have been frequently employed [46, 86–91]. In most of the realizations, the incident angle and the wavelength of the light are kept constant and a sample is interrogated by recording the reflected light intensity as a function of position. Recently, Wang et al. have achieved the label-free imaging, detection, and mass/size measurement of single viral particles with a mass detection limit of approximately 1 attogram using SPR microscopy (Fig. 9.2) [46]. The same researchers further extended the use of their SP microcopy to study the charge and Brownian motion of single particles [92].

Advancements in nanotechnology have enabled the fabrication and controlled manipulation of metallic nanoparticles of various sizes, shapes, and materials. This has led to the emergence of new detection capabilities based on the excitation of surface plasmons in metallic nanoparticles, which exhibit sharp spectral extinction peaks due to resonant excitation of their free electrons at visible and near-infrared frequencies. Contrary to the surface plasmon waves travelling at the interface of a metal–dielectric interface, plasmons excited as a result of the interaction of light with subwavelength metallic nanoscale objects or particles are highly localized and oscillate around the particle. Such plasmons are referred to as localized surface plasmons (LSPs) [93–96] (Fig. 9.3). Since LSPs are sensitive to changes in the local dielectric

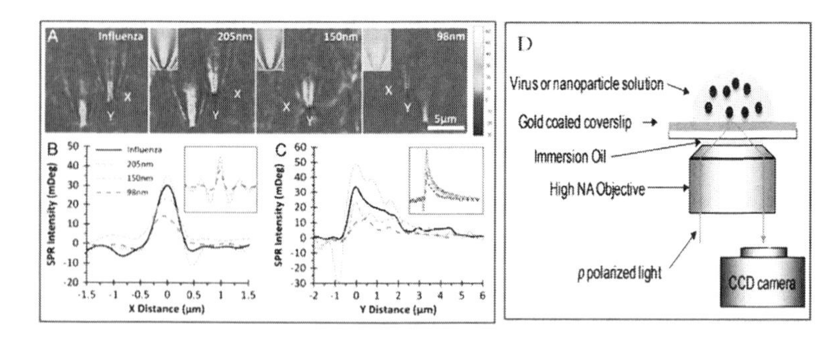

Figure 9.2 SPR image of H1N1 Influenza A virus and silica nanoparticles of various sizes in PBS buffer (A) and SPR intensity profiles along the X (B) and Y (C) directions together with the results of numerical simulations given in the insets. The schematic illustration of the SP microscopy used in the measurements. Adapted by permission from Ref. [46].

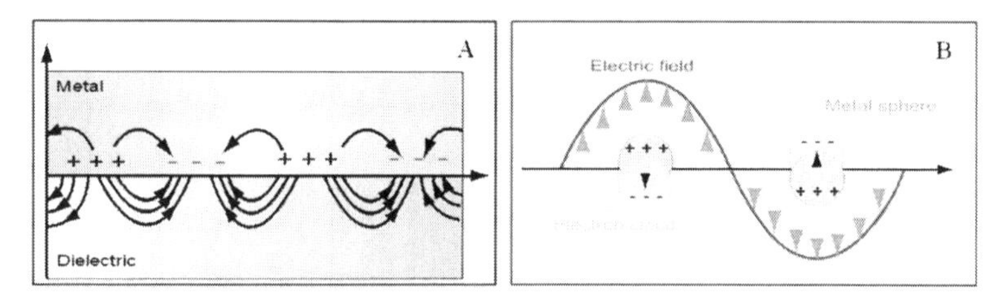

Figure 9.3 Illustration of the phyiscal mechanism for (A) surface Plasmon polariton corresponding to propagating plasmons, and (B) localized plasmons.

environment, and shape and size of the metallic particle, any change in these parameters reflects itself as a change in the localized surface Plasmon resonance (LSPR) wavelength. Various chemical and biological sensing platforms have been designed and realized using LSPR of metallic nanoparticles [97–100]. Since the LSP mode volume, which defines the extension of the localized plasmon into the surrounding, is much smaller than that of SPR, it is a better method to study the short-range perturbations, e.g., the changes due to adsorption of a molecular layer. In addition to sensing applications, LSPR phenomenon has led to the development of new technologies related to molecular rulers [101–103] and bio-imaging agents [104–107].

LSPs have been studied using nanoparticles of various shapes and geometries ranging from disks, triangles, pyramids, spheres, and cubes to crescents, stars, rings, and nanoshells [108–117]. Silver and gold nanoparticles have been shown to have refractive index sensitivities of 200 nm/RIU, and a size-, shape-, and concentration-dependent sensing volume [118, 119]. LSPR of single silver nanoparticles have been reported to undergo an LSPR shift of 40.7 nm upon adsorption of monolayer of hexadecanethiol molecules (fewer than 60,000 molecules), implying a zeptomole sensitivity [120], Larsson et al. have achieved real-time label-free monitoring of proten binding via molecular recognition and observed the LSPR shift induced by the binding of monolayers (~360 proteins) of biotin-BSA to gold nanorings. They have achieved sensitivities as high as 880 nm/RIU [121]. Dondapati et al. have used biotin-modified single gold nanostar to detect streptavidin molecules with concentrations as low as 0.1 nM, which has induced plasmon resonance shifts of approximately 2.3 nm [122].

Following the pioneering works of Haes et al. on the detection of biomarker for Alzheimer's disease using LSPR [123, 124], Vestergaard et al. have reported the detection of Alzheimer's disease biomarker, the tau protein, using an LSPR-based immunochip which was fabricated using gold-coated silica nanoparticles. They have achieved a detection limit of 10 pg/ml [125]. Mayer et al. have recently shown single-molecule LSPR detection by monitoring antibody–antigen unbinding events in the LSPR spectra of gold bipyramids [126]. In order to extract the discrete shifts of LSPR from the noisy measured spectrum, Mayer et al. performed cross-correlation between the measured data and a test function and were able to detect LSPR shifts of 0.5 nm.

It should be noted that LSPR has been the workhorse behind the recent developments in the field of surface enhanced raman spectroscopy (SERS) and other types of surface-enhanced spectroscopic methods [127–130].

9.4 Mechanical Resonators

Nanomechanical resonators have led to emerging applications in diverse fields, ranging from light modulation and quantum measure-

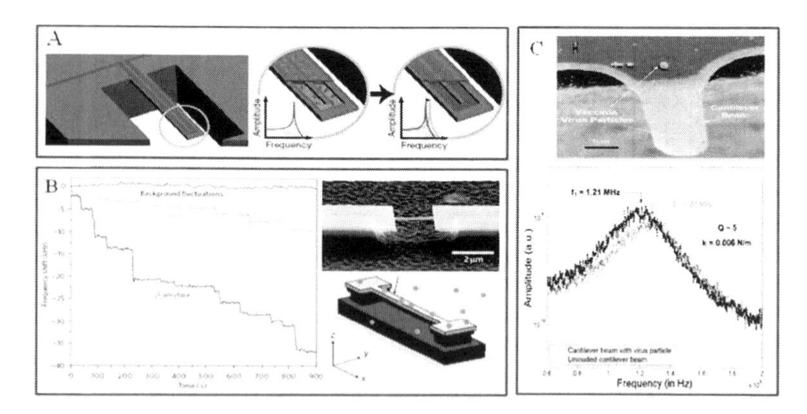

Figure 9.4 State-of-the-art nanomechanical resonators with the ability of detecting single molecules, particles, and viruses. (A) Excessive mass due to the particles brought by the fluid flowing through the suspended microchannel shifts the resonance frequency of the cantilever. The resonator has a mass of 100 ng and quality factor of 15,000 enabling highly sensitive detection of biological material in fluidic environment. Adapted by permission from Ref. [50]. (B) Real-time detection of single-molecule detection events with a doubly clamped beam nanoelectomechanical device. Each discrete jump in frequency corresponds to the binding of a single molecule. Adapted by permission from Ref. [49]. (C) A vaccinia virus deposited on the cantilever shifts the resonance frequency. Adapted by permission from Ref. [53].

ments to optical sensing and biotechnologies [94–97]. The driving force behind these important applications is the extreme sensitivity of these nanoscale devices to perturbations, which enabled the development of mass spectrometers and sensing platforms to reliably detect and analyze single cells, single molecules, and single viruses.

Nanomechanical resonators function as mass sensors with ultrahigh precision and have been able to measure the samples with masses down to 10^{-21} grams in vacuum. The most commonly used mechanical resonators are singly or doubly clamped cantilevers and beams (Fig. 9.4). The resonance frequency of a mechanical resonator is a very sensitive function of its mass. Any change of Δm in the mass m_0 of the resonator due to an adsorbed particle or absorbed chemical translate itself into a change Δf in the resonance frequency of the resonator as [50]

$$f_1 = f_0 - \delta f = \frac{1}{2\pi}\sqrt{\frac{k}{m_0 + \alpha\,\Delta m}} \quad (9.5)$$

where f_0 and f_1 are the resonance frequency of the resonator before and after the particle adsorption, respectively, k is the spring constant which is assumed to be unchanged before and after the particle adsorption, and α is a numerical constant taking into account the dependence of the process on the location of the particle on the resonator. It is clear that for a fixed $\alpha\Delta m$, the smaller the initial resonator mass is, the larger the δf is (i.e., a fixed perturbation leads to a larger frequency shift for a lighter mechanical resonator). Similarly, we see that a resonator with a higher **k** is more sensitive to mass changes than a resonator with a smaller **k**. Thus, in order to significantly improve the response of a mechanical resonator to perturbations, it is crucial to reduce the mass of the resonator while maintaining a larger spring constant, i.e., larger resonance frequency. Rearranging Eq. 9.5, one can find

$$\Delta m = \frac{k}{4\alpha\pi^2}\left(\frac{1}{f_1^2} - \frac{1}{f_0^2}\right) \quad (9.6)$$

from which the position dependent mass responsivity of mechanical resonators is clearly seen. Thus, in order to accurately estimate the mass of each adsorbed particle, one should know the resonance frequencies before and after the arrival of each particle as well as the exact location of each particle on the resonator. Returning back to Eq. 9.5 and re-arranging it, we find $f_1 = f_0/\sqrt{1 + \alpha\,\Delta m/m_0}$, which becomes $f_1 = f_0(1 - \alpha\,\Delta m/2m_0)$ for $\Delta m \ll m_0$. Consequently, we arrive at

$$\delta f = -f_0\alpha\frac{\Delta m}{2m_0} \quad (9.7)$$

Assuming that the minimum detectable resonance shift δf_{min} is of the order of the linewidth Δf of the resonance at f_0, i.e., $\delta f_{\text{min}} \approx \Delta f$, we can write $\delta f_{\text{min}} = f_0/Q$ where Q is the quality factor of the mechanical resonator. Then it is easy to see that minimum detectable mass change Δm_{min} is given by $|\Delta m_{\text{min}}| = 2m_0\,(\alpha Q)^{-1}$. We thus conclude that the lower detection limit is directly proportional to the initial mass of the resonator and inversely proportional to its quality factor. The smaller the mass of the resonator is and the higher the Q is, the smaller the lower detection limit is.

Mass estimation through monitoring the shifts in the resonance frequency of mechanical resonators has been employed to detect various particles, molecules, and viruses. A silicon nitride cantilever has been used to detect 16 *Escherichia coli* cells corresponding to a mass of $\times 10^{-12}$ g, and the detection of single vaccinia virus particles using a silicon cantilever beam has been reported [53]. Recent progress in fabrication techniques have enabled detection limits superior to these early works. Jensen et al. have reported a carbon-nanotube-based atomic resolution nanomechanical mass sensor with a mass sensitivity of 13×10^{-25} kg/$\sqrt{\text{Hz}}$ corresponding to 0.4 gold atoms /$\sqrt{\text{Hz}}$ [130]. Lassagne et al. have achieved a resolution of 1.4 zg by cooling a nanotube electromechanical resonator down to 5 K in a cryostat [131]. Naik et al. have used a suspended beam containing SiC, Al, and Ti layers to develop a single-molecule mass spectrometer which has enabled the real-time detection of individual Bovine serum albumin (BSA) and β-amylase molecules as well as gold nanoparticles with nominal radii as small as 2.5 nm [49]. Burg et al., on the other hand, achieved in weighing single gold and polystyrene nanoparticles of size $\sim$100 nm and $\sim$151 µm, single *E. coli* and *B. subtilis* bacterial cells and submonolayers of goat anti-mouse immunoglobulin-γ (IgG) bound to anti-goat IgG antibodies [101]. Burg et al. achieved this sub-femtogram resolution in water as the damping in the fluid was negligible compared to the intrinsic damping of their suspended silicon crystal resonator. Lee et al. demonstrated mass measurement in a solution with a resolution of 27 ag using suspended nanochannel resonators (SNRs), which exhibited a quality factor of approximately 8000 both in water and in dry environment [132]. They also achieved in trapping particles at the free end of the resonator using the centrifugal force caused by cantilever vibration, thus minimizing the position-dependent errors.

9.5 Photonic Crystal Resonators

Photonic crystals (PhCs) of different materials and configurations have emerged as a new class of passive and active sensors. Various sensors targeted for the detection and characterization of chemical, biological, and physical processes have been developed [133–148]. A photonic crystal is defined as structures with periodic variation of dielectric constant (refractive index) in one or more orthogonal directions. If the refractive index is varying periodically in one direction but homogeneous in the other two directions, the structure is a one-dimensional (1D) PhC. Such PhCs consist of alternating layers of two materials with different refractive indexes [149, 150]. Two-dimensional (2D) PhCs which have refractive index variation in two directions with no variation in the third direction are generally fabricated either by stacking dielectric rods in air or by drilling holes in high refractive index materials [151–153]. In 3D PhCs, refractive index varies in all three directions and they can be fabricated by orderly stacking stripes of semiconductor or dielectric materials in woodpile structures [154–156]. The periodicity creates a wavelength bandgap region through which the propagation of photons having wavelengths in this region is forbidden. If the periodicity of the structure is perturbed, for example by introducing a defect, some modes can be supported inside the band gap with a tight confinement in the defect region. Such defects can be used to build waveguides or resonators depending on their shape, size, and number. The operation wavelength or the wavelength of interaction of a PhC can be adjusted by the PhC material and the periodicity of the modulation. When the resonance condition for a specific wavelength is satisfied, light strongly couples to the PhC for a particular incidence angle, and is confined within the small mode volume, leading to the enhancement of light field within the PhC. Resonance wavelengths can be observed by illuminating the PhC at normal incidence with a source of broad spectral emission, such as light-emitting diode or white lamp, and detecting the wavelengths that are reflected back.

The ultratight confinement of the optical fields leads to nanoscale mode volumes, allowing an ultrahigh light intensity which is very sensitive to perturbations within or at the close proximity of the defect. For example, a scatterer placed in the mode volume or a molecule binding to the photonic crystal surface within the evanescent tail of the mode would effectively alter the refractive index of the PhC-environment system, leading to changes in the

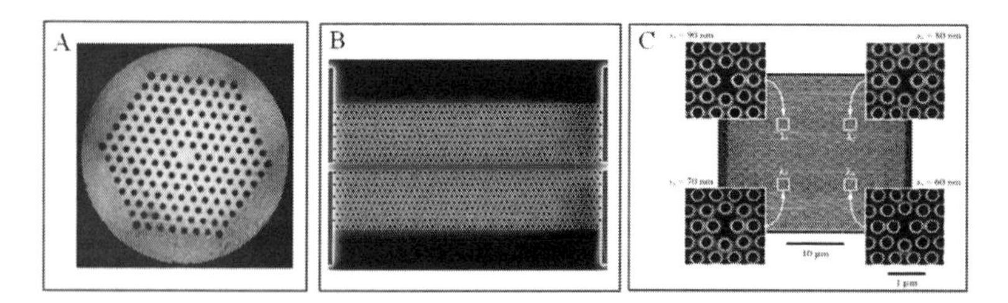

Figure 9.5 (A) End facet of a large mode area photonic crystal fiber. (B) Scanning electron microsope image of a silicon-on-insulator (SOI) planar photonic crystal waveguide, where a row of holes is removed. (C) SEM image of a 2 × 2 nanolaser array. Adapted by permission respectively from Refs. [161, 163, 172].

spectral properties of the photonic crystal. Such changes can be monitored to detect molecules, scatterers, or any other chemical or physical perturbations. Strong light confinement provided by the photonic bandgap and the flexibility of adjusting the defect mode wavelength across the photonic bandgap via fine-tuning of the structural parameters have led to increasing interests in photonic crystal based biosensors since the first demonstration of refractive index sensing using a PhC waveguide [157] and later using a PhC microcavity laser [158].

Photonic crystal structures used for biosensing applications generally involves photonic crystal fibers and slabs. PhC fibers are attractive as sensing platforms, because air holes in the fiber act as a fluidic channel to deliver biological samples, and the PhC structure ensures a strong light–matter interaction. Biomolecules, such as DNA [159] and proteins [160], have been detected with high sensitivity. Rindorf et al. have demonstrated the first detection of biomolecules using a long-period grating in a PCF (Fig. 9.5) [161]. Slab PhC sensors, on the other hand, are suitable for lab-on-chip applications. An optical biosensor based on monitoring the changes in the cut-off wavelength of a photonic crystal waveguide in response to refractive index changes was presented by Skivesen et al. [162]. Concentrations of around 0.15 μM of BSA-solution were measured with excellent signal to noise ratio.

In 2010 Jaime García-Rupérez and coworkers reported label-free antibody detection using band edge fringes in the slow-light regime near the edge of the guided band. A surface mass density detection limit below 2.1 pg/mm and a total mass detection limit below 0.2 fg were reported (Fig. 9.5) [163]. Compared to PhC microcavity based biosensors, waveguide biosensor suffers from larger sensing areas, which require larger sample volume. Lee and Fauchet demonstrated a 2D PhC microcavity biosensor that is capable of monitoring protein binding on the walls of the defect hole and quantitatively measuring the protein diameter [164]. The detection limit of this device was 2.5 fg. The authors detected the binding of glutaraldehyde and BSA by monitoring resonance wavelength, which red-shifted upon binding of the molecules. By incorporating a target delivery mechanism, the same authors improved the sensitivity down to 1 fg [165]. The performance of the device was verified by detecting single latex sphere with a diameter of 370 nm, which led to a 4 nm red-shift of the resonance [165].

Mandal et al. incorporated a 1D PhC resonator into an optofluidic biosensor and reported a detection limit on the order of 63 ag total bound mass [166]. Moreover, the authors utilized the multiplexability of their device for the concurrent detection of multiple interleukins (IL-4, IL-6, and IL-8) [166]. An optofluidic platform with a silica 1D PhC were reported by Nunes et al. with an achieved sensitivity of 480 nm/RIU and a minimum detectable refractive index change of 0.007 RIU [167]. Guo et al. have used a 1D PhC in total internal reflection geometry and demonstrated real-time detection of a wide range of molecules with molecular weights in the range of 244–150,000 Da [168]. Chan et al. incorporated photonic crystal (PC) biosensors onto standard microplates and used this platform to detect and analyze protein–DNA interactions in a high-throughput screening mode to identify compounds that prevent protein–DNA binding (Fig. 9.6) [169]. Yanik et al. have used a plasmonic nanohole array (photonic crystal like structure) in a microfluidic platform for direct detection of live viruses (vesicular stomatitis virus, pseudotyped Ebola, and Vaccinia) in biological fluid with a projected lower detection limit of approximately 10^5 PFU/mL [48].

Detection resolution of any sensor utilizing resonant structures can be improved by using high-Q resonators. This, however, causes problems in implementation as coupling light into and out of

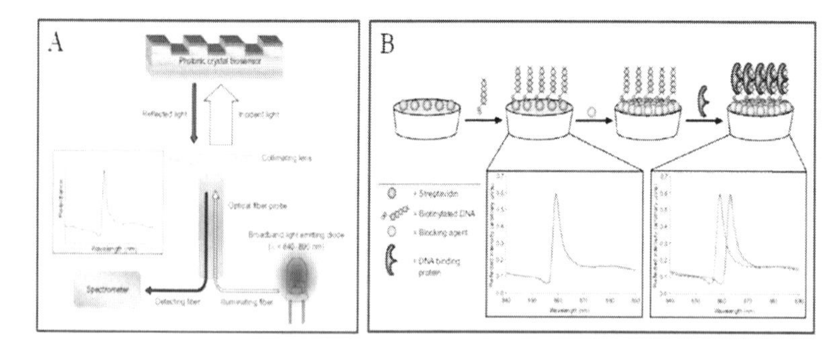

Figure 9.6 (A) Schematic of the photonic crystal biosensor developed by Chan et al. [169]. A broadband LED illuminates the biosensor, and reflected light is collected and transferred to a spectrometer. (B) Illustration showing the protein–DNA binding experiments performed with the biosensors in (A). Streptavidin-coated biosensors are used to bind biotinylated DNA oligomers. A distinct peak wavelength is observed in the reflected light. When a DNA-binding protein is introduced after the addition of starting block, the resonance wavelength peak shifts to a higher wavelength. Adapted by permission from Ref. [169].

very high-Q PhCs require high precision alignment. This can be remedied by replacing passive resonators with active ones operated in the lasing regime, creating PhC lasers which generate their own narrow linewidth emission while eliminating the need for high-precision alignment of PhCs to waveguides or tapered optical fibers. Thus, in the search for improved resolution and better detection limits, sensors, which utilize point defect nanolasers on active PhCs, were proposed and reported immediately after the early studies using passive PhC resonators for biochemical sensing. Nanocavity lasers with PhC resonators were investigated for spectroscopic tests on sample volumes down to femtoliters with a performance of roughly 1 nm spectral shift for a 0.0056 change in refractive index [158, 170]. These PhC nanolasers were then integrated in a microfluidic platform and used to detect refractive index changes as small as 0.005 [171]. A high-resolution sensor based on continuous wave PhC nanolaser with a sensitivity of 350 nm/RIU was demonstrated by Kita et al. [172]. A spectrometer-free index sensor based on the nanolaser array was also proposed (Fig. 9.5) [62]. A butt-end fiber-coupled surface emitting PhC

Γ-point band edge laser has been used as a refractive index sensor with a resolution of approximately 10^{-3} [173]. In 2010, Lu et al. proposed a PhC hetero-slab-edge microcavity design with surface mode and achieved a sensitivity of 625 nm/RIU [174]. Dündar et al. reported the sensitivity of photoluminescence from InAs quantum dots embedded PhC membrane nanocavities to refractive index changes and demonstrated a maximum sensitivity of approximately 300 nm/RIU [175].

In short, PhC passive resonators and PhC nanolasers provide an exciting tool for the manipulation of photons and have received a great deal of attention from various fields. Their high performance in detecting small changes in their lattice constants or in their effective refractive indexes and ease of integration into microfluidic platforms make PhC based systems very suitable for biological and chemical sensing.

9.6 Whispering Gallery Mode Optical Resonators

In a whispering gallery mode (WGM) resonator, light field circulates around the circular periphery of a structure with near-ideal total internal reflection (TIR) which enables extreme confinement of the field [64]. WGM resonators come in various sizes, shapes, and geometries with the common feature that there exists a curved surface which refocuses the propagating field. The propagation of the resonant light field in a WGM resonator can be intuitively understood using the arguments of geometrical optics as depicted in Fig. 9.7, where we have considered a microsphere resonator of radius R and refractive index n_{r} surrounded by an environment of refractive index $n_{\mathrm{s}} < n_{\mathrm{r}}$. For the light rays to undergo TIR at each incidence at the interface between the microsphere and the surrounding and remain trapped within the resonator, the incidence angle θ should satisfy $> \sin^{-1}(n_{\mathrm{s}}/n_{\mathrm{r}})$. If the path taken by the WGM during one roundtrip within the resonator is a multiple of the wavelength λ of the WGM light field then a resonance occurs, i.e., resonance condition is given approximately by $2\pi Rn_{\mathrm{r}} = m\lambda$ where m is an integer number. A detailed analysis of the light propagation under the TIR condition reveals that the light field trapped inside

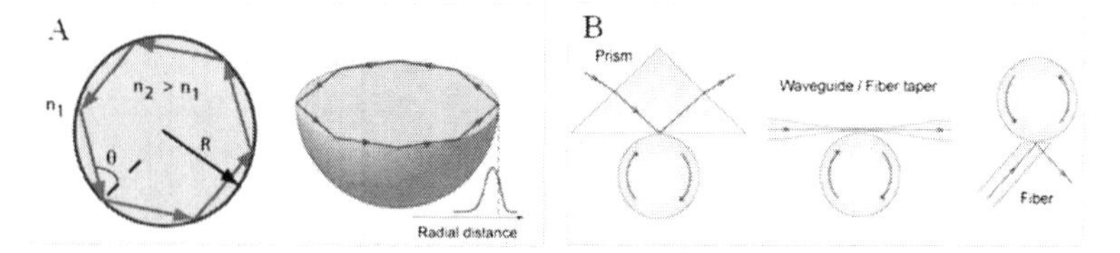

Figure 9.7 (A) Illustration of WGM in a microsphere as a light ray undergoing total internal reflection as it travels along the periphery of the dielectric sphere with radius R and refractive index n_2. The surrounding medium has refractive index of n_1 and the light ray has an incidence angle of θ at the boundary. Trapped light has an evanescent tail in the surrounding. (B) Some widely used schemes to couple light into a WGM resonator.

the resonator has an exponentially decaying evanescent tail which probes into the surrounding medium n_s. The decay length of this evanescent field is of the order of λ, making it a highly confined field. Among many various methods (e.g., prism, waveguide, fiber-taper, half-block, angle-cut fiber), waveguide and taper-fiber coupling schemes have emerged as the choices in many applications to couple light into and out of the WGM resonators (Fig. 9.7). The ideality of taper fibers for this process has been generally accepted. Coupling efficiencies as high as 99% have been achieved with taper fibers [176–178].

Evolution of the amplitude a of the field inside a WGM resonator coupled to a waveguide can be described by

$$\frac{da}{dt} = \left(i\Delta - \frac{\kappa_0 + \kappa_1}{2}\right) a - \sqrt{\kappa_1} a_{\text{in}} \tag{9.8}$$

where a_{in} is the normalized amplitude of the input field, $\kappa_0 = \omega_c / Q_0$ denotes the intrinsic damping of the resonator (i.e., due to the material, radiation, and scattering losses) with ω_c and Q_0 representing the resonance angular frequency and the intrinsic quality factor of the resonance, $\kappa_1 = \omega_c / Q_{\text{ext}}$ denotes the waveguide-resonator coupling rate with Q_{ext} being the coupling quality factor, and $\Delta = \omega - \omega_c$ is the detuning between the laser angular frequency ω and the resonance angular frequency. At steady state Eq. 9.8 becomes

$$\left(-i\Delta + \frac{\kappa_0 + \kappa_1}{2}\right) a + \sqrt{\kappa_1} a_{\text{in}} = 0 \tag{9.9}$$

Using the expression $a_{\text{out}} = a_{\text{in}} + \sqrt{\kappa_1}a$ which relates the output field a_{out} to a and a_{in}, we can write the transmission coefficient t at steady state as

$$t = 1 - \frac{2\kappa_1}{(\kappa_0 + \kappa_1) - i2\Delta} \tag{9.10}$$

from which we can express the transmission $T = |t|^2$ as

$$T = \frac{(\kappa_0 - \kappa_1)^2 + 4\Delta^2}{(\kappa_0 + \kappa_1) + 4\Delta^2} \tag{9.11}$$

Then at resonance $(\Delta = 0)$ we have

$$T = \left(\frac{1-\zeta}{1+\zeta}\right)^2 \tag{9.12}$$

with $\zeta = \kappa_1/\kappa_0$ defining a dimensionless coupling parameter. Here, $\zeta = 1$ is denoted as the critical coupling where the transmission reduces to zero, $\zeta < 1$ and $\zeta > 1$ are denoted as the under- and overcoupling, respectively. These regions are identified by the competition between different loss mechanisms [64].

At critical coupling, intrinsic losses and coupling losses are equal. In the overcoupling region, coupling losses are larger than the intrinsic losses of the cavity whereas in the undercoupling region intrinsic losses dominate. The coupling parameter is strongly dependent on the distance between the waveguide and the resonator, thus by carefully controlling this distance one can operate the waveguide-coupled resonator system in any of these coupling regions. Figure 9.8 shows the transmission versus distance between a taper-fiber and a microtoroid resonator together with the dependence of the loaded Q on the distance.

Recent developments in the fabrication technologies have made it possible to fabricate microscale WGM resonators of various shapes (Fig. 9.9) with smoother and homogeneous curved geometries, leading to significantly reduction in the number of defects and surface roughness. This, in turn, minimizes, if not eliminates, the scattering and radiation losses, leading to WGM resonators with microscale mode volumes and very high Q factors, limited mostly by the material absorption losses. The combination of highly confined field, i.e., microscale mode volume, and very high Q has made WGM resonators attractive for studying fundamental light–matter

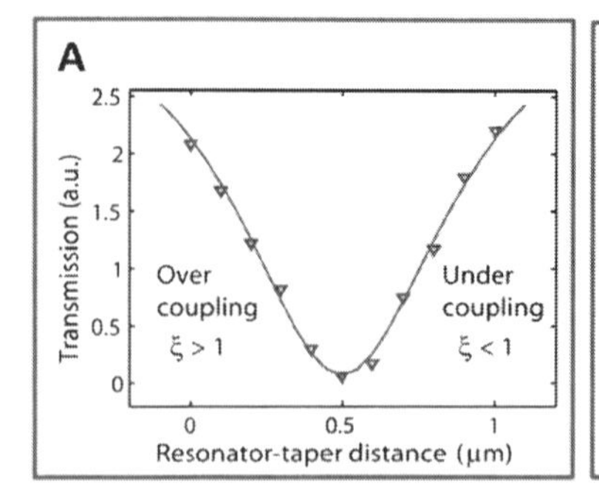

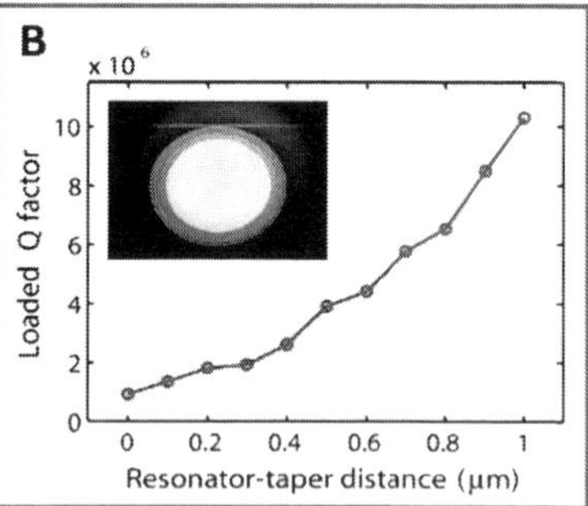

Figure 9.8 (A) Loading curve showing the dependence of transmission on the resonator-taper distance. (B) Dependence of loaded Q on the taper-resonator distance. In the overcoupling regime, coupling losses increase dominating intrinsic losses. In the deep undercoupling regime (large resonator-taper distance), coupling losses are minimized; thus, intrinsic losses of the cavity dominates. In critical coupling (i.e., cavity losses are equal to coupling losses), the transmission reduces to zero. Inset: Optical microscopy image of taper coupled microtoroid resonator.

interactions and nonlinear optics and for building optical sensing platforms with unprecedented detection capabilities.

Microdisk, microring, and microtoroid WGM resonators are fabricated on a chip using standard microfabrication technologies, and thus can be mass-produced. They have the potential to be integrated into the existing optoelectronic systems. Microtoroid has the highest reported Q value ($Q \geq 10^8$) [59, 67–69] among the on-chip resonators, and it has attracted great interest since its first report [67]. The Q values of other on-chip resonators are on the order of 10^6 in air [70–72], and degrade to 10^4 in aquatic environment. Microtoroid is fabricated from a microdisk by thermal reflow under CO_2 irradiation, which smoothens the surface eliminating the scattering losses [67]. This, in turn, allows higher Q in both air and water. In aquatic environments Q values in the range 10^6–10^8 have been demonstrated for microtoroids [57, 63, 179]. Microsphere resonators are fabricated by melting the tip of a silica fiber with flame or CO_2 irradiation [144–146]. Microspheres are formed at the tip of optical fiber stems which make them mechanically unstable. Moreover, the fabrication process is not adequate for mass fabrication. However, ease of fabrication and very

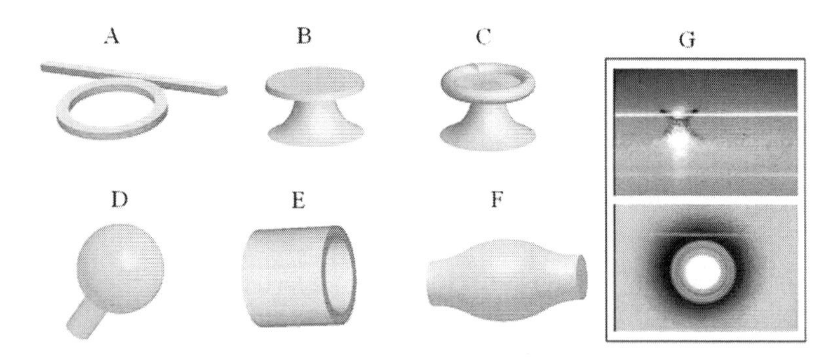

Figure 9.9 Whispering gallery mode optical resonators. (A) waveguide coupled microring, (B) microdisk, (C) microtoroid, (D) microsphere, (E) microcylinder, and (F) bottle. (G) Side and top views of a fiber-taper coupled microtoroid resonator taken with optical microscopes.

high Q values ($Q \geq 10^9$ in air and $Q \approx 10^7$ in aqueous environment) make microsphere resonators very attractive [58, 66, 180].

In biosensing applications, resonators and the couplers are immersed in water or in biological environment. The refractive indexes of these aquatic media are higher than that of the air; thus, the refractive index contrast between the resonator material and the surroundings is smaller. This causes deviation from the near-ideal TIR condition leading to increased radiation loss. The biological fluid environment also increases the material absorption losses. Thus, in liquid environment, Q is usually much smaller than its value in air. In addition, synthetic or biological particles in the surrounding attach not only to the resonator but also to the taper fiber, making the measurements difficult. Moreover, the mode volume increases in aquatic environment. Consequently Q/V, which quantifies the interaction strength between the cavity field and the matter, reduces during the transition from air to water.

Figures 9.10 and 9.11 show the effect of the surrounding and the size of the resonator on the Q and V for silica microspheres. In order to have high Q and high sensitivity in aquatic environment, the resonator should be carefully designed and fabricated. Another way of circumventing or minimizing the Q/V degradation during the transition from air to aquatic environment is to prepare the resonator from a material having refractive index higher than

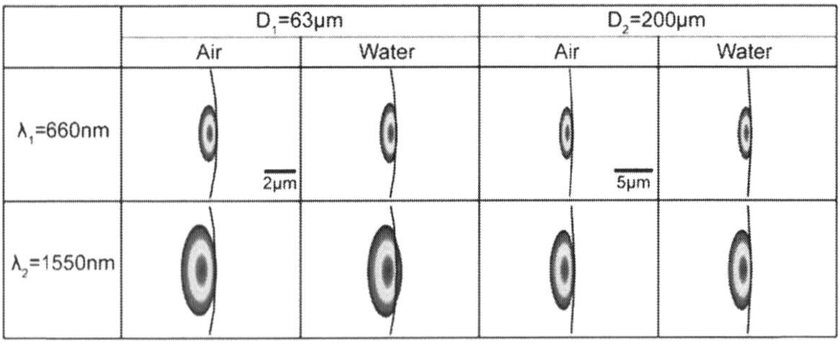

	D1=63μm		D2=200μm	
	Air	Water	Air	Water
λ1=660nm	2μm		5μm	
λ2=1550nm				

Figure 9.10 Change in mode volume at two different wavelengths λ_1 and λ_2 in air and in water for silica microspheres of two different sizes. Excerpted with permission from Ref. [180].

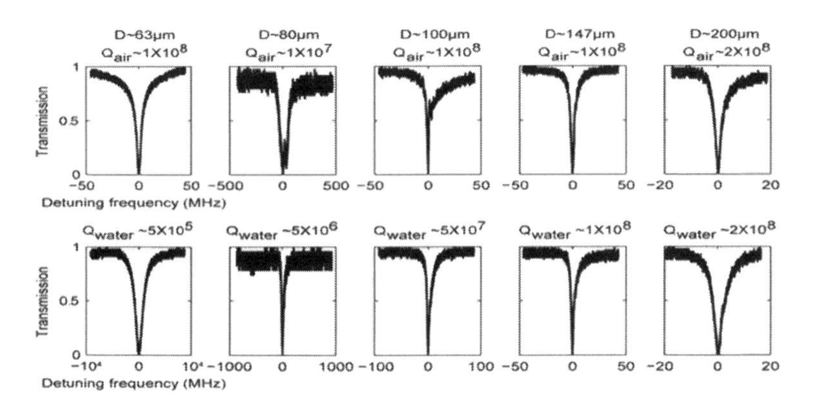

Figure 9.11 Quality factors of microspheres of various sizes in air and in water at wavelength band of 660 nm. The reduction in the quality factor for microspheres of smaller sizes is significant. Excerpted with permission from Ref. [180].

that of silica. This will allow fabricating much smaller resonators that can support WGM modes in aquatic environment. Studies on microspheres have produced some promising results, and high index material, such as chalcogenide glass, lead silicate glass, silicon, and barium titanate glass, have been used to fabricate microspheres. Quality factors greater than 2×10^4 and 2×10^6 have been reported in arsenic triselenide glass microspheres of diameters 9.2 μm [181] and larger than 55 μm [182], respectively. Resonances with *Q* factors as high as 9×10^6 have been reported in lead silicate glass microspheres of diameter $\sim$109 μm [183] Svitelskiy reported $Q = 3 \times 10^4$ for 14 μm barium titanate glass microspheres immersed in water [184].

Cylindrical liquid core resonators have emerged as promising platforms for sensing applications in aquatic environment. Such resonators, which are also referred to as the capillary-based optofluidic ring resonators, are fabricated by pulling glass preform under heat flame or CO_2 irradiation [185]. These resonators have quality factors of the order of 10^6. They have the distinct advantage that the analytes in the aquatic sensing environment flow through the core of the capillary interacting with the evanescent tail of the WGM field. Since the coupler (taper-fiber) is placed at the outer wall of the resonator (i.e., not in the aquatic environment), the system does not suffer from the problems experienced by on-chip resonators and microspheres and is more suitable for applications in aquatic environments [185–188].

Sensing with WGM resonators traditionally have relied on monitoring spectral shifts of single or multiple resonance modes upon the changes in the polarizability of the resonator-surrounding coupled system due to the arrival of analytes and synthetic or biological particles into the mode volume of the resonators [56–58, 60, 185–193]. This is the same method used in SPR and mechanical resonator based sensing platforms, and is referred to as spectral shift method (SSM) in this review. However, recently a novel sensing method which relies on a very peculiar property of WGM resonators—that is, the ability of the resonator to support frequency-degenerate counterpropagating WGM modes—has been proposed and demonstrated to detect and measure individual nanoscale objects. This sensing method, which is referred to as mode splitting method (MSM), exploits the lifting of frequency degeneracy when the coupling between the matter (analytes, particles, viruses, protein, etc.) and the WGM light field is strong enough to overcome the total loss of the system [59, 61, 62, 63, 194–202]. In both cases, as have been shown in Refs. [198, 203–205], the interaction of the cavity field with the binding molecule or particle polarizes the molecule/particle and hence changes the effective refractive index or the polarizability of the cavity system.

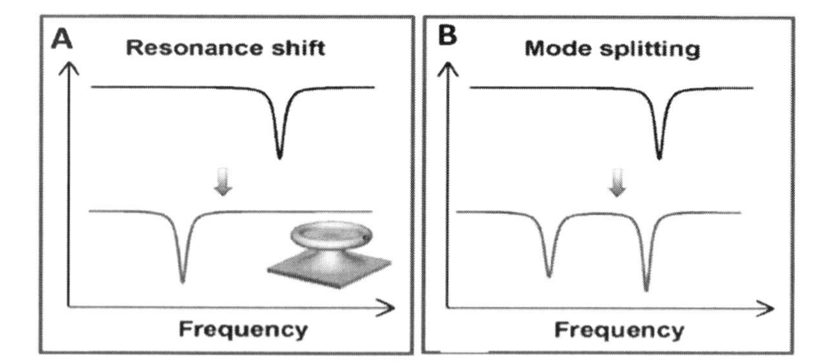

Figure 9.12 Schematic illustration of resonance shift and mode splitting based sensing methods in WGM resonators. Before the perturbation, transmission spectrum shows a single resonance line. With the perturbation, the resonance is either spectrally shifted (reactive shift) or it is split into two (mode splitting).

This change in the polarizability, similarly in the refractive index, of the cavity-surrounding system, reflects itself in the resonator transmission spectrum as a shift of the resonance frequency and/or as a broadening of the linewidth of the single resonant mode of interest or as the mode splitting. Provided that such changes in the transmission spectrum of the resonator are accurately measured, one can detect the presence of particles, count them, and measure their polarizability, size, or refractive index. Figure 9.12 depicts an illustration of sensing methodologies used in WGM based platforms.

9.7 Spectral Shift Method

In spectral shift method (Fig. 9.12), which is similar to the sensing methods employed in SPR, PhC, and NEMS resonators described in previous sections, resonant modes of a WGM resonator respond to changes occurring in or in close proximity of the resonator by undergoing spectral shift or linewidth broadening. An intuitive and simple picture to explain the resonance shift can be drawn from the resonance condition $2\pi n_e R = m\lambda$, which states that any change in R and/or in n_e (effective refractive index of the resonator taking into account the resonator material and the refractive indexes of surrounding medium as a part of the WGM field is probing the surrounding) will change the resonance wavelength. Thus, by allowing a change of Δn_e in the refractive index and ΔR in the radius of the resonator, one arrives at the expression

$$\frac{\Delta\lambda}{\lambda} = \frac{\Delta R}{R} + \frac{\Delta n_e}{n_e} \tag{9.13}$$

This expression reduces to $\Delta n_e/n_e = \Delta\lambda/\lambda$ if there is a uniform change only in the refractive index, (e.g., refractive index of the surrounding medium changes). The effective refractive index n_e can be decomposed into two parts as $n_e = \eta_1 n_r + \eta_2 n_s$, where n_r and n_s are the refractive indexes of the resonator material and the surrounding medium, respectively, and $\eta_1 + \eta_2 = 1$ with η_1 and η_2 representing the portions of the WGM field in the resonator and the surrounding, respectively. In this case, any refractive index change in the surrounding, not affecting the resonator refractive index or size, will change the resonance wavelength as $\Delta\lambda/\lambda = (\Delta n_s)/(n_s + n_r\eta_1/\eta_2)$. On the other hand, if the materials in the vicinity of the resonator have a refractive index very close to that of the resonator material and they bind to the resonator forming a uniform layer, the above equation can be approximated as $\Delta R/R = \Delta\lambda/\lambda$.

In most of the biosensing experiments, however, the small molecules in the vicinity of the resonator not only induces a change in the effective refractive index but they also bind to the resonator forming a non-uniform coating on the resonator surface. Teraoka and Arnold have shown that when these molecules enter the mode volume of the resonator and interact with the resonator field, a dipole is induced in the molecules leading to a change in the effective polarizability of the system by an amount of α_{excess} defined as the excess polarizability. This, in turn, shifts the resonance wavelength as [206–208]

$$\frac{\Delta\lambda}{\lambda} = \frac{\alpha_{\text{excess}}\sigma}{\varepsilon_0 R(n_r^2 - n_s^2)} \tag{9.14}$$

where n_r and n_s are the refractive indexes of the resonator and the surrounding, respectively, and σ is the surface density of bound molecules. In situations in which changes induce additional losses, both the resonance frequency and the resonance linewidth change. Thus by monitoring such changes one can build various sensors.

Although WGM resonators have been in use for various applications since early 20th century, the first experiments clearly showing their potential for biosensing were reported in 2002 by Kriokov et al. who reported measurement of refractive index changes as little as 10^{-4} using a WGM cylindrical resonator [209], and Vollmer et al. who, for the first time, reported the detection of streptavidin binding to biotin using a microsphere resonator (Fig. 9.13) [206]. These pioneering works were followed by the work of Vollmer et al. [210] where a two microsphere multiplexing scheme was used to detect DNA hybridization with a lower detection limit of 6 pg/mm^2 mass loading and to discriminate single-nucleotide mismatch. Demonstration of multiplexing has the potential to construct large scale detection chips in the form of arrayed microspheres. The same group later showed the detection of *E. Coli* which is a rod-like bacteria with a sensitivity of 1.2×10^2 bacteria per mm^2 corresponding to $\sim$34 pg/mm^2 dry-mass loading [191]. Using spectral shift method employed in microsphere resonators, Vollmer et al. have achieved the detection of single Influenza A (InfA) virions (Fig. 9.13), and used statistical analysis on the data obtained from multiple measurements to estimate their mean size which agrees well with the reported radii of 45–55 nm for InfA virions [58]. More recently, Shopova et al. have improved the single particle lower detection limit to 40 nm for polystyrene (PS) nanoparticles using a microsphere resonator with light from a frequency-doubled distributed feedback laser with wavelength in the band of λ = 650 nm (Fig. 9.13) [60].

Successful demonstration of biosensing applications with microsphere resonators has led to immense efforts in the direction of using other geometries of WGM resonators. In 2003, Armani et al. have reported single-molecule sensitivity using a microtoroid, and attributed this to a thermo-optic effect which enhances the amount of wavelength shift [57]. A recent theoretical work by Arnold and co-workers claims that under the experimental conditions employed in Ref. [57], the thermo-optic effect results in a smaller shift than the reactive effect and suggests that a new mechanism is necessary to explain the demonstrated single-molecule detection [211]. The aforementioned works, on single-particle and single-molecule detection, have revealed that the amount of resonance shift

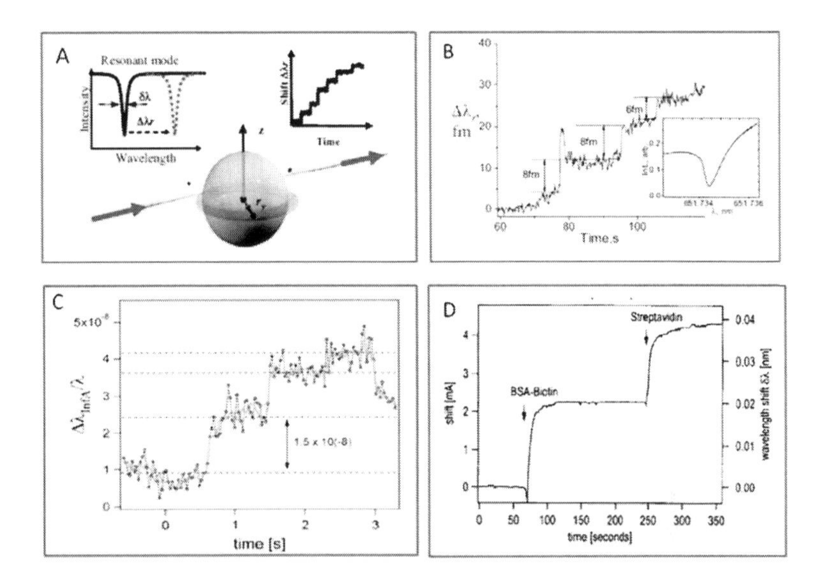

Figure 9.13 Biosensing using reactive shift method using WGM resonators. (A) Schematic illustration of the reactive shift method (i.e., spectral shift method, SSM). Particles entering the mode volume of the resonator lead to discrete jumps in the resonance wavelength. (B) Change in resonance wavelength as PS particles with average mean radii 40 nm bind to the microsphere. Each discrete jump corresponds to one particle binding event. Microsphere used in this experiment has a radius of 35 μm and $Q \approx 10^6$ (C) Response of the microsphere resonator of radius 39 μm and $Q \cong 64 \times 10^5$ to binding Influenza A virions. (D) Sensogram showing the detection of streptavidin binding to BSA biotin immobilized on the surface of a microsphere resonator with $Q \approx 2 \times 10^6$. (A) and (B) are adapted from Ref. [60], (C) from Ref. [58] and (D) from Ref. [206].

depends on the distribution of the WGM field and the location of the binding molecule or particle within the mode volume of the resonator. The shift is larger for a molecule/particle binding at a region of higher intensity. Since in practical realizations, one cannot place the binding particles always at the highest field intensity region, the shift for the same particle/molecule is different at each binding event. This is clearly seen in the results shown in Fig. 9.13. Thus, ensemble measurements are generally performed and the results are generally delivered as a frequency shift distribution. This hinders single-shot measurement of the polarizability or the size of a binding particle/molecule. A recent work by Arnold et al. showing the carousel forces acting on a particle in a liquid environment has revealed that a particle entering the WGM field of a microsphere is eventually attracted to the equatorial region where the field intensity is higher [212]. This implies the possibility of single-shot size measurement as eventually particles are likely to be brought to the region of high field intensity. However, more experimental demonstrations and both quantitative and qualitative discussions of multiple particle binding events should be carried out to confirm this possibility.

These experiments using microsphere and on-chip microtoroid resonators have shown without any doubt their unprecedented sensing capabilities. In general, microfluidic channels or micro-aquariums providing the liquid environment for biosensing are fabricated separately, and then microspheres or microtoroids are immersed into them together with the coupling fiber tapers. Special care should be taken when the tapered fibers are immersed into the same environment, as it is generally difficult to control and manipulate the taper in liquid environment, making it difficult to fine tune the gap between the taper and the resonator which is critical in most of the applications. Liquid core optical resonators have been fabricated and efficiently used in biological studies [185–190]. They provide an easy integration between photonics and microfluidics and eliminate the problems associated with immersing the fiber taper in aquatic environment. In these resonators, the taper fiber is positioned on the capillary and does not have any contact with the liquid environment flowing through the capillary. However, a drawback of these resonators compared to the microspheres and microtoroids is that up to date reported Q values are only slightly greater than 10^5. Liquid core optical resonators have been shown to provide refractive index sensitivities approaching 600 nm/RIU and been used for the detection of DNA [213], virus [214], protein [188], and cancer biomarker [215]. Despite these efforts and high sensitivities, liquid core optical resonators have not reached to the level of single-particle/molecule resolution yet.

It should be noted here that the amount of the spectral shift depends on physical/chemical properties and positions of sensing targets in the mode volume. Thus, although each arriving particle can be detected, accurate quantitative measurement of the properties of each arriving particle cannot be done. Instead, statistical analysis, such as building histograms of event probability versus spectral shift for ensembles of sequentially arriving particles of similar properties, is used to extract the required information (size, mass, or polarizability). This prevents single-shot measurement of each particle and the real-time detection capability. An equally important issue affecting them is the absence of a reference, which makes it difficult to discriminate the interactions of interest from the interfering perturbations (e.g., instrument noise, irrelevant environmental disturbances, and temperature variations). Thus, it is not easy to avoid "false signals" in the detection.

9.7.1 *Mode Splitting Method*

In addition to the spectral shift method, mode splitting demonstrated in WGM resonators has emerged as a high-performance self-referencing method for sensing applications. A WGM resonator supports two counterpropagating modes at the same resonance frequency, i.e., the transmission spectrum shows a single resonance mode (Fig. 9.14). Let us assume that only one of the modes, say the clockwise (CW) propagating mode, is excited. A subwavelength scatterer (e.g., a biomolecule, nanoparticle, or virus) in the mode volume will scatter the light. A portion of the scattering will be lost to environment as dissipation, while the rest will couple back into the resonator. A part of the backscattered light may couple into the opposite direction exciting the counterclockwise (CCW) propagating mode (Fig. 9.14). Through this process, the CW and CCW modes couple to each other. These two modes can form two standing wave modes (SWMs): a symmetric mode (SM) and an asymmetric mode (ASM). The fields of these two modes are distributed such that the SM locates the particle at its anti-node and the ASM locates the particle at its node. Consequently, the SM feels the presence of the scatterer strongly; as a result, it experiences a linewidth broadening due to increased scattering losses and a spectral shift due to the excess optical path length, which shifts the resonant wavelength. Thus, the cross-coupling between the modes due to scattering lifts the degeneracy of the WGM leading to two resonances (doublet) in

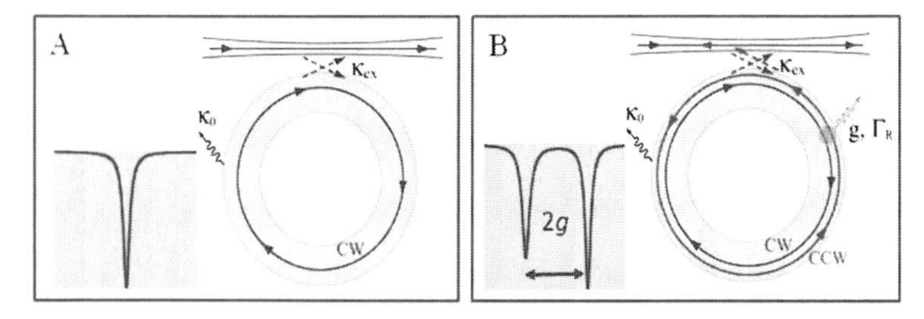

Figure 9.14 Schematic diagram explaining the mode splitting phenomenon. (A) With input light coupled in the clockwise direction (CW) only, the spectrum in the forward direction (transmission spectrum) shows single resonance, and there is no reflected light in the backward direction. The resonance is characterized with the intrinsic loss κ_0 and coupling losses κ_1. (B) A scatterer in the mode volume couples the CW input light to the counterclockwise (CCW) mode, creating two orthogonal standing wave modes spectrally shifted from each other. This is termed as mode splitting. The spectral distance between the modes and the differences in their linewidths are characterized respectively with parameters g and Γ, which depend on the polarizability of the scatterer and the field distribution at the location of the scatterer. The presence of the CCW mode leads to reflected light in the backward direction.

the transmission spectrum (Fig. 9.14). This process is referred to as the mode splitting.

Mode splitting was first reported in ultra-high-Q microspheres by Gorodetsky et al. [196], and addressed in detail by Weiss et al. [195] leading to intense theoretical and experimental investigation of intrinsic and intentionally induced mode splitting in various passive WGM resonators including microspheres, microtoroids, and microdisks [59, 194, 197]. It has also been observed in aquatic environment in microspheres and microtoroids [63, 179, 180]. Mode splitting can be detected in the transmission spectrum by linearly scanning the wavelength of a tunable laser [59] and recording the resonances or using the interferometric method introduced in Ref. [201]. For a scatterer-coupled taper-WGM cavity system, the evolution of the CW and CCW fields are given by [59, 194, 216].

$$\frac{da_{\mathrm{cw}}}{dt} = \left(i\,(\Delta - g) - \frac{\Gamma + \kappa_0 + \kappa_1}{2}\right) a_{\mathrm{cw}} - \left(ig + \frac{\Gamma}{2}\right) a_{\mathrm{ccw}} - \sqrt{\kappa_1}\, a_{\mathrm{in,cw}} \tag{9.15}$$

$$\frac{da_{\text{ccw}}}{dt} = \left(i\,(\Delta - g) - \frac{\Gamma + \kappa_0 + \kappa_1}{2}\right) a_{\text{ccw}} - \left(ig + \frac{\Gamma}{2}\right) a_{\text{cw}} - \sqrt{\kappa_1}\, a_{\text{in,ccw}} \tag{9.16}$$

where a_{cw} and a_{ccw} are the field amplitudes in the resonator, $a_{\text{in,cw}}$ and $a_{\text{in,ccw}}$ are the input fields, g represents the coupling strength between the CW and CCW modes via the scatterer, and Γ is the additional damping introduced due to the light scattered into the environment. In order to get a better understanding of the phenomenon, we switch from the CW and CCW representation to the normal mode representation with $a_+ = (a_{\text{cw}} + a_{\text{ccw}})/\sqrt{2}$ and $a_- = (a_{\text{cw}} - a_{\text{ccw}})/\sqrt{2}$ corresponding to the two SWMs. It is straightforward to show that in the steady state, the expressions in Eqs. 9.15 and 9.16 become

$$\left[-i\,(\Delta - 2g) + \frac{2\Gamma + \kappa_0 + \kappa_1}{2}\right] a_+ + \sqrt{\kappa_1}\, a_{\text{in},+} = 0 \tag{9.17}$$

$$\left[-i\Delta + \frac{\kappa_0 + \kappa_1}{2}\right] a_- + \sqrt{\kappa_1}\, a_{\text{in},-} = 0 \tag{9.18}$$

In most of the applications, there is only one input, i.e., $a_{\text{in,cw}}$ or $a_{\text{in,ccw}}$ is zero. Let us assume that $a_{\text{in,ccw}} = 0$ and $a_{\text{in}} = a_{\text{in,cw}}$ as in Fig. 9.14. Then from Eqs. 9.17 and 9.18, we find that

$$a_+ = \frac{\sqrt{2\kappa_1}\, a_{\text{in}}}{\kappa_0 + \kappa_1 + 2\Gamma - i2(\Delta - 2g)} \tag{9.19}$$

and

$$a_- = \frac{\sqrt{2\kappa_1}\, a_{\text{in}}}{\kappa_0 + \kappa_1 - i2\Delta} \tag{9.20}$$

It is seen that a_- has the similar expression as Eq. 9.10 of the degenerate case in contrast to a_+ which is detuned from it by $2g$ and has an additional damping of 2Γ. Note that even if $a_{\text{in,ccw}} = 0$, the scatterer-induced coupling creates a CCW field inside the cavity. Thus, in contrast to the degenerate case, both reflection and transmission exist and they are expressed as $a_{\text{t}} = a_{\text{cw,in}} + \sqrt{\kappa_1}\, a_{\text{cw}} = a_{\text{in}} + \sqrt{\kappa_1}\, a_{\text{cw}}$ and $a_{\text{r}} = \sqrt{\kappa_1}\, a_{\text{ccw}}$. Consequently, the mode splitting information can be extracted from both the transmission and the reflection ports (i.e., in the forward transmitted or backward reflected fields at the two outputs of the waveguide).

The amount of the spectral shift (mode splitting) $2g$ and the linewidth difference Γ between the two resonance modes are related to the polarizability α of the scatterer, the resonator mode volume V and the location $\vec{r}$ of the scatterer in the WGM field via the expressions [59, 194]

$$g = -\frac{\alpha f^2(r)\,\omega}{2V},\ \Gamma = \frac{\alpha^2 f^2(r)\,\omega^4}{6\pi\,\nu^3 V} \quad (9.21)$$

where ω is the angular resonant frequency, $f(r)$ designates normalized mode distribution and denotes the location dependence, and $\nu = c/\sqrt{\varepsilon_{\mathrm{m}}}$ with c representing the speed of light and ε_{m} denoting the electric permittivity of the surrounding. The polarizability α of the scatterer can be expressed as

$$\alpha = 4\pi R^3 \frac{\varepsilon_{\mathrm{p}} - \varepsilon_{\mathrm{m}}}{\varepsilon_{\mathrm{p}} + 2\varepsilon_{\mathrm{m}}} \quad (9.22)$$

for a spherical scatterer of radius R and electric permittivity ε_{p}. If the refractive index of the scatterer/particle is known, then its size can be estimated from the ratio of the linewidth difference and the spectral distance between the split resonances measured in the transmission spectrum. Assuming that the measurements are performed in air, this ratio is

$$\frac{\Gamma}{|g|} = \frac{8\pi^2}{3\lambda^3}\alpha \quad (9.23)$$

where we see that the location dependency shown in Eq. 9.21 is eliminated. Thus, mode splitting enables the detection and measurement of single particles regardless of where they are located within the mode volume [59, 61].

Ozdemir et al. have shown that in order to resolve the mode splitting in the transmission spectra, the coupling strength should be greater than the losses in the system [217]

$$2g > \Gamma + \frac{\omega_c}{Q} \quad (9.24)$$

This resolvability condition reveals how the properties of the scatterer and the resonator affect the mode splitting phenomenon: (i) When the cavity related losses dominate the system (i.e., $\Gamma \ll \omega_c/Q$), $2g$ should be larger than the linewidth of the resonance mode (ω_c/Q), implying that the polarizability α of the scatterer should satisfy $\alpha > f^{-2}(V/Q)$. Then, the minimum α that can lead to an observable splitting is determined by the cavity-related parameter (V/Q) and the location of the scatterer within the mode volume, which determines the intensity of the field interacting with the scatterer. Thus, larger Q, smaller V, and larger field intensity at the particle location are preferred. Experiments show that this condition is satisfied for particles smaller than 100 nm. (ii) When the scatterer related losses dominate the system (i.e., $\Gamma \gg \omega_c/Q$), $2g$ should be larger than the additional damping rate induced by the scatter, implying that $\alpha < (3/8\pi^2)\left(\lambda/\sqrt{\varepsilon_{\mathrm{m}}}\right)^3$, demonstrating the effect of the resonance wavelength and the refractive index of the surrounding on the measurable polarizability. This condition is observed for larger particles and determines the upper limit of detectable scatterer polarizability in relation with the resonance wavelength employed.

Mode splitting in a WGM microresonator has produced a new line of sensors for the detection, counting, and measurement of individual nanoscale objects. The first application of mode splitting for this purpose was reported by Zhu et al., who performed single-particle detection of KCl and PS nanoparticles down to 30 nm in air (Fig. 9.15) [59]. In this work, size measurement of the detected particles could be done with a single-shot measurement without the need for ensemble measurements or complicated statistical processing. The size of the particle could be accurately measured regardless of the location of the particle within the mode volume. It was later shown that while the detectable PS particle size is 20 nm, accurate size measurement could be performed for particles of size 30 nm or larger [61]. This discrepancy originates from the fact that detection of a single particle could be indicated by transition from a single resonance to a doublet or by any change in the splitting spectrum (e.g., amount of splitting, shift in the resonance frequencies of the doublets, or change in the linewidths of the doublets) while for accurate size measurement both the amount of splitting and the linewidth broadening information need to be accurately extracted from the splitting spectra. This becomes difficult for small particles due to the signal-processing errors and other intrinsic error sources. It is clear that the minimum detectable and measurable size of scatterers, particles, or molecules using

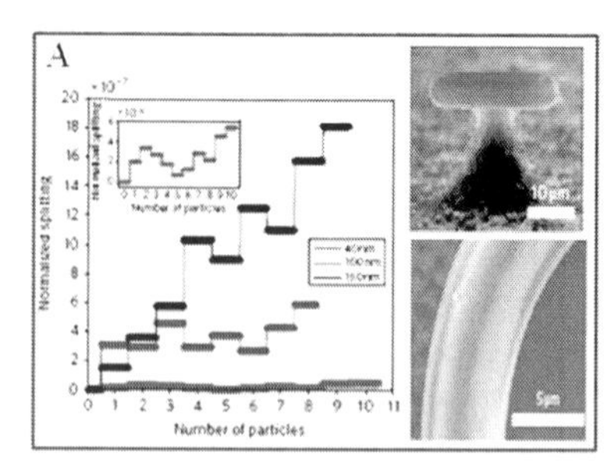

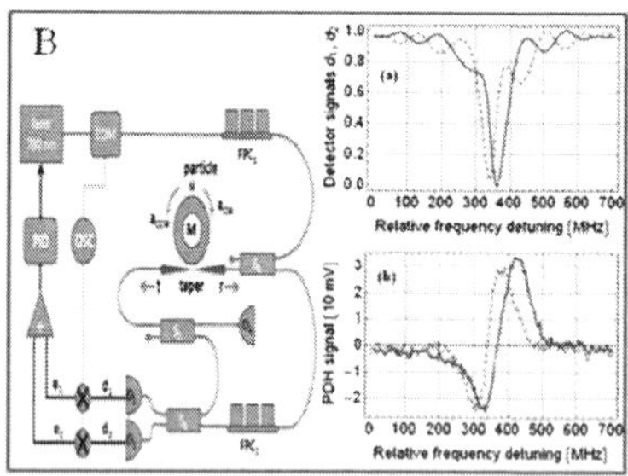

Figure 9.15 Sensing using mode splitting method in WGM resonators. (A) SEM image of a microtoroid with a single particle deposited on its ring. As the particles land in the mode volume of the resonator, discrete jumps in the mode splitting are observed. The heights and the direction of the discrete jumps depend on the location of the particle with respect to the location of previously deposited particles, if there is, within the mode volume. Splitting is determined by scanning wavelength of a tunable laser and by recording the transmission spectrum. The data shown were obtained for potassium chloride nanoparticles. Excerpted with permission from Ref. [59]. (B) An interferometric scheme using the phase difference between the reflected and transmitted fields in a WGM with split modes. The setup allows the direct detection of doublets hidden in the large linewidths of the modes without having to scan the wavelength of a tunable laser. The frequency difference of the two Pound Drever Hall (PDH) error signals is proportional to the amount of mode splitting. The experimental data shown were obtained for quantum dots attached to resonator surface. Excerpted with permission from Ref. [201].

mode splitting method strongly relies on how well the splitting spectrum is extracted from the noisy transmission or reflection spectrum.

In order to circumvent the complications related with the noise level and Q factor of the resonator, Koch et al. proposed to look at the reflection port only [218, 219]. The rationale behind this is that if the light is coupled into the resonator only in forward direction (say CW), then there will be no light in the reflection port—backward direction—in the absence of a scatterer. However, if there is a scatterer in the mode volume, a portion of the scattered CW light couples into the CCW mode resulting in a non-zero field in the backward direction. Thus, a detected light in the backward direction with forward excitation signals the presence of a scatterer in the mode volume. The authors reported the detection of gold nanoparticles of radii 50 nm using a hurricane resonator with $Q \approx 1000$ [219]. However, they could not reach single-particle resolution. In an add-drop configuration using a ring resonator of $Q \approx 500$ in the reflection mode, the same authors reported the detection of mode splitting induced by an atomic force microscopy (AFM) gold tip of diameter 250 nm. Although the reflection mode configuration allows detecting particles by power monitoring, and circumvents the mode splitting resolvability condition described by Eq. 9.24 and Ref. [217], it does not allow extracting quantitative information such as the size or refractive index of the particle or the scatterer. Knittel et al. have employed an interferometric setup which allows direct detection of the doublet even if it is hidden within the width of the resonance line (Fig. 9.15) [201]. They did this by interfering the reflected and transmitted fields at a splitter and measuring the two output ports of the splitter with two detectors which gives the normalized amplitudes of the normal modes a_+ and a_-. The authors then extracted the mode splitting by deriving a Pound–Drever–Hall error signal from each detector and showing that the frequency difference between the zero crossings of the error signals corresponds to the splitting. The authors demonstrated the effectiveness of the method by detecting CdSe quantum dots. In the proposed method, Knittel et al. eliminated the need for continuous laser scanning and curve fitting of the transmission spectra and the possible error contributions of these processes.

The measurements reported in Refs. [59, 61, 201, 217] were carried out either in air or in dry environment. For the detection of biomolecules, particles, or viruses in aquatic environment and solutions, mode splitting should be demonstrated in water environment too. Moving from air to aquatic environment hinders the measurement in various ways. First, the particle has lower polarizability in water than in air ($\alpha_{\mathrm{air}} > \alpha_{\mathrm{water}}$) due to the reduced refractive index contrast between the particle and the surrounding since $n_{\mathrm{water}} > n_{\mathrm{air}}$. Second, reduced refractive index contrast between the resonator material and the surrounding leads to deviations from the near-ideal total internal reflection, and hence the loss increases resulting in a lower Q. Kim et al. have studied mode splitting in aquatic environment in microtoroid and microsphere resonators both theoretically and experimentally [179, 180]. They have reported the observation of the first scatterer-induced mode splitting spectra in the investigated resonators. Moreover, using the mode splitting resolvability criterion introduced above, they clarified how the size and quality factor of the resonator should be chosen in order to observe mode splitting in the transmission spectra for a particle of given size and refractive index. They also discussed and clarified the reasons of why only reactive shift but not the mode splitting were observed in previous experiments performed in water and in buffer solutions.

Lu et al. employed a thermally stabilized reference interferometer to minimize the noises introduced into the transmission spectra of a microtoroid resonator, placed in aquatic environment, by spectral fluctuations of a tunable laser whose wavelength is scanned to monitor the resonance shifts and mode splitting [63]. In this scheme, the light from the tunable laser is divided into two, one of which is sent to the interferometer and the other is sent to the resonator. In this way, frequency of the scanning laser is measured accurately at the time when the resonance in the resonator is excited. The authors then use this information to reduce the frequency noise effects in the measurements. As a result they reported the detection of polystyrene beads of radii down to 12.5 nm and Influenza A virions in solution [63]. The works of Knittel et al. [201] and Lu et al. [63] are good examples of employing referencing and interferometric techniques to minimize the effects of noise in resonator-based sensing schemes and hence to improve the detection limit of mode splitting method beyond those reported in Refs. [59, 61, 179, 180]. However, these works were implemented at the expense of complicated measurement schemes. Moreover, they focus on detection and do not provide information on how the employed noise reduction techniques could be used to improve the size measurement accuracy and limit beyond that of Refs. [59, 61].

An important advantage of MSM over the SSM is its self-referencing feature which originates from the fact that the doublet reside in the same resonator and are affected similarly by the unwanted perturbations and noise sources, i.e., variations in environmental temperature and humidity. He et al. verified the self-referencing property by monitoring the effect of the temperature

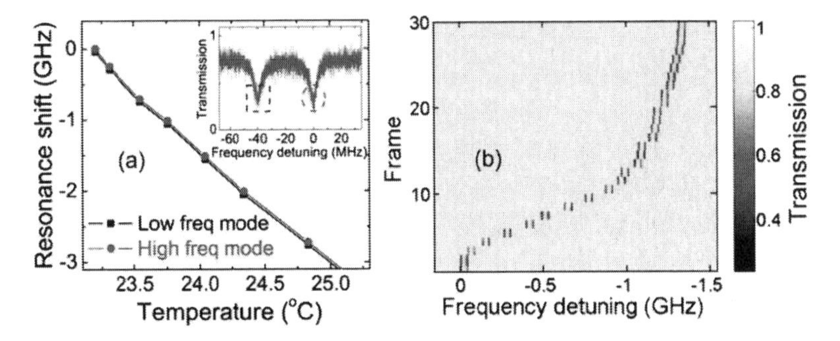

Figure 9.16 Temperature response of mode splitting in a silica micro-toroidal cavity. (a) Shift of resonance frequency as a function of temperature for the split modes. The dots and squares denote the high-frequency and low-frequency modes in the transmission spectra, respectively. (b) Intensity graph showing the evolution of the split modes as the temperature is increased. The *y* axis (Frame) denotes the time, in other words, the temperature. Adapted with permission from Ref. [199].

changes on the amount of mode splitting [199]. As seen in Fig. 9.16, when the temperature increases, the resonances of both split modes experience red shifts; however, the spectral distance between the split modes remains unchanged, thus leading to the conclusion that perturbations in the environmental temperature have a negligible influence on the mode splitting based sensor schemes. Note that in SSM, one cannot discriminate between the resonance shift due to a binding particle or molecule from spectral shift induced by thermal fluctuations occurring within or in close proximity of the resonator.

In Ref. [59], Zhu et al. have shown how the size measurement of a detected particle can be performed regardless of the location of the particle in the mode volume. A careful reader will immediately note that although in Ref. [59] tens of particles could be detected by a single resonator, only the size of the first detected particle could be measured accurately, because the theory and model introduced above did not provide a method to measure the size of each detected particle. This problem was later solved in Refs. [61, 200]. Here, we will briefly summarize the formalism for measuring size of individual nanoparticles. We start by noting that mode splitting spectra is quantified by the spectral distance and linewidth difference between the split modes. Subsequently, the frequency shift ($\Delta\omega_1^-$, $\Delta\omega_1^+$) and linewidth broadening ($\gamma_1^- \gamma_1^+$) of the split modes (doublet) with respect to the resonance frequency ω_0 and the linewidth γ_0 of the initial (pre-scatterer) WGM after the binding of the first particle are

$$\Delta\omega_1^- = \omega_1^- - \omega_0 = 2g_1, \;\; \Delta\omega_1^+ = \omega_1^+ - \omega_0 = 0 \tag{9.25}$$

$$\Delta\gamma_1^- = \gamma_1^- - \gamma_0 = 2\Gamma_1, \;\; \Delta\gamma_1^+ = \gamma_1^+ - \gamma_0 = 0 \tag{9.26}$$

where ω_1^- and ω_1^+ denote the resonance frequencies of the lower and higher frequency modes of the doublet, respectively. Then we define the mode splitting between the two modes as $\delta_1^- = \Delta\omega_1^+ - \Delta\omega_1^- = \omega_1^+ - \omega_1^- = -2g_1$ and the total frequency shift of the modes as $\delta_1^+ = \Delta\omega_1^+ + \Delta\omega_1^- = \omega_1^+ + \omega_1^- - 2\omega_0 = 2g_1$ as the total frequency shift (i.e., the sum of the frequency shift of each split mode with respect to the initial pre-scatterer WGM resonance frequency). Similarly, we define difference of the linewidths of the two modes as $\varrho_1^- = \Delta\gamma_1^+ - \Delta\gamma_1^- = \gamma_1^+ - \gamma_1^- = -2\Gamma_1$ and the sum of the linewidth-changes as $\varrho_1^+ = \Delta\gamma_1^+ + \Delta\gamma_1^- = \gamma_1^+ + \gamma_1^- - 2\gamma_0 = 2\Gamma_1$. Using these expressions, we arrive at

$$\frac{\varrho_1^-}{\delta_1^-} = \frac{\varrho_1^+}{\delta_1^+} = \frac{\Gamma_1}{g_1} \tag{9.27}$$

Substituting Eq. 9.21 in this expression, we find the polarizability of the single nanoparticle, when the surrounding medium is air, i.e., $\varepsilon_{\mathrm{m}} = 1$, independent of the particle position (r) in the mode volume as

$$\alpha_1 = -\frac{\Gamma_1}{g_1}\frac{3\lambda^3}{8\pi^2} = \frac{\varrho_1^-}{\delta_1^-}\frac{3\lambda^3}{8\pi^2} = -\frac{\varrho_1^+}{\delta_1^+}\frac{3\lambda^3}{8\pi^2} \tag{9.28}$$

We have used the expression in Eq. 9.28 to measure the size of single spherical nanoparticles down to 30 nm. Subsequent nanoparticles landing on the resonator after the first nanoparticle cause changes in the mode splitting spectra which are reflected as discrete jumps in the frequencies and linewidths of the split modes as shown in Figs. 9.15 and 9.17. The number of such discrete jumps enables counting the number of binding particles in addition to their detection with single-particle resolution.

In order to measure the polarizability of the particles binding to the resonator after the first one, we need to extend the above formalism to include the multiparticle effects. Each particle entering the mode volume of the resonator perturbs the already established SWMs which redistribute their fields such that the amount of splitting is maximized. Thus, after each binding event, the location of each individual scatterer with respect to the nodes and anti-nodes of the new SWMs will be different. Note that for the multi-scatterer case, we can no longer talk about symmetric or antisymmetric SWMs; instead, we denote them as SWM1 and SWM2. These SWMs are still orthogonal to each other and their nodes are separated by a spatial phase difference of $\pi/2$.

The amount of perturbation induced by a scatterer on the SWMs depends on how close the scatterer is located to the anti-node of the SWM. The closer is the scatterer to the anti-node of an SWM, the larger is the perturbation to that SWM. A single scatterer in the mode volume is always located at the anti-node of one of the SWMs and node of the other; thus one of the SWMs is maximally perturbed while the perturbation of the other is negligible. If we assume that there are N scatterers $(1, 2, \ldots N)$ in the mode volume of the resonator, and ϕ is the distance between the first particle and anti-node of the SWM1, and that β_i is the spatial phase distance between the first and the ith scatterer ($\beta_1 = 0$). Then we can rewrite Eqs. 9.2 and 9.3 taking into account the effect of N particles as

$$\Delta\omega_N^- = \sum_{i=1}^N 2g_i \cos^2(\phi - \beta_i),\ \Delta\omega_N^+ = \sum_{i=1}^N 2g_i \sin^2(\phi - \beta_i) \tag{9.29}$$

$$\Delta\gamma_N^- = \sum_{i=1}^N 2\Gamma_i \cos^2(\phi - \beta_i),\ \Delta\gamma_N^+ = \sum_{i=1}^N 2\Gamma_i \sin^2(\phi - \beta_i) \tag{9.30}$$

where the $\cos^2(\ldots)$ and $\sin^2(\ldots)$ terms scale the interaction strength depending on the position of the scatterer with respect to the anti-nodes of SWMs, and g_i and Γ_i correspond to the coupling coefficient and the additional damping to the system if the ith scatterer is the only scatterer in the mode volume. Then we can rewrite the amount of splitting and linewidth differences of the split modes as well as the total frequency shift and linewidth broadening after the binding of the Nth particle as

$$\delta_N^- = -2\sum_{i=1}^N g_i \cos(2\phi_N - 2\beta_i),\ \delta_N^+ = 2\sum_{i=1}^N g_i \tag{9.31}$$

$$\varrho_N^- = 2\sum_{i=1}^N \Gamma_i \cos(2\phi_N - 2\beta_i),\ \varrho_N^+ = 2\sum_{i=1}^N \Gamma_i \tag{9.32}$$

where δ_N^- and ϱ_N^- imply that each individual scatterer may increase or decrease the existing mode splitting and the linewidth difference. On the other hand, δ_N^+ and ϱ_N^+ imply that the total frequency shift and total linewidth broadening of the split modes increase with each arriving scatterer.

In practical realizations, it is impossible to know the exact values of ϕ and β_i to use δ_N^- and ϱ_N^- in above expressions for measuring the polarizability of each scatterer. However, we can use δ_N^+ and ϱ_N^+ because these depend only on g_i and Γ_i, which are directly related to the polarizability of the ith scatterer. Thus, we can write the polarizability of the Nth particle α_N as

$$\begin{aligned}\alpha_N &= -\frac{\Gamma_N}{g_N}\frac{3\lambda^3}{8\pi^2} = -\frac{3\lambda^3}{8\pi^2}\frac{\varrho_N^+ - \varrho_{N-1}^+}{\delta_N^+ - \delta_{N-1}^+} \\ &= -\frac{3\lambda^3}{8\pi^2}\frac{(\gamma_N^+ + \gamma_N^-) - (\gamma_{N-1}^+ + \gamma_{N-1}^-)}{(\omega_N^+ + \omega_N^-) - (\omega_{N-1}^+ + \omega_{N-1}^-)} \end{aligned} \tag{9.33}$$

Consequently, one can estimate the polarizability of the Nth scatterer by comparing the total frequency and linewidth of the split modes just before and just after the binding of the Nth scatterer. These observations imply that one only need to have the information (transmission spectrum) before and after a nanoparticle deposition for single-shot measurement of the particle size regardless of intrinsic splitting or the history of previous particle depositions. Zhu et al. have employed the above formulation to detect and measure InfA virions, gold and PS nanoparticles one by one (Fig. 9.17) [61]. Moreover, since the polarizability of individual particles can be measured, modality of homogenous mixtures of nanoparticles could be identified.

9.8 Sensing Using WGM Active Microresonators and Microlasers

Light–matter interaction strength quantified by the cavity parameter Q/V determines the detection limit for the WGM microresonator based sensing platforms. The quality factor Q, as discussed in the previous sections, is mainly limited by the material absorption, scattering and coupling losses. Mode volume V, on the other hand, is

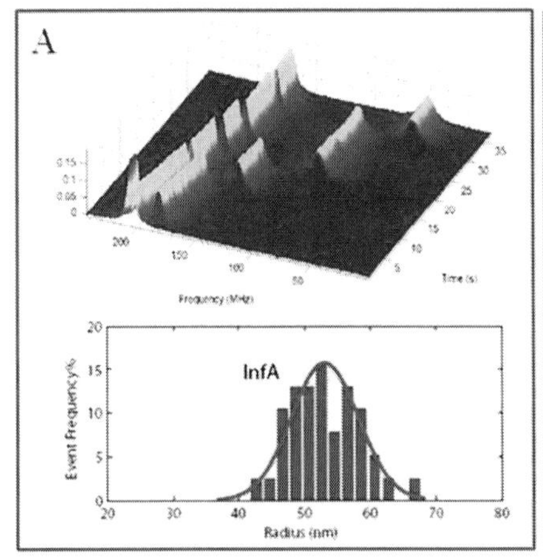

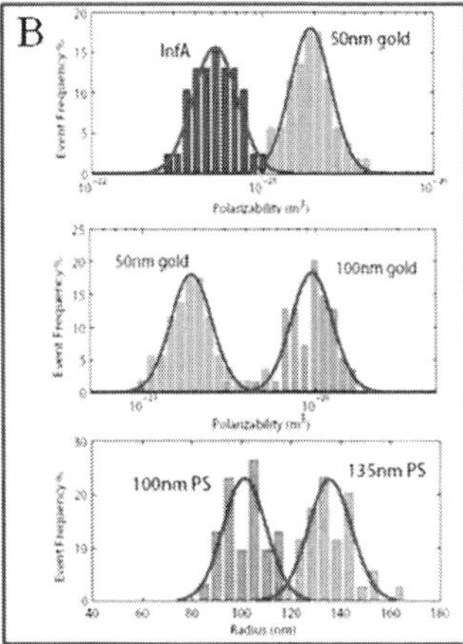

Figure 9.17 Real-time detection and size measurement of individual nanoparticles and influenza A (InfA) virions as they enter the mode volume of a microtoroid resonator one by one. (A) Spectrogram showing the change in the mode splitting as individual virions bind to the resonator. Size distribution of measured virions is given in the lower panel. (B) Histograms of size and polarizability of measured polystyrene (PS), gold, and InfA virions. Adapted with permission from Ref. [61].

determined by the size and geometry of the resonator as well as by the refractive index contrast between the resonator material and the surroundings. Although it may seem that one can arbitrarily increase Q/V by working with much smaller resonators, it is not the case as it has been demonstrated Q of a resonator significantly reduces when V becomes smaller than a critical value, which is 30 μm for a microtoroid. An alternative route is to introduce a gain medium into the resonator structure to compensate some of the losses. This will lead to an increase in Q and decrease the resonance linewidth improving the detection limit.

Resonators with a gain medium are generally referred to as "active resonators," whereas those without a gain medium are referred to as "passive resonators." The gain medium can be activated either by optical or electrical pumping depending on the type of the gain medium and the resonator material. Here, we will consider only the optically pumped resonators. The optical pumping may be on-resonant or off-resonant. As it will become clear later, if the gain provided by the active medium exceeds the total loss of the waveguide-coupled resonator, lasing takes place. Gain in WGM resonators can be provided either by doping the resonator with an active medium such as organic dyes, quantum dots (QDs), and rare earth ions, or by using the intrinsic nonlinearity of the resonator material, such as stimulated Raman scattering. Among different gain materials, rare earth ions (e.g., erbium Er^{3+}, neodymium Nd^{3+}, ytterbium Yb^{3+}), fluorescent dyes (e.g., rhodamine and Nile red) and quantum dots (e.g., CdSe, ZnSe) are popular dopants. Gain medium can be incorporated into a resonator by fabricating the resonator from active medium doped substrate [220–225], coating the resonator with light emitters [226–232], or ion implantation [233, 234]. Among all, sol–gel method provides a low-cost, fast, flexible, and convenient wet-chemical synthesis to tailor the dopants for fabricating active resonators [235, 236].

The expression for the evolution of the field inside a passive WGM resonator can be modified by the addition of a gain factor to study the properties of an active WGM resonator [237]. This then leads to

$$\frac{da}{dt} = \left(i\Delta - \frac{\kappa_0 + \kappa_1 - \xi}{2}\right) a - \sqrt{\kappa_1} a_{\text{in}} \tag{9.34}$$

where ξ denotes the round-trip energy gain and implies a reduction in the total loss of the waveguide coupled WGM resonator. Defining the effective loss, excluding the coupling losses quantified by κ_1, as $\kappa_{\text{eff}} = \kappa_0 - \xi$, we find the steady state transmission as

$$T = \frac{(\kappa_{\text{eff}} - \kappa_1)^2 + 4\Delta^2}{(\kappa_{\text{eff}} + \kappa_1) + 4\Delta^2} \tag{9.35}$$

which, at resonance ($\Delta = 0$), becomes

$$T = \left(\frac{1 - \zeta}{1 + \zeta}\right)^2 \tag{9.36}$$

with ζ now defined as $\zeta = \kappa_1/\kappa_{\text{eff}} = \kappa_1/(\kappa_0 - \xi)$. It is seen that when $\kappa_0 > \xi$, transmission at resonance becomes smaller implying attenuation ($\kappa_0 > \xi \to \zeta > 0 \to T < 1$). On the other hand, when $\kappa_0 < \xi$, transmission at resonance becomes greater than one implying an amplification ($\kappa_0 < \xi \to \zeta < 0 \to T > 1$). These imply that by tuning the gain ξ one can modify the transmission spectrum of the waveguide-coupled WGM resonator. The expression above implies five distinct operating regimes for an active resonator depending on the relation among ξ, κ_0, and κ_1. (i) No gain ($\xi = 0$). The resonator operates as a passive resonator, and the transmission spectrum is identified by a Lorentzian dip whose linewidth is determined by the total loss $\kappa_0 + \kappa_1$ of the system. (ii) Gain is smaller than the intrinsic loss of passive resonator (e.g., includes material, radiation, and scattering losses but excludes coupling loss) ($\xi < \kappa_0$), implying $\kappa_{\text{eff}} > 0$ and $\zeta > 0$. The system is still in the attenuation regime. Increasing ξ leads to decrease in effective loss κ_{eff}, implying a narrower resonance linewidth and higher Q provided that the coupling loss κ_1 is kept constant. The lower limit of resonance linewidth is ultimately determined by spontaneous emission noise. (iii) Gain is equal to passive resonator intrinsic loss ($\xi = \kappa_0$), implying $\kappa_{\text{eff}} = 0$ and $\zeta \to \infty$. This implies unit transmission; the system is at the border of attenuation and amplification regimes. (iv) Gain exceeds the intrinsic passive loss but is smaller than the total passive loss of the resonator ($\kappa_0 < \xi < \kappa_0 + \kappa_1$), implying $\zeta < 0$ and a negative effective intrinsic loss. The system is in the amplification regime; thus, the transmission spectrum is identified by a resonant peak. (v) Gain exceeds the total passive loss ($\kappa_0 + \kappa_1 < \xi$). Since all the losses are compensated in the active resonator, lasing is expected in this regime.

Incorporation of an active gain medium into the resonator structure provides a means to control the net loss in the system, allowing a narrower resonance linewidth when the system is driven below its lasing threshold and enabling generation of lasing from the system if the gain is increased to overcome the total net loss. In both of the cases, detection capabilities of the WGM resonator are significantly enhanced. The lasing region, in particular, is very attractive for detection purposes since the lasing line usually has much narrower linewidth than the resonance linewidth of the cold cavity (i.e., the gain medum is not pumped) as has been described by the famous Schawlow–Townes formula [238]

$$\Delta\nu_{\text{laser}} = \frac{\pi h\nu(\Delta\nu)^2}{P_{\text{out}}} \tag{9.37}$$

which relates the laser linewidth ($\Delta\nu_{\text{laser}}$) to the cold cavity linewidth ($\Delta\nu$) and the laser output power (P_{out}). In Eq. 9.37, ν is the laser light frequency, $h\nu$ is the photon energy, and $\Delta\nu = \nu/Q$ is the cold cavity linewidth with Q denoting the cold cavity Q factor. Then we can rewrite Eq. 9.37 for the quality factor of the laser emission as

$$Q_{\text{laser}} = \frac{\nu}{\Delta\nu_{\text{laser}}} = \frac{Q^2}{\pi h\nu^2} P_{\text{out}} \tag{9.38}$$

For Q of 10^7, operation wavelength of 1550 nm, and output laser power of 1 μW, the laser linewidth is calculated as $\Delta\nu_{\text{laser}} =$ 151 Hz, which is much narrower than the cold cavity linewidth of $\Delta\nu = 19$ MHz. This indicates that the microcavity laser as a sensing element could provide a much lower detection limit than the passive resonant cavity.

9.8.1 *Reactive Shift in Active Resonators*

The spectral shift method discussed in the previous sections is applicable for the active resonator in both the below, and above-threshold regimes. When the nanoobjects bind to the resonator surface, resonance wavelength of the cavity if the system is driven below threshold, and lasing wavelength if it is driven above threshold, undergoes spectral shift. The former corresponds to a resonator with a much narrower linewidth, and the change in the resonance wavelength is monitored in the transmission spectrum while the wavelength of a tunable laser is scanned around the resonance. In the latter, on the other hand, a tunable laser diode is no longer needed; instead a high-resolution spectrometer is required to monitor the trace amount of changes in the lasing wavelength.

Yang and Guo proposed active WGM resonators for the enhancement of sensitivity in spectral shift technique [239]. They concluded that with gain medium doped polymer microspheres it is possible to detect effective refractive index change of the order of 10^{-9} RIU. Francois and Himmelhaus have demonstrated the enhancement of sensitivity in both the spontaneous (i.e., below threshold) and stimulated emission (i.e., above threshold) regimes using Nile red doped polystyrene microspheres, reporting eightfold SNR improvement and threefold Q enhancement, which enabled monitoring of real-time adsorption kinetics of bovine serum albumin in phosphate buffered saline (PBS) [240]. Beier et al. have estimated the minimum detection limit of 260 pg/mm^2 for a minimum detectable mass of $\sim$80 fg protein [241], and Pang et al. reported the minimum detectable refractive index change of 2.5×10^{-4} RIU using CdSe/ZnS quantum dots embedded in polystyrene microspheres [242]. Detection of oligonucleotides using tetramethylrhodamine functionalized silica microspheres was reported by Nuhiji and Mulvanely [243].

9.8.2 *Mode Splitting in Active Resonators*

9.8.2.1 Below lasing threshold

Recent studies have shown that active WGM microcavities can improve the detection limit of the mode splitting technique. Assuming that the gain provided for the CW and CCW modes is the same and is equal to ξ, the expressions describing the evolution of the CW and CCW fields in an active microresonator with a coupled scatterer becomes [237]

$$\begin{aligned}\frac{da_{\text{cw}}}{dt} &= \left(i\left(\Delta - g\right) - \frac{\Gamma + \kappa_0 + \kappa_1 - \xi}{2}\right) a_{\text{cw}} \\ &\quad - \left(ig + \frac{\Gamma}{2}\right) a_{\text{ccw}} - \sqrt{\kappa_1} a_{\text{in,cw}} \end{aligned} \tag{9.39}$$

$$\begin{aligned}\frac{da_{\text{ccw}}}{dt} &= \left(i\left(\Delta - g\right) - \frac{\Gamma + \kappa_0 + \kappa_1 - \xi}{2}\right) a_{\text{ccw}} \\ &\quad - \left(ig + \frac{\Gamma}{2}\right) a_{\text{cw}} - \sqrt{\kappa_1}\, a_{\text{in,ccw}} \end{aligned} \tag{9.40}$$

In the steady state, the expressions for the normal modes are given as

$$\left[-i\left(\Delta - 2g\right) + \frac{2\Gamma + \kappa_0 + \kappa_1 - \xi}{2}\right] a_+ + \sqrt{\kappa_1} a_{\text{in},+} = 0 \tag{9.41}$$

$$\left[-i\Delta + \frac{\kappa_0 + \kappa_1 - \xi}{2}\right] a_- + \sqrt{\kappa_1} a_{\text{in},-} = 0 \tag{9.42}$$

from which we see that a_+ experiences a total loss of $\kappa_+ = 2\Gamma + \kappa_0 + \kappa_1 - \xi$ and a_- experiences a total loss of $\kappa_- = \kappa_0 + \kappa_1 - \xi$. Then the condition to resolve the mode splitting in the transmission spectrum

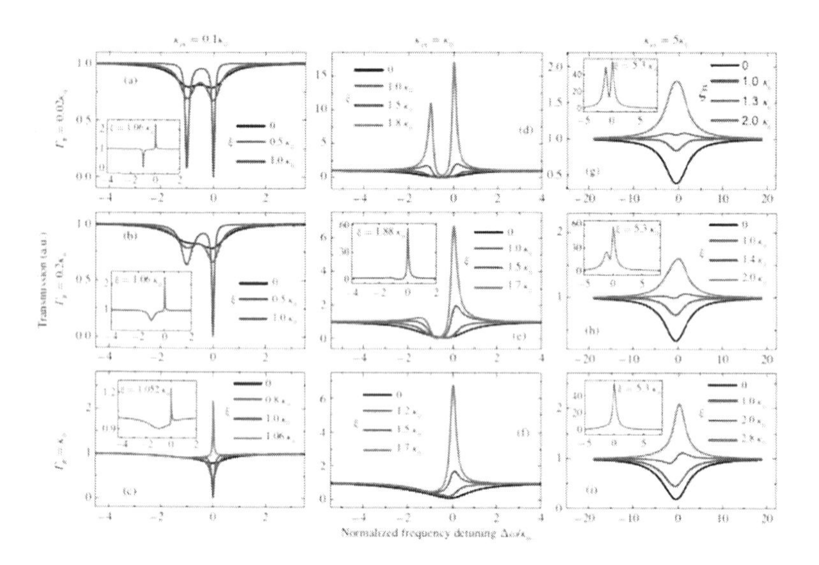

Figure 9.18 Numerical simulations showing the effects of gain and intrinsic and coupling losses on the mode splitting transmission spectra. In these simulations $\kappa_0 = 4\times10^7$ Hz and $g = -0.5\,\kappa_0$ were used in all the coupling conditions: undercoupling (a–c), critical coupling (d–f), and overcoupling (g–i). The amount of gain (ξ) and the scatterer-induced linewidth broadening (Γ) are denoted in the panels for each of the cases. Excerpted with permission from Ref. [237].

becomes

$$2\,|g| > \frac{\kappa_+ + \kappa_-}{2} = \Gamma + \kappa_0 + \kappa_1 - \xi \tag{9.43}$$

Thus, by adjusting the gain, it becomes easier to resolve the splitting in the spectra. Since the two resonance modes experience different amounts of total losses, the regions of attenuation and amplification are different for the two modes (Fig. 9.18): The a_+ mode is in the attenuation region for $\xi < 2\Gamma + \kappa_0$; in the amplification region for $\xi > 2\Gamma + \kappa_0$, and in the lasing region for $\xi > 2\Gamma + \kappa_0 + \kappa_1$. For the a_- mode, $\xi < \kappa_0$ denotes the attenuation, $\xi > \kappa_0$ the amplification, and finally $\xi > \kappa_0 + \kappa_1$ the lasing regime. This suggests that by adjusting the gain, one can drive the a_- mode into the amplification region while keeping the a_+ mode in the attenuation region. Thus, the a_- mode appears as a transmission peak while the a_+ mode appears as a transmission dip.

Assuming $a_{\text{in,ccw}} = 0$, we arrive at the following expression which gives the transmission spectrum

$$T(\Delta\omega) = 4^{-1}\left(\frac{8(\Delta\omega - 2g)^2 + 2(\kappa_+ - 2\kappa_1)^2}{4(\Delta\omega - 2g)^2 + \kappa_+^2} + \frac{8\Delta\omega^2 + 2(\kappa_- - 2\kappa_1)^2}{4\Delta\omega^2 + \kappa_-^2} - \frac{\kappa_1^2(4g^2 + \Gamma^2)}{\left[(\Delta\omega - 2g)^2 + \kappa_+^2\right]\left[4\Delta\omega^2 + \kappa_-^2\right]}\right) \quad (9.44)$$

where the first and second terms denote the contributions of a_+ and a_-, respectively, and the last term denotes interference of these two normal modes.

Figure 9.18 shows that the interplay among different loss parameters and the gain plays a crucial role in the profile of the transmission spectrum and the splitting dynamics. Three typical profiles are immediately seen [204]: (i) Both resonance modes appear as dips when $\xi < \kappa_0$, (ii) the a_+ mode appears as a dip whereas the a_- mode as a peak when $\kappa_{\text{ex}} < 2\Gamma$ and $\kappa_0 < \xi < \kappa_0 + \kappa_{\text{ex}}$ or when $\kappa_{\text{ex}} > 2\Gamma$ and $\kappa_0 < \xi < \kappa_0 + 2\Gamma$, and (iii) both modes appear as resonant peaks when $\kappa_{\text{ex}} > 2\Gamma$ and $\kappa_0 + 2\Gamma < \xi < \kappa_0 + \kappa_{\text{ex}}$. He et al. have studied the effect of gain on the mode splitting spectrum, clearly showing in the experiments that the mode splitting which could not be resolved in the transmission spectra when $\xi = 0$ becomes resolved as the amount of gain increases (Fig. 9.19) [237, 244]. In these experiments, erbium-doped microtoroids were used. A pump light in the 1460 nm wavelength band was used to excite the gain medium, while the wavelength of a probe laser in the 1550 nm wavelength band was scanned to obtain the transmission spectrum. Emission of the erbium ions into the 1550 nm wavelength band provides a mechanism to reduce the loss of the resonance modes and hence the loss experienced by the probe light, leading to a much narrower linewidth. Thus, smaller splittings could be observed.

9.8.2.2 Above the lasing threshold: WGM microlaser

When the active resonator is pumped above the lasing threshold, the laser emission coupled via the taper fiber into a single mode fiber

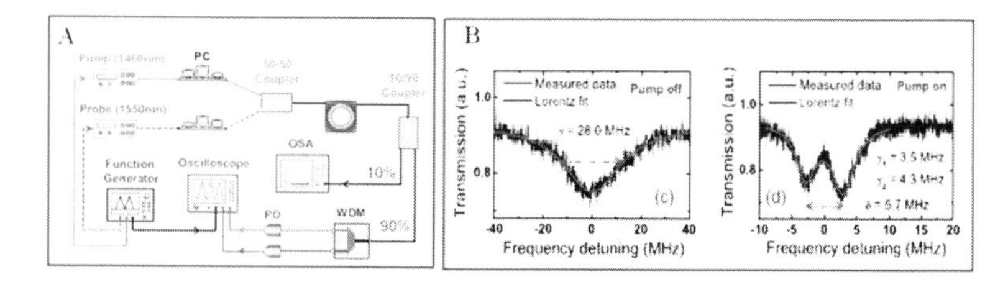

Figure 9.19 Experimental setup to study mode splitting in an active microcavity operated below the lasing threshold. Pump laser is used to excite the active medium (erbium ions) whose emission is in the wavelength band of the probe laser. When the pump is off, the transmission spectrum shows a single broad resonance mode in the probe wavelength band. As soon as the pump is turned on, emission of erbium ions into the probe band compensates a portion of the losses, leading to a narrowing of the resonance linewidths and the appearance of the split modes. Gain allows detecting the mode splitting hidden in the broad linewidth. Adapted with permission from Ref. [244].

and then detected by a photodiode leads to a sinusoidal oscillation in the detected signal. Such an oscillation is not observed for the laser used to pump the active medium. Studies have revealed that such oscillations are due to the mode splitting phenomenon, which leads to the splitting of the single laser line into two lines whose frequencies are spectrally separated from each other by the amount of mode splitting [244]. The oscillation observed in the detected signal is due to the heterodyne mixing of these two lasing lines of different frequencies at the photodiode whose bandwidth is larger than the frequency difference of the split lasing lines. The mixing process then generates a beat note signal whose frequency corresponds to the amount of frequency splitting. Any change in the surrounding or binding of target particles on the cavity surface leads to changes in the beat frequency.

He et al. have shown frequency splitting, and tested its performance for the detection of nanoscale particles, using erbium- and ytterbium-doped microtoroid resonators both in air and in water [62]. The authors used sol–gel synthesis and standard photolithography followed by etching and laser reflowing to fabricate Er- and Yb-doped microtoroidal resonators. For Er-doped microtoroids, a pump laser in the 1460 nm wavelength band was used. The emission was in the 1550 nm band. For Yb-doped microtoroids, the pump was chosen at the 980 nm band with the emission in the 1040 nm band. The emitted laser and the residual pump light were separated using wavelength division multiplexers. The laser light was sent to a photodetector with a bandwidth of 125 MHz to monitor the beat note signals from which the splitting information was extracted. The Yb-doped microlaser was preferred in water experiments because of the low absorption of water at the Yb-emission band. The authors detected gold and polystyrene particles of radius 10 nm and 15 nm, as well as individual Influenza A virions using the Er microlaser in air, and polystyrene particles of radius 30 nm using the Yb microlaser in an aquatic environment. As seen in Fig. 9.20, there are discrete step changes in the beat frequency above the noise level, indicating individual nanoparticle binding events. Since the amount of splitting induced by a particle depends on its position on the resonator surface within the mode volume as well as its location with respect to the previously adsorbed particles, the beat frequency shows discrete upward or downward jumps with varying step heights.

Microlaser-based sensing scheme provides a unique and high-performance platform for nanoscale object detection thanks to the ultranarrow linewidth of the microcavity laser which lowers the detection limit and the self-heterodyning feature, which eliminates the need for tunable laser diodes and hence reduces the costs. In addition, the tunable laser induced noises (e.g., instability of laser frequency tuning, and thermal heating induced distortion in the resonant mode) are avoided, and the response time is no longer limited by the frequency scanning speed of the laser source. It should be noted that in the case of multimode microlasers, one or more modes may experience mode splitting depending on whether the location of the particle overlaps with the field distribution of the lasing mode. In such a case, the amount of mode splitting experienced by each mode depends on the position of the particle with respect to the field maximum of each lasing mode. It may happen that while one mode does not see the particle and hence does not experience splitting, some other modes may undergo splitting induced by the same particle. This is an advantage over

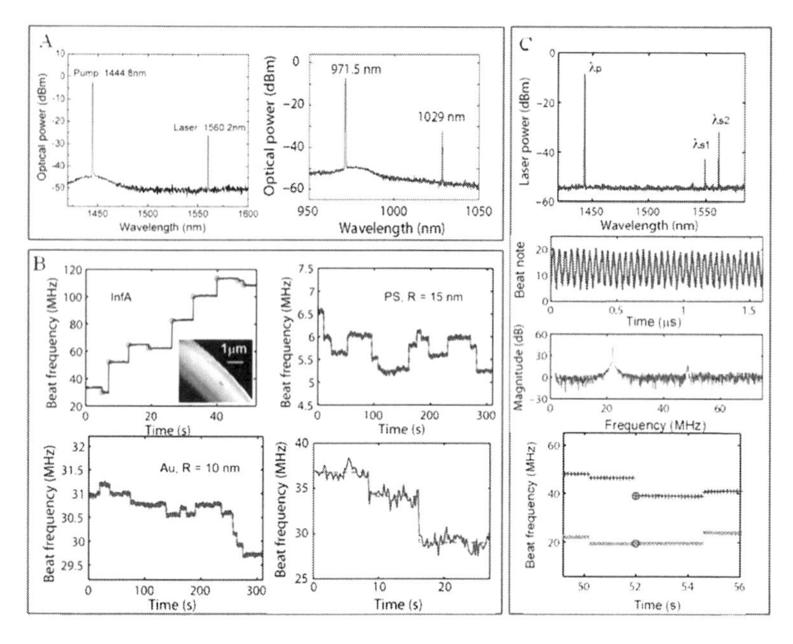

Figure 9.20 Detection of nanoparticles and influenza A virions using frequency splitting in a microcavtity laser. (A) Pump and laser emission lines for erbium- (left) and ytterbium-doped (right) toroid microcavities. Erbium ions are pumped at 1444.8 nm and laser is obtained at 1560.2 nm. Ytterbium ions are pumped at 971.5 nm and laser emission is obtained at 1029 nm. Ytterbium microcavity laser is preferred in experiments performed in water environment due to lower absorption losses in water at the pump and laser emission bands. (B) The change in beat frequency, which reflects the scatterer-induced frequency splitting in the microlaser, is given as a function of time as the scatterers enter the laser mode volume. InfA, PS (15 nm), and Au (10 nm) were detected using an erbium microcavity laser in air, whereas PS (30 nm) was detected in water with a ytterbium microcavity laser. (C) If a multimode microlaser is used, then each of the lasing modes undergoes mode splitting, the amount of which depends on the overlap of the scatterer with the mode volume of the lasing mode. The detected beat signal has contributions from each of these lines. Fourier transform of the beat signal reveals different frequency components corresponding to the amount of splitting each lasing line experiences. Using a multimode laser reduces the possibility of a binding scatterer pass undetected. In the lower panel, it is seen that the third particle does not lead to a change in the splitting frequency (beat frequency) of one of the lasing lines, while it leads to an observable change in the splitting of the other lasing line. Adapted with permission from Ref. [62].

single-mode microlaser-based sensing, as the probability of detecting a particle in a multimode microlaser is higher.

Although He et al. have improved the single-particle detection limit using microlasers beyond what was achieved in passive resonators [62], they could not measure the size of individual detected particles. Instead, they carried out a statistical analysis of the data collected from many measurements to estimate the average size for an ensemble of particles. However, one can use the size estimation method developed by Zhu et al. for mode splitting in passive resonators if a scheme capable of measuring linewidths or monitoring the changes in linewidths of each split lasing modes were present [59, 61]. Thus, to probe the ultimate limit for detecting and measuring single nanoparticles and molecules using mode splitting in microlasers, there is a need for developing schemes to accurately measure the linewidths of split modes.

9.9 Conclusions and Outlook

In this chapter, we have reviewed the underlying physics and recent progress in various sensing schemes utilizing resonant phenomena for label-free detection of molecules and particles. Among the reviewed technologies, some such as SPR-based sensors, have already matured to find their way to commercialization, while the others, such as WGM, PhC, and NEMS resonators, are still in their early development phase. However, they have shown a tremendous potential for detecting nanoscale objects at single particle/object/molecule level. Detection limits down to single nanoparticle/molecule/object have been achieved. In order to increase the interaction between the analytes and the light field in the resonant structures, fabrication processes have been improved to achieve resonators with smaller mode volumes and higher quality factors. Thus, the size of the sensors are becoming smaller and smaller. Meanwhile, there is a great demand in performing measurements with ultralow samples sizes and ultralow concentrations at single-molecule resolutions. Consequently, transporting trace amount of molecules or particles in a sample to nano- or microscale sensing area within a reasonable time have emerged as a problem to be solved for practical applications of these sensors. Thus, efficient ways of sample collection and delivery are in urgent need to improve the detection capabilities of these sensors in various environments.

Whispering gallery mode optical resonators have been the first all-optical devices to detect single molecules and particles with a label-free feature and to study molecular recognition and conformational changes. In order to achieve selectivity and specificity in sensing, various surface functionalization techniques have been developed and effectively used in PhC and WGM resonators employing the spectral shift method. Although detection and measurement of individual nanoparticles and virions have been achieved using mode splitting in WGM resonators at detection limits beyond the reach of the spectral shift method, there has been no report whether surface functionalized resonators could undergo mode splitting and enable the detection, measurement, and recognition of specific nanomaterials. This should be investigated and realized to make use of the full potential of the mode splitting technique. Currently, WGM resonators are generally fabricated from silica or silicon with a small number of reports on chalcogenide glass based resonators. Investigating and developing new materials from which WGM resonators could be fabricated will immensely help in the progress of the WGM sensing field.

In conclusion, resonant structures, especially the WGM microresonators, have the potential to provide highly sensitive detection of trace amount of disease markers for point-of-care testing, screening of drugs, and studying protein folding. Developing tools to integrate WGM resonators with microfluidic platforms without sacrificing the quality factors and mode volumes of the resonators would greatly contribute to biomolecular and biochemical sensing.

Acknowledgments

The authors would like to thank Nature Publishing Group, American Institute of Physics (AIP), Institute of Electrical and Electronics Engineers (IEEE), American Physical Society (APS), and National Academy of Sciences for authorizing the use of figures and images included in this chapter. The authors also gratefully acknowledge the support from NSF under Grant No. 0907467 and No. 0954941.

References

1. B. O'Regan and M. A. Grätzel (1991). *Nature* **353**:737–740.
2. P. Alivisatos (2004). *Nat Biotechnol* **22**:47–52.
3. N. C. Tansil and Z. Gao (2006). *Nanotoday* **1**:28–37.
4. J. M. Nam, C. S. Thaxton, and C. A. Mirkin (2003). *Science* **301**:1884–1886.
5. A. L. Simonian, T. A. Good, S.S. Wang, and J. R. Wild (2005). *Anal. Chim. Acta* **534**:69–77.
6. M. E. Davis, Z. Chen, and D. M. Shin (2008). *Nat Rev Drug Discovery* **7**:771–782.
7. J. Yang, J. You, C.-C. Chen, et al. (2011). *ACS Nano* **5**:6210–6217.
8. S. P. Sundararajan, N. K. Grady, N. Mirin, and N. J. Halas (2008). *Nano Lett.* **8**:624–630.
9. C. Eminian, F.-J. Haug, O. Cubero, X. Niquille, and C. Ballif (2011). *Prog. Photovoltaics Res. Appl.* **19**:260–265.
10. M. Danckwerts and L. Novotny (2007). *Phys. Rev. Lett.* **98**:026104.
11. N. Khlebtsov and L. Dykman (2011). *Chem. Soc. Rev.* **40**:1647–1671.
12. V. L. Colvin (2003). *Nat Biotechnol* **21**:1166–1170.
13. P. H. Hoet, I. Brüske-Hohlfeld, and O. V. Salata (2004). *J. Nanobiotechnol.* **2**:2–12.
14. N. Bellouin, O. Boucher, J. Haywood, and M. Reddy (2005). *Nature* **438**:1138–1141.
15. A. Zelenyuk, D. Imre, J.-H. Han, and S. Oatis (2008). *Anal. Chem.* **80**:1401–1407.
16. M. Brauer, C. Avila-Casado, T. I. Fortoul, S. Vedal, B. Stevens, and A. Churg, (2001). *Environ. Health Perspect.* **109**:1039–1043.
17. L. Bayer-Oglesby, L. Grize, M. Gassner, et al. (2005). *Environ. Health Perspect.* **113**:1632–1637.
18. H. Oh, H. Park, and S. Kim (2004). *Aerosol Sci. Technol.* **38**:1045–1053.
19. S. G. Agarwal, M. Mochida, Y. Kitamori, and K. Kawamura (2007). *Environ. Sci. Technol.* **41**:6920–6925.
20. S. Ghorai and A. V. Tiwanski, (2010). *Anal. Chem.* **82**:9289–9298.

21. Y. Liu, E. R. Gibson, J. P. Cain, H. Wang, V. H. Grassian, and A. Laskin (2008). *J. Phys. Chem.* **112**:1561–1571.
22. H. Geng, J. Ryu, H. J. Jung, H. Chung, K. H. Ahn, and C. U. Ro (2010). *Environ. Sci. Technol.* **44**:2348–2353.
23. M. R. Trusheim, E. R. Berndt, and F. L. Douglas (2007). *Nat. Rev. Drug Discovery* **6**:287–293.
24. M. Christ-Crain and S. M. Opal (2010). *Crit Care* **12**:203.
25. D. R. Boulware, D. B. Meya, T. L. Bergemann, et al. (2010). *PLoS Med.* 7:e1000384.
26. T. B. Sherer (2011). *Sci. Transl. Med.* **3**:79ps14.
27. S.W. Fu, L. Chen, and Yg. Man (2011). *J. Cancer* **2**:116–122.
28. L. A. Broussard (2001). *Mol. Diagn.* **6**:323–333.
29. M. R. Hillman (2002). *Vaccine* **20**:3055–3067.
30. D. Moore (2004). *Rev. Sci. Instrum.* **75**:2499–2512.
31. J. Steinfeld and J. Wormhoudt (1998). *Annu. Rev. Phys. Chem.* **49**:203–232.
32. E. Betzig, J. K. Trautmann, T. D. Harris, J. S. Weiner, and R. L. Kostelak (1991). *Science* **251**:1468–1470.
33. S. M. Nie and S. R. Emery (1997). *Science* 275:1102–1106.
34. W. Denk, J. H. Strickler, and W. W. Webb (1990). *Science* **248**: 73–76.
35. J. Prikulis, F. Svedberg, M. Kall, et al. (2004). *Nano Lett.,* **4**:115–118.
36. D. Boyer, P. Tamarat, A. Maali, B. Lounis, and M. Orrit (2002). *Science* **297**:1160–1163.
37. K. Lindfors, T. Kalkbrenner, P. Stoller, and V. Sandoghdar (2004). *Phys. Rev. Lett.* **93**:037401.
38. F. Patolsky, G. Zheng, O. Hayden, M. Lakadamyali, X. Zhuang, and C. M. Lieber (2004). *Proc. Natl. Acad. Sci. USA* **101**:14017–14022.
39. G. Zheng, F. Patolsky, Y. Cui, W. U. Wang, and C. M. Lieber (2005). *Nat. Biotechnol.* **23**:1294–1301.
40. R. G. Knollenberg (1989). *J. Aerosol Sci.* **3**:331.
41. W. W. Szymanski, A. Nagy, A. Czitrovszky, and P. Jani (2002). *Meas. Sci. Technol.* **13**:303–307.
42. A. Mitra, B Deutsch, F Ignatovich, C. Dykes, and L. Novotny (2010). *Acs Nano* **4**:1305–1312.
43. S. Person, B. Deutsch, A. Mitra, and L. Novotny (2011). *Nano Lett.* **11**:257–261.

44. E. Ozkumur, J. W. Needham, D. A. Bergstein, et al. (2008). *Proc. Natl. Acad. Sci. USA* 105:7988–7992.
45. G. G. Daaboul, A. Yurt, X. Zhang, G. M. Hwang, B. B. Goldberg, and M. S. Unlu (2010). *Nano Lett.* **10**:4727–4731.
46. S. P. Wang, X. Shan, U. Patel, et al. (2010). *Proc. Natl. Acad. Sci. USA* **107**:16028–16032.
47. C. Sonnichsen, S. Geier, N. E. Hecker, G. von Plessen, and J. Feldmann, (2000). *Appl. Phys. Lett.* **77**:2949–2951.
48. A. A. Yanik, M. Huang, O. Kamohara, et al. (2010). *Nano Lett.* **10**:4962–4969.
49. A. K. Naik, M. S. Hanay, W. K. Hiebert, X. L. Feng, and M. L. Roukes, (2009). *Nat Nanotechnol* **4**:445–450.
50. T. P. Burg, M. Godin, S. M. Knudsen, et al. (2007). *Nature* **446**:1066–1069.
51. N. V. Lavrik and P. G. Datskos (2003). *Appl. Phys. Lett.***82**:2697–2699.
52. T. Ono, X. Li, H. Miyashita, and M. Esashi (2003). *Rev. Sci. Instrum.* **74**:1240–1243.
53. A. Gupta, D. Akin, and R. Bashir (2004). *Appl. Phys. Lett.***84**:1976–1978.
54. T. P. Burg and S. R. Manalis (2003). *Appl. Phys. Lett.***83**:2698–2700.
55. T. Braun, V. Barwich, M. K. Ghatkesar, et al. (2005). *Phys. Rev. E* **72**:031907.
56. F. Vollmer and S. Arnold (2008). *Nat Methods* **5**:591–596.
57. A. M. Armani, R. P. Kulkarni, S. E. Fraser, R. C. Flagan, and K. J. Vahala (2007). *Science* **317**:783–787.
58. F. Vollmer, S. Arnold, and D. Keng (2008). *Proc. Natl. Acad. Sci. USA* **105**:20701–20704.
59. J. Zhu, S. K. Ozdemir, Y.-F. Xiao, et al. (2010). *Nat. Photon.* **4**:46–49.
60. S. I. Shopova, R. Rajmangal, Y. Nishida, and S. Arnold (2010). *Rev. Sci. Instrum.* **81**:103110.
61. J. Zhu, S. K. Ozdemir, L. He, D.-R. Chen, and L. Yang (2011). *Opt. Express* **19**:16195–16206.
62. L. He, S. K. Ozdemir, J. Zhu, W. Kim, and L. Yang (2011). *Nat. Nanotechnol.*, **6**:428–432.
63. T. Lu, H. Lee, T. Chen, et al. (2011). *Proc. Natl. Acad. Sci. USA* **108**:5976–5979.
64. K. J. Vahala (2003). Optical microcavities. *Nature* **424**:839–846.
65. D. W. Vernooy, V. S. Ilchenko, H. Mabuchi, E. W. Streed, and H. J. Kimble (1998). *Opt. Lett.* **23**:247–249.

66. M. L. Gorodetsky, A. A. Savchenkov, and V. S. Ilchenko (1996). *Opt. Lett.* **21**:453–455.
67. D. K. Armani, T. J. Kippenberg, S. M. Spillane, and K. J. Vahala (2003). *Nature***421**:925–928.
68. X. Zhang, H. S. Choi, and A. M. Armani (2010). *Appl. Phys. Lett.* **96**:153304.
69. T. J. Kippenberg, S. M. Spillane, and K. J. Vahala (2004). *Appl. Phys. Lett.***85**:6113.
70. J. Niehusmann, A. Vurckel, P. H. Bolivar, T. Wahlbrink, W. Henschel, and H. Kurz (2004). *Opt. Lett.* **29**:2861–2863.
71. I. Kiyat, A. Aydinli, and N. Dagli (2005). *Opt. Express***13**:1900–1905.
72. L.-W. Luo, G. S. Wiederhecker, J. Cardenas, C. Poitras, and M. Lipson (2011). *Opt. Express* **19**:6284–6289.
73. A.A. Savchenkov, V.S. Ilchenko, A.B. Matsko, and L. Maleki (2004). *Phys. Rev. A* **70**:051804(R).
74. Y. Takahashi, H. Hagino, Y. Tanaka, B. S. Song, T. Asano, and S. Noda (2007). *Opt. Express* **15**:17206–17213.
75. M. Notomi, E. Kuramochi, and T. Tanabe (2008). *Nat. Photon.* **2**:741.
76. M. S. Weinberg, C. E. Dube, A. Petrovich, and A. M. Zapata (2003). *J. Microelectromech. Syst.* **12**:567–576.
77. E. Kretschmann and H. Raether (1968). *Z. Naturforsch.* **23A:**2135–2136.
78. A. Otto (1968). *Z. Physik* **216:**398–410.
79. H. Raether (1988). *Surface Plasmons on Smooth and Rough Surfaces and on Gratings*. Berlin: Springer-Verlag.
80. J. Homola, S. S. Yee, and G. Gauglitz (1999). *Sens. Actuators B* **54**:3–15.
81. C. Boozer, G. Kim, S. Cong, H. Guan, and T. Londergan (2006). *Curr. Opin. Biotechnol.* **17**:400–405.
82. K. S. Phillips and Q. Cheng (2007). *Anal. Bioanal. Chem.* **387**:1831–1840.
83. J. Homola (2008). *Chem. Rev.,* **108**:462–493.
84. J. Homola (ed.) (2006). *Surface Plasmon Resonance Based Sensors*. Berlin: Springer.
85. J. Mitchell (2010). *Sensors* **10:**7323–7346.
86. B. Rothenhausler and W. Knoll (1988). *Nature***332**:615–617.
87. F.-J. Schmitt and W. Knoll (1991). *Biophys. J.* **60**:716–720.
88. B. Huang, F. Yu, and R. N. Zare (2007). *Anal Chem* **79**:2979–2983.

89. J. M. Yao, M. E. Stewart, J. Maria, et al. (2008). *Angew Chem Int Ed. Engl.* **47**:5013–5017.
90. C. T. Campbell and G. Kim (2007). *Biomaterials* **28**:2380–2392.
91. R. Thariani and P. Yager (2008). *Sens Actuators B* **130**:765–770.
92. X. Shan, S. Wang, and N. Tao (2010). *Appl. Phys. Lett.* **97**:223703.
93. B. M. Reinhard, M. Siu, H. Agarwal, A. P. Alivisatos, and J. Liphardt (2005). *Nano Lett.* **5**:2246–2252.
94. N. L. Rosi and C. A. Mirkin (2005). *Chem. Rev.* **105**:1547–1562.
95. P. K. Jain and M. A. El-Sayed, (2008). *Nano Lett.* **8**:4347–4352.
96. A.V Zayats and I. I. Smolyaninov (2003). *J. Opt. A: Pure Appl. Opt.* **5**:S16.
97. B. Sepulveda, P. C. Angelome, L. M. Lechuga, and L. M. Liz-Marzan (2009). *Nano Today* **4**:244–251.
98. A. J. Haes and R. P. Van Duyne (2002). *J. Am. Chem. Soc.***124**:10596–10604.
99. J. C. Riboh, A. J. Haes, A. D. McFarland, C. R. Yonzon, and R. P. Van Duyne (2003). *J. Phys. Chem. B* **107**:1772–1780.
100. A. J. Haes, L. Chang, W. L. Klein, and R. P. Van Duyne (2005). *J. Am. Chem. Soc.* **127**:2264–2271.
101. C. Sönnichsen, B. M. Reinhard, J. Liphardt, and A P. Alivisatos, (2005). *Nat Biotech.* **23**:741–745.
102. G. L. Liu, Y. Yin, S. Kunchakarra, et al. (2006). *Nat. Nanotechnol.***1**:47–52.
103. N. Liu, A. P. Alivisatos, M. Hentschel, T. Weiss, and H. Giessen (2011). *Science* **332**:1407–1410.
104. I. H. El-Sayed, X. Huang, and M. A. El-Sayed, (2005). *Nano Lett.* **5**:829–834.
105. J. Qian, T. Fu, Q. Zhan, and S. He (2010). *IEEE J. Sel. Top Quantum Electron.* **16**:672–684.
106. Y. Xing, Q. Chaudry, C. Shen, et al. (2007). *Nat. Protoc.* **2**:1152.
107. J. Qian, X. Li, M. Wei, X. W. Gao, Z. P. Xu, and S. L. He (2008). *Opt. Express* **16**:19568.
108. P. Hanarp, M. Kall, and D. S. Sutherland (2003). *J. Phys. Chem. B* **107**:5768–5772.
109. T. R. Jensen, G. C. Schatz, and R. P. Van Duyne (1999). *J. Phys. Chem. B* **103**:2394–2401.
110. T. Okamoto, I. Yamaguchi, and T. Kobayashi (2000). *Opt. Lett.* **25**:372–374.
111. L. J. Sherry, S. H. Chang, G. C. Schatz, R. P. Van Duyne, B. J. Wiley, and Y. N. Xia (2005). *Nano Lett.* **5**:2034–2038.

112. J. S. Shumaker-Parry, H. Rochholz, and M. Kreiter, (2005). *Adv. Mater.* **17**:2131–2134.
113. C. L. Nehl, H. W. Liao, and J. H. Hafner (2006). *Nano Lett.* **6**:683–688.
114. J. Prikulis, P. Hanarp, L. Olofsson, D. Sutherland, and M. Kall (2004). *Nano Lett.* **4**:1003–1007.
115. R. D. Averitt, D. Sarkar, and N. J. Halas (1997). *Phys. Rev. Lett.* **78**:4217–4220.
116. J. Aizpurua, P. Hanarp, D. S. Sutherland, M. Kall, G. W. Bryant, and F. J. G. de Abajo (2003). *Phys. Rev. Lett.***90**:057401.
117. H. Wang, D. W. Brandl, F. Le, P. Nordlander, and N. J. Halas (2006). *Nano Lett.***6**:827–832.
118. A. J. Haes, S. Zou, G. C. Schatz, and R. P. Van Duyne (2004). *J. Phys. Chem. B* **108**:6961.
119. A. J. Haes, S. Zou, G. C. Schatz, and R. P. Van Duyne, (2004). *J. Phys. Chem. B***108**:109.
120. A. D. McFarland and R. P. Van Duyne (2003). *Nano Lett.* **3**:1057–1062.
121. E. M. Larsson, J. Alegret, M. Käll, and D. S. Sutherland (2007). *Nano Lett.* **7**:1256–1263.
122. S. K. Dondapati, T. K. Sau, C. Hrelescu, T. A. Klar, F. D. Stefani, and J. Feldmann (2010). *ACS Nano* **4**:6318.
123. A. J. Haes, L. Chang, W. L. Klein, and R. P. Van Duyne (2005). *J. Am. Chem. Soc.* **127**:2264–2271.
124. A. J. Haes, L. Chang, W. L. Klein, and R. P. Van Duyne (2004). *Nano Lett.* **4**:1029–1034.
125. M. Vestergaard, K. Kerman, D. K. Kim, H. M. Hiep, and E. Tamiya (2008). *Talanta***74**:1038.
126. K. M. Mayer, F. Hao, S. Lee, P. Nordlander, and J. H. Hafner (2010). *Nanotechnology* **21**:255503.
127. S. A. Maier (2006). *Opt. Express* **14**:1957–1964.
128. T. Vo-Dinh, H.-N. Wang, and J. Scaffidi (2010). *J. Biophoton.* **3**:89–102.
129. P. Scaffidi, M. K. Gregas, V. Seewaldt, and T. Vo-Dinh (2009). *Anal. Bioanal. Chem.* **393**:1135–1141.
130. K. Jensen, K. Kim, and A. Zettl (2008). *Nat. Nanotech.* **3**:533–537.
131. B. Lassagne, D. Garcia-Sanchez, A. Aguasca, and A. Bachtold (2008). *Nano Lett.* **8**:3735–3738.
132. J. Lee, W. Shen, K. Payer, T. P. Burg, and S. R. Manalis (2010). *Nano Lett.* **10**:2537–2542.

133. J. T. Heeres and P. J. Hergenrother (2011). *Chem. Soc. Rev.* **40**:4398–4410.
134. S. M. Shamah and B. T. Cunningham (2011). *Analyst* **136**:1090–1102.
135. K. Lee and S. A. Asher (2000). *J. Am. Chem. Soc.* **122**:9534–9537.
136. F. Villa, L. E. Regalado, F. Ramos-Mendieta, J. Gaspar-Armenta, and T. Lopez-Rıos (2002). *Opt. Lett.* **27**:646–648.
137. V. N. Konopsky and E. V. Alieva (2007). *Anal. Chem.***79**:4729–4735.
138. E. Chow, A. Grot, L.W. Mirkarimi, M. Sigalas, and G. Girolami (2004). *Opt. Lett.* **29**:1093–1095.
139. V. Toccafondo, J. García-Rupérez, M. J. Bañuls, et al. (2010). *Opt. Lett.* **35**:3673–3675.
140. N. Skivesen, A. Tetu, M. Kristensen, J. Kjems, L. H. Frandsen, and P. I. Borel (2007). *Opt. Express* **15:**3169–3176.
141. M. Lee and P. M. Fauchet (2007). *Opt. Express* **15**:4530–4535.
142. R. A. Barry and P. Wiltzius (2006). *Langmuir* **22**:1369–1374.
143. P. A. Snow, E. K. Squire, P. St. J. Russell, and L. T. Canham (1999). *J. Appl. Phys.* **86**:1781–1784.
144. S. Chakravarty, J. Topol'ancik, P. Bhattacharya, S. Chakrabarti, Y. Kang, and M. E. Meyerhoff (2005). *Opt. Lett.* **30**:2578–2580.
145. T. Sunner, T. Stichel, S.-H. Kwon, et al. (2008). *Appl. Phys. Lett.* **92**:261112.
146. Y. Nishijima, K. Ueno, S. Juodkazis, V. Mizeikis, H. Misawa1, T. Tanimura, and K. Maeda (2007). *Opt. Express* **15**:12979–12988.
147. H. Li, L. Chang, J. Wang, L. Yang, and Y. Song (2008). *J. Mater. Chem.* **18**:5098–5103.
148. J. Ballato and A. James (1999). *J. Am. Ceram. Soc.* **82**:2273–2275.
149. M. Centini, C. Sibilia, M. Scalora, et al. (1999). *Phys. Rev. E* **60:**4891.
150. H. Inouye, M. Arakawa, J. Y. Ye, T. Hattori, H. Nakatsuka, and K. Hirao (2002). *IEEE J. Quantum Electron.* **38**:867.
151. T. F. Krauss, R. M. De La Rue, and S. Brand (1996). *Nature* **383:**699–702.
152. S. G. Johnson, S. Fan, P. R. Villeneuve, J. D. Joannopoulos, and L. A. Kolodziejski (1999). *Phys. Rev. B* **60**:5751–5758.
153. Y. Akahane, T. Asano, B.-S. Song, and S. Noda (2003). *Nature* **425**:944–947.
154. E. Ozbay, A. Abeyta, G. Tuttle, et al. (1994). *Phys. Rev. B* **50**:1945.
155. S. Noda, K. Tomoda, N. Yamamoto, and A. Chutinan (2000). *Science* **289**:604–606.

156. K. M. Ho, C. T. Chan, C. M. Soukoulis, R. Biswas, and M. Sigalas (1994). *Solid State Commun.* **89**:413–416.
157. J. Topol'ancik, P. Bhattacharya, J. Sabarinathan, and P. C. Yu (2003). *Appl. Phys. Lett.* **82**:1143.
158. M. Loncar, A. Scherer, and Y. Qiu (2003). *Appl. Phys. Lett.* **82**:4648–4650.
159. J. B. Jensen, L. H. Pedersen, P. E. Hoiby, et al. (2004). *Opt. Lett.* **29**:1974–1976.
160. J. B. Jensen, P. E. Hoiby, G. Emiliyanov, O. Bang, L. H. Pedersen, and A. Bjarklev (2005). *Opt. Express* **13**:5883–5889.
161. L. Rindorf, J. B. Jensen, M. Dufva, L. H. Pedersen, P. E. Hoiby, and O. Bang (2006). *Opt. Express* **14**:8224.
162. N. Skivesen, A. Têtu, M. Kristensen, J. Kjems, L. H. Frandsen, and P. I. Borel (2007). *Opt. Express* **15**:3169.
163. J. García-Rupérez, V. Toccafondo, M. J. Bañuls, et al. (2010). *Opt. Express* **18**:24276–24286.
164. M. R. Lee and P. M. Fauchet (2007). *Opt. Express* **15**:4530–4535.
165. M. R. Lee and P. M. Fauchet (2007). *Opt. Lett.* **32**:3284.
166. S. Mandal, J. M. Goddard, and D. A. Erickson (2009). *Lab Chip* **9**: 2924–2932.
167. N. A. M. Nunes, J. P. Kutter, and K. B. Mogensen (2008). *Opt. Lett.* **33**:1623–1625.
168. Y. Guo, J. Y. Ye, C. Divin, et al. (2010). *Anal. Chem.* **82**:5211–5218.
169. L. L. Chan, M. F. Pineda, J. Heeres, P. Hergenrother, and B. T. Cunningham (2008). *ACS Chem. Biol.* **3**:437–448.
170. M. L. Adams, G. A. DeRose, M. Loncar, and A. Scherer (2005). *J. Vac. Sci. Technol. B* **23**:3168–3173.
171. M. L. Adams, M. Loncar, A. Scherer, and Y. Qiu (2005). *IEEE J. Sel. Areas Commun.* **23**:1348–1354.
172. S. Kita, K. Nozaki, and T. Baba (2008). *Opt. Express* **16**:8174–8180.
173. S. Kim, J. Lee, H. Jeon, and H. J. Kim (2009). *Appl. Phys. Lett.* **94:**133503.
174. T.-W. Lu, Y.-H. Hsiao, W.-D. Ho, and P.-T. Lee (2010). *Opt. Lett.* **35**:1452–1454.
175. M. A. Dündar, E. C. I. Ryckebosch, R. Nötzel, F. Karouta, L. J. van IJzendoorn, and R. W. van der Heijden (2010). *Opt. Express* **18**:4049.
176. J. C. Knight, G. Cheung, F. Jacques, and T. A. Birks (1997). *Opt. Lett.* **22**:1129–1131.

177. M. Cai, O. Painter, and K. J. Vahala (2000). *Phys. Rev. Lett.* **85**:74–77.
178. S. M. Spillane, T. J. Kippenberg, O. J. Painter, and K. J. Vahala (2003). *Phys. Rev. Lett.* **91**: 043902.
179. W. Kim, S. K. Ozdemir, J. Zhu, L. He, and L. Yang (2010). *Appl. Phys. Lett.* **97**:071111.
180. W. Kim, S. K. Ozdemir, J. Zhu, and L. Yang, (2011). *Appl. Phys. Lett.* **98**:141106.
181. C. Grillet, S. N. Bian, E. C. Magi, and B. J. Eggleton (2008). *Appl. Phys. Lett.* **92**:171109.
182. D. H. Broaddus, M. A. Foster, I. H. Agha, J. T. Robinson, M. Lipson, and A. L. Gaeta (2009). *Opt. Express* **17**:5998–6003.
183. P. Wang, G. S. Murugan, T. Lee, et al. (2011). *Appl. Phys. Lett.* **98**:181105.
184. O. Svitelskiy, Y. Li, A. Darafsheh, et al. (2011). *Opt. Lett.* **36**:2862–2864.
185. I. M. White, H. Oveys, and X. Fan (2006). *Opt. Lett.* **31**:1319–1321.
186. M. Sumetsky, R. S. Windeler, Y. Dulashko, and X. Fan (2007). *Opt. Express* **15**:14376–14381.
187. I. M. White, J. Gohring, Y. Sun, G. Yang, S. Lacey, and X. Fan (2007). *Appl. Phys. Lett.* **91**: 241104.
188. H. Zhu, I. M. White, J. D. Suter, P. S. Dale, and X. Fan (2007). *Opt. Express* **15**:9139–9146.
189. H. Zhu, I. M. White, J. D. Suter, and X. Fan (2008). *Biosens. Bioelectron.* **24**:461–466.
190. H. Li and X. Fan (2010). *Appl. Phys. Lett.* **97**:011105.
191. H.-C. Ren, F. Vollmer, S. Arnold, and A. Libchaber (2007). *Opt. Express* **15**.
192. S. Arnold, M. Noto, and F. Vollmer (2005). *Frontiers of Optical Spectroscopy*. B. DiBartolo and O. Forte (eds.). Drodrecht, the Netherlands: Kluwer Academic Publishers, 337–357.
193. A. Yalçin, K. C. Popat, J. C. Aldridge, et al. (2006). *IEEE J. Sel. Top. Quantum Electron.* **12**:148–155.
194. A. Mazzei, S. Götzinger, L. de S. Menezes, G. Zumofen, O. Benson, and V. Sandoghdar (2007). *Phys. Rev. Lett.* **99**:173603.
195. D. S. Weiss, V. Sandoghdar, J. Hare, V. Lefevre-Seguin, J.-M. Raimond, and S. Haroche (1995). *Opt. Lett.* **20**:1835–1837.
196. M. L. Gorodetsky, A. D. Pryamikov, and V. S. Ilchenko (2000). *J. Opt. Soc. Am. B* **17**:1051–1057.
197. M. Borselli, T. Johnson, and O. Painter (2005). *Opt. Express* **13**:1515–1530.

198. L. Chantada, N. I. Nikolaev, A. L. Ivanov, P. Borri, and W. Langbein (2008). *J. Opt. Soc. Am. B* **25**:1312–1321.
199. L. He, S. K. Ozdemir, J. Zhu, and L. Yang (2010). *Appl. Phys. Lett.* **96**:221101.
200. J. Zhu, S. K. Ozdemir, L. He, and L. Yang (2010). *Opt. Express* **18**:23535–23543.
201. J. Knittel, T. G. McRae, K. H. Lee, and W. P. Bowen (2010). *Appl. Phys. Lett.* **97**:123704.
202. X. Yi, Y.-F. Xiao, Y. Li, et al. (2010). *Appl. Phys. Lett.* **97**:203705.
203. I. Teraoka and S. Arnold (2009). *J. Opt. Soc. Am. B* **26**:1321–1329.
204. K. R. Hiremath and V. N. Astratov (2008). *Opt. Express* **16**:5421–5426.
205. I. Teraoka and S. Arnold (2006). *J. Opt. Soc. Am. B* **23**:1381–1389.
206. F. Vollmer, D. Braun, A. Libchaber, M. Khoshsima, I. Teraoka, and S. Arnold (2002). *Appl. Phys. Lett.* **80**:4057.
207. S. Arnold, M. Khoshsima, I. Teraoka, S. Holler, and F. Vollmer (2003). *Opt. Lett.* **28**:272–274.
208. I. Teraoka and S. Arnold (2007). *J. Appl. Phys.* **102**:076109.
209. E. Krioukov, D. J. W. Klunder, A. Driessen, J. Greve, and C. Otto (2002). *Opt. Lett.* **27**:512–514.
210. F. Vollmer, S. Arnold, D. Braun, I. Teraoka, and A. Libchaber (2003). *Biophys. J.* **85**:1974–1979.
211. S. Arnold, S. Shopova, and S. Holler (2010). *Opt. Express* **18**:281–287.
212. S. Arnold, D. Keng, S. I. Shopova, S. Holler, W. Zurawsky, and F. Vollmer (2009). *Opt. Express* **17**:6230–6238.
213. J. D. Suter, I. M. White, H. Zhu, H. Shi, C. W. Caldwell, and X. Fan (2008). *Biosens Bioelectron.* **23**:1003–1009.
214. H. Zhu, I. M. White, J. D. Suter, M. Zourob, and X. Fan (2008). *Analyst* **132**:356–360.
215. J. Gohring and X. Fan (2010). *Sensors* **10**:5798–5808.
216. T. J. Kippenberg, S. M. Spillane, and K. J. Vahala (2002). *Opt. Lett.* **27**:1669–1671.
217. S. K. Ozdemir, J. Zhu, L. He, and L. Yang (2011). *Phy. Rev. A* **83**:033817.
218. B. Koch, Y. Yi, J.-Y. Zhang, S. Znameroski, and T. Smith (2009). *Appl. Phys. Lett.* **95**:201111.
219. B. Koch, L. Carson, C.-M. Guo, et al. (2010). *Sens Actuators B* **147**:573–580.
220. E. Snitzer (1966). *Appl. Opt.* **5**:1487–1499.

221. E. Snitzer and R. Woodcock (1965). *Appl. Phys. Lett.* **6**:45–46.
222. V. Sandoghdar, F. Treussart, J. Hare, V. Lefevre-Seguin, J. M. Raimond, and S. Haroche (1996). *Phys. Rev. A* **54**:R1777.
223. S. V. Frolov, M. Shkunov, Z. V. Vardeny, and K. Yoshino (1997). *Phys Rev B* **56**:R4363–R4366.
224. A. Fujii, T. Nishimura, Y. Yoshida, K. Yoshino, and M. Ozaki (2005). *Japn. J. Appl. Phys.* **44**:L1091–L1093.
225. A. Tulek, D. Akbulut, and M Bayindir (2009). *Appl. Phys. Lett.* **94**:203302.
226. L. Yang and K. J. Vahala (2003). *Opt Lett* **28**:592–594.
227. C. H. Dong, Y. F. Xiao, Z. F. Han, et al. (2008). *IEEE Photon. Technol. Lett.* **20**:342–344.
228. H. Takashima, H. Fujiwara, S. Takeuchi, K. Sasaki, and M. Takahashi (2007). *Appl. Phys. Lett.* **90**.
229. L. Yang, T. Carmon, B. Min, S. M. Spillane, and K. J. Vahala (2005). *Appl. Phys. Lett.* **86**:091114.
230. E. P. Ostby, L. Yang, and K. J. Vahala (2007). *Opt Lett.* **32**:2650–2652.
231. H. S. Hsu C. Cai, and A. M. Armani (2009). *Opt Express* **17**:23265–23271.
232. B. Min, S. Kim, K. Okamoto, et al. (2006). *Appl. Phys. Lett.* **89**191124
233. J. Kalkman A. Polman, T. J. Kippenberg, K. J. Vahala, and M. L. Brongersma (2006). *Nucl Instr Meth. B* **242**:182–185.
234. A. Polman, B. Min, J. Kalkman, T. J. Kippenberg, and K. J. Vahala (2004). *Appl. Phys. Lett.* **84**:1037–1039.
235. R. Roy (1987). *Science* **238**:1664–1669.
236. L. L. Hench and J. K. West (1990). *Chem Rev* **90**:33–72.
237. L. He, S. K. Ozdemir, Y. F. Xiao, and L. Yang (2010). *IEEE J. Quantum Electron.* **46**:1626–1633.
238. A. L. Schawlow and C. H. Townes (1958). *Phys. Rev.* **112**:1940.
239. J. Yang and L. J. Guo (2006). *IEEE J. Sel Top. Quantum Electron.* **12**:143–147.
240. A. Francois and M. Himmelhaus (2009). *Appl. Phys. Lett.* **94**:031101.
241. H. T. Beier, G. L. Cote, and K. E. Meissner (2009). *Ann Biomed Eng.* **37**:1974–1983.
242. S. Pang, R. E. Beckham, and K. E. Meissner (2008). *Appl. Phys. Lett.* **92**:221108.
243. E. Nuhiji and P. Mulvanely (2007). *Small* **3**:1408–1414.
244. L. He, S. K. Ozdemir, J. Zhu, and L. Yang (2010). *Phys. Rev. A* **82**:053810.

Chapter 10

Optical Tweezers

R. W. Bowman* and M. J. Padgett
SUPA, School of Physics and Astronomy, University of Glasgow, G12 8QQ, UK
richard.bowman@cantab.net

10.1 Introduction

Light is often used to transfer energy to biological systems, whether it is exciting a fluorophore or changing the conformational state of a photoactivated molecule. However, the momentum of light can also play an important role in biophysical experiments. Optical tweezers (OTs) exploit this to perform very sensitive force measurements and delicate manipulations. Over the four decades since Ashkin's seminal paper [1], optical micromanipulation has developed from qualitative physics experiments to provide precise tools that can be used to manipulate and measure forces and displacements on the single-molecule level.

The simplest model of an optical trap is a spring, where a trapped particle feels a force pulling it back towards the trap center. This force increases linearly with its distance from the equilibrium point. This enables optical tweezers to make sensitive

---

*Current affiliation: Nanophotonics Centre, Cavendish Laboratory, University of Cambridge, CB3 0HE, UK

Understanding Biophotonics: Fundamentals, Advances, and Applications
Edited by Kevin K. Tsia

ISBN 978-981-4411-77-6 (Hardcover), 978-981-4411-78-3 (eBook)
www.panstanford.com

Figure 10.1 Illustration of an optical trap being used to measure the force exerted by a kinesin molecule walking along a microtubule, as done in [2, 3].

force measurements, detecting forces 1000 times smaller than those accessible with atomic force microscopes (AFMs). Instead of a mechanical cantilever, the spring in optical tweezers is made of light. This confers two main benefits: firstly the spring is several orders of magnitude softer then an AFM cantilever, and secondly there is no requirement to have mechanical access to the sample, meaning that closed sample cells can be used. The soft spring has enabled groundbreaking experiments on single molecules, such as measuring the step sizes and forces exerted by single-molecular motors such as kinesins [2] and myosins [3]. The elasticity and hence conformational properties of DNA have also been probed with this technique [4].

Optical tweezers are formed with a focussed laser beam, which is deflected by the particle. It is this deflection of the light that gives rise to the forces applied to trapped objects, which can range in size from a few nanometers [5, 6] to tens or hundreds of micrometers [7]. A microscope objective is usually used to focus the laser beam onto the sample, and thus optical tweezers are generally combined with optical microscopy (tweezers systems are often based around a high-magnification microscope). In this chapter we will look at the optical systems, the forces exerted, and some applications of optical tweezers.

10.2 Optical Forces

Precise modeling of the forces and torques exerted by optical tweezers is most often done using the generalized Lorentz–Mie theory, but the simple ray-optical model (strictly valid only for the largest particles used in optical tweezers) [8] provides an excellent qualitative insight into the physics underlying optical tweezers. Fundamentally, optical tweezers function because of the momentum of light. As light passes through a trapped particle, its direction, and hence its momentum, is changed by the particle. This means that the light exerts a force on the particle, which (for particles that are stably trapped) acts to move it back toward the laser focus. The simple ray-optical model of light focussed through a sphere tells us that the light's direction is unchanged when the sphere is precisely at the laser focus, as illustrated in Fig. 10.2. When the bead is laterally displaced from the focus, it deflects the laser beam in the same direction, resulting in a force that draws it back to the optical axis. Similarly, when the bead is displaced axially, it changes the divergence of the laser beam. As we must sum over the whole beam to find its momentum, a beam with lower divergence will have a higher overall momentum than a beam with high divergence. Thus, if the particle moves behind the focus (and decreases the divergence

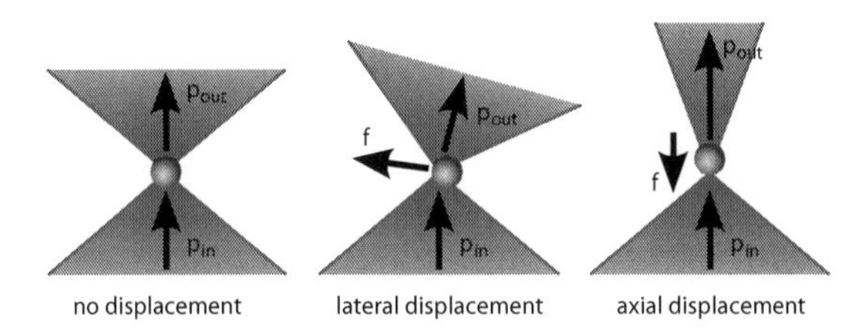

Figure 10.2 A simple model of the gradient force resulting from a laser beam being focussed through a small glass sphere. As the sphere is displaced from the focal point, it bends the light, changing its momentum. This in turn exerts a force on the particle. The laser is propagating upwards in each figure.

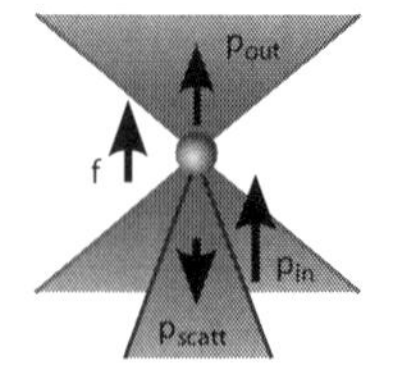

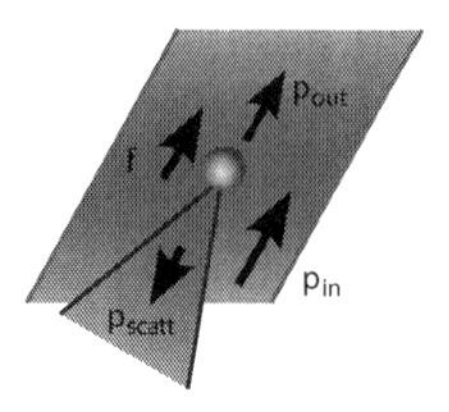

Figure 10.3 An illustration of the "scattering force" due to a partially reflective particle in an optical trap. Some of the incident light is reflected, causing a force normally in the same direction as the incident light. This displaces the equilibrium position of a bead in an optical trap axially, so it is held slightly behind the focus (left). Extended optical traps, such as line traps (right), can cause particles to move along the bright region due to the scattering force.

of the light exiting the trap), it has increased the momentum of the light in the axial direction and hence feels a force pushing it back toward the focus. Borrowing a term from Rayleigh scattering theory, this force is often referred to as the "gradient" force as it moves the particle up the intensity gradient, toward the point of maximum intensity (the laser focus).

To complete our qualitative picture of the forces, we must take into account that the bead will not perfectly refract the light; some of it will be absorbed and some reflected. This results in a small but significant force pushing the particle in the direction of propagation of the laser beam, often referred to as the "scattering" force, illustrated in Fig. 10.3. For this reason, the equilibrium position of beads in a trap does not place the center of the particle on the center of the focus, rather the particle sits behind this position (i.e., further along the direction of propagation of the laser). This is also the reason we need to use a tightly focussed beam for a single-beam trap: were it not for the axial restoring force, the particle would be moved onto the beam axis then simply pushed along the direction of propagation of the laser. In fact, our qualitative model can go further, demonstrating that the maximum force we can exert pulling a particle against the direction of propagation is limited by the divergence of the light entering the particle. The maximum possible axial momentum the light can have on exiting the particle is reached when the beam is fully collimated, i.e., the angle of the cone of light is close to zero. Thus, if we use a more weakly focussed beam it has a higher axial momentum before the particle and consequently we can reach only very small axial forces. The point where the gradient force can overcome the scattering force to create a stable trap in 3D varies depending on the size and type of particle being trapped, but it is typically at a high numerical aperture (NA), usually greater than one when working in water. It is worth noting that this corresponds to a cone angle much higher than that depicted in Fig. 10.2, often exceeding 130°.

Shaping the beam used to generate the optical traps allows some of these forces to be tuned: for example, ring-shaped beams [9, 10] can increase axial stiffness by effectively increasing the divergence of the beam before the particle (if we block out the center portion of the beam, more of the light is propagating at higher angles to the optical axis, and hence its axial momentum is lower). Laguerre–Gaussian beams focus to a ring-shaped spot, which can also increase the axial stiffness of an optical trap by concentrating more of the light at the edges of the particle.

10.2.1 *Nonconservative forces*

The gradient force corresponds to a readymade "potential," that of the intensity of the laser beam. Even in the case of larger particles, what we refer to as the gradient force will not transfer energy to a particle provided we look over a time interval where it starts and finishes in the same place. However, the scattering force is "nonconservative," meaning that it can transfer energy. For example, if a particle is pushed along the beam axis in the center of the optical trap but then moves toward the edge of the beam (where the scattering force is weaker) and then moves in the opposite direction back to its starting point, the scattering force will have done work on the particle [11]. This means that it is not strictly correct to define a potential for an optical trap where the scattering force is not zero. However, in most situations where optical traps are currently used, the nonconservative effects are sufficiently small not to have a great effect on force measurement. One notable exception is when the trapping potential is extended, so the scattering force can be used to push particles along paths, often referred to as "optical guiding"

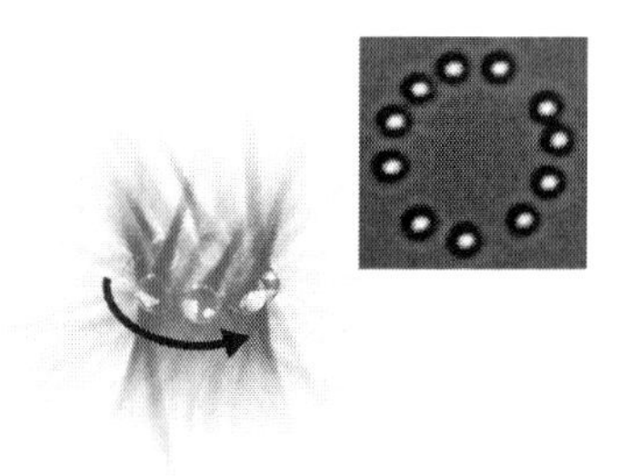

Figure 10.4 Beads circulating in a ring-shaped Laguerre–Gaussian laser beam, as used in the works by O'Neil [14] and Roichman [15] among others.

[12, 13]. An excellent demonstration of nonconservative forces is that of particles circulating around a ring-shaped trap as shown in Fig. 10.4, which has been used to study both the angular momentum of light and the colloidal dynamics of such a system [14, 15].

10.2.2 *Rayleigh Scattering: Very Small Particles*

For particles which are very much smaller than a wavelength of light (i.e., nanoparticles), we can model their interaction with the electromagnetic field of the laser beam as a single induced dipole. This is the origin of the terms *gradient force* and *scattering force*, as the equations separate quite neatly into the two different forces. Placing a polarizable particle in an electric field will induce dipole moments throughout the particle, but in the case of a very small particle it is acceptable to model this as a single point dipole at the center of the particle. As the electric field of the light is oscillating, the particle's dipole moment also oscillates, and thus it re-radiates (or "scatters") some of the light. However, this scattered light goes in all directions and not just the direction of the incident light. For this scattered light, the average momentum is zero since the different directions cancel out. This means the difference in momentum between incident and scattered light pushes it in the direction of the incident light, the "scattering force." However, there is another effect at play—namely, the interaction of an electric dipole with a gradient in the electric field. This acts to pull the particle toward the point of maximum intensity, the laser focus. The size of the induced dipole, and hence the magnitude of both gradient and scattering forces, depends not only on laser power but also on particle size. It increases very sharply with particle radius a, being proportional to volume, i.e., a^3, for very small particles [5].

10.2.3 *Mie Scattering: Intermediate-Sized Particles*

Ray-optical modeling of the forces is valid for particles much bigger than a wavelength of light (around 10 μm or larger); however, most particles used are smaller than this. Too small for ray optics and too large for approximation by a single dipole, these intermediate-sized particles are most often dealt with using a generalization of Mie theory. This is an analytic solution of Maxwell's equations for a plane wave interacting with a dielectric sphere. It is often combined with T-matrix formalism (which uses a matrix to represent the relationship between input and output light fields) to make estimates of trap potentials and stiffness values for micrometer-sized particles [16, 17]. Nonspherical particles, however, require more computationally demanding modeling such as finite-element simulations that solve Maxwell's equations directly.

10.3 System Designs for Optical Tweezers

The single-beam gradient trap is a basic design pattern which has been refined and extended in many ways; however, most systems have many elements in common. The essential feature of single-beam optical tweezers is a high-NA objective lens, to create a small focal volume which is tightly localized in three dimensions. Modern microscopes generally produce an image at infinity, which means that we use a tube lens to bring the image into focus on the camera. This also means that projecting an optical trap is as simple as coupling a collimated laser beam into the back aperture of the objective, as shown in Fig. 10.5.

Additional optics are typically needed to adjust the width of the laser beam, and to align it both in the sample and on the back aperture. If the spacing between each pair of lenses is equal to the sum of the focal lengths, tilting the beam in the Fourier plane shown in Fig. 10.6 translates the focus in the sample without

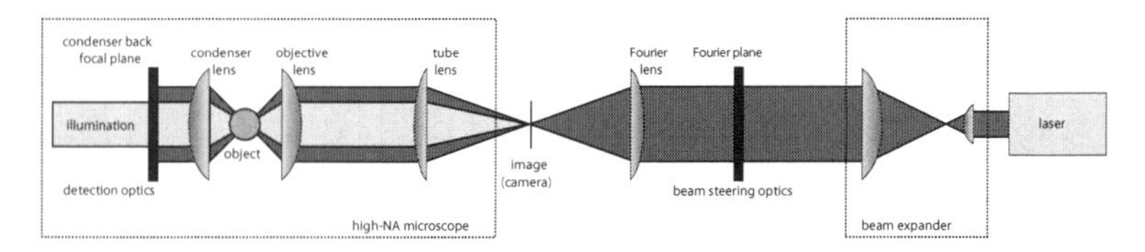

Figure 10.5 The optical system for a single-beam gradient trap. A laser beam is expanded with a telescope and then focussed by a high-NA microscope. The location of beam steering optics and detection optics for back focal plane interferometry are shown. Fold mirrors and beamsplitters are omitted for clarity, but usually the system is arranged such that the microscope is vertical, as in a standard inverted microscope. The remainder of the system is then usually laid out on an optical breadboard.

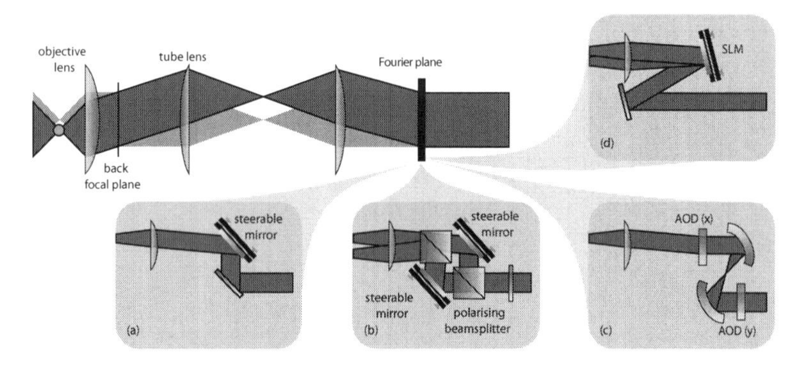

Figure 10.6 Beam steering and multiplexing techniques for movable and/or multiple optical traps: (a) steerable mirror in Fourier plane, (b) polarizing beamsplitters and two steerable mirrors, (c) acousto-optic deflectors (AODs), and (d) spatial light modulator (SLM).

changing the direction of the cone of light entering the focal point, a condition known as *telecentric imaging*. This is necessary for the trap properties (stiffness, etc.) to remain constant as the trap is moved around the sample, which is particularly important when making force measurements. Figure 10.6 shows a number of methods of moving an optical trap, ranging from a steerable mirror (for a single movable optical trap) to a spatial light modulator (SLM) which can create multiple traps at arbitrary positions in 3D [18]. Besides their cost, the simpler beam steering designs (such as a single mirror or two mirrors and beamsplitters) are often favored when forces are to be measured using an interferometric approach, which is complicated by the use of SLMs or acousto-optic deflectors (AODs).

10.3.1 *Dual Traps*

The simplest method for creating two optical traps is to divide and recombine the laser beam with two polarizing beamsplitters, as shown in Fig. 10.6. This gives rise to two orthogonally polarized traps which can be independently steered. However, any drift due to the laser will be common to both, and this has been used to cancel out drift from sensitive force measurements [19, 20]. Such setups have been used to make sensitive measurements of biomolecules such as DNA [19], as well as to explore physical concepts in bistable traps [21].

10.3.2 *Time-Shared Traps*

When more than two optical traps are required, splitting and recombining the laser with beamsplitters becomes cumbersome and inefficient. A common way around this problem is to use either a scanning mirror or a pair of AODs to move the trapping laser between a number of trap sites very rapidly [22]. If the laser is scanned much faster than the characteristic "corner frequency" of the optical traps, the fluctuations are averaged out and the system behaves as if there were multiple traps. Multiple traps created in this way do not interfere with each other (as only one trap is present at any given instant); however, they can only be scanned in 2D. More recently, multiple AODs have been used to create a scanning lens, allowing the focus to be shifted axially. This opens a number of exciting possibilities for 3D imaging, but its low optical efficiency means that this technique will need further development before it can become a viable optical trapping system [23].

10.3.3 *Holographic Optical Tweezers*

Holographic optical tweezers (HOTs) use a computer-controlled optical element to modify the phase of a laser beam [18, 24]. This allows great flexibility both in the number of traps created and in the shape of each trap. By using a spatial light modulator (SLM) to insert a hologram corresponding to the interference of several plane waves into the Fourier plane, a single beam can be multiplexed into multiple beams that produce an array of traps in 2D. However, the SLM also allows the beams to be defocussed and hence the traps can be moved in three dimensions [25]. In contrast to time-shared techniques, holographically multiplexed traps are all present simultaneously. However, this means that they can interfere with each other, producing unwanted "ghost traps" and several techniques have been developed to overcome this problem [26].

By tilting the laser beam in the Fourier plane, i.e., the back focal plane of the objective, it is possible to move the focal spot laterally. One way of achieving this would be to use a prism, but this is not easy to reconfigure. Alternatively, we can use a reprogrammable liquid crystal display or SLM to modify the phase of the laser beam in the same way that a prism does, as shown in Fig. 10.7. The phase shift from a prism is proportional to the thickness of the prism, but a phase shift of one full wavelength is equivalent to no phase shift at all. This means we can "wrap" the phase back to zero again; hence the periodic jumps from black (no phase shift) to white (one wavelength of retardation) in Fig. 10.7.

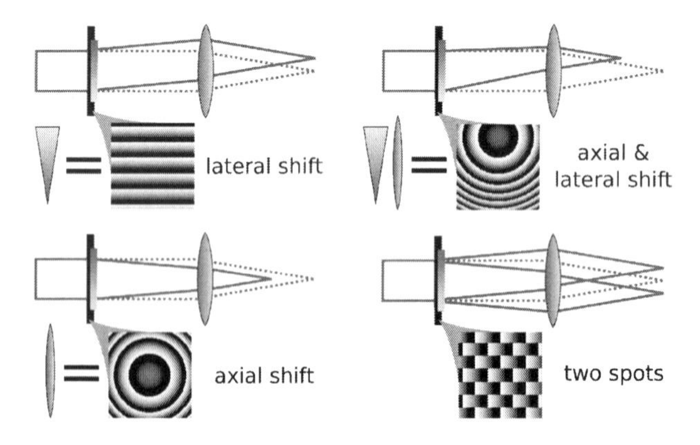

Figure 10.7 Holograms displayed on an SLM can be used to shift the optical trap laterally (by emulating a prism), axially (by displaying a lens), or any combination of the two. It is also possible to create multiple spots by displaying the interference pattern between the beams required to generate each individual spot.

It is also possible to display a virtual lens on the hologram, again having phase jumps every time the phase delay reaches a full wavelength. To shift the trap to an arbitrary position in 3D, we can combine the two phase shifts simply by adding them, as shown in Fig. 10.7. The mathematical expressions for the phase shift required to move a trap to position x, y, z as a function of u and v, the vertical and horizontal positions on the hologram, are simply

$$\phi_{\text{lateral}}(u, v) = \left(\frac{ku}{f}\right) x + \left(\frac{kv}{f}\right) y \bmod 2\pi$$

$$\phi_{\text{axial}}(u, v) = \left(\frac{k\left(u^2 + v^2\right)}{2f^2}\right) z \bmod 2\pi$$

$$\phi_{\text{3D}}(u, v) = \phi_{\text{lateral}}(u, v) + \phi_{\text{axial}}(u, v) \bmod 2\pi$$

where $k = 2\pi/\lambda$ is the wavenumber, f is the focal length of the Fourier transform lens after the SLM, and ϕ_{lateral}, ϕ_{axial}, and ϕ_{3D} are the phase shifts in radians displayed on the SLM to shift a spot laterally, axially, and in 3D, respectively.

The simplest method of creating multiple spots is to calculate the hologram $\phi_i(u, v)$ for each spot, and then to sum the light fields resulting from this [27]. To do this, we represent each light field as a complex number having a uniform amplitude and phase given by the hologram, such that the final hologram is

$$\phi_{\text{total}} = \arg\left(\sum_i \exp(\imath\phi_i)\right)$$

where $\sum_i$ denotes a sum over all the spots i, and arg() denotes taking the phase of a complex number. However, in general this method produces more than just the desired spots: we only modulate the phase of the laser beam and not the amplitude which means there is too much light on some parts of the hologram. This surplus light shows up as "ghost orders," extra spots that are produced but were not intentionally in the hologram. Ghost orders are a particular problem in highly symmetric arrays of traps.

By using iterative algorithms, it is possible to create highly efficient, uniform arrays of spots. These algorithms generally optimize the relative phase of the light fields which are summed to create the final hologram [26, 28]. The most effective algorithms iteratively transform between the spot pattern actually produced by the hologram and the hologram itself. By repeatedly optimising the intensities in the two planes, very good results can be obtained quickly. A simpler (though less effective) approach is to randomize the phase of each spot, corresponding to adding a random constant to each of the ϕ_i terms in the sum.

Simple algorithms are often favored not just because they are easier to implement, but because they require less calculation time. When manipulating particles, it is useful to be able to update the hologram as fast as the SLM can display it, usually around 60 Hz. However, even with a relatively fast multi-core CPU (e.g., a quad core 3 Ghz Pentium 4), holograms can only be calculated at the required resolution around 10 times per second, and this has been a serious limitation on holographic tweezers in the past. However, the use of consumer graphics cards has enabled much faster hologram generation [29]. This allows the tweezers system to react on the timescale of the particle's Brownian motion, which enables active feedback to increase the effective stiffness of a trap [30, 31]. As modern graphics cards have increased in power, it has even been possible to implement iterative algorithms in real time [32], though with higher latency than the simple algorithms used for active feedback.

In addition to creating multiple focussed spots and moving them around, SLMs allow very general modifications to be made to the light beam. As such they have been used to generate optical traps carrying orbital angular momentum, dubbed "optical spanners" [33, 34] and create extended line and ring traps [35, 36]. This ability to modify the shape of the focal spot in 3D has been used to trade off axial and lateral stiffness in a trap [10]. Focal spot engineering with SLMs has also been used in microscopy, for example, providing edge enhancement [37] or 3D tracking [38].

10.3.4 *Laser Sources*

One of the fundamental requirements of an optical tweezers system is the laser, which is usually the most expensive component of any trapping setup. The wavelength of laser used is important for biological experiments, as absorption of the laser causes heating and other effects and can lead to damage of the sample, or "opticution" [39]. This is of particular relevance when live cells are to be trapped directly, and a number of studies on various organisms [40, 41] have suggested that damage is minimized by using near-infrared wavelengths, although it is important to check cell viability for the particular cells and laser system used in any new experiment. The two most frequently used wavelengths are 1064 nm and around 820 nm. These coincide with local minima in the optical absorption spectrum of water, as well as corresponding to commercially available laser technologies. The former wavelength is that of Nd:YAG lasers, available as "turn-key" diode pumped solid state lasers which require no setup other than switching them on and adjusting the power level. Commercial titanium:sapphire lasers are also available, with wavelengths between about 780 nm and 830 nm. These lasers are more complicated and are often tunable and/or pulsed, but they are available in reliable, commercial systems.

Pointing stability, i.e., the drift of a laser's output beam over time, is another parameter in an optical tweezers system which has an effect on force measurement. Combined with the stability of the rest of the optical system, it sets an upper bound on the length of experiments which can be performed before drift (either of the laser or of the microscope optics) degrades the accuracy of force or position measurements. It is possible, however, to alleviate drift by taking differential measurements between two traps [19]. Power stability should also be taken into consideration, as variation in the laser power will lead to a corresponding change in the stiffness of the optical trap.

10.3.5 *Objective Lenses and Beam Shape*

Single-beam optical traps require a high numerical aperture (NA) in order for the trap to be stable in three dimensions. Combined with the small size of objects usually trapped, this leads to the use of high-magnification, high-NA objectives, typically 100× or 60× with an NA of at least 1.2. These objectives are usually designed for fluorescence microscopy, and are usually oil- or water-immersion. The use of water-immersion objectives minimizes spherical aberration, which is another limiting factor for 3D trapping; too much spherical aberration can degrade the trap's axial stiffness to the point where it is no longer stable in the axial direction [42], particularly for small or highly scattering particles [6]. Changing the refractive index of the immersion oil is one way of alleviating this problem for oil-immersion objectives, which has been used to good effect with nanoparticles [6, 43]. Other aberrations induced by misalignments or imperfections in the system, or by heterogeneities in the sample, can give rise to a misshapen focus leading to different stiffnesses in different directions or even the failure to trap particles altogether. To address this problem, a number of adaptive optics approaches have been taken, often using the same SLM which steers the beam in HOTs [44–46].

10.4 Counterpropagating Optical Traps

Some years before demonstrating the single-beam gradient trap now referred to as *optical tweezers*, Ashkin demonstrated the stable trapping of particles between the foci of two counterpropagating beams, at much lower NA [1]. This sidesteps the issue of particles being lost from the trap because the scattering forces of the two opposing beams cancel out. The particle is trapped laterally by the same mechanisms as in a single-beam trap; however, axial trapping is provided by a combination of gradient and scattering forces. Most usually, the foci of the two beams are displaced to put them outside of the particle as shown in Fig. 10.8, such that when it moves toward one focus the scattering force from that trap is higher and pushes it back toward the center. This approach frees us from the constraints of high-NA objective lenses and allows a much larger field of view. This also isolates the trapped object from the extremely high light intensity at the laser foci, which is already much reduced from the peak intensity at the focus of a high-NA lens.

Counterpropagating traps can be formed using two opposing objective lenses and still leave sufficient clearance for large sample cells [47] or even a side-looking microscope or spectroscopy system [48]. It is also possible to form a counterpropagating trap without any lenses, simply by using two optical fibers. The tips of the fibers replace the foci of the two beams in the lens-based system, and objects are stably trapped between the two. This geometry has

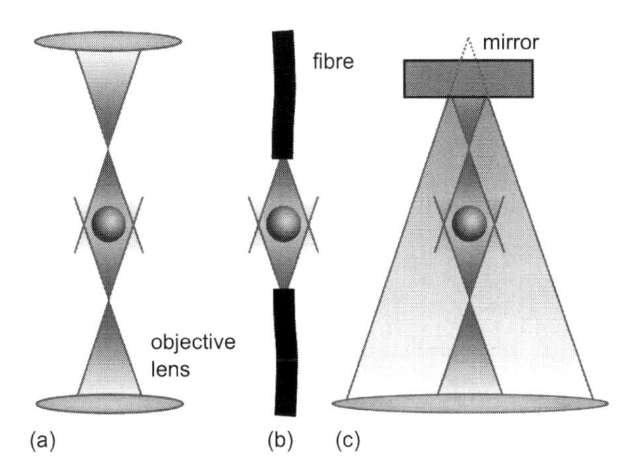

Figure 10.8 Different configurations for forming counterpropagating optical traps: (a) two opposing objective lenses, (b) two fibres, and (c) a single objective lens and a mirror, using an SLM to create two foci displaced from each other axially.

been used to trap and deform biological cells [49]. This deformation allows the mechanical properties of the cell to be probed, and this has been investigated in the context of some medical conditions, most notably cancer [50]. The two-fiber geometry does not use a microscope objective, and thus it is more flexible in that it can be integrated with a microfluidic chip [50], which has been used to allow a much higher throughput of cells than is attainable with microscope-based optical tweezers.

More recently, counterpropagating optical traps have been realized with a single objective lens [51, 52]. By using a low-NA lens in a holographic optical tweezers system, it is possible to create foci over an extended axial range. A mirror placed behind the sample can reflect some of these foci back into the sample. Thus, any beam which would focus behind the mirror actually focusses inside the sample, but is propagating in the opposite direction. This enables a single objective to produce multiple counterpropagating traps, which can be freely moved in 3D. This approach is arguably more convenient than the dual objective geometry, and has the advantage of being able to alter the focal planes of each beam individually. In contrast, most dual-objective systems alter the balance of intensity in upper and lower beams, which consequently changes the equilibrium trapping position. However, in both cases the axial stiffness can be very low, and active feedback can be used to dramatically increase the accuracy with which particles are placed in *z* [53].

10.5 Force and Position Measurement

As discussed in Section 10.2, an optical trap exerts a force on the trapped object which is modelled well (for small separations) by a Hookean spring, i.e., the force is linearly proportional to the displacment. Thus, we can measure forces by measuring the displacement of a bead and multiplying by a spring constant to find the force. Particles can typically be tracked with resolution better than a few nanometers, and spring constants are usually around $1\ \mu\mathrm{Nm}^{-1}$. Brownian motion sets a fundamental limit on the accuracy of forces which can be measured, but typically tens to hundreds of femtonewtons can be attained if data is averaged over a few seconds. The experimental hardware used to measure forces and displacements falls into two categories: laser-based detection and video camera systems.

10.5.1 *Laser Position Detection*

When a trapped object is displaced from the laser focus, it alters the distribution of light leaving the trap. In the case of a sphere, this results in a change in direction of the light which causes a translation of the intensity pattern in the back focal plane of the condenser lens, as illustrated in Fig. 10.9. The shift of the pattern is usually measured with a position sensitive detector (PSD) or a quadrant photodiode (QPD) as shown in Fig. 10.9. A PSD returns the coordinates of the center-of-mass of the light hitting its active surface, while a QPD simply returns the total intensity hitting each of the four quadrants of the detector. In the latter case, the shift of the pattern is given by the difference between the two halves of the detector, as shown in Fig. 10.9. This method of detection has been reported with precision down to 0.1 Å [54], with bandwidths of up to 1 MHz. However, it does require relatively precise alignment and

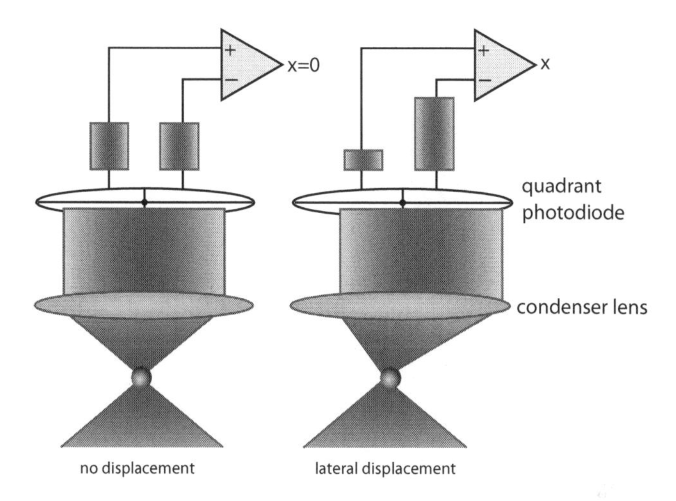

Figure 10.9 Quadrant photodiodes are an extremely sensitive way of measuring the position of a trapped object, by looking at the deflection of the beam. As the beam is bent, it is displaced in the back focal plane of the condenser. The two halves of the QPD are therefore illuminated differently, and the difference between the two halves is proportional to the particle's displacement.

calibration of the relationship between detector signal and distance. It gives the position relative to the laser, which means it will be relatively insensitive to pointing stability; however, if the laser does drift and the force is being measured relative to an object fixed to the microscope slide (e.g., a long molecule is fixed at one end to the slide and at the other end to a trapped bead), then the drift will only be detectable as an error on the force measured.

Extending this technique to multiple particles requires separating out the light which has passed through each optical trap. This is most often done using a polarising beamsplitter, where the two traps have orthogonal polarisations [19]; however, care must be taken to avoid crosstalk between the two traps when using laser-based position detection [55]. It is also possible to use a dichroic mirror in combination with separate "detection lasers" of different wavelengths, but these detection lasers must be precisely aligned with each optical trap and so very high pointing stability is required in all the lasers used to keep the system accurate over a long period of time. In the case of time-shared optical traps using an acousto-optic modulator or a scanning mirror, it is possible to separate the particles by precisely timing the readout of the detector. This has been used to track several particles in a line-shaped optical trap, made by rapidly scanning a single trap back and forth [56].

Axial tracking is also possible using a quadrant photodiode or PSD, by looking at the divergence of the light [54, 57]. Axial displacement of the particle will make the light more or less divergent and, if it is arranged to slightly overfill the detector (or the condenser has a lower NA than the objective), this will change the total intensity incident on the detector. The relationship between intensity and axial position is also approximately linear over the optical trap.

Calibration of the relationship between displacement and detector signal is generally carried out using a precision stage [57]. A particle identical to those used for force measurement is fixed to the microscope coverslip and moved in steps of a known size, and the resulting signal is recorded. This relationship is dependent on the size and composition of the particle, and the refractive index of the medium. It must therefore be repeated for each particle size and material used. The range over which such a detection scheme is linear is around the same as the region over which the optical trap has a linear force–distance relationship.

By collecting all of the light which has passed through the particle, it is in theory possible to extract the force on the particle with no calibration, by simply considering the change in momentum of the light. This has been done using counterpropagating traps with relatively low numerical aperture [58]. However, it is rarely possible to reach this level of efficiency when using a high-NA trapping objective even with oil immersion condensers, and so empirical calibration remains the norm. Also, the force extracted is the total force on the sample: if other objects distort the beam, this will introduce artefacts.

10.5.2 *Video Particle Tracking*

Position measurement using a laser is precise and provides very high bandwidth. However, it is often challenging to align and calibrate such systems, and extending them to multiple particles is difficult. It is possible to precisely locate particles in a video image, often to an accuracy of around one hundredth of a pixel [59] corresponding to a few nanometers. Particle tracking in microscopy is often used to make measurements of flow or viscosity in particle imaging velocimetry (PIV), where tracer particles are added to a system and their trajectories are analyzed. Often, these particles are the same glass or plastic microbeads used in optical tweezers experiments, imaged using either bright-field or fluorescence microscopy. PIV can be performed with a camera operating at standard video frame rates, a few tens of Hertz. In optical tweezers, it is often necessary to observe the motion of a trapped object at high speed, with a bandwidth of thousands of Hertz, as the motion of a trapped particle averages to zero over times on the order of a few tens of milliseconds. Relatively inexpensive CMOS cameras now meet this requirement, attaining frame rates of several kilohertz [60, 61] up to tens of kilohertz for more sophisticated cameras [62]. This is generally achieved by reading out only a subset of the pixels on the sensor chip, which decreases the time required to take an image and transfer it to a computer.

Tracking optically trapped particles requires relatively simple image analysis, which can now be performed on-the-fly using a typical desktop computer. When a small spherical object is slightly behind the focal plane of the microscope, it appears as a bright spot surrounded by a dark ring under bright-field illumination. After removing the background with a threshold function, simple center-of-mass algorithms can track particles in 2D with a precision of a few nanometers and a bandwidth of several kilohertz [60, 61]. Figure 10.10 shows a typical image of a 2 μm bead in an optical trap, and the image after a threshold function has been applied. The center-of-mass algorithm then operates on the background subtracted image to find the center of the particle, by considering the mean position of the intensity in the image. This is a very efficient calculation to perform, as it requires just three summations over the pixels in an image, which can be done simultaneously. If the intensity of the pixel at position (x, y) is $I_{x,y}$, we evaluate $S_I = \sum_x \sum_y I_{x,y}$, $S_x = \sum_x \sum_y x \times I_{x,y}$ and $S_y = \sum_x \sum_y y \times I_{x,y}$. The position is

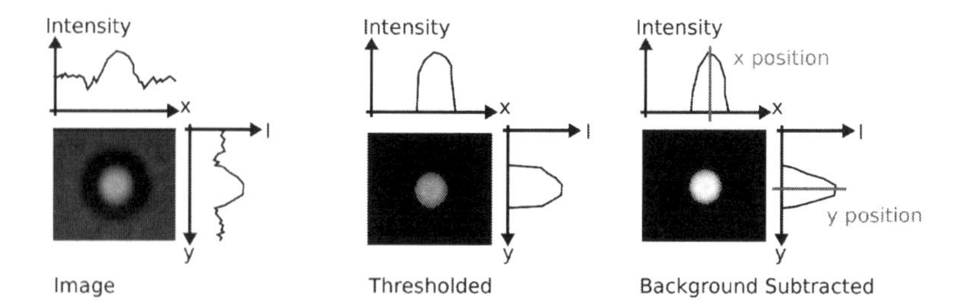

Figure 10.10 A typical image of a 2 μm silica bead in an optical tweezers system, and that same bead with a threshold function (middle) and background subtraction (right) applied. The distributions of intensity are shown beside each image. The center of mass works by finding the mean x and y position of the intensity, which can locate the particle to 1/100 of a pixel.

then given as $(S_x/S_I, S_y/S_I)$. This is analogous to calculating the mean of a dataset, where the points are weighted by their intensities. A slightly more subtle point is the difference between background subtraction and thresholding: the former subtracts a constant from the image such that the background is zero, while the latter sets pixels to zero if they are below a given value but leaves other pixels either unchanged or set to their maximum value. The latter approach means that pixels which are around the threshold value can flicker on and off, which can cause noise on the position signal in some cases.

A well-designed particle tracking algorithm can attain a resolution of order 1/100 pixel, or typically about 1 nm [63]. Tracking particles with sub-pixel accuracy enables camera-based systems to make force measurements which approach the thermal limit on accuracy set by Brownian motion. More sophisticated analysis of images allows particles to be tracked over a larger axial range, and for their axial position to be measured [64]. This requires calibration for each different bead and medium used (as changes in refractive index, etc., will change the appearance of the bead), and so a closed-loop piezo stage is usually required, as with laser-based detection schemes.

Digital holographic microscopy (DHM) has been used for particle tracking in the context of optical tweezers [65]. DHM also uses sophisticated image analysis to recover 3D position information, this time from an image taken using coherent light. It is possible to fit to an analytic model based on generalized Lorentz–Mie theory to the image of a particle. This approach recovers 3D position as well as size and refractive index; however, this is particularly demanding and only approaches video rates when accelerated with Graphics Processing Unit technology [66]. Using coherent light in imaging carries with it the inherent problem of speckle; any light scattered from outside of the focal plane (e.g., dust on lenses or out-of-focus objects in the sample) will interfere with the image. This can cause degradation of image quality, requiring careful calibration and removal of the noisy background image.

Detailed image analysis will become faster as computer power increases, but at the present time stereoscopic particle tracking [10, 67] and other techniques based on a modified imaging system [38] have the advantage of giving 3D information with only relatively simple image analysis. Stereoscopic imaging produces two images of the sample from different viewpoints, then uses parallax to find depth information. Once the pixel size of the camera and angle between the viewpoints are known, there is no need for further calibration and thus this technique is relatively robust to changes in trapped objects, etc. Typically, bright-field illumination is used to maximize the available light, but instead of illuminating on axis, two off axis sources are used. This produces a double image, where the two images of a particle converge when it is in focus. After using either color [67] or a Fourier filter [10] to separate the two images, particle tracking can be performed on each, and the particle's 3D position extracted with an accuracy of nanometers [31].

Image analysis with a camera can also measure other properties, for example the distortion of objects held in an "optical stretcher" [49]. This is one case where it is difficult to reduce the optical signal to something which can easily be picked up by a photodiode, and thus camera tracking is by far the easiest option. Cameras can be used to track heterogeneous objects such as cells, which will provide a distorted signal when measured using back-focal-plane interferometry.

10.5.3 *Particle Dynamics and Calibration*

Any particle immersed in a fluid is constantly bombarded by the atoms or molecules of the medium, as they undergo random thermal motion. This gives rise to Brownian motion, or diffusion, of the particle. By analyzing this random motion, it is possible to learn both about the optical trap and about the medium in which the trapped object is immersed. This is essential if we are to make force measurements, as we relate the force F to the displacement of the particle x using the stiffness of the optical trap κ:

$$F = \kappa x$$

The simplest approach to this comes from thermodynamics; the equipartition theorem tells us that, on average, the energy stored in each degree of freedom in a system is $k_{\mathrm{B}}T/2$, where k_{B} is Boltzmann's constant and T is the temperature. The energy stored in a linear spring (such as our optical trap) is $\kappa x^2/2$. Thus, by averaging over many measurements of the bead's position we can find $\langle x^2 \rangle$ (where the $\langle\rangle$ denote averaging), and calculate the stiffness:

$$\kappa = k_{\mathrm{B}}T/\langle x^2 \rangle$$

At the small length scales of optical tweezers, the mass of a trapped object is insignificant compared to viscous drag effects and the motion is overdamped, i.e., the particle will not keep moving due to intertia, but will stop once it is no longer being pushed along. Neglecting inertia allows us to describe the its motion with a simple equation:

$$\gamma \dot{x} + \kappa x = \zeta$$

where x is the position of the bead and $\dot{x}$ its velocity, γ is a friction coefficient (proportional to the viscosity of the fluid), and ζ is a fluctuating force term, representing the random collisions of molecules with the trapped particle. An untrapped particle follows a random walk, where the distance travelled after a time t is proportional to the square root of t. Trapped particles also move in this way, but after a short time the trap acts to pull the particle back toward the center, and limits the distance travelled. This is most often visualized by plotting the power spectral density (PSD) of the particle's position fluctuations, which has a characteristic Lorentzian shape shown in Fig. 10.11 and described by the equation

$$\mathrm{PSD} = \frac{2k_{\mathrm{B}}T/\pi\gamma}{\omega_c^2 + \omega^2}$$

where $\omega_c = \kappa/\gamma$ is the "corner frequency" [68].

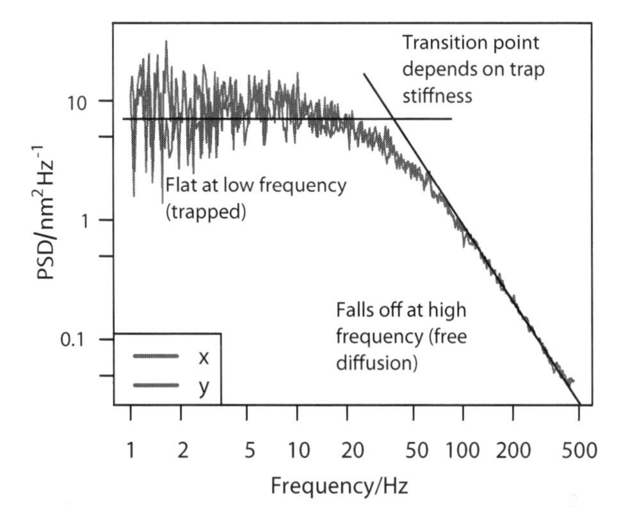

Figure 10.11 A typical power spectrum, measured by video particle tracking. It has the characteristic Lorentzian shape, with a plateau at low frequency (as the particle's diffusion is curtailed by the trap) and a tail proportional to ω^{-2} (corresponding to free diffusion over short timescales). The crossover between the two regions is determined by the ratio of trap strength to fluid viscosity.

Analysis of the power spectrum allows the trap stiffness κ to be determined in some situations where the simple equipartition expression is unreliable. It is also possible to determine other properties of the system, such as viscosity and even viscoelastic moduli, by further analyzing the motion of a trapped object [69, 70]. It is also possible to probe viscoelastic properties by actively driving the bead's motion (through moving the trap or the use of spinning, birefringent particles [71]) and observing the bead's response, which can achieve a higher signal-to-noise ratio in some circumstances.

10.6 Conclusions

Optical trapping is now a well-developed technique, able to manipulate micrometer-sized objects such as beads and cells. It is capable of applying and measuring forces from tens of piconewtons down to femtonewtons, on multiple particles simultaneously. This has been used to break new ground in research into the properties of biological molecules such as DNA and molecular motors like kinesins and myosins. As the technique continues to develop, the potential for its use in biological and biophysical experiments as well as its availability to nonspecialists is ever increasing.

In this chapter, we explored the technical aspects of constructing an optical tweezer system, employing various different technologies for control and multiplexing of the traps. These range from a simple steerable mirror to a spatial light modulator for full 3D control of many particles. Measurement of forces and positions is an important ability of optical tweezers, and both of the commonly used methods (laser-based and camera-based tracking) have been discussed. We have also briefly visited some of the physics involved in optical traps, and the expressions for trap calibration that arise from this. As optical tweezers continue to make their way out of physics laboratories to be applied to other problems, we look forward to many new developments as we further probe the world of tiny forces and displacments.

References

1. A. Ashkin, "Acceleration and trapping of particles by radiation pressure," *Phys. Rev. Lett.*, **24**, 4, 156–159, 1970.
2. K. Svoboda, C. Schmidt, B. Schnapp, and S. M. Block, "Direct observation of kinesin stepping by optical trapping interferometry," *Nature*, **365**, 6448, 721–727, 1993.
3. J. E. Molloy, J. E. Burns, J. Kendrick-jones, R. T. Tregear, and D. C. S. White, "Movement and force produced by a single myosin head," *Nature*, **378**, 6553, 209–212, 1995.
4. M. Wang, H. Yin, R. Landick, J. Gelles, and S. Block, "Stretching DNA with optical tweezers," *Biophys. J.*, **72**, 3, 1335–1346, 1997.
5. L. Bosanac, T. Aabo, P. M. Bendix, and L. B. Oddershede, "Efficient optical trapping and visualization of silver nanoparticles," *Nano Lett.*, **8**, 5, 1486–1491, 2008.
6. S. N. S. Reihani and L. B. Oddershede, "Optimizing immersion media refractive index improves optical trapping by compensating spherical aberrations," *Opt. Lett.*, **32**, 14, 1998–2000, 2007.

7. S. B. G. Thalhammer, R. Steiger and M. Ritsch-Marte, "Optical macro-tweezers: trapping of highly motile micro-organisms," *J. Optics*, **13**, 4, 044024, 2011.
8. A. Ashkin, "Forces of a single-beam gradient laser trap on a dielectric sphere in the ray optics regime," *Biophys. J.*, **61**, 2, 569–582, 1992.
9. A. T. O'Neil and M. J. Padgett, "Axial and lateral trapping efficiency of Laguerre-Gaussian modes in inverted optical tweezers," *Opt. Commun.*, **193**, 1–6, 45–50, 2001.
10. R. W. Bowman, G. Gibson, and M. Padgett, "Particle tracking stereomicroscopy in optical tweezers: Control of trap shape," *Opt. Express*, **18**, 11, 11 785–11 790, 2010.
11. Y. Roichman, B. Sun, A. Stolarski, and D. G. Grier, "Influence of nonconservative optical forces on the dynamics of optically trapped colloidal spheres: The fountain of probability," *Phys. Rev. Lett.*, **101**, 12, 128301, 2008.
12. J. Arlt, V. Garces-Chavez, W. Sibbett, and K. Dholakia, "Optical micromanipulation using a Bessel light beam," *Opt. Commun.*, **197**, 4–6, 239–245, 2001.
13. V. G. Shvedov, A. V. Rode, Y. V. Izdebskaya, A. S. Desyatnikov, W. Krolikowski, and Y. S. Kivshar, "Giant optical manipulation," *Phys. Rev. Lett.*, **105**, 11, 118103, 2010.
14. A. O'Neil, I. MacVicar, L. Allen, and M. Padgett, "Intrinsic and extrinsic nature of the orbital angular momentum of a light beam," *Phys. Rev. Lett.*, **88**, 5, 053601, 2002.
15. Y. Roichman, D. G. Grier, and G. Zaslavsky, "Anomalous collective dynamics in optically driven colloidal rings," *Phys. Rev. E*, **75**, 2, 020401, 2007.
16. T. A. Nieminen, V. L. Y. Loke, A. B. Stilgoe, G. Knoener, A. M. Branczyk, N. R. Heckenberg, and H. Rubinsztein-Dunlop, "Optical tweezers computational toolbox," *J. Opt. A*, **9**, 8, S196–S203, 2007.
17. S. Simpson and S. Hanna, "Numerical calculation of interparticle forces arising in association with holographic assembly," *J. Opt. Soc. Am. A*, **23**, 6, 1419–1431, 2006.
18. M. Reicherter, T. Haist, E. Wagemann, and H. Tiziani, "Optical particle trapping with computer-generated holograms written on a liquid-crystal display," *Opt. Lett.*, **24**, 9, 608–610, 1999.
19. J. Moffitt, Y. Chemla, D. Izhaky, and C. Bustamante, "Differential detection of dual traps improves the spatial resolution of optical tweezers," *Proc. Natl. Acad. Sci. USA*, **103**, 24, 9006–9011, 2006.

20. M. Klein, M. Andersson, O. Axner, and E. Fallman, "Dual-trap technique for reduction of low-frequency noise in force measuring optical tweezers," *Appl. Opt.*, **46**, 3, 405–412, 2007.
21. L. McCann, M. Dykman, and B. Golding, "Thermally activated transitions in a bistable three-dimensional optical trap," *Nature*, **402**, 6763, 785–787, 1999.
22. G. Brouhard, H. Schek, and A. Hunt, "Advanced optical tweezers for the study of cellular and molecular biomechanics," *IEEE Trans. Biomed. Eng.*, **50**, 1, 121–125, 2003.
23. P. A. Kirkby, K. M. N. S. Nadella, and R. A. Silver, "A compact acousto-optic lens for 2D and 3D femtosecond based 2-photon microscopy," *Opt. Express*, **18**, 13, 13 720–13 744, 2010.
24. Y. Hayasaki, M. Itoh, T. Yatagai, and N. Nishida, "Nonmechanical optical manipulation of microparticle using spatial light modulator," *Opt. Rev.*, **6**, 1, 24–27, 1999.
25. J. Liesener, M. Reicherter, T. Haist, and H. J. Tiziani, "Multi-functional optical tweezers using computer-generated holograms," *Opt. Commun.*, **185**, 77–82, 2000.
26. R. D. Leonardo, F. Ianni, and G. Ruocco, "Computer generation of optimal holograms for optical trap arrays," *Opt. Express*, **15**, 4, 1913–1922, 2007.
27. J. Leach, K. Wulff, G. Sinclair, P. Jordan, J. Courtial, L. Thomson, G. Gibson, K. Karunwi, J. Cooper, Z. J. Laczik, and M. Padgett, "Interactive approach to optical tweezers control," *Appl. Opt.*, **45**, 5, 897–903, 2006.
28. E. Dufresne, G. Spalding, M. Dearing, S. Sheets, and D. G. Grier, "Computer-generated holographic optical tweezers arrays," *Rev. Sci. Instrum.*, **72**, 1810–1816, 2001.
29. M. Reicherter, S. Zwick, T. Haist, C. Kohler, H. Tiziani, and W. Osten, "Fast digital hologram generation and adaptive force measurement in liquid-crystal-display-based holographic tweezers," *Appl. Opt.*, **45**, 5, 888–896, 2006.
30. D. Preece, R. W. Bowman, A. Linnenberger, G. Gibson, S. Serati, and M. Padgett, "Increasing trap stiffness with position clamping in holographic optical tweezers," *Opt. Express*, **17**, 25, 22 718–22 725, 2009.
31. R. W. Bowman, D. Preece, G. Gibson, and M. J. Padgett, "Stereoscopic particle tracking for 3D touch, vision and closed-loop control in optical tweezers." *J. Opt. A*, **13**, 4, 044003, 2011.
32. S. Bianchi and R. Di Leonardo, "Real-time optical micro-manipulation using optimized holograms generated on the GPU," *Comp. Phys. Commun.*, **181**, 8, 1442–1446, 2010.

33. H. He, M. E. J. Friese, N. R. Heckenberg, and H. Rubinsztein-Dunlop, "Direct observation of transfer of angular momentum to absorptive particles from a laser beam with a phase singularity," *Phys. Rev. Lett.*, **75**, 5, 826–829, 1995.

34. N. Simpson, K. Dholakia, L. Allen, and M. Padgett, "Mechanical equivalence of spin and orbital angular momentum of light: An optical spanner," *Opt. Lett.*, **22**, 1, 52–54, 1997.

35. Y. Roichman and D. G. Grier, "Projecting extended optical traps with shape-phase holography," *Opt. Lett.*, **31**, 11, 1675–1677, 2006.

36. Y. Roichman, B. Sun, Y. Roichman, J. Amato-Grill, and D. G. Grier, "Optical forces arising from phase gradients," *Phys. Rev. Lett.*, **100**, 1, 013602, 2008.

37. S. Furhapter, A. Jesacher, S. Bernet, and M. Ritsch-Marte, "Spiral phase contrast imaging in microscopy," *Opt. Express*, **13**, 3, 689–694, 2005.

38. S. R. P. Pavani, A. Greengard, and R. Piestun, "Three-dimensional localization with nanometer accuracy using a detector-limited double-helix point spread function system," *Appl. Phys. Lett.*, **95**, 2, 021103, 2009.

39. A. Ashkin, J. M. Dziedzic, and T. Yamane, "Optical trapping and manipulation of single cells using infrared laser beams," *Nature*, **330**, 6150, 769–771, 1987.

40. H. Liang, K. Vu, P. Krishnan, T. Trang, D. Shin, S. Kimel, and M. Berns, "Wavelength dependence of cell cloning efficiency after optical trapping," *Biophys. J.*, **70**, 3, 1529–1533, 1996.

41. M. Ericsson, D. Hanstorp, P. Hagberg, J. Enger, and T. Nystrom, "Sorting out bacterial viability with optical tweezers," *J. Bacteriol.*, **182**, 19, 5551–5555, 2000.

42. G. Sinclair, P. Jordan, J. Leach, M. Padgett, and J. Cooper, "Defining the trapping limits of holographical optical tweezers," *J. Mod. Opt.*, **51**, 3, 409–414, 2004.

43. E. Theofanidou, L. Wilson, W. Hossack, and J. Arlt, "Spherical aberration correction for optical tweezers," *Opt. Commun.*, **236**, 1-3, 145–150, 2004.

44. K. Wulff, D. Cole, R. Clark, R. DiLeonardo, J. Leach, J. Cooper, G. Gibson, and M. Padgett, "Aberration correction in holographic optical tweezers," *Opt. Express*, **14**, 9, 4169–4174, 2006.

45. T. Cizmar, M. Mazilu, and K. Dholakia, "In situ wavefront correction and its application to micromanipulation," *Nat. Photon.*, **4**, 6, 388–394, 2010.

46. R. W. Bowman, A. J. Wright, and M. J. Padgett, "An SLM-based Shack–Hartmann wavefront sensor for aberration correction in optical tweezers," *J. Optics*, **12**, 12, 124004, 2010.
47. P. J. Rodrigo, L. Gammelgaard, P. Bøggild, I. Perch-Nielsen, and J. Glückstad, "Actuation of microfabricated tools using multiple GPC-based counterpropagating-beam traps," *Opt. Express*, **13**, 18, 6899–6904, 2005.
48. P. Rodrigo, V. Daria, and J. Glückstad, "Four-dimensional optical manipulation of colloidal particles," *Appl. Phys. Lett.*, **86**, 7, 074103, 2005.
49. J. Guck, R. Ananthakrishnan, H. Mahmood, T. J. Moon, C. C. Cunningham, and J. Käs, "The optical stretcher: A novel laser tool to micromanipulate cells," *Biophys. J.*, **81**, 2, 767–784, 2001.
50. J. Guck, S. Schinkinger, B. Lincoln, F. Wottawah, S. Ebert, M. Romeyke, D. Lenz, H. M. Erickson, R. Ananthakrishnan, D. Mitchell, J. Käs, S. Ulvick, and C. Bilby, "Optical deformability as an inherent cell marker for testing malignant transformation and metastatic competence," *Biophys. J.*, **88**, 5, 3689–3698, 2005.
51. M. Pitzek, R. Steiger, G. Thalhammer, S. Bernet, and M. Ritsch-Marte, "Optical mirror trap with a large field of view," *Opt. Express*, **17**, 22, 19 414–19 423, 2009.
52. S. Zwick, T. Haist, Y. Miyamoto, L. He, M. Warber, A. Hermerschmidt, and W. Osten, "Holographic twin traps," *J. Opt. A*, **11**, 3, 034011, 2009.
53. S. Tauro, A. Ba nas, D. Palima, and J. Glückstad, "Dynamic axial stabilization of counter-propagating beam-traps with feedback control," *Opt. Express*, **18**, 17, 18 217–18 222, 2010.
54. A. Rohrbach, C. Tischer, D. Neumayer, E. Florin, and E. Stelzer, "Trapping and tracking a local probe with a photonic force microscope," *Rev. Sci. Instrum.*, **75**, 6, 2197–2210, 2004.
55. M. Atakhorrami, K. M. Addas, and C. F. Schmidt, "Twin optical traps for two-particle cross-correlation measurements: Eliminating cross-talk," *Rev. Sci. Instrum.*, **79**, 4, 043103, 2008.
56. M. Speidel, L. Friedrich, and A. Rohrbach, "Interferometric 3D tracking of several particles in a scanning laser focus." *Opt. Express*, **17**, 2, 1003–1015, 2009.
57. K. Neuman and S. M. Block, "Optical trapping," *Rev. Sci. Instrum.*, **75**, 9, 2787–2809, 2004.
58. W. Grange, S. Husale, H. Guntherodt, and M. Hegner, "Optical tweezers system measuring the change in light momentum flux," *Rev. Sci. Instrum.*, **73**, 6, 2308–2316, 2002.

59. J. C. Crocker and D. G. Grier, "Methods of digital video microscopy for colloidal studies," *J. Colloid. Interf. Sci.*, **179**, 1, 298–310, 1996.
60. O. Otto, C. Gutsche, F. Kremer, and U. F. Keyser, "Optical tweezers with 2.5 kHz bandwidth video detection for single-colloid electrophoresis," *Rev. Sci. Instrum.*, **79**, 2, 023710, 2008.
61. G. M. Gibson, J. Leach, S. Keen, A. J. Wright, and M. J. Padgett, "Measuring the accuracy of particle position and force in optical tweezers using high-speed video microscopy," *Opt. Express*, **16**, 19, 14 561–14 570, 2008.
62. O. Otto, F. Czerwinski, J. L. Gornall, G. Stober, L. B. Oddershede, R. Seidel, and U. F. Keyser, "Real-time particle tracking at 10,000 fps using optical fiber illumination," *Opt. Express*, **18**, 22, 22 722–22 733, 2010.
63. C. D. Saunter, "Quantifying subpixel accuracy: An experimental method for measuring accuracy in image-correlation-based, single-particle tracking," *Biophys. J.*, **98**, 8, 1566–1570, 2010.
64. Z. Zhang and C.-H. Menq, "Three-dimensional particle tracking with subnanometer resolution using off-focus images," *Appl. Opt.*, **47**, 13, 2361–2370, 2008.
65. S.-H. Lee and D. G. Grier, "Holographic microscopy of holographically trapped three-dimensional structures," *Opt. Express*, **15**, 4, 1505–1512, 2007.
66. F. C. Cheong, B. J. Krishnatreya, and D. G. Grier, "Strategies for three-dimensional particle tracking with holographic video microscopy," *Opt. Express*, **18**, 13, 13 563–13 573, 2010.
67. J. S. Dam, I. R. Perch-Nielsen, D. Palima, and J. Glückstad, "Three-dimensional imaging in three-dimensional optical multi-beam micro-manipulation," *Opt. Express*, **16**, 10, 7244–7250, 2008.
68. K. Berg-Sørensen and H. Flyvbjerg, "Power spectrum analysis for optical tweezers," *Rev. Sci. Instrum.*, **75**, 3, 594–612, 2004.
69. A. Yao, M. Tassieri, M. Padgett, and J. Cooper, "Microrheology with optical tweezers," *Lab Chip*, **9**, 17, 2568–2575, 2009.
70. I. Tolic-Norrelykke, E. Munteanu, G. Thon, L. B. Oddershede, and K. Berg-Sorensen, "Anomalous diffusion in living yeast cells," *Phys. Rev. Lett.*, **93**, 7, 078102, 2004.
71. A. Bishop, T. Nieminen, N. Heckenberg, and H. Rubinsztein-Dunlop, "Optical microrheology using rotating laser-trapped particles," *Phys. Rev. Lett.*, **92**, 19, 198104, 2004.

Chapter 11

Coherent Nonlinear Microscopy with Phase-Shaped Beams

Varun Raghunathan and Eric Olaf Potma
Department of Chemistry and Beckmann Laser Institute,
University of California Irvine, Irvine, CA 92617, USA
epotma@uci.edu

11.1 Introduction

Coherent nonlinear microscopy techniques such as coherent anti-Stokes Raman scattering (CARS), third harmonic generation (THG), and second harmonic generation (SHG) have found widespread applications in biology and nanoscience because of their high-resolution, label-free contrast, and fast imaging capabilities. These techniques are often combined in a microscope platform that enables multimodal imaging of a broad variety of samples ranging from biological tissues to individual nanostructures [1]. The popularity of these coherent nonlinear imaging techniques is exemplified by the development of various commercial multimodal microscopes, which are currently finding their way to imaging laboratories worldwide.

Understanding Biophotonics: Fundamentals, Advances, and Applications
Edited by Kevin K. Tsia

ISBN 978-981-4411-77-6 (Hardcover), 978-981-4411-78-3 (eBook)
www.panstanford.com

Unique to all these techniques is the coherent nature of the signal. To generate such a nonlinear coherent response from the sample, the incident laser pulses are focused to a tightly focused spot, where a spatially coherent nonlinear polarization is induced. This implies that the molecules in focus, which are driven by the laser pulses, are forced to oscillate in step. Hence, the subsequent radiation resulting from these molecules will mutually interfere, producing a coherent signal in the phase-matched direction of constructive interference. This phenomenon is unique to coherent microscopy and has no analog in conventional incoherent techniques such as fluorescence microscopy.

In many ways, the imaging capabilities of the coherent nonlinear microscope are dictated by the coherent nature of the signal. Understanding the mechanisms of coherent signal generation in focus not only improves our interpretation of the imaging contrast, it also offers avenues to manipulate the spatial coherences in focus for the purpose of expanding the palette of contrast mechanisms in the microscope. For instance, by tweaking the spatial coherence in focus the magnitude and phase-matched direction of the nonlinear radiation can be controlled at will. With the availability of spatial light modulators (SLM), the spatial properties of the incident beam profiles can be readily controlled, opening up a new field of flexible focus engineering.

Focus engineering with the aid of SLMs has been successfully used in SHG and THG microscopy to control the polarization state of the focal field [2, 3]. Spatially dependent control of the polarization in focus has led, for instance, to THG imaging with higher resolution. In CARS, focus engineering has been used to enhance the microscope's sensitivity to interfaces [4, 5], and mechanisms have been proposed for achieving higher-resolution CARS images [6, 7]. Although the application of focus engineering to the field of coherent microscopy is relatively new, the above examples emphasize the exciting potential that this approach may offer for improving the imaging capabilities of nonlinear microscopes.

In this chapter we will illuminate the principles of contrast control in the coherent nonlinear microscope through spatial phase shaping of the incident beams. On the basis of focus engineering examples in coherent anti-Stokes Raman scattering microscopy, we will show that additional contrast mechanisms can be achieved. These new imaging modes can highlight properties of the specimen that remain otherwise unseen, thus improving the imaging capabilities of the existing nonlinear optical microscope.

11.2 Focal Fields in a Microscope

Coherent nonlinear microscopy techniques generally utilize high NA objective lenses to focus the incoming laser beams onto the sample to a submicron spot. This tightly focused volume increases the field density and thus enhance the nonlinear optical effect in the vicinity of the focal spot. In order to appreciate the details of the coherent nonlinear process, our discussion necessarily starts with careful examination of the focal field distributions associated with these high NA objective lenses. In this section, we will briefly discuss the basics of describing the complex focal field in three dimensions and highlight the general case of the focal fields resulting from Gaussian input beams.

The complex focal fields of a high NA objective lens can be accurately modeled using the angular spectrum representation method developed by Richards and Wolf [8]. The three dimensional focal volume distributions can be written in integral form similar to diffraction integrals as

$$\vec{E}(x,y,z) = \frac{ikfe^{-ikf}}{2\pi}\sqrt{\frac{n_1}{n_2}}\int_0^{\theta_{\max}}\int_0^{2\pi} R_\phi^{-1} R_0^{-1} R_\theta \vec{E}(\theta,\phi)$$
$$\times e^{ik(x\sin\theta\cos\phi + y\sin\theta\phi + z\cos\theta)} \cdot \sqrt{\cos\theta}\sin\theta . d\phi d\theta$$
$$R_\phi = \begin{pmatrix} \cos\phi & \sin\phi & 0 \\ -\sin\phi & \cos\phi & 0 \\ 0 & 0 & 1 \end{pmatrix}, \quad R_\theta = \begin{pmatrix} \cos\theta & 0 & -\sin\theta \\ 0 & 1 & 0 \\ \sin\theta & 0 & \cos\theta \end{pmatrix}$$

Here $\vec{E}(\theta,\phi)$ is the incident laser field at the back aperture of the objective lens, f is the focal length of the lens, $\mathbf{k}$ is the wavevector of the incident light, and n_1 and n_2 are the refractive indexes of the media before and after the lens. The matrixes R_ϕ and R_θ transform the incoming light field in cylindrical coordinates, assumed to be collimated into three-dimensional

Cartesian coordinates. This also accounts of the refraction of light on the curved surface of the lens. The integral is carried out in the limits that ϕ spans the whole 0–2π range and θ is within the 0–$\theta_{\max}$ range, which corresponds to the acceptance angle of the lens.

It is difficult to solve the above integral analytically except for some simple cases such as lower-order Hermite–Gaussian (HG) beams for which semi-analytical solutions can be obtained. The reader is referred to Ref. [9] for the description of these semi-analytical forms of the focal fields. In this chapter we focus on the numerical solutions to the above integrals, which can consider any general form of input excitation. The focal field distribution is obtained by numerically solving the above integral using Simpson's 1/3 rule [5]. The only requirement is that the incoming field and the focal field be represented by a sufficient number of grid points to get good agreement with the real focal fields. For the focal field profiles to be discussed below, the volume considered is 1.5 μm × 1.5 μm × 3 μm in the X, Y, Z directions, respectively. This volume is divided into 25 nm step size to yield enough resolution to get good convergence. The NA of the lens is assumed to be 1.15, which corresponds to an acceptance angle of $\sim 60^{\circ}$.

The focal fields can also be obtained by numerical solving the Maxwell's equations as the laser beam propagates through the optics in the objective lens and the sample of interest using finite difference time domain (FDTD) algorithms. This scheme is based on appropriate boundary conditions and nonlinear effects at focus. The FDTD method has been used to obtain the focal fields of Laguerre–Gaussian beams at focus in CARS microscopy [10]. Both the diffraction integral method and FDTD yield very similar results, but the FDTD method is more time consuming, yet is more accurate in certain situations in which boundaries between media play an important role in the resultant fields.

It is typical in nonlinear microscopy to completely fill the back aperture of the objective lens with the incoming laser beam to ensure that the smallest point spread functions can be achieved at focus [9]. This can be intuitively understood from the Fourier transforming property of the lens to generate the focal fields. Figure 11.1 shows the intensity and phase of field distributions at the focal plane of the objective lens, with the X-polarized incoming

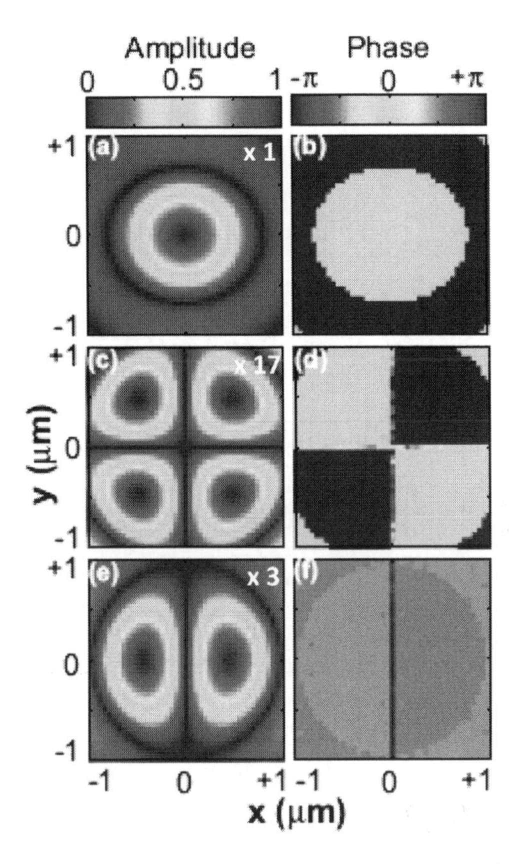

Figure 11.1 Electric field amplitude and phase profiles at the focus of a high NA objective lens at 1064 nm. *XY* profiles are shown at $Z = 0$.

beam at 1064 nm overfilling the back aperture. All the three polarization components are shown here. It is interesting to note that as the light is focused by the high NA objective, depolarization of the incident fields occurs. Hence, the focal field resulting from an input field that is linearly polarized along the X dimension exhibits Y and Z polarized components. In the case of Gaussian excitation, the Y- and Z-field components are approximately $17\times$ and $3\times$ weaker than X polarization, respectively, and does not significantly contribute to nonlinear effects studied at focus. Certain beam shapes such as radially polarized input fields create significant Z polarization at focus, which has been used to optimize field

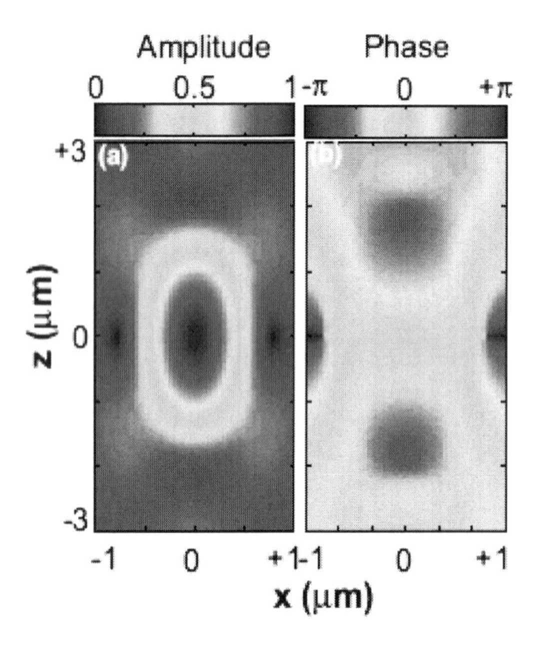

Figure 11.2 Electric field amplitude and phase profiles at the focus of a high NA objective lens at 1064 nm. The *XZ* profiles are shown at $Y = 0$.

enhancement effects at metallic tips [9]. For the X-polarized focal field the X and Y widths are slightly different with the X being wider than the Y direction. This is due to the X-polarized input excitation. The full-width half maximum (FWHM) of the point spread function achieved with the 1.15 NA lens is ~660 nm in the lateral direction. In the remainder of this chapter, we will focus mainly on the most significant field component at focus, which is along the incident polarization direction of the linearly polarized input beam.

Figure 11.2 shows the X–Z amplitude and phase profile of the X-polarized focal field at 1064 nm. The FWHM of the focal volume intensity is ~1.9 μm in the axial direction. This small spot size along with the nonlinear multiplication of the focal excitation fields forms the basis of the three-dimensional sectioning capability in nonlinear microscopy [11]. The phase profile shown in Fig. 11.2b is the resultant phase after subtracting the linear propagation phase. There exists a smooth π-phase transition as the beam crosses the focal plane. This phase reversal is called the Gouy phase shift of the field as it propagates through focus [12]. In SHG and THG techniques, the Gouy phase shift plays a very important role in controlling propagation direction of the phase-matched radiation [13, 14]. In CARS, the effect of the Gouy phase shift is small can be safely neglected for the remainder of this discussion [15]. The transverse phase profile is also changing upon propagating through focus, with the flat phase front achieved only at the focal plane. Nonetheless, close to the focal plane the transverse phase is smooth and slowly varying, and can be properly controlled with beam shaping techniques, as discussed in the next section.

11.3 Phase Shaping of Focal Fields

As emphasized in the introduction, the shaping of the phase of the focal fields provides additional degrees of control over the nonlinear emission. This control mechanism allows us to extract additional information from the sample that is otherwise hidden when using conventional Gaussian beam excitation. In this section, several different techniques of beam shaping and their corresponding focal field distributions are discussed. Beam shaping techniques can be broadly classified into three techniques, namely (i) dynamic diffractive optical elements, (ii) fixed diffractive optical elements, and (iii) beam interference. The input excitation beam can be either phase or amplitude shaped, or both. With the recent availability of phase-only shaping techniques, a popular approach is to only shape the phase of the input beam and leave the amplitude unaltered. This type of phase-only beam shaping is the topic of our discussion here.

11.3.1 *Dynamic Diffractive Optical Elements*

Dynamic diffractive optical elements, as the name implies, are reprogrammable and can generate any desired transverse phase (or amplitude) distribution. They are either based on liquid crystal arrays as in spatial light modulators (SLM) or based on microfabricated deformable mirrors (DM). DM-based spatial phase modulators rely on advanced MEMS fabrication techniques to manufacture highly reflective membranes, which can be addressed in segments and deformed to create spatial phase difference [16]. Since they rely on deformation techniques, a large net phase shift can be achieved (many multiples of 2π). However, the number of individual segments that can be addressed is rather limited, and achieving sharp phase difference in the transverse beam profile is a challenge. DM-based mirrors are extensively used in adaptive optics applications [17, 18]. Adaptive optics based phase correction techniques have also been used in CARS microscopy where aberration of the focal fields in tissues can be pre-corrected to achieve a larger depth of focus in tissues [19].

SLM-based phase modulation techniques use an array of liquid crystals (LC) which can be addressed individually to alter the refractive index and hence change the phase of incident light. One-dimensional arrays of transmission-type SLMs can be used to dress beams with a 1D phase profile, such as the phase patterns of the HG10 or HG01 modes [20]. Besides their use in spatial beam shaping applications, SLMs are also extensively used for spectral shaping of femtosecond laser pulses [21]. Two-dimensional SLMs are mainly reflective based, due to ease of fabrication of large LC arrays. Large-pixel arrays with HD resolution (1920×1080) are currently available [22]. One drawback of the reflective technology is the back panel curvature due to substrate stress, which needs to be pre-compensated before applying the desired phase pattern. Another technical challenge with SLMs is the undesired amplitude modulation when adjusting the phase, which requires polarizer-analyzer arrangements to compensate for the amplitude changes. Recently, however, new LC technology has become available that makes use of electrically controlled birefringence (ECB) [23] and can achieve phase-only modulation without additional optics. Compared to deformable mirrors, SLMs can imprint very sharp phase steps onto the beam; however, the total phase shift in each pixel is currently limited to around one cycle of the optical wave. SLMs are preferred for more complex beam shaping applications especially in situations where many different phase patterns need to be evaluated with minimal changes to the optical layout of the experiment.

A typical experimental setup that incorporates a reflective SLM for beam shaping is shown in Fig. 11.3. The SLM plane is imaged onto the back aperture of the objective lens using a 4-f imaging

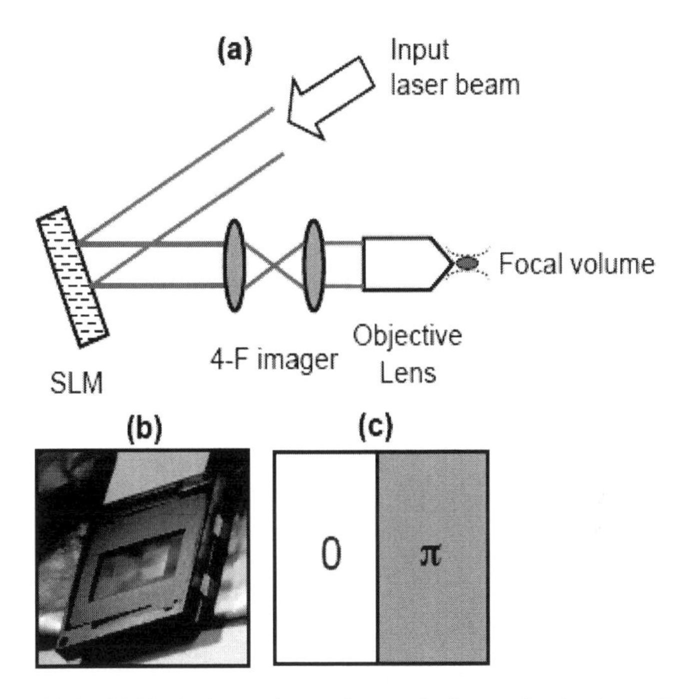

Figure 11.3 (a) Typical experimental setup for beam shaping in nonlinear microscopy. (b) Photograph of a typical SLM panel (from Holoeye). (c) Phase mask pattern to generate HG10 beam mode at focus.

system. Subsequently, the objective lens projects the phase shaped input excitation onto the focal volume. The 4-f imaging system ensures that aberrations in the beam do not accumulate as the beam propagates to the back aperture. Also shown in Fig. 11.3 is the phase pattern required to achieve HG10 mode at focus. This essentially is a horizontal 0–π division of the phase profile of the input beam, with the amplitude left unchanged. The resulting field profile in the focal plane can be approximately estimated from the Fourier transform of the input excitation. The exact 3D focal volume profile can be evaluated only by solving the integrals discussed in Section 11.2. Figure 11.4 shows the focal volume distribution of the intensity and phase distribution in the *XY* plane at focus and the *XZ* plane for 1064 nm input beam. The 0–π phase step of the input excitation results in a two-lobed focal volume with the null crossing accompanied by a phase reversal from 0 to π. This phase profile corresponds to that

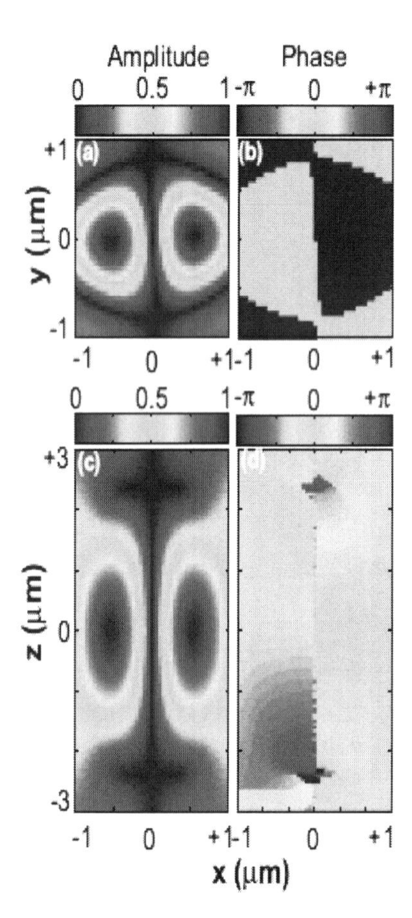

Figure 11.4 Electric field amplitude and phase profiles of HG10 beam mode at 1064 nm wavelength created using phase mask patterns as shown in Fig. 11.3c. (a) and (b) show the *XY* profiles and (c) and (d) show the *XZ* profiles. Only *X* polarization is shown.

of a focused HG10 beam. Similarly, a focal field with a HG01 phase pattern can be achieved by dividing the SLM panel into 0–π phase vertically. The nonlinear excitation volumes of these focal fields are discussed in Section 11.3. These beam shapes can be used in CARS microscopy for highlighting interface, which is further discussed in Section 11.5.

The SLM allows for flexible control of the phase compartmentalization of the focal volume. One such application of the SLM is the

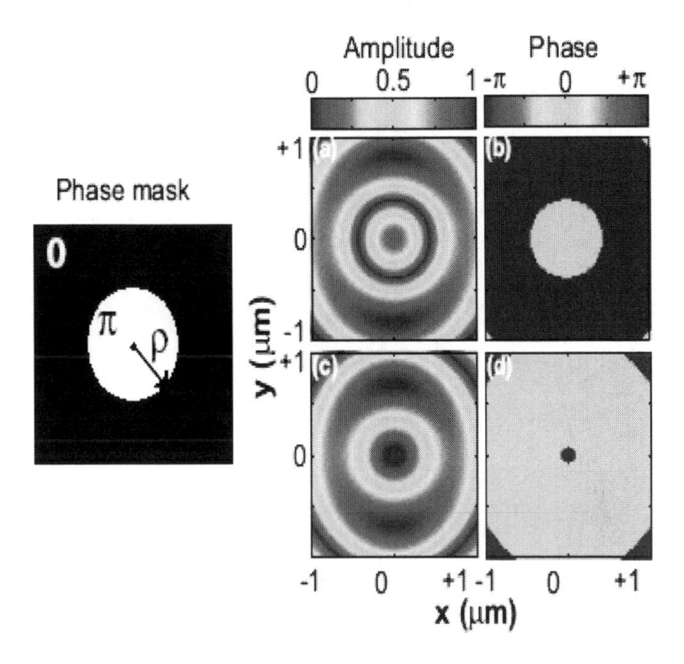

Figure 11.5 The phase mask used to create modified optical bottle beams and the simulated electric-field amplitude and phase profiles for two different normalized radius, ρ. (a) and (b) are for $\rho = 0.5$ and (c) and (d) are for $\rho = 0.6$.

generation of modified optical bottle beams. The term *optical bottle beam* (OBB) was first proposed by Padgett and Artl in their paper on shaping fields with a perfect null at focus surrounded by light in all three directions [24]. Such a field can be realized with a phase mask consisting of a circular 0–π phase pattern with a specific normalized radius ρ as shown in Fig. 11.5. Here we use the term *modified OBB* to refer to focal volume profiles generated from such a circular phase mask but without restricting the radius ρ. In this case, a perfect null is not always achieved at focus. Instead, the field strength of the light at the focal center can be conveniently varied by scaling ρ.

Figure 11.5 further shows the amplitude and phase distribution of the modified OBB profile at 1064 nm wavelength. The focal field consists of a central lobe surrounded by a concentric side-lobe, with phase difference of π between the two lobes. It can be generalized that the zero crossing of the field distribution is always accompanied

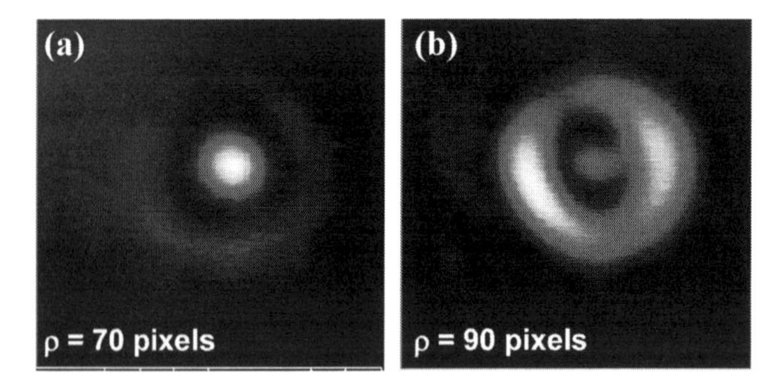

Figure 11.6 Experimental results of focal field imaging of modified optical bottle beams with increasing ρ. The images were obtained by scanning an isolated CdSe quantum dot across the focus and collecting the two-photon photoluminescence signal. Image area $\approx 2.1 \times 2.1$ μm.

by a phase reversal. The normalized radius of the central segment can be varied to change the relative strengths and size of the central-lobe with respect to side-lobes. In the simulation results shown here ρ is increased from 0.5 to 0.6. The energy lost from the central lobe is gained by the side lobes which have grown stronger. Such precise control of the annular is achievable with the recently available high-definition (HD) SLM panels [22], with suitable beam expansion. In Fig. 11.6 the experimental realization of modified OBBs is shown. These images were obtained by scanning isolated CdSe quantum dots at focus and collecting the two-photon photoluminescence (TPPL) signal in the forward direction. The input excitation was from an Nd-YVO_4 mode-locked laser at 1064 nm. The beam shaping was implemented in a setup similar to Fig. 11.3a with the phase shapes discussed above. The shrinking of the central lobe and the increase in side-lobe strength are clearly seen in these images. The application of such tunable OBBs in resolution enhancement and phase-based separation of focal volume is discussed in Sections 11.4b and 11.4c, respectively.

11.3.2 *Fixed Diffractive Optical Elements*

For certain applications of phase shaping the phase pattern applied is constant and in these situations, the use of specifically designed diffractive optical elements (DOE) or holographic patterns is preferred due to their low cost and high efficiency.

Following the discussions in Section 11.3.1 on OBB phase profiles, a perfect optical bottle beam is achieved for particular value of ρ at which the central lobe strength drops to zero. This beam is characterized by the central null at focus surrounded by light intensity in all three dimensions. These OBB profiles are used in optical tweezers to trap particles at the center [25]. For such specific applications it is better to use fixed beam shapers as there is no need to dynamically modify the focal volume. In this case holographic patterns or annular $0-\pi$ phase mask diffractive elements can be used to create such beam shapes. Figure 11.7A shows the holographic pattern which can be used to realize such OBBs at 628 nm [24]. These patterns are tailored for use with a fixed wavelength excitation. Figure 11.7B shows the OBB focal field profile in the *XZ* planes and various line scan profiles. The phase distribution (not shown here) has a sharp $0-\pi$ phase step as the beam propagates through the focal plane. It has been proposed that the sharp phase step can be used to resolve interfaces in the axial direction in CARS microscopy [26].

The OBB shapes discussed in Section 11.3.1 are a manifestation of a two-segment Toraldo pupil function [27]. In 1959 Toraldo proposed the concept of super-resolution by the use of phase-shaped annular pupil functions as a diffraction mask to create subdiffraction partition of the far-field profiles [28]. The same technique has also been extended to confocal microscopy [29, 30]. The annular phase mask profile results in concentric intensity distribution in the focal field and by properly designing the radius and number of phase segments, the subdiffraction central lobe can be sufficiently isolated from the side lobes. The apparent subdiffraction resolution capability using this scheme is due to the spatial separation of the central lobe from the strong side-lobes. Thus Toraldo pupil functions result in creation of focal fields with a reduced width of the central lobe along with a decrease in the Strehl ratio. The Strehl ratio is defined here as the ratio of the on-axis intensity achieved at focus with phase shaped beams to the theoretical maximum achievable on-axis intensity. There has been much effort devoted to fabricating special diffractive optical elements to realize Toraldo's

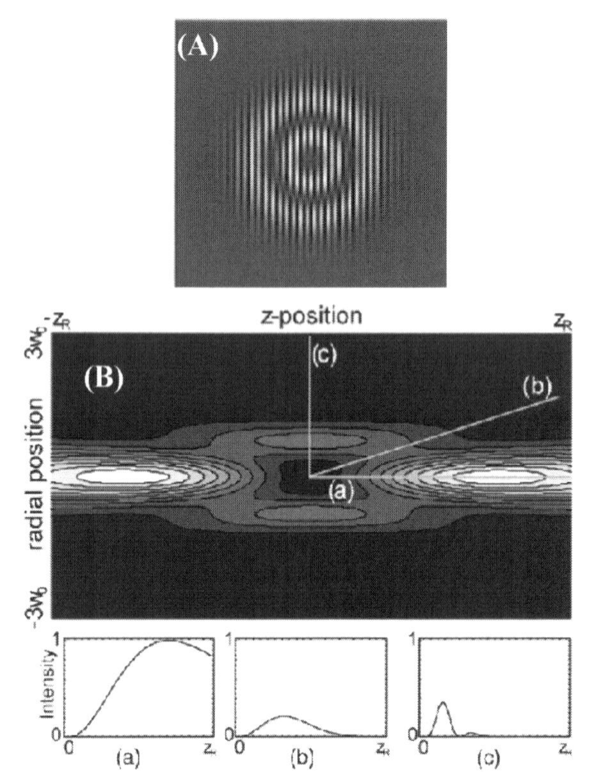

Figure 11.7 (A) Holographic pattern used to generate Optical bottle beams at 628 nm. (B) Intensity distribution of optical bottle beams along *XZ* plane at the focus. Line scan of the field profiles at different locations are also shown. Reprinted with permission from Ref. [24].

super-resolution ideas. Binary phase masks ($0-\pi$) with varying radius and number of annular rings have been designed for this purpose, although the experimental implementation of these ideas has trailed theoretical calculations. Figure 11.8 shows one such experimental implementation of a three-zone annular phase profile. Such DOEs are made possible by recent advances in microfabrication technology [31]. The fabrication procedure to manufacture such diffractive optical elements [32] is schematically shown in Fig. 11.8a. The measured intensity point spread function (PSF) distribution achieved with a three-zone phase function in comparison with the

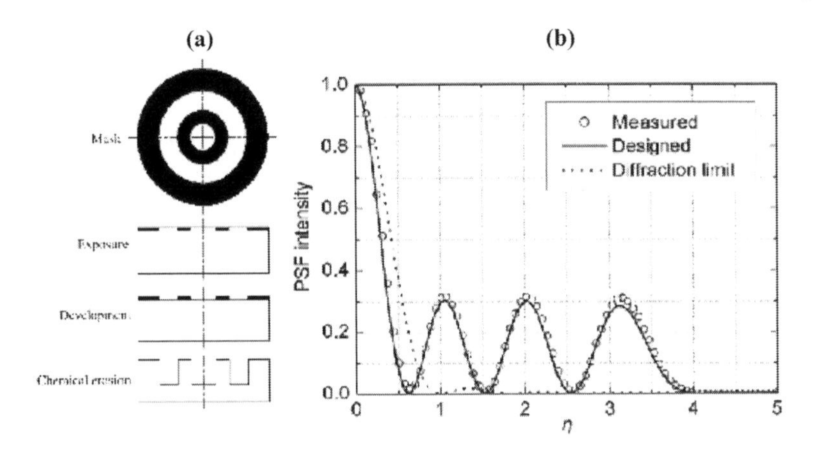

Figure 11.8 (a) The procedure to fabricate fixed diffractive optical elements for super-resolution. (b) Experimental measurement of point spread function improvement with three-zone annular pupil functions. Reprinted from Ref. [32].

Gaussian beam excitation is shown in Fig. 11.8b. Complex pupil function designs with detailed optimization algorithms have been implemented for various applications [33, 34].

Stimulated emission depletion (STED) microscopy technique uses doughnut shaped focal intensity distribution for the depletion beam to achieve super-resolution fluorescence microscopy [35]. In this case the depletion beam is phase shaped at the input to achieve the required null at focus. This can be accomplished with a Laguerre–Gaussian (LG) beam or with OBB phase shapes [36, 37]. The LG beam modes can be generated with the aid of spiral phase masks at the input. These phase masks are currently available as commercial products [38].

11.3.3 *Beam Interference*

Interference between two wave fronts can also be used to create interesting focal volume profiles. In this section we will discuss simple Gaussian beam interference and its capability to generate adjustable OBB-like profiles at focus. The addition of two **E** fields of the same wavelength but with different curvature of the phase front results in an oscillatory variation superimposed on the

Gaussian profile [39]. As explained before, the null crossing of the oscillatory terms results in phase reversal of the focal segments. By appropriately delaying the Gaussian beams relative to each other, different focal field profiles can be achieved. The interference of the two Gaussian **E** fields with different curvatures (δ_1 and δ_2) with phase difference $\Delta\varphi$ can be represented as [39]

$$E \propto \mathrm{Re}\left[\exp\left(-\frac{r^2}{2\sigma^2}\right)\exp\left(\frac{ikr^2}{2\delta_1}\right) + \exp\left(-\frac{r^2}{2\sigma^2}\right)\exp\left(\frac{ikr^2}{2\sigma_2}+i\Delta\phi\right)\right]$$

Figure 11.9a shows the experimental implementation of such a technique (reproduced from Ref. [39]). The collimated input beam goes through an interferometer setup where the beam in one of the arms is slightly diverging and delayed with respect to the beam in the other arm. The back aperture of the objective lens can be

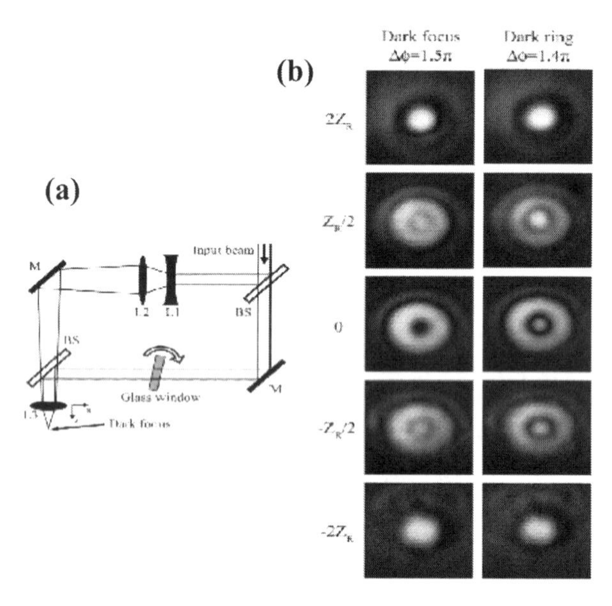

Figure 11.9 (a) Experimental setup for focal field shaping using Gaussian beam interference effect. (b) Focal field profiles achieved for two different phase delay settings between the interfering beams. Figures reproduced with permission from Ref. [39].

illuminated by the combined fields while care is taken to ensure that the spatial dimensions of the beams are comparable. Figure 11.9b shows the intensity profiles at the focus of a lens produced by this beam interference method. By varying the phase delay between the two arms it is possible to control the amount of electric field density in the central lobe.

11.4 Nonlinear Optics in a Microscope

The strong light intensities achieved in the focal spot of a high NA objective lens using short pulse excitation enable the generation of strong nonlinear optical signals [11, 40]. The third-order polarization of the material in the CARS process can be written as [41]

$$P^{(3)}(\omega_{AS}, r) = \varepsilon_o \chi^{(3)}(-\omega_{AS};\ \omega_P, -\omega_P, -\omega_S).E_P^2(\omega_P, r)E_S^*(\omega_S, r)$$

Here, the anti-Stokes wave is at frequency $\omega_{AS} = 2\omega_P - \omega_S$, and $\chi^{(3)}$ is the third-order nonlinear susceptibility tensor. Note that the CARS polarization is a multiplication of the Stokes field conjugate and the pump field squared. Thus the spatial coherence of the incoming fields is directly transferred to the nonlinear polarization induced at focus. This is to be contrasted with fluorescence techniques where the flourescence signal is incoherent and does not preserve any phase relationship with the incident excitation [11]. For the nonlinear signal to build up, the phase of the induced polarization needs to be in step with the phase of a propagating field at the anti-Stokes frequency. In other words, the phases of the polarization and the nonlinear field need to be matched. Phase matching is not always automatically fulfilled. For instance, due to the dispersion of the refractive index, and the fact that the wavevector of the participating fields depends on the refractive index, the imprinted phase by the excitation field relative to that of the signal field will diverge over extended propagation distances. This dispersion related phase-mismatching effect is very prominent in macroscopic CARS measurements. In microscopy, however, the distance during which the waves interact is of the order of an optical wavelength. Over such small distances, the effect of the material dispersion is negligible and the participating waves remain in phase during the interaction process. In addition to material dispersion, the Gouy phase shift is also a source or potential phase-mismatch. However, while significant in several other types of nonlinear coherent microscopy, we have already learned in Section 11.2 that the Gouy phase shift has no significant effect on the CARS signal generation process.

The multiplicative nature of the nonlinear signal with respect to the incoming field gives an opportunity to shape the nonlinear polarization at focus by shaping the incoming fields separately. The beam shaping techniques discussed in Section 11.3 can be applied to shape either the pump or Stokes fields, or both, and this would alter the nonlinear radiation from focus, enabling new information to be extracted from the material of interest. The applications of the beam shaping techniques are discussed in detail in Section 11.5. In this section we mainly emphasize on the nature of the nonlinear excitation fields at focus and the propagation of these focal profiles to the far field.

Figure 11.10 shows the CARS excitation profiles obtained for two different input fields. The simulated amplitude and phase profiles are shown in the first two columns, respectively. The experimentally measured focal volume profile is shown in the third column. The focal fields were experimentally mapped by scanning an isolated 30 nm-sized silicon nanoparticle through the focus and by collecting the electronic CARS signal in the epi-direction. Since the electronic four-wave mixing (CARS) signal is probed, and the electronic response from Si nanoparticles is spectrally broad, the exact wavenumber difference between pump and Stokes is not of much consequence. The pump and Stokes wavelengths used here are 817 nm and 1064 nm, respectively. The far-field radiation patterns from isolated nano-objects are discussed below. Figure 11.10a–c shows the cross-sectional profiles when both pump and Stokes are conventional Gaussian beams. In this case, the nonlinear signal at focus is characterized by a single central lobe. The diffraction limited spot size of the nonlinear signal is ~340 nm in the lateral direction, which is narrower than any of the input beams. The FWHM along the axial direction is ~950 nm, which ensures three-dimensional sectioning capability–an attractive feature of nonlinear microscopy

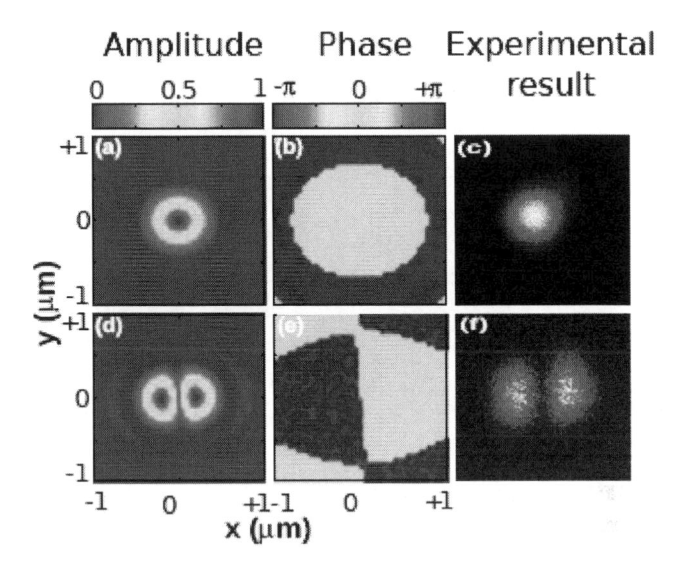

Figure 11.10 The amplitude and phase profiles of CARS excitation profiles along with the image at the focal plane mapped by scanning Si nanoparticles. For (a)–(c): the pump and Stokes excitation are both Gaussian, (d)–(f): pump is Gaussian and the Stokes is patterned with HG10 beam shape. Image size in (c) and (f) are ~2 mm × 2 mm.

for biological imaging. Figure 11.10d,e show the cross-sectional profiles for the case in which the pump beam is Gaussian and the Stokes beam is phase shaped to create a HG10 phase profile mode at focus. The nonlinear excitation field resembles a HG10 beam mode with the phase profile showing the characteristic 0–π one-dimensional phase step. For the experimental demonstrations, the phase-only SLM was patterned with 0–π phase segments, as shown in Fig. 11.3. The focal volume mapping clearly shows the two lobes in the nonlinear excitation. With phase sensitive detection techniques such as heterodyne CARS, the phase of the two lobes can also be mapped [7]. The applications of such phase shaped beams for interface highlighting and phase segmenting of focal volume are discussed in Section 11.5.

Once the coherent nonlinear dipoles are excited at focus, they radiate their energy into the far field. The far-field radiation profile can be calculated by integrating the contribution of all the dipoles at

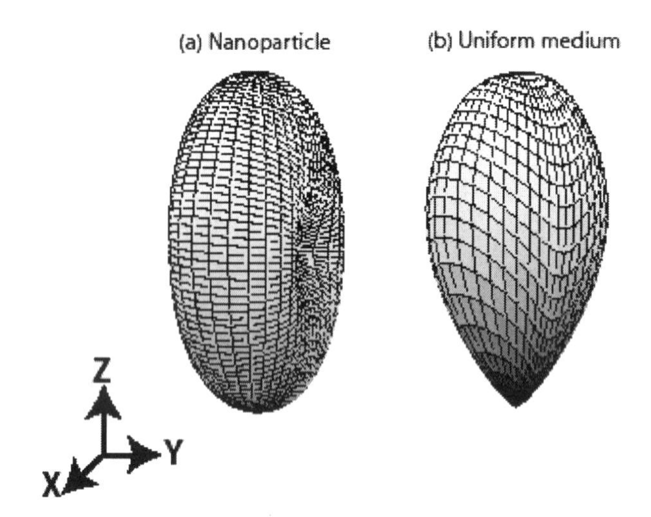

Figure 11.11 Far-field radiation patterns for (a) an isolated nanoparticle, and (b) uniform medium at focus.

focus [15]:

$$E_{\mathrm{FF}}(\omega_{\mathrm{AS}}, R) = -\int_{v} \frac{e^{ik|R-r}}{4\pi|R-r|^2}(R-r) \times [(R-r) \times P(\omega_{\mathrm{AS}}, r)]d^3r$$

The integral is carried out through the entire focal volume, v. The far-field intensity profile can be integrated on the surface of the detector, over the acceptance angle of the collection lens and detector. The acceptance angle for light collection is assumed to be 60°, unless otherwise stated.

It is instructive to consider the far-field radiation profiles from objects of specific shapes. In Fig. 11.11 we consider the coherent radiation from an isolated nanoparticles and an uniform bulk medium placed at focus. Nanoparticles are attracting considerable interest recently as nonlinear optical probes for certain biological applications [42]. Figure 11.11a shows the angularly resolved profile from an isolated nanoparticle. This radiation pattern resembles that of an dipole antenna. It is clear that these particles radiate equal portions of energy into the forward and in the epi-direction. Detecting this radiation in the epi-direction is a preferred way to image small particles as the collection of the objective lens is very efficient and this signal is not overwhelmed by a strong signal from the bulk. The radiation pattern from nonlinear emission from a uniform material is shown in Fig. 11.11b. In this case the signal is mainly concentrated along the propagation direction of the excitation beams. This is because the dipole oscillators in the focal plane can add coherently in the forward direction more efficient than in all other directions. In the case of biological tissue samples, significant amounts of the forward-propagating nonlinear signal is back-scattered and can be detected in the epi-direction as well [43].

11.5 Applications of Beam Shaping in Nonlinear Microscopy

In this section three different applications of the spatial beam shaping of excitation signal from nonlinear optical microscopy are discussed. These include (i) interface-specific imaging, (ii) resolution enhancement with annular phase functions, and (iii) nanoscale phase resolution of focal volume.

11.5.1 *Interface-Specific Imaging*

Excitation volumes that resemble HG10 and HG01 phase profiles can be used to selectively highlight interfaces in the horizontal and vertical direction [4, 5]. This can be understood as an interferometric enhancement or cancellation of dipole radiation within the acceptance angle of the detector from the two different phase segments. Figure 11.12 shows the CARS images of liquid interfaces between paraffin and d-DMSO (shown on left and right sides respectively) with HG00 and HG10 Stokes input fields. For 2899 cm^{-1} vibrational mode with HG00 excitation, paraffin is driven at a vibrational resonance and shows a strong CARS signal, while d-DMSO is nonresonant and shows a weak CARS signal. Upon exciting with a HG10 phase profile, the CARS signal from the uniform material cancels out as the dipole emitters on either side of the center are π out of phase. This cancellation effect suppresses the total signal from the bulk. At the interface, however, this cancellation effect is incomplete, because the materials on either side are different. As a consequence, the CARS signal at interfaces

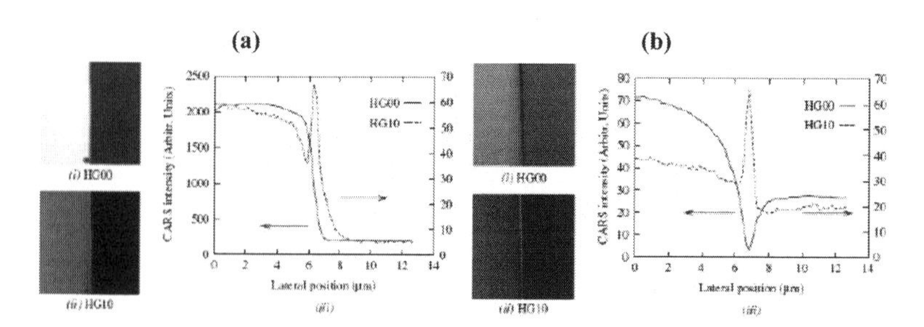

Figure 11.12 The use of focus engineering scheme to highlight interfaces selectively is shown. Interface shown here is between paraffin and d-DMSO. Images of the interface are shown at (a) 2899 cm^{-1} and (b) 2977 cm^{-1}. Excitation with HG00 and HG10 stokes beams are shown along with a line scan of the interface.

is much stronger than the CARS signal from the bulk. The contrast of the interface signal with respect to the uniform layer depends on the interplay between the spatial phase of the input beams and the spectral phase of the nonlinear susceptibility. It is found that maximum contrast is achieved in the CARS image with HG10 excitation if the Raman shift is set to at 2977cm^{-1}. This is because at this setting of the Raman shift, the spectral phase of the resonant material is $\sim\pi$ out of phase with the nonresonant material. The spectral phase difference is offset by the spatial phase difference of the HG01 mode, resulting in constructive interference at the interface. This spectral interference effect is evident even when a HG00 beam profile is used, as manifested by a dip at the interface due to the π phase difference between the two halves focal volume, resulting in destructive interference. Extending on these ideas, it has been proposed that interfaces of two dimensional structures can be imaged using Laguerre–Gaussian phase profiles [4, 10]. It has also be proposed that interfaces can be highlighted along the optic axis using optical bottle beams which are found to display a 0–π phase shift along the optical axis [26]. It should be emphasized that these techniques rely on limiting the acceptance angle of the detector to ensure good contrast. It has been found that the use of standard 0.55 NA condenser lens provides a good compromise between contrast and collection efficiency [4].

11.5.2 *Resolution Enhancement with Annular Pupils*

Subdiffraction resolution imaging in nonlinear microscopy is one of the actively pursued areas of research for resolving nanoscale features which remain out of reach when visualized with regular excitation. Unlike in fluorescence techniques, which are based on saturation effects of electronic excitations, resolution enhancement based on saturating transitions in molecules is not easily achieved in CARS microscopy. Saturation of Raman resonances involves excitation energies that are large enough to damage the samples of interest. Depleting vibrational coherences through direct infrared excitation in combination with Raman sensitive probing has been suggested as a possible route to produce tighter focal volumes in nonlinear Raman microscopy [44]. An alternative scheme based on measuring spatially varying Rabi oscillations has also been proposed [45]. Although resolution as low as 65 nm has been predicted, the need of high power for coherence depletion and the need for specialized energy states in the material for depletion may pose practical challenges for implementation. In this context, techniques for resolution enhancement requiring much lower power levels and more generic imaging conditions would be desirable. The focal volume engineering described in the previous sections can be applied to resolution enhancement to some extent. Here we discuss the use of annular pupil functions as a route to possible resolution enhancement in CARS microscopy.

In Section 11.3.2, Toraldo-style pupil functions were discussed in the context specially designed diffractive optical elements. These annular pupil functions can achieve subdiffraction focal volume segments at the center of the focus, albeit at the expense of strong side lobes. In CARS microscopy, the multiplication of the Stokes field with the square of the pump fields provides a way to filter out the side lobes to some extent and, consequently, tighter nonlinear excitation volumes can be achieved at focus. However, the side lobes cannot be suppressed indefinitely and the subdiffraction resolution has to be balanced with the side lobes increasing in strength. In general, nonlinear microscopy further extends the capability of super-resolution pupil functions by the multiplicative nature of the focal fields.

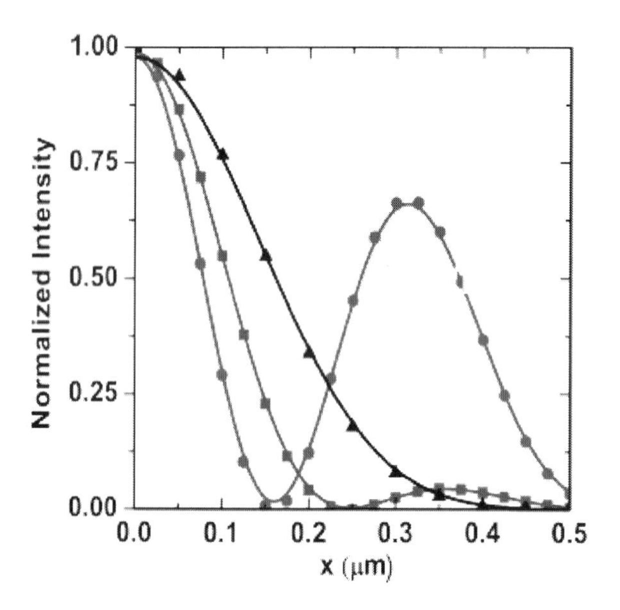

Figure 11.13 The line scan of nonlinear excitation intensity profile for modified optical bottle beams compared with Gaussian beam excitation. Gaussian profile is shown in black (triangles) trace. The modified optical bottle beam shapes include $\rho = 0.6$ and 0.62 shown by blue (square) trace and red (circle) trace, respectively.

Figure 11.13 shows the line scans of the intensity profile of CARS excitation at focus for a modified optical bottle beam with $\rho = 0.6$ and 0.62 in comparison with the Gaussian case [7]. The full-width at half maximum of the central lobe for $\rho = 0.6$ is only 210 nm along the X direction. In comparison the Gaussian excitation CARS profile has a FWHM of ~340 nm. The side lobes contribute to only 4% of the total intensity of the focal field strength. For $\rho = 0.62$, the central lobe shrinks more, but the side lobes also increase in strength making it less useful in terms of absolute resolution improvement. This is because the side lobes, although out of phase with the central lobes, contribute to the far-field signal, thus negating any resolution enhancement offered by the central lobe. This is further depicted in Fig. 11.14 in which the central lobe full width half maximum size and the side lobe to central lobe strength ratio are shown as a function of the OBB phase mask radius ρ. As shown in Fig. 11.14, the nonlinear excitation size of the central lobe goes down with increasing ρ, although at the expense of the decrease in Strehl ratio. Nonetheless, the resolution enhancement with simple binary annular functions can be up to $\sim$60% relative to the case of CARS microscopy with regular beam profiles. More complex Toraldo pupil functions with both phase and amplitude variations can be designed and optimized to separate the side lobes sufficiently from the central lobe to minimize their contribution to the nonlinear excitation field [33].

11.5.3 *Nanoscale Phase Resolution of Focal Volume*

The tunability of the OBB focal fields by dynamically changing the phase mask radius ρ was discussed previously. The focal field of the OBB as shown in Fig. 11.5 consists of a central lobe which is π phase shifted from the side lobes. The modified OBB excitation profile that uses a phase-shaped Stokes beam results in unlimited shrinking of the central lobe of the nonlinear excitation, ultimately resulting in a perfect null at the center of the focal plane. The applicability of such phase shaped beams is limited in resolution enhancement techniques due to the growth of the side-lobe strength. However, the subdiffraction sized central lobe created by the OBBs can be phase resolved using phase sensitive detection technique such as interferometric CARS, thus adding another differentiation scheme to the phase shaped focal field. Interferometric CARS has been used to reject the nonresonant signal in CARS microscopy by measuring both the amplitude and phase of the coherent nonlinear signal [46, 47]. By sending a copy of the anti-Stokes local oscillator (LO) to the microscope along with the pump and Stokes beams, the LO interferes with the generated CARS at focus and the detected anti-Stokes signal on the detector can be written as

$$I_{\text{total}} = |E_{\text{AS}} + E_{\text{LO}}|^2 = |E_{\text{AS}}|^2 + |E_{\text{LO}}|^2 + 2|E_{\text{AS}}||E_{\text{LO}}| \cdot \cos\Theta$$

Θ is the phase difference between the anti-Stokes signal generated at focus and the incident LO. By phase-modulating the LO input, lock-in detection can be used to map the phase of the nonlinear excitation as it is sensitive only to the beating term in the above equation.

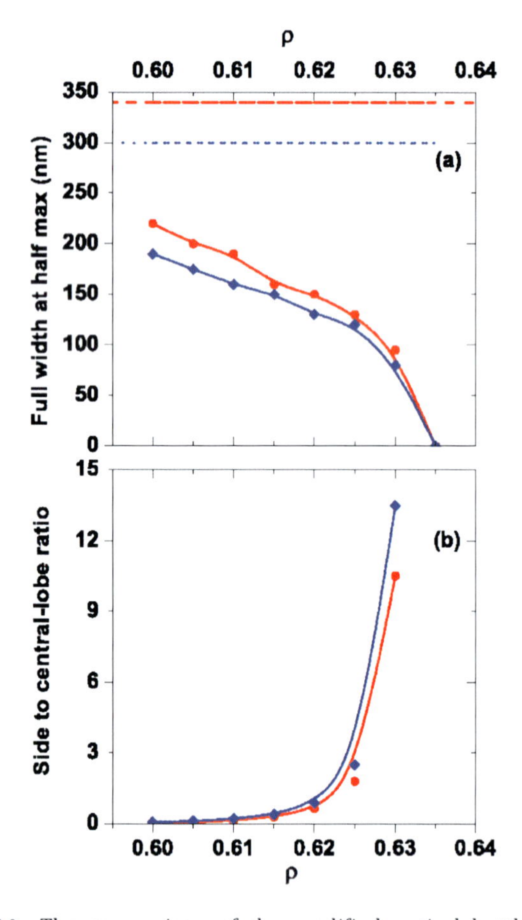

Figure 11.14 The comparison of the modified optical bottle beam (a) central-lobe size and (b) the side lobe to central lobe energy ratio as a function of the ρ parameter for the nonlinear excitation at focus. The red and blue traces correspond to the X and Y direction at the focal plane. The Gaussian case is shown by the horizontal red and blue lines in (a).

We show here simulation results of this phase sensitive detection technique [7]. In these simulations, a single 50 nm isolated nanoparticle is scanned across the focal volume and the interferometric CARS technique described above is used to detect the signal in the far field. It is assumed that the LO is modulated and the interference term is detected using a lock-in detection method. Figure 11.15 shows the line scan of the simulated lock-in signal

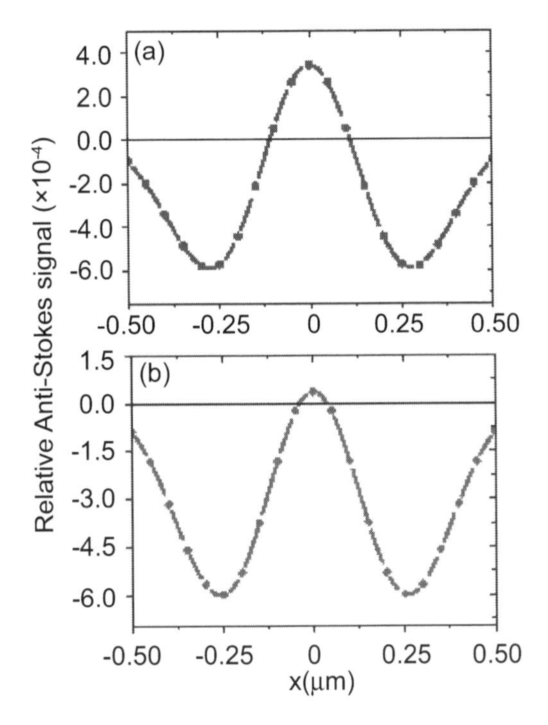

Figure 11.15 The line scan of the interferometric CARS Lock-in signal detected for an isolated nanoparticle scanned across the focal plane. Two different optical bottle beam profiles with increasing ρ are shown.

for two different OBB beam shapes. The LO is also assumed to be Gaussian beam in these simulations. The line scans show that there is a sign reversal when the object is scanned through focus. This sign reversal corresponds to a π phase step in the excitation volume, which is converted into an amplitude change by the interferometric detection method. The full width of the central lobe as measured by considering the locations of the sign reversal is ~200 nm in (a) and 100 nm in (b), respectively. The weak signal generated from the central lobe is further amplified by the use of LO in the interferometric detection scheme [46]. Thus, the combination of beam shaping and interferometric detection with phase shaping allows the identification of the subdiffraction segments of the focal volume that are separated by their distinct spatial phase.

Given that the width of the central lobe can be scaled arbitrarily small using the phase mask profiles, nanosized structures can be imaged at their natural length scales with this method. This would provide a more direct form of effective resolution improvement compared to super-resolution methods based on particle localization, which rely on some form of fitting. The calculations discussed above suggest the feasibility of visualizing individual objects with an effective resolution improvement of more than three times relative to conventional CARS imaging. This improvement is more than what is typically obtained by maximizing the optical transfer function of the imaging system. The effective resolution improvement here is achieved by extracting the information provided by both the amplitude and phase of the focal excitation volume, as opposed to using just the amplitude information contained in the focus. It should be emphasized that the results discussed here deal with only a single isolated particle scanned across the focal volume. The presence of multiple particles or a large homogeneous sample in the focal volume would result in interference of the CARS excitation signals from different segments, which would make it difficult to correctly decode the phase profile. The visualization method discussed here is thus not a general resolution enhancement method for CARS microscopy. Nonetheless, there are numerous problems in nanoscience in which the visualization of isolated nanostructures at their natural length scales can be useful. For instance, CARS excitation schemes have been used to detect the nonlinear response of plasmonic nanostructures [48, 49]. A higher resolution would be highly desirable for mapping out the spatial distribution of the surface plasmon polarizability on individual nanostructures. The CARS focal volume partitioning and interferometric detection may help to map out such details.

The ability to resolve isolated objects at their native resolution also opens up applications in particle sizing [7]. For a given nanostructure at focus, the signal from a nanostructure would increase until the size overlaps with the width of the central lobe (i.e., with increasing ρ). Beyond this point the presence of the side lobe, which is π out of phase with the central lobe, reduces the overall nonlinear signal and produces a sign reversal of the nonlinear signal. By adjusting the radius ρ on the SLM, the sign reversal can be monitored and the size of the particle can be deduced. Hence, the size of subdiffraction limited particles can be determined by using far field CARS interferometry. The presence of the phase-segmented regions at the focal plane also opens up applications in tracking moving particles. A particle moving in the focal plane can be tracked based on the phase and amplitude of the nonlinear signal detected by lock-in technique discussed previously. The sharp phase steps in focus provide reliable calibration points for the location of the particle, which are better defined relative to standard beam profiles.

11.6 Conclusions

In this chapter we studied the spatial coherences that underlie the signal generation mechanisms in coherent nonlinear imaging techniques. The spatial coherences in focus dictates the strength and propagating direction of the nonlinear radiation. While on the one hand the coherent nature of the radiation complicates the analysis of nonlinear images, it also provides additional parameters for controlling the radiation on the other hand. Radiation control is generally accomplished by shaping the transverse profiles of the excitation beams, something that has become increasingly more practical since the commercial availability of spatial light modulators. On the basis of the example of coherent anti-Stokes Raman scattering microscopy, we discussed various phase-only shaping techniques. By dressing the excitation beams with relatively simple phase profiles, the contrast of the CARS microscope can be changed in various ways, including interface highlighting and imaging with improved resolution. As implementation of beam-shaping techniques becomes increasingly more standard, we expect that such new contrast knobs on the optical microscope may find widespread use in nonlinear imaging applications.

Acknowledgments

This work is supported by the National Science Foundation (NSF), Grant No. CHE-0847097.

References

1. Chen, H., et al. (2009). A multimodal platform for nonlinear optical microscopy and microspectroscopy. *Opt. Express*, **17**:1282–1290.
2. Yew, E., and Sheppard, C. J. R. (2006). Effects of axial field components on second harmonic generation microscopy. *Opt. Express*, **14**:1167–1174.
3. Masihzadeh, O., Schlup, P., and Bartels, R. A. (2009). Enhanced spatial resolution in third-harmonic microscopy through polarization switching. *Opt. Lett.*, **34**:1240–1242.
4. Krishnamachari, V. V., and Potma, E. O. (2007). Focus-engineered coherent anti-Stokes Raman scattering: a numerical investigation. *J. Opt. Soc. Am. A*, **24**:1138–1147.
5. Krishnamachari, V. V., and Potma, E. O. (2008). Detecting lateral interfaces with focus-engineered coherent anti-Stokes Raman scattering microscopy. *J. Raman Spectrosc.*, **39**:593–598.
6. Beversluis, M. R., and Stranick, S. J. (2008). Enhanced contrast coherent anti-Stokes Raman scattering microscopy using annular phase masks. *Appl. Phys. Lett.*, **93**:231115.
7. Raghunathan, V., and Potma, E. O. (2010). Multiplicative and subtractive focal volume engineering in coherent Raman microscopy. *J. Opt. Soc. Am. A*, **27**:2365–2374.
8. Richards, B., and Wolf, E. (1959). Electromagnetic diffraction in optical systems II: Structure of the image field in an aplanatic system. *Proc. Roy. Soc. A*, **253**:358–379.
9. Novotny, L., and Hecht, B. (2006). *Principles of Nano-optics*. New York: Cambridge University Press.
10. Liu, C., and Kim, D. Y. (2007). Differential imaging in coherent anti-Stokes Raman scattering microscopy with Laguerre-Gaussian excitation beams. *Opt. Express*, **15**:10123–10134.
11. Denk, W., Strickler, J. H., and Webb, W. W. (1990). Two-photon laser scanning fluorescence microscopy. *Science*, **248**(4951):73–76.
12. Boyd, R. W. (1980). Intuitive explanation of the phase anomaly of focused light beams. *J. Opt. Soc. Am.*, **70**:877–880.
13. Cheng, J.-X., and Xie, X. S. (2002). Green's function formulation for third-harmonic generation microscopy. *J. Opt. Soc. Am. B*, **19**:1604–1610.
14. Moreaux, L., et al. (2001). Coherent scattering in multi-harmonic light microscopy. *Biophys. J.*, **80**:1568–1574.

15. Cheng, J.-X., Volkmer, A., and Xie, X. S. (2002). Theoretical and experimental characterization of coherent anti-Stokes Raman scattering microscopy. *J. Opt. Soc. Am. B,* **19**:1363–1375.
16. Bifano, T., Perreault, J., and Bierden, P. (2000). A micromachined deformable mirror to optical wavefront compensation. *Proc. SPIE,* **4124**:7–14.
17. Bifano, T. (2011). Adaptive Imaging: MEMS deformable mirrors. *Nat. Photon*ics, **5**:21–23.
18. *Imagine Optics, Adaptive Optics components (http://www.imagine-optics.com/).*
19. Wright, A. J., et al. (2007). Adaptive optics for enhanced signal in CARS microscopy. *Opt. Express,* **15**:18209–18219.
20. *Medowlark Optics, SLM Linear array (http://www.meadowlark.com/product/slmlineararray.php).*
21. Dudovich, N., Oron, D., and Silberberg, Y. (2002). Single-pulse coherently controlled nonlinear Raman spectroscopy and microscopy. *Nature,* **418**:512–514.
22. Holoeye, *HEO 1080P: High resolution LCOS phase only spatial light modulator (http://www.holoeye.com/phase_only_modulator_heo_1080p.html).*
23. Osten, S., Kruger, S., and Hermerschmidt, A. (2007). New HDTV (1920xx1080) phase-only SLM. *Proc. SPIE,* **6487**:64870X.
24. Arlt, J., and Padgett, M. J. (2000). Generation of a beam with a dark focus surrounded by regions of higher intensity: the optical bottle beam. *Opt. Lett.,* **25**:191–193.
25. Ozeri, R., Khaykovich, L., and Davidson, N. (1999). Long spin relazation time in a single beam blue detuned optical trap. *Phys. Rev. A,* **59**:R1750–R1753.
26. Krishnamachari, V. V., and Potma, E. O. (2007). Imaging chemical interfaces perpendicular to the optical axis with phase-shaped coherent anti-Stokes Raman scattering microscopy. *Chem. Phys.,* **341**:81–88.
27. Cox, I., Sheppard, C., and Wilson, T. (1982). Reapprisal of arrays of concentric annuli as superresolution filters. *J. Opt. Soc. Am. A,* **72**:1287–1291.
28. Francia, G. T. D. (1952). Nuovo pupille superrisolvente. *Atti. Fond. Giorgio Ronchi.,* **7**:366–372.
29. Hegedus, Z., and Safaris, V. (1986). Superresolving filters in confocally scanned imaging systems. *J. Opt. Soc. Am. A,* **3**:1892–1896.

30. Sheppard, C., Calvert, G., and Wheatland, M. (1998). Focal distribution for superresolving toraldo filters. *J. Opt. Soc. Am. A,* **15**:849–856.
31. Gil, D., Menon, R., and Smith, H. (2003). Fabrication of high numerical aperture phase zone plates qith a single lithography step and no etching. *J. Vac. Sci. Technol. B,* **21**:2956–2960.
32. Liu, H., Yan, Y., and Jin, G. (2006). Design and experimental test of diffractive superresolution elements. *Appl. Opt.,* **45**:95–99.
33. Sales, T., and Morris, G. (1997). Diffractive superresolution elements. *J. Opt. Soc. Am. A,* **14**:1637–1646.
34. Menon, R., Rogge, P., and Tsai, H.-Y. (2009). Design of diffractive lenses that generate optical nulls without phase singularites. *J. Opt. Soc. Am. A,* **26**:297–304.
35. Dyba, M., and Hell, S. W. (2002). Focal spots of size $\lambda/23$ open up far-field fluorescence microscopy at 33 nm axial resolution. *Phys. Rev. Lett.,* **88**:163901.
36. Willig, K., et al. (2006). SPED microscopy resolves nanoparticle assmeblies. *New J. Phys.,* **8**:106.
37. Klar, T. A., Engel, E., and Hell, S. W. (2001). Breaking Abbe's diffraction resolution limit in fluorescence microscopy with stimulated emission depletion beams of various shapes. *Phys. Rev. E,* **64**:066613.
38. Photonics, R., *Vortex phase plates (http://www.rpcphotonics.com/vortex.asp).*
39. Yelin, D., Bouma, B., and Tearney, G. (2004). Generating adjustable three dimensional dark focus. *Opt. Lett.,* **29**:661–663.
40. Zumbusch, A., Holtom, G., and Xie, X. S. (1999). Vibrational Microscopy Using Coherent Anti-Stokes Raman Scattering. *Phys. Rev. Lett.,* **82**:4142–4145.
41. Boyd, R. W. (2003). *Nonlinear Optics.* San Diego: Academic Press.
42. Wang, Y., et al. (2011). Four-wave mixing microscopy of nanostructures. *Adv. Opt. Photon.,* **3**:1–52.
43. Evans, C. L., and Xie, X. S. (2008). Coherent anti-Stokes Raman scattering microscopy: chemical imaging for biology and medicine. *Annu. Rev. Anal. Chem.,* **1**:883–909.
44. Beeker, W. P., et al. (2009). A route to sub-diffraction limited CARS microscopy. *Opt. Express,* **17**:22632–22638.
45. Beeker, W. P., et al. (2010). Spatially dependent Rabi oscillations: an approach to sub-diffraction limited coherent anti-Stokes Raman scattering microscopy. *Phys. Rev. B,* **81**:012507.

46. Potma, E. O., Evans, C. L., and Xie, X. S. (2006). Heterodyne coherent anti-Stokes Raman scattering (CARS) imaging. *Opt. Lett.,* **31**:241–243.
47. Jurna, M., et al. (2007). Shot noise limited heterodyne detection of CARS signals. *Opt. Express,* **15**:15207–15213.
48. Kim, H., et al. (2008). Spatial control of coherent anti-Stokes emission with height-modulated gold zig-zag nanowires. *Nano Lett.,* **8**:2373–2377.
49. Jung, Y., et al. (2009). Imaging gold nanorods by plasmon-resonance-enhanced four-wave mixing. *J. Phys. Chem. C,* **113**:2657–2663.

Chapter 12

Supercontinuum Sources for Biophotonic Applications

J. R. Taylor

Femtosecond Optics Group, Physics Department (Photonics), Imperial College London, Prince Consort Road, London SW7 2BW, UK

jr.taylor@ic.ac.uk

This chapter reviews the history and development of practical supercontinuum sources. As a result, only fiber-based devices will be considered. Emphasis is placed on a description of the primary nonlinear optical processes that contribute most commonly to supercontinuum generation process and how they can be controlled either through the correct choice of pump wavelength, pump pulse format, and power or by controlling the physical parameters of the host fiber such as the dispersion, mode field diameter, or associated nonlinear coefficient.

12.1 Introduction

Within a few years of the development of the laser in 1960 by Maiman, both passive and active techniques were developed to extract the energy from the laser in short pulses. Initially in

Understanding Biophotonics: Fundamentals, Advances, and Applications
Edited by Kevin K. Tsia

ISBN 978-981-4411-77-6 (Hardcover), 978-981-4411-78-3 (eBook)
www.panstanford.com

the nanosecond regime using Q switching, the picosecond regime was rapidly attained deploying the technique of mode locking. As a result, optical pulses with controlled pulse durations were readily attainable, in fact these were the shortest man-made events available. In the intervening almost 50 years, although the physics has practically remained the same, the technology has become significantly more sophisticated and pulses of only a few femtoseconds can be obtained from relatively simple laser sources. As a result, even for modest pulse energies, power densities greater than a terawatt per square centimeter can be readily achieved at the focal spot of a conventional convex lens with corresponding electric field strengths exceeding a megavolt per centimeter, resulting in a nonlinear response of the medium. In the description of the induced polarization, higher-order terms of the electric field need to be included; hence

$$P = \varepsilon_0 \left(\chi^{(1)} E + \chi^{(2)} E^2 + \chi^{(3)} E^3 + \ldots\right) \tag{12.1}$$

where P is the polarization, ε_0 is the permittivity of free space, $\chi^{(n)}$ is the nth order susceptibility, while the first term on the right hand side represents the linear response under low field strength E. Driven by a field of frequency ω, the second term gives rise to a response at frequency 2ω, second harmonic generation; however, in media exhibiting a center of symmetry this second-order term vanishes to zero and consequently in a medium such as a silica glass fiber it is the third-order term that principally contributes to the nonlinear response.

Nonlinear optical processes had been characterized prior to the invention of the laser. For example, in 1875, Kerr had investigated birefringence in isotropic media induced by a DC field, which was proportional to the intensity and so related to the third-order term in Eq. 1.1. Similarly in 1895 Pockels had shown that the birefringence induced in piezoelectric crystals was a linear function of the applied DC electric field, as is described by the second term in Eq. 1.1 above. However, in both of these examples the optical field is only one term of the overall contribution, the applied electric field and or intensity providing the other component. Only with the appearance of the pulsed laser did the optical field alone provide the sole contribution leading to the birth of modern nonlinear optics.

The first nonlinear process to be reported was second harmonic generation [1] which, however, is of little relevance to supercontinuum generation either in bulk or in fiber. More important are the processes that are a result of the third-order term in Eq. 1.1. Again, third harmonic generation originally reported by New and Ward [2] provides a relatively minor contribution to supercontinuum generation; more important are the roles played by the optical Kerr effect or intensity-dependent refractive index [3], self-focusing [4, 5], four-wave mixing [6], and stimulated Raman scattering [7].

In early laser oscillator amplifier systems self-focusing instabilities and spectral broadening were operational characteristics to be avoided; however, Jones and Stoicheff [8] utilized a modest continuum generated via anti-Stokes scattering in a cell of benzene generated by a focused giant pulse ruby laser to undertake nanosecond transient absorption measurements. Others also utilized self-focusing in cells of highly nonlinear carbon disulphide to achieve spectral broadening [9, 10] and on the basis of experimental observation Shimizu [11] theoretically demonstrated that self-phase modulation arising from intensity-dependent refractive index was the mechanism for the observed spectral broadening and interference. Although Shimizu and many others observed significant spectral broadening both intracavity with pulsed lasers or extracavity in liquid or solid samples, the most widely recognized first report of "supercontinuum generation" is that of Alfano and Shapiro [12], who reported a white light spectrum extending from 400 nm to 700 nm in a borosilicate glass sample placed at the focus of gigawatt peak power picosecond pulses from a frequency doubled Nd glass laser. The term *supercontinuum*, however, was not used to describe the spectral output, with the first reference to this nomenclature being Gersten et al. from the Alfano group in 1980 [13]. Throughout the 1970s and 1980s the supercontinuum source was extensively employed in time-resolved spectroscopic studies on picosecond or femtosecond time scales dependent on the pulse source and utilizing the same basic technique of focusing amplified pulses from various laser sources into either filled cells or jets of nonlinear liquids. Self-phase modulation was the dominant contribution to the spectral broadening mechanism, and although extremely versatile, the technique was restricted to research laboratory use and was limited by relatively low average powers, system instabilities, and alignment procedures.

12.2 Early Optical Fiber-Based Sources

By the early 1970s enormous efforts were directed toward the commercial production of low-loss single-mode optical fiber driven by the potential application in high-capacity broadband telecommunication. These developments simultaneously stimulated a new field of study: nonlinear fiber optics. The advantage of optical fiber over a bulk material is quite obvious in spite of the fact that silica has an exceedingly low nonlinear coefficient. The effective interaction length of a focused beam is limited by the confocal parameter. For a beam waist of a few micrometers, this length is of the order of a millimeter; however, if the input light can be coupled into the core of a single-mode optical fiber with the diameter of a few micrometers, the interaction length is simply limited by absorption, which for modern fibers with loss of the order of 0.2 dB/km leads to effective lengths in excess of 10 km and a consequential enhancement in fiber by a factor of 10^7–10^8 over bulk operation. Naturally, where the nonlinear processes involve phase or group velocity matching criteria then such interaction lengths may not be possible, however the substantial reduction in the power requirement to observe equivalent non linear processes in fiber over those in bulk is clear.

The first nonlinear effect to be reported in fiber was stimulated Raman scattering [14], although a hollow core fiber filled with carbon disulphide was used. Such a technique has once again been rejuvenated, like several areas in nonlinear optics, by the technological advances allowed by air core photonic bandgap fibers that can be filled with gas or liquids for nonlinear studies. As low-loss single-mode fiber of conventional structure became more readily available, all the nonlinear effects that had been previously observed in bulk configuration were observed and characterized in fiber and importantly at substantially lower power levels. The effects included stimulated Raman scattering [15], the optical Kerr effect [16], four-wave mixing [17], and self-phase modulation [18]. All of these and other effects that are vitally important in supercontinuum

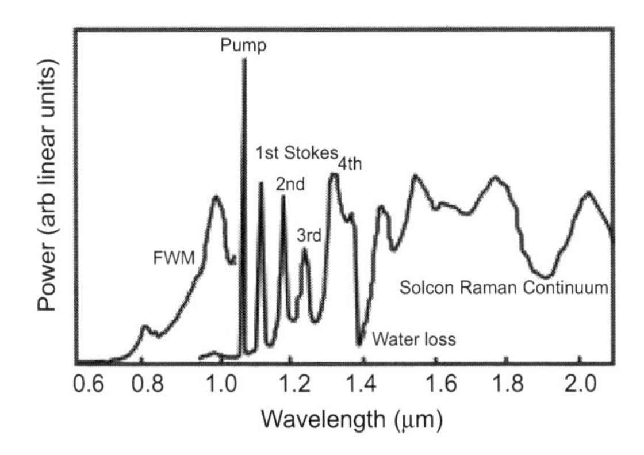

Figure 12.1 Supercontinuum spectrum generated in a 315 m length of multimode Ge-doped silica fiber at a peak pump power of 50 kW, after [21].

are concisely treated by Roger Stolen, a stalwart in the field since its inception, in a review of the early years of fiber nonlinear optics [19].

The first report of supercontinuum generation in fiber was by Lin and Stolen [20], who used various pulsed dye lasers to generate continua extending from about 392 nm to 685 nm, depending on the pump wavelength. It was recognized that stimulated Raman scattering played a dominant role in the generation process as the continua which were around 100–200 nm broad, were dominated by extension to the long wavelength side of the pump. As self-phase modulation smeared the cascaded Raman orders into a continuum, the potential application of such a source for excited state spectroscopy was recognized. The technique was extended to the near-infrared through the use of a Q-switched and mode-locked Nd:YAG laser operating at 1.06 μm to pump low-loss single-mode and multimode fibers generating the now familiar signature of cascaded Raman orders until the zero dispersion wavelength of the fiber was reached from which point on a soliton Raman continuum was obtained [21]. Figure 12.1 shows a representative supercontinuum spectrum obtained in such a manner.

The principal nonlinear processes contributing to the generated spectrum are indicated in Fig. 12.1. To the long-wavelength region of the pump at 1060 nm the cascaded Raman orders from the first at 1120 nm through to the fourth at 1305 nm are clearly identifiable, beyond which a distinct soliton Raman continuum is formed in the region of anomalous dispersion. At the time of this work, although optical solitons had been theoretically proposed [22], their existence had still not been experimentally verified and a further two years were to elapse before the classic series of experimental reports on soliton generation and characterization by Mollenauer were published [23]. A brief history of the development of optical solitons can be found in [24]. In early fibers, high water loss was present (>40 dB/km) and the result of this is identifiable as the dip in the supercontinuum in the region of 1.38 μm. For these early schemes utilizing pumping at 1060 nm, well away from the region of the zero dispersion of the fibers, four-wave mixing processes were also very inefficient as a result of poor phase matching; consequently, excursion to the short-wavelength range was limited and conversion efficiency poor. From the original, this region is plotted on a magnified intensity scale in the figure above. To achieve enhanced operation in the visible it was recommended to pump using fundamental dye laser sources [20] in the region of interest.

For high-power (~100 kW) pump pulses from a Q-switched and mode-locked Nd:YAG laser in short lengths of optical fiber, sum frequency generation between the pump radiation and longer-wavelength Stokes radiation was identified as an important process for generating radiation in the blue-green in a supercontinuum that extended over the complete window of transparency of silica fiber from 300 nm to 2100 nm [25]. The importance of operating with the pump in the region of low anomalous dispersion was also empirically identified by Washio et al. [26], who demonstrated that by using a pump at 1.34 μm the generated supercontinuum did not exhibit the distinct cascaded Raman orders characterizing Fig. 12.1 above but rather a smooth profiled supercontinuum was generated. As the physics of optical soliton generation was in its infancy, the mechanism for the smooth spectral profile—soliton Raman generation—was unknown and it was to be several years before soliton Raman effects would be both theoretically proposed [27] and experimentally demonstrated [28] as a simple yet versatile method of supercontinuum generation that plays a central role in modern compact supercontinuum sources.

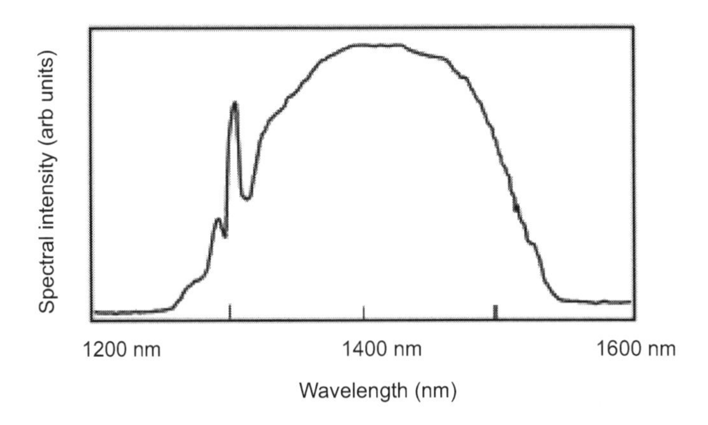

Figure 12.2 High average power soliton Raman supercontinuum [31].

By the mid 1980s, extensive investigations had been carried out on the dynamics of soliton Raman generation. It was demonstrated that this was most efficiently instigated from modulational instability [29] in the region of low anomalous dispersion and that the process could be efficiently coherently seeded by feedback of the modulational instability sideband giving rise to extended long-wavelength continua [30]. As a result, the soliton Raman continuum became a simple source of high average power in the watts level, albeit limited to the long-wavelength side of the pump wavelength. Figure 12.2 shows a typical soliton Raman supercontinuum [31].

The supercontinuum recorded in Fig. 12.2 was generated using a mode-locked Nd:YAG laser operating at 1.32 μm with an average power of 1 W in a 200 m length of single-mode fiber with a zero dispersion wavelength around 1300 nm. From the figure, the signature of modulational instability can be seen to the short-wavelength side of the pump radiation beyond which, to the long-wavelength side, lies the soliton Raman continuum which evolved from modulational instability on the long (∼100 ps) pump pulses.

12.3 Contributing Physical Processes

The production of a supercontinuum in optical fiber can be the result of numerous non linear effects taking place as the optical disturbance propagates in the fiber. Consequently, dispersion also plays an important role in the overall process, while the interplay of dispersion and nonlinearity can lead to different processes contributing to the supercontinuum at various stages of the propagation.

12.3.1 *Dispersion*

Dispersion arises as a result of the frequency dependence of the effective refractive index of the guided mode with contributions both from the material and the waveguide structure. An additional and important contribution can come from modal dispersion, and this can play an important role in the phase matching of four-wave mixing processes, where, for example, the anti-Stokes signal can propagate in a higher-order mode as compared to the pump and Stokes to achieve phase matching which would otherwise be unattainable in a purely single-mode geometry [32].

The dispersion can affect both the phase velocity and the group velocity of a signal. For high efficiency, phase matching is essential in many nonlinear interactions such as parametric generation and the nonlinear contribution to this must also be considered in most circumstances. In addition, group velocity dispersion matching is of importance, for example, in the situation of soliton-dispersive wave interactions that straddle the zero dispersion wavelength of the interaction fiber. This process as will be seen later gives rise to the important short wavelength extension of supercontinua. In conventional silica fiber structures the minimum zero dispersion wavelength achievable with the use of pure silica is 1270 nm. At wavelengths above this the dispersion is anomalous. In practical units the group delay dispersion is defined in units of ps/nm.km defined as

$$D = -\left(2\pi c/\lambda^2\right)\beta_2 \tag{12.2}$$

where β_2 is the group velocity dispersion usually written in units of s^2m^{-1}, and when β_2 is positive, the dispersion is normal and anomalous when β_2 is negative.

As solitonic effects play a major role in supercontinuum generation it is usually essential to pump fibers in a region of low anomalous dispersion. Therefore the use of conventional fibers dictates a need for pump sources above 1270 nm. In this region, only the Er-doped fiber laser provides sufficient technological sophistication for the application but also places a severe limitation on the spectral coverage of any generated supercontinuum. The impact of the introduction of photonic crystal fiber (PCF) on the field is therefore very apparent [33, 34]. Through adjustment of the pitch and diameter of the photonic crystal cladding around the core, it is possible to accurately control the dispersion, and as a consequence the zero dispersion wavelength could be shifted well into the visible, such that optimized soliton operation was possible at pump wavelength from Yb- or Nd-doped lasers around 1060 nm or mode-locked titanium sapphire systems broadly around 850 nm. In addition, the mode field diameter of the pcf could be substantially reduced, leading to an effectively higher nonlinear parameter [35] and optimized designs also ensured single transverse mode operation throughout the full spectral range of supercontinuum generation [36]

12.3.2 *Optical Solitons: Self-Phase Modulation and Dispersion*

The third-order term of Eq. 12.1 gives rise to an intensity dependent refractive index, such that the overall refractive index

$$n = n_0 + n_2 I(t) \tag{12.3}$$

where n_2 is the nonlinear refractive index, which is assumed to be instantaneous on the time scale of an incident optical pulse of intensity profile $I(t)$. The resultant modulation of the refractive index gives rise to a time-dependent phase shift which is equivalent to a frequency shift, simply proportional to the time differential of the incident intensity profile as originally shown by Shimizu [11]. This leads to a relatively linear normal frequency chirp across the central region of the pulse profile such that the frequency increases with time. Alone, this self-phase modulation does not affect the temporal profile of the pulse, but when combined with dispersion leads to temporal reshaping. In the normal regime, where longer wavelengths travel faster, the combined effects give rise to pulse broadening and chirp linearization. In the anomalous regime, dispersive broadening can also occur; however, under certain conditions, it can be envisaged that a balance can occur between the normal frequency chirp and the anomalous dispersion, as first proposed by Hasegawa [22], leading to a family of stable or period analytic solutions of the basic nonlinear Schrödinger equation, known as optical solitons that describe the system. The power required to produce a fundamental soliton that will propagate over its nonlinear length without a change in its pulse duration, in practical units, is given by

$$P_0 = \left(\frac{1.763}{2\pi}\right)^2 \frac{A_{\text{eff}}\lambda^3}{n_2 c} \frac{D}{\tau^2} \tag{12.4}$$

where A_{eff} is the effective core area of the fiber and all the other symbols have their usual meaning. It can be seen that lowering the dispersion reduces the required pulse power, while increasing the pulse length significantly reduces the power requirement; however, increased pulse length also requires increased fiber length in order to establish soliton operation. For 500 fs, 1.55 μm pulses at 1 GHz repetition rate in a standard fiber with a dispersion of 2 ps/nm.km an average power of only 15 mW is required to establish a fundamental soliton. The input power required of a pulse to form a $N = 1$ fundamental soliton does not have to exactly match that required by P_0. A pulse of power P will readjust itself to become a $N = 1$ soliton by readjusting its duration and shedding off dispersive non-solitonic radiation so long as $0.25 < P/P_0 < 2.25$.

Higher-order soliton solutions are possible given by $P_{\text{N}} = N^2 P_0$. These-higher order solitons of order N are just a nonlinear superposition of N fundamental solitons (with amplitudes A of 1, 3, 5, $\ldots 2N - 1$). Significant pulse narrowing and periodic splitting can occur with higher-order solitons on propagation which results from periodic interference between these solitons. As a result of the interference, extreme pulse compression can occur with an optimal compression factor of $4.1N$ being possible and the launch of high-power pulses near the dispersion zero and the subsequent decay of the high-order soliton was extensively researched in the 1980s as a method of ultrashort pulse generation [23, 37–39]. Associated with this extreme pulse compression is substantial spectral broadening an example of which can be seen in Fig. 12.3.

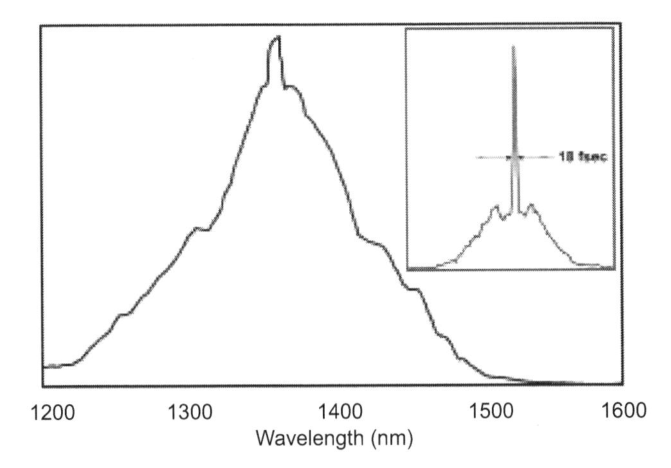

Figure 12.3 Broadband spectrum associated with high-order soliton compression to an 18 fsec pulse (inset) after [39].

The periodic breathing observed with lower-order picoseconds solitons is not observed in cases of extreme femtosecond soliton generation. Perturbations caused by higher-order dispersion or self Raman excitation, otherwise known as the soliton self-frequency shift, lead to soliton instability and fragmentation into the numerous constituent fundamental solitons. This effect, well documented both theoretically and experimentally in the 1980s, has been renamed soliton fission.

12.3.3 *Modulational Instability and Four-Wave Mixing*

Closely related to soliton generation and resulting from the interplay between anomalous dispersion and the intensity-dependent refractive index is modulational instability. Very many nonlinear systems exhibit such instabilities and the growth of modulations or perturbations from the steady state. In optical fiber, the process was first described by Hasegawa [40]. It was shown that amplitude or phase modulations on an effective continuous wave background exhibit an exponential growth rate that is accompanied by sideband evolution at a frequency separation from the carrier that is proportional to the optical pump power. For exponential growth, the sideband frequency separation from the carrier should be less than a critical frequency given by $(4\gamma P_0/\beta_2)^{1/2}$ with the maximum growth occurring at a frequency of $(2\gamma P_0/\beta_2)^{1/2}$. Modulational instability can be thought of as a four-wave mixing process phase matched through self-phase modulation, with the growth of the Stokes and anti-Stokes sidebands taking place at the expense of two photons from the carrier pump. Modulational instability is most commonly self-starting from noise at the frequency separation of the maximum gain. It is, however, possible to initiate the process, induced modulational instability, by seeding with an additional pump that lies within the gain bandwidth, and the technique has been used to enhance the long-wavelength extent of modest soliton Raman supercontinua as compared to utilizing a single pump at the same average power [30]. Cross-phase modulation can also be used to induce modulational instability on weak signals in the anomalously dispersive regime and is enhanced through group velocity matching of a pump in the normally dispersive region and the signal in the region of anomalous dispersion [41]. Modulational instability plays an important role in the initiation of short-pulse soliton formation which with subsequent amplification, self Raman interaction, and collisions forms a basis for supercontinuum generation using continuous-wave (cw), picosecond, and nanosecond pumping. With femtosecond excitation the process generally only plays a role much later in the supercontinuum development and is unnecessary for the initiation of the continuum.

12.3.4 *Soliton Instabilities and Dispersive Wave Interaction*

Although the soliton power is precisely defined for a particular optical fiber, solitons themselves are remarkably robust and in fact a pulse of any reasonable shape and with sufficient energy will, on propagation, evolve into a soliton. Solitons have been shown to emerge from noise bursts that contain sufficient energy or to be generated through the synchronous amplification of noise bursts. The energy not required to form the soliton will simply appear as dispersive radiation. For the production of supercontinua, the most common experimental situation encountered is the launch of a high-energy pulse, equivalent to a high-order solition, in the region of the zero dispersion wavelength of a nonlinear fiber. The initial part of the evolution is identical to that described above, in that the high-order soliton will temporally compress accompanied by associated spectral broadening. It is at this stage that perturbations come into play and lead to a breakup of the soliton structure into a collection of lower-amplitude "colored" soliton and non soliton structures. The perturbation to the soliton can arise through high-order dispersion terms which are of particular importance once a high-order soliton has compressed generating an extensive bandwidth. The other effect also associated with the broad bandwidth of a femtosecond-scale pulse is self-Raman interaction [28] or the so-called soliton self frequency shift [42, 43]. The peak of the Raman gain bandwidth of silica is at a frequency shift of approximately 440 cm^{-1} from the pump and as a result of the glassy structure the gain bandwidth is exceptionally broad ($\sim$800 cm^{-1}) [15]. Even for frequency shifts of 100 cm^{-1} the Raman gain is approximately 20% that of the peak. Consequently for pulses of the order of 100 fs significant gain can be achieved on the long-wavelength side of the pulse spectrum provided by pumping by components of the short-wavelength side of the pulse spectrum. On propagation, ultrashort pulse solitons will experience a continuous red shift. However, the soliton power is dependent on the dispersion (see Eq. 12.4) as well as cubic power of the wavelength. With the increasing wavelength and most commonly with an increasing dispersion with wavelength the power demands for a fundamental soliton increase, resulting in an adiabatic pulse broadening. Since the soliton self-frequency shift is proportional to τ^{-4} [43], where τ is the soliton pulse width, the process becomes self-terminating. Naturally, the process can be enhanced through fiber design, either by deploying dispersion flattened or dispersion decreasing fiber formats.

For high-power pulses launched in the region of, or straddling, the zero dispersion, it has been predicted that solitons will emerge from pulses of any arbitrary shape and amplitude [44]. It was proposed that with increasing amplitude the solitons would frequency down shift and that the non-solitonic component in the normally dispersive regime, dispersive waves, would correspondingly frequency up shift, which was experimentally verified [45]. Earlier it had been experimentally reported that in the decay of high-order femtosecond solitons, the self-frequency shifted solitons trapped a dispersive wave component that moved to shorter wavelengths coincident with the long-wavelength shifting soliton. The authors demonstrated that group velocity matching was essential for the process to take place [46]. A more recent and more precise series of measurements have been undertaken by Nishizawa and Goto mapping the spatio-temporal evolution of a supercontinuum showing clear evidence of soliton–dispersive wave trapping [47, 48], the process which is vital for the short-wavelength extension of supercontinua.

12.4 Modern Supercontinuum Sources

Although Alfano and Shapiro made their first report of extensive supercontinuum generation in a bulk glass sample in 1970 [12], the technique changed little over more than 30 years. Impressive and highly applicable spectral coverage was obtainable, yet the experimental configurations detracted from widespread application, particularly in the area of biophotonics. There are several factors that contributed to this. Of primary concern was that low-repetition rate pulsed lasers were used as the pump source; consequently extremely low average powers and low spectral power densities were obtained, limiting sensitivity. The matching of large-frame pulsed lasers originally with bulk samples and latterly with conventional single-mode fibers, introduced long-term instability and unreliability of the sources, requiring source attention and realignment, when what was needed was a "plug and play" solution for real-world applications. Finally, earlier fiber configurations relied upon conventional fiber structures where soliton effects were restricted to the wavelength range above 1300 nm, which did not match with any efficient pump sources. As a consequence, the generated supercontinua were severely restricted in their wavelength coverage in particular in the near-UV and visible, regions that are of particular importance for biophotonic application. A typical supercontinuum spectrum pumped by a Nd:YAG laser is that of Fig. 12.1. Full integration with fiber laser pumps was possible and relatively high-power operation was achieved with very compact packages. Chernikov et al. [49] reported a novel all-fiber Q-switched Yb-fiber laser pumped supercontinuum source based upon conventional single-mode fiber with an average output power of up to 1.2 W, with a supercontinuum dominated by cascaded Raman scattering and soliton Raman generation, which consequently limited operation to the range 1060–2300 nm with an output reminiscent of Fig. 12.1. Four-wave mixing extended the short wavelength extent but with less than 10% of the efficiency of the generation of the infrared continuum. Generation of the second and third harmonics ensured that the complete transmission window of silica was covered, although relatively inefficiently in the visible. Peak powers were such that the infrared components of the continuum could be externally frequency-doubled to generate the visible frequency but with limited conversion efficiency. The inability to efficiently generate visible components using the more common pump laser sources restricted supercontinuum applications; however, this was to dramatically change with the introduction of photonic crystal fibers [33, 34].

12.4.1 *Femtosecond Pulse Pumped Supercontinua*

In 2000 Ranka et al. made the first report of supercontinuum generation in photonic crystal fiber using the 100 fs pulses from a mode-locked Ti:sapphire laser at 790 nm to pump a 75 cm length of photonic crystal fiber with a zero dispersion wavelength of 770 nm. With a peak launch power of approximately 8 kW in the anomalously dispersive regime, the pump was a high-order soliton with the generated supercontinuum extending from approximately 400 nm to 1600 nm as represented in Fig. 12.4.

The most striking features of this supercontinuum are its relative flatness, the extended coverage of the visible, and the complete saturation of the pump. With the introduction of photonic crystal fibers, the ability to manipulate the air-silica microstructure led to an extremely strong weighting of the waveguide dispersion contribution, as a result of huge effective core-cladding refractive index difference. Consequently, the zero dispersion could be manufactured by design to any required wavelength, allowing soliton generation with the most common laser sources [51, 52]. In addition with small effective core areas the nonlinearity coefficient could be

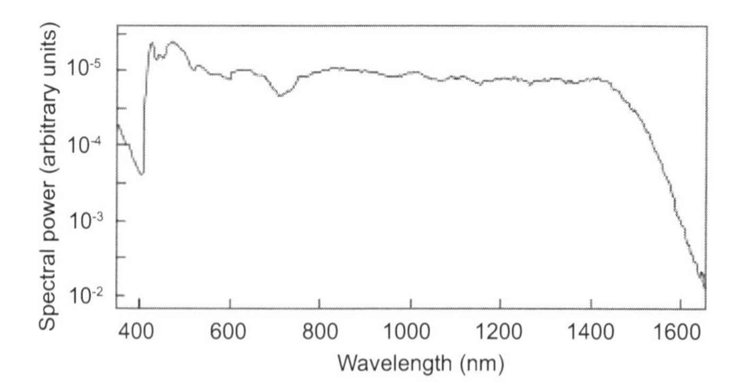

Figure 12.4 Characteristic supercontinuum generated in a photonic crystal fiber with a zero dispersion wavelength at 770 nm pumped by 100 fs, 8 kW pulses at 795 nm, after [50].

significantly increased, again enhancing nonlinearity and allowing soliton generation at lower peak powers. The role played by solitons was recognized by Ranka et al. [50], and although no new physical processes resulted from the introduction of photonic crystal fiber, it was to play a pivotal role in the rejuvenation of supercontinuum sources ultimately leading to not just the scientific but commercial success of the device.

The principal processes involved in the generation of a supercontinnum under high peak power femtosecond pulse pumping in the anomalously dispersive regime are well understood and have been extensively modeled and experimentally characterized [53–56]. Key to the mechanism is the rapid pulse compression of the high-order soliton to its minimum pulsewidth with associated spectral broadening [57] over a fiber length scale given approximately by $\tau_0^2/N\beta_2$, where N is the soliton order. As described above, for pulses of several femtoseconds, inherent system perturbations such as high-order dispersion and the soliton self-frequency shift prohibit the simple theoretically predicted behavior of high-order soliton breathing, with the high-order soliton breaking up, shedding off numerous fundamental solitons and dispersive radiation. In the region of the zero dispersion wavelength, if the bandwidth of the soliton overlaps the normal dispersion region, dispersive radiation feeds off the soliton. Cross-phase modulation leads to interaction between the solitons and the dispersive radiation in the normal regime such that as the solitons experience the red shifting of the soliton self-frequency shift, the induced phase modulation slows the dispersive waves, matching the group velocities of the red-shifting solitons and the blue-shifting dispersive wave components [58]. This interaction leads to an enhancement of the spectral extremes of the continuum.

Beyond the point of maximum compression as the input pulse fragments into various colored solitons leading to instability, noise-driven processes influence modulational instability evolution in the developing supercontinuum spectrum. This role of noise is quite often overlooked since generated supercontinua most often exhibit exceedingly smooth profiles, as in Fig. 12.4 for example; however, this is quite misleading and is simply a result of the averaging of numerous spectra that do indeed exhibit very poor shot to shot reproducibility, stability, and coherence. The effect of noise provides the principal contribution to the observation of "rogue waves" in supercontinuum generation [59, 60]. This can be simply thought of as manifesting itself through the process of soliton self-frequency shift. Any large intensity spike or intense noise feature on the input signal will through amplification, via modulational instability, evolve into a soliton. The more intense, the more the feature will experience self-Raman interaction and so the greater the red shift. Consequently, rogue wave behavior in supercontinua will be characterized by extreme red shifting of the spectrum, and also since these events are rare, they will exhibit an "L-shaped" statistical distribution in the number of events against the spectral shift, simply indicative of the noise contribution to the process and the rarity of the extreme intensity spike above the noise floor. Consequently, if noise reduction and high stability are essential to the application of the supercontinuum, such as in metrology, then it is probably better to use pump pulses with durations of less than 50 fs and with as high peak powers as possible while employing fiber lengths such that the point of the most extreme pulse compression is just at the point of exit. Alternatively, purely self-phase modulation could be utilized by operating solely in the normal dispersion regime; however, spectral coverage would tend to be reduced from that obtainable using solitonic effects at equivalent power levels.

A disadvantage of the mode-locked Ti:sapphire lasers as a femtosecond pump source for supercontinnum generation is the requirement of bulk optical coupling to the fibers and the associated alignment problems and instability associated with this, which detracts from ease of use and hands-free operation, essential for widespread application. Femtosecond pumping in fiber-integrated packages has been reported using Er oscillator-amplifier configurations and highly nonlinear conventional fiber [61]. Although octave spanning supercontinua can be achieved utilizing femtosecond pumping around 1550 nm both in photonic crystal and conventional fiber structures, the short-wavelength extent rarely extends below 800 nm; however, by using a hybrid, cascaded fiber assembly of initial generation in a highly nonlinear fiber (HNLF), followed by excitation of the photonic crystal fiber by the supercontinuum generated in the initial HNLF visible generation has been possible [62].

12.4.2 *Picosecond Pulse Pumped Supercontinua*

For many applications in biophotonics, wide spectral coverage of both the visible and the near-infrared is required, also high spectral power density and consequently high average powers. Despite impressive spectral coverage employing femtosecond Ti:sapphire laser pumping, average power levels are typically in the range of 10s of mW with associated spectral power densities of a few 10s of μW/nm. The power scaling of picosecond-based lasers allows much higher power levels to be achieved in oscillator-amplifier assemblies before nonlinear distortion is encountered and therefore present greater opportunity for increased spectral power density to be achieved. Rulkov et al. [63] reported the first all-fiber-integrated supercontinuum source utilizing a picosecond Yb-MOPFA (master oscillator power fiber amplifier) coupled with a photonic crystal fiber with a zero dispersion at 1040 nm to generate a supercontinuum operating at the watt level average power and allowing for the first time spectral power densities of greater than 1 mW/nm to be achieved over a spectral range from about 500 nm to 2000 nm. Since then, continuous development, primarily based on the power scaling, has resulted in average

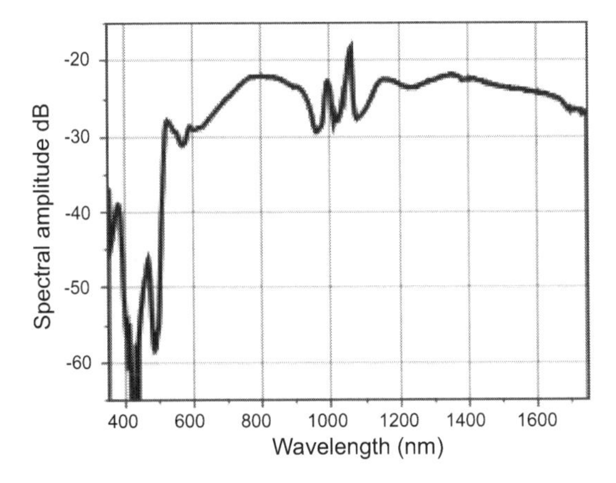

Figure 12.5 Spectrum of high average power supercontinuum obtained in a 35 m length of photonic crystal fiber, with a zero dispersion at 1040 nm, pumped by 1.5 W average power (16 kW peak) picosecond pulses from a Yb fiber MOPFA system after [63].

powers of up to 39 W from piciosecond Yb-pumped supercontinuum sources [64].

It is interesting to note that irrespective of fiber length and pump power beyond a length of about 10 m in the photonic crystal fiber used, Rulkov et al. noted that spectral extension below 500 nm was not possible. A typical supercontinuum obtained is shown in Fig. 12.5.

With picosecond pulse pumping in the anomalously dispersive region, high-order soliton dynamics does not play an important role in the formation of the supercontinuum. From Eq. 12.4 it may be thought that because of the increased pulse duration, the soliton power of a fundamental soliton would be substantially reduced as compared to the femtosecond regime and that consequently picosecond pumping would lead to exceedingly high-order soliton operation and extreme pulse compression. The characteristic nonlinear or soliton length, however, also scales proportionally to the square of the pulse duration and so picosecond soliton effects consequently are not observed over the relatively short lengths of fiber employed for supercontinuum generation. In the picosecond regime, for pump pulses in the region of the zero dispersion

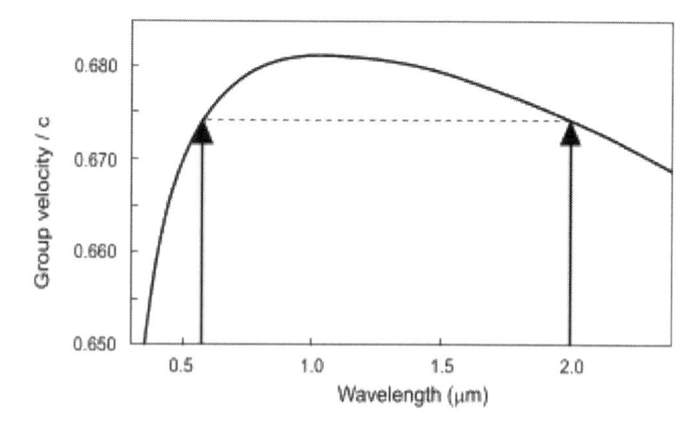

Figure 12.6 Variation of the group velocity with wavelength for a photonic crystal fiber with a zero dispersion at 1040 nm. The group velocity matching of a soliton at 2 μm and a dispersive wave around 0.56 μm is indicated by the arrows.

wavelength modulational instability and four-wave mixing are the processes that initiate continuum generation. Modulational instability leads to the rapid temporal breakup of the long-input picosecond pulses into femtosecond-scaled subpulses that are amplified to soliton powers and give rise to spectral broadening through the self frequency shift and soliton collision effects, while soliton-dispersive wave interactions as described in the section above contribute dominantly to the short wavelength extension.

The short wavelength restriction on the operation of the supercontinuum is a result of the need for soliton dynamics to initiate and in fact play the major role in supercontinuum generation and shaping. It is somewhat surprising that it is the long-wavelength characteristics of the fiber that affect the short-wavelength operation and extent of the supercontinnum. However, this is best explained with reference to Fig. 12.6 below, which shows the wavelength dependence of the group velocity of a photonic crystal fiber with a zero dispersion wavelength at 1040 nm.

If one considers the long wavelength extent of the soliton self-frequency shift and in the example above this is chosen at 2 μm, the corresponding group velocity matched wavelength permitting soliton-dispersive wave trapping in the normally dispersive region is

0.56 μm. For shorter wavelength extension, in this fiber it is essential for the solitons to further wavelength shift above 2 μm. This is possible, however, in most fibers beyond about 2.3 μm waveguiding is poor and the fibers are lossy and so the short wavelength would extend little below 500 nm.

There have been several approaches to resolve this. Travers et al. [65] first used the technique of cascading two photonic crystal fibers, such that the first fiber had a zero dispersion wavelength substantially longer than the zero dispersion wavelength of the second. The first fiber was 0.7 m long and had a zero dispersion at 1040 nm with a characteristic group velocity similar to that of Fig. 12.6 and allowed modulational instability to initiate supercontinuum generation via solitonic effects. The supercontinuum generated in the first fiber, for an average pump power of approximately 3.5 W, produced a spectrum that only extended to 600 nm on the short wavelength of the continuum. The crux of the technique was to use this continuum to pump the second photonic crystal fiber which had a zero dispersion at 780 nm. As a consequence of the shifted dispersion zero, the group velocity curve of this second fiber was shifted to the short-wavelength side relative to Fig. 12.6 above, although the material dispersion does dominate the response in this spectral region. As a result, however, the wavelengths at the short side of the initial continuum, generated in the first fiber, pump the second fiber and giving rise to solitons and dispersive wave radiation that can be group velocity matched in the second fiber efficiently down to 400 nm. A supercontinuum with an average power in excess of 1 W was obtained extending to 1800 nm.

The concept of the concatenated, dispersion-decreasing fiber configuration was further developed in the form of a continuously dispersion decreasing photonic crystal fiber—a long length taper. Even though relatively low loss splicing (<1 dB) is possible between differing photonic crystal fibers, if several of these have to be assembled, there will be a significant loss. An equivalent low loss option was elegantly produced at the University of Bath through the production of long lengths of a continuously tapering photonic crystal fiber pulled directly from a preform at the pulling tower, generating a fiber with a precisely controlled dispersion shifting profile. Initial fibers were produced with a constant air-hole-to-pitch

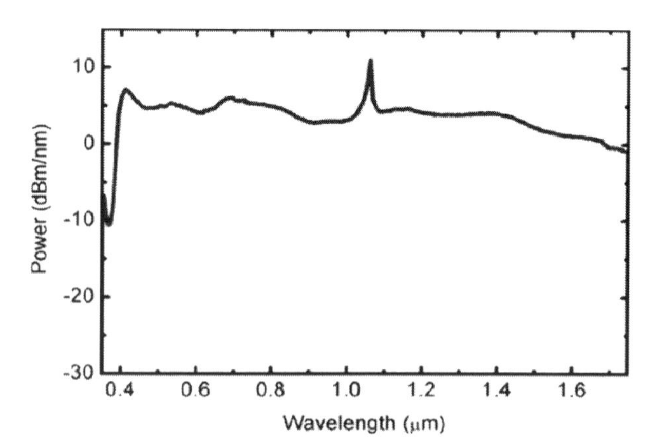

Figure 12.7 Supercontinuum, with an average power of 3.5 W produced in a 1.5 m length of tapered photonic crystal fiber, with an input zero dispersion wavelength of 1040 nm, pumped by picosecond pulses at 1064 nm.

ratio of 0.7 with the diameter of the pitch reducing from 5 μm to 1 μm over a 10 m length of fiber with the associated zero dispersion wavelength shifting from 1047 nm to 690 nm respectively. It should be noted that as the dimension of the effective core of the photonic crystal fiber reduces and, for example, with the output dimension of the PCF described above, a second zero dispersion wavelength was also exhibited at 1390 nm. Using this technique a supercontinuum with an average power of 3.5 W was produced, extending from 320 nm to 2300 nm, effectively the complete window of transparency of silica fiber. In the spectral range 400–800 nm, a spectral power density in excess of 2 mW/nm was obtained in the initial demonstration [66]. The technique was also successfully demonstrated with nanosecond pumping where the dynamics of the supercontinuum, initiated by modulational instability, are similar to picosecond pulse pumping. A representative supercontinuum spectrum obtained in a tapered PCF structure is shown in Fig. 12.7.

As the pitch is reduced in a tapered PCF, for a given soliton wavelength in the anomalously dispersive region, the corresponding group velocity matched wavelength in the normal dispersion region reduces. It may be thought that a PCF with a fixed diameter and structure identical to the output end of the fiber would be

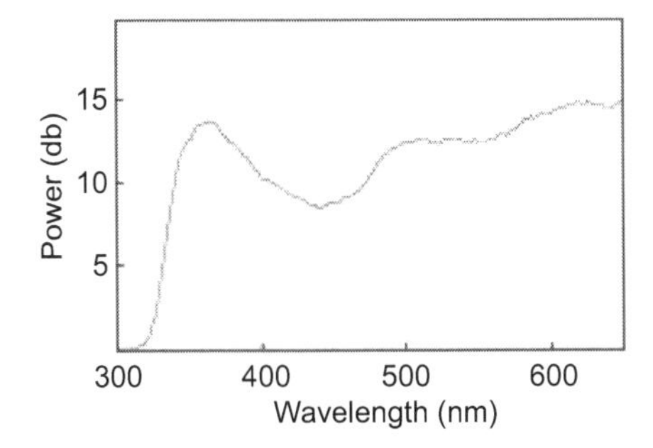

Figure 12.8 Short wavelength extension and enhancement achieved in a tapered photonic crystal fibre pumped at 1060 nm.

adequate to achieve extended short-wavelength operation of the supercontinuum. It must be remembered, however, that such a fiber would have a zero dispersion wavelength that is shifted to around 785 nm. For a pump wavelength around 1060 nm, the initiation of modulational instability and soliton formation would be too difficult and power demanding. Such a fiber would be appropriately and efficiently pumped by a frequency-doubled Er fiber laser or Ti:sapphire laser. Consequently, the tapered structure is a requirement if pumping at 1060 nm is employed.

The use of tapered PCF is also important with regard to the soliton–dispersive wave trapping process. For a fixed or constant core PCF the group velocity matching is well defined by a single profile (see Fig. 12.6). For a tapered construction, the group velocity critera is continuously changing along the fiber length. In addition, the tapering leads to a shifting of the zero dispersion wavelength to shorter wavelengths on propagation; hence for a fixed wavelength the group velocity continuously decreases with length. This is the equivalent of the soliton self-frequency shift process where for a fixed core PCF the soliton shifts to longer wavelengths, experiencing deceleration and a lower group velocity leading to a trapping of the short-wavelength dispersive waves. Consequently, in a taper trapping can take place without even the need for the soliton self-frequency shift, so enhancing the process. As a result, tapers can enhance the short-wavelength side of the supercontinuum by as much as 300 nm as compared to operation in a fixed core fiber.

Figure 12.8 shows the enhanced short-wavelength region obtained in a 1.5 m long taper pumped at 1060 nm.

In the UV, up to 2 mW/nm has been achieved using 1060 nm pumping of various tapered photonic crystal fiber structures of length scales of a few meters.

Group index matching to shorter wavelengths for a given wavelength in the infrared can be achieved by making the PCF structure more like a few micrometer diameter strand of silica surrounded by air. This can be achieved by approximating the structure with a PCF with a large pitch and a high air filling fraction. Optimization of this structure has resulted in supercontinua pumped by sub-nanosecond pulses at 1064 nm generating impressive spectral coverage with a continuum that extended from 400 nm to 2500 nm in a fiber with a constant core diameter [67].

12.4.3 *CW Pumped Supercontinua*

With the commercial development of high-power fiber lasers far outpacing academic research and development in the area, average powers of 10 kW are now available from a single-mode Yb fiber laser [68]. As a result, nonlinear optics in fiber is directly accessible with cw pump sources operating at more modest power levels, permitting increased scalability of the spectral power density achievable in continuum generation, as well as providing simplicity of the all-fiber configuration.

The first report of high-power cw all-fiber supercontinuum sources employed both highly nonlinear fibers and photonic crystal fibers pumped in the region of their zero dispersion wavelength and in the anomalously dispersive regime [69]. Efficient supercontinuum generation requires the pump to experience anomalous dispersion, which leads to modulational instability and the adiabatic evolution and amplification of optical solitons. Clearly the linewidth of the pump is an important consideration. Since the process is effectively a self-seeded noise-driven process, if the linewidth is too narrow, the inverse of the linewidth would infer noise fluctuations relatively long in time. The length scale for these to evolve into solitons is long and leads to longer soliton structures that do not self-frequency-shift as efficiently. On the other hand, for exceedingly broad linewidths, the temporal format of the noise is very short and the power requirement for soliton formation too high. Therefore empirically it can be seen that there is an optimal median linewidth requirement for the cw pump source. Once solitons have evolved from the cw pump signal, they shift to longer wavelength through the mechanism of the self-frequency shift as described above, while soliton collisions in the presence of Raman amplification also contribute to the red shifting and evolution of the continuum. Primarily cw pumped supercontinua are characterized by the dominant red shift as a result of the self-frequency shift. For this to efficiently take place, it is important that the dispersion does not increase significantly with wavelength, otherwise the power demands (see Eq. 12.4) placed on the soliton will lead to a self-termination of the shifting process. Similarly any increase in the effective mode area with increased wavelength will reduce the overall nonlinear coefficient and increase the soliton power requirement. Loss too should be minimized since cw pumped systems generally will require proportionally longer fiber lengths to exhibit comparable nonlinearity; hence distributed loss will lead to increased soliton durations with length and self-termination of the self-frequency shift. In early PCF-based schemes pumped at 1060 nm, water loss at the air hole interface limited the upper wavelength extent to the region around 1380 nm [70], although this was not a problem in conventional highly nonlinear fiber structures [69]. An additional factor which can affect the long wavelength extent of a supercontinuum source is the presence of a second zero dispersion wavelength. This too inhibits extension of the soliton self-frequency shift and leads to dispersive wave generation in the long-wavelength normal dispersion region. Consequently considerable effort has been directed toward the optimization of photonic crystal fiber design for cw pumped supercontinuum [71, 72]. Figure 12.9 shows a 29 W average power supercontinuum pumped at 1060 nm. The photonic crystal fiber used was 20 m long and exhibited double zero dispersion wavelengths at 810 nm and 1730 nm. More than 50mW/nm was obtained over the spectral range 1060–1380 nm [73]. Above 1380 nm the supercontinuum spectrum rolls off in intensity as a result of the water loss described above.

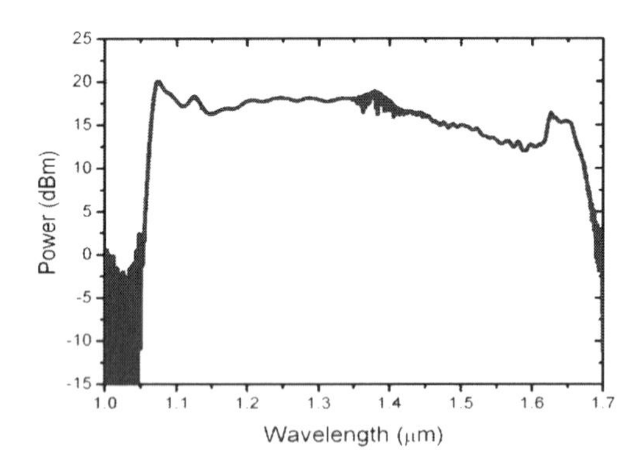

Figure 12.9 29 W average power, cw pumped supercontinuum pumped at 1060 nm after [73].

Cw pumped systems have allowed the highest spectral power densities to be achieved. Using an industrial scale Yb fiber laser operating at 1060 nm to pump a 20 m length of PCF with a single zero dispersion at 840 nm a supercontinuum extending from 1060 nm to beyond 2200 nm was obtained with spectral powers of greater than 100 mW/nm from 1100 nm to 1400 nm and more than 50 mW/nm from 1060 nm to 1700 nm. However, like the majority of cw pumped systems, these supercontinua are characterized by an absence of spectral components to the short-wavelength side of the pump.

For short-wavelength generation it is essential that the pump lies in the anomalous dispersion region close to the dispersion zero such that with the evolution of modulational instability and subsequent soliton generation that there is spectral overlap of the evolving solitons in the normal dispersion regime. In this way, the process of soliton-dispersive wave trapping, as discussed above, takes place, which is key to short-wavelength evolution of the supercontinuum. The process is aided by low values of dispersion and high nonlinearity and the group velocity matching mechanism between self-frequency-shifted soliton and dispersive wave determines the spectral extent of the supercontinuum and consequently optimization of the fiber geometry can enhance the processes involved. With the improved understanding of the processes involved in cw supercontinuum generation, together with

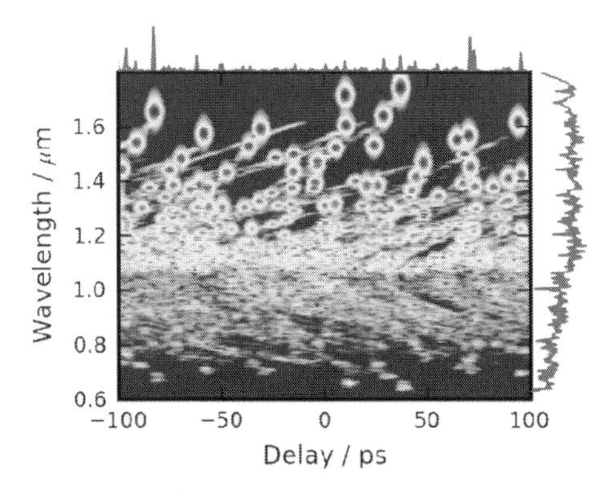

Figure 12.10 Simulated spectrogram of a 1060 nm, 170 W cw pumped supercontinuum after 25 m of fiber with a zero dispersion at 1050 nm. The associated spectrum is shown to the right hand side of the figure, after [74].

advances in technology in the design of specific PCF structures, the performance of cw pumped supercontinuum sources have been comprehensively modeled and characterized. Figure 12.10 shows a theoretical prediction of supercontinuum generation for an average pump power of 170 W in a 25 m length of a photonic crystal fiber with a hole pitch of 3.4 μm and a hole diameter to pitch ratio of 0.47 giving a zero dispersion at 1050 nm. In the spectrogram the one to one correspondence of the optical solitons in the infrared and the trapped dispersive radiation in the visible region is clearly seen. The vast number of solitons involved gives rise to a relatively smooth spectrum as can be seen from the associated spectral profile shown to the right hand side of the spectrogram. Since the origin of the supercontinuum is essentially noise driven, the integration of many of these single shots essentially gives rise to integration and further smoothing of the overall spectral profile.

This has been experimentally realized [74] and the generated cw pumped supercontinuum with extension to the visible, with a minimum wavelength of 600 nm achieved, is shown in Fig. 12.11, for an equivalent pump power of 230 W. As can be seen total depletion of the pump was achieved. The dip in the supercontinuum spectrum occurs at the zero dispersion wavelength. More than 50 mW/nm was achieved in the infrared and around 3 mW/nm in the visible. Again, the smooth spectral profile of the continuum arises through

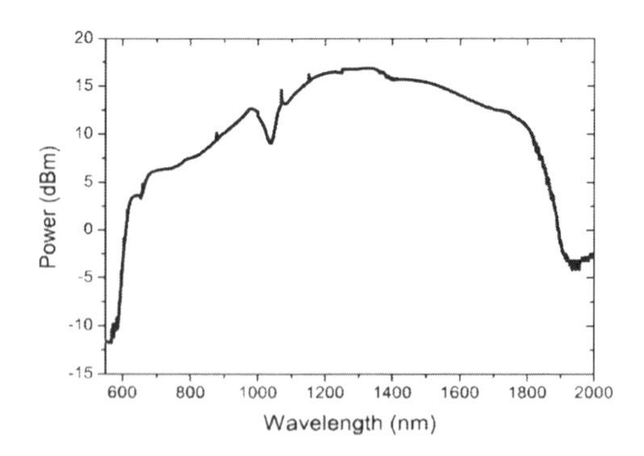

Figure 12.11 Experimentally measured cw pumped supercontinuum spectrum in 50 m of PCF with a zero dispersion at 1050 nm pumped by 230 W at 1060 nm.

the integration of large numbers of randomly distributed temporal structures.

Through improvements in fiber design and the use of high GeO_2 doping which allows increased nonlinear response, a 180 m long fiber with a 130 m tapered end section allowed wavelength as short as 470 nm to be achieved for a cw pump power of only 40 W at 1060 nm [72]. However, one should be aware that the transmission of high average powers in the visible should be avoided in fibers with high Germania doping as this leads to photodarkening of the fibers through color center production and an associated drop in transmission efficiency.

12.5 Application and Future Developments

The combination of the femtosecond Kerr lens mode-locked Ti:sapphire laser in conjunction with photonic crystal fiber, albeit in a bulk optically coupled geometry, with its spectral coverage from 400 nm to 1600nm and beyond, see for example Fig. 12.4, has revolutionized supercontinuum applications particularly in the biomedical area. Further developments through the integration of high average power nanosecond and picosecond Yb fiber lasers with photonic crystal fiber has produced highly versatile, hands-free supercontinuum sources that cover the complete transmission window of silica fiber, in user-friendly commercial packages that have extended the capabilities of the supercontinuum source and led to its rapid deployment in routine measurement.

Biophotonic applications, from spectroscopy to microscopy and too numerous to mention all, have relied upon the extended spectral coverage, the high spectral power densities, and the ability to select tunable femtosecond or picosecond pulses. Broadband near-infrared radiation has been utilized in confocal reflectance microscopy [75] achieving an axial resolution of about a micrometer. In confocal microscopy the first proposal of the application of a supercontinuum source was reported in 2001 [76]. The pulsed nature of the supercontinuum and the ability to spectrally select excitation wavelengths from the near-UV throughout the visible has found wide application in fluorescence lifetime imaging microscopy [77]. Multiphoton fluorescence microscopy has also utilized the broadband tunability together with the femtosecond pulse capability of the supercontinuum source [78]. With the increasing operational power levels of commercial supercontinuum sources, spectral selection within the continuum appears a likely application to provide tunability of both the excitation and depletion beams [79] for application to stimulated emission depletion microscopy, surpassing the diffraction-limited resolution of confocal microscopes and allowing resolution of a few tens of nanometers.

Optical coherence tomography is another extremely active area where the broad bandwidth of the supercontinuum can be widely applied [80, 81] to improve axial resolution, but the source would become more attractive if the cost of commercial units could be substantially reduced to compete with broadband semiconductor diode sources, which may not provide the resolution but are cheaper. Supercontinuum sources have also been utilized in time-resolved diffuse optical spectroscopy [82] and a compact device has been developed incorporating a commercial supercontinuum source [83].

In silica-based supercontinuum sources the spectral range 320–2400 nm is readily obtained with future developments mainly directed at the power scaling of commercial devices. However, many potential biophotonic applications demand wavelengths outside these spectral ranges. In the UV, loss and problems of multiphoton ionization leading to the formation of color centers will most likely inhibit operation below current short-wavelength limits and it is more likely that practical vacuum ultraviolet (VUV) and VUV sources will be based upon the high power femtosecond pulsed pumping of gas-filled hollow-core photonic bandgap fibers [84]. Although the experimental configurations are not as compact and integrated as conventional commercial supercontinuum sources, relatively high conversion efficiencies with tunability in the 200–300 nm range can be achieved.

In the mid-infrared the greatest potential is afforded by the so-called soft glasses. Although at present devices based on these materials have lower power handling capabilities, the substantially higher nonlinear coefficients of many of the glasses means that significantly shorter lengths of fibers and lower power levels are required to obtain significant spectral broadening and nonlinearity in the samples. Mid-infrared supercontinum generation has been reviewed in microstructured optical fibers constructed from a variety of materials [85]. The ability to simplify the manufacture of the microstructured fibers through the use of extrusion techniques, as has been demonstrated with tellurite fibers, coupled with the requirement for relatively short lengths of the samples is particularly attractive, and extremely impressive spectral coverage, extending over more than 4000 nm, from 500 nm to 4500 nm with pumping at 1550 nm, has been reported in a 8 mm sample of tellurite photonic crystal fiber [86].

Fluoride fibers also enable low loss transmission in the mid-infrared and following the first reports of relatively low power supercontinuum generation using Er fiber laser pump technology at 1550 nm which produced a 5 mW average power continuum that extended beyond 3000 nm [87], the technology has been refined such that average powers in excess of 10 W have been reported [88] and a continuum extending to 6280 nm has been achieved in ZBLAN fluoride fiber [89], which has the potential to support radiation to beyond 8000 nm.

It can therefore be seen that at present all the processes contributing to supercontinuum generation are very well understood and that with this understanding, theoretical prediction agrees remarkably well with experimental realization. In addition, the technology currently exists that allows the supercontinuum source to be routinely deployed in compact, user-friendly packages as a hands-free diagnostic. Both the high-power fiber laser and the photonic crystal fiber have contributed to the commercial advancement. With power scaling and the move to new materials, supercontinuum sources will routinely operate from the extreme UV to the mid-infrared and will underpin future applications in the biophotonics arena.

Acknowledgements

I would like to thank and acknowledge the research of the members of the Femtosecond Optics Group at Imperial College who have contributed to the development of high-power and versatile supercontinuum sources over many years, and in particular Dr. Sergei Popov, Dr. John Travers, Dr. Andrei Rulkov, Dr. Burley Cumberland, Dr. Pierre Champert, Mr. Ed Kelleher, and Mr. Ben Chapman. I also acknowledge the support of Professor Valentin Gapontsev and the IPG Group of companies and the Royal Society London.

References

1. Franken, P. A., Hill, A. E., Peters, C. W., and Weinreich, G. (1961). Generation of optical harmonics. *Phys. Rev. Lett.*, **7**:118–119.
2. New, G. H. C., and Ward, J. F. (1967). Optical third harmonic generation in gases. *Phys. Rev. Lett.*, **19**:556–559.
3. Maker, P. D., Terhune, R. W., and Savage, C. M. (1964). Intensity-dependent changes in the refractive index of liquids. *Phys. Rev. Lett.*, **12**:507–509.
4. Askaryan, G. A. (1962). Effects of the gradient of a strong electromagnetic beam on electrons and atoms. *JETP*, **15**:1088–1090.
5. Shen, Y. R., and Shaham, Y. J. (1965). Beam deterioration and stimulated Raman effects. *Phys. Rev. Lett.*, **15**:1008–1010.
6. Carman, R. L., Chiao, R. Y., and Kelley, P. L. (1966). Observation of degenerate stimulated four-photon interaction and four-wave parametric amplification. *Phys. Rev. Lett.*, **17**:1281–1283.

7. Woodbury, E. J., and Ng, W. K. (1962). Ruby laser operation in the near IR. *Proc. IRE*, **50**:2367.
8. Jones, W. J., and Stoicheff, B. P. (1964). Inverse Raman spectra induced absorption at optical frequencies. *Phys. Rev. Lett.*, **13**:657–659.
9. Ueda, Y., and Shimoda, K. (1967). Observation of stimulated Raman emission and stimulated Rayleigh-wing scattering from self-trapped filaments of a laser beam. *Jap. J. App. Phys.*, **6**:628–633.
10. Brewer, R. G. (1967). Frequency shifts in self-focused light. *Phys. Rev. Lett.*, **19**:8–10.
11. Shimizu, F. (1967). Frequency broadening in liquids by a short light pulse. *Phys. Rev. Lett.*, **19**:1097–1100.
12. Alfano, R. R., and Shapiro, S. L. (1970). Emission in the region 4000 Å to 7000 Å via four-photon coupling in glass. *Phys. Rev. Lett.*, **24**:584–587.
13. Gersten, J., Alfano, R. R., and Belic, M. (1980). Combined stimulated Raman scattering and continuum self phase modulation. *Phys Rev. A*, **21**:1222–1224.
14. Ippen, E. P. (1970). Low-power quasi-cw Raman oscillator. *App. Phys. Lett.*, **16**:303–305.
15. Stolen, R. H., Ippen, E. P., and Tynes, A. R. (1972). Raman oscillation in glass optical waveguides. *App. Phys. Lett.*, **20**:62–64.
16. Stolen, R. H., and Ashkin, A. (1973). Optical Kerr effect in glass waveguide. *App. Phys. Lett.*, **22**:294–296.
17. Stolen, R. H., Bjorkholm, J. E., and Ashkin, A. (1974). Phase-matched three wave mixing in silica fiber optical waveguides. *App. Phys. Lett.*, **24**:308–310.
18. Stolen, R. H., and Lin, C. (1978). Self phase modulation in silica optical fibers. *Phys. Rev. A*, **17**:1448–1453.
19. Stolen, R. H. (2008). The early years of fiber nonlinear optics. *J. Lightwave Tech.*, **26**:1021–1031.
20. Lin, C., and Stolen, R. H. (1976). New nanosecond continuum for excited-state spectroscopy. *App. Phys. Lett.*, **28**:216–218.
21. Lin, C., Nguyen, V. T., and French, W. G. (1978). Wideband near IR continuum (0.7–2.1 μm) generated in low loss optical fibres. *Electron. Lett.*, **14**:822–823.
22. Hasegawa, A., and Tappert, F. (1973). Transmission of stationary nonlinear optical pulses in dispersive dielectric fibers. *App. Phys. Lett.*, **23**:142–144.

23. Mollenauer, L. F., Stolen, R. H., Gordon, J. P., and Tomlinson, W. J. (1980). Experimental observation of picosecond pulse narrowing and solitons in optical fibers. *Phys. Rev. Lett.*, **45**:1095–1098.
24. Mollenauer, L. F., and Gordon, J. P. (2006). *Solitons in Optical Fibers – Fundamentals and Applications* (Elsevier Academic Press) ISBN-13:978-0-12-504190-4.
25. Fuji, Y, Kawasaki, B. S., Hill, K. O., and Johnson, D. C. (1980). Sum-frequency light generation in optical fibers. *Opt. Lett.*, **5**:48–50.
26. Washio, K., Inoue, K., and Tanigawa, T. (1980). Efficient generation of near i.r. stimulated light scattering in optical fibres pumped in low-dispersion region at 1.3 μm. *Electron. Lett.*, **16**:331–333.
27. Vysloukh, V. N., and Serkin, V. N. (1983). Generation of high-energy solitons of stimulated Raman radiation in fiber light guides. *JETP Lett.*, **38**:199–202.
28. Dianov, E. M., Karasik, A. Ya., Mamyshev, P. V., Prokhorov, A. M., Serkin, V. N., Stelmakh, M. F., and Fomichev, A. A. (1985). Stimulated Raman conversion of multisoliton pulses in quartz optical fibers. *JETP Lett.*, **41**:294–297.
29. Gouveia-Neto, A. S., Faldon, M. E., and Taylor, J. R. Spectral and temporal study of the evolution from modulational instability to solitary wave. *Opt. Commun.*, **69**:325–328.
30. Gouveia-Neto, A. S., Faldon, M. E., and Taylor, J. R. (1989). Raman amplification of modulational instability and solitary wave formation. *Opt. Lett.*, **13**:1029–1031.
31. Gouveia-Neto, A. S., Gomes, A. S. L., and Taylor, J. R. (1988). Femtosecond Soliton Raman Generation. *IEEE J. Quantum Electron.*, **24**:332–340.
32. Stolen, R. H. (1975). Phase-matched stimulated-four-wave-mixing in silica fiber waveguides. *IEEE J. Quantum Electron.*, **11**:100–103.
33. Knight, J. C., Birks, T. A., Russell, P. St. J., and Atkin, D. M. (1996). All-silica single-mode optical fiber with photonic crystal cladding. *Opt. Lett.*, **21**:1547–1549.
34. Russell, P. St. J. (2006). Photonic-crystal fibers. *J. Lightwave Tech.*, **24**:4729–4749
35. Broderick, N. G. R., Monro, T. M., Bennett, P. J., and Richardson, D. J. (1999). Nonlinearity in holey optical fibers: measurement and future opportunities. *Opt. Lett.*, **24**:1395–1387.
36. Birks, T. A., Knight, J. C., and Russell, P. St. J. (1997). Endlessly single-mode photonic crystal fiber. *Opt. Lett.*, **22**:961–963.

37. Tai, K., and Tomita, A. (1986). 50x optical fiber pulse compression at 1.319 μm. *App. Phys. Lett.,* **56**:135–138.

38. Grudinin, A. B., Dianov, E. M., Korobkin, D. V., Prokhorov, A. M., Serkin, V. N., and Kaidarov, D. V. (1987). Stimulated-Raman scattering of 18fs pulses in 1.6 μm region during pumping of a single mode optical fibre by the beam from a Nd:YAG laser ($\lambda = 1.06$ μm). *JETP Lett.,* **45**:260–2263.

39. Gouveia-Neto, A. S., Gomes, A. S. L., and Taylor, J. R. (1988). Pulses of four optical cycles from an optimized optical/fibre grating pair/soliton pulse compressor. *J. Mod. Opt.,* **35**:7–10.

40. Hasegawa, A., and Brinkman, F. (1980). Tunable coherent IR and FIR sources utilizing modulational instability. *IEEE J. Quantum Electron.,* **16**.:694–699.

41. Gouveia-Neto, A. S., Faldon, M. E., Sombra, A. B., Wigley, P. G. J., and Taylor, J. R. (1988). Subpicosecond-pulse generation through cross phase modulation induced modulational instability in optical fibers. *Opt. Lett.,* **13**:901–903.

42. Mitschke, F. M., and Mollenauer, L. F. (1986). Discovery of the soliton self-frequency shift. *Opt. Lett.,* **11**:659–661.

43. Gordon, J. P. (1986). Theory of the soliton self-frequency shift. *Opt. Lett.,* **11**:662–664.

44. Wai, P. K. A., Menyuk, C. R., Lee, C., and Chen, H. H. (1987). Soliton at the zero-group dispersion wavelength of a single mode fiber. *Opt. Lett.,* **12**:628–630.

45. Gouveia-Neto, A. S., Faldon, M. E., and Taylor, J. R. (1988). Temporal and spectral evolution of femtosecond solitons in the region of the zero-group velocity dispersion of a single mode optical fibre. *Opt. Commun.,* **69**:173–176.

46. Beaud, P., Hodel, W., Zysset, B., and Weber, H. P. (1987). Ultrashort pulse propagation, pulse breakup and fundamental soliton formation in a single mode optical fiber. *IEEE J. Quantum Electron.,* **23**:1938–1946.

47. Nishizawa, N., and Goto, T. (2001). Widely broadened supercontinuum generation using highly nonlinear dispersion shifted fibers and femttosecond fiber lasers. *Jap. J. App. Phys.,* **40**:L365–367.

48. Nishizawa, N., and Goto, T. (2002). Characteristics of pulse trapping by use of ultrashort soliton pulses in optical fibers across the zero-dispersion wavelength. *Opt. Express,* **10**:1151–1159.

49. Chernikov, S. V., Zhu, Y., Taylor, J. R., and Gapontsev, V. P. (1997). Supercontinuum self-Q-switched ytterbium fiber laser. *Opt. Lett.,* **22**:298–300.

50. Ranka, J. K., Windeler, R. S., and Stentz, A. J. (2000). Visible continuum generation in air-silica microstructure optical fibers with anomalous dispersion at 800 nm. *Opt. Lett.*, **25**:25–27.
51. Wadsworth, W. J., Knight, J. C., Ortigosa-Blanch, A., Arriaga, J., Silvestre, E., and Russell, P. St. J. (2000). Soliton effects in photonic crystal fibres at 600 nm. *Electron. Lett.*, **36**:53–55.
52. Price, J. H. V., Belardi, W., Monro, T. M., Malinowski, A., Piper, A., and Richardson, D. J. (2002). Soliton transmission and supercontinuum generation in holey fiber using a diode pumped Ytterbium fiber source. *Opt. Express*, **10**:382–387.
53. Dudley, J. M., Genty, G., and Coen, S. (2006). Supercontinuum generation in photonic crystal fiber. *Rev. Mod. Phys.*, **78**:1135–1184.
54. G. Genty, G., Coen, S., and Dudley, J. M. (2007). Fiber supercontinuum sources. *J. Opt. Soc. Am.*, 24:1171–1785.
55. Herrmann, J., griebner, U., Zhavoronkov, N., Husakou, A., Nickel, D., Knight, J. C., Wadsworth, W. J., Russell, P. St. J., and Korn, G. (2002). Experimental evidence for supercontinuum generation by fission of higher-order-solitons in photonic fibers. *Phys. Rev. Lett.*, **88**:173901-1–4.
56. Dudley, J. M., Gu, X., Xu, L., Kimmel, M., Zeek, E., O'Shea, P., Trebino, R., Coen, S., and Windeler, R. S. (2002). Cross-correlation frequency resolved optical gating analysis of broadband continuum generation in photonic crystal fiber: simulations and experiment. *Opt. Express*, **10**:1215–1221.
57. Dianov, E. M., Nikonova, Z. S., Prokhorov, A. M., and Serkin, V. N. (1986). Optimal compression of multi-soliton pulses in optical fibers. *Sov. Tech. Phys. Lett.* **12**:311–313.
58. Gorbach, A. V., and Skyrabin, D. V. (2007). Light trapping in gravity-like potentials and expansion of supercontinuum spectra in photonic-crystal fibres. *Nat. Photonics*, **1**:653–657.
59. Solli, D. R., Ropers, C., Koonath, P., and Jalali, B. (2007). Optical rogue waves. *Nature*, **450**:1054–1058.
60. Dudley, J. M., Genty, G., and Eggleton, B. J. (2008). Harnessing and control of optical rogue waves in supercontinuum generation. *Opt. Express*, **16**:3644–3651.
61. Nicholson, J. W., Yablon, A. D., Westbrook, P. S., Feder, K. S., and Yan, M. E. (2004). High power, single mode,all-fiber source of femtosecond pulses at 1550 nm and its use in supercontinuum generation. *Opt. Express*, **12**:3025–3034.

62. Nicholson, J. W., Bise, R., Alonzo, J., Stockert, T., Trevor, D. J., Dimarcello, E., Monberg, E., Fini, J. M., Westbrook, P. S., Feder, K., and Grüner-Nielsen, L. (2008). *Opt. Lett.,* **33**:28–30.
63. Rulkov, A. B., Vyatkin, M. V., Popov, S. V., Taylor, J. R., and Gapontsev, V. P. (2005). High brightness picosecond all-fiber generation in 525-1800 nm range with picosecond Yb pumping. *Opt. Express,* **13**:2377–2381.
64. Chen, K. K., Alam, S-u., Price, J. H. V., Hayes, J. R., Lin, D., Malinowski, A., Codemard, C., Ghosh, D., Pal, M., Bhadra, S. K., and Richardson, D. J. (2010). Picosecond fiber MOPA pumped supercontinuum source with 39W output power. *Opt. Express,* **18**:5426–5432.
65. Travers, J. C., Popov, S. V., and Taylor, J. R. (2005). Extended blue supercontinuum generation in cascaded holey fibers. *Opt. Express,* **30**:3132–3134.
66. Kudlinski, A., George, A. K., Knight, J. C., Travers, J. C., Rulkov, A. B., Popov, S. V., and Taylor, J. R. (2006). Zero-dispersion wavelength decreasing photonic crystal fibers for ultraviolet-extended supercontinuum generation. *Opt. Express,* **14**:5715–5722.
67. Stone, J. M., and Knight, J. C. (2008). Visibly "white" light generation in uniform photonic crystal fiber using a microchip laser. *Opt. Express,* **16**:2670–2675.
68. See for example www.ipgphotonics.com
69. Popov, S. V., Champert, P. A., Solodyankin, M. A., and Taylor, J. R. (2002). Seeded fibre amplifiers and multi-watt average power continuum generation in holey fibres. Paper WKK2, Proceeding of the Optical Society of America Annual meeting, 117.
70. Avdokhin, A. V., Popov, S. V., and Taylor, J. R. (2003). Continuous wave, high-power, Raman continuum generation in holey fibres. *Opt. Lett.,* **28**:1353–1355.
71. Kudlinski, A., Bouwmans, G., Douay, M., Taki, M., and Mussot, A. (2009). Dispersion-engineered photonic crystal fibers for cw-pumped supercontinuum sources. *J. Lightwave Tech.,* 27:1556–1564.
72. Kudlinski, A., Bouwmans, G., Quiquempois, Y., LeRouge, A., Bigot, L., Mélin, G., and Mussot, A. (2009). White-light cw-pumped supercontinuum generation in highly GeO_2-doped–core photonic crystal fibers. *Opt. Lett.,* **34**:3631–3633.
73. Cumberland, B. A., Travers, J. C., Popov, S. V., and Taylor, J. R. (2008). 29W high power cw supercontinuum source. *Opt. Express,* **16**:5954–5962.

74. Travers, J. C., Rulkov, A. B., Cumberland, B. A., Popov, S. V., and Taylor, J. R. (2008). Visible supercontinuum generation in photonic crystal fibers with a 400W continuous wave fiber laser. *Opt. Express*, **16**:14435–14447.
75. Garzon, J., Meneses, J., Tribillion, G., Gharbi, T., and Plata, A. (2004). Chromatic confocal microscopy by means of continuum light generated through a standard single mode fibre. *J. Opt. A: Pure Appl. Opt.*, **6**:544–548.
76. Birk, H., and Storz, R. (2001) Illuminating device and microscope. United States, Leica Microsystems; Heidelberg GmbH, Germany.
77. Owen, D. M., Auksorius, E., Manning, H. B., Talbot, C. B., De Beule, P. A. A., Dunsby, C., Neil, M. A. A., and French, P. M. W. (2007) Excitation-resolved hyperspectral fluorescence lifetime imaging using a UV-extended supercontinuum source. *Opt. Lett.*, **32**:3408–3410.
78. Jureller, J. E., Scherer, N. F., Birks, T. A., wadsorth, W. J., and Russell, P. St. J. (2003) Widely tunable femtosecond pulses from a tapered fiber for ultrafast microscopy and multiphoton applications. In Miller, R. J. D., Murnane, M. M., Scherer, N. F., and Weiner, A. M. (eds.), *Ultrafast Phenomena XIII*, Berlin, Springer-Verlag.
79. Wildanger, D., Rittweger, E., Kastrup, L., and Hell, S. W. (2008). STED microscopy with a supercontinuum laser source. *Opt. Express*, **16**:9614–9621.
80. Hsiung, P. L., Chen, Y., Ko, T. H., Fujimoto, J. G., deMatos, C. J. S., Popov, S. V., Gapontsev, V. P., and Taylor, J. R. (2004). *Opt. Express*, **12**:5287–5295.
81. Bizheva, K., Pflug, B., Ermann, B., Povazay, B., Sattmann, H., Qiu, P., Anger, E., Reitsamer, H., Popov, S. V., Taylor, J. R., Unterbuber, A., Ahnelt, P., and Drexler, W. (2006). Optophysiology: depth-resolved probing of retinal physiology with functional ultrahigh-resolution optical coherence tomography. *Proc. Natl. Acad. Sci. USA*, **103**:5066–5071.
82. Bassi, A., Swartling, J., D'Andrea, C., Pifferi, A., Torricelli, A., and Cubeddu, R. (2004). Time-resolved spectrometer for turbid media based upon supercontinuum generation in a photonic crystal fiber. *Opt. Lett.*, **29**:2405–2407.
83. Bassi, A., Farina, A., D'Andrea, C., Pifferi, A., valentine, G., and Cubeddu, R. (2007). Portable, large-bandwidth time-resolved system for diffuse optical spectroscopy. *Opt. Express*, **15**:14482–14487.

84. Nold, J., Hölzen, P., Joly, N. Y., Wong, G. K. L., Nazarkin, A., Podlipensky, A., Scharrer, M., and Russell, P. St. J. (2010). Pressure-controlled phase matching to third harmonic in Ar-filled hollow core photonic crystal fiber. *Opt. Lett.*, **35**:2922–2924.
85. Price, J. H. V., Monro, T. M., Ebendorff-Heidepriem, H., Poletti, F., Horak, P., Finazzi, V., Leong, J. Y. Y., Petropoulos, P., Flanagan, J. C., Brambilla, G., feng, X., and Richardson, D. J. (2007). Mid-IR supercontinuum generation from nonsilica microstructured optical fibers. *IEEE J. Sel. Top. Quantum Electron.*, **13**:738–749.
86. Domachuk, P., Wolchover, N. A., Cronin-Golomb, M., Wang, A., George, A. K., Cordeiro, C. M. B., Knight, J. C., and Omenetto, F. G. (2008). Over 4000nm bandwidth of mid-IR supercontinuum generation in sub-centimeter segments of highly nonlinear Tellurite PCFs. *Opt. Express*, **16**:7161–7168.
87. Hagen, C. L., Walewski, J. W., and Sanders, S. T. (2006). Generation of a continuum extending to the midinfrared by pumping ZBLAN fiber with an ultrafast 1550-nm source. *IEEE Photonic Technol. Lett.*, **18**:91–93.
88. Xia, C., Xu, Z., Islam, M. N., Terry Jr, F. L., Freeman, M. J., Zakel, A., and Mauricio, J. (2009). 10.5 watts time-averaged power mid-infrared supercontinuum generation extending beyond 4μm with direct pulse pattern modulation. *IEEE J. Sel. Top. Quantum Electron.*, **15**:422–434.
89. Qin, G., Yan, X., Kito, C., Liao, M., Chaudhari, C., Suzuki, T., and Ohishi, Y. (2009). Ultrabroadband supercontinuum generation from ultraviolet to 6.28 μm in a fluoride fiber. *App. Phys. Lett.*, **95**:161103–1–3.

Chapter 13

Novel Sources for Optical Coherence Tomography Imaging and Nonlinear Optical Microscopy

Kenneth Kin-Yip Wong

Department of Electrical and Electronic Engineering, University of Hong Kong, Chow Yei Ching Building, Pokfulam Road, Hong Kong

kywong@eee.hku.hk

The ever-growing demands of life-science and healthcare services and solutions call for continuous substantial advancements in various biophotonic applications, in particular in label-free imaging and diagnostics. The powerful knowledge base acquired in telecommunications and its well-developed photonic components can be leveraged for the stringent requirements in biophotonic applications.

In this chapter, we will explore some novel sources, especially fiber-based optical parametric oscillator (OPO), to provide a fundamental platform for two promising biophotonic applications, namely optical coherence tomography (OCT) and nonlinear optical microscopy (such as coherent anti-Stokes Raman scattering (CARS) microscopy). Each of them has significantly different requirements for the laser sources due to their different mechanisms. OCT is

Understanding Biophotonics: Fundamentals, Advances, and Applications

Edited by Kevin K. Tsia

ISBN 978-981-4411-77-6 (Hardcover), 978-981-4411-78-3 (eBook)

www.panstanford.com

an optical interferometric imaging technique that complements other non-invasive lower-resolution (100 μm to 1 mm) imaging techniques such as X-ray radiography, magnetic resonance imaging (MRI), computer tomography (CT), and ultrasound imaging. CARS on the other hand allows for chemically specific imaging without the need for fluorescent labels because the contrast is derived from specific molecular vibrations, which are interrogated by two laser pulses (*pump* and *Stokes*). We will explore such kind of fiber-based source as core technology to meet some of the critical challenges among next-generation multimodal optical imaging systems, in this chapter. Such challenges include the high-speed imaging capabilities with enhanced axial resolution, versatile working wavelength window, robustness of the system, etc. First, we review the basic principles of OCT and CARS systems. Second, we explore some enabling optical amplifier technologies to generate the respective laser sources for OCT and CARS applications. Third, we focus some state-of-the-art technologies by highlighting few illustrations at the end of the chapter.

13.1 Introduction

Biophotonics, the interdisciplinary research of the photonics and biology, is an emerging frontier in dealing with interactions between light and biological matter. It provides challenges for fundamental research and opportunities for new technologies. Two of the most promising technologies are optical coherence tomography (OCT) and coherent anti-Stokes Raman scattering (CARS) microscopy. However, the outlook for its development into practical applications still faces the following few key issues:

1. High-speed imaging capabilities with enhanced axial resolution in OCT systems are essential among these issues, as they can enable the rapid acquisition rates that are often necessary to reduce artifacts due to patient motion. In addition, these high-speed imaging capabilities enable the generation of 3D volumetric images within reasonable time constraints. Conventional swept-source technologies have been constrained by mechanically moving parts or piezo-tuning devices.

2. Different wavelengths have been utilized as the laser sources for the OCT systems with different requirements. For example, shorter wavelengths (850/1050, nm [1, 2]) work better for retinal imaging, but optical scattering in tissues is reduced and the penetration depth is increased at longer wavelengths [3, 4]. Current laser sources are generally optimized for certain wavelengths; typical tunable sources are limited in their operating wavelengths and power output. The outlook for the development of conventional tunable sources to cover the wide wavelength window (850–1700 nm) is deemed as not promising.
3. Though some of the current OCT and CARS systems have already leveraged on the advancements in the fiber optic industry, most of them are still confined to standard off-the-shelf components (passive components, optical amplifiers etc.). Furthermore, conventional light sources in CARS setups still rely on bulky and costly Ti: Sapphire lasers. Significant improvements can be expected by utilization of state-of-art fiber optic components and technologies.

13.2 Background

13.2.1 *Optical Coherence Tomography*

Optical coherence tomography (OCT) has been recognized as an "optical biopsy" of different organs in recent decades. It can provide depth information (optical biopsy), and it avoids the physical cutting of samples so that in vivo non-invasive imaging is made possible. OCT can obtain ~10 μm resolutions and 2~3 mm imaging depths in highly scattering biological tissues based on low-coherence interferometery and fiber optic technology [5]. The most basic form is called time-domain OCT (TD-OCT), which relies on a Michelson-type interference pattern generated from a focused sample arm beam and a lateral-scanning mechanism. The origins of OCT can be traced to the field of optical coherence reflectometry, a 1D distance mapping technique that was originally developed to identify the faults from reflections in optical fiber networks [6–7] and was almost immediately applied to biological applications [8]. It also explains the ties of OCT to the telecommunications industry and how the OCT community enjoys the existing and low-cost hardware for biomedical applications OCT was first demonstrated for cross sectional retinal imaging in 1991 [9]. Since then, it has become a clinically viable diagnostic technique in the ophthalmological community. Its technologies enjoyed substantial advancement in the past decade due to the various promising development in the telecommunications. As a result, it is now well positioned for wide deployment in various clinical and research applications [10, 11]. OCT imaging can achieve at least one order of magnitude higher spatial resolution compared with commonly used ultrasound imaging (~50 μm). As another comparison, MRI has a typical resolution of 0.1 mm, whereas CT normally provides ~1 mm resolutions.

For better diagnostic ability required in the new frontiers of OCT application, it is essential to achieve rapid acquisition rates to reduce artifacts due to patient motion, to capture fast dynamic and to generate 3D volumetric images within reasonable time frames. OCT imaging speed increases have been achieved with Fourier domain detection (FD-OCT) technique [12, 13]. It measures axial (axial line, A-line) back-reflection signals in the frequency domain (i.e., the wave-number domain), in contrast to the conventional TD-OCT technique, which measures an interferometric signal in the time domain (or delay length domain). In other words, the FD-OCT systems speed up the detection process in obtaining the axial scattering information since the obtained raw data are converted into the time/length domain using discrete Fourier transformations (DFT) with digital signal processors. Recent studies have shown that FD-OCT detection techniques (either frequency swept sources or spectrometers as explained in the next section) can dramatically improve imaging speed or detection sensitivity compared to standard TD-OCT detection [14–15].

13.2.2 *Fourier Domain Optical Coherence Tomography (FD-OCT)*

The first type of FD-OCT implementation is known as spectral-domain OCT (SD-OCT). In SD-OCT, the interfered signal is detected using a spectral-domain detection system that is comprised of a collimator, grating, and a linear CCD or multi-channel photodetector array so that the signals are captured in a parallel manner [16, 17]. The second type of FD-OCT implementation is called optical frequency-domain imaging (OFDI) or swept-source OCT (SS-OCT), which primarily relies on the frequency swept sources. It then avoids the need for high performance spectrometers and CCDs which are used in SD-OCT detection [18–19]. Furthermore, it can leverage the advantage of dual balancing detection to mitigate the effect of the intrinsic relative intensity noise (RIN) of a light source, which attracts interests in the development of frequency swept sources for OCT.

Recent efforts to achieve sweep rates with tens of kHz such as Fourier domain mode-locked (FDML) laser [20, 21]. However, some require tunable filters, such as rotatable diffraction grating [22], polygonal mirror [23] or piezo Fabry-Perot filter [20, 21], with limited tuning speeds. Alternative high-speed swept-sources which utilize fiber dispersion such as dispersion tuning [24], change of the modulation frequency and chromatic dispersion in the laser cavity [25], though the tunable range is limited due to the rational modelocking [26] and cavity length. Another approach is based on using a stretched pulse supercontinuum (SC) source. However, the low sensitivity and narrow depth range inherited from the significant intensity noise of the SC source limits its application [27].

13.2.3 *Alternative OCT Systems*

Some more recent developments further enrich OCT community. For example, phase-conjugate OCT (PC-OCT) uses phase conjugation to invert the spectral phase to compensate even-order dispersion [28] and provides the axial resolution improvement (factor of 2), which has been demonstrated recently [29]. Another development called dual-band OCT, which improves the classification of different tissues types by the analysis of spectroscopic properties to bring additional imaging contrast. A recent demonstration by implementing SD-OCT in two wavelength bands (740 and 1300 nm) simultaneously, which shown to be able to permit medical in vivo diagnostics with high resolution and improved spectroscopic imaging contrast with low computational costs, though it required different scan cameras at their respective wavelengths [30].

13.3 Coherent Anti-Stokes Raman Scattering (CARS) Systems

On the other hand, microscopic imaging based on coherent anti-Stokes Raman scattering (CARS) allows for chemically specific imaging of cells, tissues, and drug delivery systems, whereas the contrast is derived from molecular vibrations. CARS is a third-order nonlinear optical process involving three laser beams: a pump beam of frequency ω_P, a Stokes beam of frequency ω_S, and a probe beam at frequency ω_{Pr} (sometimes equals to the pump frequency for some configurations). These beams interact with the sample and generate a coherent optical signal at the anti-Stokes frequency ($\omega_{Pr} + \omega_P - \omega_S$). The CARS phenomenon was reported for the first time in 1965 but the technique has achieved a major development since 1999 [31]. The employed pump and Stokes pulses can both be picosecond [32] or picosecond and femtosecond [33]. They drive Raman-active vibrations at the pump-Stokes frequency, ω_P-ω_S, generating a blue-shifted anti-Stokes signal (ω_{CARS}). The CARS signal is enhanced, when ω_P-ω_S is equal to the frequency of a Raman active vibration. It has been demonstrated successfully in the non-invasive imaging of living cells [34] and lipid membranes [33]. Furthermore, CARS microscopy can combine with other linear and nonlinear imaging modalities such as confocal Raman microscopy and multiphoton fluorescence microscopy to become a single microscopy platform [35, 36]. In summary, CARS provide a few unique advantages: (a) they permit nondestructive molecular imaging without any labeling; (b) the coherent nature of CARS signal result in a highly directional output, which greatly facilitate the signal collection; (c) they also allow high-speed vibrational imaging; (d) CARS signal generate at a wavelength, spectrally separated from the one-photon fluorescence background, also simplify the detection process since they are shorter than the excitation wavelengths.

13.3.1 *Fiber-Based CARS Systems*

However, the practical real-time application of CARS imaging and spectroscopy outside the laboratory environment is primarily limited by the lack of compact, high-performance light sources with sufficient power and stability. Conventional sources such as femtosecond solid-state lasers (e.g., Ti: sapphire laser) are regularly utilized inside the laboratory for CARS imaging and spectroscopy, but they are not practical to use for general applications outside the laboratory because they are widely known to be expensive, difficult to handle and fairly unstable. On the other hand, fiber lasers offer attractive properties for this application such as stability, good beam quality, permanent optical alignment, and potentially low cost. A fiber-based source for coherent Raman microspectroscopy was reported recently [37], which relies on the valuable property of being environmentally stable provided by the fiber-based picosecond and femtosecond sources. It is more desirable to use picosecond sources since it maximizes spectral resolution and contrast. Another recent effort [38] utilizes an integrated fiber source of high-energy transform-limited picosecond pulses. A frequency-doubled fiber source can pump a bulk optical parametric oscillator (OPO), which provides tunable two pulse trains that can be used for CARS imaging. Even though it is not an all-fiber source yet (since OPO is generated by LBO crystal), it is still a promising step toward an all-fiber-based CARS system [39]. In short, pulse energy and pulse duration which yields comparable peak powers as high as the conventional solid-state systems are still the primarily design issues since some of the schemes may not be scalable to high powers.

13.4 Enabling Optical Amplifier Technologies

The conventional EDFA used in telecommunications (C-band) earned unprecedented success as optical amplifier. However, its bandwidth limitation and wavelength window (1530–1562 nm) motivated significant amounts of research interest in alternative types of optical amplifiers. This interest can be attributed to

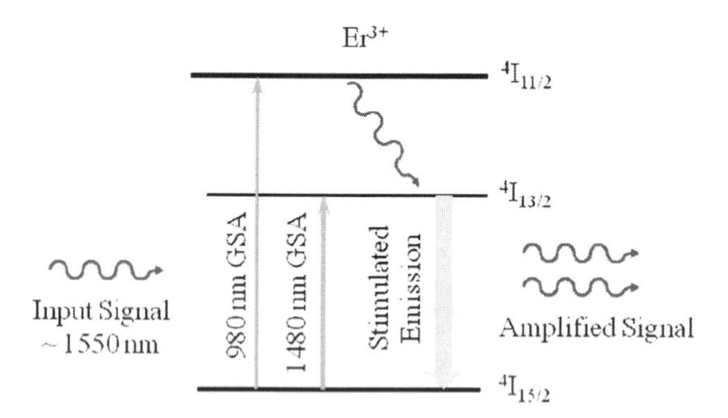

Figure 13.1 Energy levels of erbium ions. GSA: ground state absorption.

the different requirements for optical amplifiers from system to system.

This section introduces some of the most common optical amplifiers, such as erbium-doped fiber amplifiers (EDFAs), fiber Raman amplifiers (FRAs), semiconductor optical amplifiers (SOAs), and fiber-optical parametric amplifiers (OPAs). The aim of this section is to give an update of the progress in optical amplifiers as a background to the current trend in the development of optical amplifiers for biophotonic applications [40].

13.4.1 *Erbium-Doped Fiber Amplifier (EDFA)*

Several rare earth elements showed absorption spectrum in the low-attenuation window of silica fiber which implied their possible use as optical amplifiers. The most successful one to date is erbium ions (Er^{3+}), which are used in EDFAs. EDFAs have a similar operating principle as lasers, except that there is no oscillating cavity inside EDFAs. A simplified version of the erbium ion energy levels are shown in Fig. 13.4: the ions are optically pumped to the excited state by shorter wavelength light (commonly-used pump wavelengths are 980 nm or 1480 nm), and amplify light near 1550 nm, which is right in the lowest-loss transmission window of optical fiber. The quantum efficiency of a 980 nm pump is poorer due to greater mismatch of energies of the pump photons, as compared to signal photons and a significant amount of excited state absorption

(ESA). An example of an EDFA configuration is shown in Fig. 13.2, with the isolators preventing any leakage of pump (co- or counter-propagating) into the signal input or output ends. Variations include two-stage configurations, with different pumping schemes to meet different requirements.

Since the first germanosilicate-based conventional (C-band) EDFA was demonstrated, many different kinds of systems were investigated. For example, fluorozirconate fibers based on ZBLAN (Zr-Ba-La-Al-Na) were initially proposed as a means of broadening EDFAs without the need for filtering. The ultimate exhaust of C-band gain bandwidth naturally drove the emergence of L-band EDFA (1570–1605 nm [41]), which is essentially the same as C-band EDFA but utilizes longer EDF with a lower inversion level (~40%) [42]. Alternative L-band tellurite EDFA, which utilizes tellurite glass as the host material, can achieve 50-nm bandwidth from 1560 to 1610 nm [43]. With this technology, an over-80 nm EDFA combining C- and L-band has been demonstrated [44]. However, even L-band has been fully utilized, and research focus has been shifted to S-band (1450 to 1520 nm). Different candidates include TDFA (thulium-doped fiber amplifier) for 1450 to 1480 nm, and gain-shifted TDFA (GS-TDFA) for 1480 to 1510 nm [45].

Most of these developments show promising results, but in order to employ true wideband amplifier under practical conditions (e.g., power conversion efficiency (PCE) or interface with silica-based fibers), further improvements have to be achieved. EDFA has occasionally been used for OCT applications [46], which will be explored further in the later section.

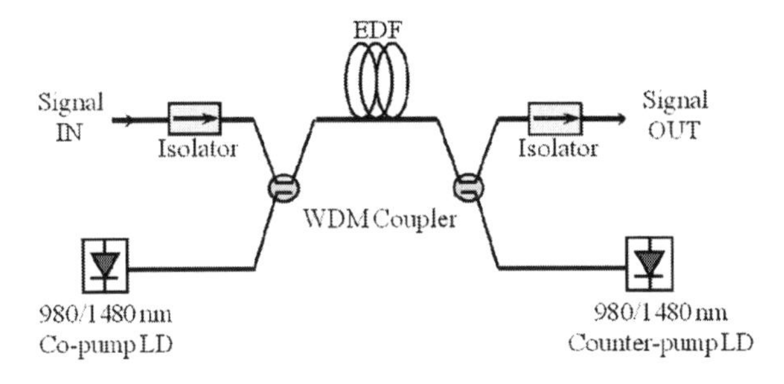

Figure 13.2 Example of an EDFA configuration.

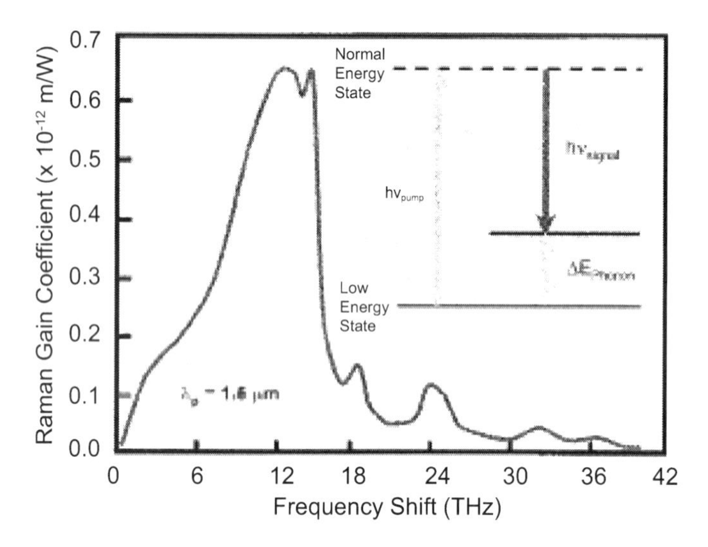

Figure 13.3 Measured Raman-gain spectrum for fused silica at a pump wavelength of 1:5 μm [48].

13.4.2 *Fiber Raman Amplifier (FRA)*

In contrast to EDFA, fiber Raman amplifiers (FRAs) are based on a fiber nonlinear phenomenon called stimulated Raman scattering (SRS). SRS is the interaction between an incident photon and an optical phonon. The incident photon is scattered by a molecule to a lower-frequency photon, and the molecule itself makes a transition between its two vibrational states [47]. In other words, Raman gain arises from the transfer of power from one optical beam to another that is downshifted in frequency by the energy of an optical phonon. The Raman gain spectrum in fused silica fibers is illustrated in Fig. 13.3 [48]. The gain bandwidth can be as wide as 40 THz, with the dominant peak near 13.2 THz. Note that the gain band shifts with the pump spectrum, while the peak value of the gain coefficient, g_R, is inversely proportional to the pump wavelength. For example, it is around 13.2 THz, which corresponds to approximately 100 nm in the telecommunication bands around 1550 nm.

Figure 13.4 shows an example of an FRA configuration. It is similar to the EDFA configuration, but is polarization sensitive due to the fact that FRA is based on a $\chi^{(3)}$ nonlinear phenomenon.

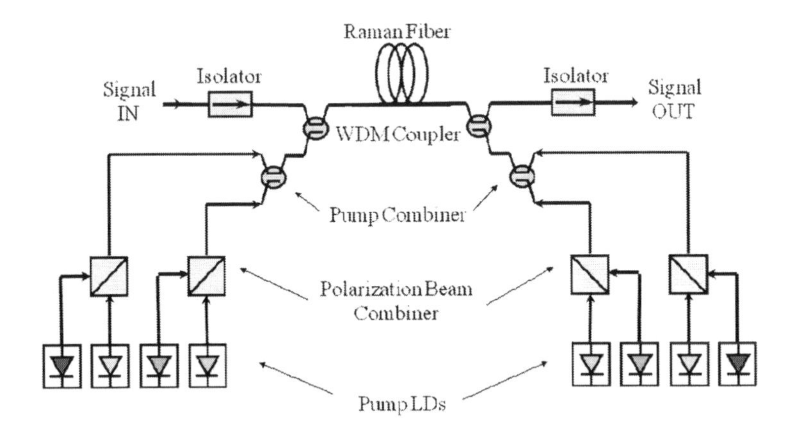

Figure 13.4 A typical configuration of FRA.

Therefore, some techniques such as pumping by orthogonal polarizations, or recently polarization scrambling, were proposed [49], to reduce polarization dependence. In addition, the gain bandwidth of FRAs is generally narrow [48]. Therefore, in order to achieve a large bandwidth, multiple pumps with different wavelengths are necessary [50].

Compared with EDFA, FRAs have several advantages. First of all, the center wavelength of the amplification spectrum of an FRA depends on the pump wavelength, while the EDFA is restricted by the rare-earth element, doping etc. Therefore, FRAs can provide amplifiers in other bands, such as the 1.3 μm or S-band [50], provided that they are properly pumped. We can also make broadband Raman amplifiers with multiple pumps because the gain spectra of individual pumps overlap [51].

Although FRAs have many good features, appropriate pump sources have always been a problem because of the weak Raman effect (or small Raman gain coefficient) of silica glass. FRAs are generally less efficient than EDFA, and therefore they require higher pump power (up to ~1 or 2 W). Recently, high pump power laser diodes have become available [52]. Another critical issue is due to the short response time of Raman gain (can be as short as 3 to 6 fs): it leads to a coupling of pump fluctuations to the signal. The usual way of avoiding this deleterious coupling is to make the pump

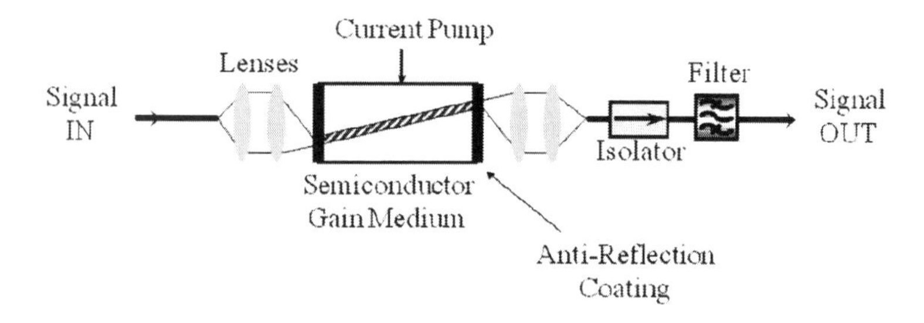

Figure 13.5 A typical configuration of SOA [55].

and signal counter-propagate. However, this will degrade the noise figure of the Raman amplifier. If co-propagating pumps and signals are to be used, then the pump lasers must have a very low relative intensity noise (RIN) [53]. Such developments in high quality pump lasers and broadband Raman amplifiers with multiple pumps have revived interest in Raman amplifiers, such as its application in OCT applications [54].

13.4.3 *Semiconductor Optical Amplifiers (SOAs)*

SOAs are one of the most widely used gain media for biophotonic applications. A typical configuration of a SOA is shown in Fig. 13.5. When semiconductor gain material (commonly with a direct bandgap) is pumped by injecting electric current, a population inversion of electrons and holes can occur in the valence and conduction bands [55]. Upon passing a light signal through the semiconductor, amplification takes place through stimulated emission, as shown in Fig. 13.6.

The most attractive feature of the SOA is that it is a compact device that can be fabricated efficiently in high volume using an IC production process. Compactness and low cost go well together to enable use of a component in many places in the practical systems, particularly in the point-of-care locations, which are size- and cost-sensitive. By a proper choice of material, the center wavelength of an SOA can be customized anywhere between 1 μm and 1.6 μm. It also makes SOAs a candidate amplifier outside the EDFA band for biophotonic applications.

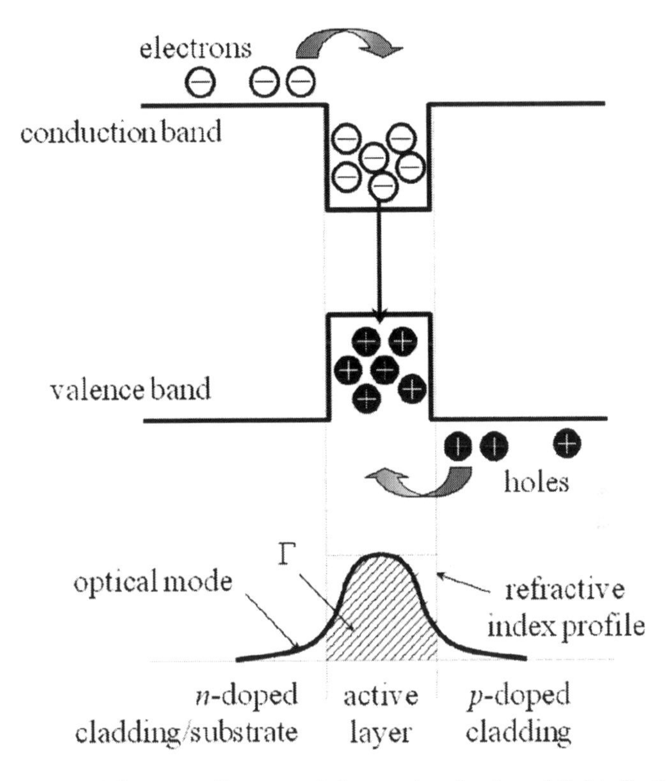

Figure 13.6 Schematic diagram of the carrier (top) and light (bottom) confinement in the active layer of a SOA. (Top) The smaller bandgap confines the carriers (electrons and holes); (bottom) while the large refractive index introduces optical waveguiding. The fraction of the mode inside the active layer is denoted by the confinement factor Γ [55].

One problem of SOA is the polarization sensitivity due to the asymmetry of its basic configuration. This polarization dependent gain (PDG) can range from a few to tens of dB, which is clearly not acceptable in any application in which the input signal polarization is not controlled. Several methods have been proposed and the PDG of a SOA can be reduced to a few tenths of a dB [56]. Furthermore, coupling at the fiber-waveguide interface also poses another problem. The coupling loss increases the SOAs noise figure, and the SOA gain is also sensitive to optical reflections at the end facets, especially in the high-gain region (~30 dB). Researchers have proposed several methods, including angled waveguide stripes (as shown in Fig. 13.5), antireflection coating, and window regions to make the reflections lower than 10^{-6}, and the gain ripple $\sim$0.1 dB for a chip gain of 30 dB. Improvements are still being made [57].

13.4.4 *OPAs as Amplifiers*

Fiber OPAs are based on the third-order nonlinear susceptibility $\chi^{(3)}$ in fiber. Ref. [58] shows a generic approach to dealing with the nonlinear propagation equations. It can be shown that the nonlinear polarization density, $P_{\mathrm{NL}}(z;t)$, is given by

$$P_{NL}(z,t) = \chi^{(3)} E^3(z,t) \qquad (13.1)$$

where

$$E(z,t) = E\cos(\omega t - \beta z) \qquad (13.2)$$

ω is the angular frequency and β is the propagation constant of the signal. The nonlinear polarization density causes the refractive index to become intensity dependent, which is the root cause of many nonlinear effects, namely self-phase modulation (SPM), cross-phase modulation (XPM) and four-wave mixing (FWM). Note that the refractive index n can be expressed as follows:

$$n = n_0 + n_2 I \qquad (13.3)$$

where n_0 is the linear refractive index and n_2 is the nonlinear refractive index and I is the optical field intensity. Measured values of n_2 for silica fibers vary in the range 2.2–3.4 $\times$ 10^{-20} m^2/W [47]. The numerical value of n_2 depends not only on the core composition but also on whether the input polarization is preserved inside the fiber or not. If the core of an optical fiber is doped with germania, n_2 is enhanced because n_2 for GeO_2 is larger by about a factor of 3 compared with its value for SiO_2. That is why the core of highly-nonlinear dispersion-shifted fiber (HNLF) is doped with GeO_2.

Let us now consider an example: assume there are two signals with angular frequencies ω_1 and ω_2, respectively. The input field is

$$E(z,t) = E_1\cos(\omega_1 t - \beta_1 z) + E_2\cos(\omega_2 t - \beta_2 z) \qquad (13.4)$$

where β_1 and β_2 are the propagation constants of the two signals, respectively. Using Eqs. 13.4 and 13.5, the nonlinear polarization density is given by

$$
\begin{aligned}
P_{NL}(z,t) &= \chi^{(3)}\left[E_1\cos(\omega_1 t-\beta_1 z)+E_2\cos(\omega_2 t-\beta_2 z)\right]^3 \\
&= \chi^{(3)}\left\{\left(\frac{3E_1^3}{4}+\frac{3E_2^2E_1}{2}\right)\cos(\omega_1 t-\beta_1 z)\right. \\
&+\left(\frac{3E_2^3}{4}+\frac{3E_1^2E_2}{2}\right)\cos(\omega_2 t-\beta_2 z) \\
&+\frac{3E_1^2E_2}{4}\cos\left[(2\omega_1-\omega_2)t-(2\beta_1-\beta_2)z\right] \\
&+\frac{3E_2^2E_1}{4}\cos\left[(2\omega_2-\omega_1)t-(2\beta_2-\beta_1)z\right] \\
&+\frac{3E_1^2E_2}{4}\cos\left[(2\omega_1+\omega_2)t-(2\beta_1+\beta_2)z\right] \\
&+\frac{3E_2^2E_1}{4}\cos\left[(2\omega_2+\omega_1)t-(2\beta_2+\beta_1)z\right] \\
&\left.+\frac{E_1^3}{4}\cos(3\omega_1 t-3\beta_1 z)+\frac{E_2^3}{4}\cos(3\omega_2 t-3\beta_2 z)\right\}
\end{aligned}
\tag{13.5}
$$

The terms at $2\omega_1+\omega_2$, $2\omega_2+\omega_1$, $3\omega_1$, and $3\omega_2$ can be neglected since the phase-matching condition will not be satisfied for these terms due to the fiber chromatic dispersion. New frequency components appear at $2\omega_1$-ω_2 and $2\omega_2$-ω_1, are shown in Fig. 13.7. This phenomenon is generally referred to as four-wave mixing (FWM).

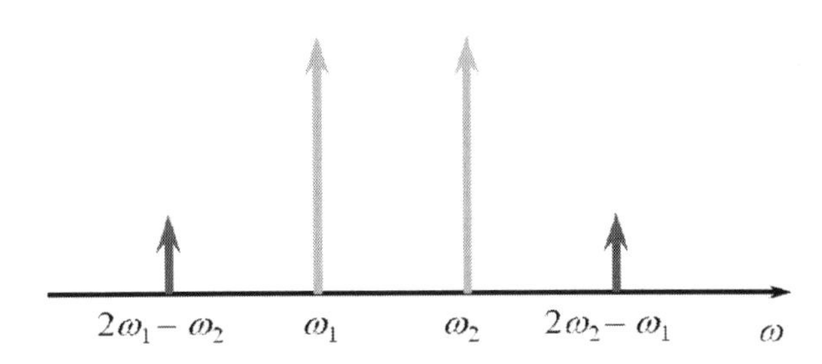

Figure 13.7 Illustration of four-wave mixing.

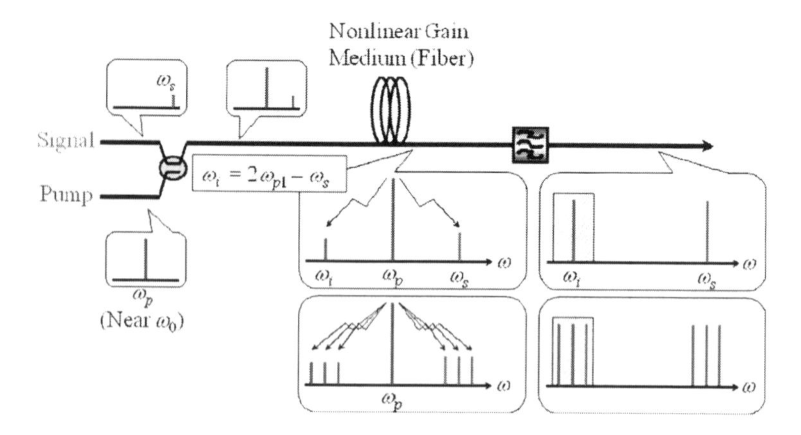

Figure 13.8 A typical schematics of one-pump OPA.

When the input signals are strong (which will then be called pumps), FWM can provide gain for other co-propagating weak signals. In such cases, a more complicated OPA mechanism must be invoked. Interested readers can refer to the literature [59–60] for detailed explanations.

Fiber OPAs can be operated in either one- or two-pump configuration. The schematic of a one-pump OPA is shown in Fig. 13.8. It is similar to other fiber-based amplifiers (EDFA, FRA etc.) as the input signal is amplified, but a new wavelength called idler is also generated at the same time. The quantum mechanical picture illustrates this: for every two photons lost from the pump, two new photons are generated, one at the signal frequency and the other at the idler frequency. Therefore, according to the principle of energy conservation, the idler frequency (ω_i) has to satisfy the frequency relationship: $\omega_i = 2\omega_p - \omega_s$. A filter is required after the fiber to filter out residual pump power, ASE noise, and the idler (or the signal if using as wavelength converter instead of optical amplifier).

The gain spectrum and bandwidth of this parametric process strongly depends on the pump wavelength. The pump is usually operated slightly above the zero-dispersion wavelength (λ_0) of the fiber in order to fulfill the phase matching condition, as shown in Fig. 13.8 [61].

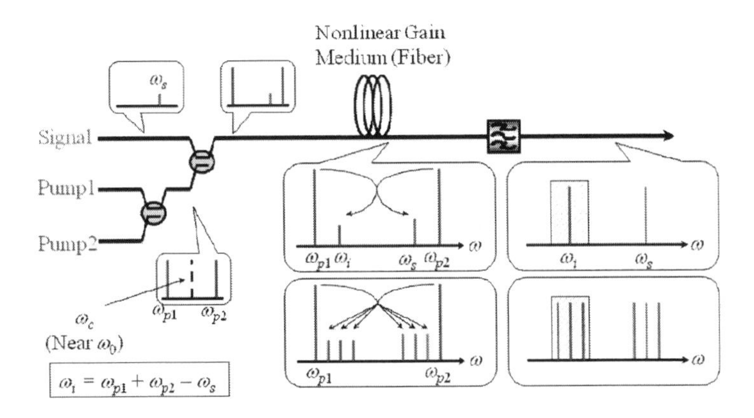

Figure 13.9 A typical schematics of two-pump OPA.

One can also operate OPA in a two-pump configuration as illustrated in Fig. 13.9. In contrast to the one-pump configuration, the two pumps are operated far from the signal band, but the average frequency (also called the center frequency, ω_c), is maintained near the zero-dispersion frequency. Idler wavelength will also be generated in two-pump OPA, except that it is determined by the frequency relationship: $\omega_i = \omega_{p1} + \omega_{p1} - \omega_s$. Even though the two-pump configuration is more complicated than the one-pump configuration, it provides extra degree of freedom to optimize the bandwidth, gain flatness [60] and polarization independence.

The advantages of the OPA include low insertion loss, high gain, and large amplification bandwidth. Experiments showed that the gain bandwidth can be as wide as 400 nm [62], while the CW signal gain can be as high as 70 dB [63]. Similar to the FRAs, the center of the OPA's gain bandwidth depends on the pump wavelength, thus the OPA can be used to amplify wavelengths outside the conventional EDFA range (C-band). In contrast to FRA, the OPA also has some other unique features, for example, its gain spectrum is also pump-wavelength dependent and gain flatness can be achieved by a two-pump configuration. Last but not least, OPA also provides wavelength conversion (phase-conjugated idler) with gain, which can be utilized in the phase-conjugated OCT (PC-OCT) [64].

13.4.4.1 Fiber-optical parametric oscillator (FOPO)

The basic principle of fiber optical parametric oscillator (FOPO) is similar to other kind of oscillating cavity except that the gain medium is obtained from FOPA. However, since the nonlinearity of commonly available fibers (e.g., single-mode fiber [SMF]) is fairly low (<1/W/km), obtaining sufficient parametric gain to form an oscillating cavity with such fibers generally requires watt-level pump powers, which can best be provided by pulsed lasers. As a result, until recently there have been few reports of fiber OPOs, and most of them have been operated with pulsed pumps [65, 66]. Their close relatives, modulation-instability lasers [67], have been operated with a CW pump [68]; their output, however, consists of solitons instead of CW output.

More recently, with the availability of highly-nonlinear fibers (HNLFs) and high-power CW pumps, it is becoming feasible to implement a fiber OPOs in a CW regime [69, 70], up to watt-level output power and over 100-nm of tuning range [71]. In this section, we review the operation of such CW fiber OPOs by investigating a specific demonstration in details [69]. They were consisted of HNLFs, in length of hundred of meters, with a Fabry-Perot cavity formed by two fiber Bragg gratings (FBGs), singly-resonant at the signal wavelength; the idler constitutes the useful output, which is illustrated in Fig. 13.10. For this specific configuration, because the

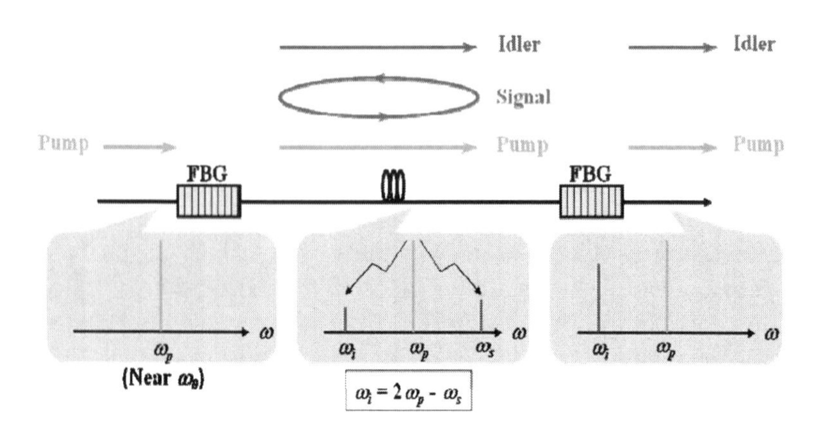

Figure 13.10 One configuration of optical parametric oscillator.

idler is completely coupled out as output and the signal is reflected completely by the FBG after each forward pass, only signal and pump are present at the input FBG. Therefore, the signal gain and the idler conversion efficiency are independent of the initial phases of the signal and pump. This phase independence is the same as found in OPAs with no idler at the input, such as in [72], is also known as phase-insensitive OPA. Such kind of configuration can be applied to either a one-pump OPO, or a two-pump OPO.

It can be illustrated by some numerical examples corresponding to the experimental parameters. For example, we implemented a FOPO by a 100 m-long HNLF, and FBGs written directly onto it, we obtained a threshold pump power of 240 mW, but low output power. On the other configuration, we implemented another FOPO by a 1 km long HNLF with FBGs attached by means of fiber connectors we have extracted up to 100 mW of idler power at 1566 nm, for 700 mW of pump power at 1563 nm. Excluding coupling losses, this corresponds to an internal conversion efficiency of 30%, compared to the maximum theoretical efficiency of 50%.

In a one-pump fiber OPA, the small-signal gain for a signal wavelength close to the pump can be obtained as follows [47, 73]:

$$G_s = \exp\left(-\alpha L\right)\left[1 + \left(\gamma P_0 L_{\mathrm{eff}}\right)^2\right] \tag{13.6}$$

where γ is the fiber nonlinearity coefficient, P_0 is the pump input power; $L_{\mathrm{eff}} = (1 - e^{-\alpha L})/\alpha$ is the effective length, α is the fiber attenuation coefficient, and L is the fiber length. Let T denote the round-trip transmittance of the cold cavity without the distributed fiber loss. Then the threshold condition is $G_s T e^{-\alpha L} = 1$. For a given L, we can solve for P_0 required to reach threshold, by means of the equation for $X = e^{-\alpha L}$. Eq. 13.7 becomes

$$T X^2 = \left[1 + r^2 (1 - X)^2\right] = 1 \tag{13.7}$$

where $r = \gamma P_0/\alpha$. This leads to the threshold pump power

$$\begin{aligned} r &= \frac{\gamma P_{\mathrm{th}}}{\alpha} = \frac{\sqrt{\frac{1}{TX^2} - 1}}{1 - X} \\ P_{\mathrm{th}} &= \frac{\alpha}{\gamma}\frac{\sqrt{1/T - X^2}}{X\,(1 - X)} \end{aligned} \tag{13.8}$$

For the Sumitomo HNLF used in our experiments, $\gamma = 17\ \mathrm{W}^{-1}\,\mathrm{km}^{-1}$, and $\alpha = 0.2$ Nepers/km. If we consider a short cavity with $L = 100$

m, and integral FBGs, so that $T \approx 1$, we find that $P_{\text{th}} = 0.12$ W. In the other configuration of a 1 km long cavity with FBGs attached through lossy connectors, so that $T \approx 1/20$, we obtain $P_{\text{th}} = 0.35$ W. Therefore, it is practical to reach threshold with a few hundred milliwatts of pump power, or even lower with the improvement of the gain medium and cavity loss, in either type of cavity.

In the later section, we will explore different configurations of FOPO which can be utilized in biophotonic applications.

13.5 Recent Research Efforts

13.5.1 *OCT Applications*

In this section, we will focus more the novel swept-source design required in SS-OCT. Those who are not too familiar with the principle of frequency-domain ranging and OCT system instrumentations are encouraged to refer to Chapter 3, "OCT Imaging." Swept-source (SS) lasers have advanced rapidly over the past few years, primarily driven by demanding applications in OCT. A recent coverage for various OCT technologies can be referred to [11], while we will focus more on the fiber-based solutions in this section.

13.5.1.1 Fourier domain mode locking (FDML)

SS laser has been employed as a standard driving technology for sensing applications until recently [74, 75], which has also been explored in other application such as the field of biomedical imaging, especially after the popular realizations of OCT in the form of Fourier domain OCT (FD-OCT) [76, 77]. In order to enhance the axial and transverse resolutions, improve the signal-to-noise ratio (SNR), and increase the A-line scan rate to acquire faster and larger three-dimensional image sets [78, 79] or even video-rate imaging, several OCT techniques have been developed. One of the most promising is optical frequency-domain imaging (OFDI) or swept-source OCT (SS-OCT), whereas the wavelength of a laser source is swept linearly in time before illuminating onto the sample. Recently, a new SS technology has been introduced and attracted wide recognition in the research community is called Fourier domain modelocking

(FDML) developed by Huber et al. [21]. In summary, it matches the fiber laser ring cavity length with the fundamental or harmonic of the round-trip time of the photons circulating in the cavity such that when one wavelength arrives at the tunable filter, it coincides with the same particular wavelength that is being generated through the gain medium (such as the seminconductor optical amplifier (SOA)). Therefore, sweeping wavelengths can be continuously generated inside the cavity with tens of milliwatts of output power. Optical filters are commonly used to sweep the wavelengths such as the piezoelectric driven tunable Fabry-Perot interferometer (TFPI) filters [21] and polygon-grating filters [77]. Comparing with the traditional SS, FDML laser can generate equivalent or even higher optical output power at a higher sweeping speed, which enables a high imaging speed and large imaging depths.

13.5.1.2 Hybrid FDML

It is conventional to use SOA as the gain medium to construct FDML laser due to its wide availability (e.g., in the 1050 nm regime [2]), compact size and relatively wide gain bandwidth ($\sim$100 nm) as mentioned in the previous sections. However, there are several disadvantages of SOA comparing with other kinds of optical amplifiers, such as its relatively higher noise figure (typically 7 dB or so) and lower gain comparing with EDFA. Therefore, some previous studies have demonstrated the use of Raman amplifier [54] and EDFA [46] as gain media in the 1550nm regime, most of these were off-the-shelf telecom components. However, the results had shown limited bandwidth and power, which inherited from the C-band EDFA and Raman amplifier, when comparing with the SOA counterparts. Another study demonstrated a parallel configuration of two SOAs with different gain spectra [80], where the overall bandwidth was increased. However, it was challenging to combine the two paths leading to the two different SOAs with low coupling loss and equalize the power level of the two paths. In addition, it is essential to exactly match the length in the two paths. Otherwise the overall the wavelength sweeping and thereby the FDML operation would be compromised.

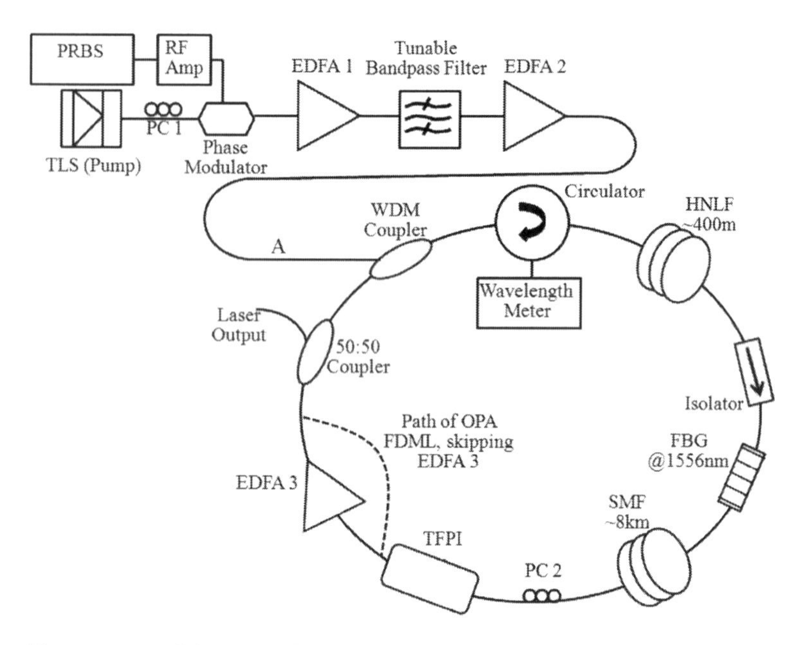

Figure 13.11 Schematic diagram of the hybrid FDML. The dotted line bypasses the EDFA3, which corresponds to the OPA FDML configuration [81].

Therefore, we proposed and demonstrated a hybrid FDML configuration to leverage the advantages of different kind of optical amplifiers [81]. The design concept related to the fiber OPA has already been explained in details in the previous section. The setup of the hybrid FDML configuration is illustrated in Fig. 13.11. First, the pump (of the OPA) wavelength was located at ~1555 nm by the tunable laser source (TLS) which has the linewidth of 20 MHz and power of 5 dBm. The specific choice of the wavelength fulfilled the phase-matching condition required by the OPA. It was then phase-dithered via a phase modulator which was driven by a pseudo-random binary sequence (PRBS) to suppress the stimulated Brillouin scattering (SBS) [82]. Otherwise, the maximum pump power that could be launched into the nonlinear gain medium (i.e., HNLF) would be limited by the SBS. Such kind of SBS was also monitored by the wavelength meter. After that, the pump passed through a two-stage EDFA with the ASE noise suppressed by a tunable bandpass filter in-between. The parametric amplification was realized by the HNLF, with a zero-dispersion wavelength (ZDW) of 1554 nm, dispersion slope of 0.02ps/nm^2/km, and non-linear coefficient of 14 W^{-1} km^{-1}. A FBG and an isolator were used to attenuate the pump power in the cavity in order to protect the following fiber components from optical damage. A fiber-based TFPI filter was driven sinusoidally for continuously filtering within FDML configuration, which preceded the EDFA3. This specific EDPA would be bypassed during the OPA FDML configuration as shown in the Fig. 13.11 to illustrate the better performance of the hybrid FDML versus its OPA FDML counterpart. The ~8 km SMF fiber extended the cavity length to match the cavity round-trip time with the two periods of the driving frequency of the TFPI filter, which was 39.5 kHz. It also corresponded to the second harmonic frequency of the fundamental FDML frequency, which was limited by the piezoelectric specification of the TFPI filter used in this experiment. A higher sweeping rate and thereby the imaging speed can be achieved by a TFPI filter that can be driven with higher frequency and shorter SMF (i.e., significantly less than 8 km) or buffering technique [79]. The FDML effect could still be observed by matching the sweeping period of the TFPI filter with half round-trip time in the cavity.

Figure 13.12 shows the FDML spectrum obtained when both the one-pump OPA and C-band EDFA were utilized to generate the

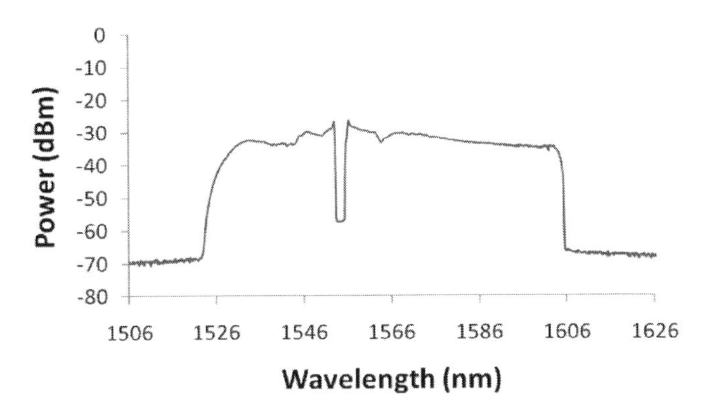

Figure 13.12 Hybrid FDML spectrum (Resolution: 0.06 nm) [81].

hybrid FDML laser. The spectrum was solely limited by the gain spectrum of the OPA, which can be further increased by stronger pump power [58]. In the hybrid FDML operation both the red and blue wavelength sides of the OPA gain spectrum compensated the non-uniform spectrum of the EDFA, while the dip in the M-shaped one-pump OPA gain spectrum was partially compensated by the EDFA gain spectrum. As a result, the hybrid FDML configuration illustrates the principle of utilizing different characteristics of the optical amplifiers.

This proof-of-principle study only utilized a one-pump OPA and EDFA. Two-pump OPAs are yet to be investigated since it could be used to provide a uniform flattened spectrum [62] and polarization independent operation [58]. Other than that, the versatile gain spectrum at arbitrary wavelength regions of OPA can be incorporated into any kind of hybrid FDML configuration with other complementary type of optical amplifiers (e.g., SOA or Raman amplifiers), which further enhances the performance of the existing FDML laser systems.

13.5.1.3 Swept-pump FOPO/FOPA

In addition to utilizing SOA, EDFA, and Raman amplifier as the gain medium, fiber optical parametric amplifier (OPA) by itself has also been demonstrated to be one of the most promising candidates in generating a wide-tunable SS. A recent report on utilizing narrowband, widely tunable fiber optical parametric process as the gain and the wavelength selection mechanism in a ring laser topology, simultaneous access to two distinct bands (NIR and SWIR) over 329 nm has been demonstrated. A nonmechanical cavity-tuning mechanism allowed for ultrafast sweep rate, limited primarily by the speed of the tunable pump source confined to a conventional 1550 nm band. Using a tunable pumping source sweep across 23 nm, the wavelength-swept FOPO device generated light in two contiguous bands in 1300 nm NIR band and 1800-nm SWIR range [83]. We also demonstrated a similar technique but utilizing a chirped pump instead of a tunable pump in FOPA [84]. The high-speed swept pump is achieved with the swept rate as high as 78 MHz by a technique called dispersive Fourier transformation (DFT), which circumvents the fundamental speed limitation appeared in the conventional SS based on the cavity configurations. Based on such swept pump FOPA, the idler can be generated with a wavelength range twice as the pump bandwidth. Such all-optical approach offers order-of-magnitude higher swept rate and thus lends itself to many applications such as high-speed signal processing and optical imaging such as OCT.

13.5.1.4 FOPA in phase-conjugated OCT (PC-OCT)

Another novel application of FOPA in the alternative OCT system is in the form of phase-conjugated OCT (PC-OCT) as mentioned in the previous section. A typical PC-OCT system requires two key components: a source, which generates two phase-conjugated beams, and a phase conjugator to compensate even-order dispersion. In principle, both processes can be completed in a FOPA setup instead of the spontaneous parametric down-conversion (SPDC) with fiber OPA [29]. We reported a strongly amplified high speed wavelength-swept light source for PC-OCT. By introducing a bidirectional fiber OPA (BD-FOPA) and a FDML laser, the source achieves an output power on the order of milliwatts, which is several orders of magnitude higher than the SPDC source. Improvement in source power and speed should further strengthen the potential of PC-OCT as a practical imaging method [64].

Others have also studied the possibility of using nonlinear wavelength conversion to extend the bandwidth coverage of the traditional swept-source [85]. This, together with our efforts, demonstrates that fiber nonlinear effects, particularly FOPA, can potentially be utilized to construct next-generation swept-sources in advancing OCT technologies.

13.5.2 *CARS Applications*

First-generation CARS microscopes used two temporally synchronized femtosecond (fs) or picosecond (ps) pulsed lasers [34]. Second-generation CARS systems used optical parametric oscillators (OPOs) pumped with fs or ps solid state lasers to generate multiple inherently synchronized pulse trains without the need for electronic synchronization [86, 87]. Thanks to the ultrafast response, wide-gain bandwidth [62], high gain [63], and large detune from the pump [62] of the FOPA, efficient short pulse generation is possible at nonconventional wavelengths by using an OPO configuration.

A conventional OPO utilizing parametric oscillation is based on the $\chi^{(2)}$ nonlinear effect of crystals [88]. Its outstanding performance attracted spectacular applications in optical research and development. However, this approach usually requires a precise free-space alignment for optimal performance. On the other hand, FOPO is based on the $\chi^{(3)}$ nonlinear effect of optical fiber, which eliminates the requirement of precise alignment and allow potential integration with fiber components [89] or even CMOS-compatible [90]. It has been demonstrated CW [69, 89] and nanosecond [92] pulse generations with a wide wavelength tuning range using various FOPO configurations. In the picosecond and femtosecond regimes, since the walk-off between the pump and the signal and idler is significant, it is required to use short fiber. For example, a microstructure fiber (MF) based FOPO has been demonstrated to have a tuning range mostly in the visible regime [93, 94]. However, most of the biophotonic applications such as CARS require sources in the spectral region around 800–1000 nm or even longer wavelengths. Some other MF based FOPOs [95], dispersion-shifted fiber (DSF) based FOPOs [66], and HNLF based FOPOs [96] were designed to operated in 1550 nm regime, but usually with limited tuning range. Thus, it is desirable to build a widely tunable FOPO operated in versatile wavelength regimes.

In this section, we illustrate the principle by demonstrating a fully fiber-integrated picosecond FOPO with a wavelength tuning range of 440 nm pumped by a simple intensity-modulated pump [97]. Tuning is achieved by changing the pump wavelength from 1532 to 1549 nm. A 40-m dispersion-shifted fiber (DSF) is employed inside a cavity as the parametric gain medium as shown in Fig. 13.13. Moreover, in the wavelength region from 1320 to 1395 nm and 1720 to 1820 nm, the linewidth of the FOPO output is less than 1 nm without the use of any wavelength-selective element inside the cavity. Within the narrow linewidth region, pulses with pulsewidths from 7 to 17 ps are generated. This scheme has the potential to be a

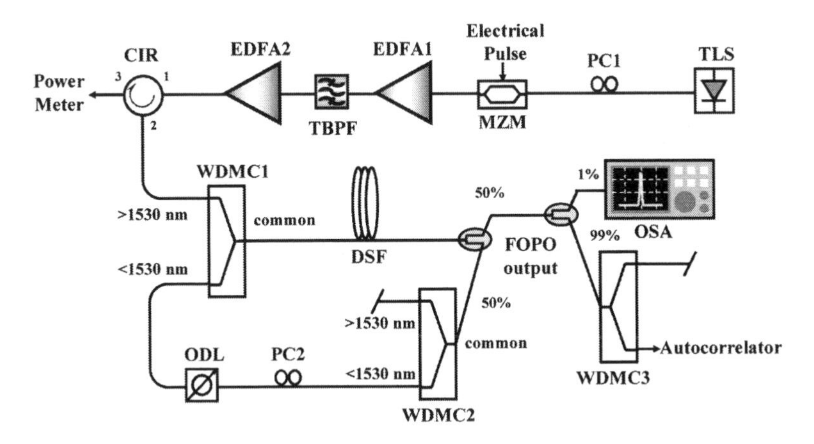

Figure 13.13 Experimental setup of the FOPO [97].

cost-effective source in generating short pulse for nonlinear optical microscopy applications such as CARS.

When the parametric amplification is achieved through the gain medium (e.g., DSF) inside a cavity, the noise in either of the sidebands (signal and idler) will be amplified and then fed back to the DSF as a seed. After multiple paths there will be significant signal and idler components at the output.

A phase-matching condition is required to generate signal and idler with substantial output power. The phase-matching condition is

$$\Delta\beta + 2\gamma P = 0, \tag{13.9}$$

where γ is the fiber nonlinear coefficient and P is the incident pump power. $\Delta\beta$ is the total phase mismatch, which can be approximated by the following equation [98]:

$$\Delta\beta = \beta^{(2)}\,(\Delta\omega)^2 + \beta^{(4)}(\Delta\omega)^4/12, \tag{13.10}$$

where $\Delta\omega$ is the frequency detune of the signal (and the idler as well) from the pump; $\beta^{(\mathrm{m})}$ is the mth order derivative of the propagation constant.

The shape of the gain spectrum is determined by the fiber dispersion profile as shown in Fig. 13.14 [62]. For example, when $\beta^{(2)} < 0$ and $\beta^{(4)} < 0$, it will operate as a conventional FOPA (i.e., anomalous-dispersion pumping), which has a continuous gain

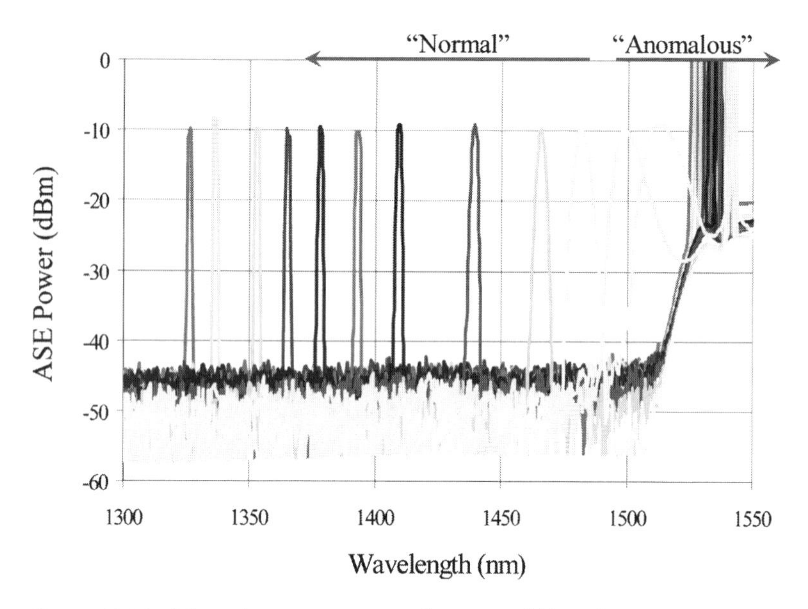

Figure 13.14 The gain spectrum as a function of fiber dispersion profile: (right) anomalous-dispersion pumping, i.e. $\beta^{(2)} < 0$; (right), i.e. normal-dispersion pump, $\beta^{(2)} > 0$.

spectrum around the pump. One the contrary, when $\beta^{(2)} > 0$ and $\beta^{(4)} < 0$, idler and signal will be generated with frequency detune much larger than the bandwidth of the conventional FOPA (i.e., normal-dispersion pumping). Such narrowband gain spectrum of FOPA provides an extra degree of freedom in controlling the FOPA and therefore the corresponding FOPO. For a given fiber, we can vary $\beta^{(2)}$ over a large range by tuning the pump wavelength as shown in Fig. 13.14, thus making the widely tunable FOPO feasible.

The experimentally measured linewidths of the sidebands are plotted as a function of wavelength in Fig. 13.15 (green rectangles). The linewidth of the sideband is relatively broad when the pump wavelength is close to the ZDW ($\sim$10 nm when pump at 1548 nm) but becomes narrower quickly as the pump is tuned away from the ZDW in the normal dispersion region. This trend is in agreement with the analytical expression for the gain bandwidth of the FOPA versus sideband wavelength using the DSF parameters (black solid

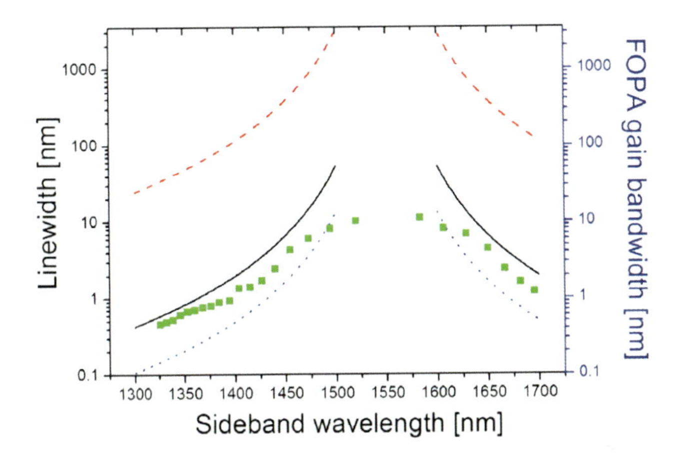

Figure 13.15 Linewidth of the sideband versus wavelength (green rectangles). Theoretical FOPA gain bandwidth versus sideband wavelength when using current DSF (black solid line), HNL-DSF (red dashed line), and a DSF with a larger $\beta^{(4)}$ (blue dotted line) [97].

line), which can be calculated by the following equation [62]:

$$\delta\lambda \approx \frac{24}{C^4}\left|\frac{\gamma P_0}{\beta^{(4)}\,(\Delta\lambda)^3}\right| \qquad (13.11)$$

where is the FOPA gain bandwidth, is the wavelength separation of between the pump and the signal, is the peak power of the pump, is the nonlinear coefficient, and C $\approx 7.85 \times 10^{20}$ $s^{-1}m^{-1}$. Between 1320 and 1395 nm, the linewidths of the sidebands are below 1 nm. Since signal and idler at each sideband should be symmetrical in frequency, thus approximately symmetric in wavelength, the linewidth of the sideband between 1720 and 1820 nm is also expected to be less than 1 nm. The linewidth of the output signal or idler is reasonably narrow ($\sim$1 nm), given it requires no additional bandpass filters inside the cavity to select the wavelengths. It is a highly desirable feature for this configuration because of the availability of filter at certain wavelength may limit the tuning range of the FOPO. This linewidth is narrower than our previous work in [98], where we used an HNLF as a gain medium. The fairly large nonlinear coefficient (14 $W^{-1}km^{-1}$) and small $\beta^{(4)}(-5.8 \times 10^{-56}$ s^4 $m^{-1})$ of the HNLF we used will result in a much broader gain bandwidth, which is shown in the red dashed line of Fig. 13.15.

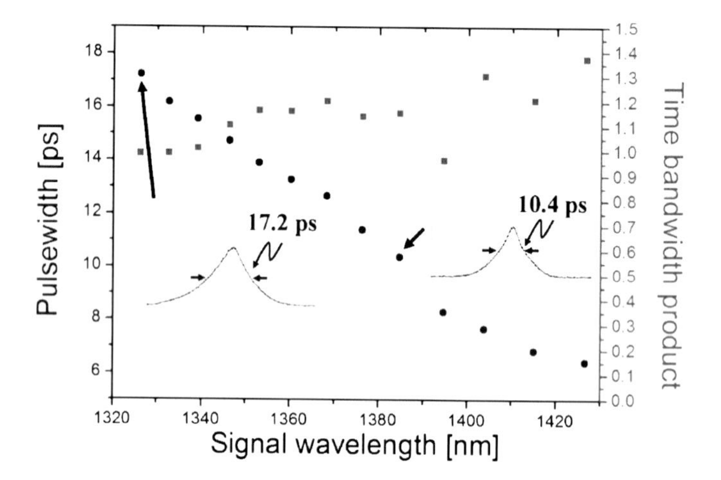

Figure 13.16 Signal pulsewidth versus wavelength (black circles). The blue rectangles are signal time bandwidth products (TBP) as a function of wavelength. Insets are autocorrelation traces at some wavelengths [97].

Furthermore, if we can use a DSF with a larger $\beta^{(4)}$ (four times larger than that of the DSF used in this experiment), the gain bandwidth can be reduced further (blue dotted line). Thus, this scheme has the potential to generate pulses with narrower linewidths.

Figure 13.16 shows the output pulsewidths (circles) and signal TBPs (rectangles) as a function of signal wavelength when the linewidth of the signal is narrower than 1 nm, which is the wavelength region of our primary interest. Insets are autocorrelation traces at some wavelengths measured by the autocorrelator. The real full-width at half-maximum (FWHM) pulsewidth is calculated by assuming a sech^2 pulse shape, multiplied by the FWHM correlation width using a deconvolution factor of 0.648. The pulsewidth increases from 7 to 17 ps when the signal detunes further from the pump, thus the walk-off between the signal and the pump becomes larger which broaden the signal pulsewidth. The output pulses are narrower than the pump pulses (100 ps) because of the pulse narrowing effect of FOPA [99]. The TBP is calculated to be around 1, which is larger than that of the transform-limited soliton pulse, 0.315. The fairly large TBP is primarily due to the cross-phase modulation (XPM) between the pump and the signal. The peak power of the output signal at 1320 nm is measured to be 970 mW. Thus, the conversion efficiency of the FOPO is calculated to be 3.88%. The conversion efficiency is perceived to be limited by the walk-off between the signal and the pump. We also use a digital communication analyzer at the FOPO output to measure the pulse shape of the signal; there is no observable pulse shape variation over 1 hour. This indicates that the FOPO output is reasonably stable. More recent effort has been demonstrated a multimodal CARS microscopy of using FOPO [39].

This chapter shows a glimpse of the utilization of the powerful knowledge base acquired in telecommunications and its well-developed photonic components, which can be leveraged for the stringent requirements in biophotonic applications.

References

1. Srinivasan, V. J., Huber, R., Gorczynska, I., Fujimoto, J. G., Jiang, J. Y., Reisen, P., and Cable, A. E. (2007). High-speed, high-resolution optical coherence tomography retinal imaging with a frequency-swept laser at 850 nm. *Opt. Lett.*, **32**:361–363.
2. Huber, R., Adler, D. C., Srinivasan, V. J., and Fujimoto, J. G. (2007). Fourier domain mode locking at 1050 nm for ultra-high-speed optical coherence tomography of the human retina at 236,000 axial scans per second. *Opt. Lett.*, **32**:2049–2051.
3. Kakuma, H., Ohbayashi, K., and Arakawa, Y. (2008). Optical imaging of hard and soft dental tissues using discretely swept optical frequency domain reflectometry optical coherence tomography at wavelengths from 1560 to 1600 nm. *J. Biomed. Opt.*, **13**:014012.
4. Nishizawa, N., Chen, Y., Hsiung, P., Ippen, E. P., and Fujimoto, J. G. (2004). Real-time, ultrahigh-resolution, optical coherence tomography with an all-fiber, femtosecond fiber laser continuum at 1.5 μm. *Opt. Lett.*, **29**:2846–2848.
5. Zysk, A. M., Nguyen, F. T., Oldenburg, A. L., Marks, D. L., and Boppart, S. A. (2007). Optical coherence tomography: a review of clinical development from bench to bedside. *J. Biomed. Opt.*, **12**:051403.
6. Takada, K., Yokohama, I., Chida, K., and Noda, J. (1987). New measurement system for fault location in optical waveguide devices based on an interferometric technique. *Appl. Opt.*, **26**:1603–1606.

7. Youngquist, R. C., Carr, S., and Davies, D. E. N. (1987). Optical coherence domain reflectometry: A new optical evaluation technique. *Opt. Lett.*, **12:**158–160.
8. Fercher, A. F., Mengedoht, K., and Werner, W. (1988). Eye-length measurement by interferometry with partially coherent light. *Opt. Lett.*, **13**:186–188.
9. Huang, D., Swanson, E. A., Lin, C. P., Schuman, J. S., Stinson, W. G., Chang, W., Hee, M. R., Flotte, T., Gregory, K., Puliafito, C. A., and Fujimoto, J. G. (1991). Optical coherence tomography. *Science*, **254**(5035):1178–1181.
10. Brezinski, M. (2006). Optical Coherence Tomography: Principles and Applications, Academic Press, London.
11. Drexler, W., and Fujimoto, J. G. Eds., (2010). Optical Coherence Tomography: Technology and Applications (Springer).
12. Leitgeb, R. A., Drexler, W., Unterhuber, A., Hermann, B., Bajraszewski, T., Le, T., Stingl, A., and Fercher, A. F. (2004). Ultrahigh resolution Fourier domain optical coherence tomography. *Opt. Lett.*, **12**:2156–2158.
13. Yun, S., Tearney, G., de Boer, J., Iftimia, N., and Bouma, B. (2003). High-speed optical frequency-domain imaging. *Opt. Express*, **11**:2953–2963.
14. de Boer, J. F., Cense, B., Park, B. H., Pierce, M. C., Tearney, G. J., and Bouma, B. E. (2003). Improved signal-to-noise ratio in spectral-domain compared with time-domain optical coherence tomography. *Opt. Lett.*, **28**:2067–2069.
15. Choma, M. A., Sarunic, M. V., Yang, C., and Izatt, J. (2003). Sensitivity advantage of swept source and Fourier domain optical coherence tomography. *Opt. Express*, **11**:2183–2189.
16. Yun, S. H., Tearney, G. J., Bouma, B. E., Park, B. H., and de Boer, J. F. (2003). High-speed spectral-domain optical coherence tomography at 1.3 μm wavelength. *Opt. Express*, **11**:3598–2165.
17. Park, B., Pierce, M. C., Cense, B., Yun, S., Mujat, M., Tearney, G., Bouma, B., and de Boer, J. (2005). Real-time fiber-based multi-functional spectral-domain optical coherence tomography at 1.3 μm. *Opt. Express*, **13**:3931–3944.
18. Golubovic, B., Bouma, B. E., Tearney, G. J., and Fujimoto, J. G. (1997). Optical frequency-domain reflectrometry using rapid wavelength tuning of a Cr4+:forsterite laser. *Opt. Lett.*, **22**:1704–1706.
19. Chinn, S. R., Swanson, E. A., and Fujimoto, J. G. (1997). Optical coherence tomography using a frequency tunable optical source. *Opt. Lett.*, **22**:340–342.

20. Huber, R., Wojtkowski, M., Taira, K., Fujimoto, J. G., and Hsu, K. (2005). Amplified, frequency swept lasers for frequency domain reflectometry and OCT imaging: design and scaling principles. *Opt. Express*, **13**:3513–3528.

21. Huber, R., Wojtkowski, M., and Fujimoto, J. G. (2006). Fourier Domain Mode Locking (FDML): A new laser operating regime and applications for optical coherence tomography. *Opt. Express*, **14**:3225–3237.

22. Yun, S. H., Boudoux, C., Pierce, M. C., de Boer, J. F., Tearney, G. J., and Bouma, B. E. (2004). Extended-cavity semiconductor wavelength-swept laser for biomedical imaging. *IEEE Photonics Technol. Lett.*, **16**:293–295.

23. Yun, S. H., Boudoux, C., Tearney, G. J., and Bouma, B. E. (2003). High-speed wavelength-swept semiconductor laser with a polygon-scanner-based wavelength filter. *Opt. Lett.*, **28**:1981–1983.

24. Li, S., and Chan, K. T. (1998). Electrical wavelength tunable and multiwavelength actively mode-locked fiber ring laser. *Appl. Phys. Lett.*, **72**:1954–1956.

25. Yamashita, S., and Asano, M. (2006). Wide and fast wavelength-tunable mode-locked fiber laser based on dispersion tuning. *Opt. Express*, **14**:9299–9306.

26. Vlachos, K., Bintjas, C., Pleros, N., and Avramopoulos, H. (2004). Ultrafast semiconductor-based fiber laser sources. *IEEE J. Sel. Top. Quantum Electron.*, **10**:147–154.

27. Moon, S., and Kim, D. Y. (2006). Ultra-high-speed optical coherence tomography with a stretched pulse supercontinuum source. *Opt. Express*, **14**:11575–11584.

28. Erkmen, B. I., and Shapiro, J. H. (2006). Phase-conjugate optical coherence tomography. *Phys. Rev. A*, **74**:041601(R).

29. Le Gouet, J., Venkatraman, D., Wong, F. N. C., and Shapiro, J. H. (2009). Classical low-coherence interferometry based on broadband parametric fluorescence and amplification. *Opt. Express*, **17**:17874–17887.

30. Kray, S., Spöler, F., Först, M., and Kurz, H. (2009). High-resolution simultaneous dual-band spectral domain optical coherence tomography. *Opt. Lett.*, **34**:1970–1972.

31. Zumbusch, A., Holtom, G. R., and Xie, X. S. (1999). Three-dimensional vibrational imaging by Coherent Anti-Stokes Raman Scattering. *Phys. Rev. Lett.*, **82**:4142–4145.

32. Potma, E. O., Jones, D. J., Cheng, J.-X., Xie, X. S., and Ye, J. (2002). High-sensitivity coherent anti-Stokes Raman scattering microscopy with two tightly synchronized picosecond lasers. *Opt. Lett.*, **27**:1168–1170.

33. Müller, M., and Schins, J. M. (2002). Imaging the thermodynamic state of Lipid membranes with Multiplex CARS Microscopy. *J. Phys. Chem. B*, **106**:3715–3723.
34. Cheng, J.-X., and Xie, X. S. (2004). Coherent Anti-Stokes Raman Scattering Microscopy: Instrumentation, Theory, and Applications. *J. Phys. Chem.*, **108**:827–840.
35. Wang, H.-W., Le, T. T., and Cheng, J.-X. (2008). Label-free imaging of arterial cells and extracellular matrix using a multimodal CARS microscope. *Opt. Commun.*, **281**(7):1813–1822.
36. Chen, H., Wang, H., Slipchenko, M. N., Jung, Y. K., Shi, Y., Zhu, J., Buhman, K. K., and Cheng, J.-X. (2009). A multimodal platform for nonlinear optical microscopy and microspectroscopy. *Opt. Express*, **17**:1282–1290.
37. Andresen, E. R., Nielsen, C. K., Thøgersen, J., and Keiding, S. R. (2007). Fiber laser-based light source for coherent anti-Stokes Raman scattering microspectroscopy. *Opt. Express*, **15**:4848–4856.
38. Kieu, K., Saar, B. G., Holtom, G. R., Xie, X. S., and Wise, F. W. (2009). High-power picosecond fiber source for coherent Raman microscopy. *Opt. Lett.*, **34**:2051–2053.
39. Zhai, Y. H., Goulart, C., Sharping, J. E., Wei, H., Chen, S., Tong, W., Slipchenko, M. N., Zhang, D., and Cheng, J.-X. (2011). Multimodal coherent anti-Stokes Raman spectroscopic imaging with a fiber optical parametric oscillator *Appl. Phys. Lett.*, **98**:191106.
40. Wong, K. K. Y. (2003). Toward Practical Application of Fiber Optical Parametric Amplifiers in Optical Communication Systems. PhD thesis, Stanford University.
41. Chung, H. S., Lee, M. S., Lee, D., Park, N., and DiGiovanni, D. J. (1999). Low noise, high efficiency L-band EDFA with 980 nm pumping. *IEEE Electron. Lett.*, **35**(13):1099–1100.
42. Massicott, J. F., Wyatt, R., and Ainslie, B. J. (1992). Low noise operation of Er3+ doped silica fibre amplifier around 1.6^{-1} m. *IEEE Electron. Lett.*, **28**(20):1924–1925.
43. Ono, H., Mori, A., Shikano, K., and Shimizu, M. (2002). A low-noise and broad-band erbium-doped tellurite fiber amplifier with a seamless amplification band in the C- and L-bands. *IEEE Photonics Technol. Lett.*, **14**(8):1073–1075.
44. Sun, Y., Sulhoff, J. W., Srivastava, A. K., Zyskind, J. L., Strasser, T. A., Pedrazzani, J. R., Wolf, C., Zhou, J., Judkins, J. B., Espindola, R. P., and Vengsarkar, A. M. (1997). 80 nm ultra-wideband erbium-doped silica fibre amplifier. *IEEE Electron. Lett.*, **33**(23):1965–1967.

45. Ng, L. N., Taylor, E. R., and Nilsson, J. (2002). 795 nm and 1064 nm dual pump thulium-doped tellurite fibre for S-band amplification. *IEEE Electron. Lett.*, **38**(21):1246–1247.
46. Lee, H. S., Jung, E. J., Son, S. N., Jeong, M. Y., Kim, C. S. FDML wavelength-swept fiber laser based on EDF gain medium. Presented in *IEEE 14th OptoElectronics and Communications Conference (OECC)*, paper FA5, (2009).
47. Agrawal, G. P. (1995). *Nonlinear Fiber Optics*, 2nd ed. Academic Press, San Diego, CA.
48. Stolen, R. H., and Ippen, E. P. (1973). Raman gain in glass optical waveguides. *Appl. Phys. Lett.*, **22**(6):276–281.
49. Lin, Q., and Agrawal, G. P. (2003). Vector theory of stimulated Raman scattering and its application to fiber-based Raman amplifiers. *J. Opt. Soc. Am. B*, **20**:1616–1631.
50. Islam, M. N. (2002). Raman amplifiers for telecommunications. *IEEE Sel. Top. Quantum Electron.*, **8**(3):548–559.
51. Namiki, S., and Emori, Y. (2001). Ultrabroad-band Raman amplifiers pumped and gain-equalized by wavelength-division-multiplexed high-power laser diodes. *IEEE Sel. Top. Quantum Electron.*, **7**(1):3–16.
52. Garbuzov, D., Menna, R., Komissarov, A., Maiorov, M., Khalfin, V., Tsekoun, A., Todorov, S., and Connolly, I. 1400–1480 nm ridge-waveguide pump lasers with 1 Watt CW output power for EDFA and Raman amplification. In *Optical Fiber Communication Conference*, March 2001. PD18.
53. Espindola, R. P., Bacher, K. L., Kojima, K., Chand, N., Srinivasan, S., Cho, G. C., Jin, F., Fuchs, C., Milner, V., and Dautremont-Smith, W. (2002). Penalty-free 10 Gbit/s single-channel co-pumped distributed Raman amplification using low RIN 14xx nm DFB pump. *IEEE Electron. Lett.*, **38**(3):113–115.
54. Klein, T., Wieser, W., Biedermann, B. R., Eigenwillig, C. M., Palte, G., and Huber, R. (2008). Raman-pumped Fourier-domain mode-locked laser: analysis of operation and application for optical coherence tomography. Opt. Lett., **33**(23):2815–2817.
55. Kaminow, I. P., and Li, T. Y. (2002). Optical Fiber Telecommunications IV-A (Academic Press).
56. Itoh, M., Shibata, Y., Kakitsuka, T., Kadota, Y., and Tohmori, Y. (2002). Polarization-insensitive SOA with a strained bulk active layer for network device application. *IEEE Photonics Technol. Lett.*, **14**(6):765–767.

57. Durhuus, T., Mikkelsen, B., Joergensen, C., Danielsen, S. L., Stubkjaer, K. E., van den Hoven, G. N., and Tiemeijer, L. F. (1996). All-optical wavelength conversion by semiconductor optical amplifiers. *IEEE/OSA J. Lighwave Technol.*, **14**:942–954.

58. Marhic, M. E. (2007). Fiber Optical Parametric Amplifiers, Oscillators and Related Devices (Cambridge University Press).

59. Marhic, M. E., Kagi, N., Chiang, T.-K., and Kazovsky, L. G. (1996). Broadband fiber optical parametric amplifiers. *Opt. Lett.*, **21**(8):573–575.

60. Marhic, M. E., Park, Y., Yang, F. S., and Kazovsky, L. G. (1996). Broadband fiber-optical parametric amplifiers and wavelength converters with low-ripple Chebyshev gain spectra. *Opt. Lett.*, **21**(17):1354–1356.

61. Yamamoto, T., and Nakazawa, M. (1997). Highly efficient four-wave mixing in an optical fiber with intensity dependent phase matching. *IEEE Photonics Technol. Lett.*, **9**(3):327–329.

62. Marhic, M. E., Wong, K. K.-Y., and Kazovsky, L. G. (2004). Wide-band tuning of the gain spectra of one-pump fiber optical parametric amplifiers *IEEE J. Sel. Top. Quantum Electron.*, **10**:1133.

63. Torounidis, T., Andrekson, P. A., and Olsson, B.-E. (2006). Fiber-optical parametric amplifier with 70-dB gain. *IEEE Photonics Technol. Lett.*, **18**:1194.

64. Zhu, R., Cheng, K. H., Lam, E. Y., Wong, F. N., and Wong, K. K. (2010). Power Enhanced and Fast Swept Source for Phase Conjugate Optical Coherence Tomography. Presented in *Frontiers in Optics*, paper FTuY1.

65. Hill, K. O., Kawasaki, B. S., Fujii, Y., and Johnson, D. C. (1980). Efficient sequence frequency generation in a parametric fiber-optic oscillator. *Appl. Phys. Lett.*, **36**(11):888–890.

66. Serkland, D. K., and Kumar, P. (1999). Tunable fiber-optic parametric oscillator. *Opt. Lett.*, **24**(2):92–94.

67. Nakazawa, M., Suzuki, K., and Haus, H. A. (1989). The modulation instability laser. *IEEE J Quantum Electron.*, **25**(9):2036–2045.

68. Coen, S., and Haelterman, M. (2001). Continuous-wave ultrahigh-repetition-rate pulsetrain generation through modulational instability in a passive fiber cavity. *Opt. Lett.*, **26**(1):39–41.

69. Marhic, M. E., Wong, K. K.-Y., Kazovsky, L. G., and Tsai, T. E. (2002). Continuous-wave fiber optical parametric oscillator. *Opt. Lett.*, **27**:1439–1441.

70. Xu, Y. Q., Murdoch, S. G., Leonhardt, R., and Harvey, J. D. (2009). Raman-assisted continuous-wave tunable all-fiber optical parametric oscillator. *J. Opt. Soc. Am. B*, **26**:1351.

71. Malik, R., and Marhic, M. E. (2010). Tunable Continuous-Wave Fiber Optical Parametric Oscillator with 1 W Output Power. *Optical Fiber Communication (OFC)*, paper JWA18.

72. Marhic, M. E., Wong, K. K. Y., Ho, M. C., and Kazovsky, L. G. (2001). 92% pump depletion in a continuous-wave one-pump fiber optical parametric amplifier. *Opt. Lett.*, **26**(9):620–622.

73. Stolen, R. H., and Bjorkholm, J. E. (1982). Parametric amplification and frequency conversion in optical fibers. *IEEE J Quantum Electron.*, **18**(7):1062–1072.

74. Jung, E. J., Kim, C.-S., Jeong, M. Y., Kim, M. K., Jeon, M. Y., Jung, W., and Chen, Z. (2008). Characterization of FBG sensor interrogation based on a FDML wavelength swept laser. *Opt. Express*, **16**:16552–16560.

75. Kranendonk, L. A., Walewski, J. W., Sanders, S. T., Huber, R. J., and Fujimoto, J. G. (2006). Measurements of Gas Temperature in an HCCI Engine by Use of a Fourier-Domain Mode-Locking Laser. *Laser Applilcations to Chemical, Security and Environmental Analysis*, Technical Digest (Optical Society of America), paper TuB2.

76. Vakoc, B. J., Shishko, M., Yun, S. H., Oh, W. Y., Suter, M. J., Desjardins, A. E., Evans, J. A., Nishioka, N. S., Tearney, G. J., and Bouma, B. E. (2007). Comprehensive esophageal microscopy by using optical frequency-domain imaging (with video). *Gastrointest. Endosc.*, **65**(6):898–905.

77. Mariampillai, A., Standish, B. A., Moriyama, E. H., Khurana, M., Munce, N. R., Leung, M. K. K., Jiang, J., Cable, A., Wilson, B. C., Vitkin, I. A., and Yang, V. X. D. (2008). Speckle variance detection of microvasculature using swept source optical coherence tomography. *Opt. Lett.*, **33**(13):1530–1532.

78. Schmidt-Erfurth, U., Leitgeb, R. A., Michels, S., Povazay, B., Sacu, S., Hermann, B., Ahlers, C., Sattmann, H., Scholda, C., Fercher, A. F., and Drexler, W. (2005). Three-dimensional ultrahigh-resolution optical coherence tomography of macular diseases. *Invest. Ophthalmol. Vis. Sci.*, **46**(9):3393–3402.

79. Adler, D. C., Huber, R., and Fujimoto, J. G. (2007). Phase-sensitive optical coherence tomography at up to 370,000 lines per second using buffered Fourier domain mode-locked lasers. *Opt. Lett.*, **32**(6):626–628.

80. Jeon, M. Y., Zhang, J., Wang, Q., and Chen, Z. (2008). High-speed and wide bandwidth Fourier domain mode-locked wavelength swept laser with multiple SOAs. *Opt. Express,* **16**(4):2547–2554.
81. Cheng, H. Y., Standish, B. A., Yang, V. X. D., Cheung, K. K. Y., Gu, X., Lam, E., and Wong, K. K. Y. (2010). Wavelength-swept spectral and pulse shaping utilizing hybrid Fourier domain modelocking by fiber optical parametric and erbium-doped fiber amplifiers. *Opt. Express,* **18**(3):1909–1915.
82. Korotky, S. K., Hansen, P. B., Eskildsen, L., and Veselka, J. J. (1995). Efficient phase modulation scheme for suppressing stimulated Brillouin scattering. Technical Digest International Conference Integrated Optics and Optical Fiber Communications (Hong Kong), pp. 110–111.
83. Kuo, B. P.-P., Alic, N., Wysocki, P. F., and Radic, S. (2011). Simultaneous Wavelength-Swept Generation in NIR and SWIR Bands Over Combined 329-nm Band Using Swept-Pump Fiber Optical Parametric Oscillator. *J. Lightwave Technol.,* **29**:410–416.
84. Zhang, C., Cheung, K. K. Y., Chui, P. C., Tsia, K. K., Wong, K. K. Y. (2011). Fiber Optical Parametric Amplifier with High-Speed Swept Pump. *IEEE Photonics Technol. Lett.,* **23**:1022–1024.
85. Leonhardt, R., Biedermann, B. R., Wieser, W., and Huber, R. (2009). Nonlinear optical frequency conversion of an amplified Fourier Domain Mode Locked (FDML) laser. *Opt. Express,* **17**(19):16801–16808.
86. Evans, C. L., Potma, E. O., Puoris'haag, M., Cote, D., Lin, C. P., and Xie, X. S. (2005). Chemical imaging of tissue *in vivo* with video-rate coherent anti-Stokes Raman scattering microscopy. *Proc. Natl. Acad. Sci. USA,* **102**:16807.
87. Ganikhanov, F., Carrasco, S., Xie, X. S., Katz, M., Seitz, W., and Kopf, D. (2006). Broadly tunable dual-wavelength light source for coherent anti-Stokes Raman scattering microscopy. *Opt. Lett.,* **31**:1292–1294.
88. Dunn, M. H., and Ebrahimzadeh, M. (1999). Parametric Generation of Tunable Light from Continuous-Wave to Femtosecond Pulses. *Science,* **286**:1513.
89. Sharping, J. E. (2008). Microstructure Fiber Based Optical Parametric Oscillators. *J. Lightwave Technol.,* **26**:2184–2191.
90. Razzari, L., Duchesne, D., Ferrera, M., Morandotti, R., Chu, S., Little, B. E., and Moss, D. J. (2010). CMOS-compatible integrated optical hyper-parametric oscillator. *Nat. Photonics,* **4**:41–45.
91. de Matos, C. J. S., Taylor, J. R., and Hansen, K. P. (2004). Continuous-wave, totally fiber integrated optical parametric oscillator using holey fiber. *Opt. Lett.,* **29**:983–985.

92. Wong, G. K. L., Murdoch, S. G., Leonhardt, R., Harvey, J. D., and Marie, V. (2007). High-conversion-efficiency widely-tunable all-fiber optical parametric oscillator. *Opt. Express,* **15**:2947–2952.
93. Deng, Y., Lin, Q., Lu, F., Agrawal, G., and Knox, W. (2005). Broadly tunable femtosecond parametric oscillator using a photonic crystal fiber. *Opt. Lett.,* **30**:1234–1236.
94. Sharping, J. E., Fiorentino, M., Kumar, P., and Windeler, R. S. (2002). Optical-parametric oscillator based on four-wave mixing in microstructure fiber. *Opt. Lett.,* **27**:1675–1677.
95. Lasri, J., Devgan, P., Tang, R., Sharping, J. E., and Kumar, P. (2003). A microstructure-fiber-based 10-GHz synchronized tunable optical parametric oscillator in the 1550-nm regime. *IEEE Photonics Technol. Lett.,* **15**:1058.
96. Devgan, P., Lasri, J., Tang, R., Grigoryan, V., Kath, W., and Kumar, P. (2005). 10-GHz dispersion-managed soliton fiber-optical parametric oscillator using regenerative mode-locking. *Opt. Lett.,* **30**:528–530.
97. Zhou, Y., Cheung, K. K. Y., Yang, S., Chui, P. C., and Wong, K. K. Y. (2010). Ultra-Widely Tunable, Narrow Linewidth Picosecond Fiber Optical Parametric Oscillator. *IEEE Photonics Technol. Lett.,* **22**:1756–1758.
98. Zhou, Y., Cheung, K. K. Y., Yang, S., Chui, P. C., and Wong, K. K. Y. (2009). Widely-tunable picosecond optical parametric oscillator using highly-nonlinear fiber. *Opt. Lett.,* **34**:989–991.
99. Torounidis, T., Karlsson, M., and Andrekson, P. A. (2005). Fiber Optical Parametric Amplifier Pulse Source: Theory and Experiments. *J. Lightwave Technol.,* **23**:4067–4073.

Chapter 14

Optical Scanning Holography and Sectional Image Reconstruction for Biophotonics

Edmund Y. Lam
Imaging Systems Laboratory, Department of Electrical and Electronic Engineering, The University of Hong Kong, Pokfulam, Hong Kong
elam@eee.hku.hk

Digital holographic microscopy is an interferometric technique that records the wavefront information of an object as a hologram. The recording is achieved with digital image sensors such as a CCD or CMOS camera; computations are then needed to reconstruct the object. In this chapter, we focus on a particular form called the optical scanning holography (OSH). With a single scan in the x–y directions, OSH allows us to capture the entire three-dimensional volume in a digital hologram. The data can then be processed to obtain the individual sections, commonly known as sectional image reconstruction. Here, we introduce the basic principles and modeling of the OSH system, followed by several reconstruction algorithms that vary in terms of sophistication. The application of

Understanding Biophotonics: Fundamentals, Advances, and Applications
Edited by Kevin K. Tsia

ISBN 978-981-4411-77-6 (Hardcover), 978-981-4411-78-3 (eBook)
www.panstanford.com

this system in biophotonics is demonstrated through imaging of biological specimens.

14.1 Digital Holographic Microscopy

A high-resolution and high-speed capture of small volumetric objects, particular dynamic biological specimens, can greatly enhance a scientist's perception of the microscopic world. However, these two goals are often conflicting [1]. For instance, confocal microscopy provides excellent two-dimensional images with high lateral resolution at the expense of a shallow depth due to the inherent physics. A single spot on the sample is illuminated at any given time through a pinhole. The light reflected from the sample is then imaged by the objective back to the pinhole, giving excellent rejection of out-of-focus information, and in turn leading to high-quality images. To obtain a volume imagery, however, sequential acquisition is needed at different axial planes, which is a slow process and unsuitable for objects that might move during the scan [2].

On the other hand, holographic imaging, although currently providing images with a lower lateral resolution, can capture the entire three-dimensional volume during a single x–y scan in a digital hologram. The data can then be processed to obtain the individual two-dimensional sections for viewing and analysis. High-resolution volume imagery of dynamic objects is therefore possible with holographic imaging. Indeed, the first intended application for holograms was in microscopy [3, 4], and their use in three-dimensional microscopy of living biological specimens have long been demonstrated [5].

The advent of digital holography has further facilitated the development of holographic microscopy, as it makes the recording and the processing of the holograms much more flexible [6]. Digital holography can be traced to the work of Goodman and Lawrence in 1967, who reported the first instance of detecting a hologram and reconstructing an image using only electronic means [7, 8]. They processed a 256 $\times$ 256 image recorded on vidicon using a PDP-6 computer, with a fast Fourier transform based reconstruction algorithm that took about five minutes of computation. Since then, several factors have contributed to the phenomenal growth of digital holography: the potential in three-dimensional imaging, the exponential gain in computational power due to Moore's law, the advancement in signal processing techniques, particularly in image recovery and synthesis, and the prevalent use of electronic sensors. Digital holographic systems now have different architectures, such as off-axis hologram [9, 10], in-line hologram with digital filtering [11], and phase-shift hologram [12], and various applications ranging from shape measurement, remote sensing, to encryption [13], but imaging remains the predominant theme [6].

In digital holographic imaging, data capture at only a single two-dimensional plane is necessary. It therefore holds a significant advantage in terms of capturing speed, and is particularly appropriate for imaging fast-moving *in vivo* objects. After the three-dimensional information has been acquired and stored, however, we often want to view individual two-dimensional planes of the object. This step is called sectioning [14]. For confocal microscopy this is a trivial process, as data are originally captured per individual section. For holographic imaging, however, this often involves significant computation using techniques such as deconvolution, which is challenging because of the ill-posed nature of the problem [15, 16]. Furthermore, in reconstructing any given plane, the hologram can be viewed as containing defocused light and many spurious structures from the other sections, reducing the signal-to-noise ratio (SNR) of the resulting sectional image [17]. Depending on the architecture, holographic imaging may also suffer from speckle noise due to the coherent nature of its recording [18].

14.2 Optical Scanning Holography

In this chapter we focus on a specific form of digital holography called the optical scanning holography (OSH), which is a two-pupil system consisting of active optical scanning and optical

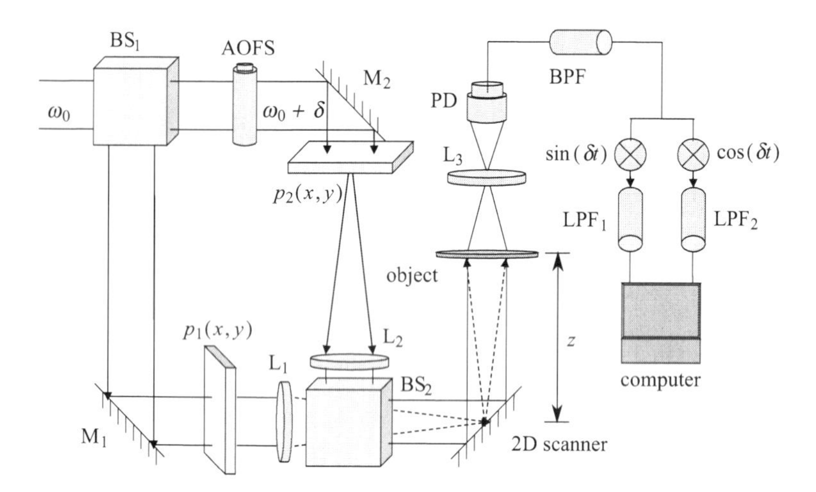

Figure 14.1 The OSH system architecture [21]. All figures reprinted with permission from the Optical Society of America.

heterodyning [19]. Unlike other holographic microscopes, OSH can capture the information of fluorescent specimens in three dimensions, making it a unique candidate for certain biophotonics applications [20].

The operating principle of OSH has been described in detail, for example, in Ref. [20]. Here, our objective is to provide only sufficient description for the readers to appreciate the data-recording power of OSH and the challenges of sectional image reconstruction that necessarily follows the processing of the hologram.

14.2.1 *The Physical System*

Figure 14.1 shows The OSH system architecture used in digital holographic microscopy. Two coherent beams illuminate the system. They are at different temporal frequencies, one at ω_0 (with a corresponding wavelength λ_0) and the other one at $\omega_0 + \delta$, where $\delta \ll \omega_0$. The first beam passes through a pupil represented by $p_1(x, y)$ and then a lens L_1 with focal length f; similarly, the second beam passes through a pupil represented by $p_2(x, y)$ and then a lens L_2 also with focal length f. A beam splitter (denoted BS_2 in the diagram) then combines the light, and a two-dimensional scanner projects it onto an object at a distance z away. Alternatively, the sample rather than the projected pattern can be scanned. A photodetector (PD) then collects the transmitted and scattered light from the object and converts it to an electronic signal. The signal then goes through demodulation, which includes a bandpass filter (BPF) centered at frequency δ, electronic multipliers, and low-pass filters (LPF_1 and LPF_2), resulting in two quadrature electronic holograms.

14.2.2 *Mathematical Representation*

We can model the OSH as a linear space-invariant (LSI) system by fixing the object distance z. The resulting optical transfer function (OTF) is denoted by $\mathcal{H}(k_x, k_y; z)$, where k_x and k_y are the transverse spatial frequency coordinates. Under paraxial approximation, it can be expressed as

$$\begin{aligned}\mathcal{H}(k_x, k_y; z) &= \exp\left\{j\frac{z}{2k_0}\left(k_x^2 + k_y^2\right)\right\} \\ &\times \iint_{-\infty}^{\infty} p_1^*(\hat{x}, \hat{y})\, p_2\left(\hat{x} + \frac{f}{k_0}k_x, \hat{y} + \frac{f}{k_0}k_y\right) \\ &\times \exp\left\{j\frac{z}{f}(\hat{x}k_x + \hat{y}k_y)\right\} d\hat{x}\, d\hat{y},\end{aligned} \tag{14.1}$$

where k_0 is the wave number given by $k_0 = 2\pi/\lambda_0$ [22]. The detailed derivation can be found in Ref. [20]. The inverse Fourier transform of the OTF is the spatial impulse response of the system, which is also known as the point spread function (PSF). It is given by

$$h(x, y; z) = \frac{1}{4\pi^2}\iint_{-\infty}^{\infty} \mathcal{H}(k_x, k_y; z) \exp\left\{j(k_x x + k_y y)\right\} dk_x\, dk_y. \tag{14.2}$$

While Eq. (14.1) is a general expression valid for different pupil functions, we can consider two specific choices:

(1) Let $p_1(x, y) = 1$ and $p_2(x, y) = \delta(x, y)$. The OTF becomes

$$\begin{aligned}\mathcal{H}(k_x, k_y; z) &= \exp\left\{j\frac{z}{2k_0}\left(k_x^2 + k_y^2\right)\right\} \\ &\quad\times \exp\left\{j\frac{z}{f}\left(-\frac{f}{k_0}k_x^2 - \frac{f}{k_0}k_y^2\right)\right\} \\ &= \exp\left\{j\frac{z}{2k_0}\left(k_x^2 + k_y^2\right)\right\}\exp\left\{-j\frac{z}{k_0}\left(k_x^2 + k_y^2\right)\right\} \\ &= \exp\left\{-j\frac{z}{2k_0}\left(k_x^2 + k_y^2\right)\right\}. \qquad (14.3)\end{aligned}$$

This creates an impulse response whose phase is a quadratic function of x and y, known as a Fresnel zone pattern (FZP).

(2) Let $p_1(x, y) = \delta(x, y)$ and $p_2(x, y) = 1$. The OTF becomes

$$\begin{aligned}\mathcal{H}(k_x, k_y; z) &= \exp\left\{j\frac{z}{2k_0}\left(k_x^2 + k_y^2\right)\right\}\exp\left\{j\frac{z}{f}(0 \cdot k_x + 0 \cdot k_y)\right\} \\ &= \exp\left\{j\frac{z}{2k_0}\left(k_x^2 + k_y^2\right)\right\}. \qquad (14.4)\end{aligned}$$

The OTF is again a quadratic phase factor, but with an opposite sign. This is reasonable considering the symmetry between the two paths passing through $p_1(x, y)$ and $p_2(x, y)$, and that the former is at a frequency δ smaller than the latter.

In what follows, assume we take the case (1) above. The free-space spatial impulse response is then given by

$$\begin{aligned}h(x, y; z) &= \frac{1}{4\pi^2}\iint_{-\infty}^{\infty} \exp\left\{-j\frac{z}{2k_0}\left(k_x^2 + k_y^2\right)\right\} \\ &\quad\times \exp\left\{j(k_x x + k_y y)\right\} dk_x\, dk_y \\ &= \frac{1}{4\pi^2}\exp\left\{j\frac{k_0}{2z}(x^2 + y^2)\right\}\iint_{-\infty}^{\infty} \exp\left\{-j\left(\frac{z}{2k_0}k_x^2 - k_x x\right.\right. \\ &\quad\left.\left. + \frac{k_0}{2z}x^2 + \frac{z}{2k_0}k_y^2 - k_y y + \frac{k_0}{2z}y^2\right)\right\} dk_x\, dk_y \\ &= \frac{1}{4\pi^2}\exp\left\{j\frac{k_0}{2z}(x^2 + y^2)\right\} \\ &\quad\times \iint_{-\infty}^{\infty} \exp\left\{-j\frac{z}{2k_0}\left(\tilde{k}_x^2 + \tilde{k}_y^2\right)\right\} d\tilde{k}_x\, d\tilde{k}_y. \qquad (14.5)\end{aligned}$$

The last expression is obtained when we let $\tilde{k}_x$ and $\tilde{k}_y$ be the shifted version of k_x and k_y, respectively. This can be further simplified by noting that

$$\int_{-\infty}^{\infty} \exp\left\{-j\frac{z}{2k_0}\tilde{k}_x^2\right\} d\tilde{k}_x = \int_{-\infty}^{\infty} \cos\left(\frac{z}{2k_0}\tilde{k}_x^2\right) d\tilde{k}_x - j \times \int_{-\infty}^{\infty} \sin\left(\frac{z}{2k_0}\tilde{k}_x^2\right) d\tilde{k}_x,$$

and we can have a closed form because [23]

$$\int_{-\infty}^{\infty} \cos\left(\frac{z}{2k_0}\tilde{k}_x^2\right) d\tilde{k}_x = \int_{-\infty}^{\infty} \sin\left(\frac{z}{2k_0}\tilde{k}_x^2\right) d\tilde{k}_x = \sqrt{\frac{2k_0\pi}{z}} \int_{-\infty}^{\infty} \cos\pi\hat{k}_x^2\, d\hat{k}_x = \sqrt{\frac{k_0\pi}{z}}.$$

Therefore, the impulse response given by Eq. (14.5) can be simplifed to

$$\begin{aligned} h(x, y; z) &= \frac{1}{4\pi^2}\exp\left\{j\frac{k_0}{2z}(x^2+y^2)\right\}\left(\sqrt{\frac{k_0\pi}{z}} - j\sqrt{\frac{k_0\pi}{z}}\right)^2 \\ &= \frac{1}{4\pi^2}\exp\left\{j\frac{k_0}{2z}(x^2+y^2)\right\}\left(-2j\frac{k_0\pi}{z}\right) \\ &= -j\frac{k_0}{2\pi z}\exp\left\{j\frac{k_0}{2z}(x^2+y^2)\right\}, \end{aligned} \qquad (14.6)$$

which is quadratic in x and y.

We are now ready to express the hologram formation mathematically. Assume that we have an object with complex amplitude $a(x, y; z)$. The complex hologram is then given by [20]

$$g(x, y) = \int_{-\infty}^{\infty} \left(|a(x, y; z)|^2 * h(x, y; z)\right) dz. \qquad (14.7)$$

The symbol $*$ denotes a two-dimensional convolution. Alternatively, we can express the right hand side in the spatial frequency domain. Let the Fourier transform of the object's squared intensity be

$\mathcal{A}(k_x, k_y; z) = \mathcal{F}_{xy}\{|a(x, y; z)|^2\}$. Another way to write Eq. (14.7) is therefore

$$g(x, y) = \int_{-\infty}^{\infty} \mathcal{F}_{xy}^{-1}\left\{\mathcal{A}(k_x, k_y; z)\mathcal{H}\left(k_x, k_y; z\right)\right\} dz. \tag{14.8}$$

In numerical computation, it is often necessary to discretize the image formation, hence Eq. (14.7) is also written as

$$g(x, y) \approx \sum_{i=1}^{N} \left(|a(x, y; z_i)|^2 * h(x, y; z_i)\right), \tag{14.9}$$

where the object is assumed to have N sections denoted by $z_1, z_2, \ldots, z_N$.

14.2.3 *Diffraction Tomography*

We can gain further insight into the nature of the data captured by OSH by relating it to diffraction tomography [21]. We first note that the wave number k_0 is related to k_x, k_y, and k_z, which are the x, y and z components of the propagation vector, by

$$k_0 = \sqrt{k_x^2 + k_y^2 + k_z^2}. \tag{14.10}$$

However, $k_x^2 + k_y^2 \ll k_0^2$, because in OSH we assume that the waves are forward propagating along the positive z direction, resulting in much smaller x and y components. Invoking this paraxial approximation, we have

$$k_z = k_0\left(1 - \frac{k_x^2 + k_y^2}{k_0^2}\right)^{1/2} \approx k_0\left(1 - \frac{1}{2}\frac{k_x^2 + k_y^2}{k_0^2}\right) = k_0 - \frac{k_x^2 + k_y^2}{2k_0}. \tag{14.11}$$

Substituting the approximation into Eq. (14.3), we can express the OTF as [24]

$$\mathcal{H}\left(k_x, k_y; z\right) = \exp\left\{-j\frac{z}{2k_0}\left(k_x^2 + k_y^2\right)\right\} \approx \exp\left\{-jz(k_0 - k_z)\right\}. \tag{14.12}$$

We can now express the hologram formation by means of Eq. (14.8). Writing out the inverse Fourier transform explicitly, we get

$$g(x, y) \approx \iiint_{-\infty}^{\infty} \mathcal{A}(k_x, k_y; z) \exp\{-jz(k_0 - k_z)\}$$
$$\times \exp\left\{j(xk_x + yk_y)\right\} dk_x\, dk_y\, dz$$
$$= \iiint_{-\infty}^{\infty} \left(\mathcal{A}(k_x, k_y; z) \exp\{-jzk_0\}\right)$$
$$\times \exp\left\{j(xk_x + yk_y + k_z z)\right\} dk_x\, dk_y\, dz. \quad (14.13)$$

Thus, the complex hologram formation consists of the following steps:

(1) Compute the two-dimensional Fourier transform of the object's squared intensity, i.e., $\mathcal{A}(k_x, k_y; z) = \mathcal{F}_{xy}\{|a(x, y; z)|^2\}$.
(2) Multiply the above by a linear phase factor, $\exp\{-jzk_0\}$.
(3) Compute the three-dimensional inverse Fourier transform of the result above, taking k_x, k_y, and z as the three variables to be integrated.

It is also instructive to consider the Fourier transform of the complex hologram, denoted as $G(k_x, k_y)$. Taking the two-dimensional Fourier transform on both sides of Eq. (14.8), we have

$$G(k_x, k_y) = \int_{-\infty}^{\infty} \mathcal{A}(k_x, k_y; z)\mathcal{H}\left(k_x, k_y; z\right) dz$$
$$\approx \int_{-\infty}^{\infty} \mathcal{F}_{xy}\left\{|a(x, y; z)|^2\right\} \exp\{-jz(k_0 - k_z)\}\, dz$$
$$= \iiint_{-\infty}^{\infty} |a(x, y; z)|^2 \exp\left\{-j\left[xk_x + yk_y + z(k_0 - k_z)\right]\right\}$$
$$dx\, dy\, dz. \quad (14.14)$$

Letting $\widehat{\mathcal{A}}(k_x, k_y, k_z) = \mathcal{F}_{xyz}\left\{|a(x, y; z)|^2\right\}$, i.e., the three-dimensional Fourier transform of the squared magnitude of the object, we can further simplify $G(k_x, k_y)$ as

$$G(k_x, k_y) = \widehat{\mathcal{A}}(k_x, k_y, k_0 - k_z) = \widehat{\mathcal{A}}\left(k_x, k_y, k_0 - \sqrt{k_0^2 - k_x^2 - k_y^2}\right). \quad (14.15)$$

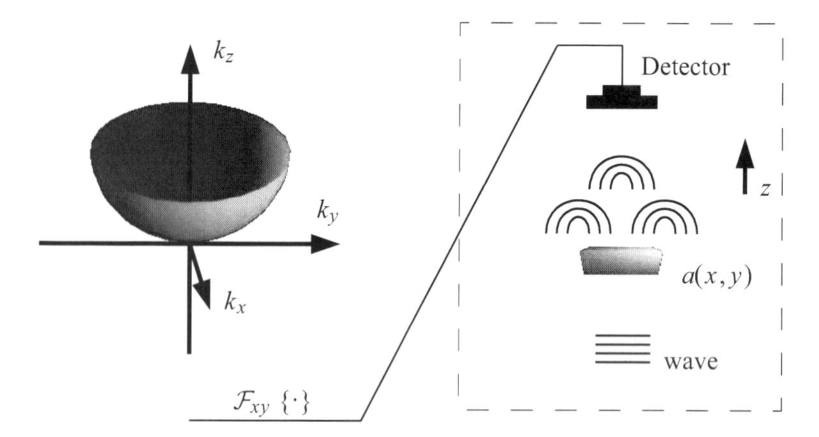

Figure 14.2 A view of the spatial frequency values captured in the complex hologram [21].

Thus, $G(k_x, k_y)$ consists of points of $\widehat{\mathcal{A}}(k_x, k_y, k_0 - k_z)$ equidistant from the coordinates $(0, 0, k_0)$ in the three-dimensional spatial frequency domain. This is a semi-spherical surface depicted in Fig. 14.2.

We can interpret Eq. (14.15) in light of the Fourier diffraction theorem commonly used in diffraction tomography. OSH records a diffracted propagation through the object by a collimated light. The theorem states that when an object is illuminated by a plane wave, the Fourier transform of the diffracted field produces the Fourier transform of the object along the semi-spherical surface in the three-dimensional spatial frequency domain [25]. This generalizes the projection-slice theorem useful for cases such as computed tomography (CT) [23], by taking into account the wave nature of light propagation and therefore incorporating the diffraction effect. However, unlike in CT where multiple slices are acquired and methods such as filtered back-projection can be employed to reconstruct the volumetric data, in OSH we only receive a single semi-spherical slice of data in the spatial frequency domain. The reconstruction problem is therefore severely underdetermined, which explains why regularization is necessary to cope with its ill-posed nature [26], an essential element in the inverse imaging approach to be described in Section 14.3.6.

14.3 Sectional Image Reconstruction

The complex hologram contains the three-dimensional information of an object, but such information is not trivial to interpret visually. Frequently, we would like to reconstruct a cross-section of the object, i.e., $|a(x, y; z)|$ for a specific value of z. This is commonly known as sectional image reconstruction. Below we describe several techniques that have been applied to OSH to date. However, first let us investigate how we can estimate the impulse response of the system from the observed data.

14.3.1 *Identifying the Impulse Response*

The theoretical free-space impulse response, given by Eq. (14.6), is dependent on the depth z. For experimental data, the depth locations where there are objects of interest may be unknown. In that case, the reconstruction of $|a(x, y, z)|$ given $g(x, y)$ in Eq. (14.7) is an instance of blind reconstruction because of the lack of information about the impulse response [27]. A separate identification method, such as the edge-based technique in Ref. [28], may be needed.

Yet, even if the useful values of z are known, another issue is that Eq. (14.6) represents only an ideal impulse response. For instance, we have taken $p_1(x, y) = 1$ and $p_2(x, y) = \delta(x, y)$ in its derivation, but in practice, we have to take into account the finite size of the pupil. In reality, we often use a broad and a narrow Gaussian expressions for these pupil functions respectively [29]. One practical way to estimate the real impulse response is to measure the output for various inputs that resemble an impulse. Here, such an input would be a point in the three-dimensional space [21].

14.3.2 *Conventional Method in Sectioning*

Let us take the discrete version of the complex hologram formation given by Eq. (14.9). Assume that among the N sections in the object $a(x, y; z_i)$, we would like to reconstruct the plane at z_1 from $g(x, y)$. We can make use of the conjugate of the impulse response, where

$$h^*(x, y; z) = h(x, y; -z) = j\frac{k_0}{2\pi z}\exp\left\{-j\frac{k_0}{2z}(x^2 + y^2)\right\}, \quad (14.16)$$

and note that $h(x, y; z) * h^*(x, y; z) = \delta(x, y)$ because

$$\mathcal{F}\left\{h(x, y; z) * h^*(x, y; z)\right\} = \mathcal{H}\left(k_x, k_y; z\right) \mathcal{H}^*\left(-k_x, -k_y; z\right) = 1.$$

Therefore, $h^*(x, y; z)$ is a matched filter of the impulse response. By convolving it with the complex hologram, we have

$$\begin{aligned} g(x, y) * h^*(x, y; z_1) &= \sum_{i=1}^{N} \left(|a(x, y; z_i)|^2 * h(x, y; z_i)\right) * h^*(x, y; z_1) \\ &= |a(x, y; z_1)|^2 + \sum_{i=2}^{N} |a(x, y; z_i)|^2 * h(x, y; z_i) \\ &\quad * h^*(x, y; z_1). \end{aligned} \tag{14.17}$$

The first term above is the target of our reconstruction. The subsequent terms are called the defocus noise, contributed by signals at other sections. We call this reconstruction scheme the conventional method, being the first technique suggested for the sectional image reconstruction [20]. Note also that the first term above is real, while the rest may be complex. The imaginary part of the reconstructed signal is commonly discarded as there is no signal content.

14.3.3 *Sectioning Using the Wiener Filter*

The Wiener filter approach to sectional image reconstruction is designed as an improvement to the conventional method [17]. The major change is that, instead of simply discarding the imaginary part of Eq. (14.17), both the real and the imaginary parts of $g(x, y) * h^*(x, y; z_1)$ are recorded. The square of their magnitudes, after smoothing with rectangular pulses, are labeled $\mathcal{P}_{\mathrm{real}}(k_x, k_y)$ and $\mathcal{P}_{\mathrm{imag}}(k_x, k_y)$ respectively, which are used to estimate the power spectra of the signal and noise. A Wiener filter using the OTF $\mathcal{H}\left(k_x, k_y; z_1\right)$ and these power spectra is given by

$$W(k_x, k_y) = \frac{\mathcal{H}^*\left(k_x, k_y; z_1\right)}{|\mathcal{H}\left(k_x, k_y; z_1\right)|^2 + \dfrac{\mathcal{P}_{\mathrm{imag}}(k_x, k_y)}{\mathcal{P}_{\mathrm{real}}(k_x, k_y) - \mathcal{P}_{\mathrm{imag}}(k_x, k_y)}}. \tag{14.18}$$

This is used to filter the complex hologram $g(x, y)$, giving the reconstructed sectional image $\mathcal{F}_{xy}^{-1}\{G(k_x, k_y)W(k_x, k_y)\}$.

This method has been demonstrated on synthetic data with two sections [17]. However, the computation of the power spectra relies on idealized assumptions (such as the absence of random noise) and the reconstruction results would have significant errors in practical scenarios.

14.3.4 *Sectioning Using the Wigner Distribution*

Time-frequency analysis presents an alternative approach to separate the signal and the defocus noise in Eq. (14.17). Assume that there are only two sections. The Wigner distribution function (WDF), which is a standard time-frequency analysis tool, of the complex hologram consists of three terms: the WDF of the focused signal, the WDF of the defocus signal, and a cross term. Properly chosen filters can then be designed to get rid of or at least suppress the latter two [30].

There are, however, several drawbacks related to the WDF approach. First, the computation load is significant, as time-frequency analysis doubles the dimension of the underlying signal. Second, the cross term causes degradation to the reconstruction quality. This is a common problem in time–frequency analysis, especially for WDF, although alternative time–frequency analysis tools exist that can suppress it better [31]. Nevertheless, the number of cross terms grows with the number of sections (where for N sections, there are $\binom{N}{2}$ cross terms), making them more and more difficult to suppress or ignore. Third, as with the Wiener filter technique, random noise is ignored in the problem formulation, resulting in an important source of error in the practical use of this technique. So far, WDF has only been demonstrated feasible for synthetic two-section objects under ideal conditions.

14.3.5 *Sectioning Using a Random-Phase Pupil*

The preceding methods are purely postprocessing in nature. There is also an attempt for sectional image reconstruction by putting in a random-phase pupil and capturing the hologram twice [32]. Consider, for simplicity, that the object only contains information at a particular depth z_c; the complex hologram, following Eq. (14.8), is given by

$$g_c(x, y) = \mathcal{F}_{xy}^{-1}\left\{\mathcal{A}(k_x, k_y; z)\exp\left\{-j\frac{z_c}{2k_0}\left(k_x^2 + k_y^2\right)\right\}\right\}, \quad (14.19)$$

where we have used the pupil functions $p_1(x, y) = 1$ and $p_2(x, y) = \delta(x, y)$ as done earlier. This is referred to as the coding part.

Now suppose we have a second object, which is a pinhole at depth z_d, and mathematically expressed as $\delta(x, y; z_d)$. This is put into the OSH system, considered the decoding part of the algorithm, but we reverse the two pupils so that $p_1(x, y) = \delta(x, y)$ and $p_2(x, y) = 1$, resulting in an OTF given by Eq. (14.4). The complex hologram is therefore

$$g_d(x, y) = \mathcal{F}_{xy}^{-1}\left\{\exp\left\{j\frac{z_d}{2k_0}\left(k_x^2 + k_y^2\right)\right\}\right\}. \quad (14.20)$$

Convolving the two holograms gives

$$\begin{aligned} g(x, y) &= g_c(x, y) * g_d(x, y) \\ &= \mathcal{F}_{xy}^{-1}\left\{\mathcal{A}(k_x, k_y; z)\exp\left\{j\frac{z_d - z_c}{2k_0}\left(k_x^2 + k_y^2\right)\right\}\right\}. \end{aligned} \quad (14.21)$$

Thus, the single-section object can be recovered if $z_d = z_c$ and that no noise is present. When the object has multiple sections, this method will contain defocus noise. An attempt has been made to transfer the defocus images into "specklelike noise" [32]. In the encoding part, we let

$$p_2(x, y) = \exp\{j2\pi s(x, y)\}, \quad (14.22)$$

where $s(x, y)$ is an independent random function uniformly distributed in the interval [0, 1]. Details of the mathematics are given in Ref. [32].

14.3.6 *Sectioning Using Inverse Imaging*

The inverse imaging approach attempts to recover all sections together by considering them as the "input" to the forward problem, and the observed hologram as the "output" under a known system. It gives possibly the best reconstruction results to date [21, 26].

Using the discrete hologram formation equation in Eq. (14.9), we convert the two-dimensional hologram $g(x, y)$, of size $N_x \times N_y$, into a length $N_x N_y$ vector $\boldsymbol{g}$ by lexicographic ordering. Similarly, $|a(x, y, z_i)|^2$ becomes $\boldsymbol{a}_i$. An $N_x N_y \times N_x N_y$ matrix H_i is formed from $h(x, y; z_i)$ such that $\mathrm{H}_i \boldsymbol{a}_i$ is equivalent to $|a(x, y, z_i)|^2 * h(x, y; z_i)$ after lexicographic ordering.

With these quantities, the complex hologram formation can be written as

$$\boldsymbol{g} = \sum_{i=1}^{N} \mathrm{H}_i \boldsymbol{a}_i + \boldsymbol{e}, \tag{14.23}$$

where $\boldsymbol{e}$ is a vector of length $N_x N_y$ representing the Gaussian random noise. This can be further simplified to

$$\boldsymbol{g} = \mathrm{H}\boldsymbol{a} + \boldsymbol{e}, \tag{14.24}$$

where

$$\mathrm{H} = \left[\mathrm{H}_1\ \mathrm{H}_2 \ldots \mathrm{H}_N\right] \qquad \text{and} \qquad \boldsymbol{a} = \begin{bmatrix} \boldsymbol{a}_1 \\ \boldsymbol{a}_2 \\ \vdots \\ \boldsymbol{a}_N \end{bmatrix}.$$

Since H is known from the system and $\boldsymbol{g}$ is the observed data, the inverse problem of finding $\boldsymbol{a}$ in the above equation can be solved by minimizing the function

$$J(\boldsymbol{a}) = \|\mathrm{H}\boldsymbol{a} - \boldsymbol{g}\|^2 + \mu\|\boldsymbol{a}\|^2, \tag{14.25}$$

where $\|\cdot\|$ denotes the ℓ_2 norm of a vector. The term $\mu\|\boldsymbol{a}\|^2$ represents the regularization to the ill-posed inverse problem, where μ is called the Lagrangian. This minimization has an analytic solution given by

$$\hat{\boldsymbol{a}} = \mathrm{H}^T\left(\mathrm{H}\mathrm{H}^T + \mu\mathrm{I}\right)^{-1}\boldsymbol{g}, \tag{14.26}$$

where I is the identity matrix. Note that the above formulation imposes smoothness on the whole reconstructed image, and therefore a noticeable drawback is that the reconstructed objects often show blurred edges.

It is also possible to use other norms, particularly for the purpose of preserving edge sharpness [21]. We can instead minimize the function

$$J_{\mathrm{TV}}(\boldsymbol{a}) = \|\mathrm{H}\boldsymbol{a} - \boldsymbol{g}\|^2 + \mu\|\boldsymbol{a}\|_{\mathrm{TV}}, \tag{14.27}$$

where $\|\cdot\|_{\mathrm{TV}}$ stands for the total variation (TV) regularization of the underlying two-dimensional image. This is given by

$$\|\boldsymbol{a}\|_{\mathrm{TV}} = \iint_{-\infty}^{\infty} |\nabla a(x, y)|\, dx\, dy,$$

and $\nabla a(x, y) = (\partial a/\partial x, \partial a/\partial y)$ stands for the gradient of $a(x, y)$.

A further refinement to Eq. (14.3.6) is also possible, by noting that $\boldsymbol{a}_i$ contains entries in $|a(x, y; z_i)|^2$, and therefore it must be nonnegative for all pixels. Thus, we can minimize

$$J_{\mathrm{TV}}(\boldsymbol{a}) = \|\mathrm{H}\boldsymbol{a} - \boldsymbol{g}\|^2 + \mu\|\boldsymbol{a}\|_{\mathrm{TV}} \qquad \text{subject to} \qquad \boldsymbol{a} \geq 0, \quad (14.28)$$

where the inequality is taken to be componentwise.

Both Eq. (14.27) and Eq. (14.28) are convex optimization problems [33], and can be solved efficiently. For details, the readers are referred to Ref. [21].

14.4 Experiments

Several OSH sectional image reconstruction results have been reported in the research literature. Here, we highlight two experiments, one with a synthetic object and another one with a biological specimen, and demonstrate the results of the inverse imaging approach.

14.4.1 *Reconstruction Experiment with a Synthetic Object*

A synthetic object as drawn in Fig. 14.3 is being imaged by an OSH system. This object contains two sections of interest: one is a hollow "S" at $z_1 = 87$ cm and another is a hollow "H" at $z_2 = 107$ cm. The difference, Δz, is then 20 cm. The light sources have frequencies 40 MHz and 40.01 MHz. The diameter of the collimated beam is 25 mm and the focal length of the lens is 50 cm, resulting in a small numerical aperture (NA) around 0.025.

The captured complex hologram is shown in Fig. 14.4(a) and (b), where the former is the real part and the latter is the imaginary part. If we apply the conventional reconstruction described in Section 14.3.2, the results at the two sections contain significant

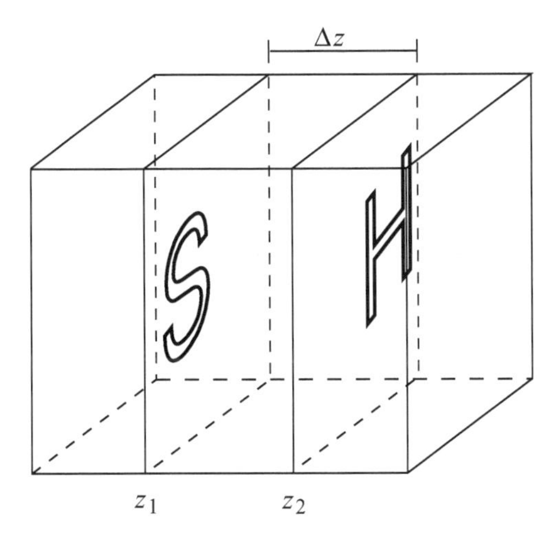

Figure 14.3 A synthetic two-section object for use in an OSH system [21].

defocus noise, as shown in (c) and (d). The noise is significantly suppressed with the more sophisticated inverse imaging sectional reconstruction scheme. The results of such reconstruction are given in (e) and (f).

14.4.2 *Reconstruction Experiment with a Biological Specimen*

We also obtain the complex hologram of a three-dimensional object consisting of a slide of fluorescent beads (DukeR0200, 2 μm in diameter, excitation around 542 nm, emission around 612 nm). The beads tend to stick either to the top or the bottom surface, giving us a test sample with two dominant sections. In this sample, the two planes are separated by around 35 μm [34]. An emission filter and a fluorescence detector are attached to the system to measure the emitted light from the beads.

We need to estimate experimentally the values z_1 and z_2 that correspond to the two sections. This is done by reconstructing the object using all the impulse responses from 70 μm to 125 μm. At this preprocessing stage, the conventional method is used for a fast reconstruction. Figure 14.5 shows the image for two particular

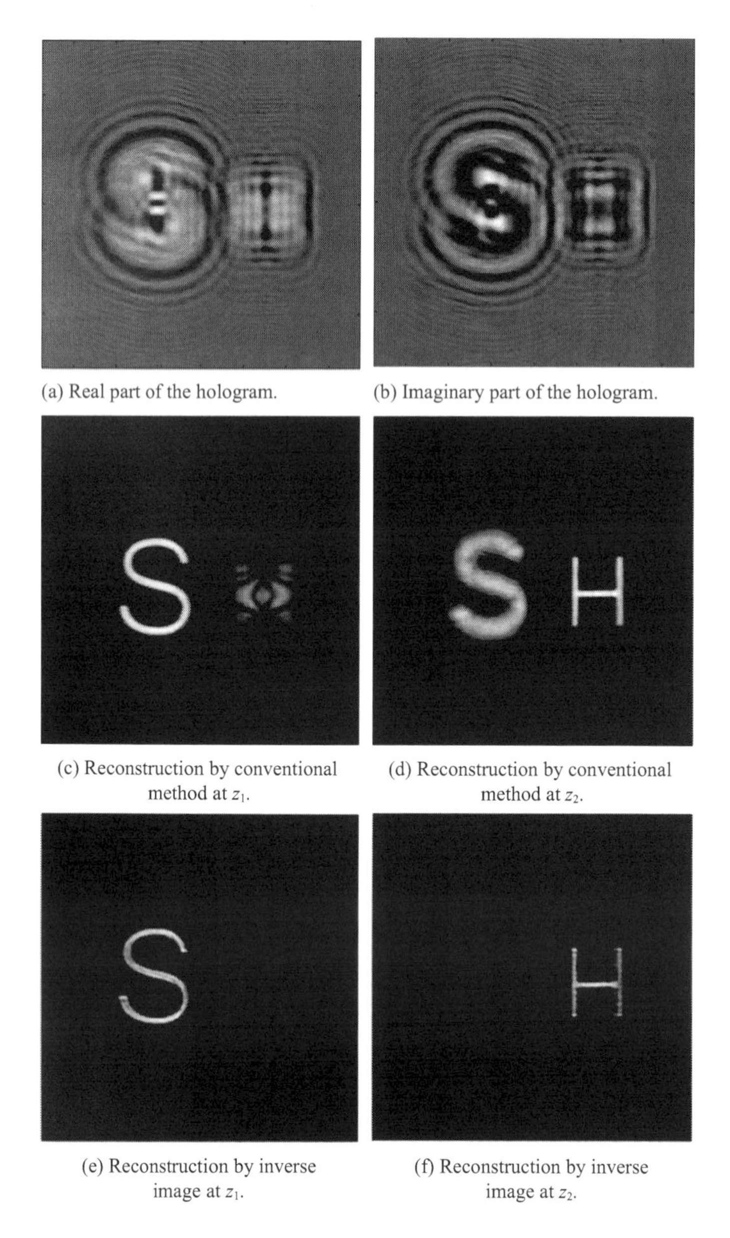

(a) Real part of the hologram.

(b) Imaginary part of the hologram.

(c) Reconstruction by conventional method at z_1.

(d) Reconstruction by conventional method at z_2.

(e) Reconstruction by inverse image at z_1.

(f) Reconstruction by inverse image at z_2.

Figure 14.4 Complex hologram and image reconstruction for a synthetic two-sectional image [21, 28]. The orientation is consistent with the drawing in Fig. 14.3.

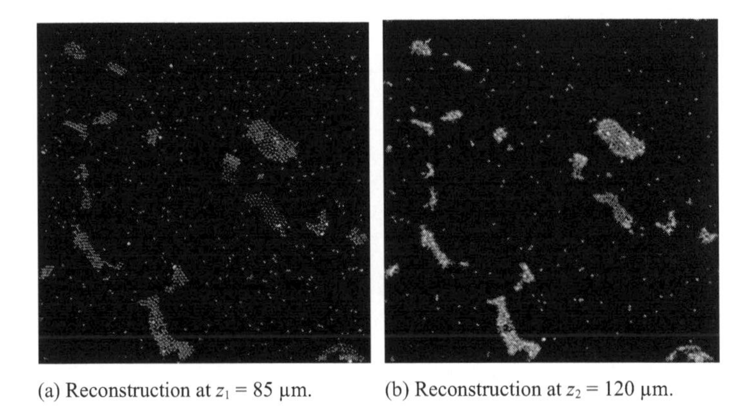

(a) Reconstruction at $z_1 = 85$ μm. (b) Reconstruction at $z_2 = 120$ μm.

Figure 14.5 Reconstructed sections by the conventional method at the two estimated planes [1].

depths, 85 μm and 120 μm, which have the most visible signals in spite of the defocus noise. These are then considered the planes of interest, and at the actual reconstruction step we apply the inverse imaging technique at these planes. The results are shown in Fig. 14.6.

It is observed that defocus noise is present in both sections by the conventional reconstruction method, and is particularly significant at 120 μm. In comparison, it is suppressed in the reconstruction by inverse imaging, and we can observe the beads with good resolution.

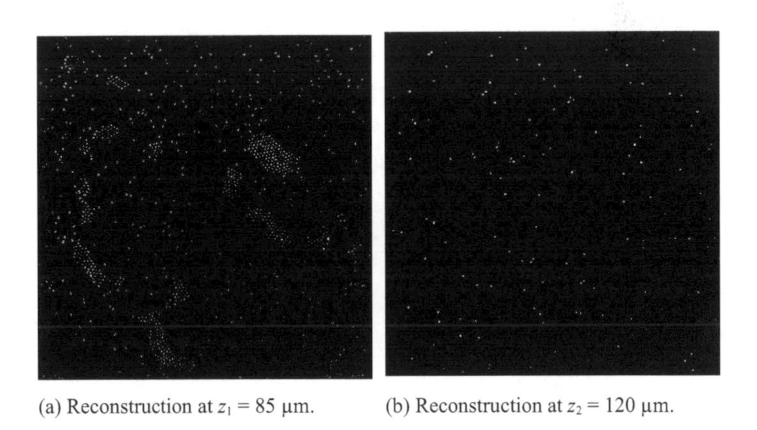

(a) Reconstruction at $z_1 = 85$ μm. (b) Reconstruction at $z_2 = 120$ μm.

Figure 14.6 Reconstructed sections by inverse imaging at the two estimated planes [1].

In fact, although not should in this figure (but interested readers can refer to the online dataset present at Ref. [1]), the beads are present not just at $z_1 = 85$ μm and $z_2 = 120$ μm but also at several other sections. There are mainly two reasons for this phenomenon. The first is that the beads are spherical, not flat, and therefore extend to other sections. The other is that the beads are not completely assembled at the top and bottom surfaces. A few beads may be moving to the surfaces or just staying at certain sections.

14.5 Concluding Remarks

We have shown in this chapter the basic working principle and modeling of the OSH system, and the mathematical algorithms used in sectional image reconstruction. Given its strength in holographic imaging of fluorescent objects, we expect there will be further demonstrations of OSH for biophotonics in the future.

Acknowledgments

This chapter is substantially based on published articles including Refs. [1, 21, 26, 28], and OSA's permission to reuse part of the contents and figures is gratefully acknowledged. The author also wishes to thank Prof. Ting-Chung Poon for a lot of helpful discussions, and Prof. Xin Zhang, who carried out many of the studies described here while a student in my group. This work was supported in part by the Research Grants Council of the Hong Kong Special Administrative Region, China, under Projects HKU 7138/11E, 7131/12E, and N_HKU714/13.

References

1. E. Y. Lam, X. Zhang, H. Vo, T.-C. Poon, and G. Indebetouw. Three-dimensional microscopy and sectional image reconstruction using optical scanning holography, *Applied Optics*. **48**(34), H113–H119 (December 2009).

2. T. R. Corle and G. S. Kino. *Confocal Scanning Optical Microscopy and Related Imaging Systems.* (Academic Press, 1996).
3. D. Gabor. A new microscope principle, *Nature.* **161**, 777 (1948).
4. J. W. Goodman. *Introduction to Fourier Optics,* 3rd ed. (Roberts and Company, 2004).
5. C. Knox. Holographic microscopy as a technique for recording dynamic microscopic subjects, *Science.* **153**, 989 (1966).
6. M. K. Kim. Principles and techniques of digital holographic microscopy, *SPIE Review.* **1**, 018005 (May 2010).
7. J. W. Goodman and R. W. Lawrence. Digital image formation from electronically detected holograms, *Applied Physics Letters.* **11**(3), 77–79 (August 1967).
8. J. W. Goodman. Digital image formation from holograms: early motivations and modern capabilities. In *OSA Topical Meeting in Digital Holography and Three-Dimensional Imaging,* p. JMA1 (June 2007).
9. E. N. Leith and J. Upatnieks. Wavefront reconstruction with diffused illumination and three-dimensional objects, *Journal of the Optical Society of America.* **54**(11), 1295–1301 (November 1964).
10. E. Cuche, P. Marquet, and C. Depeursinge. Simultaneous amplitude-contrast and quantitative phase-contrast microscopy by numerical reconstruction of Fresnel off-axis holograms, *Applied Optics.* **38**(34), 6994–7001 (December 1999).
11. L. Onural and P. D. Scott. Digital decoding of in-line holograms, *Optical Engineering.* **26**, 1124–1132 (November 1987).
12. I. Yamaguchi and T. Zhang. Phase-shifting digital holography, *Optics Letters.* **22**(16), 1268–1270 (August 1997).
13. H. Di, K. Zheng, X. Zhang, E. Y. Lam, T. Kim, Y. S. Kim, T.-C. Poon, and C. Zhou. Multiple-image encryption by compressive holography, *Applied Optics.* **51**(7), 1000–1009 (March 2012).
14. E. N. Leith, W.-C. Chien, K. D. Mills, B. D. Athey, and D. S. Dilworth. Optical sectioning by holographic coherence imaging: A generalized analysis, *Journal of the Optical Society of America A.* **20**(2), 380–387 (February 2003).
15. E. Y. Lam and J. W. Goodman. Iterative statistical approach to blind image deconvolution, *Journal of the Optical Society of America A.* **17**(7), 1177–1184 (July 2000).
16. J. Ke, T.-C. Poon, and E. Y. Lam. Depth resolution enhancement in optical scanning holography with a dual-wavelength laser source, *Applied Optics.* **50**(34), H285–H296 (December 2011).

17. T. Kim. Optical sectioning by optical scanning holography and a Wiener filter, *Applied Optics*. **45**(5), 872–879 (February 2006).
18. J. W. Goodman. *Speckle Phenomena in Optics: Theory and Applications*. (Roberts & Company, Englewood, Colorado, 2007).
19. A. W. Lohmann and W. T. Rhodes. Two-pupil synthesis of optical transfer functions, *Applied Optics*. **17**(7), 1141–1151 (April 1978).
20. T.-C. Poon. *Optical Scanning Holography with MATLAB* (Springer, 2007).
21. X. Zhang and E. Y. Lam. Edge-preserving sectional image reconstruction in optical scanning holography, *Journal of the Optical Society of America A*. **27**(7), 1630–1637 (July 2010).
22. M. Born and E. Wolf. *Principles of Optics*, 7th ed. (Cambridge University Press, 1999).
23. R. N. Bracewell. *The Fourier Transform and Its Applications*, 3rd ed. (McGraw-Hill, 2000).
24. G. Indebetouw and W. Zhong. Scanning holographic microscopy of three-dimensional fluorescent specimens, *Journal of the Optical Society of America A*. **23**(7), 1699–1707 (July 2006).
25. E. Wolf. Three-dimensional structure determination of semi-transparent objects from holographic data, *Optics Communications*. **1**(4), 153–156 (April 1969).
26. X. Zhang, E. Y. Lam, and T.-C. Poon. Reconstruction of sectional images in holography using inverse imaging, *Optics Express*. **16**(22), 17215–17226 (October 2008).
27. Z. Xu and E. Y. Lam. Maximum a posteriori blind image deconvolution with Huber-Markov random-field regularization, *Optics Letters*. **34**(9), 1453–1455 (May 2009).
28. X. Zhang, E. Y. Lam, T. Kim, Y. S. Kim, and T.-C. Poon. Blind sectional image reconstruction for optical scanning holography, *Optics Letters*. **34**(20), 3098–3100 (October 2009).
29. B. D. Duncan and T.-C. Poon. Gaussian beam analysis of optical scanning holography, *Journal of the Optical Society of America A*. **9**(2), 229–236 (February 1992).
30. H. Kim, S.-W. Min, B. Lee, and T.-C. Poon. Optical sectioning for optical scanning holography using phase-space filtering with Wigner distribution functions, *Applied Optics*. **47**(19), 164–175 (April 2008).
31. L. Cohen. *Time Frequency Analysis: Theory and Applications* (Springer-Verlag, 2007).

32. Z. Xin, K. Dobson, Y. Shinoda, and T.-C. Poon. Sectional image reconstruction in optical scanning holography using a random-phase pupil, *Optics Letters*. **35**(17), 2934–2936 (September 2010).
33. S. Boyd and L. Vandenberghe. *Convex Optimization* (Cambridge University Press, 2004).
34. G. Indebetouw. Scanning holographic microscopy with spatially incoherent sources: reconciling the holographic advantage with the sectioning advantage, *Journal of the Optical Society of America A*. **26**(2), 252–258 (February 2009).

Chapter 15

Subcellular Optical Nanosurgery of Chromosomes

Linda Shi,[a] Veronica Gomez-Godinez,[c] Norman Baker,[b] and Michael W. Berns[a,c]

[a] *Institute of Engineering in Medicine and Department of Bioengineering, University of California, CA, USA*
[b] *Department of Electrical and Computer Engineering, University of California, San Diego, La Jolla, CA 92093, USA*
[c] *Beckman Laser Institute and Department of Biomedical Engineering, University of California, Irvine, CA 92697, USA*
zhixiashi@gmail.com

We describe two laser microirradiation optical platforms to study: (i) DNA damage responses following selective damage to a single mitotic chromosome arm and (ii) inhibition of cytokinesis resulting from damage to single and/or multiple chromosome tips (presumptive telomeres) of cells in the anaphase stage of mitosis. For studies on DNA damage a 200 femtosecond near-IR laser and a 12 nanosecond 532 nm second harmonic Nd:YVO_4 laser were used, and for microirradiation of chromosome tips a 532 nm 12 picosecond second harmonic Nd:YAG laser was used. The DNA damage studies showed that the structure of the alteration was the same for both lasers, and very specific chemical sensing and repair pathways were activated in mitosis. This was evidenced by the accumulation of these proteins at the irradiation sites. Irradiation of the chromosome

Understanding Biophotonics: Fundamentals, Advances, and Applications
Edited by Kevin K. Tsia

ISBN 978-981-4411-77-6 (Hardcover), 978-981-4411-78-3 (eBook)
www.panstanford.com

tips resulted in an inhibition or delay in cytokinesis in a significant percentage of cells. This result suggests that the chromosome tip, the location of telomere, may have a regulatory role in the final stage of cell division.

15.1 Introduction

This chapter will focus on recent studies that use microscope-focused laser light to alter submicrometer regions of chromosomes in living cells. The ability to use a focused laser to alter and/or ablate a submicron region of a pre-selected single chromosome [3, 5] or numerous other cell organelles [6, 7] has been the subject of many research articles, reviews, and three books [4, 8, 15, 25].

The fundamental physics of the ablation process is not well understood in single cell biological systems. It can be due to a linear absorption process such as heating or photochemistry, or it could be due to a non-linear process resulting in a micro-plasma, including shock waves, cavitation bubbles, and high electric fields. Or the ablation process may involve a combination of linear and non-linear processes [10]. Though there are good reviews on the subject [24], the physics of the ablation process of an organelle or structure inside a single living cell is not well characterized. Despite the lack of a precise understanding of the physics of subcellular ablation, laser micro- and nanosurgery have been used to alter and study virtually every cell organelle that can be visually resolved with the light microscope. This includes fluorescence using standard fluorophores or genetically expressed fusion proteins, the latter providing the ability to visualize and ablate structures such as a single microtubule. With a 25 nm diameter, a microtubule is well below the resolution of the optical microscope, yet it can be easily cut with a focused laser beam [11, 23].

In this chapter we describe two laser microirradiation optical platforms to study: (i) DNA repair following selective damage to a single mitotic chromosome arm and (ii) inhibition of cytokinesis resulting from damage to single and/or multiple chromosome tips (presumptive telomeres) of cells in the anaphase stage of mitosis.

15.2 Robotic Laser Microscope Systems Design

We have developed several robotic laser microscope systems with remote operational capabilities [10]. Each system is comprised of four key component systems: (i) the microscope and associated imaging systems, (ii) the laser sources, (iii) the array of controlling optical elements, and (iv) the computers and software–hardware interfaces. Here we mainly describe a single laser ablation system using two different laser sources.

15.2.1 *Robotic Laser Microscope Using a Picosecond Green Laser (RoboLase II)*

RoboLase II is comprised of an inverted microscope (Zeiss Axiovert 200M, Germany), external optics to direct the laser beam into the microscope, a CCD digital camera, and hardware-software for control of laser power, specimen stage and microscope stand focus and illumination. The optical setup is shown in Fig. 15.1 and the hardware block diagram is shown in Fig. 15.2. For ablation experiments, either a 40× oil NA 1.3 or 63 × 1.4 plan-apochromat PH3 oil objective is used. Optics external to the microscope guide the lasers into the microscope and provide automated laser power

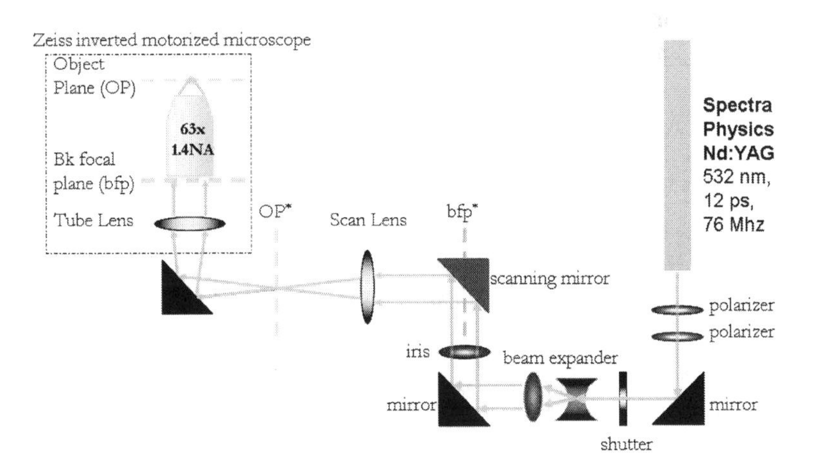

Figure 15.1 Optical design of RoboLase II.

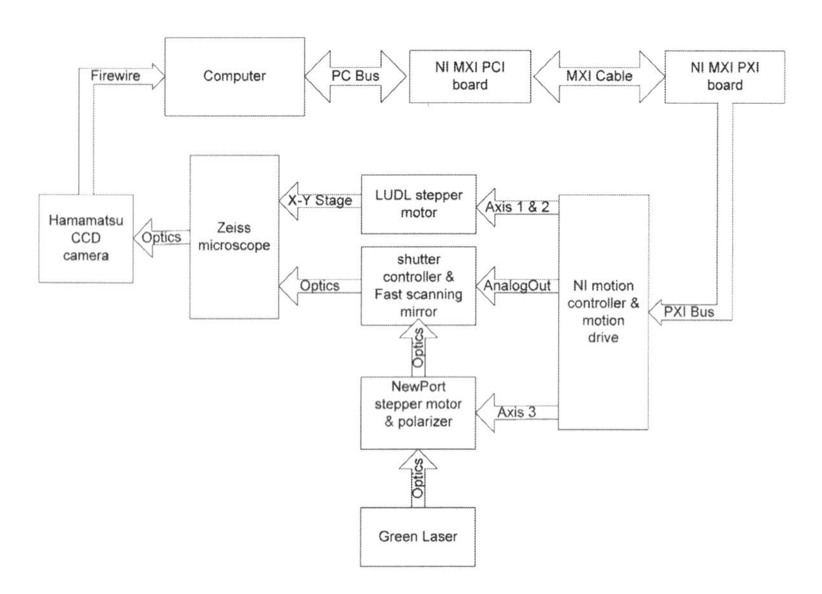

Figure 15.2 Hardware block diagram of RoboLase II.

control, laser shuttering, and laser power monitoring. The ablation laser is a diode-pumped Spectra-Physics Vanguard Nd:YAG with a second harmonic generator (SHG) providing a TEM00 mode 532 nm laser beam that is linearly polarized with 100:1 purity. The laser head emits 12 ps pulses at 76 MHz with a 2 W average power. The laser beam is transmitted through the first of two stationary glan polarizers (CLPA-12.0-425-675, CVI laser, LLC, Albuquerque, NM) in front of the laser cavity. The laser power is controlled by rotating the second polarizer mounted in a motorized rotational mount driven by an open-loop two-phase stepper motor with a 0.05 degree accuracy (PR50PP, Newport Corp, Irvine, CA). The stepper motor rotates the polarizer from its vertical orientation with maximum transmission to its horizontal orientation with minimum transmission well below the required threshold for ablation of cell organelles such as microtubules or chromosomes. The stepper motor is controlled via the flexmotion board in the PXI chassis. A mechanical shutter (Vincent Associates, Rochester, NY) gates the main laser beam to provide "short" (in our studies 30 ms duty cycle) bursts of macro-pulses to the microscope. The laser beam is then expanded using an adjustable-beam expander (2-8×, 633/780/803 nm correction, Rodenstock, Germany) and lowered to a height just above the optical table by two additional mirrors. Telecentric beam steering is achieved by placing a single dual-axis fast scanning mirror (Newport Corp.) at an image plane conjugate to the back focal plane of the microscope objective. This image plane is formed by a 250 mm biconvex lens positioned with its front focal plane at the image plane of the under the microscope *Keller–Berns* port and with its back focal plane at the fast scanning mirror surface. To access the *Keller–Berns* port, the microscope is raised 70 mm above the table via custom-machined metal alloy posts to leave room for a 45 degree mirror, which vertically redirects incident laser light running parallel to the table. Once inside the microscope, the laser light passes through the tube lens and through an empty slot of the reflector turret before entering the back of the objective lens. All external mirrors in the ablation laser light path are virtually loss-less dielectric mirrors optimized for 45 degree reflections of 532 nm S-polarized light (Y2-1025-45-S, CVI Laser LLC). The fluorescence filter wheel contains filter cubes (Chroma Technology Corp., Bellows Falls, VT) that span the range for most commonly used fluorophores. The cubes are removable and can have excitation/absorption/bandpass filter combinations designed for any chromophore that a collaborator might use.

Specimens are mounted in an X-Y stepper stage (Ludl Electronic Products, Hawthrone, NY) controlled with a PXI-7344 stepper motor controller (National Instruments, Austin, TX) and an MID-7604 power drive (National Instruments). The motion board is mounted in a PXI chassis (National Instruments), connected to the host computer through two MXI4 boards (one in the PXI chassis, the other in the host computer) through the MXI-3 fiber-optic cable (National Instruments).

A high quantum efficiency digital camera is used to capture transmitted and fluorescent images. RoboLase implements a Hamamatsu ORCA-AG deep-cooled 1,344 X 1,024 pixel 12-bit digital CCD camera with digital (firewire) output. The ORCA can read out sub regions of the chip for increased frame rates, bin pixels for increased signal to noise, and adjust gain and exposure time to trade off between signal to noise characteristics and arc lamp exposure times.

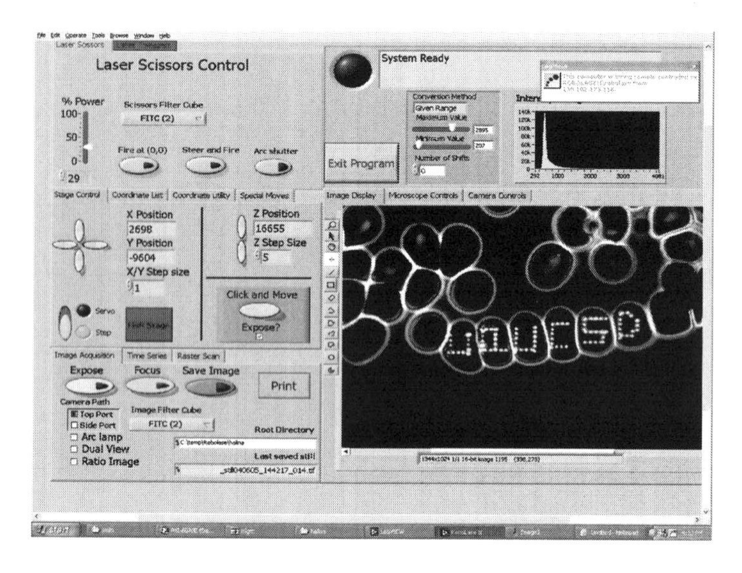

Figure 15.3 Front panel of RoboLase II.

The control software for this system is programmed in the LabVIEW (National Instruments) language and is responsible for control of the microscope, cameras, laser ablation and external light paths. The control software also manages image and measurement file storage. It communicates with the user through the graphic user interface or the "front panel" in LabVIEW. The front panel receives user input and displays images and measurements. The control software interprets commands sent by the user into appropriate hardware calls and returns the results of that action to the front panel and/or computer's hard drive. Emphasis is placed on the design of the front panel, such that it is easy to learn while providing the features needed to search for a cell of interest and then to perform cellular manipulation on that cell. Figure 15.3 shows an image of the front panel of RoboLase II.

15.2.2 *Robotic Laser Microscope with a Femtosecond Near-Infrared (NIR) Laser*

This laser system is built around a Ti:sapphire femtosecond near infrared (NIR) laser (Fig. 15.4). The NIR Ti:sapphire laser is a 76 MHz

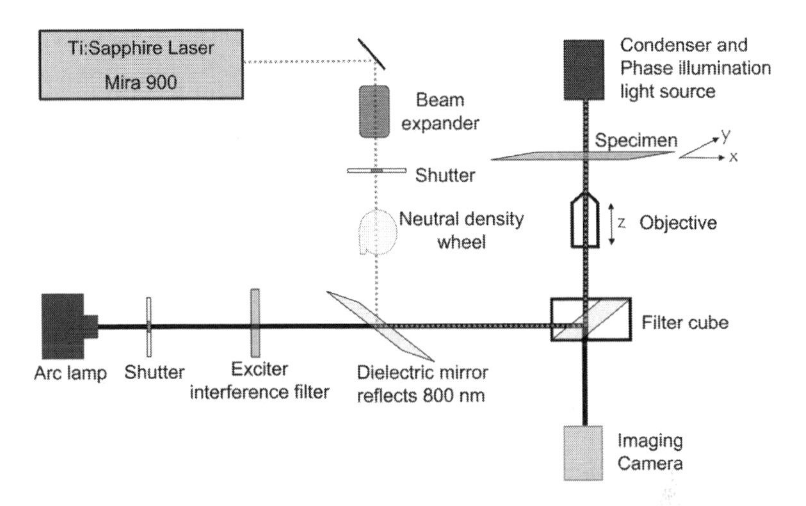

Figure 15.4 Optical design of RoboLase with femtosecond IR laser system [14].

~200 fs beam tunable between 720 and 850 nm (Coherent Mira Model Coherent Inc., Santa Clara, CA). The beam is coupled to the side port of a Zeiss inverted microscope (Axiovert 200M) via use of a series of highly reflective coated mirrors. A single dual-axis fast scanning mirror (Newport Corp.) in the beam path enables precise scanning of the beam in the microscope optical field. The beam is focused via a 63× (1.4 NA) microscope objective to a diffraction-limited spot. For 800 nm, the transmission factor of the Zeiss Plan-Apochromat 63× /1.4 NA objective was determined to be 0.6, using the dual objective method [14]. The pulse energy at the focused spot is varied by a control on the orientation of a Glan-Thompson polarizer mounted on a motorized rotational stage in the beam path. The scanning mirror is controlled by in-house developed software on a LabView platform to create a pattern of any desired geometry in the target cell or organelle. The macropulse exposure time (usually 20–30 milliseconds) of the focused spot is controlled by use of a computer-controlled mechanical shutter (Uniblitz). Since the laser is generating 200 femtosecond pulses at a rate of 76 MHZ, the total number of short micropulses exposing the target is considerable. A Hamamatsu ORCA-AG digital CCD camera with digital (firewire) output is used to capture transmitted and fluorescent images.

The system uses Labview to communicate and control the ORCA camera controller. Software for computer control of all hardware and image acquisition is similar to that described in the previous section.

15.3 Chromosome Studies

15.3.1 *DNA Damage Responses of Mitotic Chromosomes*

The natural occurrence of DNA damage within cells through endogenous metabolic by-products, DNA replication errors and environmental factors such as the sun's harmful UV rays has resulted in organisms evolving the ability to recognize and repair DNA damage. Unrepaired DNA damage can lead to an accumulation of mutations that can be deleterious to tissues through cell senescence, apoptosis or even up-regulated cell divisions. Understanding the responses to DNA damage is critical to understanding aging, developmental defects, and diseases such as cancer. The induction of DNA damage in a specified sub-micron linear region of interphase nuclei using microscope-focused laser beams is becoming a standard method to study cellular responses to DNA damage [9, 13, 16, 25]. However, only a few of these studies have looked at the DNA damage response in mitotic cells [1, 12, 19]. This is primarily due to the inherent difficulties associated with the rounding and eventual detachment of mitotic cells from their growing substrates. The fact that marsupial Ptk2 cells (*Potorous tridactylus)* remain very flat and adhere to the growing surface during mitosis, and that they have only 12 chromosomes, facilitates selective laser targeting [17].

Our recent studies have focused on two aspects of laser induced DNA damage/response in mitotic cells: (i) ultrastructural transmission electron microscope (TEM) characterization of the chromatin alteration produced either by the femtosecond near-IR (NIR), or the nanosecond green lasers, and (ii) immunostaining for the double stand break marker (γH2AX) as well as other proteins associated with (a) sensing of double strand breaks (DSBs) such as Nbs1 and Rad50 and (b) the actual repair of the damaged DNA, by the protein Ku.

Chromosomes observed 1–10 seconds post laser by phase contrast microscopy initially demonstrate paling (a change in refractive index) at the irradiation site (Fig. 15.5a,d). To examine the ultrastructural nature of this change, chromosomes in live cells were exposed to either the femtosecond near-IR (Fig. 15.5a–c)

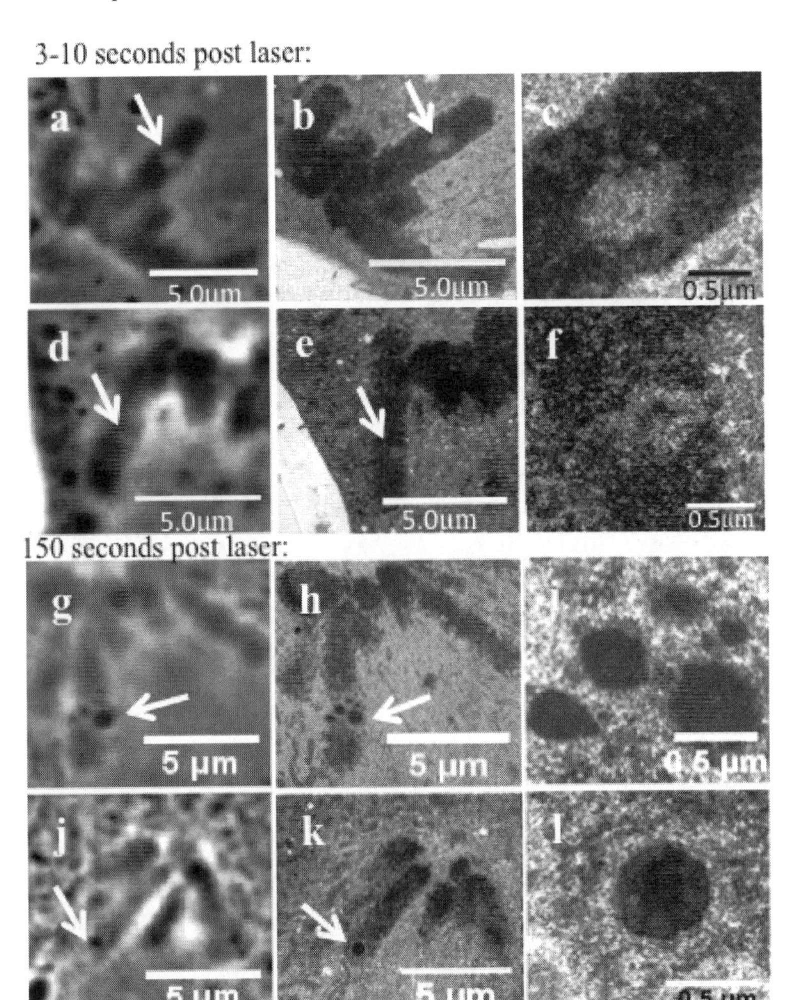

Figure 15.5 Phase (a, d, g, j) and transmission electron micrographs of cells damaged by the femtosecond near-IR (a–c and g–j) and nanosecond lasers (d–f and j–l) [14].

or nanosecond green lasers (Fig. 15.5d–f). They were fixed for TEM 3–10 seconds post laser, serial thin-sectioned, and examined using established TEM procedures. A phase image of an anaphase chromosome (Fig. 15.5a, arrow) shows the paling observed 3 seconds after laser exposure. Low and high magnification electron micrographs of the paled region demonstrate reduced electron density in that region (Fig. 15.5b,c). The diameter and shape of the paled area observed by phase contrast light microscopy is virtually the same as the TEM region showing diminished electron density. The diameter of the paled area in Fig. 15.5a is 0.8 μm, and in the corresponding electron micrograph it is 0.87 μm. An anaphase chromosome damaged by the green nanosecond laser also causes phase paling at the laser irradiation site (Fig. 15.5d, arrow). The corresponding low- and high-magnification electron micrographs (Fig. 15.5e,f) similarly reveal reduction in electron density corresponding to the phase paled region. The diameter of the light microscope phase paling is 0.7 μm (Fig. 15.5d) and for the TEM electron diminished zone it is 0.72 μm (Fig. 15.5e,f). Therefore, it is concluded that the chromosome alterations created by the femtosecond NIR and nanosecond green lasers are similar at the energy and power densities used in these studies: (i) phase-lightened (paled) regions when observed using phase contrast microscopy, and (ii) a diminution of electron density in the same region when observed by TEM [14].

Analysis of the changes occurring in the irradiation sites using immunostaining reveal that both lasers are capable of inducing a DNA DSB response (Fig. 15.6). This is evidenced by positive staining for the DSB marker γH2AX as well as the accumulation of proteins involved in DSB sensing, such as Nbs1 and Rad50. Fig. 15.6h shows a phase image of an anaphase chromosome fixed eight minutes post laser. Immunostaining for Nbs1 (Fig. 15.6j) shows strong fluorescence that appears to co-localize with the dark material when both images are overlapped (Fig. 15.6h). Figure 15.6k demonstrates fluorescence for γH2AX which extends beyond the actual lesion area. However, this is expected as γH2AX is known to spread beyond the damage site resulting in an amplification of the DSB signal [20]. Overlapping the γH2AX fluorescence image with the Nbs1 fluorescence image (Fig. 15.6n) and the phase dark

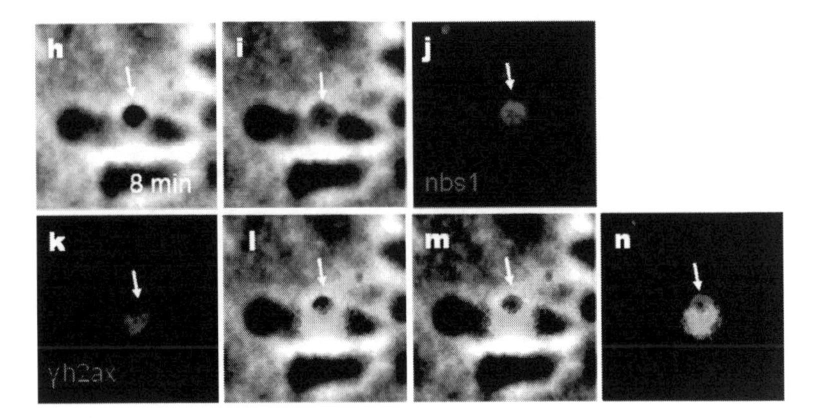

Figure 15.6 The femtosecond and near-IR lasers elicit a double strand break response as demonstrated by yH2AX and Nbs1 immuno-staining. Dark material is observed to co-localize with Nbs1. (h) An anaphase chromosome fixed eight minutes post laser. An arrow depicts the dark material that formed after laser irradiation. (i) Overlay between (j), Nbs1 fluorescence and (h). (j) Nbs1 fluorescence image. (k) γH2AX fluorescence image. (l) Overlay between (h) and (k). (m) Overlap of Nbs1 and γH2AX fluorescence with phase image. (n) Nbs1 and γH2AX fluorescence [14].

material (Fig. 15.6l) demonstrates that γH2AX does not co-localize with either Nbs1 or the phase dark material. Similar results were obtained when the green nanosecond laser was used [14].

To determine whether the DSB response in mitosis involves activation of a repair pathway in addition to DSB sensing, cells were immunostained for Ku (Fig. 15.7b), a heterodimer that is part of the non-homologues end joining (NHEJ) repair pathway. The results show co-localization between Ku and the phase dark material when metaphase cells are fixed and stained three minutes post laser (Fig. 15.7e). These results demonstrate that the near-IR and green nanosecond lasers can be used to elicit a DSB response which results in accumulation of a repair pathway protein in mitotic cells.

The formation of the dark material was identified to be a result of the accumulation of DNA damage response proteins as demonstrated by GFP-Nbs1 accumulation and immunostaining for Nbs1, Rad50, and Ku [14, 18]. The accumulation of the fluorescent fusion protein GFP-Nbs1 to mitotic chromosomes occurred in a

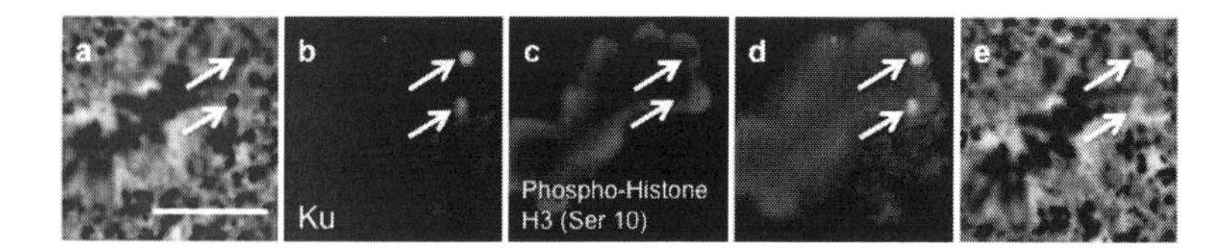

Figure 15.7 Ku, a heterodimer from the non-homologous end joining pathway, can recruit to mitotic chromosomes damaged by the laser. This example is from chromosomes damaged with the near-IR laser; however, the green can elicit the same response. Staining for a mitotic marker, histone H3 phosphorylated on Ser 10, demonstrates a lack of histone protein at the damaged area [14].

similar manner as the formation of the dark material. It increased in diameter and intensity over time. Further, visualization of Nbs1 through immunofluorescence was not observed prior to the formation of dark material. The co-localization of Nbs1 and Ku to the phase dark material supports the TEM observations that the dark material is an aggregation of electron dense material.

In addition to following chromosomes within mitosis, cells with laser-induced mitotic chromosome alterations have been followed into G1 [14]. The phase dark material persists into G1 and there is positive immunostaining for Nbs1 and Ku associated with this material. This result suggests that at least some of the DNA damage produced in mitosis is still being sensed, and repaired in G1 [14].

In summary, the studies reviewed here demonstrate that both the femtosecond near-IR and nanosecond green lasers are capable of creating damage that is in the submicron range which initially appears as a change in phase contrast (phase paling) and is subsequently characterized by the formation of phase dense dark material with a distinct ultrastructural character suggestive of accumulation of material. Fluorescence immunostaining of the irradiated region reveals DNA DSB response proteins, NBS1 and γH2AX, as well as the homologous end joining repair pathway protein Ku. The laser-induced damage produced on mitotic chromosomes was found to persist into G1 where γH2AX, Nbs1, and Ku continued to be associated with the phase dark material initiated in mitosis [14].

15.3.2 *Chromosome Tips (Telomeres) Regulate Cytokinesis*

Genome maintenance is crucial to survival of species and the maintenance of healthy tissues and organs throughout life. Lesions in DNA and subsequent repair pathways have been intensively studied, particularly in G1, S, and G2 [21]. As seen in the previous section, DNA damage response mechanisms during mitosis have only recently been studied [14].

Most of the DNA damage response studies in mitosis have focused on the period of before anaphase onset, prior to the chromosomes initialing movement to respective poles. In those studies, coupling DNA damage and repair to the spindle assembly checkpoint has been suggested [19], however it is not absolutely clear whether these mechanisms are actually related to the sensing of DNA damage [22].

In our recent studies we have continued to use PTK cells because they remain flat during cell division, and they have a relatively small number of large chromosomes (12) whereas most other cell types round-up during mitosis, and the chromosome numbers are generally much larger. Because of the large size and small number of chromosomes, these cells have been used in a significant number of studies employing laser-mediated damage [4, 5].

To investigate the response to DNA damage in mitosis after anaphase is initiated (separation of chromosomes) laser energy from the green picosecond RoboLase II system (see Section 15.2.1, Fig. 15.1) was applied to the chromosome arms, chromosome tips, or cytoplasm (Fig. 15.8A–C respectively) to evaluate the effects of disrupting these structures on progression through mitosis [2].

Laser ablation of the cytoplasmic region distal from the midzone resulted in no discernable morphological changes (Fig. 15.8C). Targeting of chromosome arms resulted in either a severing of the arm and production of a chromatin fragment free from the motion of the chromosome body (Fig. 15.8A), or an optical phase-contrast "paling" (i.e., change in refractive index) in the irradiated region of the chromosome without distinct severing of the chromosome arm was observed (see Fig. 15.8A,D in the previous section). Irradiation of the chromosome tips result loss of chromosome tip structure,

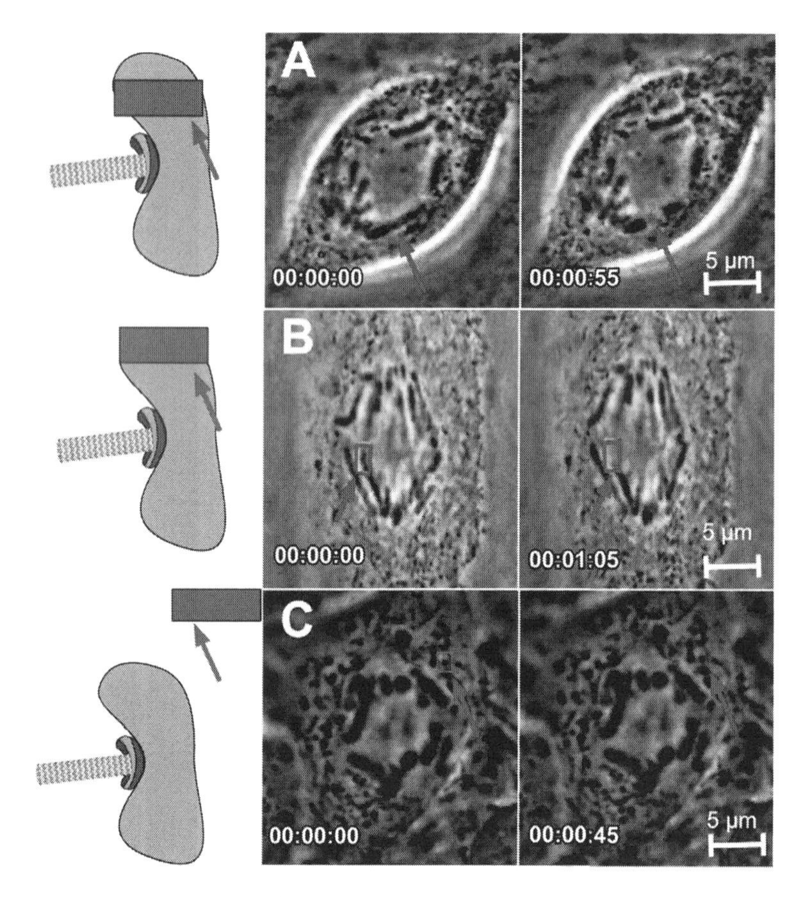

Figure 15.8 Examples of chromosome and cytoplasmic ablations. Left side is a cartoon of different irradiation sites (non-tip and tip chromosome ablation and non-chromosome -cytoplasm) ablation. (A) Arm chromosome ablation in the mid-region between chromosome tip and centromere preserving the distal remnant of the chromosome. (B) Chromosome tip ablation eliminates the distal region of the chromosome with no observable chromosome fragment remaining. (C) Cytoplasmic ablation avoids chromosomes but may be targeted within or outside the mitotic spindle [2].

also indicated by a distinct phase-lightening at the irradiation site (Fig. 15.8B).

Different effects were observed on the progression of anaphase depending on the location of the laser damage (see Table 15.1).

Table 15.1 Timing of mitotic exit for anaphase laser ablation to cytoplasm, chromosome arm and chromosome tips [2]

Region of ablation	Outcome	*n*	% total	Mean time (min)	STDEV
Cytoplasm distal to midzone	Cell divides	38	100	24	4
Cytoplasm midzone	Cell divides	33	100	49	11
Chromosome (non-tip)	Cell divides	76	100	26	7
Chromosome tip (single)	No furrow	16	16		
	Furrow regression	12	12	88	25
	Delay in division	39	40	70	15
	Cell divides	32	32	27	5
	Total cells	99			
Chromosome tip (multiple)	No furrow	32	37		
	Furrow regression	16	19	101	32
	Delay in division	21	24	79	14
	Cell divides	17	20	30	4
	Total cells	86			

In cells with selective focal laser damage to non-tips of the chromosomes (non-telomere regions) normal timing of cell division including completion of cytokinesis was observed ($N = 95$) (Fig. 15.9b). However, targeting of one chromosome tip resulted in 68% of cells with disrupted mitotic progression. Only 32% of the tip-irradiated cells underwent cytokinesis similar to control non-tip-abated cells. The remainder of tip-irradiated cells were divided into three categories: (1) cells that did not initiate a furrow after more than two hours post-irradiation (16%), (2) cells that had furrow regression after initiating cytokinesis (12%), or (3) cells with delayed furrow formation followed by cytokinesis (30%).

Chromosome tip ablation also impacted the duration of cell division (Fig. 15.9A). While >90% of control cells including non-irradiated and cells subjected to cytoplasmic ablation distal from the midzone or chromosome arms completed cytokinesis within 30 minutes, 68% of cells in which chromosome tips were ablated did not complete cytokinesis within this time frame. Of the cells that were delayed, approximately half exhibited an extended delay, particularly in the earliest stages of furrow initiation (Fig. 15.9B, blue). These observations indicate that, unlike damage to a

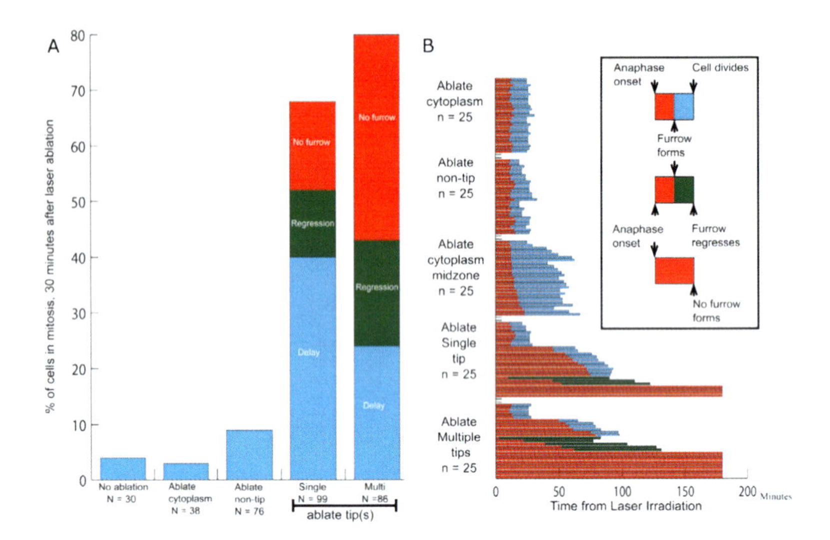

Figure 15.9 (A) Chromosome tip ablation delays exit from mitosis. Control mitotic cells or cytoplasm and arm laser ablation treated cells predominantly exit mitosis within 30 minutes of anaphase onset. However, single or multiple chromosome tip ablations result in a dramatic increase in delayed mitotic exit (red, no furrow formed; green, furrow regression; and blue, normal cytokinesis exit). (B) Timeline for control and chromosome tip ablation. Time histograms of 25 representative cells each for cytoplasmic ablation distal from the midzone, chromosome arm ablation, cytoplasmic midzone ablation, and single and multiple chromosome tip ablation. The inset-rectangle on the right side of the figure defines the transitions as well as the beginning and end-points of the cell data in the figure. Transitions are based on morphological criteria of anaphase onset, furrow formation, furrow regression and successful cell division [2].

chromosome arm, damage to the tip, the presumptive telomere, negatively impacts progression through cytokinesis.

To examine whether there is a quantitative relationship between the number of tips (telomeres) damaged and the inhibition of cytokinesis, experiments were conducted in which more than one tip was irradiated. In these experiments 56% of the cells subjected to multiple (two or three) tip ablations (Table 15.1) either did not form a furrow or had furrow regression after it was formed. Since this is more than a two-fold increase compared to single tip ablation, the results demonstrate that multiple tip ablations increase the proportion of cells with disrupted cytokinesis. Notwithstanding, this increase of cytokinetic defects could be due to inaccuracy in laser targeting, whereas multiple laser exposures increases the probability of successfully hitting a tip. On the other hand, multiple tip ablations may result in a cumulative effect, resulting in an increase in the frequency of cytokinesis defects.

In summary, the results of laser tip ablation implicate telomeres as regulatory sites for cytokinesis and suggest a telomere-based signaling pathway that couples post-segregation chromosome damage to completion of cell division. This pathway is possibly linked to DNA repair; however, the existence of other telomeric-specific protein damage/repair pathways cannot be excluded.

15.4 Conclusions

This review has focused on the genetic-bearing structure, the chromosome, because it has been studied more than any other cell structure using laser microirradiation. Recent advances in understanding DNA damage and repair have been possible using this approach and will likely expand in the future. In addition, the studies on ablation of chromosome tips are particularly relevant in light of the recent high interest in the function of telomeres and their relation to both ageing and oncogenesis. Finally, for the bioengineer and the biologist, we have shown that laser nanosurgery is not necessarily laser-specific. In other words, with carefully chosen irradiation parameters, such as wavelength, pulse duration, and energy delivery, similar biological and structural effects can be produced. However, it is important for all investigators to provide details on the exact lasers and irradiation parameters used so that studies can be clearly understood, compared, and replicated.

Acknowledgements

The authors would like particularly to thank Kyoko Yokomori, Jagesh Shah, and Samantha Zeitlin for their contributions to the original work described in this review. They provided key insight and support for those studies. We also acknowledge the Beckman laser Institute Foundation for the financial support to conduct much of this work.

References

1. Anantha, R., Sokolova, E., and Borowiec, J. (2008). RPA phosphorylation facilitates mitotic exit in response to mitotic DNA damage, *Proc. Natl. Acad. Sci.,* **105**(35), pp. 12903–12908.
2. Baker, N. M., Zeitlin, S. G., Shi, L. Z., Shah, J., and Berns, M. W. (2010). Chromosome tips damaged in anaphase inhibit cytokinesis, *PLoS One,* **5**(8), pp. e12398.
3. Berns, M. W., Olson, R. S., and Rounds, D. E. (1969). *In vitro* production of chromosomal lesions using an argon laser microbeam, *Nature,* **221**, pp. 74–75.
4. Berns, M. W. (1974). *Biological Microirradiation*, Prentice Hall Series on Biological Techniques, Englewood Cliffs, USA.
5. Berns, M. W. (1974). Directed chromosome loss by laser microirradiation, *Science*, **186**, pp. 700–705.
6. Berns, M. W., Aist, J., Edwards, J., Strahs, K., Girton, J., McNeill, P., Rattner, J. B., Kitzes, M., Hammer-Wilson, M., Liaw, L. H., Siemens, A., Koonce, M., Peterson, S., Brenner, S., Burt, J., Walter, R., van Dyk, D., Coulombe, J., Cahill, T., and Berns, G. S. (1981). Laser microsurgery in cell and developmental biology, *Science*, **213**, pp. 505–513.
7. Berns, M. W., Wang, Z., Dunn, A., Wallace V., and Venugopalan, V. (2000). Gene inactivation by multiphoton-targeted photochemistry. *Proc. Natl. Acad. Sci.*, **97**, pp. 9504–9507.
8. Berns, M. W. (2007). A history of laser scissors (microbeams), *Laser Manipulation of Cells and Tissues*, ed. Berns, M. W., and Greulich, K. O., Academic Press, San Diego, Methods in Cell Biology, **82**, pp. 3–58.
9. Botchway, S. W., Reynolds, P., Parker, A. W., and O'Neill, P. (2010). Use of near infrared femtosecond lasers as sub-micron radiation microbeam for cell DNA damage and repair studies, *Mutat. Res.*, **704**, pp. 38–44.
10. Botvinick, E. L., Berns, M. W. (2005). Internet-based robotic laser scissors and tweezers microscopy, *Microsc. Res. Tech.*, **68**(2), pp. 65–74.
11. Colombelli, J., Reynaud, E. G., Rietdorf, J., Pepperkok, R., and Stelzer, E. H. (2005). In vivo selective cytoskeleton dynamics quantification in

interphase cells induced by pulsed ultraviolet laser nanosurgery, *Traffic (Oxford, U.K.)* **6**(12), pp. 1093–1102.

12. Giunta, S., Belotserkovskaya, R., and Jackson, S. P. (2010). DNA damage signaling in response to double-strand breaks during mitosis, *J. Cell Biol.*, **190**(2), pp. 197–207.
13. Gomez-Godinez, V., Wakida, N. M., Dvornikov, A. S., Yokomori, K., Berns, M. W. (2007). Recruitment of DNA damage recognition and repair pathway proteins following near-IR femtosecond laser irradiation of cells, *J. Biomed. Opt.*, **12**(2), 020505.
14. Gomez-Godinez, V., Wu, T., Sherman, A. J., Lee, C. S., Liaw, L. H., Zhongsheng, Y., Yokomori, K., and Berns, M. W. (2010). Analysis of DNA double-strand break response and chromatin structure in mitosis using laser microirradiation, *Nucleic Acids Res.*, **38**, e202.
15. Greulich, K. A. (1999). *Micromanipulation by Light in Biology and Medicine*, Birkhausr Verlag, Basel.
16. Kong, X., Mohanty, S. K., Stephens, J., Heale, J. T., Gomez-Godinez, V., Shi, L. Z., Kim, J. S., Yokomori, K., and Berns, M. W. (2009). Comparative analysis of different laser systems to study cellular responses to DNA damage in mammalian cells, *Nucleic Acids Res.*, **37**, e68.
17. Liang, H., and Berns, M. W. (1983). Establishment of nucleolar deficient sublines of PtK2 (Potorous tridactylis) by ultraviolet laser microirradiation, *Exp Cell Res.* **144**, pp. 234–240.
18. Mari, P. O., Florea, B. I., Persengiev, S. P., Verkaik, N. S., Bruggenwirth, M., Modesti, M., Giglia-Mari, G., Bezstarosti, K., Demmers, J. A., Luider, T. M., Houtsmuller, A. B., and van Gent, D. C. (2006). Dynamic assembly of end-joining complexes requires interaction between Ku70/80 and XRCC4, *Proc. Natl. Acad. Sci.*, **103**, pp. 18597–18602.
19. Mikhailov A., Cole R. W., Rieder, C. L. (2002). DNA damage during mitosis in human cells delays the metaphase/anaphase transition via the spindle-assembly checkpoint, *Curr. Biol.*, **12**, pp. 1797–1806.
20. Rogakou, E. P., Boon, C., Redon, C., and Bonner, W. M. (1999). Megabase chromatin domains involved in DNA double-strand breaks in vivo, *J. Cell Biol.*, **146**, pp. 905–916.
21. Shrivastav, N., Li, D., and Essigmann, J. M. (2010). Chemical biology of mutagenesis and DNA repair: cellular responses to DNA alkylation. *Carcinogenesis,* **31**, pp. 59–70.
22. Skoufias, D., Lacroix, F., Andreassen, P., Wilson, L., and Margolis, R. (2004). Inhibition of DNA decatenation, but not DNA damage, arrests cells at metaphase, *Mol Cell*, **15**, pp. 977–990.

23. Wakida, N. M., Lee, C. S., Botvinick, E. L., Shi, L. Z., Dvornikov, A., and Berns, M. W. (2007). Laser nanosurgery of single microtubules reveals location dependent depolymerization rates, *J. Biomed. Opt.*, **12**, pp. 1–8.
24. Vogel, A., Noack, J., Huttman, G., and Paltauf, G. (2005). Mechanisms of femtosecond laser nanosurgery of cells and tissue, *Appl. Phys. B*, **81**(8), pp. 1015–1047.
25. Yokomori, K., and Berns, M. W. (2010). Analysis of DNA double-strand break response and chromatin structure in mitosis using laser microirradiation, *Nucleic Acids Res.*, **38**, e202.

Chapter 16

Optical Transfection of Mammalian Cells

Maciej Antkowiak,[a] Kishan Dholakia,[b] and Frank Gunn-Moore[c]
[a]*School of Medicine, University of St. Andrews, St. Andrews, UK*
[b]*School of Physics and Astronomy, University of St. Andrews, St. Andrews, UK*
[c]*School of Biology, University of St. Andrews, St. Andrews, UK*
ma81@st-andrews.ac.uk

16.1 Introduction

Mammalian cells are studied for two reasons: firstly to comprehend the complex and myriad different biochemical processes that occur within cells and, in so doing develop a deep understanding of these processes so that we correct and protect these events from disease or malfunction, and secondly to modify them in order to produce biological relevant compounds for industry. In both cases, biologists very often wish to introduce nucleic acids to modify the expression of genes (transfection) or introduce new chemical compounds and, as is becoming increasingly more common, physical structures (injection) into mammalian cells. Many differing techniques have been developed over the years to achieve this with various advantages and disadvantages [24, 76]. These

Understanding Biophotonics: Fundamentals, Advances, and Applications
Edited by Kevin K. Tsia

ISBN 978-981-4411-77-6 (Hardcover), 978-981-4411-78-3 (eBook)
www.panstanford.com

methods range from the use of chemicals that either envelop or bind to the membrane impermeable compound and transport the subsequent complex across the membrane, or use a modified viral vector, or more direct physical methods such as electroporation or microinjection. The technique that is used depends on a number of biological parameters such as the nature of the impermeable compound to be introduced into the cell, whether many cells of the same cell type or a select few within a population of different cells are required to be modified, and whether it is possible to do this in a living organism. It is of note that, as recently highlighted, many of these techniques have proven to be limited to a capability of introducing DNA into cells, and other large macromolecules have proven to be difficult [68].

Optical methods that would show these capabilities have been gaining recent interest with a flurry of imaginative approaches. In this chapter we describe many of these new methodologies, which rely upon the inherent advantages of utilizing light: it is eminently controllable both in its energy and shape, it is sterile, provides unparalleled precision, and can be coupled to other optical techniques such as the manipulation of objects.

16.2 Biophysical Mechanisms of Laser-Mediated Membrane Permeabilization and Cell Transfection

16.2.1 *Laser-Mediated Membrane Poration*

In the next sections of this chapter we will show that the term *optical transfection* covers a wide range of techniques in which a light field is used, either directly or indirectly, to cause increased permeability of a cellular membrane. Many types of lasers have been successfully used to disrupt the plasma membrane and it is understood that various light–matter interactions can mediate this interaction. Depending upon the irradiation parameters (wavelength, energy density, pulse duration) and properties of the irradiated molecules (absorption cross section), photons can be absorbed either in a linear or nonlinear process.

Importantly, in biological samples there are no strong endogenous biological absorbers, except for melanin and hemoglobin, that would provide linear absorption in the 350–1100 nm wavelength range efficient enough to mediate a direct microsurgical (vaporization) effect at typical irradiation parameters [64]. However, absorption of photons may result in the excitation of molecules, leading in turn to temperature increase through "electron–phonon" interactions. If energy is deposited at a time scale shorter than the thermal diffusion time (which is typically in the range of 60 ns–1.5 μs), "thermal confinement" can be achieved, leading to thermal injury in the irradiated area. Also, a thermoelastic stress can result from the thermal expansion of the irradiated volume when energy is deposited in pulses shorter than 100 ps [64]. In the UV part of the spectrum the absorbed photons may additionally lead to direct molecular scission resulting in photochemical damage [93]. All these processes may disturb the plasma membrane and increase its permeability.

When a pulsed irradiation beam is used or the beam is tightly focused through a high numerical aperture lens, the energy density at the sample may be high enough to result in enhanced nonlinear absorption processes. First, the multiphoton absorption may significantly enhance absorption at longer, mainly NIR, wavelengths. In this mechanism *n* photons are simultaneously absorbed in a process equivalent to the absorption of a single photon at the *n* times shorter wavelength. Second, at high irradiances the intrinsic "single photon" absorption coefficient may be increased through medium ionization and generation of quasi-free electrons which contribute to photon absorption. If the critical density of quasi-free electrons $\sim 10^{23}$ cm^{-3} is reached, plasma formation occurs resulting in optical breakdown of the medium. A detailed description of multiphoton ionization, quantum tunneling, and avalanche ionization involved in optical breakdown can be found in the review paper by Quinto-Su et al. [64]. It is important to note that different mechanisms dominate depending on the pulse duration and repetition rate of the laser beam. In the nanosecond regime, plasma is formed through avalanche ionization initiated by seed electrons resulting from multiphoton ionization, while in the case of ultrashort (femtosecond) pulses, plasma formation results directly from multiphoton ionization and quantum tunneling.

The relatively long duration of nanosecond pulses enables efficient avalanche ionization resulting in significant heating of the plasma and a high volumetric energy density. Consequently, even when pulse energies close to the ablation threshold are used, the resulting damage from cavitation bubble and related shock wave may be lethal to the cells in vicinity of irradiation spot. In the next section we will show in more detail how this type of pulsed lasers can be used for optical transfection.

A much more gentle and localized cavitation bubble can be obtained with picosecond and, in particular, femtosecond pulsed lasers. While picosecond pulses still to some degree lead to avalanche ionization, femtosecond pulses give a much better control over the density of free electrons and volumetric energy density in the focal volume. Above the breakdown threshold the disruptive mechanical effects in the cellular membrane are highly localized with the cavitation bubbles smaller than 300 nm for a typical 100 fs NIR beam focused through a NA = 0.9 microscope objective [91].

Even more important for optical transfection is the ability of focused femtosecond pulsed irradiation to sustain the creation of free electrons even below the optical breakdown threshold. In this "low-density plasma" regime the highly reactive free electrons cause photochemical damage of biological molecules through generation of reactive oxygen species and photodissociation [92]. This gentle effect is typically accumulated over thousands or millions of pulses by using a high repetition (80 MHz) laser. Figure 16.1 shows the variety of experiments performed with femtosecond lasers in different regimes. It is worth noting that cell transfection is achieved at a fraction (10–30%) of the optical breakdown threshold energy.

16.2.2 *Mechanisms of Optoinjection and Phototransfection*

The cellular plasma membrane is a thin (4–8 nm) lipid bilayer which surrounds the cytoplasm and constitutes a barrier between the cell's interior and the external environment. Although small objects can be endocytosed, larger molecules such as RNA, DNA,

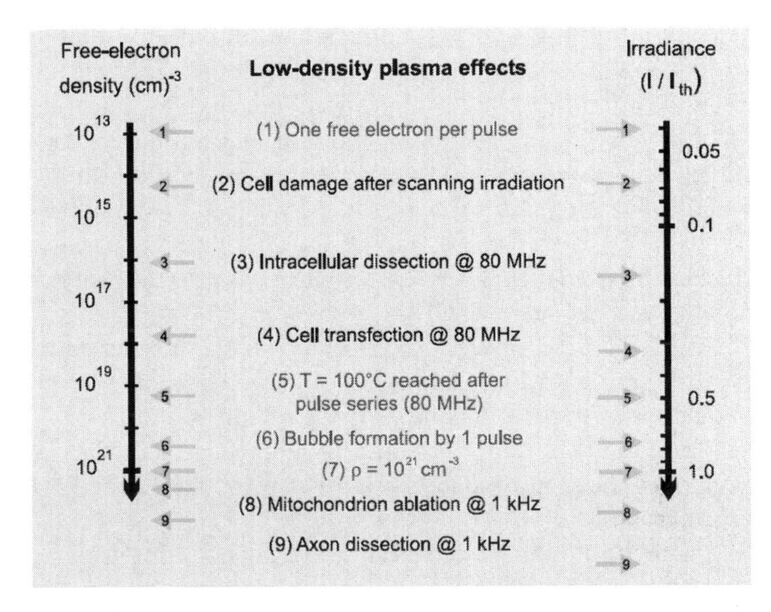

Figure 16.1 A summary of physical breakdown phenomena induced by femtosecond laser pulses. The irradiance values are normalized to the optical breakdown threshold I_{th} defined by a critical electron density of $\rho_{\text{cr}} = 10^{21}$ cm^{-3}. All data refer to plasma formation in water with femtosecond pulses of about 100 fs duration and 800 nm wavelength [92]. Notably, the irradiance used in cell transfection experiments at high pulse repetition rate is on the order of 10–30% of the optical breakdown threshold.

and most fluorophores are excluded. Specifically, as nucleic acids are negatively charged, they are repelled from the negatively charged phospholipids that constitute the plasma membrane. Therefore, for a trans-membrane transition to occur it is necessary to either enclose the DNA into cationic containers, such as lipoplex or polyplex, or permeabalize the membrane by changing its biochemical properties or simply creating physical pores. In this section we explore the optical techniques that enable penetration of naked nucleic acids and other membrane impermeable molecules into the cell's cytoplasm.

Optoinjection of fluorophores and other small molecules is believed to be based on passive diffusion through the opening into the cell cytoplasm in a manner similar to electroporation.

Depending on the size of the molecules, extracellular substances diffuse through a pore in the membrane into the whole cell within a few seconds. The number of optoinjected molecules will thus depend on their extracellular concentration, diffusion coefficient, but more importantly, on the number, lifetime, and size of the membrane pores. As a result, the number of optoinjected molecules is expected to be less controllable than with needle-based microinjection. Indeed this capability was directly measured with the use of femtosecond laser poration by Baumgart et al. [9], where the authors used patch-clamp loading to calibrate the light intensity emitted by a known concentration of the fluorescent dye Lucifer yellow (LY) within the cell's cytoplasm. This allowed the quantification of the concentration of optoinjected molecules, which was found to be approximately 40% of the extracellular concentration.

In addition to passive diffusion, active volume exchange may also occur upon membrane poration if the intra- and extracellular osmolarities are different [34]. Most cells will use a mechanism of regulatory volume increase (RVI) or decrease (RVD) when exposed to a hypo- or hypertonic medium respectively. In such situations, poration of the membrane will result in an efflux or influx of water through the pore, bringing the cell volume to its original value. While efflux of water from the cell should counteract loading, influx of the extracellular medium into the cell can be expected to increase the amount of injected molecules. This was demonstrated by Kohli et al. [34], who measured loading efficiency of 72.3% in Madin-Darby Canine Kidney cells when exposed to an 0.2 M sucrose solution prior to poration. A similar technique involving a hypertonic poration medium was also used in optical transfection of embryonic calli cells of rice (*Oryza sativa*) [26].

In contrast to optoinjection of small molecules, phototransfection with plasmid DNA is a more complex phenomenon. The size of plasmid DNA prevents its efficient diffusion within the cytoplasm. It was shown that DNA molecules larger than 1–2 kbp are practically immobile inside the cell, which is also the case for other large molecules such as dextrans $>$1000 kDa [45, 90]. At the same time, using direct microinjection, it has been estimated that on average only 1 DNA molecule strand out of a 1000, delivered to the cytoplasm, finds its way to the nucleus and can be subsequently transcribed into mRNA [61]. However, only a few (<10) plasmids delivered to the nucleus are necessary for successful gene expression [61].

Although no direct study has been published on the mechanism of plasmid DNA phototransfection, a substantial amount of work has been done on plasmid DNA transfer with electroporation, which also uses naked supercoiled strands of DNA. Golzio et al. [27] used fluorescently tagged DNA to show that within 1 second after membrane electroporation, the DNA strands interact with the outside of the lipid layer resulting in a formation of a localized DNA-membrane aggregate. These aggregates are strongly attached to the membrane and cannot be disrupted by subsequent electroporation with pulses of reversed polarity. The DNA remains embedded within the membrane for up to 30 minutes after which time it starts to diffuse in the cytoplasm. This demonstrates that the mechanism behind DNA phototransfection is most probably not based on a simple passive diffusion into the cell cytoplasm but is multi-step and strongly dependent on the biochemical DNA–pore interaction [25]. A deeper understanding of this process is still required to fully optimize the efficiency and realize the full potential of this transfection technique.

While most of the transfection techniques focus on crossing the cellular membrane, the second important barrier in the plasmid DNA transfection is the nuclear membrane. In contrast to optoinjection of fluorophores and RNA transfection, plasmid DNA has to be ultimately delivered to the cell nucleus for the gene to be transcribed into RNA and expressed. In intact cells even if plasmid DNA is successfully delivered to the cytoplasm it is unlikely to passively cross the nuclear membrane [13] unless the cell undergoes mitosis during which the nuclear membrane is being opened and reorganized. This is reflected in the very low transfection efficiency of post-mitotic nondividing cells using non-viral transfection techniques. In dividing cells it has been shown that the appropriate timing of transfection in the cell cycle may significantly enhance transfection efficiency [14, 44], including when using femtosecond laser optical transfection [51] with S/G2 being the optimal phase for transfection. This is attributed to the fact that the delivered plasmid DNA degrades with time, and if delivery occurs within a few hours before mitosis, this increases the probability of an intact strand of DNA entering the nucleus and being correctly transcribed. Recently to improve transfection efficiency, Praveen et al. [63] performed optical transfection with the addition of a synthetic peptide, Nupherin-neuron (Biomol Research Laboratory), containing a DNA binding and nuclear localization sequences that enhance DNA import into the nucleus. They achieved a threefold enhancement in the transfection efficiency of nonsynchronized adherent CHO-K1 cells and reported, for the first time, successful optical transfection of recently trypsinised cells which are typically put into the G0 phase of the cell cycle.

Another method to improve plasmid DNA uptake into the nucleus would be to porate directly the nuclear membrane. Indeed, a sequential irradiation of both the cellular and nuclear membrane was recently found to improve transfection efficiency in femtosecond laser optical transfection [30]. However, this process has yet to be studied in detail and optimized.

16.3 Optical Transfection with Various Types of Lasers

One of the main advantages of almost all types of optical transfection is their easy integration with standard microscopes. Typically, the beam is combined with the imaging path using a dichroic mirror and delivered to the sample through a microscope objective, as shown in Fig. 16.2.

In approaches that require a tightly focused beam a beam expansion telescope is used to slightly overfill the back aperture of the microscope objective, guaranteeing a diffraction limited focal spot in the sample plane. Alternatively, this telescope can also be combined with a beam shaping element, such as an axicon [20], to sculpt the shape of the irradiation light at the sample. The steering mirror (SM) is relayed to the back focal plane of the objective by a 4f relay telescope, in a configuration similar to holographic optical tweezers [22, 74], and provides aberration free lateral beam steering capabilities. In more sophisticated designs it can be replaced by a fast scanning mirror [21] or a spatial light modulator [3].

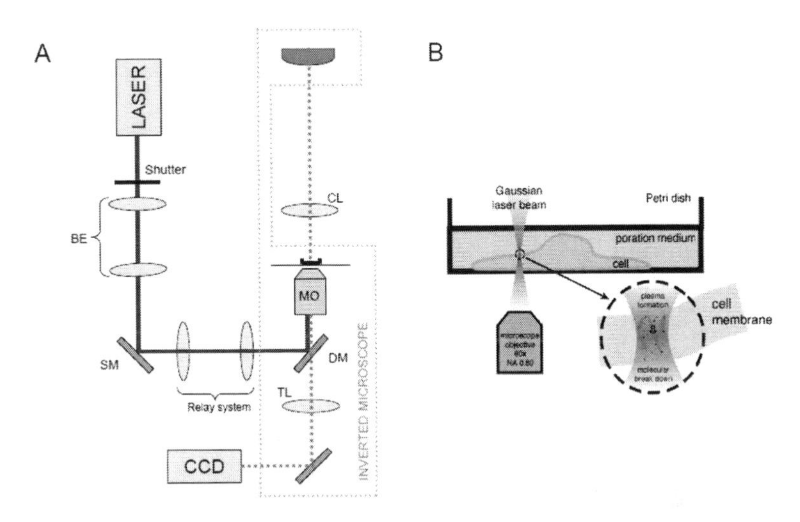

Figure 16.2 Optical transfection techniques are usually easy to integrate with microscopic imaging. (A) The most common configuration for integration of a poration laser beam with an inverted microscope. BM, beam expanding telescope; SM, steering mirror; DM, dichroic mirror; CL, condenser lens; TL, tube lens. (B) Typically the laser beam is focused on the plasma membrane of a targeted cell, In this example a NIR femtosecond beam is focused through a high numerical aperture objective to cause plasma formation. The genetic material or molecules to be optoinjected are present in the extracellular medium.

The microscope based configuration shown in Fig. 16.2 facilitates continuous imaging of the sample using a chosen microscopy modality before, during and after the experiment. In many cases it also enables image guided targeting of selected cells. The commonly used fluorescent imaging, both wide-field and scanned, gives good insight into the transfection process and can be used to assay the efficiency of optoinjection and gene transfection. The typical assays for optoinjection include Lucifer yellow and propidium iodide, with the latter being particularly convenient due to its significant fluorescence enhancement upon binding to intracellular nucleic acids. At the same time viability of irradiated cells is usually verified by using Calcein AM, which becomes fluorescent in healthy cells, and Annexin V conjugates, which indicate early stage apoptosis. Figure 16.3 shows an example of optoinjection experiment in which

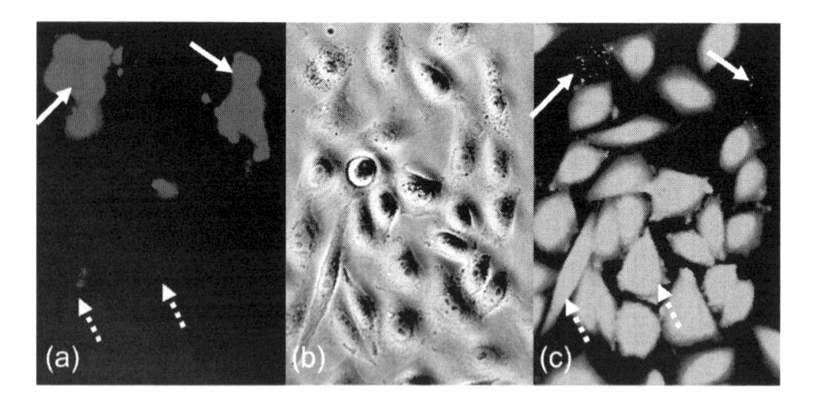

Figure 16.3 Typical images acquired in an optoinjection efficiency and viability study: (a) PI signal 5 minutes after irradiation, (b) phase contrast image 5 minutes after irradiation, (c) Calcein AM 90 minutes after irradiation. Examples of necrotic (solid arrows) and viable cells (dashed arrow) are shown. Reprinted with permission from [3]. Copyright 2010 Wiley-VCH Verlag GmbH & Co. KGaA, Weinheim.

the combination of propidium iodide and Calcein AM was used to quantify optoinjection efficiency and cell viability, respectively. It was also demonstrated that non-fluorescent, label-free techniques such as digital holography microscopy (DHM) can be used to asses cellular response to laser irradiation and predict viability [2, 101].

Transient plasmid DNA transfection is usually assayed by expression of a fluorescent protein, e.g., GFP or dsRED, 24–48 hours post irradiation. These types of plasmid DNA are typically added to the transfection medium at a concentration of around 10 μg/ml and their expression is easy to detect with standard fluorescent microscopes. It is important to note that in dividing cell types, such as the most popular CHO, HEK, or HeLa cells, the population of cells proliferates during this time and in order to accurately assess the number of successfully transfected cells, the number of fluorescing cells after 24–48 hours has to be corrected by the proliferation rate of the culture. For example, let's consider an experiment in which 100 cells were irradiated in a culture with population doubling time of 24 hours. 48 hours post irradiation 50 cells are found to fluoresce giving $50/100*100\% = 50\%$ *uncorrected* transfection efficiency but most probably only 50/(100*2*2)*100% = 12.5% of irradiated cells were successfully transfected. Unfortunately, in many studies it is not clearly stated whether the transfection efficiency has been corrected for the culture proliferation, which makes a direct quantitative comparison between techniques very difficult.

In recent years many types of laser irradiation have been tried on a wide range of cell types and tissue samples. Many authors have investigated how various continuous-wave (CW) and pulsed lasers, working from UV up to NIR wavelengths, interact with biological materials and how this interaction could be used in cell nanosurgery, ablation and which is of particular relevance to optical transfection, plasma membrane poration and permeabilization. A complete review of this field is not within the scope of this chapter; however, there is extensive literature on that subject [19, 92, 93]. In this section we describe the most successful and promising optical transfection approaches with various kinds of lasers. Since the power density of the laser beam defines the balance between linear and nonlinear absorption processes we group the techniques according to the delivered pulse duration.

16.3.1 *Continuous-Wave Lasers*

Historically, the first demonstration of transfection with a continuous-wave light source was provided in 1996 by Palumbo et al. [56]. They have used an argon-ion 488 nm laser tightly focused (100×, NA > 1) to a submicrometer-sized spot on the membrane of NIH-3T3 murine fibroblast cells. This technique has been further studied by Schneckenburger et al. [71], who used a lower NA objective (40×, NA = 0.6) achieving good transient transfection of CHO-K1 cells. A few years later the same laser and a standard confocal scanning head was used by Nikolskaya et al. for spatially defined optoinjection and transfection of plasmid DNA into neonatal rat cardiac myocytes [54]. It is worth noting that the 488 nm argon-ion laser is widely available as a standard element of the confocal scanning systems as the excitation source for the popular FITC/GFP group of fluorophores. This makes this technique an attractive and accessible choice. However, when this laser is not readily available in a confocal system its cost and relatively large footprint may discourage potential users. It was recently shown that a similar system based on a compact and less expensive violet (405 nm) diode laser can be used to produce both stable [57] and transient transfection [84] of plasmid DNA in a range of standard cell lines. With the optimized irradiation time on the order of 1 second and the power at the sample as low as 2 mW the efficiency of transient plasmid DNA transfection may reach 40%, which is comparable with other optical transfection techniques. This diode laser was also used to optically transfect siRNA to obtain cell selective gene knockdown [84]. Importantly, just as in the case of the 488 nm laser, the 405 nm source is a common part of confocal scanning systems used to excite DAPI/Hoechst33258 family of stains.

The mechanism behind the CW laser membrane poration is debatable and various explanations have been proposed. In all published studies authors use media with a high ($\pm$40 μM) concentration of phenol red, a substance which is a standard component of cell culture media used as a pH sensor. It is highly absorptive in the violet-blue part of the spectrum and its presence in the transfection medium has proven to be required to enable cell damage or membrane poration. It was proposed [56] that highly efficient absorption of light results in local heating of the membrane, which may lead to its increased permeability. This hypothesis was backed by Schneckenburger et al. [71] in a study in which the authors used a fluorescent membrane marker 6-dodecanoyl-2-dimethylamino-naphthalene (laurdan) to directly observe the phase transition of the lipids in the membrane from gel phase to a liquid crystalline phase. They observed a ratiometric change in the emission spectrum of the fluorophore which can be related to liquidity of the membrane lipids. However, an independent numerical model suggests [84] that the possible increase in temperature is on the order of 0.02 K, which would not suffice to introduce any thermal effect in the lipid layer. As an alternative explanation, it has been proposed that reactive oxygen species (ROS) produced in a photochemical process can enhance the permeability of the membrane [84]. It seems that at this stage it is difficult to conclusively describe the role of phenol red and the methodology is still open to discussion and further optimization.

16.3.2 *Nanosecond Pulsed Lasers*

In the previous section we described how the linear absorption of light delivered by a CW laser can mediate membrane poration and cell transfection. While this approach is attractive due to a wide availability of light sources and a relatively simple system configuration, it has one significant drawback: the irradiation time is on the order of seconds which significantly limits a potential throughput. To address this issue, many research groups have investigated the applicability of pulsed laser sources to cell nanosurgery and membrane poration. Pulsed lasers are capable of delivering large numbers of photons in short bursts of light, which greatly enhances the efficiency of nonlinear light–matter interactions and enables a variety of damage mechanisms, such as heat deposition and molecule ionization, as described in Section 16.2.

Fundamentally, there are two different mechanisms behind membrane poration with nanosecond lasers. In the most straightforward approach the laser is focused directly on the membrane through a high numerical aperture microscope objective, in a manner similar to femtosecond laser poration described in the next section. However, due the relatively long pulse duration and high pulse energy, the membrane disruption created is much larger and very often results in low cell viability. Using this approach usually only the targeted cell is affected, making this technique truly cell selective. Alternatively, the laser can be focused in a water-based medium, particles in the medium, or glass–medium interface in the proximity of a cell and the shock wave and sheer stress of fluid flow created by medium breakdown and resulting cavitation, permeabilizes the membrane of all neighboring cells.

The most popular and accessible nanosecond pulsed laser source is the Q-switched Nd:YAG emitting at the fundamental wavelength of 1064 nm. Very often the second (532 nm) and third (355 nm) harmonic are used as their absorption coefficients are much higher in biological samples. The very first optical transfection experiment with a nanosecond laser, and at the same time the first laser-mediated gene transfection ever, was reported by Tsukakoshi et al. in 1984 [88], with follow-up work by Kurata et al. [39]. The authors used focused single pulses with a pulse energy of around 1 mJ at

355 nm wavelength to obtain both stable and transient transfection of an Eco-gpt gene in normal rat kidney cells. Importantly, only cells irradiated in the nuclear area expressed the gene with the success rate of around 10% for transient and 0.066% for stable transfection. Stable transfection using this kind of system was also reported by Tao et al. [79], while Shirahata et al. [72] combined it with optical tweezers achieving transfection of non-adherent cells. In addition, plant cells can also be optically transfected with UV nanosecond pulses, which was demonstrated on embryonic calli of rice [26] and wheat [5] (in this case a 308 nm 6 ns pulsed excimer laser was used).

In a similar manner, Mohanty et al. [50] used the fundamental 1064 nm wavelength of an Nd:YAG laser. The beam (pulse energy 150–250 μJ at 1–10 Hz repetition rate) was tightly focused with a high numerical aperture (NA $>$ 1) objective to obtain energy density of 2.4 – 4 $\times$ 10^4 J/cm^2 at the sample. Both fluorophore optoinjection and GFP encoding plasmid DNA transient transfection were achieved.

As mentioned above, a nanosecond laser beam does not necessarily have to be focused directly on the cellular membrane. Soughayer et al. [73] tightly focused single pulses (10 μJ) of the second harmonic (532 nm) beam from a pulsed Nd:YAG laser at the interface of the glass coverslip and the aqueous buffer medium, thereby creating a stress wave capable of permeabilizing surrounding cells. A characteristic pattern of cellular response was observed with cells within a 30 μm radius dead or detached, and a ring of viably optoinjected cells at a distance of around 60 μm from the irradiation spot. This demonstrates the significant gradient of the shear stress inflicted on the cells. A very similar, however much larger, pattern was observed by Rau et al. [66], who focused a 6 ns 532 nm beam in the medium 10 μm above the glass coverslip. Using time-resolved imaging they found that the extent of cell lysis and optoinjection zones depends on the dynamics of the cavitation bubble, which is determined by the relation of the pulse energy to the plasma formation threshold of the medium [65]. As this threshold is relatively high in case of water, the created bubbles and the correlated damage zone cannot be completely minimized [29]. This problem can be alleviated by choosing a different material with a lower breakdown threshold. Arita et al. [4] used 500 nm polystyrene particles which were optically trapped at the desired height above the cell monolayer and ablated by focused 1 μJ pulses. This technique provides both optoinjection as well as transient gene transfection capability with significantly minimized cell lysis and necrosis, being an attractive alternative to laser-induced water breakdown.

The mechanical stress induced by nanosecond laser pulse absorption has also been used by Terakawa et al. [80]. In their approach the nanosecond laser is focused on a black rubber sheet placed in close proximity of the cells to be transfected. High-energy 1 ns pulses of 120 mJ were focused to a spot 3 mm in diameter giving energy density of 1.7 J/cm^2. A relatively low gene transfection efficiency of around 2% was reported; however, this is compensated by the large number of cells affected at once.

Finally, it was shown that to interact with cells a nanosecond beam does not have to be tightly focused so that the energy density can reach the ablation threshold. The second harmonic (532 nm) from a pulsed Nd:YAG laser was successfully used in conjunction with an automated high-throughput cell imaging and processing LEAP™ (Cyntellect) to obtain transfection and optoinjection at rates exceeding 1000 cells/second [21, 67]. The system used a loosely focused (focal spot between and 14 and 56 μm) 0.5 ns pulse at the optimized energy density of 23–60 $nJ/\mu m^2$, and delivered between 1 and 30 pulses to the cells in either a cell-targeted or raster-scanned irradiation mode. A wide range of substances injected into numerous cell types included ions, small molecules, dextrans, siRNA, plasmid DNA, proteins, and semiconductor nanocrystals. The mechanism behind membrane permeabilization in this study is not fully understood as the laser was not tightly focused on the membrane and the energy density was below the photo-ablation threshold, which was confirmed by the absence of bubble formation or shockwave propagation.

In a similar manner, although with a much larger irradiated area, a Ho:YAG laser emitting nanosecond pulses at 2080 nm was used [33, 69], providing a transfection efficiency of up to 41.3%. The beam from this laser, which is popular in endourology, was delivered through a 220 μm fiber, making it potentially attractive in in vivo endoscopic applications.

16.3.3 *Femtosecond Pulsed Lasers*

Femtosecond lasers have been shown to provide the most localized and gentle membrane poration with relatively limited damage and cytotoxicity, in particular when compared to focused nanosecond lasers [102]. A typical system configuration is based on a Ti:Sapphire laser operating at around 800 nm wavelength giving 80–200 fs pulses at 76–80 MHz repetition rate. Since the poration mechanism depends on a multiphoton process, the laser beam is usually focused by a relatively high numerical aperture objective (NA = 0.8–1.3). At a high repetition rate (on the order of MHz) the photochemical damage caused by low density plasma formation is cumulative between the pulses with the typical irradiation time being on the order of 2–100 ms, resulting in a delivery of a train of 10^5–10^6 pulses. Depending on the irradiation time, the mean power of 30–100 mW is usually used corresponding to 0.4–1.2 nJ pulse energy. However, it was demonstrated [89] that a much shorter, sub-20 fs, pulse duration may significantly enhance the efficiency of nonlinear effects, enabling even more delicate poration at a much lower mean power of around 7 mW. Alternative configurations include NIR lasers at low repetition rate on the order of 1 kHz [37], in which case a single pulse is used to porate the cellular membrane. However, due to the high single pulse energy they are thought to be excessively damaging and less suitable for viable cell poration than high repetition rate lasers [37]. Also, in one study a 1554 nm fiber femtosecond laser with a 20 MHz repetition rate was successfully used for both optoinjection and transfection [28]. However, in this case a much longer irradiation time of 5–20 s and the power of 100 mW had to be used to mediate poration, suggesting a lower multiphoton efficiency most probably resulting from a lower energy of single photons at the longer wavelength. Assuming that the low-density plasma formation depends on the ionization of water molecules with an ionization energy of 6.5 eV, absorption of nine photons at 1554 nm is required as compared to five photons at 800 nm.

Studies on the cellular reaction to femtosecond infrared irradiation revealed that, apart from the potential thermomechanical damage related to the poration and gas bubble formation, there are significant photochemical effects within the cell. In particular, creation of reactive oxygen species (ROS) [82], such as H_2O_2, may trigger Ca^{2+} waves [31, 32], activate membrane ion channels and result in hyperpolarization or depolarization of the cellular membrane [1]. Since this type of stress is likely to activate stress pathways in the cell and compromise its long term viability, it should necessarily be taken into account when an experiment is designed. Baumgart et al. showed that addition of an antioxidant, e.g., ascorbic acid, can considerably limit this problem [10]. They also showed that repeated irradiation, even after a few minutes, may notably aggravate ROS production, especially when a laser with kHz repetition rate is used. Importantly, the same group showed that the presence of a fluorophore may provide seed electrons for plasma formation, making the cell damage more severe [38], which means that the laser power may have to be reduced in experiments that involve cell staining or optoinjection of fluorescent markers.

The popularity of femtosecond lasers in optical transfection, in spite of their relatively high cost, was fueled by their wide use in two-photon scanning microscopy, which made them ubiquitous in molecular biology laboratories. Moreover, in many cases the scanning mirrors can be used to precisely position and park the laser spot on the sample, enabling simultaneous fluorescent imaging and phototransfection with the existing hardware and software.

A wide variety of cell types and molecules have been successfully used in phototransfection experiments with femtosecond lasers. Popular fluorophores and molecules like propidium iodide [3, 9, 28, 41, 48, 58, 82], ethidium bromide [89], Lucifer yellow [7, 9, 82], FITC labeled dextran [77], sucrose [34], trypan blue [12], and cascade blue [77] have all been cell-selectively loaded into cell cytoplasm. Mthunzi et al. [51] optoinjected mouse embryonic stem cells with the Gata-6 transcription factor triggering their differentiation into extraembryonic endoderm. With the assistance of optical tweezers McDougall et al. also demonstrated optoinjection of single 100 nm gold nanoparticles [49]. At the same time multiple groups achieved transient transfection of plasmid DNA encoding fluorescent proteins [9, 28, 51, 75, 77, 81, 85-86, 89, 103] and luciferase [103].

The vast majority of experiments were performed on dividing cell lines. This not only greatly simplifies cell culturing and preparation, but also typically enhances the efficiency of transfection as the mitotic cells are more likely allow the plasmid DNA into the nucleus (cf. Section 16.2.2). The only examples of femtosecond laser phototransfection of post-mitotic cells, namely primary rat hippocampal neurons, come from the Eberwine group. First, they successfully phototransfected GFP encoding mRNA, observing expression as early as 30 minutes after poration [7]. Next, they used the subcellular selectivity of phototransfection to introduce mRNA encoding nuclear transcription factor *Elk1* into specific regions of the cells. This enabled them to prove that *Elk1* injected in the dendrites leads to cell death while the cells remained viable when *Elk1* was injected into the soma. In their next work, Sul et al. used femtosecond phototransfection to inject mRNA harvested from an astrocyte, into a post-mitotic neuron obtaining a change in the phenotype of cell into one with astrocytic characteristics [78]. This technique, named transcriptome-induced phenotype remodeling (TIPeR), is a perfect example of a new approach in molecular biology at the single-cell level enabled by optical transfection. In both experiments 16 randomly chosen sites on the membrane were irradiated at a low dose ($P = 35$ mW, $T = 5$ ms), which as supported by the numerical simulation of the transport efficiency through the pores, resulted in enhanced mRNA transfection. It is noteworthy that all these experiments were performed with mRNA and not plasmid DNA, thus avoiding the barrier of the nuclear membrane in the transfection process. Plasmid DNA phototransfection of post-mitotic cells with femtosecond laser remains yet to be shown.

16.3.4 *Nanoparticle-Mediated Techniques*

An interesting recent innovation in optical transfection came with the realization that the laser light does not necessarily have to interact with the cellular membrane directly but can instead be absorbed by nano- or microparticles. Depending on the material properties, the absorbed photons are usually turned into heat, often leading to microscopic cavitation when a pulsed laser is used [43], or their electric field is enhanced by many orders of magnitude by a plasmonic resonance, which is the case for metallic nanoparticles [11]. Usually, this strong but highly localized response to irradiation alleviates the problem of light focusing on the membrane as the position of the particles defines the affected area. Also, typically it is no longer necessary to focus the laser beam since a much lower energy density suffices to mediate membrane poration.

Historically, the first attempts to use nanoparticles for membrane permeabilization were inspired by the idea behind photodynamic therapy in which irradiation of a highly absorbing photosensitizer leads to photochemical damage and death of pre-stained cells. Using cell-type specific antibodies, the nanoparticles can be targeted to the chosen cells [18, 42], e.g., tumor tissue [23], resulting in a cell-type selective therapy even when a nonselective irradiation is applied. Pitsillides et al. [59] demonstrated that 20–30 nm gold nanoparticles and 830 nm iron oxide-doped silica particles can not only kill cells when irradiated, but they also cause viable permeabilization of the cellular membrane to 10 kDa FTIC-dextran. The authors used an unfocused 565 nm laser beam delivering 20 ns pulses at an energy density of 0.35 J/cm^2. Yao et al. optimized this technique further in terms of particle size and irradiation parameters, showing good viable injection with a similar ns pulsed laser but at an energy density of 30 mJ/cm^2 [99, 100]. Recently, the same group reported GFP encoding plasmid DNA transfection using this technique; however, the efficiency was only around 1% [98].

As an extension of this technique, Wu et al. [95] proposed to embed the gold nanoparticles into the glass substrate on which the monolayer of adherent cells is cultured, rather than to add the nanoparticles to the medium. This provides a more controlled and uniform distribution of particles and prevents their endocytosis. In contrast to earlier studies, the authors replaced a scanning laser beam with a patterned illumination defined by a shadow mask. Although the laser and energy density used were similar to previous experiments, their system provided more flexibility and control over the irradiated area.

All the above described experiments suggest that the irradiation of gold nanoparticles with a nanosecond laser creates membrane pores that favor injection of small molecules, e.g., 10 kDa dextran, but does not provide a good efficiency when larger molecules or plasmid DNA are to be injected. Recently, Chakravarty et al. showed that carbon black may be a promising alternative to gold nanoparticles [16]. When activated by a defocused 800 nm femtosecond laser pulses (1 kHz repetition rate, 1 mJ pulse energy) at the energy density of 5 mJ/cm^2, carbon particles with the average diameter of 25 nm can facilitate viable delivery of calcein molecules to over 90% of cells. Importantly, over 20% of irradiated cells also showed uptake of plasmid DNA, which was confirmed both by direct fluorescent tagging of DNA as well as expression of luciferase gene. The only significant drawback of this approach lies in the very long irradiation time of around 5–10 minutes. This severely limits the throughput of this technique.

Another implementation of the concept of creating localized cavitation by pulsed irradiation of absorbing objects was described by Wu et al. [96, 97] in the form of a photothermal blade. They coated the 2 μm tip of a microcapillary pipette with a 100 nm thick layer of titanium. This tip was brought into light contact with the cellular membrane and illuminated with a 6 ns pulse at 532 nm and an optimized energy density of 180 mJ/cm^2, resulting in the creation of small cavitation bubbles and poration of the cellular membrane. At the same time, the cargo was delivered through the capillary directly into the created pores. The authors succeeded to inject large molecules such as RNA, DNA, 200 μm beads and even 2 μm long bacteria. Although not purely optical and non-contact, this technique is an appealing combination of traditional microinjection with the advantages of optical transfection.

It is worth mentioning that nanoparticles based optical transfection does not necessarily have to use their absorption properties to permeabilize the cellular membrane. It is well established that many small nanoparticles, including gold spheres and rods, are endocytosed by cells when added to the culture medium. Chen et al. [17] used DNA–gold nanorod conjugates to deliver DNA into cultured HeLa cells and showed that femtosecond pulsed 800 nm laser can selectively release strands of plasmid DNA from the particles surface upon irradiation. Selective transfection with EGFP expressing plasmids in the irradiated area, was observed with an efficiency of around 20%. A similar experiment was later demonstrated with ssDNA-nanoshell conjugates [6], gold nanospheres with double-stranded interfering oligonucleotides [40], gold nanospheres irradiated by a green nanosecond laser [62], and with independent release of DNA from two types of nanorods with two different resonant wavelengths [94]. All these experiments take advantage of the fact that a single nanoparticle can carry a few hundred oligonucleotides providing good loading efficiency. However, the fact that the gold nanoparticles remain in the treated cells and cannot be removed after the procedure may be a significant limiting factor in many experiments.

16.4 Technical Advancements Toward High-Throughput, in vivo, and Clinical Applications

In the previous sections, we discussed the advantages brought by various techniques developed for laser-mediated membrane poration and cell transfection. While the recent advancements make these techniques more efficient and reliable, the field is yet to be implemented on a wider scale in the routine practices of molecular biology laboratories. In particular, aspects such as reliability, throughput, biocompatibility, and ease of use are crucial for this technique to be widely adopted in bio-industry and research labs, not to mention clinical applications such as in vivo genetic therapy. In this section, we look at recently demonstrated optical transfection experiments in which the above-mentioned issues have been addressed.

16.4.1 *Reliability, Ease of Use, and Throughput*

As we explained in the previous sections, femtosecond pulsed NIR lasers are the most versatile solution for gentle cell optoinjection and transfection. As a consequence, most of the research effort has been put into systems based on this type of lasers.

Since it is necessary to tightly focus the femtosecond beam on the cellular membrane, this technique suffers from an intrinsic limitation on the achievable throughput. With the typical irradiation time on an order of tens of milliseconds, a single beam can only dose a few tens of cells per second. Moreover, this can only be achieved when each cell is only dosed once and there is no significant delay between doses, which requires additional engineering solutions.

Below, we show how beam shaping, beam steering, and image processing have been used to make the femtosecond laser based systems more robust and easier to use.

The first issue that had to be addressed was the precise membrane targeting with a tightly focused beam. As the poration mechanism depends on a multiphoton, highly nonlinear process of light absorption on the cellular membrane, it is crucial to create a possibly high power density of light in the region of the membrane, which is only 4–8 nm thick. This is typically achieved with a high numerical aperture microscope objective which provides a diffraction limited Gaussian focal spot of around 400–600 nm. This tight lateral confinement of the focused light enhances efficiency of multiphoton effects in the focal plane of the objective. However, at the same time it means that the length of the focal spot is drastically decreased along the axial direction, with the typical confocal parameter of 2–4 μm. This has been visualized by Niioka et al. [53], who by fixing irradiated cells and using transmission electron microscopy (TEM), directly measured the dimensions of the region ablated by a tightly focused femtosecond beam. They showed that the length of the ablated region is between 1 and 2 μm, depending on the applied power. This clearly illustrates how sensitive to axial misalignment this technique can be and that beam positioning is a key to successful, efficient phototransfection.

An interesting solution to the above problem has been proposed by Tsampoula et al. [85], who used a "non-diffracting" zero-order Bessel beam instead of a Gaussian beam. This beam, due to its propagation invariant nature over a long distance, acts as a rod of light, or an "optical syringe" that can extend over tens or hundreds of micrometers. The high intensity central rod provides good power density to mediate multiphoton absorption over its whole lengths, while the low intensity outer rings do not affect the cell noticeably. Importantly, the peak transfection efficiency is comparable with that achieved with a Gaussian beam, while the useful axial range is shown to be around 20 times larger. The authors also used the "self-healing" property of the Bessel beam to demonstrate successful transfection through a turbid layer. The only drawback of using the Bessel beam compared to the Gaussian beam is the total power required to create the rod of light. In their experiment Tsampoula et al. [85] had to use 628 mW to obtain around 70 mW in the central core.

The idea of using a Bessel beam has been further extended by Čižmár et al. [20], who used a spatial light modulator (SLM) to create multiple Bessel beams and steer their three-dimensional position at will. This gave birth to the concept of a phototransfection workstation, on which a user would simply "point and shoot" cells at will with a software and diffractive optics steering the beam to a desired cell. In that manner, although still not automated, the transfection process would be greatly simplified and accelerated. A further work of Antkowiak et al. [3] showed that when a Bessel beam cannot be used a Gaussian beam can be steered by an SLM to multiple axial positions in order to enhance the probability of irradiating the membrane. This strategy significantly enhances the poration efficiency without compromising cell viability. However, it comes at the price of serial multiple doses, which could slow the procedure.

A further step toward an automated phototransfection of adherent cells has been made by Cappelleri et al. [15], who used phase contrast images to automatically target primary neurons in a Petri dish. This development, combined with long-term imaging allowed them to greatly simplify the mRNA transfection experiments discussed in Section 3.3 [7, 78]. However, the high complexity of phase contrast images required a sophisticated and laborious (computation time of 30 s per single cell) numerical algorithm for image segmentation and cell identification. This effectively limited the throughput to 120 cells per hour, which was still estimated to give a sixfold improvement over a fully manual procedure. A further acceleration of the process is expected when the concentration of cells increases and the software is optimized. This work illustrates the major issue of the most suitable imaging modality to accompany the photoporation process. Various groups have applied different approaches to membrane targeting and cell identification. Figure 16.4 shows the same colony of adherent CHO-K1 cells visualized using some of the common microscopy techniques, namely phase contrast (PC), differential interference contrast (DIC), and wide-field epi-fluorescence. It is clear that each of these techniques highlights different details in the cell morphology and poses

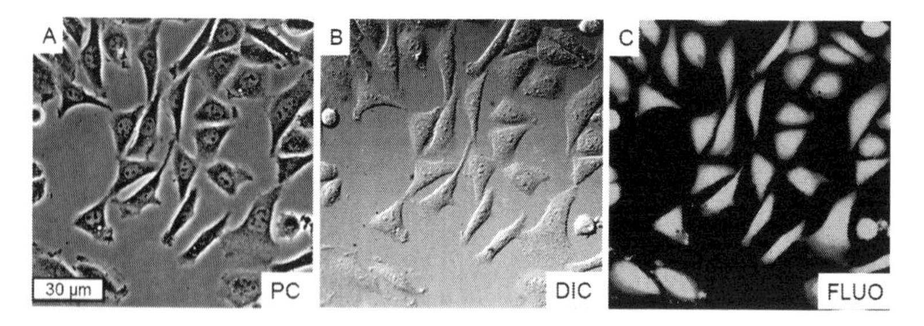

Figure 16.4 Comparison of popular imaging techniques that can be used for image-guided targeted optical transfection: (A) phase contrast (PC), (B) differential interference contrast (DIC), and (C) wide-field fluorescence (calcein AM).

different challenges to the image-processing algorithms. It seems that cells uniformly stained with the cytoplasm stain Calcein-AM (Fig. 16.4c) are the least demanding on the image segmentation procedures at the cost of additional, invasive sample preparation step and cell modification which may not be compatible with other aspects of the experiment. PC and DIC images offer relatively low contrast between the cell and background. Moreover, they suffer from artifacts that have to be compensated for in the software. It seems that efficient image processing is one of the major challenges in automated phototransfection of adherent cells and probably limits the throughput more significantly than the speed of beam steering and targeting.

An automated targeted and untargeted laser transfection of adherent cells has also been incorporated into a commercial cell culture processing system produced by Cyntellect (San Diego, CA). As mentioned in Section 16.3.2, a nanosecond laser (532 nm, 0.5 ns pulse duration) was integrated into their laser-enabled analysis and processing LEAP^{TM} system and has been demonstrated to optoinject a variety of substances into a range of cell types [21, 67]. The laser was defocused to a 14 µm diameter spot and either targeted to a specific cell or raster scanned over a large area of the cell monolayer. The achievable throughput was estimated to be up to 500 cells/second in the cell targeted mode and 2000–5000 cells/second depending on plating density in the untargeted operation. Interestingly, it was determined that multiple, sequential low radiant exposures produced better results than a single high radiant exposure.

As we already discussed in the earlier sections, both nanosecond and femtosecond beams can be used in conjunction with nanoparticles to permeabilize large populations of cells with a defocused, scanned beam [16, 100] or simply wide-field illumination [95]. Both gold [100] and carbon [16] nanoparticles have been shown to mediate membrane poration and lead to gene transfection at very low laser fluences (cf. Section 3.4). While this technique does not require a precisely positioned tightly focused beam, which facilitates *en masse* transfection of large groups of cells, the introduction of nanoparticles may cause limiting side-effects. In particular endocytosed particles may alter the properties of studied cells and limit their usefulness in further experiments. Also, gold nanoparticles embedded in the substrate [95] may inhibit cell adhesion and proliferation in some cell types. Further studies into biocompatibility of these techniques with the most demanding cell types (e.g., primary cells) are required to prove their applicability. However, they definitely constitute one of the most promising developments in optical transfection.

An alternative approach to phototransfection of large numbers of cells has been investigated by Marchington et al. [48]. They used the "lab-on-a-chip" approach to flow non-adherent or recently trypsinized cells through a microfluidic chip. In this approach, hydrodynamics focusing was used to confine the flowing cells within a 20 μm wide area in the center of the channel, where a stationary femtosecond beam is parked. While this system did not allow for cell selective irradiation it can be easily modified toward a shape- or fluorescence-activated system using a feedback loop from an integrated imaging system. The authors demonstrated efficient (42 ± 8%) optoinjection of propidium iodide into trypsinized HEK293 cell with good cell viability (28 ± 4% of cells were viably optoinjected). The achieved throughput was on the order of 1 cell/second, which is comparable to the throughput achieved by Antkowiak et al. [3] in the raster scanned irradiation of adherent CHO-K1 cells.

16.4.2 *Towards in vivo Applications*

The in vitro studies on established cell lines and ex vivo harvested primary cells have provided a tremendous amount of breakthrough results in molecular biology. However, there is an ongoing debate about the difference between in vitro and in vivo behavior of cells and tissues. The extracellular matrix and presence of other cells and biochemicals produced by them may significantly influence the biophysical properties of investigated cells. In particular, gene expression levels and activity of signaling pathways are strongly dependent on the environmental factors. Therefore, in many molecular biology studies it is necessary to validate the in vitro results in an animal model.

At the same time, the very promising initial results of genetic therapy of tissues and organs in living organisms increased the interest in new transfection techniques. In light of the recently identified safety issues with in vivo viral gene delivery [8, 52], new reliable non-viral naked DNA and drug delivery methods are of great interest. In this section we show how phototransfection has been used for drug and gene delivery in animal models, both adult and at the embryonic stage. We also look at engineering advances toward a miniaturized endoscopic optical fiber-based phototransfection probe.

To date there has not been much work performed with laser-mediated transfection in animal models. Satoh et al. used the nanosecond pulsed laser-induced stress waves (LISW) [80] mentioned in Section 16.3.2 for gene transfer into the mouse central nervous system [70]. They showed an enhanced expression of the EGFP gene injected directly into tissue. In a similar experiment Ogura et al. demonstrated transfection of rat skin in vivo [55]. While a large of population of cells around the irradiation site at depths of a few millimeters can be transfected, these experiments used a mechanical stress rather light–tissue interaction to permeate the membrane, and as such were inherently site but not cell selective. Also, it seems that this technique may be somewhat limited in vivo by the need to place a rubber sheet on the tissue surface.

A purely optical technique of tissue transfection in animal model was used by Zeira et al. to transfect tibial muscle in mice [103].

The injection of naked DNA into the tissue was followed by scanned irradiation of a $95 \times 95\ \mu m^2$ area at a depth of 1–3 mm with a loosely focused (objective NA = 0.5) 800 nm femtosecond laser beam at moderate powers (10–30 mW at focus). The process was optimized in terms of irradiation dose and beam position within tissue. Interestingly, since the beam was scanned over a relatively large area over the period of 5–20 seconds, the irradiation time in any single spot was on the order of tens of microseconds, which is three orders of magnitude shorter than the typically used in in vitro studies with similar beams. In the next step, the group replaced the microscope objective-based system with a miniaturized lens on an articulated arm, which made the system easier to use in a clinical in vivo setting. Both these experiments demonstrated a long lasting expression of the transfected gene (>7 months). They also showed that laser mediated genetic vaccination with plasmid DNA containing the hepatitis B virus (HBV) surface antigen (HBsAg) induces high titers of HBsAg-specific antibodies [104] proving the potential use in genetic therapy and prevention.

More recently a different NIR femtosecond laser beam ($\lambda = 1043$ nm, $T_P = 500$ fs, repetition rate = 200 kHz) without any focusing (beam diameter $\approx$4 mm) was used by Tsen et al. to enhance transfection of tumors in mice with intradermally injected luciferase-encoding plasmid DNA [87]. The required irradiation time of 80 seconds and the power density of 0.04 GW/cm^2 were significantly larger than in Zeira's experiments. However, a much larger area could be irradiated at once, making this technique an interesting alternative for large scale bulk transfection.

A second exciting in vivo application of phototransfection is in developmental biology. The ability to selectively inject and transfect single blastomere cells at the earliest stages of embryo development is crucial in studies involving cell lineage mapping and genetic modification of subsections of developing organisms. Although up to date no mammalian embryos have been successfully optoinjected or phototransfected, a very promising experiment on zebra fish embryos was demonstrated by Kohli et al. [35, 36]. The authors used a tightly focused (NA = 1.0) 800 nm beam to optically induce transient pores in both chorionated and dechorionated early to mid cleavage stage (2-cell to 8/16-cell) embryos. The irradiated blastomere cells were successfully loaded with a fluorescent dye (FITC) and Streptavidin-conjugated quantum dots. The authors also obtained selective phototransfection with plasmid DNA (Simian-CMV-EGFP) and the fluorescent protein was expressed in the developed larvae. The long-term viability of irradiated embryos was close to 100% and no morphological differences were observed in developed animals when compared to the control population.

A similar experiment was performed by Torres-Mapa et al. [83] on early-stage embryos of *Pomatoceros lamarckii*. These embryos are typically only 60 μm in diameter and, due to their small size, are extremely difficult to viably microinject. Using a femtosecond 800 nm laser the authors demonstrated cell-selective optoinjection of fluorescent markers into cells deep within the embryo without any damage to the cells in outer layers, enabling cell lineage and fate mapping in the developing organism. They also combined the optoinjection modality of their system with holographic optical trapping by switching the Ti:sapphire laser to a CW mode so that the embryos could also be rotated, translated, and positioned in three dimensions, thus creating a complete all-optical non-contact manipulation workstation. This is yet another example of an easy integration with a microscopy platform and another optical manipulation technique, which is one of the most important advantages of optical transfection.

Both these studies show that femtosecond laser-mediated injection and transfection of developing embryos may be an attractive alternative to the traditionally used microinjection and electroporation, in particular in high-throughput lab-on-a-chip applications.

One of the main engineering issues in the in vivo setting is the beam delivery to the tissue of interest. In particular, the use of high NA microscope objectives limits this technique to the more easily accessible outer layers of the body, e.g., skin, muscles, and outer layers of the brain. A miniaturized endoscopic probe capable of delivering the phototransfection beam would greatly enhance the applicability of the technique deep within the living organism in an almost non-invasive way. This would be particularly interesting when combined with endoscopic imaging leading to a cell selective transfection within a living animal.

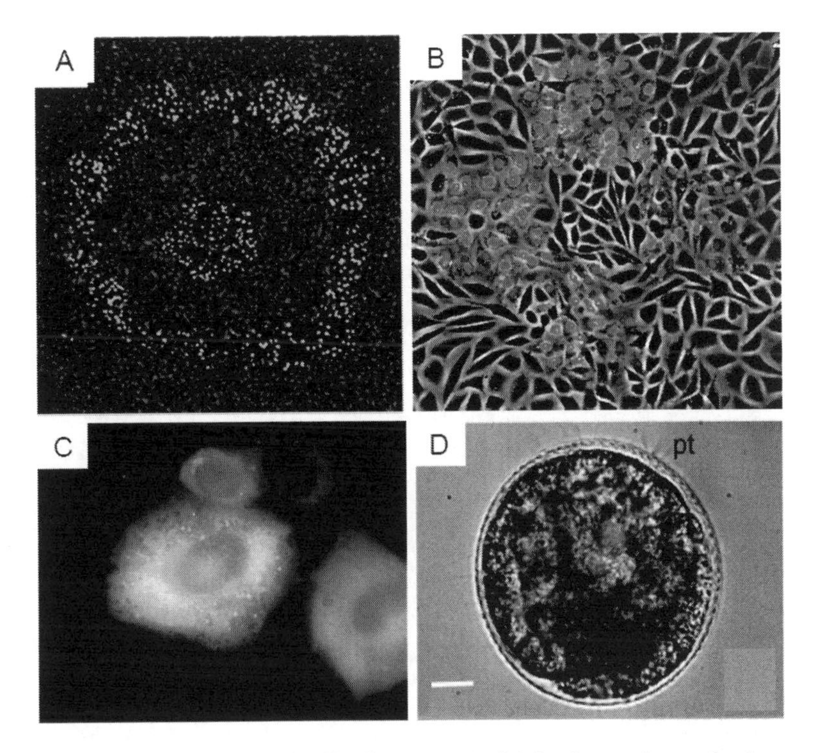

Figure 16.5 Examples of cell selective optical injection and transfection: (A) sequential patterned optoinjection of multiple fluorophores using a nanosecond laser [21]. (B) Patterned optoinjection of PI using a femtosecond laser with fast beam steering capabilities. (C) Cell-selective transfection of dsRed fluorescent protein [3]. (D) Cell-selective optoinjection of PI into single cells within inner layers of a developing embryo of *Pomatoceros lamarckii* [83].

The first work in this direction was done by Tsampoula et al. [86], who showed that a standard single mode optical fiber with a conical tip can deliver NIR ultrashort pulses ($\sim$800 fs @ 800 nm) to the cells at its distal end. The conical axicon tip provided moderate elongated focusing of the beam (focal spot $\approx$ 2–3 μm) at a working distance of around 13 μm. With the optimized irradiation time and dose, this system provided transfection efficiencies comparable with objective-based system on CHO cells in vitro (30–50%). In the next step Ma et al. [46] replaced the conical tip with a microlens obtaining a longer working distance (20 μm). They also integrated

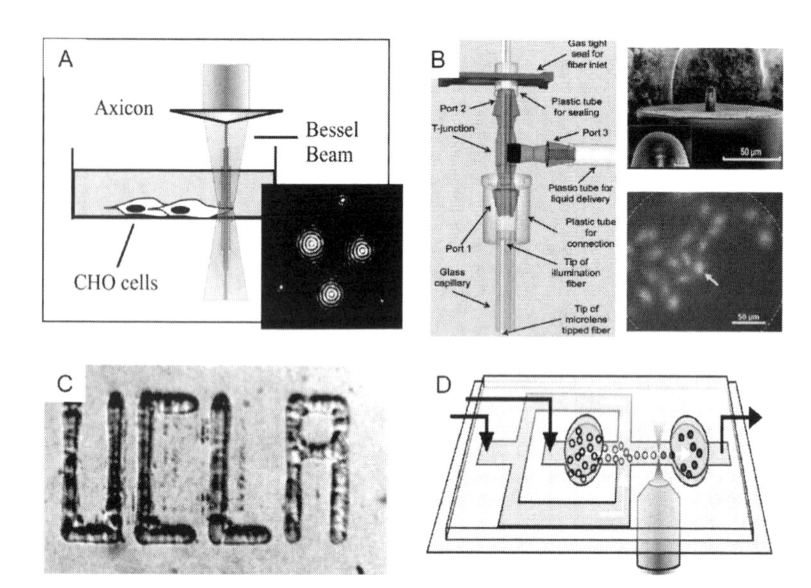

Figure 16.6 Advanced optical transfection techniques: (A) "Non-diffracting" Bessel beam forms an optical syringe which facilitates membrane targeting [85]. Inset shows multiplexed Bessel beam for simultaneous targeted poration of multiple cells [20]. (B) optical transfection through a fiber-based endoscopic probe—an integrated system with microfluidic DNA delivery [46], optimized microlens tip [46], and image-guided endoscopic transfection through an imaging fiber bundle [47]. (C) Patterned wide-field illumination of gold nanoparticles with nanosecond pulsed laser enables high-throughput permeabilization of multiple cells (image shows cavitation bubbles 120 ns after patterned irradiation) [95]. (D) Microfluidic format enables high-throughput optoinjection of nonadherent cells [48].

the optical fiber with a capillary delivering the genetic material to the vicinity of the fiber tip creating the first system ready for a truly endoscopic operation. Later on, Ma et al. [47] used a coherent fiber bundle to realize parallel fluorescent imaging and phototransfection through its multiple cores. This enabled cell selective endoscopic phototransfection using a 300 μm diameter probe. Looking at the pace of technological developments it is highly likely that the nearest future will bring a variety of in vivo phototransfection experiments using this kind of probes.

16.5 Conclusions

In recent years we have witnessed a rapid development and maturing of a plethora of optical transfection techniques. In this chapter we discussed the various approaches to laser-mediated cell permeabilization and highlighted their advantages and drawbacks in the context of fundamental biophysical research as well as their potential use in in vivo setting and high-throughput applications. These characteristics are summarized in Table 16.1.

It is remarkable that a 4–8 nm wide layer has caused so many difficulties for scientists. The use of light to create a permeable "pore" across this structure is an easy concept but it is apparent that we have already moved away from this rather simplistic view and that, as described above, a range of subtle and ingenious ways have been developed for allowing impermeable biomolecules and structures across the plasma membrane. So what for the future? This can be broadly seen in three approaches: (i) The ability to inject a wider range of bio-relevant compounds and structures, e.g., the rise of the use quantum dots in imaging, makes them an attractive structure that could be introduced into cells, along with the ability to insert specific antibodies where it might be possible to immuno-stain cells with no detergent permeabilization. With the described ability to combine optical tweezing with optoinjection, this now becomes a possibility. (ii) At one stage the ability to transfect optically cells in living organisms would have been considered too difficult, but as described above the first forays into this capability have already been made, and coupling this with the advances in fiber imaging make this a particularly intriguing possibility. (iii) Increasing high throughput would be of major interest to the pharmaceutical and bio-manufacturing industries. Again we have seen the first tentative steps toward this in the form of microfluidic platforms, which might form the basis of this technology. However, last but not the least is the realization that much of this truly interdisciplinary research has been pioneered by optical physics, and experience would suggest that as this field continues to evolve, more applications will become apparent; for example, the ability to shape and control light is

Table 16.1 Comparison of optical transfection techniques

Type of optical transfection technique	Single cell selectivity	Throughput	Applicability in vivo	Cost/accessibility	Ease of use
CW	+	Low due to a long irradiation time	Not demonstrated yet	Low / light sources popular in confocal scanning microscopy	Easy to align and operate
Ns pulsed	+/−	moderate	Not demonstrated yet (only LISW on outer layer of tissue)	Moderate/rare in biology labs	Moderately difficult to set up
Fs pulsed	+	moderate	Yes, both through objective and an endoscopic probe	High cost but NIR fs lasers are widely used in 2 photon scanning microscopy	Requires precise targeting and alignment
Particle mediated	+/−	High, but plasmid DNA transfection efficiency tends to be low	Difficult (particles need to be introduced to cells prior to transfection)	moderate/particles commonly available but a pulsed laser is necessary (cf. ns and fs)	Easy alignment but delivery of particles to cells may be a problem (requires pre-incubation or special substrate)

exemplified by the rise of the Bessel beam, which at one time was considered to be a physical oddity but has now been applied not just in optical transfection but also in imaging [60]. Therefore, the future is bright.

References

1. Ando, J., Smith, N. I., Fujita, K., and Kawata, S. (2009). Photogeneration of membrane potential hyperpolarization and depolarization in non-excitable cells. *Eur. Biophys. J.*, **38**:255–262.
2. Antkowiak, M., Torres-Mapa, M. L., Dholakia, K., and Gunn-Moore, F. J. (2010). Quantitative phase study of the dynamic cellular response in femtosecond laser photoporation. *Biomed. Opt. Express*, **1**:414–424.
3. Antkowiak, M., Torres-Mapa, M. L., Gunn-Moore, F., and Dholakia, K. (2010). Application of dynamic diffractive optics for enhanced femtosecond laser based cell transfection. *J. Biophotonics*, **3**:696–705.
4. Arita, Y., Torres-Mapa, M. L., Lee, W. M., Cizmar, T., Campbell, P., Gunn-Moore, F. J., and Dholakia, K. (2011). Spatially optimized gene transfection by laser-induced breakdown of optically trapped nanoparticles. *Appl. Phys. Lett.*, **98**:093702–093703.
5. Badr, Y. A., Kereim, M. A., Yehia, M. A., Fouad, O. O., and Bahieldin, A. (2005). Production of fertile transgenic wheat plants by laser micropuncture. *Photochem. Photobiol. Sci.*, **4**:803–807.
6. Barhoumi, A., Huschka, R., Bardhan, R., Knight, M. W., and Halas, N. J. (2009). Light-induced release of DNA from plasmon-resonant nanoparticles: Towards light-controlled gene therapy. *Chem. Phys. Lett.*, **482**:171–179.
7. Barrett, L. E., Sul, J. Y., Takano, H., Van Bockstaele, E. J., Haydon, P. G., and Eberwine, J. H. (2006). Region-directed phototransfection reveals the functional significance of a dendritically synthesized transcription factor. *Nat. Methods*, **3**:455–460.
8. Baum, C., Düllmann, J., Li, Z., Fehse, B., Meyer, J., Williams, D. A., and von Kalle, C. (2003). Side effects of retroviral gene transfer into hematopoietic stem cells. *Blood*, **101**:2099–2113.
9. Baumgart, J., Bintig, W., Ngezahayo, A., Willenbrock, S., Murua Escobar, H., Ertmer, W., Lubatschowski, H., and Heisterkamp, A. (2008). Quantified femtosecond laser based opto-perforation of living GFSHR-17 and MTH53 a cells. *Opt. Express*, **16**:3021–3031.

10. Baumgart, J., Kuetemeyer, K., Bintig, W., Ngezahayo, A., Ertmer, W., Lubatschowski, H., and Heisterkamp, A. (2009). Repetition rate dependency of reactive oxygen species formation during femtosecond laser-based cell surgery. *J. Biomed. Opt.*, **14**:054040.
11. Ben-Yakar, A., Eversole, D., and Ekici, O. (2007). Spherical and anisotropic gold nanomaterials in plasmonic laser phototherapy of cancer. *Nanotechnologies for the Life Sciences* (Wiley-VCH Verlag GmbH & Co. KGaA).
12. Brown, C. T. A., Stevenson, D. J., Tsampoula, X., McDougall, C., Lagatsky, A. A., Sibbett, W., Gunn-Moore, F., and Dholakia, K. (2008). Enhanced operation of femtosecond lasers and applications in cell transfection. *J. Biophotonics*, **1**:183–199.
13. Brown, M. D., Schatzlein, A. G., and Uchegbu, I. F. (2001). Gene delivery with synthetic (non viral) carriers. *Int. J. Pharm.*, **229**:1–21.
14. Brunner, S., Sauer, T., Carotta, S., Cotten, M., Saltik, M., and Wagner, E. (2000). Cell cycle dependence of gene transfer by lipoplex, polyplex and recombinant adenovirus. *Gene Ther.*, **7**:401–407.
15. Cappelleri, D. J., Halasz, A., Sul, J. Y., Kim, T. K., Eberwine, J., and Kumar, V. (2010). Towards a fully automated high-throughput phototransfection system. *JALA Charlottesv Va*, **15**:329–341.
16. Chakravarty, P., Qian, W., El-Sayed, M. A., and Prausnitz, M. R. (2010). Delivery of molecules into cells using carbon nanoparticles activated by femtosecond laser pulses. *Nat. Nanotechnol.*, **5**:607–611.
17. Chen, C.-C., Lin, Y.-P., Wang, C.-W., Tzeng, H.-C., Wu, C.-H., Chen, Y.-C., Chen, C.-P., Chen, L.-C., and Wu, Y.-C. (2006). DNA–gold nanorod conjugates for remote control of localized gene expression by near infrared irradiation. *J. Am. Chem. Soc.*, **128**:3709–3715.
18. Cherukuri, P., Glazer, E. S., and Curley, S. A. (2010). Targeted hyperthermia using metal nanoparticles. *Adv. Drug Delivery Rev.*, **62**:339–345.
19. Chung, S. H., and Mazur, E. (2009). Surgical applications of femtosecond lasers. *J. Biophotonics*, **2**:557–572.
20. Cizmar, T., Kollarova, V., Tsampoula, X., Gunn-Moore, F., Sibbett, W., Bouchal, Z., and Dholakia, K. (2008). Generation of multiple Bessel beams for a biophotonics workstation. *Opt. Express*, **16**:14024–14035.
21. Clark, I. B., Hanania, E. G., Stevens, J., Gallina, M., Fieck, A., Brandes, R., Palsson, B. O., and Koller, M. R. (2006). Optoinjection for efficient targeted delivery of a broad range of compounds and macromolecules into diverse cell types. *J. Biomed. Opt.*, **11:** 014034.

22. Curtis, J. E., Koss, B. A., and Grier, D. G. (2002). Dynamic holographic optical tweezers. *Opt. Commun.*, **207**:169–175.

23. El-Sayed, I. H., Huang, X., and El-Sayed, M. A. (2006). Selective laser photo-thermal therapy of epithelial carcinoma using anti-EGFR antibody conjugated gold nanoparticles. *Cancer Lett.*, **239**:129–135.

24. Escoffre, J. M., Teissie, J., and Rols, M. P. (2010). Gene transfer: how can the biological barriers be overcome? *J. Membr. Biol.*, **236**:61–74.

25. Faurie, C., Rebersek, M., Golzio, M., Kanduser, M., Escoffre, J. M., Pavlin, M., Teissie, J., Miklavcic, D., and Rols, M. P. (2010). Electro-mediated gene transfer and expression are controlled by the life-time of DNA/membrane complex formation. *J. Gene Med.*, **12**:117–125.

26. Gao, Y., Liang, H., and Berns, M. W. (1995). Laser-mediated gene transfer in rice. *Physiol. Plant.*, **93**:19–24.

27. Golzio, M., Teissie, J., and Rols, M. P. (2002). Direct visualization at the single-cell level of electrically mediated gene delivery. *Proc. Natl. Acad. Sci. USA*, **99**:1292–1297.

28. He, H., Kong, S. K., Lee, R. K., Suen, Y. K., and Chan, K. T. (2008). Targeted photoporation and transfection in human HepG2 cells by a fiber femtosecond laser at 1554 nm. *Opt. Lett.*, **33**:2961–2963.

29. Hellman, A. N., Rau, K. R., Yoon, H. H., and Venugopalan, V. (2008). Biophysical response to pulsed laser microbeam-induced cell lysis and molecular delivery. *J. Biophotonics*, **1**:24–35.

30. Hosokawa, Y., Iguchi, S., Yasukuni, R., Hiraki, Y., Shukunami, C., and Masuhara, H. (2009). Gene delivery process in a single animal cell after femtosecond laser microinjection. *Appl. Surf. Sci.*, **255**:9880–9884.

31. Iwanaga, S., Kaneko, T., Fujita, K., Smith, N., Nakamura, O., Takamatsu, T., and Kawata, S. (2006). Location-dependent photogeneration of calcium waves in HeLa cells. *Cell Biochem. Biophys.*, **45**:167–176.

32. Iwanaga, S., Smith, N. I., Fujita, K., and Kawata, S. (2006). Slow Ca(2+) wave stimulation using low repetition rate femtosecond pulsed irradiation. *Opt. Express*, **14**:717–725.

33. Knoll, T., Trojan, L., Langbein, S., Sagi, S., Alken, P., and Michel, M. S. (2004). Impact of holmium : YAG and neodymium : YAG lasers on the efficacy of DNA delivery in transitional cell carcinoma. *Lasers Med. Sci.*, **19**:33–36.

34. Kohli, V., Acker, J. P., and Elezzabi, A. Y. (2005). Reversible permeabilization using high-intensity femtosecond laser pulses: Applications to biopreservation. *Biotechnol. Bioeng.*, **92**:889–899.

35. Kohli, V., and Elezzabi, A. Y. (2008). Laser surgery of zebrafish (Danio rerio) embryos using femtosecond laser pulses: optimal parameters for exogenous material delivery, and the laser's effect on short- and long-term development. *BMC Biotechnol.*, **8**:7.
36. Kohli, V., Robles, V., Cancela, M. L., Acker, J., Waskiewicz, A. J., and Elezzabi, A. Y. (2007). An alternative method for delivering exogenous material into developing zebrafish embryos. *Biotechnol. Bioeng.*, **98**:1230–1241.
37. Kuetemeyer, K., Baumgart, J., Lubatschowski, H., and Heisterkamp, A. (2009). Repetition rate dependency of low-density plasma effects during femtosecond-laser-based surgery of biological tissue. *Appl. Phys. B: Lasers Opt.*, **97**:695–699.
38. Kuetemeyer, K., Rezgui, R., Lubatschowski, H., and Heisterkamp, A. (2010). Influence of laser parameters and staining on femtosecond laser-based intracellular nanosurgery. *Biomed. Opt. Express*, **1**:587–597.
39. Kurata, S., Tsukakoshi, M., Kasuya, T., and Ikawa, Y. (1986). The Laser Method for Efficient Introduction of Foreign DNA into Cultured-Cells. *Exp. Cell Res.*, **162**:372–378.
40. Lee, S. E., Liu, G. L., Kim, F., and Lee, L. P. (2009). Remote Optical Switch for Localized and Selective Control of Gene Interference. *Nano Lett.*, **9**:562–570.
41. Lei, M., Xu, H., Yang, H., and Yao, B. (2008). Femtosecond laser-assisted microinjection into living neurons. *J. Neurosci. Methods*, **174**:215–218.
42. Lévy, R., Shaheen, U., Cesbron, Y., and Sée, V. (2010). Gold nanoparticles delivery in mammalian live cells: a critical review. *Nano Rev.*, **1.** doi: 10.3402/nano.v1i0.4889. Epub 2010 Feb 22.
43. Liu, C., Li, Z., and Zhang, Z. (2009). Mechanisms of laser nanoparticle-based techniques for gene transfection-a calculation study. *J. Biol. Phys.*, **35**:175–183.
44. Lloyd, D. R., Leelavatcharamas, V., Emery, A. N., and Al-Rubeai, M. (1999). The role of the cell cycle in determining gene expression and productivity in CHO cells. *Cytotechnology*, **30**:49–57.
45. Lukacs, G. L., Haggie, P., Seksek, O., Lechardeur, D., Freedman, N., and Verkman, A. S. (2000). Size-dependent DNA mobility in cytoplasm and nucleus. *J. Biolog. Chem.*, **275**:1625–1629.
46. Ma, N., Ashok, P. C., Stevenson, D. J., Gunn-Moore, F. J., and Dholakia, K. (2010). Integrated optical transfection system using a microlens fiber combined with microfluidic gene delivery. *Biomed. Opt. Express*, **1**:694–705.

47. Ma, N., Gunn-Moore, F., and Dholakia, K. (2011). Optical transfection using an endoscope-like system. *J. Biomed. Opt.*, **16**:028002.

48. Marchington, R. F., Arita, Y., Tsampoula, X., Gunn-Moore, F. J., and Dholakia, K. (2010). Optical injection of mammalian cells using a microfluidic platform. *Biomed. Opt. Express*, **1**:527–536.

49. McDougall, C., Stevenson, D. J., Brown, C. T. A., Gunn-Moore, F., and Dholakia, K. (2009). Targeted optical injection of gold nanoparticles into single mammalian cells. *J. Biophotonics*, **2**:736–743.

50. Mohanty, S. K., Sharma, M., and Gupta, P. K. (2003). Laser-assisted microinjection into targeted animal cells. *Biotechnol. Lett.*, **25**:895–899.

51. Mthunzi, P., Dholakia, K., and Gunn-Moore, F. (2010). Photo-transfection of mammalian cells using femtosecond laser pulses: optimisation and applicability to stem cell differentiation. *J. Biomed. Opt.*, **15**:041507.

52. Nienhuis, A. W., Dunbar, C. E., and Sorrentino, B. P. (2006). Genotoxicity of Retroviral Integration In Hematopoietic Cells. *Mol. Ther.*, **13**:1031–1049.

53. Niioka, H., Smith, N. I., Fujita, K., Inouye, Y., and Kawata, S. (2008). Femtosecond laser nano-ablation in fixed and non-fixed cultured cells. *Opt. Express*, **16**:14476–14495.

54. Nikolskaya, A. V., Nikolski, V. P., and Efimov, I. R. (2006). Gene printer: Laser-scanning targeted transfection of cultured cardiac neonatal rat cells. *Cell Commun. Adhes.*, **13**:217–222.

55. Ogura, M., Sato, S., Nakanishi, K., Uenoyama, M., Kiyozumi, T., Saitoh, D., Ikeda, T., Ashida, H., and Obara, M. (2004). In vivo targeted gene transfer in skin by the use of laser-induced stress waves. *Lasers Surg. Med.*, **34**:242–248.

56. Palumbo, G., Caruso, M., Crescenzi, E., Tecce, M. F., Roberti, G., and Colasanti, A. (1996). Targeted gene transfer in eucaryotic cells by dye-assisted laser optoporation. *J. Photochem. Photobiol. B-Biol.*, **36**: 41–46.

57. Paterson, L., Agate, B., Comrie, M., Ferguson, R., Lake, T. K., Morris, J. E., Carruthers, A. E., Brown, C. T. A., Sibbett, W., Bryant, P. E., Gunn-Moore, F., Riches, A. C., and Dholakia, K. (2005). Photoporation and cell transfection using a violet diode laser. *Opt. Express*, **13**:595–600.

58. Peng, C., Palazzo, R. E., and Wilke, I. (2007). Laser intensity dependence of femtosecond near-infrared optoinjection. *Phys. Rev. E: Stat. Nonlinear Soft Matter Phys.*, **75**:041903,041901–041908.

59. Pitsillides, C. M., Joe, E. K., Wei, X. B., Anderson, R. R., and Lin, C. P. (2003). Selective cell targeting with light-absorbing microparticles and nanoparticles. *Biophys. J.*, **84**:4023–4032.
60. Planchon, T. A., Gao, L., Milkie, D. E., Davidson, M. W., Galbraith, J. A., Galbraith, C. G., and Betzig, E. (2011). Rapid three-dimensional isotropic imaging of living cells using Bessel beam plane illumination. *Nat. Methods*, **8**:417–423.
61. Pollard, H., Remy, J. S., Loussouarn, G., Demolombe, S., Behr, J. P., and Escande, D. (1998). Polyethylenimine but not cationic lipids promotes transgene delivery to the nucleus in mammalian cells. *J. biolog. chem.*, **273**:7507–7511.
62. Poon, L., Zandberg, W., Hsiao, D., Erno, Z., Sen, D., Gates, B. D., and Branda, N. R. (2010). Photothermal release of single-stranded DNA from the surface of gold nanoparticles through controlled denaturating and Au–S bond breaking. *ACS Nano*, **4**:6395–6403.
63. Praveen, B. B., Stevenson, D. J., Antkowiak, M., Dholakia, K., and Gunn-Moore, F. J. (2011). Enhancement and optimization of plasmid expression in femtosecond optical transfection. *J. Biophotonics*, **4**:229–235.
64. Quinto-Su, P. A., and Venugopalan, V. (2007). Mechanisms of laser cellular microsurgery. *Methods Cell Biol.*, **82**:113–151.
65. Rau, K. R., III, A. G., Vogel, A., and Venugopalan, V. (2004). Investigation of laser-induced cell lysis using time-resolved imaging. *Appl. Phys. Lett.*, **84**:2940–2942.
66. Rau, K. R., Quinto-Su, P. A., Hellman, A. N., and Venugopalan, V. (2006). Pulsed laser microbeam-induced cell lysis: time-resolved imaging and analysis of hydrodynamic effects. *Biophys. J.*, **91**:317–329.
67. Rhodes, K., Clark, I., Zatcoff, M., Eustaquio, T., Hoyte, K. L., Koller, M. R., Michael, W. B., and Greulich, K. O. (2007). Cellular Laserfection. *Methods Cell Biol.*, **82**:309–333.
68. Rusk, N. (2011). Seamless delivery. *Nat Methods*, **8**:44.
69. Sagi, S., Knoll, T., Trojan, L., Schaaf, A., Alken, P., and Michel, M. S. (2003). Gene delivery into prostate cancer cells by holmium laser application. *Prostate Cancer Prostatic Dis.*, **6**:127–130.
70. Satoh, Y., Kanda, Y., Terakawa, M., Obara, M., Mizuno, K., Watanabe, Y., Endo, S., Ooigawa, H., Nawashiro, H., Sato, S., and Takishima, K. (2005). Targeted DNA transfection into the mouse central nervous system using laser-induced stress waves. *J. Biomed. Opt.*, **10**:060501.

71. Schneckenburger, H., Hendinger, A., Sailer, R., Strauss, W. S. L., and Schmidtt, M. (2002). Laser-assisted optoporation of single cells. *J. Biomed. Opt.*, **7**:410–416.

72. Shirahata, Y., Ohkohchi, N., Itagak, H., and Satomi, S. (2001). New technique for gene transfection using laser irradiation. *J. Investig. Med.*, **49**:184–190.

73. Soughayer, J. S., Krasieva, T., Jacobson, S. C., Ramsey, J. M., Tromberg, B. J., and Allbritton, N. L. (2000). Characterization of cellular optoporation with distance. *Anal. Chem.*, **72**:1342–1347.

74. Spalding, G. C., Courtial, J., and Di Leonardo, R. (2008). Holographic optical trapping, in *Structured Light and Its Applications: An Introduction to Phase-Structured Beams and Nanoscale Optical Forces*, edited by D. L. Andrews, Elsevier.

75. Stevenson, D., Agate, B., Tsampoula, X., Fischer, P., Brown, C. T. A., Sibbett, W., Riches, A., Gunn-Moore, F., and Dholakia, K. (2006). Femtosecond optical transfection of cells: viability and efficiency. *Opt. Express*, **14**:7125–7133.

76. Stevenson, D. J., Gunn-Moore, F. J., Campbell, P., and Dholakia, K. (2010). Single cell optical transfection. *J. R. Soc. Interface*, **7**:863–871.

77. Stracke, F., Rieman, I., and Konig, K. (2005). Optical nanoinjection of macromolecules into vital cells. *J. Photochem. Photobiol. B-Biol.*, **81**:136–142.

78. Sul, J. Y., Wu, C. W., Zeng, F., Jochems, J., Lee, M. T., Kim, T. K., Peritz, T., Buckley, P., Cappelleri, D. J., Maronski, M., Kim, M., Kumar, V., Meaney, D., Kim, J., and Eberwine, J. (2009). Transcriptome transfer produces a predictable cellular phenotype. *Proc. Natl. Acad. Sci. USA*, **106**:7624–7629.

79. Tao, W., Wilkinson, J., Stanbridge, E. J., and Berns, M. W. (1987). Direct Gene-Transfer into Human Cultured-Cells Facilitated by Laser Micropuncture of the Cell-Membrane. *Proc. Natl. Acad. Sci. USA*, **84**:4180–4184.

80. Terakawa, M., Ogura, M., Sato, S., Wakisaka, H., Ashida, H., Uenoyama, M., Masaki, Y., and Obara, M. (2004). Gene transfer into mammalian cells by use of a nanosecond pulsed laser-induced stress wave. *Opt. Lett.*, **29**:1227–1229.

81. Tirlapur, U. K. and Konig, K. (2002). Targeted transfection by femtosecond laser. *Nature*, **418**:290–291.

82. Tirlapur, U. K., Konig, K., Peuckert, C., Krieg, R., and Halbhuber, K. J. (2001). Femtosecond near-infrared laser pulses elicit generation of

reactive oxygen species in mammalian cells leading to apoptosis-like death. *Exp. Cell Res.*, **263**:88–97.

83. Torres-Mapa, M. L., Antkowiak, M., Cizmarova, H., Ferrier, D. E. K., Dholakia, K., and Gunn-Moore, F. J. (2011). Integrated holographic system for all-optical manipulation of developing embryos. *Biomed. Opt. Express*, **2**:1564–1575.
84. Torres, A., Lorenzo, V., Hernandez, D., Rodriguez, J. C., Concepcion, M. T., Rodriguez, A. P., Hernandez, A., de Bonis, E., Darias, E., Gonzalez-Posada, J. M., and et al. (1995). Bone disease in predialysis, hemodialysis, and CAPD patients: evidence of a better bone response to PTH. *Kidney Int.*, **47**:1434–1442.
85. Tsampoula, X., Garces-Chavez, V., Comrie, M., Stevenson, D. J., Agate, B., Brown, C. T. A., Gunn-Moore, F., and Dholakia, K. (2007). Femtosecond cellular transfection using a nondiffracting light beam. *Appl. Phys. Lett.*, **91**:053902,053901–053903.
86. Tsampoula, X., Taguchi, K., Cizmar, T., Garces-Chavez, V., Ma, N., Mohanty, S., Mohanty, K., Gunn-Moore, F., and Dholakia, K. (2008). Fibre based cellular transfection. *Opt. Express*, **16**:17007–17013.
87. Tsen, S. W., Wu, C. Y., Meneshian, A., Pai, S. I., Hung, C. F., and Wu, T. C. (2009). Femtosecond laser treatment enhances DNA transfection efficiency in vivo. *J. Biomed. Sci.*, **16**:36.
88. Tsukakoshi, M., Kurata, S., Nomiya, Y., Ikawa, Y., and Kasuya, T. (1984). A novel method of DNA transfection by laser microbeam cell surgery. *Appl. Phys. B: Photophys. Laser Chem.*, **35**:135–140.
89. Uchugonova, A., Konig, K., Bueckle, R., Isemann, A., and Tempea, G. (2008). Targeted transfection of stem cells with sub-20 femtosecond laser pulses. *Opt. Express*, **16**:9357–9364.
90. Verkman, A. S. (2002). Solute and macromolecule diffusion in cellular aqueous compartments. *Trends Biochem. Sci.*, **27**:27–33.
91. Vogel, A., Linz, N., Freidank, S., and Paltauf, G. (2008). Femtosecond-laser-induced nanocavitation in water: implications for optical breakdown threshold and cell surgery. *Phys. Rev. Lett.*, **100**:038102.
92. Vogel, A., Noack, J., Huttman, G., and Paltauf, G. (2005). Mechanisms of femtosecond laser nanosurgery of cells and tissues. *Appl. Phys. B: Lasers Opt.*, **81**:1015–1047.
93. Vogel, A. and Venugopalan, V. (2003). Mechanisms of pulsed laser ablation of biological tissues. *Chem. Rev.*, **103**:2079–2079.
94. Wijaya, A., Schaffer, S. B., Pallares, I. G., and Hamad-Schifferli, K. (2008). Selective release of multiple DNA oligonucleotides from gold nanorods. *ACS Nano*, **3**:80–86.

95. Wu, T. H., Kalim, S., Callahan, C., Teitell, M. A., and Chiou, P. Y. (2010). Image patterned molecular delivery into live cells using gold particle coated substrates. *Opt. Express*, **18**:938–946.
96. Wu, T. H., Teslaa, T., Kalim, S., French, C. T., Moghadam, S., Wall, R., Miller, J. F., Witte, O. N., Teitell, M. A., and Chiou, P. Y. (2011). Photothermal nanoblade for large cargo delivery into mammalian cells. *Anal. Chem.*, **83**:1321–1327.
97. Wu, T. H., Teslaa, T., Teitell, M. A., and Chiou, P. Y. (2010). Photothermal nanoblade for patterned cell membrane cutting. *Opt. Express*, **18**:23153–23160.
98. Yao, C., Qu, X., and Zhang, Z. (2009). Laser-based transfection with conjugated gold nanoparticles. *Chin. Opt. Lett.*, **7**:898–900.
99. Yao, C., Qu, X., Zhang, Z., Huttmann, G., and Rahmanzadeh, R. (2009). Influence of laser parameters on nanoparticle-induced membrane permeabilization. *J. Biomed. Opt.*, **14**:054034.
100. Yao, C. P., Rahmanzadeh, R., Endl, E., Zhang, Z. X., Gerdes, J., and Huttmann, G. (2005). Elevation of plasma membrane permeability by laser irradiation of selectively bound nanoparticles. *J. Biomed. Opt.*, **10:**064012.
101. Yu, L., Mohanty, S., Zhang, J., Genc, S., Kim, M. K., Berns, M. W., and Chen, Z. (2009). Digital holographic microscopy for quantitative cell dynamic evaluation during laser microsurgery. *Opt. Express*, **17**:12031–12038.
102. Zeigler, M. B. and Chiu, D. T. (2009). Laser selection significantly affects cell viability following single-cell nanosurgery. *Photochem. Photobiol.*, **85**:1218–1224.
103. Zeira, E., Manevitch, A., Khatchatouriants, A., Pappo, O., Hyam, E., Darash-Yahana, M., Tavor, E., Honigman, A., Lewis, A., and Galun, E. (2003). Femtosecond infrared laser: an efficient and safe in vivo gene delivery system for prolonged expression. *Mol. Ther.*, **8**:342–350.
104. Zeira, E., Manevitch, A., Manevitch, Z., Kedar, E., Gropp, M., Daudi, N., Barsuk, R., Harati, M., Yotvat, H., Troilo, P. J., Griffiths, T. G., 2nd, Pacchione, S. J., Roden, D. F., Niu, Z., Nussbaum, O., Zamir, G., Papo, O., Hemo, I., Lewis, A., and Galun, E. (2007). Femtosecond laser: a new intradermal DNA delivery method for efficient, long-term gene expression and genetic immunization. *FASEB J.*, **21**:3522–3533.

Index

ablation 166, 177–184, 187, 676, 686–687, 703
  plasma-induced 184–187
absorption
  multi-photon 146–147
  two-photon 146, 200
acousto-optic crystal 315–316, 318–319, 322
acousto-optic deflector (AOD) 310, 312–318, 320–321, 323, 514–515
action potentials (APs) 310, 313, 494
active resonators 483–487, 489
AFM, *see* atomic-force microscope
anaphase onset 685, 688
anti-Stokes Raman scattering 265
AOD, *see* acousto-optic deflector
AOD scanners 317, 320–323, 328
AOD two-photon microscope design 313
APs, *see* action potentials
ArF excimer laser 180
atomic-force microscope (AFM) 410, 477, 508
atomic transitions 108–109, 111, 146, 151
autofluorescence 126, 128, 212, 269, 273, 278–279, 284

beam delivery medium 347
beam shaping techniques 543, 554
beam splitter 26–27, 210, 312, 314, 652
BFP, *see* blue fluorescent protein
biological fluorescence nanoscopy 221–222, 224, 226, 228, 230, 232, 234, 236, 238, 240, 242, 244, 246, 248, 250
biological particles 463, 465
biological tissues 2, 5, 58, 82, 85–86, 93–95, 97–99, 130, 135–136, 159, 174, 181–182, 337, 365, 537
biology, molecular 404, 443, 710, 718
biomolecular interactions 386, 402
biophotonics 1–2, 18, 29, 57–58, 96–97, 156, 187, 385, 584, 588, 610, 649–650, 668
biosensors 422, 442, 457–458
birefringence 18, 63, 65, 67, 69, 572
blue fluorescent protein (BFP) 134
bovine serum albumin (BSA) 420–421, 454, 457, 486
brain imaging 200
BSA, *see* bovine serum albumin

cancer 126, 172, 199, 217, 291, 293, 424, 521, 680
cancer cells 291, 412–413

cancer detection 200, 291, 368
CARS, *see* coherent anti-Stokes Raman scattering
CARS microscopy 164–165, 347, 367, 540, 544, 546, 549, 559, 561, 564, 614
CCD, *see* charge-coupled device
cells
  mitotic 680, 683, 710
  thalassemic 300
CFP, *see* cyan fluorescent protein
charge-coupled device (CCD) 44, 278, 297, 402, 613, 649, 675
chromophores 71–72, 94, 99, 124, 167, 169, 176, 182, 677
chromosome tip ablation 686–689
chromosomes, anaphase 682–683
CL, *see* condenser lens
CLSM, *see* confocal laser scanning microscopy
coherent anti-Stokes Raman scattering (CARS) 154–155, 157–158, 162, 164–165, 301, 335–338, 368, 372, 537–538, 543, 554, 609–610, 614, 634–635
coherent nonlinear microscopy 537–538, 540, 542, 544, 546, 548, 550, 552, 554, 556, 558, 560, 562, 564, 566
coherent Raman scattering (CRS) 154, 268, 301
collagen 67, 122, 126, 139, 161, 163, 181, 199–200, 203, 205, 207–209
condenser lens (CL) 312, 369, 514, 522–523, 701
condenser lens, dark-field 402
confocal detection 274, 276, 278, 297
confocal laser scanning microscopy (CLSM) 228
conventional biophotonics 385, 387, 389
cornea 5, 179, 185
corneal tissue 180
CRS, *see* coherent Raman scattering
cyan fluorescent protein (CFP) 122, 134
cytokinesis 281, 674, 687–689
cytoplasm 212, 272, 404, 685–688, 696, 698–699
cytotoxicity 128–129, 133, 170, 708

DCF, *see* double-cladding fiber
DCPCF, *see* double-cladding photonic crystal fiber
DCU, *see* dispersion compensation unit
deep-tissue imaging 7, 159, 177
DFG, *see* difference-frequency generation
DHM, *see* digital holographic microscopy
DIC, *see* differential interference contrast
dichroic mirror (DM) 158, 204–205, 210, 215, 312, 365, 523, 543, 700–701
difference-frequency generation (DFG) 139–141
differential interference contrast (DIC) 385, 715–716
diffraction tomography 656, 658
diffractive optical elements (DOE) 543, 548–551, 559
diffuse optical tomography (DOT) 99, 101
digital holographic microscopy (DHM) 22, 29, 526, 649–650, 652, 702
diode laser 704
dispersion compensation unit (DCU) 314, 321

dispersion precompensation 206, 364
dispersion-shifted fiber (DSF) 634–635, 637–638
DM, *see* dichroic mirror
DNA damage 673, 680, 684–685
DOE, *see* diffractive optical elements
DOT, *see* diffuse optical tomography
double-clad fiber 363–365, 369
double-cladding fiber (DCF) 214–215, 369
double-cladding photonic crystal fiber (DCPCF) 215–216
DSF, *see* dispersion-shifted fiber

EDFA, *see* erbium-doped fiber amplifiers
electroporation 694, 697, 699, 720
emission, fluorescent 117–118, 250, 397
erbium-doped fiber amplifiers (EDFA) 616–619, 624, 629–632

far-field diffraction 31–32
FBFL, *see* fiber-based femtosecond laser
FBGs, *see* fiber Bragg gratings
FDML, *see* Fourier domain mode-locked
FDML configuration, hybrid 630, 632
femtosecond fiber lasers 205, 217
femtosecond pumping 588
fiber
  elastin 203, 207–208
  hollow-core 345
  optical glass 50
fiber-based femtosecond laser (FBFL) 368–369
fiber Bragg gratings (FBGs) 626–628, 631
fiber cantilever 350, 358
fiber-cantilever piezotube scanners 358
fiber lasers 205, 295, 615, 629
  high-power 594, 601
fiber-optical parametric amplifier 616
fiber-optical parametric oscillator (FOPO) 626–628, 634–635, 637, 639
fiber optics 335, 372
fiber Raman amplifiers (FRAs) 616, 618–619, 624–625
fiber rotary joint (FRJ) 348, 360
fluoresceins 122–123, 126–127, 129–130, 216
fluorescence
  metal-enhanced 389, 405–406
  one-photon 201–202
  single-photon 159–160
  two-photon 158, 160, 162, 201–202
fluorescence cell imaging 418
fluorescence excitation, single-photon 162
fluorescence microscopy 40, 115, 121, 129–130, 135–136, 153, 221, 261, 301, 386, 443, 519, 525, 538
  multiphoton 159, 599, 614
  single-photon 159–160, 162
  two-photon 159–160
fluorescence recovery after photobleaching (FRAP) 124
fluorescent proteins 125, 132, 134–135, 148, 231, 233, 242–244, 249–250, 405, 702, 709, 720
  photoswitchable 242–243

fluorophores 122, 125–130, 147, 163, 224–225, 230–232, 234, 237, 243, 245, 405–406, 697, 699, 703–704, 709
  synthetic 242–243
FOPO, *see* fiber-optical parametric oscillator
Förster resonance energy transfer (FRET) 135, 401, 403
four-wave mixing (FWM) 142–143, 156, 160, 342–343, 573–575, 581, 585, 590, 622–624
Fourier domain mode-locked (FDML) 613, 628–629
Fourier optics 33–34
FRAP, *see* fluorescence recovery after photobleaching
FRAs, *see* fiber Raman amplifiers
frequency-resolved optical gating (FROG) 27–28
FRET, *see* Förster resonance energy transfer
FRJ, *see* fiber rotary joint
FROG, *see* frequency-resolved optical gating
FWM, *see* four-wave mixing

gaussian beam optics 42–43, 45
GDD, *see* group-delay dispersion
gene delivery 415–416, 718
gene expression 386, 390, 415, 417, 424–425, 693, 699
gene transfer 718
GFP, *see* green fluorescent protein
gold nanoparticles 199, 390–392, 397, 425, 442, 451, 454, 477, 709, 711–713, 717, 722
green fluorescent protein (GFP) 122, 125, 134, 159, 247, 279, 416, 702
ground state depletion (GSD) 229, 231–232, 234, 238
group-delay dispersion (GDD) 322
group velocity dispersion (GVD) 80–81, 342–343, 578
GSD, *see* ground state depletion
GVD, *see* group velocity dispersion

HeLa cells 232, 240, 416, 702
hemoglobin 75, 161, 176–177, 299, 695
holographic imaging 650–651, 668
holographic microscopy, digital 22, 29, 526, 649–650, 652
holographic optical tweezers (HOTs) 515, 520–521, 700
HOTs, *see* holographic optical tweezers

instrument response function (IRF) 120
interface, metal–dielectric 446–447, 449
interference microscopy 228–229
intravascular imaging 359, 369–370
intravital endomicroscopy 335–372
IR photoablation 181–183
IRF, *see* instrument response function

laser-induced stress waves (LISW) 718
laser-induced stress waves, pulsed 718
laser irradiation 175, 416–417, 683, 702–703
laser microirradiation 673–674, 689

laser trap 295
laser tweezers Raman spectroscopy (LTRS) 289–291, 293–295, 297, 299
light polarization 10, 14, 17–18, 63, 66, 137
light scattering 53, 55, 58, 82–83, 85, 87–88, 93, 130, 147, 393–394, 398, 444
  inelastic 259–260, 262, 264
light–matter interactions 2, 10, 14, 55, 58, 101, 108, 441
LISW, *see* laser-induced stress waves
localized surface plasmon resonance (LSPR) 387–388, 391–392, 413, 419, 427–428, 450–451
LSPR, *see* localized surface plasmon resonance
LTRS, *see* laser tweezers Raman spectroscopy

magnetic resonance imaging (MRI) 262, 336, 610, 612
Maxwell's equations 6–7, 398, 513, 540
MEF, *see* metal-enhanced fluorescence
metal-enhanced fluorescence (MEF) 389, 405–406
microlasers 492–493
  multimode 491–493
microscopy, fluorescence photoactivation localization 238
Mie scattering 85, 87, 393, 513
mitotic chromosomes 680, 683–684
mode splitting method (MSM) 465, 471, 476, 478
MPM, *see* multiphoton microscopy
MPM endoscope 216–217, 338, 361
MRI, *see* magnetic resonance imaging
MSM, *see* mode splitting method
multiphoton endomicroscopy 214–215
multiphoton ionization 184, 599, 695–696
multiphoton microscopy (MPM) 81, 157, 199–212, 214, 216–217, 336–339, 344–345, 361–362, 368

nanoparticles
  metallic 388, 397, 406, 409, 413, 449–450, 710
  plasmonic 392, 399, 414–416, 423
nanoplasmonic biophotonics 385–418
nanoplasmonic structures 392, 415
nanoplasmonic trapping 389–390, 418–420
nanoplasmonics 388–393, 395, 397, 399, 401–402, 417, 421
  fundamentals of 390–391, 393, 395, 397, 399
nonlinear microscopy 540, 542, 545, 554, 557, 559, 561, 563
nonlinear optical microscope 157
nonlinear optical microscopy 135–136, 157–160, 162, 557, 609
nonlinear optics 135–137, 139, 141, 143, 145, 147, 149, 151, 153, 155–157, 159, 161, 163, 553, 574–575

OBB, *see* optical bottle beam

OCT
  see optical coherence tomography
  phase-conjugated 625, 633
OFDI, *see* optical frequency-domain imaging
OPA
  see optical parametric amplifiers
  one-pump 624, 631–632
  two-pump 625
OPL, *see* optical path length
OPO, *see* optical parametric oscillator
optical bottle beam (OBB) 547–550, 558, 560–563
optical coherence tomography (OCT) 21, 27–29, 201, 209, 211–213, 335–336, 338–339, 345, 347–348, 359, 368–372, 599, 609–613, 628, 633
optical fibers, single-mode 574
optical frequency-domain imaging (OFDI) 613, 628
optical Kerr effect 144–146, 573–574
optical microscopes, nonlinear 158, 539
optical parametric amplifiers (OPA) 616, 622, 625, 627, 630, 632
optical parametric oscillator 615, 626
optical parametric oscillator (OPO) 161, 609, 615, 626, 633
optical path length (OPL) 25–26, 211, 426, 471
optical scanning holography (OSH) 649, 651–653, 655–659, 668
optical transfection 694, 696, 698–701, 703, 705, 707, 709–712, 716–717, 720, 722, 725
optical transfer function (OTF) 564, 653–654, 656, 660, 662
optical trapping 53, 290, 298, 301, 419, 428, 529, 720
  conventional 390, 419–420, 427
optical tweezers 53, 290, 294–295, 297–298, 386, 418–419, 507–510, 512–522, 524–526, 528, 530, 549, 700, 706, 709
optical waveguides 46–47, 49, 302
OSH, *see* optical scanning holography
OTF, *see* optical transfer function

particle imaging velocimetry (PIV) 525
PBF, *see* photonic bandgap fiber
PCA, *see* principal component analysis
PCF, *see* photonic crystal fibers
PDG, *see* polarization dependent gain
phase contrast microscopy 29–30, 681
phase-shaped beams 537–538, 540, 542, 544, 546, 548, 550, 552, 554, 556, 558, 560, 562, 564, 566
photochemical decomposition 183
photodiode 490, 527
photodynamic therapy 73, 167, 169–170, 386, 711
photomultiplier tube 204–205, 210, 365, 369
photonic bandgap fiber (PBF) 215, 574, 600
photonic crystal 454–457, 588

photonic crystal fibers (PCF) 217, 344–345, 362–363, 367, 456, 579, 585–586, 588–592, 594–598, 601
photonic crystal resonators 454–457
photosensitizer 169–171, 173
PIV, *see* particle imaging velocimetry
plasmon resonance 389–390, 397
  localized surface 387–388, 391–392, 450
plasmon resonance energy transfer (PRET) 389, 403–404
plasmonic enhanced photothermal effect 413–417, 427
point spread function (PSF) 36, 38–39, 222–224, 228–229, 232, 237, 239–240, 249, 251, 540, 542, 550–551, 653
polarization
  electronic 59–60, 64, 388
  molecular 59–60, 64
polarization dependent gain (PDG) 621
polarization microscopy 18
position sensitive detector (PSD) 522, 524, 528
PRET, *see* plasmon resonance energy transfer
principal component analysis (PCA) 272, 286–289, 291
PSD, *see* position sensitive detector
PSF, *see* point spread function

QDs, *see* quantum dots
quadrant photodiodes 522–524
quantum dots (QDs) 122, 126, 131–133, 199, 242, 386, 405, 413, 459, 476–477, 484, 487, 548, 720, 723

Raman amplifiers 620, 629, 632
Raman analysis of biological cells 276, 297
Raman scattering 152, 262–265, 267–271, 273–275, 277, 279, 281, 283, 301, 389, 407–409
  coherent anti-Stokes 301, 537–538, 609–610, 614
  principles of 262–273
  spontaneous 149, 151, 155, 280
Raman spectroscopy 150–154, 259–262, 268, 274–275, 280, 287, 291, 298, 300–302, 385, 443
random-phase pupil 661
ray-tracing techniques 37–38
Rayleigh range 44–46
Rayleigh scattering 85–87, 150, 152, 262, 267, 512
RBCs, *see* red blood cells
reactive oxygen species (ROS) 133, 168–170, 696, 704, 709
red blood cells (RBCs) 75, 176, 296–300
relative intensity noise (RIN) 613, 620
resonance Raman scattering (RRS) 268
RIN, *see* relative intensity noise
ROS, *see* reactive oxygen species
RRS, *see* resonance Raman scattering

sapphire 5, 63, 112, 204, 210, 215, 312
sapphire lasers 97, 141, 147, 200, 205, 209, 215, 343, 363, 585, 588, 593, 598, 611, 615
saturated pattern excitation microscopy (SPEM) 235
saturated structural illumination microscopy (SSIM) 229, 235

SBS, *see* stimulated Brillouin scattering
Schrödinger equation 102–103, 105
second-harmonic generation (SHG) 139–144, 157–158, 160–164, 200, 202–205, 208–212, 217, 335–338, 343, 368, 372, 537, 676
selective photothermolysis 176
self-phase modulation (SPM) 144–145, 156, 160, 342–343, 362, 367, 573–575, 579, 582, 587, 622
semiconductor optical amplifiers (SOAs) 616, 620–621, 629, 632
SERDS, *see* shifted excitation Raman difference spectroscopy
SERS, *see* surface-enhanced Raman scattering
SFG, *see* sum-frequency generation
SHG, *see* second-harmonic generation
shifted excitation Raman difference spectroscopy (SERDS) 286
SIM, *see* structured illumination microscopy
single-cell analysis 260, 274–275, 277, 279, 281
single-mode fibers (SMF) 341, 361, 367, 574, 577, 626
SLMs, *see* spatial light modulator
SMF, *see* single-mode fibers
SNR, *see* suspended nanochannel resonators
SOAs, *see* semiconductor optical amplifiers
spatial light modulator (SLMs) 290, 301, 514–518, 520–521, 530, 538, 543–544, 546, 555, 564–565, 700, 715
spectral shift method (SSM) 465–467, 469, 471, 473, 475, 477–479, 481, 486, 494
SPEM, *see* saturated pattern excitation microscopy
SPM, *see* self-phase modulation
SPR, *see* surface plasmon resonance
SRS, *see* stimulated Raman scattering
SSIM, *see* saturated structural illumination microscopy
SSM, *see* spectral shift method
standing wave modes (SWMs) 471, 473, 481
stimulated Brillouin scattering (SBS) 342, 630
stimulated Raman scattering (SRS) 89, 154–155, 157–158, 162, 164, 301, 342, 367, 484, 573–575, 618
Stokes waves 150, 154, 156
structured illumination microscopy (SIM) 235–237
sum-frequency generation (SFG) 139–141
super-resolution ensemble nanoscopy 229–235
super-resolution imaging techniques 237, 242
super-resolution single molecule nanoscopy 237–249
supercontinuum 165, 574, 576–578, 584–585, 587–593, 596–597, 599–600
supercontinuum sources 571, 573, 586, 589, 595, 599, 601
surface-enhanced Raman scattering (SERS) 268, 389, 408–412, 451
surface-enhanced Raman spectroscopy 407, 409, 411

surface plasmon resonance (SPR) 391, 444, 447–448, 450, 465–466
suspended nanochannel resonators (SNR) 310–311, 328, 454, 628, 651
SWMs, *see* standing wave modes

TEM, *see* transmission electron microscope
TERS, *see* tip-enhanced Raman scattering
THG, *see* third-harmonic generation
third-harmonic generation (THG) 139, 142–143, 160, 342, 537
tip-enhanced Raman scattering (TERS) 410
TIR, *see* total internal reflection
TIRFM, *see* total internal reflection fluorescence microscopy
tissue ablation 179
tissue absorption 159, 168
tissues
  absorptive 179, 184
  coagulated 172, 175
  functional imaging of 209, 368
TLS, *see* tunable laser source
tomography 97–98, 176, 178
total internal reflection (TIR) 3, 46–48, 226–227, 341, 459–460
total internal reflection fluorescence microscopy (TIRFM) 225, 227
TPA, *see* two-photon absorption
TPEF, *see* two-photon excited fluorescence
TPFM, *see* two-photon fluorescence microscopy
transfection efficiency 699–700, 703, 707, 710, 721
transmission electron microscope (TEM) 317, 676, 680, 682, 714
tube lens 158, 312, 314, 513–514, 677, 701
tumors 167–168, 176, 199, 412–413, 719
tunable laser source (TLS) 630
two-photon absorption (TPA) 146, 156, 200
two-photon beam splitter 312, 314
two-photon excited fluorescence (TPEF) 160, 200, 202–205, 207–213, 217, 335–336, 338, 343, 347–348, 368, 372
two-photon fluorescence excitation 162
two-photon fluorescence microscopy (TPFM) 159–160, 162–163, 200

ultrasensitive biomolecular absorption spectroscopy 403
UV photoablation 182–183

virions 468–469, 478, 483, 491–492, 494
viruses 135, 443–444, 449, 452, 454, 465, 477

waveguide confined Raman spectroscopy (WCRS) 302
waveguide dispersion 342
wavelength converters 624
wave–particle duality 53, 102
waves
  acoustic 177, 316, 318
  electromagnetic 4–5, 398
  photoacoustic 176
  ultrasonic 176–177

WCRS, *see* waveguide confined Raman spectroscopy
WDF, *see* Wigner distribution function
WGM, *see* whispering-gallery-mode
WGM optical resonators 459, 461, 463, 465
WGM resonators 459–461, 465–466, 468, 471, 476, 484–485, 494
whispering-gallery-mode (WGM) 445, 459–460, 466, 471, 476, 480, 493–494
wide-field Raman imaging 282–283
Wigner distribution function (WDF) 661
wound healing 207, 209, 212, 214, 217, 368

yeast cells 279
yellow fluorescent protein (YFP) 122, 134, 148, 233
YFP, *see* yellow fluorescent protein
ytterbium microcavity laser 492

ZDW, *see* zero-dispersion wavelength
zero-dispersion wavelength (ZDW) 575, 577–579, 582, 585–586, 590–594, 597, 624, 631, 636